Stochastic Damage Mechanics

随机损伤力学

李　杰　任晓丹　著

Jie Li, Xiaodan Ren

同济大学出版社
TONGJI UNIVERSITY PRESS
·上海·

图书在版编目(CIP)数据

随机损伤力学：Stochastic Damage Mechanics：英文/李杰，任晓丹著．—上海：同济大学出版社，2022.10

ISBN 978-7-5765-0291-6

Ⅰ．①随… Ⅱ．①李… ②任… Ⅲ．①损伤力学—英文 Ⅳ．①O346

中国版本图书馆 CIP 数据核字(2022)第 130705 号

Stochastic Damage Mechanics

随机损伤力学

李 杰 任晓丹 **著**

责任编辑 宋 立 **责任校对** 徐春莲 **封面设计** 唐思雯

出版发行 同济大学出版社 www.tongjipress.com.cn
(地址：上海市四平路 1239 号 邮编：200092 电话：021-65985622)

经 销 全国各地新华书店

排 版 南京文脉图文设计制作有限公司

印 刷 江阴市机关印刷服务有限公司

开 本 787 mm×1092 mm 1/16

印 张 29.75

字 数 743 000

版 次 2022 年 10 月第 1 版

印 次 2022 年 10 月第 1 次印刷

书 号 ISBN 978-7-5765-0291-6

定 价 218.00 元

李杰，工学博士，丹麦奥尔堡大学荣誉博士，中国科学院院士，同济大学特聘讲座教授，上海防灾救灾研究所所长。

长期在结构工程与工程防灾领域从事研究工作。在随机力学、工程结构可靠性与生命线工程研究中取得了具有国际影响的研究成果。包括：发展了随机损伤力学基本理论；建立了随机系统分析的概率密度演化理论，解决了复杂结构整体抗灾可靠度分析问题；建立了大规模工程网络抗震可靠性分析与优化设计理论等。领衔获得国家自然科学二等奖(2016)、国家科技进步三等奖(1997)各 1 项、部省级科技奖励一等奖 5 项。

1998 年获得国家杰出青年科学基金，1999 年入选“长江学者奖励计划”首批特聘教授。2014 年，因在概率密度演化理论与大规模基础设施系统可靠性方面的学术成就，被美国土木工程师学会(ASCE)授予领域最高学术成就奖——Alfred M. Freudenthal 奖章。2017 年当选国际结构安全性与可靠性学会主席。

任晓丹，工学博士，同济大学教授，博士生导师。兼任国际期刊 *Engineering Failure Analysis* 土木工程主题编辑、上海力学学会结构振动专业委员会秘书长、中国建筑学会建筑结构分会青年理事。

长期从事混凝土随机损伤理论与结构非线性分析方法的研究和开发工作。在国内外学术期刊及国际学术会议发表论文 110 余篇，其中 SCI 收录国际期刊论文 70 余篇，出版学术专著 1 部。曾获上海市优秀博士学位论文奖(2012)，世界华人计算力学学会优秀青年科学家奖(2018)和上海市科技进步一等奖、二等奖(2019、2020)。在混凝土损伤力学方面的研究成果已列入我国《混凝土结构设计规范》(GB 50010—2010)。主持国家自然科学基金 3 项、教育部博士点基金 1 项、专用项目 2 项，作为骨干成员先后参与国家自然科学基金重大研究计划集成项目以及国家自然科学基金重大国际合作项目的研究。

Jie Li holds a Ph. D. in Structural Engineering and an honorary doctorate of Aalborg University in Denmark. He is an academician of the Chinese Academy of Sciences and the distinguished chair professor of Tongji University. He also serves as the director of Shanghai Institute for Disaster Prevention and Relief.

Professor Jie Li's research interests focus on the structural engineering and disaster prevention. His research findings are of international impacts in the field of stochastic mechanics, reliability of engineering structures and lifeline engineering. He developed the fundamental theory of stochastic damage mechanics and established the probability density evolution theory for stochastic analysis and global reliability analysis of complex structures. In addition, Professor Li developed the theory of aseismic reliability analysis and optimization design for large-scale engineering networks. He has been awarded the Second Class National Natural Science Prize (2016), the Third Class National Science and Technology Progress Prize (1997) and five One Class Ministerial-Provincial Science and Technology Prizes.

Professor Jie Li was awarded the National Science Fund for Distinguished Young Scholars of China in 1998. He was selected as the Cheung Kong Scholar Distinguished Professors in 1999. In the year 2014, he received the Alfred M. Freudenthal Medal from the American Society of Civil Engineers (ASCE) for his significant contributions of the probability density evolution theory and reliability of large-scale infrastructure systems. In 2017, Prof. Li was elected as the president of the International Association for Structural Safety and Reliability (IASSR).

Xiaodan Ren, Ph. D. , is professor of structural engineering at Tongji University, in the College of Civil Engineering. He is the subject editor of civil engineering for the international journal *Engineering Failure Analysis*. He serves as the secretary of Committee of Vibration Mechanics, Shanghai Society of Theoretical and Applied Mechanics and young member of council of Association of Building Structures, the Architecture Society of China.

Prof. Ren has been working on stochastic damage mechanics and nonlinear analysis of structures for nearly 20 years. He has published 110 papers (70 international journal papers) and 1 book. He received the award of Excellent Doctoral Dissertation of Shanghai in 2012, the ICACM young investigator award in 2018. He also won the First and Second Prizes for Science & Technology Achievement of Shanghai in 2019 and 2020, respectively. His contributions in damage mechanics were included in the Chinese Code for design of concrete structures (GB 50010—2010). He is also PI of three research projects granted by the National Natural Science Foundation of China and one granted by the Ministry of Education of the People's Republic of China, respectively. Prof. Ren also took part in the major project and the international collaboration project of NSFC.

Contents

Part 1 Stochastic Damage Mechanics

Part 2 Continuum Damage Mechanics

Part 3 Numerical Methods for Damage Mechanics

Part 4 Application of Damage Mechanics

Part 1

Stochastic Damage Mechanics

Indentation tests based multi-scale random media modeling of concrete[①]

Hankun Liu, Xiaodan Ren, Jie Li

School of Civil Engineering, Tongji University, 1239 Siping Road, Shanghai 200092, China

Abstract In the present paper, a multi-scale random media model is proposed based on the indentation tests for each constituent of concrete including hydrated cement paste(HCP), aggregate and interfacial transition zone(ITZ). Firstly, systematic indentation tests are performed for each constituent of concrete at the nano- and micro-scales. Following the random field theory, each constituent of concrete materials is modeled as a random field. The scale of fluctuation is investigated based on the results of indentation tests. At the nano-scale, the scales of fluctuation of HCP and ITZ both turn to be roughly 20 μm, but that of aggregate is much larger. At the micro-scale, the scales of fluctuation of HCP and aggregate stay from 167 to 569 μm. Then the pointwise parameter estimation and model verification are performed for each constituent of concrete, and the one-dimensional (1-D) probabilistic density function (PDF) of the random field is obtained based on the proposed maximum possibility criterion. The probability distributions of the indentation modulus and hardness for each constituent of concrete are identified based on the statistical analysis. By introducing the reconstruction technique, concrete materials could be reconstructed as the random medium at the nano- and micro-scales. With the local averaging theory, the reduce factor for mechanical properties between the nano- and micro-scales is studied and it is shown that the experimental reduce factor agrees well with the theoretical one.

Keywords Concrete; Indentation test; Random field; Random media; Reduce factor

1 Introduction

It is generally believed that the properties such as elastic modulus, strength and fracture energy, are governed by the nanoscale properties[1], which are not intriguing with the development of indentation techniques. In recent years, indentation tests have been widely used to investigate the properties for cementitious materials. In early years, the microhardness tests were used to study the bulk properties of cement paste with a maximum penetration depth in the level of 10^{-5} m[2-3]. With the significant advances for the conventional Vickers microhardness testing technique, a novel microindentation technique was developed to investigate the elastic modulus and the microstrength of ITZ in reinforced concrete[4-6]. To obtain the submicron-scale mechanical properties of cement paste, the nanoindentation technique was emerged and applied to determine the first indentation

① Originally published in *Construction and Building Materials*, 2018, 168: 209-220.

results of individual constituents of Portland cement clinker with the penetration depth around $(0.3\sim0.5)\times10^{-6}$ m[7]. With the help of nanoindentation technique, the existence of two types of calcium-silicate-hydrates(C-S-H) with the corresponding volume fraction was found in Refs.[8-9]. Based on the work of Refs.[8-9], a statistical analysis of massive nanoindentation tests, namely the grid indentation technique, was proposed to allow for the information on the phase mechanical properties and the volume fraction[10-11]. Later, the previous work was extended to the bone and the shale[12], C-S-H at elevated temperatures[13], the ultra-high performance concrete[14], and so on. As to ITZ, several works have been done to provide better insight into the mechanical properties in cement paste[15], steel fiber reinforced mortar[16], and recycled aggregate concrete[17].

The statistical information(e.g. the PDF and the relative volume fraction) of cement paste can be provided by the previous work. However, in the field of structural engineering, ones always pay more attention to the probabilistic and statistical knowledge about concrete instead of mortar and cement paste. Inspired by the previous researches, a statistical way is proposed in the present paper to model concrete at the nano- and micro-scales.

This study is organized as follows. Section 2 presents the experimental details for the nano- and micro-indentations. In Section 3, we conduct the probabilistic and statistical analysis on each constituent of concrete, e.g. HCP, ITZ and aggregate, and obtain the corresponding 1-D PDF of the random field by a "maximum possibility criterion". After that, a random media model of concrete is proposed by synthesizing the PDFs of the three constituents in Section 4. Section 5 develops the intrinsic connection between the nano- and micro-scales using the local averaging theory. Numbers of conclusions are given in Section 6.

2 Materials and methods

The material prepared herein was the regular concrete, which was casted into steel molds at the water : cement : sand : aggregate ratio of 0.4 : 1 : 2 : 5, to form bars with the dimensions of 0.1 m×0.1 m×0.3 m. The specimens were hydrated with a humidity of 95% at the room temperature and kept in the standard conditions for 3 months.

The specimens were then cut into several small sample blocks with the size of 20 mm× 20 mm× 20 mm. To meet the requirements of the nano- and micro-testing, small specimens with the approximate dimensions of 20 mm×20 mm×5 mm were made up using a diamond saw. The small samples were embedded into the epoxy resin firstly. In the second step, according to the previous polishing technique[15, 17], the small specimens were ground and polished with silicon carbide papers down to 6 μm, then the diamond lapping films of gradations 6, 3, 1, and 0.5 μm were used to polish the samples. To avoid the effect of surface roughness on the indentation results, the measured root-mean-square roughness of the sample surface should be lower than one third of the average indentation depth[18]. Finally, the small samples were ultrasonically cleaned in water for 1 min, in order to remove the polishing debris.

The Nano Test Vantage was used to measure the elastic properties of materials at the nano- and micro-scales, by using electromagnetic force application and capacitive depth measurement. A Berkovich tip was used for the indentation. According to Oliver and Pharr method, the mechanical properties where the indentation hardness is defined as the maximum load divided by the contact area and the modulus is obtained from the final unloading curve can be obtained from the loading-unloading curves[19]. The indentation law for the nano- and micro-indentations was shown in Table 1. A series of nano- and micro-tests were conducted on the specimens shown in Fig. 1. To acquire the properties of each constituent of concrete, the preselecting testing zones were made on the specimens with regard to each constituent. For the simplicity, the aggregate was abbreviated as the capital letter "A", HCP as "H" and ITZ as "I". Each indentation area and its corresponding indentation lattice were listed in Table 2. For the tests at the nanoscale, the dimensions of the indent lattice for HCP, ITZ and aggregate were all 25×20. And for the micro-testing, the indent lattice dimensions for HCP and aggregate were both 14×20.

Table 1. Indentation law

Control factor	Nano-test	Micro-test
Maximum depth	300 nm	10 μm
Loading and unloading rate	0.2 mN/s	100 mN/s
Hold time	15 s	15 s
Grid space	seeing Table 2 and Fig. 2	

Fig. 1. Specimens for testing

Taking the indent areas $A_{n,1}$ and $A_{m,1}$ for instance, the schematic diagrams are shown in Fig. 2. For a relatively large indentation depth, due to the restrain of the stress field and the physical interaction of the penetrating indenter and the rigid inclusion, the penetration near a rigid inclusion surface would be reduced[20]. In addition, as for the micro-tests, the indentations operated to a depth of h_{max}, would activate material situated on a surface radius of $4h_{max}$[21], which may be larger than the width of ITZ. To avoid this argument, the indents on the ITZ phase were not considered in the tests at the microscale, as detailed in Table 2.

Table 2. Indent area and the corresponding lattice

	Constituent	Indent area							Total
Nanoscale	Aggregate	Indent area	$A_{n,1}$	$A_{n,2}$	$A_{n,3}$	$A_{n,4}$	$A_{n,5}$	—	5
		Lattice	3×20	3×20	8×20	1×20	10×20	—	25×20
	HCP	Indent area	$H_{n,1}$	$H_{n,2}$	$H_{n,3}$	$H_{n,4}$	$H_{n,5}$	—	5
		Lattice	4×20	7×20	4×20	4×20	6×20	—	25×20
	ITZ	Indent area	$I_{n,1}$	$I_{n,2}$	$I_{n,3}$	$I_{n,4}$	$I_{n,5}$	$I_{n,5}$	6
		Lattice	5×20	4×20	4×20	4×20	3×20	5×20	25×20
Microscale	Aggregate	Indent area	$A_{m,1}$	$A_{m,2}$	$A_{m,3}$	—	—	—	3
		Lattice	4×20	5×20	5×20	—	—		14×20
	HCP	Indent area	$H_{m,1}$	$H_{m,2}$	$H_{m,3}$	—	—	—	3
		Lattice	4×20	5×20	14×20	—	—	—	14×20

Note: $A_{n,1}$, $H_{n,1}$ and $I_{n,1}$ stand for the 1st nano-test area on the aggregate, HCP and ITZ specimens, respectively; $A_{m,1}$ and $H_{m,1}$ stand for the 1st micro-test area on the aggregate and HCP specimens, respectively.

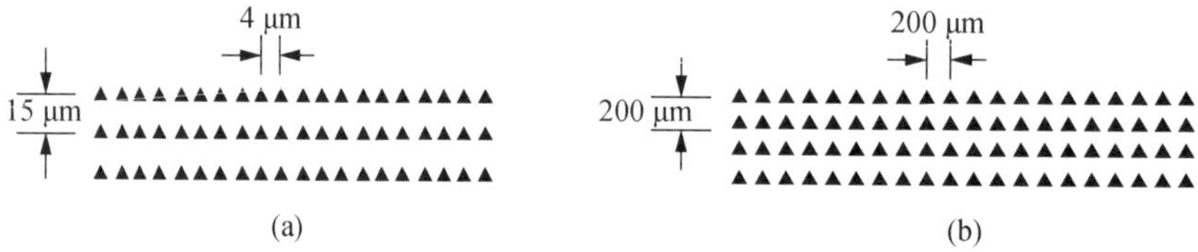

Fig. 2. Indent lattice: (a) 3×20 (nanoscale) and (b) 4×20 (microscale)

3 Probabilistic analysis on each constituent

3.1 Probabilistic homogenization

In reality, concrete with its attribute displaying sufficient large degree of disorder should be described by a probabilistic rather than the deterministic model. Particularly, the random field theory is born to model the complex distributed disordered system[22]. In the following, each constituent of concrete is treated as a random field.

To derive stochastic model, each sample(random series) with 20 observations from the nano- and micro-tests is divided into 4 sub-samples consisted of 6 observations(Fig. 3). Totally, the sample number for HCP is 100, that for ITZ is 100, and that for aggregate is 100 at the nanoscale, and that for HCP and aggregate both are 84 at the microscale.

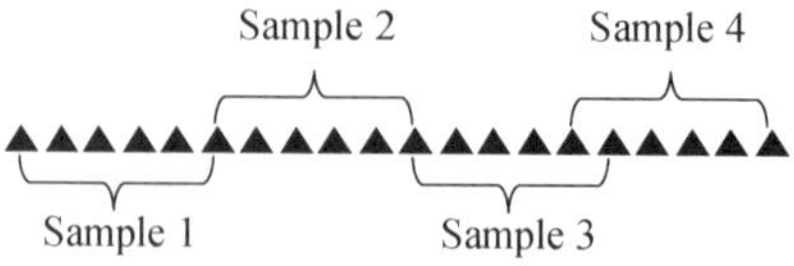

Fig. 3. Segmentation for random series

At the nanoscale, Fig. 4 displays the sample data, the related first-order and second-order statistics of the nano-hardness for HCP. At the microscale, the statistics of the micro-hardness for HCP are shown in Fig. 5.

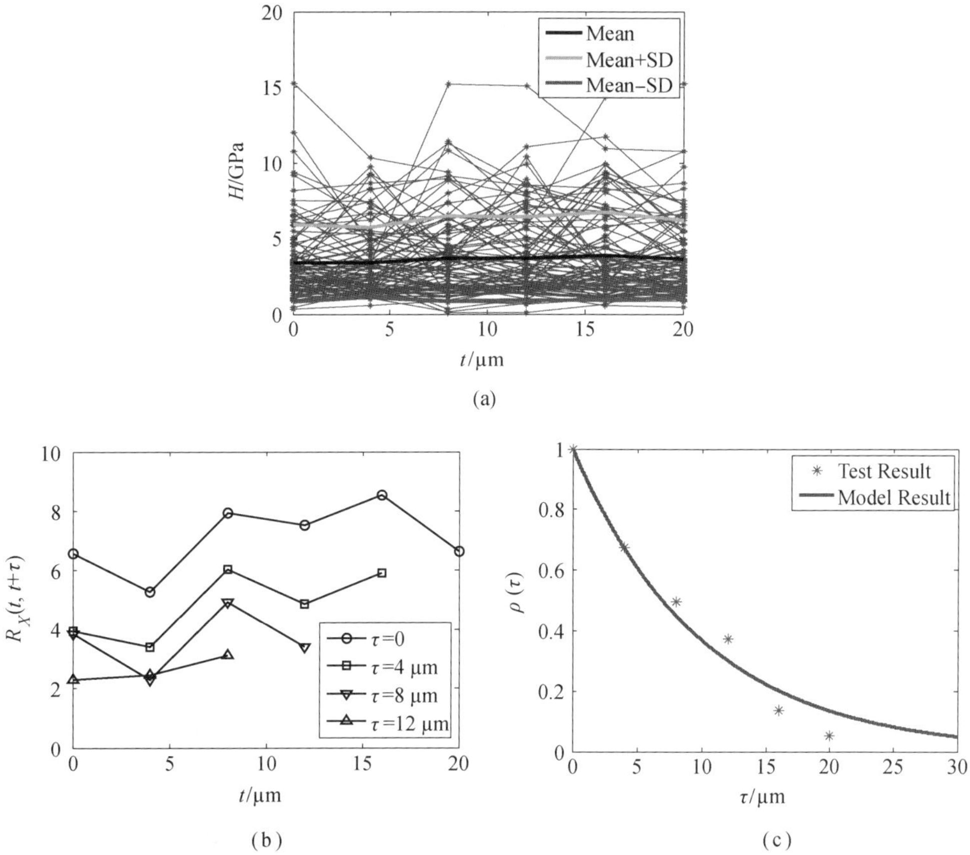

Fig. 4. Nano-hardness of HCP (100 samples): (a) random samples and its statistics ('Mean' and 'SD' stand for the mean value and the standard deviation respectively); (b) autocorrelation function curve and (c) normalized autocorrelation curve fitted using Eq. (1)

Figs. 4a and 5a confirm that most of the values fall within the range from the mean value minus the standard deviation (denoted by "Mean − SD") to "Mean + SD". As shown in Figs. 4a, b and 5a, b, the mean values, the standard deviations and the autocorrelation results for the nano- and micro-hardness remain constant with the variation of space. Moreover, the probability characteristics only depend on the relative locations of the points (the relative distance here $\tau=0$, 4, 8 and 12 μm at the nano-scale, and $\tau=0$, 0.2, 0.4, and 0.6 mm at the micro-scale). From the properties listed above, the 1-D random fields for the nano- and micro-hardness of HCP could be considered to be homogeneous. With regard to the nano- and micro-modulus for HCP and the nano- and micro-properties (indentation hardness and modulus) for aggregate and ITZ, the same probabilistic analysis repeatedly applied to the relative random field, it turns out that all the random field could be called

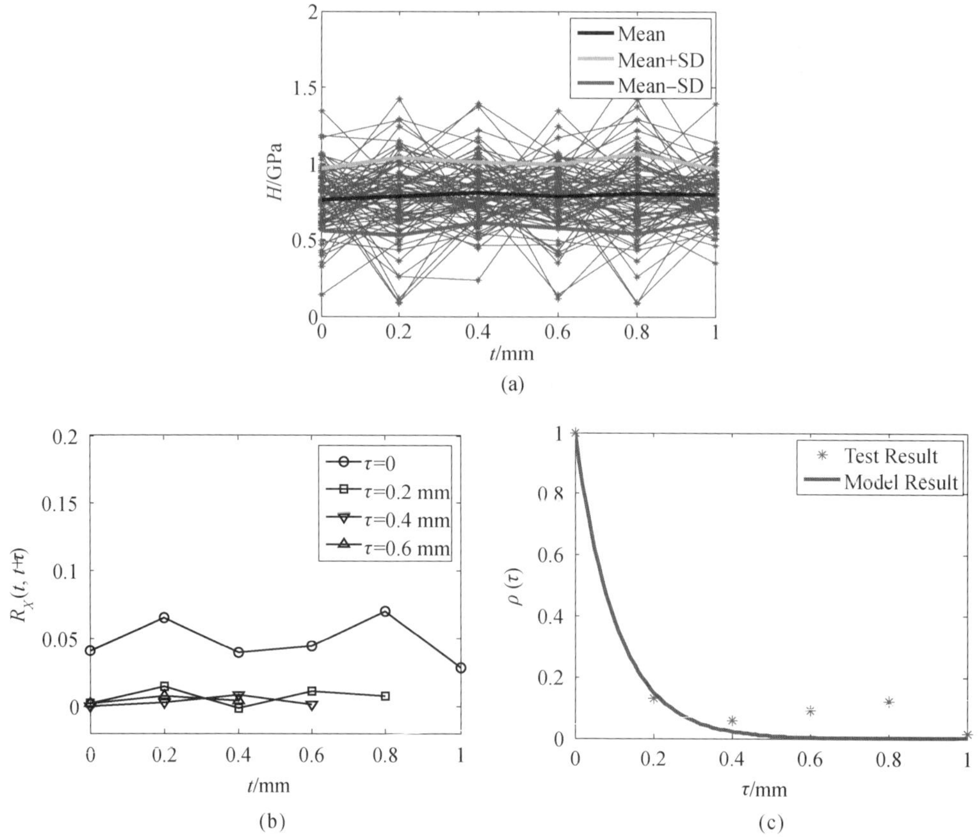

Fig. 5. Micro-hardness of HCP(84 samples): (a) random samples and its statistics; (b) autocorrelation function curve and (c) normalized autocorrelation curve fitted using Eq.(1)

homogeneous(stationary). Although the intrinsic complexity in the microstructure and the physical and mechanical properties has not been well understood yet, the concrete constituents can also be regarded as a kind of probabilistic homogenous materials in the current stage of investigation.

3.2 Scale of fluctuation θ

To focus on the scale of fluctuation, consider the correlation function expressed as Eq.(1), based on Vanmarcke' work[22], the scale of fluctuation θ can be derived as Eq.(2).

$$\rho(\tau) = \exp(-a\tau) \tag{1}$$

$$\theta = 2\int_0^{\infty} \rho(\tau)\mathrm{d}\tau = \frac{2}{a} \tag{2}$$

where $\rho(\tau)$ is the correlation function.

Following Eq. (1), the parameter a in the correlation function for the nano-and micro-properties of concrete constituents can be obtained by the fitting of the observed data plotted in Figs. 4c and 5c. And the scale of fluctuation θ can also be calculated from Eq.(2), as shown in Table 3.

Table 3. Scale of fluctuation for each constituent

	Constituents	Properties	a	$\theta = 2/a$ /μm
Nano-scale	HCP	H	0.100 1	19.985
		M	0.101 4	19.714
	ITZ	H	0.111 9	17.875
		M	0.084 1	23.785
	Aggregate	H	0.040 9	48.898
		M	0.059 7	33.508
Micro-scale	HCP	H	9.504 6	210
		M	3.514 4	569
	Aggregate	H	12.007 5	167
		M	0.000 3	6 137 000

At the nanoscale, the scale of fluctuation for the nano-properties of HCP and ITZ are both around 20 μm, and that of aggregate is much larger. At the microscale, that for the micro-properties of HCP and the micro-hardness of aggregate stay from 167 to 569 μm. It is worth noting that the scale of fluctuation depends on the microstructure of materials. And according to Jennings's qualitative and quantitative study[23-24], the amorphous colloidal structure of the C-S-H is organized in 'globules' composed of basic building blocks and globules porosity. Nevertheless, the microstructure of aggregate is highly complex. That is why the scale of fluctuation of the modulus for aggregate exhibits inconceivable to some extent.

3.3 Parameter estimation and model verification

With regard to each constituent(HCP, ITZ or aggregate), a number of observed random series(as described in the "Probabilistic homogenous" section) can be obtained. In the following, it is necessary to investigate the point distribution and to verify it by Kolmogorov-Smirnov test(referred as K-S test)[25].

Both of the point estimation and the interval estimation are firstly conducted to estimate the distribution parameters with a hypothesized PDF. Then, K-S test is performed to measure the maximum difference between the observed PDF and the hypothesized PDF estimated for the nano- and micro-tests. Here, the hypothesized distribution types, including normal distribution(ND), lognormal distribution(LD), Weibull distribution (WD) and gamma distribution(GD), are evaluated by the P-value(the probability for the maximum difference greater than the critical value), one of which is chosen and regarded as the "good" or "best" PDF. Finally, it is hopeful to determine the best estimated 1-D PDF for the random field of each constituent if not only are the P-values of each point larger than the specified significance level(herein 5%), but also is the mean P-value of six points relatively large. Tables 4 and 5 summarize the "best estimate" PDF and the relevant P-values by the application of K-S test for the nano- and micro-tests.

Hence, as shown in Table 4, the nano-hardness and modulus of HCP, ITZ and aggregate follow the lognormal distribution, the gamma distribution and the normal distribution, respectively. In the meanwhile, the micro-hardness and modulus of HCP and aggregate follow the normal distribution. Evidently, the P-values in the tables above are greater than 5% except for that of micr-omodulus for aggregate, due to the complexity and heterogeneity of aggregate.

Table 4. Pointwise P-values from K-S test for the nano-testing

Materials type		Distribution type	P-values for 6 sections of 1D random field						
			1	2	3	4	5	6	Mean value
HCP	H	LD	0.964	0.766	0.570	0.336	0.748	0.846	0.705
	M	LD	0.283	0.965	0.167	0.276	0.917	0.967	0.596
ITZ	H	GD	0.947	0.827	0.618	0.189	0.119	0.904	0.601
	M	GD	0.452	0.733	0.555	0.788	0.398	0.758	0.614
Aggregate	H	ND	0.557	0.674	0.539	0.071	0.327	0.723	0.482
	M	ND	0.379	0.760	0.377	0.548	0.583	0.770	0.570

Table 5. Pointwise P-values from K-S test for the micro-testing

Materials type		Distribution type	P-values for 6 sections of 1D random field						
			1	2	3	4	5	6	Mean value
HCP	H	ND	0.898	0.529	0.584	0.632	0.573	0.771	0.665
	M	ND	0.556	0.269	0.461	0.432	0.341	0.712	0.462
Aggregate	H	ND	0.864	0.147	0.607	0.845	0.452	0.487	0.567
	M	ND	0.020	0.011	0.058	0.068	0.018	0.023	0.033

3.4 1-D PDF for random field

To model each constituent of concrete accurately, it is necessary to acquire the knowledge of 1-D PDF of the random field. Considering that the field belongs to homogenous field, a "maximum possibility criterion" is proposed to capture the biggest possibility matching the reality based on the P-values for the obtained data, which is as follows

$$P = \max\left(\frac{1}{6}\sum_{i=1}^{6} P_i\right) \tag{3}$$

$$P_i (i = 1, 2, \cdots, 6) > \alpha \tag{4}$$

Following the simplex method[26-27], the optimal values meeting the requirements of "maximum possibility criterion" and the relevant P-values are shown in Tables 6 and 7. Figs. 6 and 7 show the histograms of the nano- and micro-properties(indentation hardness and modulus) for HCP together with the theoretical PDFs using the optimal parameters in Table 6. It is worth noting that for each constituent at each scale only one out of six points on the random process is herein exhibited.

Table 6. Optimal parameters for 1-D random field

Constituents		Properties	Distribution type	Optimal parameters	
Nanoscale	HCP	H	LD	1.03	0.72
		M	LD	4.01	0.44
	ITZ	H	GD	2.74	0.75
		M	GD	3.70	12.01
	Aggregate	H	ND	11.71	3.52
		M	ND	98.81	16.32
Microscale	HCP	H	ND	0.80	0.19
		M	ND	34.16	4.88
	Aggregate	H	ND	3.94	1.81
		M	ND	56.21	13.52

Note: for instance, optimal parameters are the mean value and the standard deviation for ND.

Table 7. P-values for 1-D random field

Constituent		Properties	P-values for 6 points					
Nanoscale	HCP	H	0.483	0.811	0.409	0.565	0.396	0.938
		M	0.614	0.762	0.931	0.942	0.857	0.896
	ITZ	H	0.614	0.841	0.634	0.691	0.542	0.735
		M	0.513	0.893	0.949	0.532	0.643	0.904
	Aggregate	H	0.478	0.695	0.606	0.339	0.761	0.821
		M	0.311	0.941	0.778	0.777	0.514	0.885
Microscale	HCP	H	0.211	0.723	0.810	0.988	0.795	0.455
		M	0.567	0.066	0.266	0.932	0.274	0.647
	Aggregate	H	0.829	0.044	0.717	0.920	0.211	0.132
		M	0.004	0.201	0.147	0.454	0.040	0.085

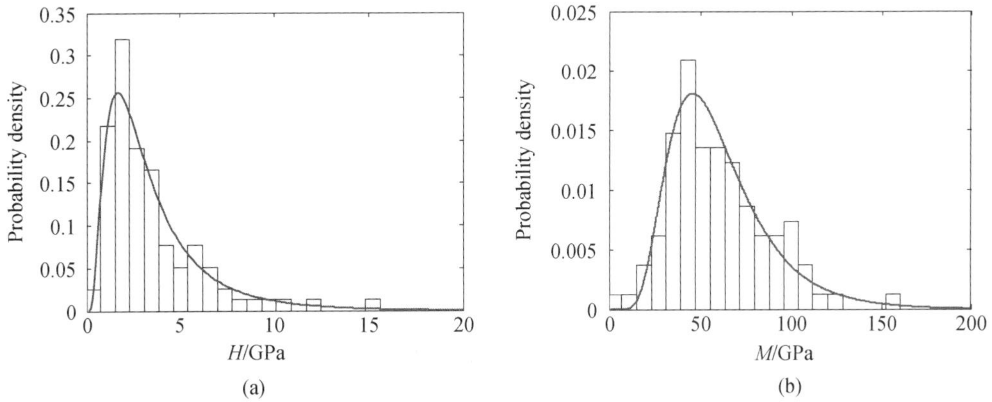

Fig. 6. Histograms and theoretical PDF curves for HCP: (a) nano-hardness; (b) nano-modulus

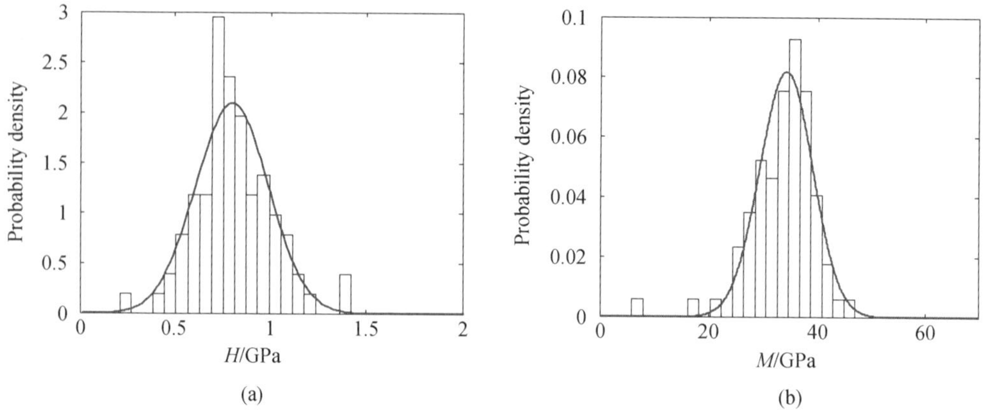

Fig. 7. Histograms and theoretical PDF curves for HCP: (a) micro-hardness; (b) micro-modulus

Following the similar statistical analysis, the comparison between the histograms and the theoretical PDFs for the nano- and micro-hardness and modulus of each concrete constituent show better agreement other than that of micro-properties for aggregate. Moreover, it shows only one peak in the histogram of HCP and ITZ. In other words, when statistically modeling HCP, ITZ and aggregate, the distribution listed in Table 6 may be a good choice. In addition, the nano-properties of HCP follow the lognormal distribution and that of ITZ the Gamma distribution, which implies that there are different statistical characteristics between HCP and ITZ, although they have almost the same constituents (C-S-H and portlandite). And the nano-properties for aggregate and micro-properties for HCP and aggregate all follow the normal distribution.

Nevertheless, the bad estimating results for the micro-modulus of aggregate in Table 5 and Table 7 are highly consistent with the bad fitting results for that in Table 3. Also the histogram compares worse with the theoretical distribution for the micro-modulus of aggregate exhibited in Fig. 8. For aggregate, it lacks statistical regularity and shows at least two peaks in the histograms, which can be attributed to the heterogeneity within the materials.

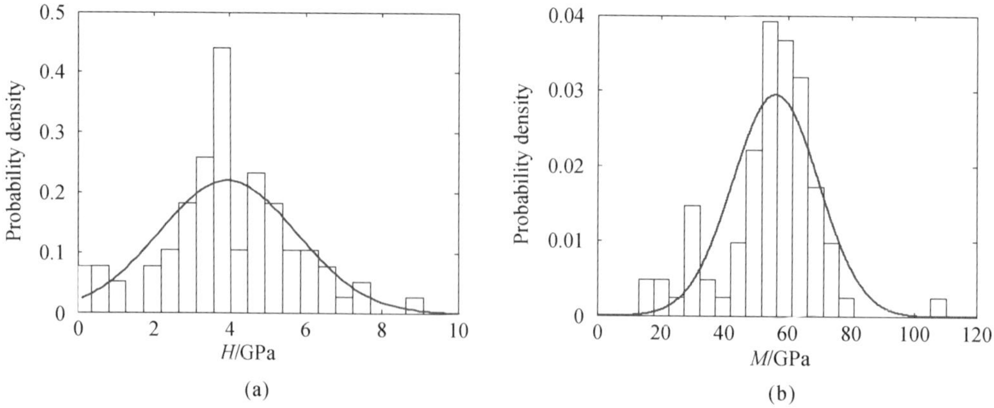

Fig. 8. Histograms and theoretical PDF curves for aggregate: (a) micro-hardness; (b) micro-modulus

4 Random media modeling

A previous study has demonstrated that the constituents of HCP — the low stiffness/hardness phase(MP), the low density C-S-H phase(LD C-S-H), the high density C-S-H phase (HD C-S-H) and the portlandite(CH) and anhydrated clinker phases, all follow the normal distribution[11]. It is evidently to find that the mechanical properties obey different distributions not only for the different constituents but also at the different scales(i.e. nano-, micro-, meso- and macro-level). In the present paper, the focus is to reconstruct concrete materials at the nano- and micro-scales by combining HCP, ITZ and aggregate, and less attention is paid to the detailed components of HCP, ITZ and aggregate.

On the basis of the similarity of constitutive relationships for each concrete constituent and the additivity of probability, it is reasonable to use the following equation to reconstruct the concrete materials.

$$\phi(x)=\sum_{i=1}^{n} f_i\phi_i(x) \quad i=1, 2, 3 \tag{5}$$

where x is the mechanical property H or M, f_i is the volume fraction occupied by HCP, ITZ or aggregate, $\phi_i(x)$ is the relative PDF, and $\phi(x)$ is the reconstruction PDF of concrete.

Coarse aggregate is incorporation with continuous graded of diameter 10 ~ 25 mm. Furthermore, the particle size r and the sieving rate m follow the relationship as follows.

$$r(m)=15m+10 \tag{6}$$

Assuming that the aggregate is round, the volume of ITZ around that is approximately expressed as

$$\frac{f_{ITZ}}{f_A}=\int_0^1 \frac{V_{ITZ}}{V_A}dm=\int_0^1\left[\left(1+\frac{D}{r(m)}\right)^3-1\right]dm \tag{7}$$

where f_{ITZ} and V_{ITZ} are the volume fraction and the volume of ITZ; f_A and V_A are that of aggregate; D is the thickness of ITZ.

The ITZ width which has been discussed and disputed until now, should be given. According to Refs.[28-30], the ITZ of the order of 10~50 μm in width is reported; however, the ITZ cannot be observed in Ref.[31]. As displayed in Fig. 9, the indentation results in this paper show four types of ITZ profiles coinciding with Ref.[20]. Fig. 9 displays the properties of ITZ higher than (Type 1), equal to (Type 2), lower than (Type 3) and much lower than (Type 4) that of the bulk paste. The profiles in Type 1 and Type 2 may result from the well bond at the interface; however, the poor bond leads to the profiles in Type 3 and Type 4. From the statistical perspective, the mean observation series shows the trend of increasing indentation properties with increasing distance apart from the interface, and the ITZ width could be measured approximately of 25 μm.

Calculating by Eq.(7) with $D=25$ μm, a volume fraction ratio of HCP : aggregate : ITZ is about 0.611 : 0.387 : 0.002.

Using Eq.(5) and the estimated distributions in Table. 6, the random media distributions of concrete are given in Fig. 10. There are two peaks dominating the PDF, which are attributed to the material regions dominated by the HCP phase and the aggregate phase. It is worth mentioning that ITZ governs the initial damage of concrete, although it takes up a very small proportion in concrete.

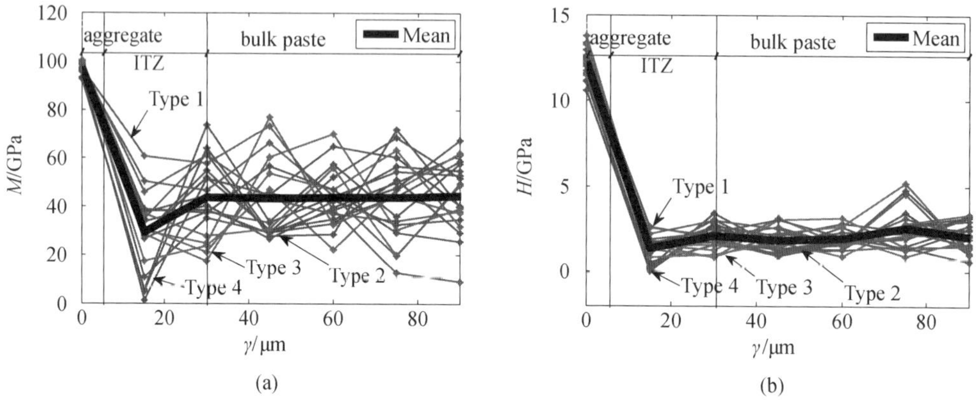

Fig. 9. The ITZ properties plotted against distance from the aggregate: (a) modulus; (b) hardness

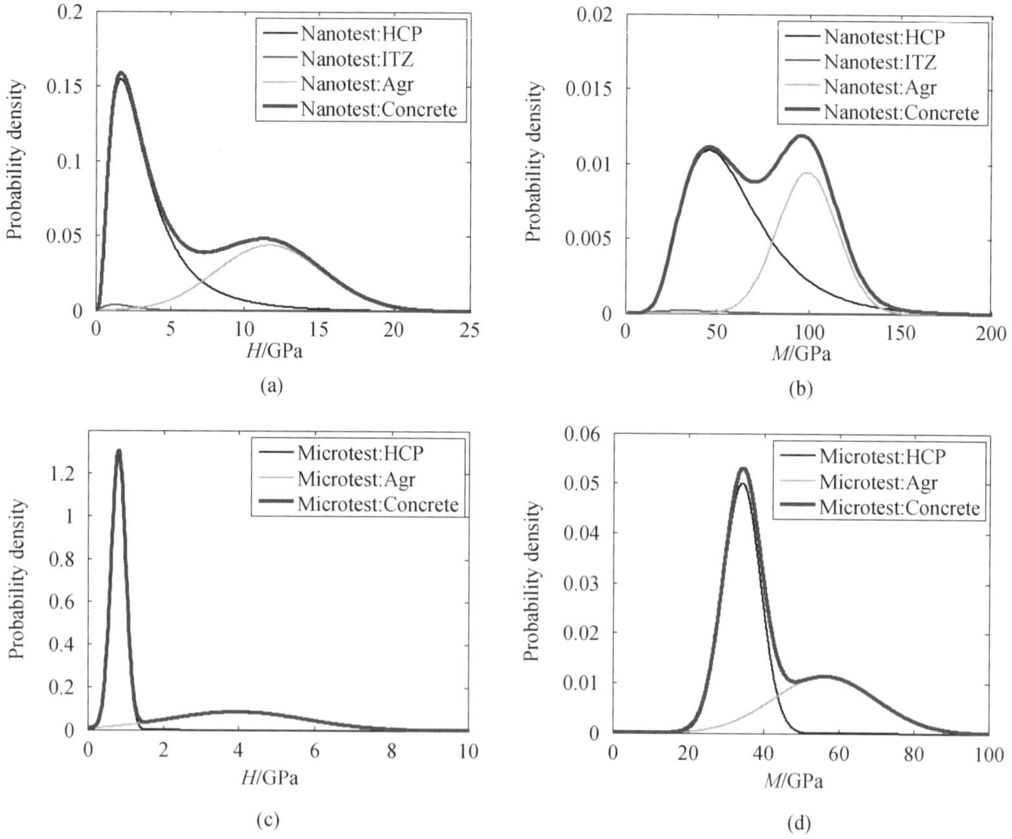

Fig. 10. Random media distribution for concrete: (a) nano-hardness; (b) nano-modulus; (c) micro-hardness and (d) micro-modulus

In Ref. [11], the de-convolution technique is used to extract different phases in HCP materials. Inspired by this enlightening work, a simple reconstruction technique is adopted to directly reconstruct the concrete materials, which has a very clear boundary between its individual constituents. By the reconstruction technique, it provides a new perspective that the properties of concrete and its constituents exhibit statistical and probabilistic characteristics. Traditionally, the properties of each constituent may be referring to the mean values corresponding to the peak in the PDF shown in Fig. 10 in the engineering application. Moreover, it also updates the traditional point of view that concrete constitute is always regarded as a kind of materials with uniform properties.

As shown in Fig. 11, the comparisons between the histogram and the reconstructed PDFs for the properties of concrete show well agreement. Moreover, the histograms of observed results could also be a verification for the reconstruction technique proposed in this paper. Furthermore, a reasonable conjecture could be made that the first peak attributed to the HCP phase may move to the right side even go over the second peak mainly due to the increasing column of HD C–S–H when the high-strength or super high-strength concrete is concerned. On the contrary, the first peak would depart from the second peak mainly due to the increasing volume of LD C–S–H[31].

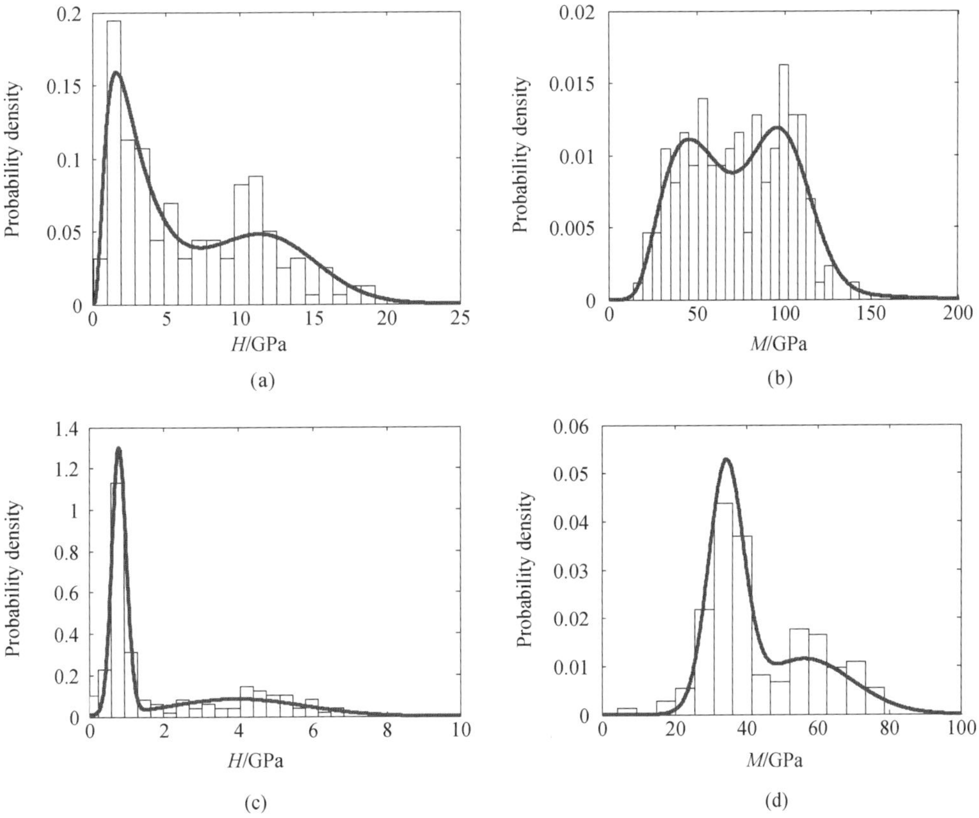

Fig. 11. Histograms and theoretical PDF curves for concrete: (a) nano-hardness; (b) nano-modulus; (c) micro-hardness and (d) micro-modulus

5 Local averaging from nano- to micro-scale

5.1 Gedanken experiment and local averaging process

Consider that the concrete materials are indented at two different scales (Fig. 12). The collective outcome with a small indentation depth comprising a stationary random process is denoted by $X(t)$ with mean m and variance σ^2. When an indentation test performed to a large depth, each "big" indentation can be regarded as a local averaging over the small indentations. Thus, a family of moving average processes $X_T(t)$, proposed by Vanmarcke[22], can be expressed as follows

$$X_T(t) = \frac{1}{T}\int_{t-T/2}^{t+T/2} X(u)\mathrm{d}u \tag{8}$$

where T denotes the averaging space.

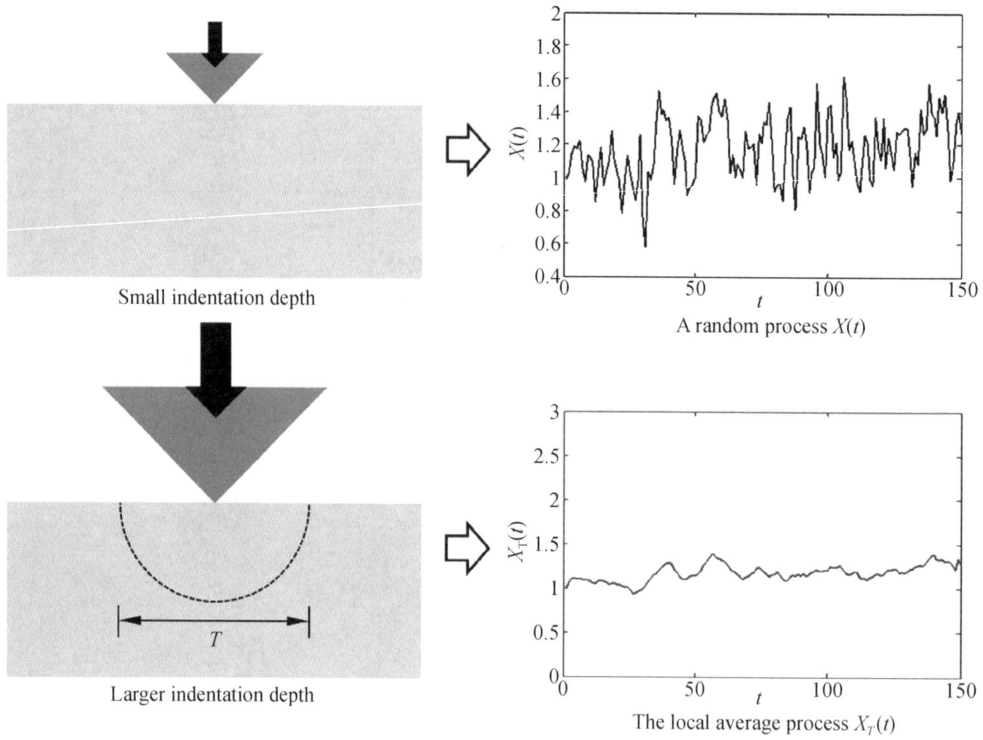

Fig. 12. Schematic representation of a local averaging process for indentation

5.2 "Theoretical reduce factor"

The mean of the local average process $X_T(t)$ is not affected by the local averaging operation while the variance of $X_T(t)$ may be expressed as follows[22]

$$Var[X_T(t)] \equiv \sigma_T^2 = \gamma(T)\sigma^2 \tag{9}$$

where $\gamma(T)$ measures the reduction of the point variance σ^2 under local averaging, and its square root $[\gamma(T)]^{1/2}$ is a "reduce factor" to be applied to the standard deviation σ, which is called "theoretical reduce factor" in this paper. The "theoretical reduce factor" can be calculated from the variance function with exponential correlation function aforementioned, expressed as follows[22]

$$\gamma(T)=2\left(\frac{\theta}{T}\right)^{2}\left(\frac{2T}{\theta}-1+e^{-2T/\theta}\right) \tag{10}$$

According to Ulm et al's research[19], the radius of the activated area around the indentations is about 4 times indentation depth. Consequently, the dimension of the activated area for the nano-tests can be on the order of $T=8\times0.3=2.4$ μm, and that for the micro-test $T=8\times10=80$ μm. In addition, the scale of fluctuation θ has been listed in Table 3.

5.3 "Experimental reduce factor"

Assume that a simple linear relationship is followed for either the hardness or the modulus between the nano- and micro-scale as follows

$$Y_H=a_H X_H+b_H \tag{11}$$

$$Y_M=a_M X_M+b_M \tag{12}$$

where X_H and Y_H are the micro- and nano-scale hardness respectively, and X_M and Y_M are the micro- and nano-scale modulus respectively, which all follow a typical distribution aforementioned. It is obvious that the ratio of standard deviation at the micro- and nano-scales, σ_X/σ_Y, is $1/a_H$ for the hardness and $1/a_M$ for the modulus. To keep the consistence with the "reduce factor" in local averaging theory above, we called the parameters $1/a_H$ and $1/a_M$ the "experimental reduce factor".

With the assist of relative entropy[32], the parameters a_H, b_H, a_M and b_M can be identified when the smallest relative entropy occurs[Figs. 13(a), 14(a)]. And the identification results are shown in Eqs.(13) and (14). Consequently, the "experimental reduce factor" of concrete can be calculated as displayed in Table 8. With the parameters identified, two PDFs at two scales coincide basically[Figs. 13(b), 14(b)] for the indentation hardness and indentation modulus, respectively, which reveals the consistency of two scales to a certain extent.

$$Y_H=2.8X_H-0.3 \tag{13}$$

$$Y_M=3.4X_M-63 \tag{14}$$

Similarly, the reduce factors of HCP and aggregate(phase material) can also be calculated using the relative entropy analysis, which are seen in Table 8.

As shown in Table 8, the theoretical reduce factors are 0.446~0.657, which are within the range of the experimental results: 0.180~0.828. It is thus not surprising that the local averaging theory can explain the standard deviation variation from the nano- to micro-scale, although there may be size effect existing in the scaling process[33]. In addition, small indentation depths(nanoscale) lead to the local properties, while large indentation depths (microscale) lead to the homogenized material properties. The homogenization from the

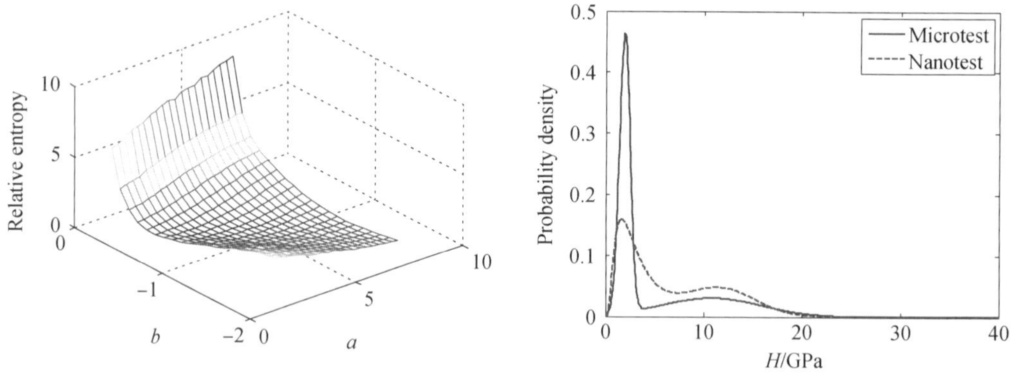

(a) (b)

Fig. 13.　Parameters identification: (a) the relative entropy and (b) the corresponding PDFs at two scales for the hardness

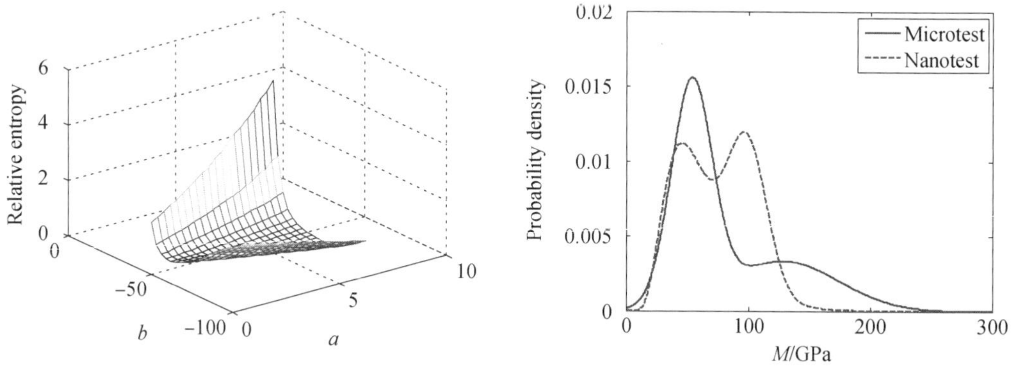

(a) (b)

Fig. 14.　Parameters identification: (a) the relative entropy and (b) the corresponding PDFs at two scales for the modulus

nano- to micro-scale results in the point variance reduction, which is quantitatively investigated by the reduce factor in the present paper.

Table 8.　Experimental and theoretical reduce factors

		Reduce factors	
		Experimental result	Theoretical result
Concrete	Hardness	0.357	—
	Modulus	0.294	—
HCP	Hardness	0.180	0.468
	Modulus	0.317	0.465
Aggregate	Hardness	0.514	0.657
	Modulus	0.828	0.576
ITZ	Hardness	—	0.446
	Modulus	—	0.503

6 Conclusions

A comprehensive nano- and micro-tests have been under conduction on each constitute of concrete(i.e. HCP, ITZ and aggregate). Random field theory and statistics theory based analysis on the testing results show some probabilistic and statistical information that: (1) the randomness of each constituent follows the homogeneous random field; (2) there is an exponential correlation structure for the nano- and micro-properties of aggregate, HCP and ITZ, and the scale of fluctuation for HCP and ITZ is around 20 μm; (3) the properties of each constituent follow a specific distribution for the different constituents at the different levels. Additionally, the "maximum possibility criterion" proposed has proved to be a valuable tool in the assessment of 1-D PDF of the random field.

With the reconstruction technique, the concrete PDFs for the indentation properties are reconstructed by the combination of that of HCP, ITZ and aggregate with the relevant volume fraction, which leads to the random media modeling of concrete in a statistical sense at the nano- and micro-scales. Thus, the random media distribution of concrete can serve as an input for the already existing multiscale modeling.

Indentation tests from the nano- to micro-scale show mathematically a local averaging process and mechanically homogenization, which reveals that local averaging theory is an effective tool to discover the statistical variation between the different scales and to measure the reduction of the point variance during the averaging process. It is expected that a new insight could be provided to deal with up scaling problem.

Acknowledgements

Financial supports from the National Science Foundation of China (Grant No. 51261120374, 91315301, 51678439) are greatly appreciated. The indentation tests were carried out in Southeast University, with the help of Dr. Wei Huang, Dr. Weiwei Zhu and Prof. Yamei Zhang. The authors would like to thank the anonymous reviewers for the valuable suggestions.

References

[1] CORR D, SHAH S P. Concrete materials science at the nanoscale[C]// Applications of Nanotechnology in Concrete Design: Proceedings of the International Conference held at the University of Dundee, Scotland, UK on 7 July 2005. Thomas Telford Publishing, 2005: 2-12.

[2] FELDMAN R F, CHENG-YI H. Properties of Portland cement-silica fume pastes Ⅱ. Mechanical properties[J]. Cement and Concrete Research, 1985, 15(6): 943-952.

[3] IGARASHI S, BENTUR A, MINDESS S. Characterization of the microstructure and strength of cement paste by microhardness testing[J]. Advances in Cement Research, 1996, 8(30): 87-92.

[4] ZHU W, BARTOS P J M. Assessment of interfacial microstructure and bond properties in aged GRC using a novel microindentation method[J]. Cement and Concrete Research, 1997, 27(11): 1701-1711.

[5] ZHU W, BARTOS P J M. Application of depth-sensing microindentation testing to study of interfacial transition zone in reinforced concrete[J]. Cement and Concrete Research, 2000, 30(8): 1299-1304.
[6] TRTIK P, REEVES C M, BARTOS P J M. Use of focused ion beam (FIB) for advanced interpretation of microindentation test results applied to cementitious composites[J]. Materials and Structures, 2000, 33(3): 189-193.
[7] VELEZ K, MAXIMILIEN S, DAMIDOT D, et al. Determination by nanoindentation of elastic modulus and hardness of pure constituents of Portland cement clinker[J]. Cement and Concrete Research, 2001, 31(4): 555-561.
[8] CONSTANTINIDES G, ULM F J, VAN VLIET K. On the use of nanoindentation for cementitious materials[J]. Materials and Structures, 2003, 36(3): 191-196.
[9] CONSTANTINIDES G, ULM F J. The effect of two types of CSH on the elasticity of cement-based materials: Results from nanoindentation and micromechanical modeling[J]. Cement and Concrete Research, 2004, 34(1): 67-80.
[10] CONSTANTINIDES G, CHANDRAN K S R, ULM F J, et al. Grid indentation analysis of composite microstructure and mechanics: Principles and validation[J]. Materials Science and Engineering: A, 2006, 430(1-2): 189-202.
[11] CONSTANTINIDES G, ULM F J. The nanogranular nature of C-S-H[J]. Journal of the Mechanics and Physics of Solids, 2007, 55(1): 64-90.
[12] ULM F J, VANDAMME M, BOBKO C, et al. Statistical indentation techniques for hydrated nanocomposites: concrete, bone, and shale[J]. Journal of the American Ceramic Society, 2007, 90(9): 2677-2692.
[13] DEJONG M J, ULM F J. The nanogranular behavior of CSH at elevated temperatures (up to 700 C)[J]. Cement and Concrete Research, 2007, 37(1): 1-12.
[14] SORELLI L, CONSTANTINIDES G, ULM F J, et al. The nano-mechanical signature of ultra high performance concrete by statistical nanoindentation techniques[J]. Cement and Concrete Research, 2008, 38(12): 1447-1456.
[15] MONDAL P, SHAH S P, MARKS L D. Nanoscale characterization of cementitious materials[J]. ACI Materials Journal, 2008, 105(2): 174.
[16] WANG X H, JACOBSEN S, HE J Y, et al. Application of nanoindentation testing to study of the interfacial transition zone in steel fiber reinforced mortar[J]. Cement and Concrete Research, 2009, 39(8): 701-715.
[17] LI W, XIAO J, SUN Z, et al. Interfacial transition zones in recycled aggregate concrete with different mixing approaches[J]. Construction and Building Materials, 2012, 35: 1045-1055.
[18] SAKULICH A R, LI V C. Nanoscale characterization of engineered cementitious composites (ECC)[J]. Cement and Concrete Research, 2011, 41(2): 169-175.
[19] OLIVER W C, PHARR G M. An improved technique for determining hardness and elastic modulus using load and displacement sensing indentation experiments[J]. Journal of Materials Research, 1992, 7(6): 1564-1583.
[20] IGARASHI S, BENTUR A, MINDESS S. Microhardness testing of cementitious materials[J]. Advanced Cement Based Materials, 1996, 4(2): 48-57.
[21] ULM F J, VANDAMME M, JENNINGS H M, et al. Does microstructure matter for statistical nanoindentation techniques?[J]. Cement and Concrete Composites, 2010, 32(1): 92-99.
[22] VANMARCKE E. Random fields: analysis and synthesis[M]. World Scientific, 2010.

[23] TENNIS P D, JENNINGS H M. A model for two types of calcium silicate hydrate in the microstructure of Portland cement pastes[J]. Cement and Concrete Research, 2000, 30(6): 855-863.

[24] JENNINGS H M. A model for the microstructure of calcium silicate hydrate in cement paste[J]. Cement and Concrete Research, 2000, 30(1): 101-116.

[25] SOONG T T. Fundamentals of probability and statistics for engineers[M]. John Wiley & Sons, 2004.

[26] NELDER J A, MEAD R. A simplex method for function minimization[J]. The Computer Journal, 1965, 7(4): 308-313.

[27] MATHEWS J H, FINK K D. Numerical methods using MATLAB[M]. Upper Saddle River, NJ: Pearson prentice hall, 2004.

[28] LI G, ZHAO Y, PANG S S. Four-phase sphere modeling of effective bulk modulus of concrete[J]. Cement and Concrete Research, 1999, 29(6): 839-845.

[29] SUN Z, GARBOCZI E J, SHAH S P. Modeling the elastic properties of concrete composites: Experiment, differential effective medium theory, and numerical simulation[J]. Cement and Concrete Composites, 2007, 29(1): 22-38.

[30] SCRIVENER K L, CRUMBIE A K, LAUGESEN P. The interfacial transition zone (ITZ) between cement paste and aggregate in concrete[J]. Interface Science, 2004, 12(4): 411-421.

[31] P. MONDAL. Nanomechanical properties of cementitious materials[M], Dissertations & Theses — Gradworks, 2008.

[32] SOBEZYK K, TRĘBICKI J. Maximum entropy principle in stochastic dynamics[J]. Probabilistic Engineering Mechanics, 1990, 5(3): 102-110.

[33] NĚMEČEK J, KOPECKÝ L, BITTNAR Z. Size effect in nanoindentation of cement paste: Key Paper [C]//Applications of Nanotechnology in Concrete Design: Proceedings of the International Conference held at the University of Dundee, Scotland, UK on 7 July 2005. Thomas Telford Publishing, 2005: 47-53.

Physical mechanism of concrete damage under compression①

Hankun Liu[1], Xiaodan Ren[2], Shixue Liang[3], Jie Li[2]

1 *Sichuan Institute of Building Research, Chengdu 610081, China; lhksibr@foxmail.com*

2 *School of Civil Engineering, Tongji University, 1239 Siping Road, Shanghai 200092, China; lijie@tongji.edu.cn*

3 *School of Civil Engineering and Architecture, Zhejiang Sci-Tech University, Hangzhou 310018, China; liangshixue0716@126.com*

Abstract Although considerable effort has been taken regarding concrete damage, the physical mechanism of concrete damage under compression remains unknown. This paper presents, for the first time, the physical reality of the damage of concrete under compression in the view of statistical and probabilistic information (SPI) at the mesoscale. To investigate the mesoscale compressive fracture, the confined force chain buckling model is proposed; using which the mesoscale parameters concerned could be directly from nanoindentation by random field theory. Then, the mesoscale parameters could also be identified from macro-testing using the stochastic damage model. In addition, the link between these two mesoscale parameters could be established by the relative entropy. A good agreement between them from nano- and macro-testing when the constraint factor approaches around 33, indicates that the mesoscale parameters in the stochastic damage model could be verified through the present research. Our results suggest that concrete damage is strongly dependent on the mesoscale random failure, where meso-randomness originates from intrinsic meso-inhomogeneity and meso-fracture arises physically from the buckling of the confined force chain system. The mesoscale random buckling of the confined force chain system above tends to constitute the physical mechanism of concrete damage under compression.

Keywords Concrete; Damage; Compression; Random field; Nanoindentation; Multiscale

1 Introduction

Concrete, a mixture of Portland cement, water, sand and aggregate, hydrated to form cementitious material with micro-crack, void and inhomogeneous[1], exhibits nonlinearity and randomness of mechanical properties. Under loading, concrete and its properties suffer from deterioration, which could be regarded as damage. The concept of damage is firstly developed by Kachanov[2], introduced into concrete material by Dougill soon afterwards[3]. Over decades, concrete nonlinearity, especially the characterization of properties softening, has been modeled by damage mechanics, whose branches include continuum

① Originally published in *Materials*, 2019, 12(20): 3295.

damage mechanics[4-5] and stochastic damage mechanics[6-7]. The former one regards macroscopic homogeneity, however, the latter one pays more attention to mesoscopic inhomogeneity and the progressive transition between different levels. Mazars modeled the degradation of concrete by bi-scalar model[4], based on this, Wu and Li proposed a plastic damage model by introducing the elastoplastic damage release rate[5]. Moreover, to discover the physical mechanism of concrete damage, the latter one, idealized as a parallel fiber bundle including two levels, has been always applied due to its simplicity in reproducing randomness and nonlinearity. Actually, concrete could be always regarded as a set of parallel small concrete rods connected on two ends and deform compatibly. At the macroscale, concrete could be modeled as a fiber bundle, giving a smooth curve reflecting the averaging response. While at the mesoscale, a small concrete rod could be considered as a fiber, exhibiting a kind of elastic-brittle relationship, referring to the previous researches[6-8]. Up to now, the fiber bundle model has been used for modeling concrete damage including tensile and compressive damage, under the static loading[7] and the dynamic loading[8]. In addition, Bazant and Pang also systematically investigated the size effect of concrete materials based on the bundle model[9]. It is noted that the physical mechanism for concrete tensile damage could be easily disclosed by the bundle model; however, the physical mechanism for concrete compressive damage remains a mystery.

To disclose the physical mechanism, the physical experiment could always be considered as a direct and effective way. Although it is generally realized that the nonlinearity and randomness of the mechanical properties for concrete are directly related to the fracture of the meso-element and its accumulation, the direct access to the knowledge regarding the local properties has never been provided until the 1970s. Hereafter, Beaudoin and Feldman systematically studied the properties of the autoclaved calcium silicate systems including elastic modulus, micro-strength, micro-hardness and the relationships between the properties by micro-hardness testing, and additionally, suggested the linear relationship between microscale compressive strength and micro-hardness[10]. From Igarashi's research, the microscale compressive strength and the micro-hardness both increased with increasing curing age and decreasing water to cement ratio, meanwhile, the same relationship as Beaudoin and Feldman was also verified[11]. Zhu and Bartos proposed a novel microindentation method continuously monitoring load and displacement, to assess the elastic modulus and microhardness of the interfacial zone of reinforced concrete[12]. According to the research of Buckle and Durst et al.[13-14], small indentation depths lead to mechanical phase properties, while greater indentation depths result in homogenized material properties. Georgios and Ulm proposed a novel method by means of grid indentation, using deconvolution technique to identify in situ two calcium-silicate-hydrates (C-S-H) phase(low-density C-S-H and high-density C-S-H)[15-17]. To anticipate the nature of strength, the dual-nanoindentation technique was proposed to assess the strength of bone and cohesive-frictional materials, suggesting the nanogranular friction responsible for the increased intrinsic resistance in compression[17-19]. Vandamme et al. also investigated the nanogranular origin of concrete creep by nanoindentation[20]. Liu et al. combined nanoindentation with random field modeling to study the probabilistic and statistical properties of concrete materials, suggested the multiscale SPI of concrete in a comprehensive manner and proposed a multi-scale random media model for concrete[21]. In addition, Mondal et al. studied the topological structure of cementitious materials using nanoindentation[22]. In short, microhardness testing[10-11], microindentation[12] and the nanoindentation technique[13-22] appeared successfully, using that which researchers have investigated; the fundamental knowledge including topological structure, mechanical and physical information, SPI, as well as their interrelationships. With the development of material science and technology, the materials genome could be discovered. Through the usage of this, the elementary physical properties and fundamental structural characteristics could also be predicted[23]. Usually, for concrete, a rather complex system, it is crucial and feasible to use the SPI of the intrinsic structure and properties at the nanoscale or mesoscale to investigate macroscopic properties. Unfortunately, the physical mechanism of concrete damage or constitutive relationship under compression remains unknown, despite the great achievements on continuum damage mechanics which fails to reveal the mechanism, and on stochastic damage mechanics (the bundle model, etc.) which has failed to be verified by the physical experiment until nowadays. On the basis of the recent progress on nanoindentation, stochastic mechanics and random media modeling for concrete, the paper focuses on the link between concrete damage and materials SPI at the lower scale, and on the origin of concrete damage.

In this research, the damage of concrete is investigated when subjected to compressive loading. Based on the previous research achievement, it is hypothesized that the fracture of "concrete fiber" originates at the mesoscale. On the one hand, nanoindentation tests are conducted on each constituent of concrete: hardened cement paste (HCP), interfacial transition zone (ITZ) and aggregate. By application of the confined force chain buckling model proposed in this paper, reconstruction technique[21] and random field theory, the SPI of fracture behavior could be obtained for concrete fiber at the mesoscale. On the other hand, the SPI of meso-parameters of concrete could also be recognized from macro-testing employing the stochastic damage model. Thus, this hypothesis could be proven, as long as the SPI of the meso-fiber from the nanoindentation coincides with that recognized from the macro-testing.

2 Materials and Methods

2.1 Materials and Preparation

The material prepared here was ordinary concrete with water : cement : sand : with an aggregate ratio of 0.4 : 1 : 2 : 5. The bars were made measuring 0.1 m×0.1 m×0.3 m, and hydrated with the humidity of 95% at the room temperature for three months, which were used for uniaxial compression testing at the macroscale. Then, the prisms were sliced into small specimenc with approximate dimensions of 0.02 m × 0.02 m × 0.005 m, which were prepared for nanoindentation. After embedding into the epoxy resin, grinding and polishing with silicon carbide papers. and ultrasonically cleaning, the samples for nanoindentation (Fig. 1) were prepared. Details of sample preparation for nanoindentation have been described in the previous study[21].

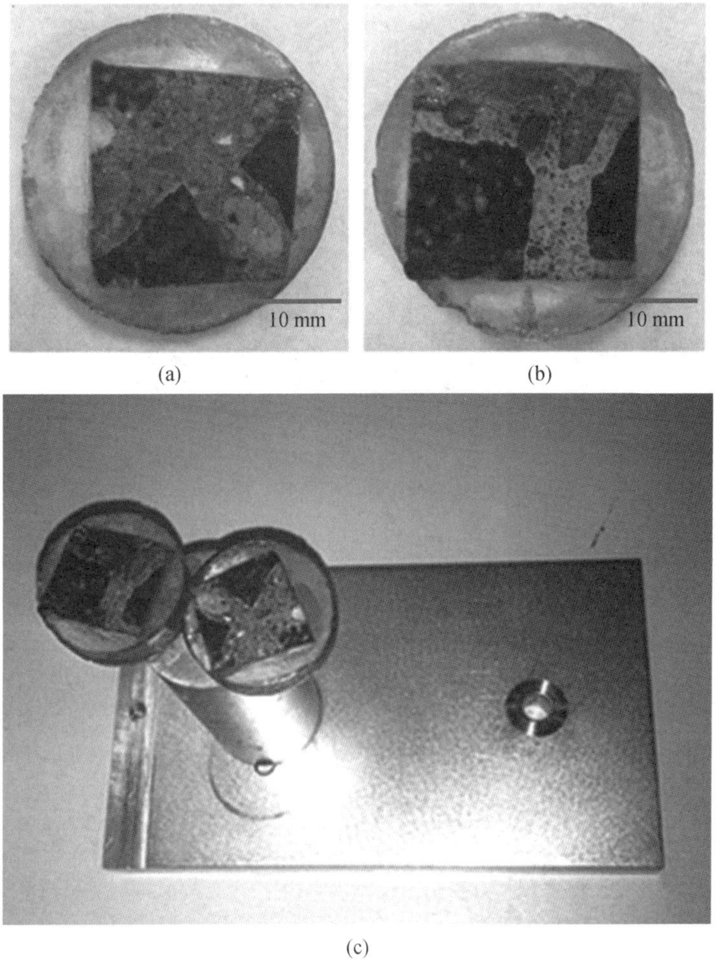

Fig. 1. Samples for testing in nanoindenter: (a) specimen 1; (b) specimen 2; and (c) specimens installed

2.2 Nanoindentation

The equipment conducting nanoindentation in the present paper is the NanoTest Vantage system (Fig. 2, from Micro Materials Limited in Wrexham, UK) to offer nanomechanical and nanotribologcal tests, with electromagnetic load application, with a maximum load of 500 mN, load resolution of 3 nN and displacement resolution of 0.002 nN. A series of nanoindentation tests were conducted with a Berkovich tip, with a maximum depth of 300 nm, loading and unloading rate of 0.2 mN/s and holding time of 15 s. According to the approach of Oliver and Pharr[24], indentation hardness H and indentation modular M were obtained from loading-unloading curves. Furthermore, Young's modulus linking to the elastic constants of specimen and indenter[25], could be expressed as

$$\frac{1}{M}=\frac{1-\nu^2}{E}+\frac{1-\nu_i^2}{E_i} \tag{1}$$

where E_i and ν_i are Young's modulus and Poisson's ratio of the indenter employed with a given value of 1 140 GPa and 0.07; E and ν are that of the tested materials, and the Poisson's ratio is 0.2.

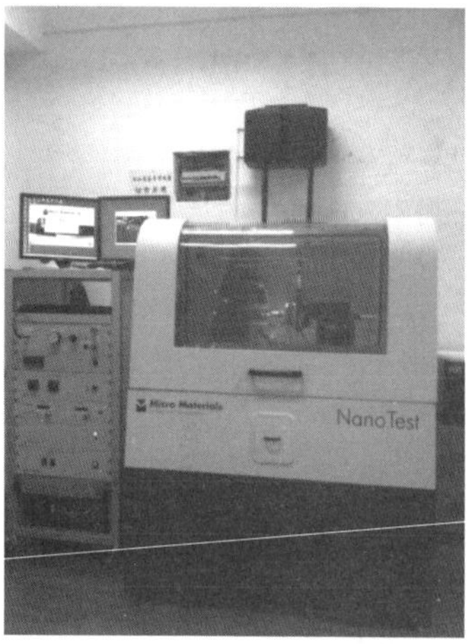

(a)

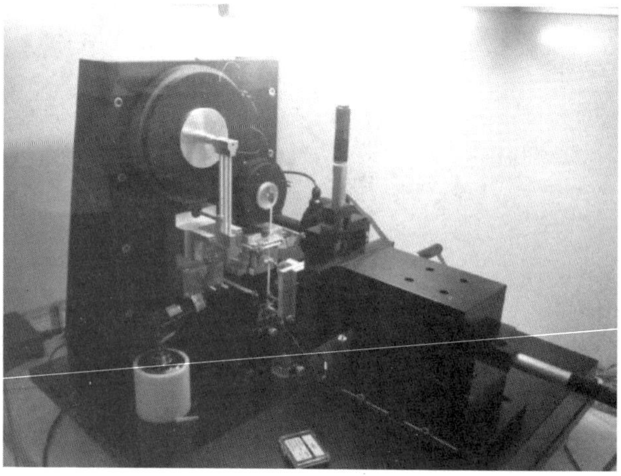

(b)

Fig. 2. NanoTest Vantage testing system: (a) appearance; and (b) internal details

Nanoindentation on each constituent of concrete was conducted in the present paper, and the dimensions of the indent lattice for HCP, ITZ and aggregate were all 25×20.

Fig. 3. Macro-testing system of uniaxial loading

2.3 Macro-Testing

A total of seven prism specimens subjected to uniaxial compressive loading were investigated by an electro-hydraulic servo-controlled concrete testing system from MTS Systems Corporation in Eden Prairie, MN, USA (Fig. 3). The hinge on the top and the scale marks on the bottom are used to guarantee the axial compression and the accurate centration of the tested samples. The stiffness of 1.1×10^{10} N/m is enough to provide the closed-loop controlled compression and the data collection accuracy. A pair of extensometers were installed on opposite sides of the specimen shown in Figure 3, which collected the axial displacement data. And the strain rate was 10^{-5} which guaranteed a static loading.

3 Theoretical, Experimental and Numerical Approach

3.1 Shear Fracture Strain(SFS) from Nanoindentation

3.1.1 Force Chain Based Modeling for Hardness

Due to the nanogranular nature of C-S-H[26], the aggregative particles of C-S-H could be assumed to yield the confined three-particle force chain buckling mechanism which was firstly proposed by Tordesillas and Muthuswamy[27]. The schematic is shown in Figure 4. Based on the force chain theory, a connection could be established from nanoscale to mesoscale.

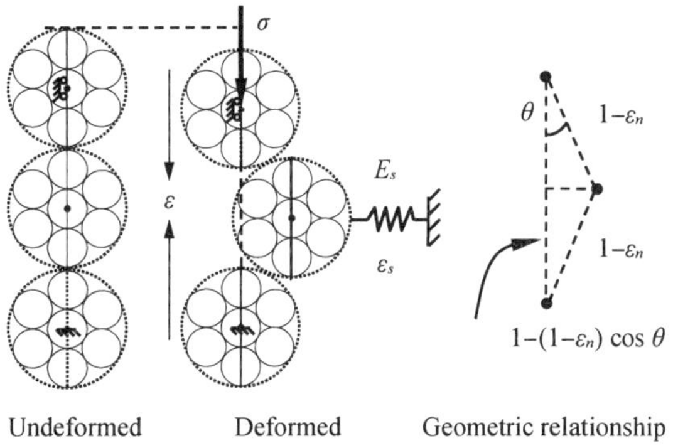

Fig. 4. Three-particle force chain model

For simplicity and possibility, only the contact force between particles and the lateral supporting force from the surrounding weak force chains are assumed for C-S-H in the present paper. The assumption could be extended to cement paste, ITZ and aggregate, because of the similar characteristics among the constituents of concrete. Next, the contact law is introduced to describe the key features in the force chain model.

1. Contact Models

The homogenized normal stress and tangential stress between particles are respectively

$$\sigma_n = E_n \varepsilon_n \tag{2}$$

$$\sigma_t = E_t \varepsilon_t \tag{3}$$

where E_n and E_t denote the normal and tangential modulus respectively; ε_n and ε_t are the normal and tangential strain.

To reflect the constraint effect of the weak force chain network, the applied confined pressure could be expressed as follows

$$\sigma_s = E_s \varepsilon_s \tag{4}$$

where E_s and ε_s denote the confined modulus and the corresponding stain.

2. Averaged Potential Energy Density

Considering a certain region (at the mesoscale) of cement paste in the surrounding area of the indenter, when applying compression by a strain ε, the averaged potential energy density e_p could be expressed as follows

$$e_p = u - w = \frac{1}{2}E_n\varepsilon_n^2 + \frac{1}{2}E_t\varepsilon_t^2 + \frac{1}{2}E_s\varepsilon_s^2 - \sigma\varepsilon \tag{5}$$

where u is the stored energy density; w is the work density with the stress σ. As shown in Figure 4, the strain ε_n, ε_t and ε_s could be rewritten in terms of θ and ε_n as follows

$$|\varepsilon_t| = (1-\varepsilon_n)\sin\theta \approx \sin\theta \tag{6}$$

$$\varepsilon_s = (1-\varepsilon_n)\sin\theta \approx \sin\theta \tag{7}$$

$$\varepsilon = 1-(1-\varepsilon_n)\cos\theta \tag{8}$$

Substituting Eqs.(6)—(8) into Eq.(5), one obtains the potential energy density

$$e_p = \frac{1}{2}E_n(\varepsilon_n)^2 + \frac{1}{2}E_t(\sin\theta)^2 + \frac{1}{2}E_s(\sin\theta)^2 - \sigma[1-(1-\varepsilon_n)\cos\theta] \tag{9}$$

It is evidently observed that the normal strain and the tangential strain could be decoupled.

3. Critical Load

According to the principle of resident potential energy, the partial derivative of the averaged potential energy density with respect to each degree of freedom should be zero, we get

$$\frac{\partial e_p}{\partial\theta} = 0,\ \frac{\partial e_p}{\partial\varepsilon_n} = 0 \tag{10}$$

Substituting Eq.(9) into Eq.(10), and solving Eq.(10), the relationship between the stress and the normal strain yields Eq.(11)

$$\sigma = \frac{(E_t + E_s)\cos\theta}{1-\varepsilon_n},\ \varepsilon_n = \frac{\sigma\cos\theta}{E_n} \tag{11}$$

From Eq.(11), an apparently stable path for the confined force chain could be achieved when the force chain structure keeps straight from the beginning of loading, which corresponds to

$$\sigma = E_n\varepsilon_n \tag{12}$$

Solving Eq.(11), σ, ε_n and ε could be obtained, which yields

$$\sigma = \frac{E_n}{2\cos\theta}\left(1-\sqrt{1-4\cos\theta\frac{(E_t+E_s)}{E_n}}\right) \tag{13}$$

$$\varepsilon_n = \frac{1}{2}\left(1-\sqrt{1-4\cos\theta\frac{(E_t+E_s)}{E_n}}\right) \tag{14}$$

$$\varepsilon = 1-\frac{1}{2}\left(1+\sqrt{1-4\cos\theta\frac{(E_t+E_s)}{E_n}}\right)\cos\theta \tag{15}$$

Another stable path could be acquired when θ is zero and this critical point with the stress and strain of Eqs.(16) and (17) corresponds to a buckling load of the force chain system.

$$\sigma_{cr}=\frac{E_n}{2}\left(1-\sqrt{1-4\frac{(E_t+E_s)}{E_n}}\right) \tag{16}$$

$$\varepsilon_{cr}=\varepsilon_n=\frac{1}{2}\left(1-\sqrt{1-4\frac{(E_t+E_s)}{E_n}}\right) \tag{17}$$

To conduct Taylor expansion for Eq.(16) and Eq.(17) at $E_t+E_s/E_n=0$, the results obtained are as follows

$$\varepsilon_{cr}=\frac{1}{2}\left(1-\sqrt{1-4\frac{(E_t+E_s)}{E_n}}\right)\approx\frac{E_t+E_s}{E_n} \tag{18}$$

$$\sigma_{cr}=E_n\varepsilon_{cr}\approx E_t+E_s \tag{19}$$

As easily seen from Eq.(19), a mixed-mode of shearing (E_t) and constraint (E_s) is included in the critical load, which means that the critical compressive load of the meso-element is attributed to the meso-element itself and the surrounding materials. With the framework of hardness theory, the hardness value H is the pressure at a limiting condition where the pressure keeps constant with increasing load[28]. Meanwhile, for indentation on cement paste, there is also a mixed fracture mode by cutting and hydrostatic pressure under the indenter at the limiting state[11]. Based on the statement above, the indentation hardness could be equivalent to the constrained compressive strength σ_c of the meso-element confined by the surrounding material around the indenter

$$\sigma_c=\sigma_{cr}\approx E_t+E_s=H \tag{20}$$

3.1.2 Constrained SFS

Over decades, researchers have investigated the relationship between the strength and the micro-hardness, with the conclusion that for cementitious materials(frictional materials), the hardness and the yield stress relationship of the form H/Y were reported on the order of 20～30[29]. Similarly, in Ref.[11], the ratio was found to be 2.7～3 more order of magnitude than that of metals, which was in the range of 30～60. As known from early on, the ratio of metal discussed above, named "the constraint factor", was evidently less than that of cementitious materials, due to the frictional effect on the hardness within the cohesive-frictional materials.

Based on the previous researches[11, 29] and Eq.(20), the relationship between the hardness H and the compressive strength σ_y was followed by introducing a constraint factor C

$$C=\frac{H}{\sigma_y} \tag{21}$$

With the assumption of elastic-brittle property of mesoscale element aforementioned[6-8], the SFS $\Delta_{1,s}(x)$ from nanoindentation could be rewritten as

$$\Delta_{1,s}(x)=\frac{\sigma_y}{E}=\frac{H}{CE}=\frac{\Delta_{con}}{C} \tag{22}$$

where Δ_{con} is defined as the constrained SFS. Combining Eqs.(1) and (22), the constrained

SFS could be rewritten as

$$\Delta_{con}=\frac{H}{E}=\frac{H}{1-\nu^2}\left(\frac{1}{M}-\frac{1-\nu_i^2}{E_i}\right) \tag{23}$$

3.1.3 Random Field Modeling and Statistical Modeling

Due to the attribute of a sufficient large degree of disorder, the knowledge of the probabilistic characteristics of concrete is fundamental to understanding the intrinsic random microstructure, even the nanostructure. Generally, random field theory deals effectively with the complex distributed disordered system[30]. As the random heterogeneous materials, Torquato has made considerable progress on 2D and 3D microstructure characterization, also on the relationship between mechanical properties and microstructure[31]. However, it is very difficult to use 2D or 3D modeling to investigate the physical mechanism of concrete damage, due to more parameters and complex simulation execution. To make it convenient and effective, the 1D random field is still adopted in the present paper.

The random field could be defined as homogeneous, as long as the mean $m_X(t_j)$ and the covariance $R_X(t_j, t_j+\tau)$ as follows keep constant with the variance of space, in other words, these values only depend on the relative distance τ.

$$m_X(t_j)=E[X(t_j)] \tag{24}$$

$$R_X(t_j, t_j+\tau)=E[X(t_j)X(t_j+\tau)] \tag{25}$$

where $E[\cdot]$ is the expected operator, $X(t_j)$ is the observation on the random series with respect to t_j.

To investigate the statistical characteristics, Kolmogorov-Smirnov test (referred to as the K-S test) could be adopted to acquire the probabilistic density function (PDF) of each point of the random series (containing six sections). To execute the K-S test, the main procedure is outlined as follows:

(1) Choose a sample X_i from the population X and rearrange sample values x_i in increasing order of magnitude.

(2) Compute the observed cumulative distribution function (CDF) $F_n(x_i)$ at each ordinal sample value.

(3) Estimate the parameters of the hypothesized distribution as described below based on the observed data and determine the theoretical CDF $F(x_i)$ at the same sample value above using the hypothesized distribution.

(4) Form the differences $|F_n(x_i)-F(x_i)|$, and calculate the statistics:

$$D=\max_{i=1}^{n}\{|F_n(x_i)-F(x_i)|\} \tag{26}$$

(5) Select a value of α and determine the critical value D_α.

(6) Accept or reject the testing hypothesis by comparing D and D_α.

The procedure stated above is the classic K-S test process. However, since the critical value is approximate, the null hypothesis is usually rejected or accepted by comparing the returned P value and the significance level α.

In this study, the hypothesized PDF commonly used in civil engineering could be made including normal distribution, lognormal distribution, Weibull distribution and gamma distribution. Then, the estimated PDF could be acquired by executing the K-S test. According to Ref. [21], there is only a little difference between PDFs using the mean estimated parameters of six sections and the parameters given by a maximum possibility criterion proposed by the authors. Therefore, in this paper, the PDF with the mean estimated parameters of six points could be referred to as the best estimate.

By conducting the probabilistic and statistical modeling on the constrained SFS, the 1D PDF of constrained SFS for each constituent of concrete could be obtained.

3.2 SFS from the Macro-Testing

Generally, complex global behaviors could be captured on the basis of the fiber bundle model whose individual element is endowed with a simple response (elastic-brittle prosperity shown in Figure 5). Under compressive loading, one of these fibers would fail when the overall strain exceeds the random SFS denoted by the random variable $\Delta_{2,s}(x)$, which could be considered to be a homogenous lognormal random field with the mean λ and the standard deviation ζ in Refs.[6-8]. According to Refs.[7 - 8], the damage of the fiber bundle represented by $d(\varepsilon)$ could be defined as follows

$$d(\varepsilon)=\int_0^1 H[\varepsilon-\Delta_{2,s}(x)]\mathrm{d}x \tag{27}$$

where $H[\cdot]$ is the Heaviside function, ε is the elastic strain, $\Delta_{2,s}(x)$ is the 1D random field for fracture strain, and x is the spatial coordinate of the meso-fiber. In the present work, the focus would be on the expected value of the damage variable $d(\varepsilon)$ given as

$$E[d]=F(\varepsilon) \tag{28}$$

where $F(\varepsilon)$ denotes the first-order cumulative distribution function of $\Delta_{2,s}(x)$.

The 1D expected stress and strain relationship could be expressed as follows

$$E[\sigma]=[1-F(\varepsilon)]E_0\varepsilon \tag{29}$$

where σ denotes the stress, and E_0 denotes the initial elastic modulus.

3.3 Multiscale Approach

Fig. 5 shows the topological structure, physical model and mechanical properties at three scales. At the nanoscale, a confined force chain buckling model was established for force-chain based materials, which could be extended to the mesoscale using the principle of resident potential energy, to investigate the relationship between the compressive strength and the hardness for the mesoscale concrete element. Meanwhile, at the mesoscale, the mesoscale parameters for concrete damage under compression could be not only from the identification combining model results with macro testing results but also from the direct experiment (nanoindentation). Then, further than that, the former one could be regarded as a traditional method: the parameters at the lower scale could be recognized provided that a reasonable model and the macroscale experiment results are given. The latter pays more attention to the verifiability of the model and the development of the microscale testing techniques. Actually, the fiber bundle model would be an optimal and convenient model to connect these three scales.

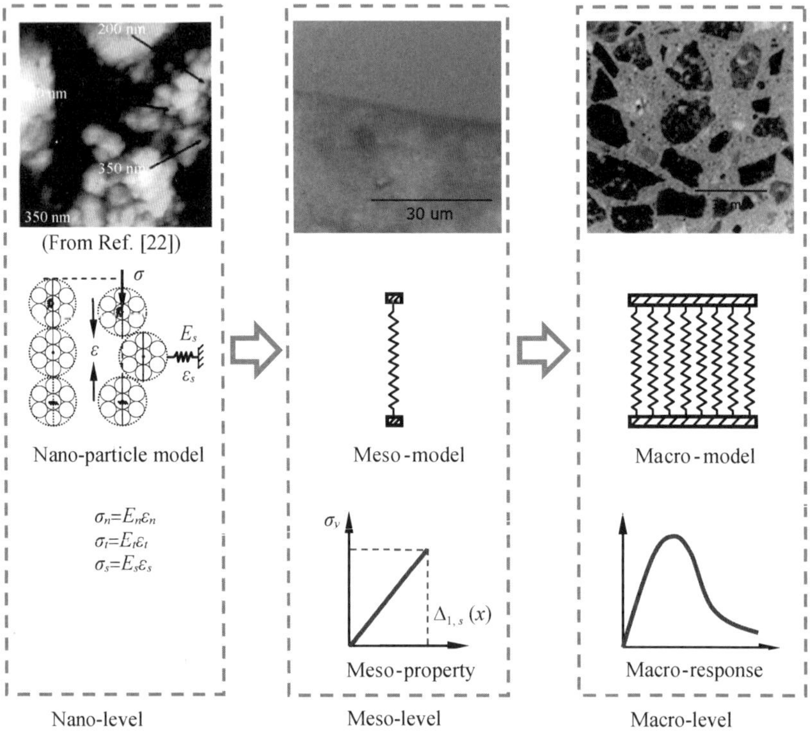

Fig. 5. Concrete modeling at the nano-, meso- and macro-scales

On the one hand, each concrete constituent (HCP, ITZ and aggregate) could be modeled as a random field using the observed values from nanoindentation tests. Performing "maximum possibility criterion"[21], the knowledge of PDF for constrained SFS Δ_{con} could be provided, with which SFS $\Delta_{1,s}(x)$ could also be obtained using Eq.(22)(in Section 3.1). Then, employing stochastic damage theory, the statistical characteristics of SFS $\Delta_{2,s}(x)$ could also be obtained including the mean and the standard deviation of a homogenous lognormal random field, by comparing the macroscale experimental stress and strain curves with the model results(in Section 3.2). Finally, the constraint factor in Eq.(22) could be recognized by comparing $\Delta_{1,s}(x)$ with $\Delta_{2,s}(x)$ through the relative entropy(in Section 4.3).

4 Results and Discussion

4.1 Concrete Damage SPI from Nanoindentation

Specifically, to investigate the SPI of concrete damage, the phase SPI could be generated firstly due to the obvious bound among each constituent of concrete. Combined with the phase SPI, concrete SPI could be reproduced by the reproduction technique[21].

4.1.1 Phase SPI and Random Field Theory

To characterize the probability distribution of properties from nanoindentation, three zones including HCP, ITZ and aggregate were selected randomly shown in Fig. 6(a) before nanoindentation. Following random field modeling of nanoindentation results, it leads to a sample number of 100 for HCP, aggregate and ITZ, respectively, with respect to the constrained SFSΔ_{con}. In Figure 6(b)—(d), they show the samples of the random field, and the corresponding mean and standard deviation for HCP, ITZ and aggregate, respectively. From the first and second-order statistical characteristics, the mean value and the standard deviation calculated both keep in constant. It is clear that the 1D random filed for the concrete constituent herein is belonging to a stationary random process. Notably, concrete constituents could be considered to be homogeneous from the probabilistic and statistical standpoint, although concrete is well-known as a kind of inhomogeneous material. Actually, concrete could also be regarded as homogenous, provided that one considers the complex materials as random media.

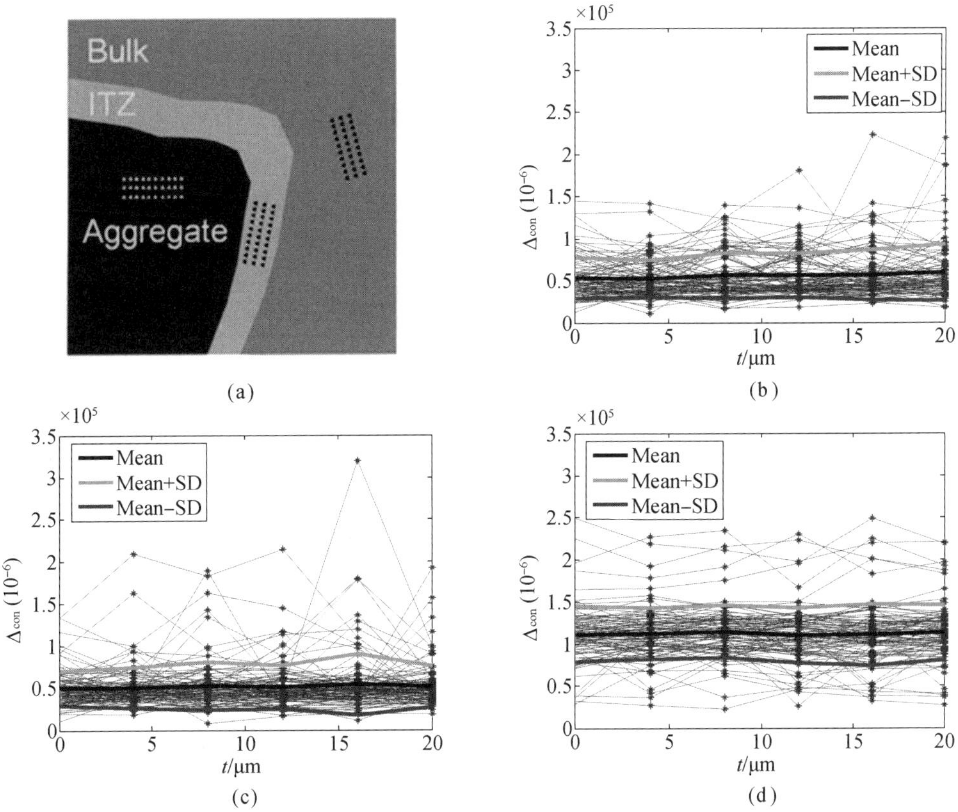

Fig. 6. 1D random filed observations: (a) the indent zones; the samples, the mean and the standard deviation for the constrained SFS of (b) HCP, (c) ITZ and (d) aggregate

By performing "maximum possibility criterion"[21], the 1D PDF for the random field could be obtained. It shows the optimal parameters in Table 1 for the 1D PDF of the random field by the simplex method[32-33]. Fig. 7(a)—(c) gives the histograms of the constrained SFS for each concrete constituent together with the theoretical PDF using the optimal parameters in Table 1. The comparisons between the histograms and the theoretical PDFs show better agreement.

Table 1. Optimal parameters for the 1D PDF

Constituents	Properties	Distribution Type	Optimal Parameters	
			Mean Value	Standard Deviation
HCP	Δ_{con}	Lognormal distribution	10.82	0.43
ITZ	Δ_{con}	Lognormal distribution	10.72	0.29
Agr	Δ_{con}	Normal distribution	111 730	23 910

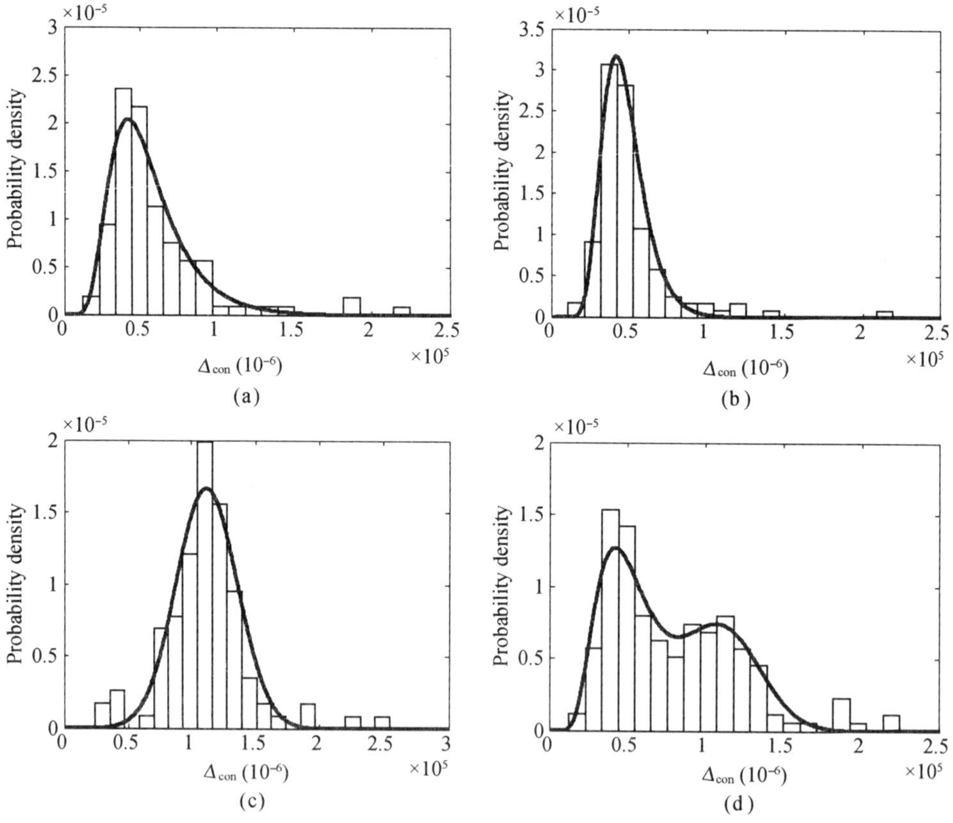

Fig. 7. Reconstruction of the constrained SFS for concrete: histograms and theoretical probability density curves for (a) HCP, (b) ITZ, (c) aggregate, and (d) concrete reconstructed with Equation(30)

4.1.2 Synthesis Technique of Concrete SPI

As mentioned above, each constituent of concrete materials exhibits randomness, which constitutes the complex, distributed disordered system. The PDF of concrete, regarded as a random media[21], could be directly obtained by adding the individual PDF, expressed as follows

$$\phi(x)=\sum_{i=1}^{n} f_i \phi_i(x) \quad i=1, 2, 3 \tag{30}$$

where x denotes the property of concrete or its constituent; f_i denotes the volume fraction of HCP, ITZ or aggregate; $\phi_i(x)$ denotes the PDF of the concrete constituent; $\phi(x)$ denotes the reconstructed concrete PDF. According to the previously reported result[21], a volume fraction ratio of HCP : aggregate : ITZ is around 0.611 : 0.387 : 0.002, and the corresponding random media distributions of concrete are displayed in Fig. 7(d). It is evident that the histograms and the theoretical PDFs of the constrained SFS are in close agreement with each other.

One may argue that the optimal distributions listed in Table 1 remain subjective. However, the theoretical distribution obtained could really model the primary characteristics for each constituent, even concrete. In the meantime, it is really a choice for engineers and researchers to statistically and probabilistically model concrete from a practical point of view. Especially, the distribution of concrete shown in Fig. 7(d) provides the physical reality for concrete damage under compression in Section 4.2.

4.2 Concrete Damage SPI from Macro-testing

To validate the PDF of SFS from nanoindentation, the behaviors of the total seven prism specimens measuring 0.1 m×0.1 m×0.3 m subjected to uniaxial compressive loading were investigated by MTS in this research. Fig. 8(a) shows the uniaxial test results of the concrete specimens at the macroscale. It is observed from Fig. 8(a) that all the stress-strain responses are displayed, and they appear randomly. Taking expectation of the stress with respect to the strain, the experimental mean stress and strain relationship could be also plotted. By comparing the theoretical mean stress-strain result from Eq. (29) with the

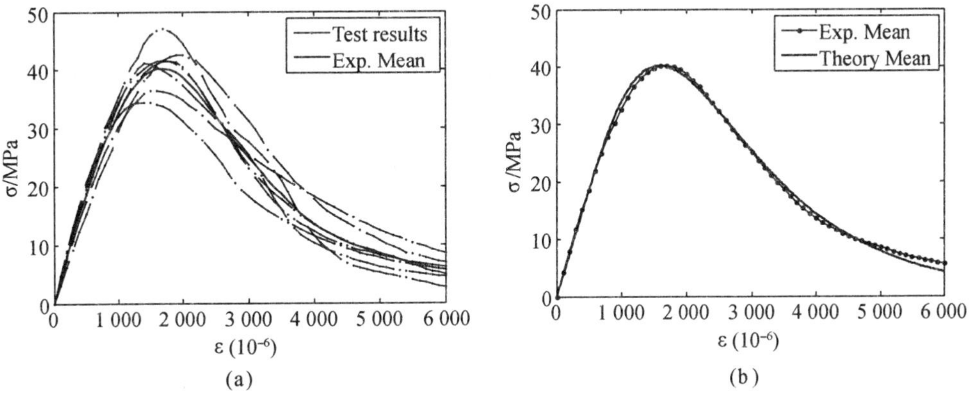

Fig. 8. Stress-strain curves under uniaxial compression: (a) the stress-strain curves of seven prisms and a mean stress vs. strain curve; and (b) comparison of expected experimental and theoretical results with the best-fit parameters

experimental result, the mesoscale parameters are found to be $E=37.00$ GPa, $\lambda=7.62$ and $\xi=0.52$. The theoretical and experimental results are displayed in Fig. 8 (b), which indicates that the theoretical result agrees well with the experimental one. From the failure patterns, the classic failure pattern with a major diagonal crack could be observed shown in Fig. 9.

Fig. 9. Photos of failed specimens

In the present paper, the focus would be on the mean stress and strain relationship, from which the parameters including the mean value and the standard deviation of the distribution could be recognized. However, the standard deviation stress and strain relationship could also be plotted in the figure, which would lead to the relative length of the random field for concrete materials. That is also a key point deserving further research.

4.3 Constraint Factor: Linking $\Delta_{1,s}$ with $\Delta_{2,s}$

For establishing the link between SFS $\Delta_{1,s}$ (with an unknown constraint factor C) and $\Delta_{2,s}$, the relative entropy theory could be employed to study on the similarity between these two distributions of $\Delta_{1,s}$ and $\Delta_{2,s}$. In statistics, the relative entropy could be regarded as a measure of the distinguishability between two probability density distributions, which is also called Kullback-Leibler divergence[34]. For the probability distributions of two discrete random variables P and Q, the relative entropy $D_{KL}(P \parallel Q)$ could be expressed as

$$D_{KL}(P \parallel Q)=\sum_i P(i)\ln\frac{P(i)}{Q(i)} \tag{31}$$

Assume that the random field in the stochastic damage model following a lognormal distribution is represented by P, and that from the nanoindentation by Q. Apparently, a distinct relative entropy $D_{KL}(P \parallel Q)$ would be calculated with different constraint factors, the smallest one of which means a minimum difference between P and Q. In other words, the constraint factor C meeting the minimum $D_{KL}(P \parallel Q)$ is an optimal value for concrete. Fig. 10(a) shows the relative entropy with different constraint factors. It is clear that when the constraint factor is very small, the result of Eq. (31) would approach zero. In other words, the constraint factor less than five is not of any meaning for concrete materials, despite the pseudo smaller relative entropy. While the relative entropy corresponding to the constraint factor $C = 33.12$ approaches the minimum, giving the knowledge that the discrepancy between these two PDFs is the minimum, namely, they are in good agreement. Moreover, the constraint factor identified agrees well with the previously reported results[11, 24]. Fig. 10(b) shows the detailed PDFs of concrete constituent together with the PDFs of $\Delta_{1,s}$ and $\Delta_{2,s}$ obtained above, It is easy to see that the distribution recognized from the macro-test is an ideal lognormal distribution, while the distribution reconstructed from the nano-test is a two-peak distribution, where the first peak is attributed to HCP and the second one is ascribed to the aggregate. It is interesting to note that the difference between probability density at around $\Delta = 3\ 400$ (the second peak) reaches a larger value, which is attributed to HCP and ITZ playing a more important role than the aggregate during damage evolution. From the failure patterns (Fig. 9), through the main crack it is also observed that a small amount of coarse aggregate was broken apart.

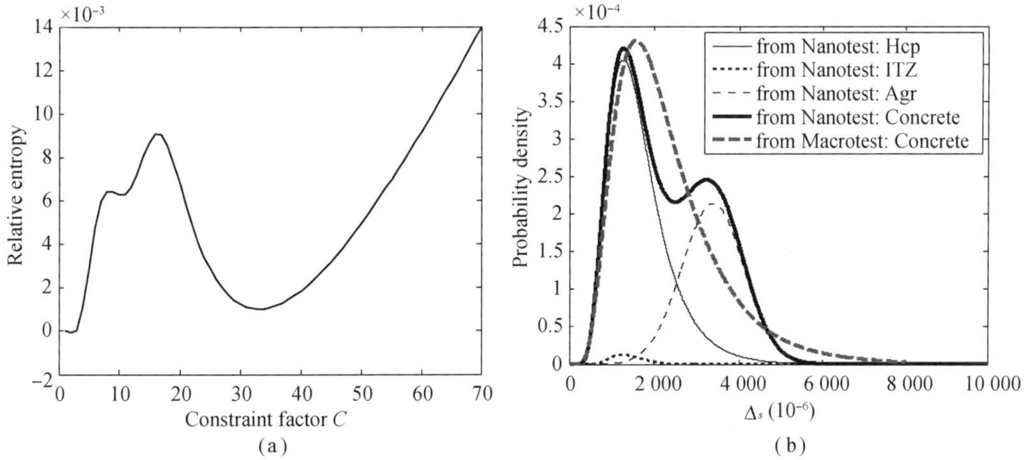

Fig. 10. SFS from the nano- and macro-testing: (a) the relative entropy vs. the constraint factor. (b) PDFs from two different scales corresponding to the smallest relative entropy

From the nanoscale to mesoscale, SFS could be generated based on the nanoindentation, which shows that the meso-fracture arises physically from the bulking of confined force chain system at the mesoscale. From the mesoscale to macroscale, SFS could be recognized based on the macro-testing, which provides the information that the damage of concrete is strongly dependent on the random fracture of mesoscale concrete element. In a word, the concrete damage under compression results from the random buckling of confined force chain system at the mesoscale.

5 Conclusions

In summary, this study sets a framework to investigate the damage of concrete under compression based on the SPI. Base on the multiscale research including experimental results, theoretical derivation and numerical analysis, the following conclusions could be drawn:

(1) The confined force chain buckling model proposed indicates the relationship between mesoscale strength and mesoscale hardness. The indentation hardness could be equivalent to the constrained compressive strength confined by the surrounding material around the indenter. The nanoindentation combined with the proposed model and random field theory provides direct access to the SPI of mesoscale fracture behavior of concrete. Meanwhile, the mesoscale fracture behavior of each constituent follows the homogeneous random field.

(2) Nanoindentation combined with macro-testing under compression could lead to the constraint factor linking two scales from mesoscale to macroscale, by comparing the difference between distributions of mesoscale fracture behavior from nano- and macro-testing. This multiscale method provides an effective way to investigate concrete damage under compression, to offer the physical reality of concrete damage evolution, and to estimate the effect of concrete constituents on damage evolution. Up to now, it is interesting to see that the nature of mechanical properties, e.g., strength, creep and damage, could be anticipated based on the SPI by using nanoindentation[17-20].

(3) At the mesoscale, the meso-fracture arises physically from the bulking of the confined force chain system. At the macroscale, the concrete damage is strongly dependent on the random fracture at the mesoscale. From mesoscale to macroscale, the accumulation of mesoscopic fracture results in the macroscopic damage. Notably, the mesoscale inhomogeneity and the mesoscale confined bulking are intrinsic to concrete, which may constitute the physical mechanism controlling concrete damage subjected to compression.

(4) Our investigation provides the possibility to control damage and to strengthen cementitious materials. Additionally, evaluating the macroscopic properties based on the SPI at the lower scales could be a feasible option. However, in the present paper, only one concrete mix was studied. The concrete materials with higher and smaller water to cement ratios still deserve further investigation and thus the effect of concrete constituents, especially the aggregate on damage evolution could be systematically investigated in the future. Reasonable speculation could be given that the greater the concrete strength, the smaller difference in the second peak in Figure 10b, and vice versa.

Funding:

This research was funded by National Key R&D Program of China(grant No. 2017TFC0702900), National Science Foundation of China for Major Project (grant No. 51261120374,

91315301), National Science Foundation of China (grant No. 51808499) and Science Foundation of Zhejiang Province of China(grant No. LQ18E080009).

Acknowledgments

The authors would like to acknowledge the kindly help of Southeast University in Nanjing, China, and the helpful suggestion of Yamei Zhang, Weiwei Zhu and Wei Huang. The authors also thank the anonymous reviewers for the constructive suggestions.

References

[1] BREYSSE D. Probabilistic formulation of damage-evolution law of cementitious composites[J]. Journal of Engineering Mechanics, 1990, 116(7): 1489-1510.

[2] KACHANOV L M. Rupture time under creep conditions[J]. International Journal of Fracture, 1999, 97(1): 11-18.

[3] DOUGILL J W. On stable progressively fracturing solids[J]. Zeitschrift für angewandte Mathematik und Physik ZAMP, 1976, 27(4): 423-437.

[4] MAZARS J. A description of micro-and macroscale damage of concrete structures[J]. Engineering Fracture Mechanics, 1986, 25(5-6): 729-737.

[5] WU J Y, LI J, FARIA R. An energy release rate-based plastic-damage model for concrete[J]. International Journal of Solids and Structures, 2006, 43(3-4): 583-612.

[6] KANDARPA S, KIRKNER D J, SPENCER Jr B F. Stochastic damage model for brittle materials subjected to monotonic loading[J]. Journal of Engineering Mechanics, 1996, 122(8): 788-795.

[7] LI J, REN X D. Stochastic damage model for concrete based on energy equivalent strain[J]. International Journal of Solids and Structures, 2009, 46(11-12): 2407-2419.

[8] REN X D, LI J. A unified dynamic model for concrete considering viscoplasticity and rate-dependent damage[J]. International Journal of Damage Mechanics, 2013, 22(4): 530-555.

[9] BAŽANT Z P, PANG S D. Activation energy based extreme value statistics and size effect in brittle and quasibrittle fracture[J]. Journal of the Mechanics and Physics of Solids, 2007, 55(1): 91-131.

[10] FELDMAN R F, HUANG C Y. Properties of Portland cement-silica fume pastes II [J]. Mechanical Properties. Cement and Concrete Research, 1985, 15: 943-952.

[11] IGARASHI S, BENTUR A, MINDESS S. Characterization of the microstructure and strength of cement paste by microhardness testing[J]. Advances in Cement Research, 1996, 8(30): 87-92.

[12] ZHU W, BARTOS P J M. Assessment of interfacial microstructure and bond properties in aged GRC using a novel microindentation method[J]. Cement and Concrete Research, 1997, 27(11): 1701-1711.

[13] WESTBROOK J H. The science of hardness testing and its research applications[J]. American Society for Metals: Metal Park, OH, USA, 1973: 453-491.

[14] DURST K, GÖKEN M, VEHOFF H. Finite element study for nanoindentation measurements on two-phase materials[J]. Journal of Materials Research, 2004, 19(1): 85-93.

[15] CONSTANTINIDES G, ULM F J, VAN VLIET K. On the use of nanoindentation for cementitious materials[J]. Materials and Structures, 2003, 36(3): 191-196.

[16] DEJONG M J, ULM F J. The nanogranular behavior of CSH at elevated temperatures (up to 700 C) [J]. Cement and Concrete Research, 2007, 37(1): 1-12.

[17] ULM F J, VANDAMME M, BOBKO C, et al. Statistical indentation techniques for hydrated nanocomposites: concrete, bone, and shale[J]. Journal of the American Ceramic Society, 2007, 90(9): 2677-2692.

[18] TAI K, ULM F J, ORTIZ C. Nanogranular origins of the strength of bone[J]. Nano Letters, 2006, 6(11): 2520-2525.

[19] GANNEAU F P, CONSTANTINIDES G, ULM F J. Dual-indentation technique for the assessment of strength properties of cohesive-frictional materials[J]. International Journal of Solids and Structures, 2006, 43(6): 1727-1745.

[20] VANDAMME M, ULM F J. Nanogranular origin of concrete creep[J]. Proceedings of the National Academy of Sciences, 2009, 106(26): 10552-10557.

[21] LIU H, REN X D, LI J. Indentation tests based multi-scale random media modeling of concrete[J]. Construction and Building Materials, 2018, 168: 209-220.

[22] MONDAL P, SHAH S P, MARKS L D. Nanoscale characterization of cementitious materials[J]. ACI Materials Journal, 2008, 105(2): 174.

[23] ALKHATEB H, AL-OSTAZ A, CHENG A H D, et al. Materials genome for graphene-cement nanocomposites[J]. Journal of Nanomechanics and Micromechanics, 2013, 3(3): 67-77.

[24] OLIVER W C, PHARR G M. An improved technique for determining hardness and elastic modulus using load and displacement sensing indentation experiments[J]. Journal of Materials Research, 1992, 7(6): 1564-1583.

[25] PHARR G M, OLIVER W C, BROTZEN F R. On the generality of the relationship among contact stiffness, contact area, and elastic modulus during indentation[J]. Journal of Materials Research, 1992, 7(3): 613-617.

[26] CONSTANTINIDES G, ULM F J. The nanogranular nature of C-S-H[J]. Journal of the Mechanics and Physics of Solids, 2007, 55(1): 64-90.

[27] TORDESILLAS A, MUTHUSWAMY M. On the modeling of confined buckling of force chains[J]. Journal of the Mechanics and Physics of Solids, 2009, 57(4): 706-727.

[28] FISCHER-CRIPPS A C. Nanoindentation[M]. 2nd ed. Springer: New York, USA, 2011.

[29] KHOLMYANSKY M, KOGAN E, KOVLER K. On the hardness determination of fine grained concrete[J]. Materials and Structures, 1994, 27(10): 584-587.

[30] VANMARCKE E. Random Fields: Analysis and Synthesis[M]. The MIT Press: London, UK, 2010.

[31] TORQUATO S. Random Heterogeneous Materials: Microstructure and Macroscopic Properties[M]. Springer: New York, NY, USA, 2002.

[32] NELDER J A, MEAD R. A simplex method for function minimization[J]. The Computer Journal, 1965, 7(4): 308-313.

[33] MATHEWS J H, FINK K D. Numerical methods using MATLAB[M]. Upper Saddle River, NJ: Pearson Prentice Hall, 2004.

[34] SOBEZYK K, TRĘBICKI J. Maximum entropy principle in stochastic dynamics[J]. Probabilistic Engineering Mechanics, 1990, 5(3): 102-110.

Stochastic damage model for concrete based on energy equivalent strain①

Jie Li[1,2], **Xiaodan Ren**[1]

1 School of Civil Engineering, Tongji University, 1239 Siping Road, Shanghai 200092, China
2 The State Key Laboratory on Disaster Reduction in Civil Engineering, Tongji University, Shanghai 200092, China

Abstract Starting with the framework of conventional elastoplastic damage mechanics, a class of stochastic damage constitutive model is derived based on the concept of energy equivalent strain. The stochastic damage model derived from the parallel element model is adopted to develop the uniaxial damage evolution function. Based on the expressions of damage energy release rates (DERRs) conjugated to the damage variables thermodynamically, the concept and its tensor formulations of energy equivalent strain is proposed to bridge the gap between the uniaxial and the multiaxial constitutive models. Furthermore, a simplified coupling model is proposed to consider the evolution of plastic strain. And the analytical expressions of the constitutive model in 2-D are established from the abstract tensor expression. Several numerical simulations are presented against the biaxial loading test results of concrete, demonstrating that the proposed models can reflect the salient features for concrete under uniaxial and biaxial loading conditions.
Keywords Damage mechanics; Stochastic model; Energy equivalent strain; Concrete; Plastic model

1 Introduction

Numerical simulation of concrete structures has attracted interests for over four decades. Meanwhile almost all the breakthroughs of solid mechanics have been adopted to construct or reconstruct analytical models of concrete materials. However, except for several remarkable progress and achievements, the constitutive modeling and nonlinear analysis of concrete structures are still quite challenging problems. The elaborate modeling of concrete performances, which is able to reproduce the complete range of mechanical properties of concrete subjected to external effects in a consistent manner, remains a somewhat controversial subject.

In the recent 20 years, the formulation of continuum damage mechanics (CDM) has been introduced to nonlinear constitutive modeling of concrete materials. Due to its solid foundation of irreversible thermodynamics and relevant consideration of physical mechanisms, CDM provides a powerful framework for the construction of constitutive models for concrete. Furthermore, in order to capture the essential characteristics of

---

① Originally published in *International Journal of Solids and Structures*, 2009, 46(11-12): 2407-2419.

concrete bearing damage and failure, e.g. the different behaviors under tension and compression, the unilateral effect, the residual strain after unloading, the strain stiffening in tension and stress softening in compression, several creative techniques were introduced to reform the classical framework of CDM. For example, Mazars[1-2] proposed a bi-scalar model to represent the degradation of concrete under tensile and compressive loading, respectively. Ju[3] developed a coupled plastic damage model to describe the strengthening of concrete multiaxial compression and remnant strain in repeated loading. Faria et al.[4] proposed the split of the effective stress tensor to account for different nonlinear performances of concrete under tension and compression in cyclic loading. And Wu et al.[5] and Wu and Li[6] introduced the formulation of elasto-plastic damage release rate and developed a framework of plastic damage model for concrete. However, CDM just provides the theoretical framework of constitutive modeling for a wide range of softening materials. For a specific material like concrete, its damage evolution function could not be proposed within the framework of CDM. The principle of irreversible thermodynamics, which is the foundation of CDM, just specifies the necessary condition of damage evolution by using inequality. Hence the empirical damage evolution functions are widely adopted in most of the CDM based material models, in which many empirical parameters without clear physical meanings are required.

Another possible approach for development of damage evolution function is the micro-mechanical models, which take into account the micro-flaws and cracks in the solids. Actually, concrete is a kind of complex composite material made up of water, cements, aggregates and admixtures. During its solidification process, randomly scattered micro-cracks and voids are nucleated due to the shrinkage of cement matrix and evaporation of pore water. When subjected to external loads the evolution and the propagation of the initial damage result in the nonlinearity of the stress-strain relationship of concrete. On the other hand, due to the random distribution of the aggregates and micro-flaws, the stochastic damage evolution for concrete should be considered and accordingly the stochastic damage model for concrete should be developed through the micro-mechanical approach. The most widely used stochastic micro-mechanical model is the parallel element model. In the early literatures,[7-10] this model was proposed to describe the progressive damage propagation of a brittle rod subjected to uniaxial loading. According to this model, it is assumed that the stiffness of all the elements is identical whereas the rupture strengths are independent and identically distributed(i.i.d.) random variables. The monotonic load-displacement behavior of the brittle materials can then be calculated through the stochastic averaging approaches. In 1996, Kandarpa et al.[11] proposed an extended model to represent the randomness of strength and the load-displacement relationship of brittle materials. In this model, the failure strengths of the micro-elements are modeled as a 1-D random field. With proper selection of the distribution for the rupture strength and the spatial correlation for the random field, the statistics including the mean value and variance of the load-displacement relationships are analytically derived in the macro scale. Li and Zhang[12] and Li et al.[13] defined the fracture strain of the element as a 1-D random field, based on which the uniaxial stochastic damage model for concrete was developed. However, under the multi-axial loading condition, it is hard to derive the compact analytical expressions of damage evolution within the frame of micromechanical theory.

The present paper aims at providing a novel stochastic damage model of concrete that is established based on the framework of CDM with the damage evolution function developed through the stochastic micro-mechanical approach. The present model could not only reflect the nonlinear behaviors but also capture the stochastic features of concrete. Moreover, it fits well with the experimental observations including the nonlinear stress-strain curves and stochastic fluctuation measures.

In Section 2, an energy release rate based plastic damage model is adopted as the continuum framework. According to this model, a tensile damage variable and a shear damage variable leading to a fourth-order damage tensor are chosen to represent the degradation of the mechanical properties of concrete. A decomposition of the effective stress tensor is presented thereafter to define an elastic Helmholtz free energy (HFE), after which the plastic-damage constitutive relation with internal variables is derived. Additionally, the theoretical as well as empirical evolution laws of plastic strain are proposed. In Section 3, the stochastic damage evolution function is developed based on the parallel element model. Both the mean value and standard deviation for damage evolution are derived. Section 4 is devoted to developing the expressions of energy equivalent strain to bridge the gap between the continuum damage framework and the micro-mechanics based stochastic damage evolution function. In order to improve the practical applicability of this model, analytical expressions are derived from the abstract tensor representation in Section 5. The numerical results in Section 6 reveal that the proposed stochastic damage model in this paper could reflect the multi-dimensional nonlinearity and the randomness of concrete as well as fit well with the experimental observations.

2 Continuum damage framework

2.1 Damage variables and their expressions

According to the principle of strain equivalence[3, 14], the effective stress tensor $\bar{\boldsymbol{\sigma}}$ in damaged material may be described as

$$\bar{\boldsymbol{\sigma}} = \mathbf{C}_0 : \boldsymbol{\varepsilon}^e = \mathbf{C}_0 : (\boldsymbol{\varepsilon} - \boldsymbol{\varepsilon}^p) \tag{1}$$

where $\mathbf{C}_0$ denotes the usual fourth order isotropic linear-elastic stiffness; $\boldsymbol{\varepsilon}$, $\boldsymbol{\varepsilon}^e$ and $\boldsymbol{\varepsilon}^p$ are rank-two tensors, denoting the strain tensor and its elastic and plastic components, respectively.

To account for the unilateral effect, the effective strain tensor is decomposed as follows:

$$\bar{\boldsymbol{\sigma}} = \bar{\boldsymbol{\sigma}}^+ + \bar{\boldsymbol{\sigma}}^- \tag{2}$$

$$\bar{\boldsymbol{\sigma}}^{+}=\mathbf{P}^{+}:\bar{\boldsymbol{\sigma}} \tag{3}$$

$$\bar{\boldsymbol{\sigma}}^{-}=\bar{\boldsymbol{\sigma}}-\bar{\boldsymbol{\sigma}}^{+}=\mathbf{P}^{-}:\bar{\boldsymbol{\sigma}} \tag{4}$$

where the fourth-order projection tensors $\mathbf{P}^{+}$ and $\mathbf{P}^{-}$ are expressed as[4]

$$\mathbf{P}^{+}=\sum_{i}H(\hat{\bar{\sigma}}_{i})(p_{i}\otimes p_{i}\otimes p_{i}\otimes p_{i}) \tag{5}$$

$$\mathbf{P}^{-}=\mathbf{I}-\mathbf{P}^{+} \tag{6}$$

in which $\mathbf{I}$ is the forth order identity tensor; $\hat{\bar{\sigma}}_{i}$ and p_{i} denote the ith eigenvalue and eigenvector of the effective stress tensor $\bar{\boldsymbol{\sigma}}$, respectively; $H(\cdot)$ is the Heaviside function, which is defined as

$$H(x)=\begin{cases}0 & x<0\\ 1 & x\geqslant 0\end{cases} \tag{7}$$

To establish the intended constitutive law, an elastic Helmholtz free energy (HFE for short) potential should be introduced as function of the free and internal variables. The initial elastic HFE potential ψ_{0}^{e} is defined as the elastic strain energy. It can be written as the summation of its positive and negative components ($\psi_{0}^{\mathrm{e}+}$, $\psi_{0}^{\mathrm{e}-}$) considering the decomposition of the effective stress tensor in Eqs. (2)—(4)

$$\psi_{0}^{\mathrm{e}}=\frac{1}{2}\bar{\boldsymbol{\sigma}}:\boldsymbol{\varepsilon}^{\mathrm{e}}=\frac{1}{2}\bar{\boldsymbol{\sigma}}^{+}:\boldsymbol{\varepsilon}^{\mathrm{e}}+\frac{1}{2}\bar{\boldsymbol{\sigma}}^{-}:\boldsymbol{\varepsilon}^{\mathrm{e}}=\psi_{0}^{\mathrm{e}+}+\psi_{0}^{\mathrm{e}-} \tag{8}$$

where the superscript "e" refers to "elastic" whereas the subscript "0" refers to "initial" states. Considering the tensile and shear mechanisms for degradation of the macro-mechanical properties under tensile and compressive loading conditions, two damage scalars, d^{+} and d^{-}, are adopted here and therefore the elastic HFE with the form

$$\psi^{\mathrm{e}}(\boldsymbol{\varepsilon}^{\mathrm{e}}, d^{+}, d^{-})=\psi^{\mathrm{e}+}(\boldsymbol{\varepsilon}^{\mathrm{e}}, d^{+})+\psi^{\mathrm{e}-}(\boldsymbol{\varepsilon}^{\mathrm{e}}, d^{-}) \tag{9}$$

can be postulated, where $\psi^{\mathrm{e}+}$ and $\psi^{\mathrm{e}-}$ are defined as

$$\psi^{\mathrm{e}\pm}(\boldsymbol{\varepsilon}^{\mathrm{e}}, d^{\pm})=(1-d^{\pm})\psi_{0}^{\mathrm{e}\pm} \tag{10}$$

with symbol "$\pm$" denoting "$+$" or "$-$" as appropriate.

The total elasto-plastic HFE potential could be defined as the sum of the elastic component ψ^{e} and plastic component ψ^{p}, [5, 15] that is

$$\psi(\boldsymbol{\varepsilon}^{\mathrm{e}}, \boldsymbol{\kappa}, d^{+}, d^{-})=\psi^{\mathrm{e}}(\boldsymbol{\varepsilon}^{\mathrm{e}}, d^{+}, d^{-})+\psi^{\mathrm{p}}(\boldsymbol{\varepsilon}^{\mathrm{e}}, \boldsymbol{\kappa}, d^{+}, d^{-}) \tag{11}$$

where $\boldsymbol{\kappa}$ denotes a suitable set of plastic variables; the plastic HFE potential ψ^{p} is defined as

$$\psi^{\mathrm{p}}(\boldsymbol{\varepsilon}^{\mathrm{e}}, \boldsymbol{\kappa}, d^{+}, d^{-})=\psi^{\mathrm{p}}(\boldsymbol{\varepsilon}^{\mathrm{e}}, \boldsymbol{\kappa}, d^{-})=(1-d^{-})\psi_{0}^{\mathrm{p}}=(1-d^{-})\int_{0}^{\boldsymbol{\varepsilon}^{\mathrm{p}}}\bar{\boldsymbol{\sigma}}^{-}:\mathrm{d}\boldsymbol{\varepsilon}^{\mathrm{p}} \tag{12}$$

According to the second principle of thermodynamics, any arbitrary irreversible process satisfies the Clausius-Duheim inequality, of which the reduced form is

$$\dot{\gamma} = -\dot{\psi} + \boldsymbol{\sigma} : \dot{\boldsymbol{\varepsilon}} \geqslant 0 \tag{13}$$

Differentiating both sides of Eq.(11) with respect to time yields

$$\dot{\psi} = \frac{\partial \psi^{e}}{\partial \boldsymbol{\varepsilon}^{e}} : \dot{\boldsymbol{\varepsilon}}^{e} + \frac{\partial \psi}{\partial d^{+}} \dot{d}^{+} + \frac{\partial \psi}{\partial d^{-}} \dot{d}^{-} + \frac{\partial \psi^{p}}{\partial \boldsymbol{\kappa}} \cdot \bar{\boldsymbol{\kappa}} \tag{14}$$

Taking the assumption that the damage and plastic unloading are elastic processes, for any admissible process the following conditions have to be fulfilled:

$$\boldsymbol{\sigma} = \frac{\partial \psi^{e}}{\partial \boldsymbol{\varepsilon}^{e}} \tag{15}$$

$$\dot{\gamma}^{d} = -\left(\frac{\partial \psi}{\partial d^{+}} \dot{d}^{+} + \frac{\partial \psi}{\partial d^{-}} \dot{d}^{-} \right) \geqslant 0 \tag{16}$$

$$\dot{\gamma}^{p} = \boldsymbol{\sigma} : \dot{\boldsymbol{\varepsilon}}^{p} - \frac{\partial \psi^{p}}{\partial \boldsymbol{\kappa}} \cdot \dot{\boldsymbol{\kappa}} \geqslant 0 \tag{17}$$

From Eq. (15) it can be clearly seen that here the Cauchy stress tensor is only dependent on the elastic HFE potential, which is a variant to Faria et al.[4] where the total one are considered. Substituting the elastic HFE potential in Eqs.(8)—(11) in Eq. (15) leads to

$$\boldsymbol{\sigma} = (1 - d^{+})\mathbf{P}^{+} : \bar{\boldsymbol{\sigma}} + (1 - d^{-})\mathbf{P}^{-} : \bar{\boldsymbol{\sigma}} = (\mathbf{I} - d^{+}\mathbf{P}^{+} - d^{-}\mathbf{P}^{-}) : \bar{\boldsymbol{\sigma}} \tag{18}$$

It is then possible to obtain a final form for the constitutive law, which is a very visual expression for the Cauchy stress tensor $\boldsymbol{\sigma}$ [15]

$$\boldsymbol{\sigma} = (\mathbf{I} - \mathbf{D}) : \bar{\boldsymbol{\sigma}} = (\mathbf{I} - \mathbf{D}) : \mathbf{C}_0 : \boldsymbol{\varepsilon}^{e} = (\mathbf{I} - \mathbf{D}) : \mathbf{C}_0 : (\boldsymbol{\varepsilon} - \boldsymbol{\varepsilon}^{p}) \tag{19}$$

where the fourth-order damage tensor $\mathbf{D}$ is given by

$$\mathbf{D} = d^{+}\mathbf{P}^{+} + d^{-}\mathbf{P}^{-} \tag{20}$$

From Eq. (16), the tensile and the shear damage energy release rate (DERR) Y^{+} and Y^{-}, conjugated to the corresponding damage variables, can be expressed as

$$Y^{\pm} = -\frac{\partial \psi}{\partial d^{\pm}} \tag{21}$$

Eq. (21) shows that the DERRs depend on the total elasto-plastic HFE potential, not just on the elastic one as in the classical damage model of Lemaitre[14]. Further, substituting Eqs. (10)—(12) in Eq.(21), one obtains

$$Y^{+} = \psi_0^{e+} \tag{22}$$

$$Y^{-} = \psi_0^{e-} + \psi_0^{p} \tag{23}$$

Thus the coupled plastic damage model can be proposed based on definition of elasto-plastic DERRs.

2.2 Evolution model for plastic strain

Different types of plasticity models combined with damage expressions have been proposed in the literatures. Among them one family of models relies on the stress-based plasticity formulated in the Cauchy stress space, say Simo and Ju[16], Lubliner et al.[17] and Carol et al.[18] However, the Cauchy stress may descend due to the softening property of concrete, which will result in local concave and global shrinkage of the plastic potential function. It is difficult to conduct reliable numerical simulation under this situation. Another family of models is based on the plasticity formulated in the effective stress space, say Ju, Lee and Fenves, Faria et al., Li and Wu and Jason et al.[3-4, 15, 19-20] The effective stress represents the micro-stress on the undamaged material. Hence the plastic potential function will keep expanding throughout the loading process and thus it is convenient to develop reliable numerical algorithms for this model. According to the "effective stress space plasticity", the evolution law of plastic strain is expressed as follows:

$$\dot{\boldsymbol{\varepsilon}}^{\mathrm{p}}=\dot{\lambda}^{\mathrm{p}}\frac{\partial F^{\mathrm{p}}}{\partial\bar{\boldsymbol{\sigma}}} \tag{24}$$

$$\dot{\boldsymbol{\kappa}}=\dot{\lambda}^{\mathrm{p}}\mathbf{H} \tag{25}$$

$$F^{\mathrm{p}}\leqslant 0,\ \dot{\lambda}^{\mathrm{p}}\geqslant 0,\ \dot{\lambda}^{\mathrm{p}}F^{\mathrm{p}}\leqslant 0 \tag{26}$$

where F^{p} is the plastic potential, which is the yield function in the associated flow rule; λ^{p} is the plastic flow parameter; **H** denotes the vectorial hardening function.

The Drucker-Prager function is adopted as the plastic potential function as follows:

$$F^{\mathrm{p}}=\alpha\bar{I}_1+\sqrt{3\bar{J}_2} \tag{27}$$

where $\bar{I}_1$ is the first invariants of $\bar{\boldsymbol{\sigma}}$; $\bar{J}_2$ is the second invariants of $\bar{\boldsymbol{s}}$, the deviatoric components of $\bar{\boldsymbol{\sigma}}$; α is material parameter.

Considering Eq. (24), the initial plastic HFE potential could be expressed as

$$\psi_0^{\mathrm{p}}=\int_0^{\varepsilon^{\mathrm{p}}}\bar{\boldsymbol{\sigma}}^{-}:\mathrm{d}\boldsymbol{\varepsilon}^{\mathrm{p}}=\int_0^{\lambda^{\mathrm{p}}}\bar{\boldsymbol{\sigma}}^{-}:\frac{\partial F^{\mathrm{p}}}{\partial\bar{\boldsymbol{\sigma}}}\mathrm{d}\lambda^{\mathrm{p}} \tag{28}$$

Then substituting Eq. (27) in Eq. (28) yields

$$\begin{aligned}\psi_0^{\mathrm{p}}&=\int_0^{\lambda^{\mathrm{p}}}\bar{\boldsymbol{\sigma}}^{-}:\frac{\partial(\alpha\bar{I}_1+\sqrt{3\bar{J}_2})}{\partial\bar{\boldsymbol{\sigma}}}\mathrm{d}\lambda^{\mathrm{p}}\\&=\frac{b}{2E_0}\left(3\bar{J}_2^{-}+\alpha\bar{I}_1^{-}\cdot\sqrt{3\bar{J}_2}-\frac{1}{2}\bar{I}_1^{+}\bar{I}_1^{-}\right)\geqslant 0\end{aligned} \tag{29}$$

where $b=\frac{4}{3}\lambda^{\mathrm{p}}E_0/\sqrt{\bar{\boldsymbol{s}}:\bar{\boldsymbol{s}}}\geqslant 0$ is defined as the material parameter.

Wu et al.[5] proposed a vectorial hardening function as follows

$$\mathbf{H}=\mathrm{diag}[w,-(1-w)]\left\{\frac{\partial F^{\mathrm{p}}}{\partial\bar{\boldsymbol{\sigma}}}\right\} \tag{30}$$

where diag [•] is a diagonal matrix; w is the weight factor expressed as

$$w=\frac{\sum_{i=1}^{3}\langle\bar{\sigma}_i\rangle}{\sum_{i=1}^{3}|\bar{\sigma}_i|} \tag{31}$$

Here $\bar{\sigma}_i$ is the ith eigenvalue of the effective stress $\bar{\boldsymbol{\sigma}}$; the symbol $\langle \cdot \rangle$ is the Macaulay bracket defined as

$$\langle x \rangle = (x + |x|)/2 \tag{32}$$

Although the "effective stress space plasticity" provides a strict framework to represent the evolution of plastic strain, its numerical implementation is sometimes time consuming and unstable during iterative solving process. Therefore, the above theoretical plastic model is usually just adopted to establish the expression of plastic HFE potential. In practical applications, on the other hand, some empirical plastic models which were usually more tractable were adopted to calculate the numerical values of plastic strain. For example, Dahlblom and Ottosen[21] introduced a fraction δ of the maximum developed principal strain. The plastic strain takes the form

$$\boldsymbol{\varepsilon}^{\mathrm{p}} = \delta \boldsymbol{\varepsilon}_{\max} \tag{33}$$

Faria et al.[4] proposed an evolution law for the plastic strain

$$\dot{\boldsymbol{\varepsilon}}^{\mathrm{p}} = \beta E_0 H(\dot{d}^-) \frac{\langle \bar{\boldsymbol{\sigma}} : \dot{\boldsymbol{\varepsilon}} \rangle}{\bar{\boldsymbol{\sigma}} : \bar{\boldsymbol{\sigma}}} : \mathbf{C}_0^{-1} : \bar{\boldsymbol{\sigma}} \tag{34}$$

where besides the Young's modulus E_0, a material parameter $\beta > 0$ is introduced in order to control the rate intensity of plastic deformation. The Macaulay brackets enable one to set a non-negative value for the product $\bar{\boldsymbol{\sigma}} : \dot{\boldsymbol{\varepsilon}}$.

Considering the coupling of damage and plasticity, we propose a practical plastic evolution model taking the form

$$\boldsymbol{\varepsilon}^{\mathrm{p}} = \mathbf{C}_0^{-1} : \mathbf{F}(\mathbf{D}) : \mathbf{C}_0 : \boldsymbol{\varepsilon} \tag{35}$$

where $\mathbf{F}(\mathbf{D})$ is a forth-order tensor function of damage scalars and takes the form

$$\mathbf{F}(\mathbf{D}) = f_{\mathrm{p}}^{+}(d^{+})\mathbf{P}^{+} + f_{\mathrm{p}}^{-}(d^{-})\mathbf{P}^{-} \tag{36}$$

$f_{\mathrm{p}}^{\pm}(d^{\pm})$ are the scalar functions for the corresponding damage parameter, and meet the conditions of

$$\begin{cases} f_{\mathrm{p}}^{\pm}(x) = 0 & \text{if} \quad x = 0 \\ f_{\mathrm{p}}^{\pm}(x) < 1 & \text{if} \quad x = 1 \\ f_{\mathrm{p}}^{\pm}(x_1) \geqslant f_{\mathrm{p}}^{\pm}(x_2) & \text{if} \quad x_1 > x_2 \end{cases} \tag{37}$$

In the present paper a linear expression for the scalar function $f_{\mathrm{p}}^{\pm}(d^{\pm})$ is adopted as follows

$$f_{\mathrm{p}}^{\pm}(d^{\pm}) = \xi_{\mathrm{p}}^{\pm} d^{\pm} \tag{38}$$

where the plastic parameter $\xi_{\mathrm{p}}^{\pm}$ matches the criteria

$$0 \leqslant \xi_{\mathrm{p}}^{\pm} < 1. \tag{39}$$

3 Damage evolution function

Although the above macro-continuum damage mechanics framework could provide a rational procedure to describe the mechanical behavior of concrete, however, the model cannot explain the damage evolution rule rationally. Especially, the framework cannot answer such a problem: Why does the curve of damage evolution always exhibit some kind of nonlinearity? In order to recognize the nonlinearity in the damage evolution process, detailed analysis based on the micro-level investigation is required. In this section, the micro-parallel element model is introduced firstly. By introducing the micro-fracture strain as a basic random variable, a type of stochastic damage evolution function is developed.

3.1 Uniaxial stochastic damage model

According to Kandarpa et al[11]. and Li and Zhang[12], a structural element in uniaxial loading condition can be idealized as a series of micro-elements jointed in parallel (Fig. 1). The elements are linked with rigid bar on the ends so that each of them bears uniform deformation during the loading process. It is observed that the model has two scales: ① the micro-scale and, ② the macro or structural scale. The individual element represents the micro-properties of the material and the element system describes the macro response. Therefore, complex macro-material behaviors can be obtained based on the parallel system in which the individual element is endowed with simple material properties.

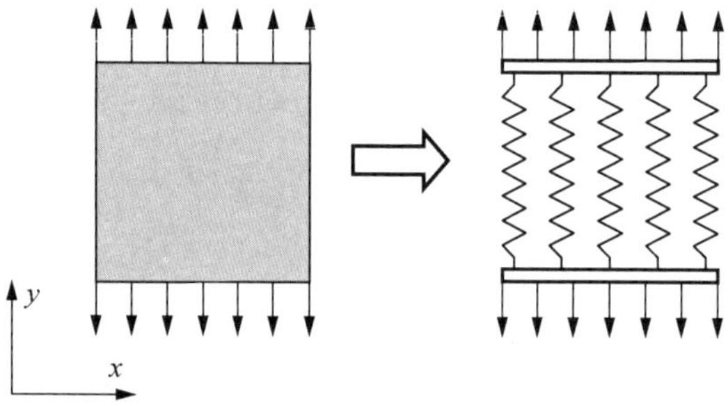

Fig. 1. Parallel element model

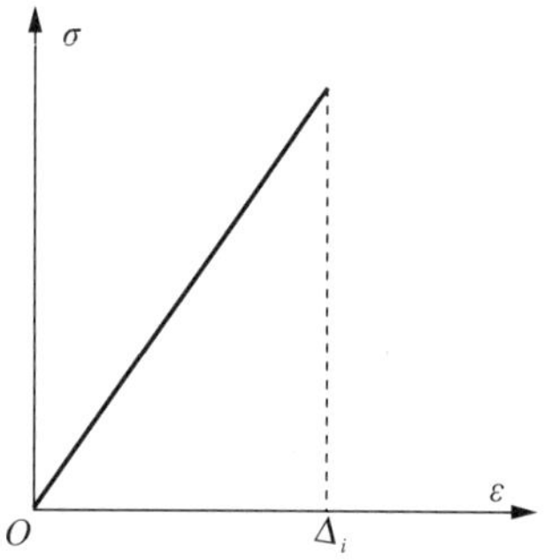

Fig. 2. Stress-strain relationships of micro-elements

Let N denote the total number of micro-elements. The spatial coordinate of the ith element is $i/N(i=1, 2, \cdots, N-1)$. The stress-strain relationship is assumed as perfect elasto-brittle type with random fracture strain Δ_i (Fig. 2).

In this model, the nucleation and propagation of micro-cracks inside concrete are simulated by sequential fracture of micro-elements throughout the entire loading process. Fig. 3 illustrates the stress-strain responses of the discrete bundle. It is observed that the bundle exhibits linear stress-strain behavior at the very beginning. Then the fracture of a micro-element leads to the local dropdown of stress-strain curve. Before the next fracture, the stress-strain curve remains linear with a degraded slope, which could be defined as the secant stiffness. Finally, a saw-toothed stress-strain curve is developed for discrete bundle due to sequential but finite fracture of micro-elements.

Call for the classic definition of damage in Robotnov[22]

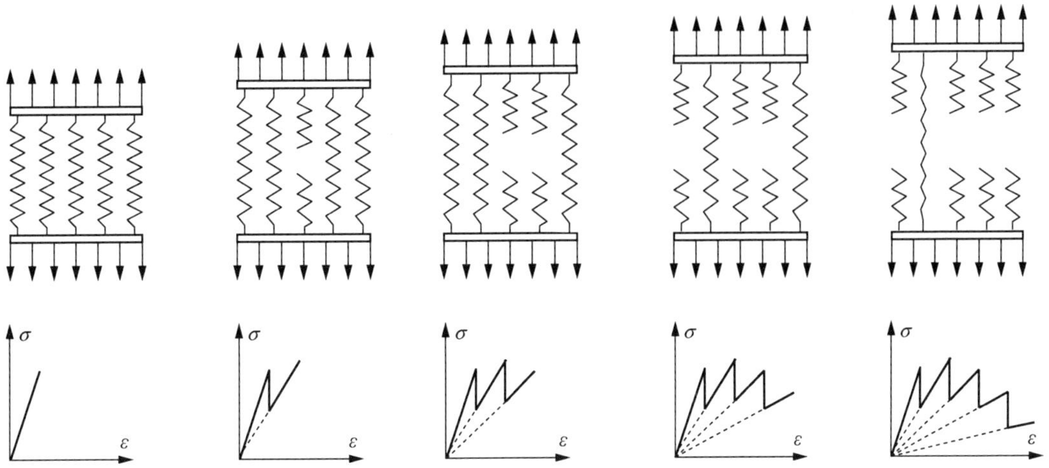

Fig. 3. Stress-strain responses of discrete bundle

$$d = \frac{A_d}{A} \tag{40}$$

where A_d and A are the cracked area and the initial undamaged area, respectively. The damage variable for the parallel element model can be defined as

$$d(\varepsilon^e) = \frac{1}{N}\sum_{i=0}^{N-1} H(\varepsilon^e - \Delta_i) \tag{41}$$

where $H(\cdot)$ is the Heaviside function as in Eq. (7). Taking the limit of Eq. (41) as N approaches infinity and accounting for the definition of stochastic integral, one obtains

$$d(\varepsilon^e) = \int_0^1 H[\varepsilon^e - \Delta(x)]\mathrm{d}x \triangleq g(\varepsilon^e) \tag{42}$$

where $\Delta(x)$ is the 1-D micro-fracture strain random field; x denotes the spatial coordinate of the micro-element; and $\triangleq$ denotes "defining as". When the number of the micro-element N approaches infinity, the discrete parallel element bundle will be translated into a continuum bundle. It is shown in Fig. 4 that the smooth stress-strain curve is obtained for the continuum bundle.

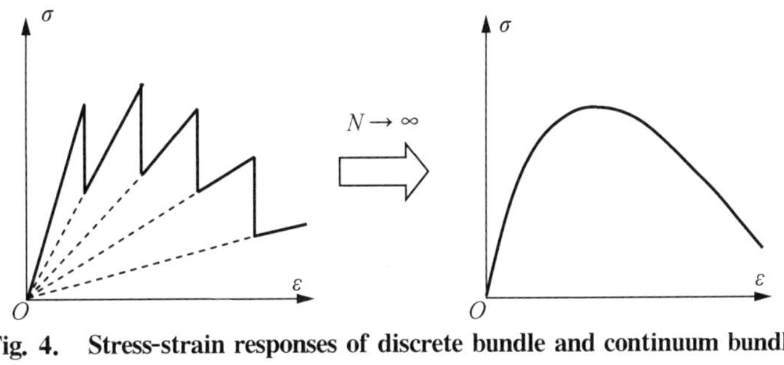

Fig. 4. Stress-strain responses of discrete bundle and continuum bundle

Eq. (42) is actually a macro-stochastic damage evolution law. Suppose $\Delta(x)$ is a homogenous random field with the first-order and second-order probability density functions

$$f(\Delta;\ x)=f(\Delta) \tag{43}$$

$$f(\Delta_1,\ \Delta_2;\ x_1,\ x_2)=f(\Delta_1,\ \Delta_2;\ |x_1-x_2|) \tag{44}$$

Let $\varphi(x)=H[\varepsilon^e-\Delta(x)]$ be a generating random field of $\Delta(x)$. Obviously, $\varphi(x)$ obeys the(0, 1) distribution. That is

$$\begin{aligned}P[\varphi(x)=1]&=P\{H[\varepsilon^e-\Delta(x)]=1\}=P\{[\varepsilon^e-\Delta(x)]\geqslant 0\}\\&=\int_0^{\varepsilon^e} f(\Delta)d\Delta \mathrel{\hat{=}} F(\varepsilon^e)\end{aligned} \tag{45}$$

$$P[\varphi(x)=0]=1-F(\varepsilon^e) \tag{46}$$

where $F(\varepsilon^e)$ is the first-order cumulative distribution function of $\Delta(x)$. Thus the ensemble average of $\varphi(x)$ is

$$\mu_\varphi(x)=E[\varphi(x)]=1\times F(\varepsilon^e)+0\times[1-F(\varepsilon^e)]=F(\varepsilon^e) \tag{47}$$

where $E[\cdot]$ denotes the expectation operator.

The two dimensional distribution of $\varphi(x)$ could be obtained as follows

$$\begin{aligned}P[\phi(x_1,\ x_2)=1]&=P[\varphi(x_1)\varphi(x_2)=1]=P[\varphi(x_1)=1\cap\varphi(x_2)=1]\\&=P[\Delta_1(x)\leqslant\varepsilon^e\cap\Delta_2(x)\leqslant\varepsilon^e]\\&=\int_0^{\varepsilon^e}\int_0^{\varepsilon^e} f(\Delta_1,\ \Delta_2;\ |x_1-x_2|)d\Delta_1 d\Delta_2\\&\mathrel{\hat{=}} F(\varepsilon^e,\ \varepsilon^e;\ |x_1-x_2|)\end{aligned} \tag{48}$$

$$P[\phi(x_1,\ x_2)=0]=1-F(\varepsilon^e,\ \varepsilon^e;\ |x_1-x_2|) \tag{49}$$

where $F(\varepsilon^e,\ \varepsilon^e;\ |x_1-x_2|)$ is the second-order cumulative distribution function of $\Delta(x)$. Therefore the ensemble average function of $\phi(x_1,\ x_2)$ as well as the autocorrelation function of $\varphi(x)$ is

$$R_\varphi(x_1,\ x_2)=E[\varphi(x_1)\varphi(x_2)]=E[\phi(x_1,\ x_2)]=F(\varepsilon^e,\ \varepsilon^e;\ |x_1-x_2|) \tag{50}$$

It is concluded that the generated field $\varphi(x)$ is homogenous random field in a wide sense. Performing expectation operator on Eq. (42) and considering Eq. (47), we obtain the expectation of the damage scalar $d(\varepsilon^e)$

$$\mu_d(\varepsilon^e)=E\left[\int_0^1\varphi(x)\mathrm{d}x\right]=\int_0^1 E[\varphi(x)]\mathrm{d}x=\int_0^1 F(\varepsilon^e)\mathrm{d}x=F(\varepsilon^e) \tag{51}$$

The variance of $d(\varepsilon^e)$ is given by

$$V_d^2=E[d(\varepsilon^e)]^2-[\mu_d(\varepsilon^e)]^2=E\left[\int_0^1\varphi(x)\mathrm{d}x\right]^2-[\mu_d(\varepsilon^e)]^2 \tag{52}$$

Using the properties of mean square stochastic integral and double integral, one gets

$$\begin{aligned}E\left[\int_0^1\varphi(x)\mathrm{d}x\right]^2&=\int_0^1\int_0^1 R_\varphi(x_1,\ x_2)\mathrm{d}x_1\mathrm{d}x_2=\int_0^1\int_0^1 F(\varepsilon^e,\ \varepsilon^e;\ |x_1-x_2|)\mathrm{d}x_1\mathrm{d}x_2\\&=2\int_0^1(1-\gamma)F(\varepsilon^e,\ \varepsilon^e;\ \gamma)\mathrm{d}\gamma\end{aligned} \tag{53}$$

where $\gamma=|x_1-x_2|$.

Therefore, Eq. (52) could be rewritten as

$$V_d^2=E[d(\varepsilon^e)]^2-[\mu_d(\varepsilon^e)]^2=2\int_0^1(1-\gamma)F(\varepsilon^e,\ \varepsilon^e;\ \gamma)\mathrm{d}\gamma-F(\varepsilon^e)^2 \tag{54}$$

If $\Delta(x)$ is an independent random field, then

$$F(\varepsilon^e,\ \varepsilon^e;\ \gamma)=[F(\varepsilon^e)]^2 \tag{55}$$

Hence the variance function is

$$V_d^2=2\int_0^1(1-\gamma)F(\varepsilon^e)^2\mathrm{d}\gamma-F(\varepsilon^e)^2=0 \tag{56}$$

This means that ignoring the spatial correlation of the micro-fracture strain field will lead to deterministic damage evolution model. Therefore, the spatial correlation of $\Delta(x)$ should be taken into account. Let μ_Δ and σ_Δ be the first-order expectation and standard deviation of $\Delta(x)$, respectively. Suppose $Z(x)=\ln\Delta(x)$ be a normal random field with the first-order parameter $(\lambda,\ \zeta)$. Then the relations among parameters can be given as follows

$$\lambda=E[\ln\Delta(x)]=\ln\left(\frac{\mu_\Delta}{\sqrt{1+\sigma_\Delta^2/\mu_\Delta^2}}\right) \tag{57}$$

$$\zeta^2=\mathrm{Var}[\ln\Delta(x)]=\ln(1+\sigma_\Delta^2/\mu_\Delta^2) \tag{58}$$

It is deduced that $\Delta(x)$ is a lognormal random field with the first-order cumulative distribution given by

$$F(\varepsilon^e)=\Phi\left[\frac{\ln\varepsilon^e-\lambda}{\zeta}\right]\triangleq\Phi(\alpha) \tag{59}$$

where $\Phi(\cdot)$ is the cumulative distribution function of a standard normal distribution. Consider exponential auto-correlation coefficient function for $Z(x)$ as follows

$$\rho_z(\gamma)=\exp(-\xi\gamma) \tag{60}$$

Then the second-order cumulative distributed function of $\Delta(x)$ is

$$F(\varepsilon^e,\ \varepsilon^e;\ \gamma)=\Phi\left(\frac{\ln\varepsilon^e-\lambda}{\zeta},\ \frac{\ln\varepsilon^e-\lambda}{\zeta}\bigg|\rho_z\right)\triangleq\Phi(\alpha,\ \alpha|\rho_z) \tag{61}$$

where $\Phi(y_1,\ y_2|\rho)$ is the standard expression of a second-order cumulative distributed function for two-dimensional normal distribution, which is defined as double integral of its second-order probability density function. In addition, the 2-D function $\Phi(\alpha,\ \alpha|\rho_z)$ could be reduced to a 1-D integral expression by introducing[23]

$$\Phi(\alpha,\ \alpha|\rho_z)=\Phi(\alpha)-\frac{1}{\pi}\int_0^\beta\frac{1}{1+t^2}\exp\left[-\frac{\alpha^2}{2}(1+t^2)\right]\mathrm{d}t \tag{62}$$

where the upper limit of integral is given by

$$\beta=\sqrt{\frac{1-\rho_z}{1+\rho_z}} \tag{63}$$

Three parameters $(\lambda,\ \zeta,\ \xi)$ are introduced in the parallel element model to represent the stochastic damage evolution under uniaxial loading. As is well known, the performance of concrete depends to a large degree on the loading condition. Thus it is reasonable that two groups of parameters are introduced to describe the damage evolutions under tension and compression, respectively. In the present paper, $(\lambda^+, \zeta^+, \xi^+)$ denotes the tensile material parameter triple related to the evolution of d^+, whereas $(\lambda^-, \zeta^-, \xi^-)$ denotes the compressive material parameter triple related to the evolution of d^-. By the way, damage and failure under uniaxial compression are mainly due to the shear damage mechanism, hence the triple $(\lambda^-, \zeta^-, \xi^-)$ is also referred as the shear material parameters.

3.2 Parameter identification

As we known, there are considerable experimental difficulties in determining the full process multiaxial performance of concrete. Nonetheless, the material parameters introduced in the above model could be identified through the results of uniaxial tests. Since the formulation system of the present model exhibits high nonlinear behaviors even during the uniaxial loading, direct parameter identification algorithms such as the least square method are not reliable for the subsequent structural simulation. Therefore, the optimizing algorithm is adopted to identify the two groups of material parameters $(E_0, \lambda^\pm, \zeta^\pm, \xi^\pm, \xi_p^\pm)$ based on the results of uniaxial tensile and compressive tests, respectively. The stochastic modeling principle could be expressed as follows

$$J_m = [E(\bar{x}) - E(x)]^T [E(\bar{x}) - E(x)] \rightarrow \min \tag{64}$$

$$J_v = [\mathrm{Var}(\bar{x}) - \mathrm{Var}(x)]^T [\mathrm{Var}(\bar{x}) - \mathrm{Var}(x)] \rightarrow \min \tag{65}$$

where J_m and J_v are the optimizing objective functions based on the mean value and variance, respectively; $E(\cdot)$ is the expectation operator; $\mathrm{Var}(\cdot)$ is the variance operator; $\bar{x}$ denotes the experimental result; and x is the simulation results based on certain values of parameters.

4 Energy equivalent strain

In order to bridge the gap between the micro-model, i.e. the one-dimensional model, and the macro-model, i.e. the multi-dimensional CDM model, described as above two sections, the concept of energy equivalent strain is introduced which could be derived from the damage energy release rate (DERR) described in Section 2.

According to Eqs. (8), (22), (23) and (29), the DERRs are proposed as[5, 15]

$$Y^+ = \psi_0^{e+} = \frac{1}{2E_0}\left[\frac{2(1+v_0)}{3} 3\bar{J}_2^+ + \frac{1-2v_0}{3}(\bar{I}_1^+)^2 - v_0 \bar{I}_1^+ \bar{I}_1^-\right] \tag{66}$$

$$Y^- = \psi_0^- = \psi_0^{e-} + \psi_0^{p-} \approx b_0 [\alpha \bar{I}_1^- + \sqrt{3\bar{J}_2^-}]^2 \tag{67}$$

where $\bar{I}_1^\pm$ are the first invariants of $\bar{\boldsymbol{\sigma}}^\pm$; $\bar{J}_2^\pm$ are the second invariants of $\bar{\boldsymbol{s}}^\pm$, the deviatoric components of $\bar{\boldsymbol{\sigma}}^\pm$; v_0 denotes the Poisson's ratio; b_0 and α are material parameters determined from multiaxial experimental data.

According to the thermo-dynamic theory, the damage evolution function could be defined as[5]

$$d^{\pm}=g_{Y}^{\pm}(Y^{\pm}) \tag{68}$$

However, the traditional thermo-dynamics theory cannot specify the explicit form of the function $g(\cdot)$. In most of the CDM based models, the empirical damage evolution functions are widely adopted. On the other hand, noticing the monotonic increasing characteristic of damage, we could determine that there exists a one to one map between g and Y. This means that for two stress states, if the initial damage states are equal, then for the same increased damage released energy in the damage process, the corresponding damage will be the same no matter the materials are in one-dimensional stress state or in multi-dimensional stress state. We call this background as the damage consistent condition.
Actually, the DRREs are scalar functions of effect stress and elastic strain as follows

$$Y^{\pm}=Y^{\pm}(\bar{\boldsymbol{\sigma}})=Y^{\pm}(\boldsymbol{\varepsilon}^{e})=Y^{\pm}(\varepsilon_1^e, \varepsilon_2^e, \varepsilon_3^e) \tag{69}$$

Combining Eqs. (68) and (69), it is easy to derive a multivariate function between the damage scalar $d^{\pm}$ and the principal elastic strains $(\varepsilon_1^e, \varepsilon_2^e, \varepsilon_3^e)$.
According to the above section, one-dimensional damage evolution rule could be expressed as

$$d^{\pm}=g^{\pm}(\varepsilon^{e\pm})=\int_0^1 H[\varepsilon^{e\pm}-\Delta^{\pm}(x)]\mathrm{d}x \tag{70}$$

Their evolution law in the sense of the mean value and standard deviation could be demonstrated through Eqs. (51) and (54).
On the other hand, it is easy to understand that the damage scalars $d^{\pm}$ in the 3-D loading condition should be determined towards a given set of principal elastic strains $(\varepsilon_1^e, \varepsilon_2^e, \varepsilon_3^e)$, along an arbitrary loading path Γ. This means that a 4-D damage evolution surface should be established based on the 3-D damage evolution surface within coordinate space. For the sake of simplicity, Fig. 5 only illustrates the damage evolution manifold in the 2-D strain condition. According to the damage consistent condition, there must exist an iso-line along

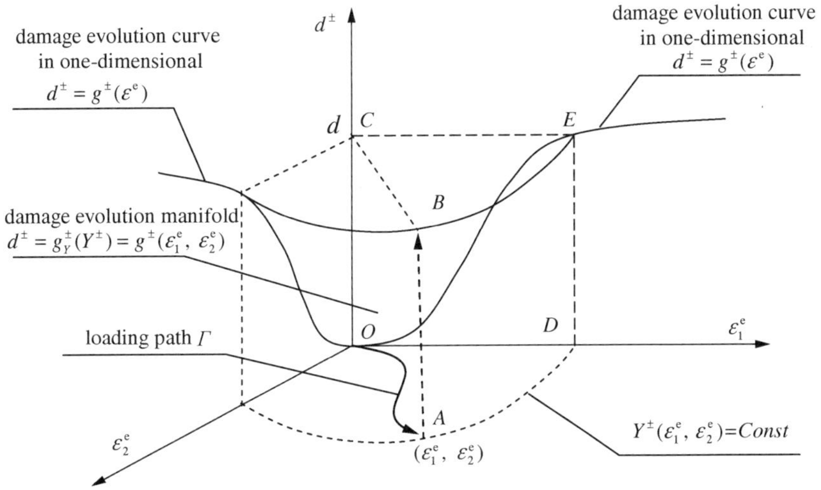

Fig. 5. Geometry structure of damage evolution

which the damage energy release rate and the corresponding damage scalar keep invariant. Noticing Eq. (68), the iso-line could be derived by solving the following equation

$$Y^{\pm}(\varepsilon_1^e, \varepsilon_2^e) = \text{const.} \tag{71}$$

It is clearly that along the curve determined by Eq. (71), the damage scalar $d^{\pm}$ remains constant.

For a given pair of principal elastic strains $(\varepsilon_1^e, \varepsilon_2^e)$, an equivalent uniaxial strain $\bar{\varepsilon}^e$ is easily obtained. We define this equivalent uniaxial strain as "energy equivalent strain".

Actually, for the uniaxial loading $(\bar{\sigma}_2 = 0, \bar{\sigma}_3 = 0)$, the foregoing Eqs. (66) and (67) can be rewritten as

$$y^{+}(\varepsilon^{e+}) = \frac{E_0}{2}(\varepsilon^{e+})^2 \tag{72}$$

$$y^{-}(\varepsilon^{e-}) = b_0[(\alpha - 1)E_0\varepsilon^{e-}]^2 \tag{73}$$

where $y^{\pm}$ are the DERRs expressions related to the uniaxial loading. According to the damage consistent condition, the energy equivalent strain could be solved based on the following equation

$$Y^{\pm}(\varepsilon_1^e, \varepsilon_2^e, \varepsilon_3^e) = \text{const.} = y^{\pm}(\bar{\varepsilon}^e) \tag{74}$$

Substituting Eqs. (72) and (73) in (74), the energy equivalent elastic strains become

$$\bar{\varepsilon}^{e+} = \sqrt{\frac{2Y^{+}}{E_0}} \tag{75}$$

$$\bar{\varepsilon}^{e-} = \frac{1}{(\alpha - 1)}\sqrt{\frac{Y^{-}}{b_0}} \tag{76}$$

Noticing Eqs. (35)—(39), we obtain the expressions of energy equivalent strain as follows

$$\bar{\varepsilon}^{\pm} = \frac{\bar{\varepsilon}^{e\pm}}{1 - \xi_p^{\pm} d^{\pm}} \tag{77}$$

Substituting Eqs. (75) and (76) in (42), the multi-dimensional damage evolution functions are established as

$$d^{\pm} = g^{\pm}(\bar{\varepsilon}^{e\pm}) = \int_0^1 H[\bar{\varepsilon}^{e\pm} - \Delta^{\pm}(x)]\mathrm{d}x \tag{78}$$

5 Model verification

Usually, the numerical results in various loading conditions should be presented to illustrate the applicability and effectiveness of the proposed model. Simultaneously, the results obtained by the suggested model should be compared with the corresponding experimental results to verify its performance. However, due to lack of stress-strain experimental data under multiaxal loading condition, the multi-dimensional responses of the proposed material model often verifies in the structural scale. It is clear that the gap between the material scale and the structural scale significantly degrades the reliability of verification conclusions. In other words, the local stress strain responses simulated by the theoretical model may be far from accurate, even though perfect agreement is obtained in the force-deformation behaviors in certain points of structures. Thus the proposed constitutive law is examined using the "elementary" test data in the present paper.

5.1 Algorithmic aspects

In contrast to the damage constitutive equations which are explicit once the elastic strains are given, Eq. (35) is implicit in the plastic part of the model and requires an iterative process in the solution. Fortunately, the total strain expressions for plastic strain evolution are more robust and reliable than the differential models in numerical implementation procedure. With Eqs. (19), (35), (36) and (38) in mind, the expressions for the total strain $\boldsymbol{\varepsilon}$ and the elastic strain $\boldsymbol{\varepsilon}^{e}$ is deduced as

$$\boldsymbol{\varepsilon} = \{\mathbf{I} - \mathbf{C}_0^{-1} : [\xi_p^{+} d^{+}(\boldsymbol{\varepsilon}^{e})\mathbf{P}^{+} + \xi_p^{-} d^{-}(\boldsymbol{\varepsilon}^{e})\mathbf{P}^{-}] : \mathbf{C}_0\} : \boldsymbol{\varepsilon}^{e} \tag{79}$$

For a given total strain tensor $\boldsymbol{\varepsilon}$, appropriate numerical algorithms, such as the Newton-Raphson method, the modified Newton method, etc. could be used to solve the nonlinear implicit Eq. (79) for the elastic strain $\boldsymbol{\varepsilon}^{e}$. Actually, the mean values of the damage scalars $\mu_{d^{\pm}}$ are calculated during the iterative process. After that the variance $V_{d^{\pm}}$ could be evaluated through the definition of energy equivalent strain. Finally, the mean value and variance of the Cauchy stress tensor $\boldsymbol{\sigma}$ are figured out based on $\boldsymbol{\varepsilon}$, $\boldsymbol{\varepsilon}^{e}$, $\mu_{d^{\pm}}$ and $V_{d^{\pm}}$.

5.2 Brief introduction to biaxial experiment

Because the high-performance concrete (HPC) is superior to ordinary concrete in strength, toughness, workability, durability, etc., which makes HPC particularly suitable for construction of tall buildings, long-span bridges and other significant structures such as the nuclear reactor containment, thus considerable attention has been attracted to relevant fields. A systematic experimental research was carried out by the authors to investigate the mechanical properties of HPC.[24] The behaviors of HPC specimens, with the dimension of 150 mm × 150 mm × 50 mm, subjected to uniaxial and biaxial loading were investigated in this research. Total 91 plate specimens were tested in four different loading conditions including the uniaxial tension, uniaxial compression, biaxial tension and compression on the Instron 8506 close-loop testing machine (Fig. 6). Uniaxial and biaxial complete stress-strain curves were obtained under strain control loading scheme. During the biaxial loading process, the ratio between global strains measured in different loading directions maintained a constant value. For each biaxial loading case, three different biaxial strain ratios $\alpha(=\varepsilon_1/\varepsilon_2)$ were chosen, i.e. $\alpha=1$, 0.4, 0.1 for compression-compression loading. And the loading rates adopted in uniaxial as well as biaxial tests were very low to ensure the stability of crack propagation and avoid strain rate effect. The ultimate strength envelopes in both stress and strain space were developed through parameter identification of the complete stress-strain curves.

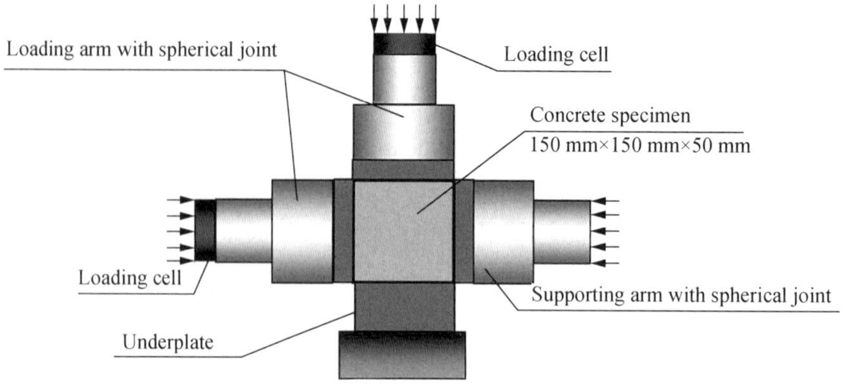

Fig. 6. Experimental setup of biaxial loading

5.3 Monotonic uniaxial tests

(1) *Uniaxial* tension: The matcrial parameters identified based on test results are: $E_0=$ 37 559 MPa, $\lambda^+=4.92$, $\zeta^+=0.30$, $\xi^+=40$ and $\xi_p^+=0.70$. It is observed form Fig. 7(a) and (b) that the simulation results agree well with the testing data both in the sense of the mean value and standard deviation (STD).

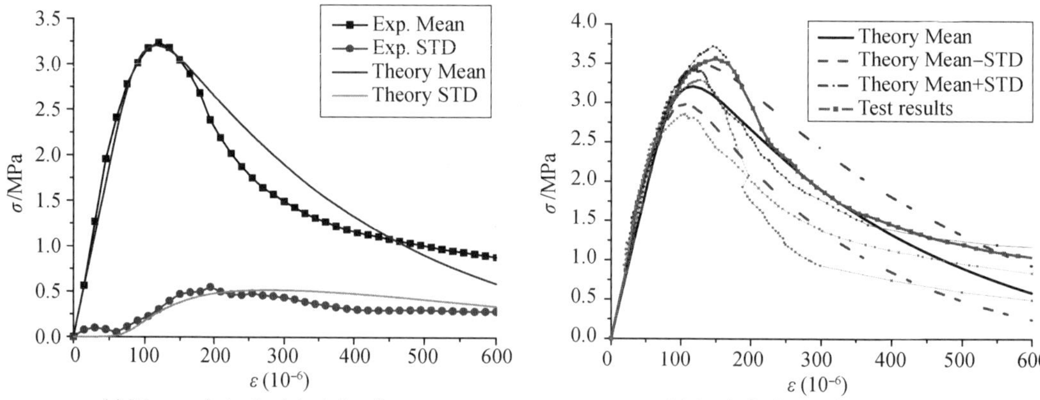

(a) Mean and standard deviation for stresses

(b) Analytically predicted variability

Fig. 7. Monotonic uniaxial tension tests

(2) *Uniaxial* compression: The parameters of material properties identified for the simulations are: $E_0=37\ 559$ MPa, $\lambda^-=7.77$, $\zeta^-=0.37$, $\xi^-=50$ and $\xi_p^-=0.20$. It is found that good agreement with experimental results is obtained. Both the mean value and the STD are well reproduced (Fig. 8).

5.4 Monotonic biaxial tests

The proposed model is also verified with our experimental results of biaxial compression ($\sigma_3=0$). Based on the material parameters identified through the former uniaxial experimental data, predicted biaxial stress-strain curves in sense of mean value and standard deviation are simulated and shown in Figs. 9—11.

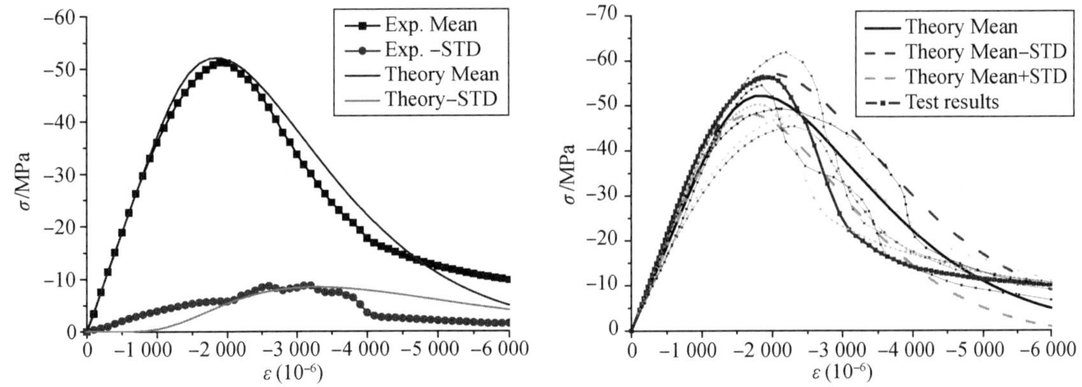

(a) Mean and standard deviation for stresses

(b) Analytically predicted variability

Fig. 8. Monotonic uniaxial compression tests

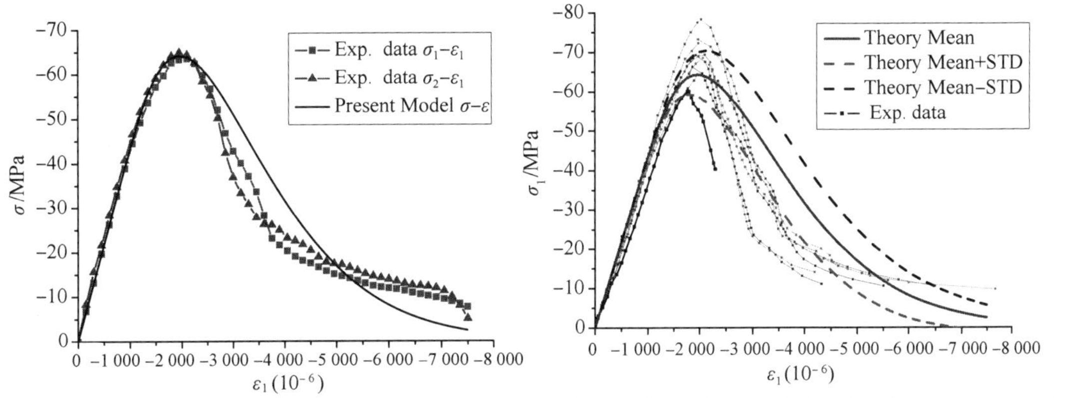

(a) Biaxial stress-strain curves

(b) Analytically predicted variability in primary loading direction

Fig. 9. Monotonic biaxial tests ($\varepsilon_2/\varepsilon_1=-1/-1$)

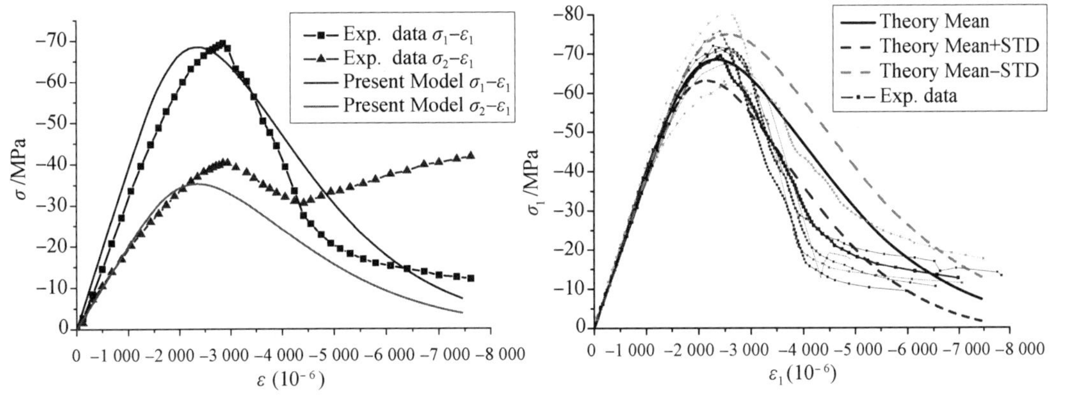

(a) Biaxial stress-strain curves

(b) Analytically predicted variability in primary loading direction

Fig. 10. Monotonic biaxial tests ($\varepsilon_2/\varepsilon_1=-0.4/-1$)

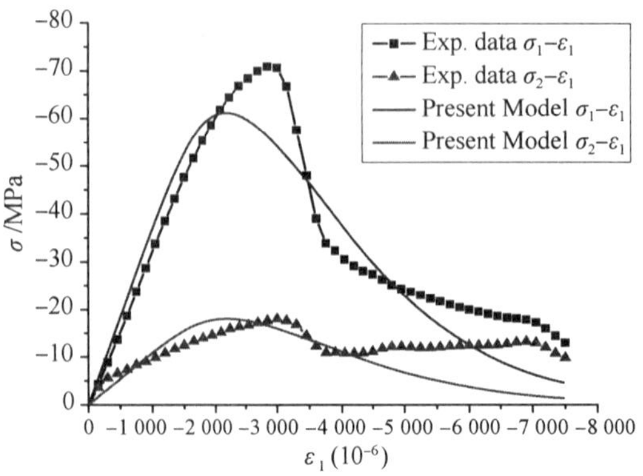

(a) Biaxial stress-strain curves

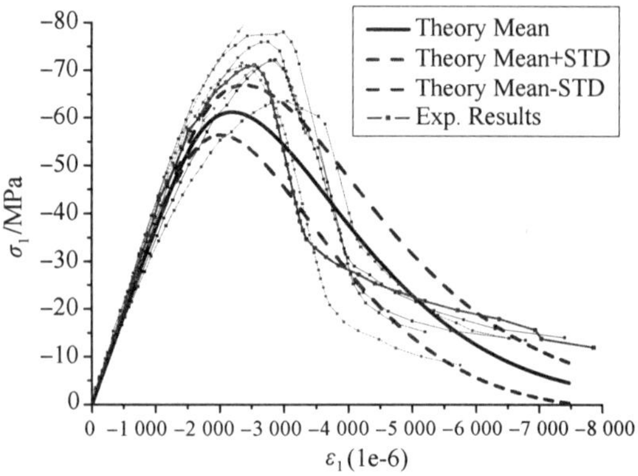

(b) Analytically predicted variability in primary loading direction

Fig. 11. Monotonic biaxial tests($\varepsilon_2/\varepsilon_1=-0.1/-1$)

It is noted that the simulated curves in the primary loading direction agree well with the experimental data both in the sense of the mean value and STD. However, the rehardening of concrete after entering the softening stage in the secondary loading direction, which is observed in our experiment, is difficult to simulate precisely using the present model. In our opinion, the combined action between the shear dilation and the compressive deformation lead to the reascension. Hence if we intend to simulate such phenomenon through damage approach, the shear dilation of concrete material should be carefully taken into account.

5.5 Biaxial envelope

Based on the parameters identified in preceding section, the biaxial peak stress envelopes predicted by the present model are plotted in Fig. 12, together with the testing results including Kupfer et al., Tasuji et al., Li and Guo, Lee et al. and our own experiments.[25-28] It is shown that most of the testing points are located in the domain between Mean + STD and Mean − STD curves. Therefore, it can be concluded that the proposed model is applicable in the analysis of structures bearing multi-dimensional excitations and damages.

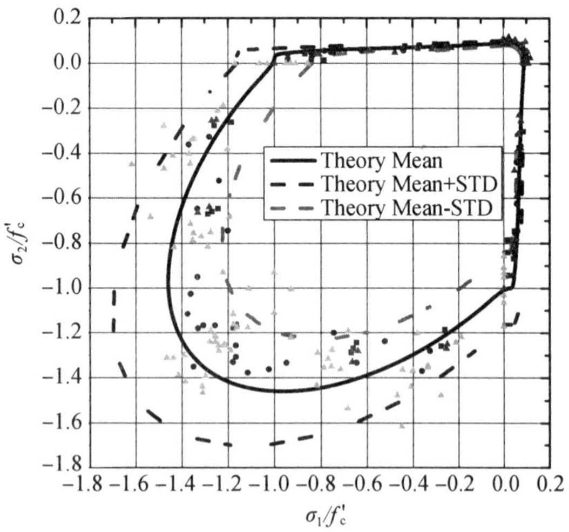

Fig. 12. Biaxial peak stress envelopes

6 Conclusions

Considering the nonlinearity and randomness of concrete materials simultaneously, a stochastic damage constitutive model is proposed in the present paper. Within the framework of damage mechanics, two damage scalars are adopted to describe the degradation of the macro-mechanical properties of concrete. A uniaxial stochastic damage model is adopted to represent the coupling between the nonlinearity and randomness, where an empirical model is employed to describe the evolution of plasticity. Based on the damage energy release rates (DERRs), the energy equivalent strain is defined to bridge the gap between the one-dimensional and multi-dimensional constitutive law. Finally, the numerical predictions are verified by comparison with the testing results. Based on the formulations and simulations, several conclusions could be drawn:

(1) The bi-scalar CDM framework proposed in this paper could represent the nonlinear behaviors of concrete through a unified theoretical approach.

(2) Based on the definition of energy equivalent strain, the multi-dimensional stochastic damage evolution could be established. The agreement between simulated biaxial envelopes and the experimental data shows that the proposed model is able to reproduce the multi-dimensional stochastic nonlinear behaviors of concrete.

(3) The randomness of multiaxial strength predicted using the present model is of great importance to predict the reliability of concrete structure under multiaxial loading condition.

(4) Combining with appropriate method for stochastic response analysis of nonlinear structures, the present model could simulate the subtle stochastic response of concrete structures.

Acknowledgments

Financial supports from the National Science Foundation of China for Innovative Research Groups (Grant Nos. 50621062 and 50321803) and for key Project (Grant Nos. 90715033) are greatly appreciated.

Appendix A Two-dimensional stochastic damage formulations

It is clear that the tensor formulations are extremely convenient to develop the multi-dimensional constitutive model in a unified theoretical representation. But in practical application, the tensor based programming is not only complicated in data structure but also time and storage consuming in program execution. Therefore the vector and matrix representations of constitutive model are usually adopted to optimize the coding structure and efficiency in the finite element programming. In this section, the 2-D closed-form analytical formulations of stochastic damage models are developed. Curves in Fig. A.0 are helpful to establish the direct physical understanding of the entities rather than rigorous mathematical definitions

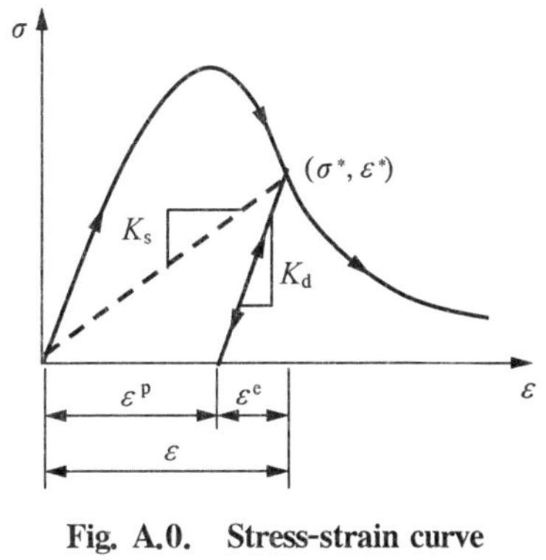

Fig. A.0. Stress-strain curve

in multi-dimensional loading condition.

The 2-D stochastic damage model could be expressed by the following two equations

Loading equation

$$\begin{Bmatrix}\sigma_1\\ \sigma_2\end{Bmatrix}=\mathbf{K}_s\begin{Bmatrix}\varepsilon_1\\ \varepsilon_2\end{Bmatrix} \tag{A.1}$$

where $\mathbf{K}_s$ is a 2×2 tangential stiffness matrix.

Unloading/reloading equation

$$\begin{Bmatrix}\sigma_1-\sigma_1^*\\ \sigma_2-\sigma_2^*\end{Bmatrix}=\mathbf{K}_d\begin{Bmatrix}\varepsilon_1-\varepsilon_1^*\\ \varepsilon_2-\varepsilon_2^*\end{Bmatrix} \tag{A.2}$$

where $\mathbf{K}_d$ is a 2×2 unloading/reloading stiffness matrix; $\{\sigma_1^*, \sigma_2^*\}^T$ is the unloading/reloading stress vector and $\{\varepsilon_1^*, \varepsilon_2^*\}^T$ is the unloading/reloading strain vector.

As is mentioned in Section 2.1, split of effective stress tensor is adopted to account for the unilateral effect of concrete in different tensile and compressive loading conditions. Thus the unified tensor formulae are developed using the Heaviside function. But their expanded analytical formulas could not be expressed in a consistent way due to its discontinuity crossing tensioncompression boundary. Hence different analytical formulae are developed, respectively, considering different tension-compression combinations.

A.1 T-T region($\bar{\sigma}_1>0$, $\bar{\sigma}_2\geqslant0$)

In the T-T region of the biaxial loading condition, the value of the Heaviside function in Eq. (5) could be determined. Then substituting Eqs. (20) and (36) in the damage constitutive relationship Eq. (19) and considering the definitions of stiffness matrices in Eqs. (A.1) and (A.2) will yield the following expressions:

Tangential stiffness matrix

$$\mathbf{K}_s=(1-d^+)(1-\xi_p^+d^+)\frac{E_0}{1-v_0^2}\begin{bmatrix}1 & v_0\\ v_0 & 1\end{bmatrix} \tag{A.3}$$

Unloading/reloading stiffness matrix

$$\mathbf{K}_d=(1-d^+)\frac{E_0}{1-v_0^2}\begin{bmatrix}1 & v_0\\ v_0 & 1\end{bmatrix} \tag{A.4}$$

Substituting Eqs. (3) and (4) in Eqs. (66) and (67), one obtains DERRs

$$\begin{aligned}Y^+&=\frac{1}{2E_0}\left\{\frac{2(1+v_0)}{3}(\bar{\sigma}_1^2+\bar{\sigma}_2^2-\bar{\sigma}_1\bar{\sigma}_2)+\frac{1-2v_0}{3}(\bar{\sigma}_1+\bar{\sigma}_2)^2\right\}\\ &=\frac{E_0}{2(1-v_0^2)}[(\varepsilon_1^e)^2+(\varepsilon_2^e)^2+2v_0\varepsilon_1^e\varepsilon_2^e]\end{aligned} \tag{A.5}$$

$$Y^-=0 \tag{A.6}$$

Then substituting Eqs. (A.5) and (A.6) in Eqs. (75) and (76), the expressions of energy equivalent strain in the T-T region are derived as follows

Energy equivalent strain

$$\bar{\varepsilon}^{e+}=\sqrt{\frac{1}{1-v_0^2}[(\varepsilon_1^e)^2+(\varepsilon_2^e)^2+2v_0\varepsilon_1^e\varepsilon_2^e]} \tag{A.7}$$

$$\bar{\varepsilon}^{e-}=0 \tag{A.8}$$

Noticing the stochastic damage evolution function (42) and the damage loading/unloading conditions, the damage evolution in the T-T region could be represented as follows

$$d^+=g^+(r_n^+)=\int_0^1 H[r_n^+-\Delta^+(x)]\mathrm{d}x d^-=0 \tag{A.9}$$

$$r_n^+=\max_{\tau\in[0,n]}\bar{\varepsilon}^{e+} \tag{A.10}$$

The derivations of stochastic damage formulations in the C-C and T-C regions are similar to that in the T-T region.

A.2 C-C region($\bar{\sigma}_1\leqslant 0$, $\bar{\sigma}_2<0$)

Tangential stiffness matrix

$$\mathbf{K}_s=(1-d^-)(1-\xi_p^-d^-)\frac{E_0}{1-v_0^2}\begin{bmatrix}1 & v_0\\ v_0 & 1\end{bmatrix} \tag{A.11}$$

Unloading/reloading stiffness matrix

$$\mathbf{K}_d=(1-d^-)\frac{E_0}{1-v_0^2}\begin{bmatrix}1 & v_0\\ v_0 & 1\end{bmatrix} \tag{A.12}$$

Energy equivalent strain

$$\bar{\varepsilon}^{e+}=0 \tag{A.13}$$

$$\bar{\varepsilon}^{e-}=\frac{1}{(1-v^2)(\alpha-1)}\{\alpha(1+v_0)(\varepsilon_1^e+\varepsilon_2^e)+\sqrt{[(\varepsilon_1^e+v_0\varepsilon_2^e)^2+(\varepsilon_2^e+v_0\varepsilon_1^e)^2-(\varepsilon_1^e+v_0\varepsilon_2^e)(\varepsilon_2^e+v_0\varepsilon_1^e)]}\} \tag{A.14}$$

Damage evolution

$$d^+=0\quad d^-=g^-(r_n^-)=\int_0^1 H[r_n^--\Delta^-(x)]\mathrm{d}x \tag{A.15}$$

$$r_n^-=\min_{\tau\in[0,n]}\bar{\varepsilon}^{e-} \tag{A.16}$$

A.3 T-C region($\bar{\sigma}_1\geqslant 0$, $\bar{\sigma}_2<0$)

Tangential stiffness matrix

$$\mathbf{K}_s=\frac{E_0}{1-v_0^2}\begin{bmatrix}(1-d^+)(1-\xi_p^+d^+) & (1-d^+)(1-\xi_p^+d^+)v_0\\ (1-d^-)(1-\xi_p^-d^-)v_0 & (1-d^-)(1-\xi_p^-d^-)\end{bmatrix} \tag{A.17}$$

Unloading/reloading stiffness matrix

$$\mathbf{K}_d=\frac{E_0}{1-v_0^2}\begin{bmatrix}(1-d^+) & (1-d^+)v_0\\ (1-d^-)v_0 & (1-d^-)\end{bmatrix} \tag{A.18}$$

Energy equivalent strain

$$\bar{\varepsilon}^{e+} = \sqrt{\frac{1}{(1-v^2)}\varepsilon_1^e(\varepsilon_1^e + v_0\varepsilon_2^e)} \tag{A.19}$$

$$\bar{\varepsilon}^{e-} = \frac{1}{(1-v^2)(\alpha-1)}\{\alpha(1+v_0)(\varepsilon_1^e+\varepsilon_2^e)+ \sqrt{[(\varepsilon_1^e+v_0\varepsilon_2^e)^2+(\varepsilon_2^e+v_0\varepsilon_1^e)^2-(\varepsilon_1^e+v_0\varepsilon_2^e)(\varepsilon_2^e+v_0\varepsilon_1^e)]}\} \tag{A.20}$$

Damage evolution

$$d^+=g^+(r_n^+)=\int_0^1 H[r_n^+-\Delta^+(x)]\mathrm{d}x \quad d^-=g^-(r_n^-)=\int_0^1 H[r_n^--\Delta^-(x)]\mathrm{d}x \tag{A.21}$$

$$r_n^+ = \max_{\tau\in[0,n]} \bar{\varepsilon}^{e+},\ r_n^- = \min_{\tau\in[0,n]} \bar{\varepsilon}^{e-} \tag{A.22}$$

It is observed from Eqs. (A.3), (A.4) and (A.11) and (A.12) that isotropic stiffness matrices are developed in the T-T or C-C region because a single mode of scalar damage evolution is activated in each loading condition. On the other hand, tensile and compressive damage evolutions are activated by the effective stresses in corresponding directions. Thus the anisotropic unsymmetrical stiffness matrices [Eqs. (A.17) and (A.18)] are established in the T-C region.

Appendix B　Mean and variance

Notice that the damage scalars d^+ and d^- in Appendix I are random variables. Hence the stochastic properties of stiffness matrices and stresses should be discussed. Considering the 2-D expressions developed as above, one observes that the random evolution of stiffness matrices depend on the following two entities

$$\Psi_s=(1-d^\pm)(1-\xi_p^\pm d^\pm) \tag{B.1}$$

$$\Psi_d=1-d^\pm \tag{B.2}$$

Since the expectation and variance of the damage scalars $d^\pm$ could be calculated through Eqs. (51) anfd (54), the expectation and variance of Ψ_s and Ψ_d could be expressed in terms of $\mu_{d^\pm}$ and $V_{d^\pm}$.

According to the probability theory, one easily gets

$$E(\Psi_d)=1-\mu_{d^\pm} \tag{B.3}$$

$$\mathrm{Var}(\Psi_d)=V_{d^\pm}^2 \tag{B.4}$$

The following two methods are proposed to evaluate the mean value and variance of Ψ_s:

(1) Direct stochastic uncoupling method (DSUM). It is assumed that the relation between the $\boldsymbol{\varepsilon}^p$ and $\boldsymbol{\varepsilon}$ is deterministic, then the damage variables $d^\pm$ in Eqs. (35)—(38) should be replaced by their mean value function $\mu_{d^\pm}$. Thus Ψ_s could be expressed as

$$\Psi_s=(1-d^\pm)(1-\xi_p^\pm\mu_{d^\pm}) \tag{B.5}$$

Performing expectation and variance operators on Eq. (70) gives

$$E(\Psi_s) = (1 - \mu_{d^\pm})(1 - \xi_p^\pm \mu_{d^\pm}) \tag{B.6}$$

$$V(\Psi_s) = V_{d^\pm}(1 - \xi_p^\pm \mu_{d^\pm})^2 \tag{B.7}$$

(2) Stochastic truncation method (STM). According to this approach, Taylor expansions are employed where the third and higher order terms are omitted in the expectation evaluation procedure. Then we have

$$\begin{aligned} E(\Psi_s) &= 1 - (1 + \xi_p^\pm)E(d^\pm) + \xi_p^\pm E[(d^\pm)^2] \\ &= 1 - (1 + \xi_p^\pm)\mu_{d^\pm} + \xi_p^\pm [V_{d^\pm}^2 + (\mu_{d^\pm})^2] \end{aligned} \tag{B.8}$$

$$\begin{aligned} E(\Psi_s)^2 &\approx 1 - 2(1 + \xi_p^\pm)E(d^\pm) + [1 + 4\xi_p^\pm + (\xi_p^\pm)^2]E[(d^\pm)^2] \\ &= 1 - 2(1 + \xi_p^\pm)\mu_{d^\pm} + [1 + 4\xi_p^\pm + (\xi_p^\pm)^2][V_{d^\pm}^2 + (\mu_{d^\pm})^2] \end{aligned} \tag{B.9}$$

$$\mathrm{Var}(\Psi_s) = E(\Psi_s)^2 - [E(\Psi_s)]^2 \tag{B.10}$$

Generally speaking, both the DSUM and STM are approximation methods for statistical estimation. The former is more efficient in computing whereas the latter gets higher order precision in theory. Therefore a proper approach should be adopted considering the balance between efficiency and precision. DSUM is adopted in the former numerical simulations.

References

[1] MAZARS J. Application de la mecanique de l'endommangement au comportement non-lineaire et a la rupture du beton de structure[M]. These de Doctorate d'Etat, L.M.T., Universite Paris, France, 1984.

[2] MAZARS J. A description of micro- and macro-scale damage of concrete structures[J]. Engineering Fracture Mechanics, 1986, 25: 729-737.

[3] JU J W. On energy-based coupled elastoplastic damage theories: constitutive modeling and computational aspects[J]. International Journal of Solids Structures, 1989, 25(7): 803-833.

[4] FARIA R, OLIVER J, CERVERA M. A strain-based plastic viscous-damage model for massive concrete structures[J]. International Journal of Solids Structures, 1998, 35(14): 1533-1558.

[5] WU J Y, LI J, FARIA R. An energy release rate-based plastic-damage model for concrete[J]. International Journal of Solids and Structures, 2006, 43(3-4): 583-612.

[6] WU J Y, LI J. Unified plastic-damage model for concrete and its applications to dynamic nonlinear analysis of structure[J]. Structural Engineering and Mechanics, 2007, 25(5): 519-540.

[7] PEIRCE F T. Tensile test for cotton Yarns-the weakest link[J]. Journal of Texture Institute, 1926, 17: 355-370.

[8] DANIELS H E. The statistic theory of the strength of bundles of threads. I [J]. Proceedings of the Royal Society of London A, 1945, 183: 405-435.

[9] KRAJCINOVIC D, SILVA M A G. Statistical aspects of the continuous damage theory [J]. International Journal of Solids and Structures, 1982, 18(17): 557-562.

[10] KRAJCINOVIC D. Damage Mechanics[M]. second ed. Elsevier B.V., Amsterdam, 1996.

[11] KANDARPA S, KIRKNER D J, SPENCER B F. Stochastic damage model for brittle materiel subjected to monotonic loading[J]. Journal of Engineering Mechanics, 1996, 126(8): 788-795.

[12] LI J, ZHANG Q Y. Study of stochastic damage constitutive relationship for concrete material[J]. Journal of Tongji University, 2001, 29(10): 1135-1141(in Chinese).

[13] LI J, LU Z H, ZHANG Q Y. Research on the stochastic damage constitutive model of concrete[C]//

Advances in Concrete and Structures (Proceedings of the International Conference ICACS 2003), 2003: 44-52.

[14] LEMAITRE J. Evaluation of dissipation and damage in metals submitted to dynamic loading[C]// Proceeding of ICAM-1, Japan, 1971.

[15] LI J, WU J Y. Energy-based CDM model for nonlinear analysis of confined concrete structures[C]// Proceedings of the ISCC-2004 (No. Key-9), Changsha, China, 2004: 209-221.

[16] SIMO J C, JU J W. Strain- and stress-based continuum damage model-I formulation[J]. International Journal of Solids and Structures, 1987, 23(7): 821-840.

[17] LUBLINER J, OLIVER J, OLIVER S, et al. A plastic-damage model for concrete[J]. International Journal of Solids Structures, 1989, 25(3): 299-326.

[18] CAROL I, RIZZI E, WILLAM K. On the formulation of anisotropic elastic degradation. Ⅰ. Theory based on a pseudo-logarithmic damage tensor rate; Ⅱ. Generalized pseudo-Rankine model for tensile damage[J]. International Journal of Solids and Structures, 2001, 38(4): 491-546.

[19] LEE J, FENVES G L. Plastic-damage model for cyclic loading of concrete structures[J]. Journal of Engineering Mechanics Division, ASCE 124, 1998: 892-900.

[20] JASON L, HUERTA A, PIJAUDIER-CABOT G, et al. An elastic-plastic damage formulation for concrete: application to elementary tests and comparison with an isotropic damage model [J]. Computational Modelling of Concrete, 2006, 195(52): 7077-7092.

[21] DAHLBLOM O, OTTOSEN N S. Smeared crack analysis using generalized fictious model[J]. Journal of Engineering Mechanics, ASCE 116, 1990: 55-76.

[22] ROBOTNOV Y N. Creep rupture. Applied Mechanics[C]//Proceedings of ICAM-12, 1968: 342-349.

[23] ZHANG X T, FANG K T. Introduction to Multivariate Statistical Analysis [M]. Science Press, Beijing: 1982 (in Chinese).

[24] REN X D, YANG W Z, ZHOU Y, et al. Behavior of high-performance concrete under uniaxial and biaxial loading[J]. ACI Materials Journal, 2008, 105(6): 548-557.

[25] KUPFER H, HILSDORF H K, RÜSCH H. Behavior of concrete under biaxial stresses[J]. ACI Journal, 1969, 66(8): 656-666.

[26] TASUJI M E, SLATE F O, NILSON A H. Stress-strain response and fracture of concrete in biaxial loading[J]. ACI Journal, 1978, 75(7): 306-312.

[27] LI W Z, GUO Z H. Experimental research for strength and deformation of concrete under biaxial tension-compression loading[J]. Journal of Hydraulic Engineering, 1991, 8: 51-56.

[28] LEE S K, SONG Y C., HAN S H. Biaxial behavior of plain concrete of nuclear containment building [J]. Nuclear Engineering and Design, 2003, 227: 143-153.

A rate-dependent stochastic damage-plasticity model for quasi-brittle materials①

Xiaodan Ren[1], Shajie Zeng[2], Jie Li[1]

1 School of Civil Engineering, Tongji University, 1239 Siping Road, Shanghai 200092, China
2 Shanghai Jianke Engineering Consulting Co., Ltd., 75 South Wanping Road, Shanghai 200032, China

Abstract In this work, a rate-dependent model for the simulation of quasi-brittle materials experiencing damage and randomness is proposed. The bi-scalar plastic damage model is developed as the theoretical framework with the damage and the plasticity opening for further developments. The governing physical reason of the material rate-dependency under relatively low strain rates, which is defined as the Strain Delay Effect, is modeled by a differential system. Then the description of damage is established by further implementing the rate-dependent differential system into the random damage evolution. To reproduce the evolution of plasticity under a variety of stress conditions, a multi-variable phenomenological plastic model is proposed and the description of plasticity is then formulated. An explicit integration algorithm is developed to implement the proposed model in the structural simulation. The model results are validated by a series of numerical tests that cover a wide variety of stress conditions and loading rates. The proposed model and algorithm offer a package solution for the nonlinear dynamic structural simulations.

Keywords Damage-plasticity model; Strain delay effect; Stochastic damage; Multi-variable plasticity; Dynamic loading

1 Introduction

The quasi-brittle materials like concrete and rocks usually exhibit highly complicated behaviors when experiencing stresses. Indicated by the intrinsic heterogeneity of the material meso-structure, the typical mechanical behaviors of concrete could be attributed into three aspects: the nonlinearity, the randomness and the rate-dependency. After decades of studies, one of the most used descriptions of the quasi-brittle nonlinearity at present are the damage-plasticity models[11, 15-16, 21, 38, 47]. Both the damage variables and the plastic deformations are included in these models to describe the material softening and the residual strains, respectively. By introducing the framework of thermodynamics and the energy based representations, damage and plasticity could be well organized in a class of unified theories. Currently the damage-plasticity models have been used as the standard tools for the nonlinear numerical simulations of concrete structures.

The other two aspects, e.g. randomness and rate-dependency, have not been well

① Originally published in *Computational Mechanics*, 2015, 55(2): 267-285.

considered in the existing theories. Experiences indicate that the tested stress-strain curves might be quite different even for the concrete specimens prepared based on the same mixture ratios and the same curing conditions. In many cases the randomness of concrete is able to deviate the nonlinear responses dramatically so that it deserves careful considerations in the analytical model. Krajcinovic[20] firstly introduced the conventional bundle model to the modeling of random damage behavior of concrete, although their results were restricted to the mean value evolution of damage. Kandarpa et al.[17] investigated the stochastic damage model and developed the standard deviation of damage evolution. Li and Ren[23] further improved the results of[17] and implemented the stochastic damage model into the energy based multiaxial damage-plasticity framework.

It is well known that most of the engineering materials are sensitive to the strain rates. Simultaneously, relatively high strain rates could be detected for the engineering structures subjected to the dynamic loads such as impact, explosion and earthquake. In most cases, the difference becomes significant when the rate changes[4]. The fact that concrete and other materials are sensitive to the rate of loading has been investigated by decades of experiments[1, 4, 13, 19, 25, 35, 43-45, 49] and investigations[9-12, 16, 24, 30-32, 48]. The physical mechanisms of the strain-rate effect can be attributed into two governing aspects: the viscous effect and the inertial effect. The rate-dependency is governed by the inertial effect at very high loading rate ($\dot{\varepsilon} \geqslant 100\ \mathrm{s}^{-1}$). For the low and moderate strain rates ($\dot{\varepsilon} < 100\ \mathrm{s}^{-1}$), the viscous effect plays governing role. And the latter case has been concerned more and more by civil engineers and researchers in recent years. The milestone of the rate-dependent material models was settled by Perzyna[30], who defined the viscoplastic evolutions by the over-stress function. Starting from this celebrated idea, the classical viscoplastic theory was well developed. Following the Perzyna's idea, the rate-dependent damage models could be also extended from the inviscid damage model. And the recently developed dynamic damage models[9, 11, 16, 32] are mainly following this idea.

Based on the analysis of literatures, we have reached the fact that the pragmatic rate-dependent constitutive models with the appropriate description of the material randomness and rate-dependency deserve further investigation. Therefore, the present work concentrates on the rate-dependency of concrete relying on the stochastic damage-plasticity framework. In Chapter 2, the energy based framework of damage-plasticity is developed, with the damage evolution as well as the plastic evolution opening for further development. Chapter 3 proposes a physics based model for the rate-dependent damage evolutions. The material rate-dependency is defined as the Strain Delay Effect(SDE) and described by the proposed differential system. The plasticity of concrete is discussed in Chapter 4. A multi-variable description of plasticity is proposed to consider the difference between the residual strain evolutions under the tensile and the compressive stress conditions, respectively. And the evolutionary functions of the plastic strains are defined with the corresponding damage variables involved. In Chapter 5, an explicit numerical scheme is proposed for the present constitutive model. Within the proposed scheme, the differential system defining the rate-dependent damage evolution is simulated by the finite difference method and the global convergence is defined by the drift of damage criteria. The stabilities of the physical system as well as the numerical system are analysed in Chapter 6. Chapter 7 validates the proposed model and algorithms by a series of numerical tests. Finally, a number of conclusions are draw in Chapter 8.

2 Representation of damage-plasticity

The material damage and plasticity could be unified in the following form as the starting point of the plastic damage constitutive model[47].

$$\boldsymbol{\sigma} = (\mathbb{I} - \mathbb{D}) : \mathbb{E}_0 : (\boldsymbol{\epsilon} - \boldsymbol{\epsilon}^{\mathrm{p}}) \tag{1}$$

where $\boldsymbol{\sigma}$ and $\boldsymbol{\epsilon}$ are the second order stress tensor and strain tensor, respectively; $\mathbb{E}_0$ is a fourth order tensor denoting the initial undamaged elastic stiffness; $\mathbb{D}$ is the fourth order damage tensor; and $\boldsymbol{\epsilon}^{\mathrm{p}}$ is a second order tensor denoting the plastic strain.

The proposed model is developed based on the strain equivalence hypothesis[22]: *the strain associated with a damaged state under the applied stress is equivalent to the strain associated with its undamaged state under the effective stress.*

The undamaged state undergoing the same strain as the damage state defines the effective stress tensor as follows

$$\bar{\boldsymbol{\sigma}} = \mathbb{E}_0 : (\boldsymbol{\epsilon} - \boldsymbol{\epsilon}^{\mathrm{p}}) \tag{2}$$

The strain equivalence hypothesis decouples the plastic damage model shown in Eq.(1) and defines the solution of variables, such as the plastic strain, in the undamaged state. Eq.(2) actually defines the plastic constitutive relationships for the undamaged state. Hence the evolution of the plastic strain, which is equivalent to the plastic strain undergoing damage, should be governed by the stress of the undamaged state, that is, the effective stress. Define the yield condition and the evolution potential based on the effective stress, one obtains the following effective stress space plasticity(ESP).

$$\begin{cases} \dot{\boldsymbol{\epsilon}}^{\mathrm{p}} = \dot{\lambda} \dfrac{\partial \tilde{F}(\bar{\boldsymbol{\sigma}}, \boldsymbol{\kappa})}{\partial \bar{\boldsymbol{\sigma}}} \\ \dot{\boldsymbol{\kappa}} = \dot{\lambda} \left[\mathbf{h} \cdot \dfrac{\partial \tilde{F}(\bar{\boldsymbol{\sigma}}, \boldsymbol{\kappa})}{\partial \bar{\boldsymbol{\kappa}}} \right] \\ F(\bar{\boldsymbol{\sigma}}, \boldsymbol{\kappa}) \geqslant 0, \ \dot{\lambda} \leqslant 0, \ \dot{\lambda} F(\bar{\boldsymbol{\sigma}}, \boldsymbol{\kappa}) = 0 \end{cases} \tag{3}$$

where F and $\tilde{F}$ are the yield function and the plastic potential, respectively; λ and $\boldsymbol{\kappa}$ are the plastic flow parameter and the hardening parameter; and $\mathbf{h}$ denotes the vectorial hardening function.

Based on the trivial derivations in plasticity theory, we could obtain the rate form of the constitutive law as follows

$$\dot{\bar{\boldsymbol{\sigma}}} = \mathbb{E}^{\mathrm{ep}} : \dot{\boldsymbol{\epsilon}} \tag{4}$$

where $\mathbb{E}^{\mathrm{ep}}$ is the elastoplastic tangential stiffness.

Eqs.(2)—(4) define the effective stress space plasticity by analog with the framework of the classical plastic theory. This framework is often referred as the theoretical effective stress space plasticity (T-ESP). T-ESP offers a rigorous and complete framework to describe the plastic evolution so that it is often adopted in the theoretical development. However, numbers of deficiencies may happen in its applications. The physical background of the yield function F and the plastic potential $\tilde{F}$ defined in the effective stress space are not very clear. Expressed by the effective stress, F and $\tilde{F}$ are also difficult to be measured experimentally. Moreover, the numerical implementation of the T-ESP is also rather intricate due to its mathematical structure. The return-mapping algorithm[37] may be performed with numbers of iterations at each integration points during each time-step, which may be time-consuming during the simulation of large scale structures. Therefore, the phenomenological plastic models are also proposed and widely used to describe the plasticity of concrete in a simple way but with sound experimental support. Chapter 4 of the present paper works on this side.

Substituting Eq.(2) into Eq.(1) yields the damage constitutive law as follows

$$\boldsymbol{\sigma} = (\mathbb{I} - \mathbb{D}) : \bar{\boldsymbol{\sigma}} \tag{5}$$

where the effective stress $\bar{\boldsymbol{\sigma}}$ has been solved in the undamaged state. The fourth order damage tensor $\mathbb{D}$ degrades the effective stress applied on the undamaged material. To solve $\mathbb{D}$, two aspects should be carefully considered. First of all, the general fourth order tensor is too complex and unnecessary, so that the structure of the damage tensor should be well simplified. Second, the damage criterion which governs the evolution of damage tensor should be pre-defined.

Despite the anisptropic damage represented by the second order tensor, the present paper adopts the bi-scalar damage representation as follows

$$\mathbb{D} = D^+\mathbb{P}^+ + D^-\mathbb{P}^- \tag{6}$$

where D^+ and D^- are the tensile and the compressive damage scalars, respectively; $\mathbb{P}^+$ and $\mathbb{P}^-$ are the fourth order projective tensors, which project D^+ and D^- to the corresponding tensile and compressive stress states, respectively. As shown in Eq.(6), the damage tensor is split into two components. Thus the evolution of damage should be governed by the evolutions of the corresponding eigen-damages as well as the rotation of the basis defined by the projective tensor.

Technically, damage is the continuum measure of material degradation induced by cracks and defects in the sub-scale. As the cracks and defects are initiated and driven by the applied stress, damage should be governed by the stress also. However, stress is not an appropriate variable to drive the damage evolution because of the strain softening even under monotonic loading. To address this issue and also to remain the form of stress in the damage criterion, the effective stress is usually adopted. To consider the tensile damage and the compressive damage in a separate way, the effective stress split[11, 47] is introduced as follows

$$\bar{\boldsymbol{\sigma}} = \bar{\boldsymbol{\sigma}}^{+} + \bar{\boldsymbol{\sigma}}^{-} \tag{7}$$

where the tensile effective stress and the compressive effective stress are

$$\begin{cases} \bar{\boldsymbol{\sigma}}^{+} = \mathbb{P}^{+} : \bar{\boldsymbol{\sigma}} = \sum_{t} H(\hat{\bar{\sigma}}_t)\hat{\bar{\sigma}}_t \boldsymbol{p}^{(t)} \otimes \boldsymbol{p}^{(t)} \\ \bar{\boldsymbol{\sigma}}^{-} = \mathbb{P}^{-} : \bar{\boldsymbol{\sigma}} = \sum_{t} [1 - H(\hat{\bar{\sigma}}_t)]\hat{\bar{\sigma}}_t \boldsymbol{p}^{(t)} \otimes \boldsymbol{p}^{(t)} \end{cases} \tag{8}$$

and the projective tensors

$$\begin{cases} \mathbb{P}^{+} = \sum_{t} H(\hat{\bar{\sigma}}_t) \boldsymbol{p}^{(t)} \otimes \boldsymbol{p}^{(t)} \otimes \boldsymbol{p}^{(t)} \otimes \boldsymbol{p}^{(t)} \\ \mathbb{P}^{-} = \mathbb{I} - \mathbb{P}^{+} \end{cases} \tag{9}$$

where $\hat{\bar{\sigma}}_t$ and $\boldsymbol{p}^{(t)}$ are the t-th eigenvalue and eigenvector of the effective stress tensor, respectively. And the Heaviside function reads

$$H(x) = \begin{cases} 0 & x \leqslant 0 \\ 1 & x > 0 \end{cases} \tag{10}$$

As shown in Fig. 1, the eigen-based effective stress split defines the tensile stress state and the compressive stress state respectively. The projective tensors $\mathbb{P}^{+}$ and $\mathbb{P}^{-}$, which define the eigen-tensors of the fourth order damage tensor, are expressed by the principal directions of effective stress. Thus the proposed damage model is a kind of equivalent form to the rotating angle shear model[14] because the rotation of the principal effective stress yields the rotation of the principal damage.

The evolutions of the tensile damage D^{+} and the compressive damage D^{-} should be governed by the corresponding stress states. Meanwhile, the irreversible thermodynamics suggests the energy conjugated force as the driven force (criterion) of the internal variable. Thus by proposing the expression of Helmholtz free energy(HFE)

$$\Psi = (1 - D^{+})\Psi_0^{+} + (1 - D^{-})\Psi_0^{-} \tag{11}$$

the energy conjugated quantities of the tensile and the compressive damages, which are defined as the damage release rates, are

$$\begin{cases} Y^{+} = -\dfrac{\partial \Psi}{\partial D^{+}} = \Psi_0^{+} \\ Y^{-} = -\dfrac{\partial \Psi}{\partial D^{-}} = \Psi_0^{-} \end{cases} \tag{12}$$

The initial HFEs Ψ_0^{+} and Ψ_0^{-} could be further split into the elastic part and the plastic part. We have

$$\begin{cases} \Psi_0^{+} = \Psi_0^{e+} + \Psi_0^{p+} \\ \Psi_0^{-} = \Psi_0^{e-} + \Psi_0^{p-} \end{cases} \tag{13}$$

The elastic parts

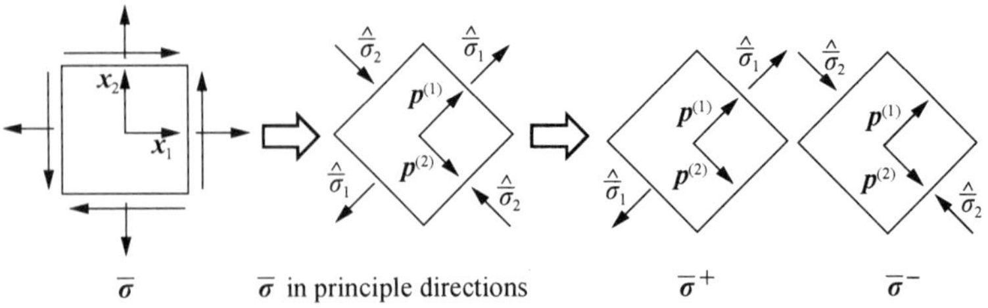

Fig. 1. Effective stress decomposition

$$\Psi_0^{e\pm} = \frac{1}{2}\bar{\boldsymbol{\sigma}}^{\pm} : \boldsymbol{\epsilon}^e = \frac{1}{2}\bar{\boldsymbol{\sigma}}^{\pm} : \mathbb{E}_0^{-1} : \bar{\boldsymbol{\sigma}} \tag{14}$$

and the plastic parts[16]

$$\Psi_0^{p\pm} \approx \int \bar{\boldsymbol{\sigma}}^{\pm} : d\boldsymbol{\epsilon}^p \tag{15}$$

Wu et al.[47] neglected the tensile plastic HFE Ψ_0^{p+} and introduced the Drucker-Prager plastic model for the compressive HFE Ψ_0^{p-}. They derived the approximate explicit expressions of the damage release rates as follows

$$\begin{cases} Y^+ \approx \dfrac{1}{2E_0}\left[\dfrac{2(1+v_0)}{3}3\bar{J}_2^+ + \dfrac{1-2v_0}{3}(\bar{I}_1^+)^2 - v_0\bar{I}_1^+\bar{I}_1^-\right] \\ Y^- \approx b_0(\alpha\bar{I}_1^- + \sqrt{3\bar{J}_2^-})^2 \end{cases} \tag{16}$$

where $\bar{I}_1^{\pm}$ are the first invariants of the tensile and the compressive effective stresses $\bar{\boldsymbol{\sigma}}^{\pm}$, respectively; and $\bar{J}_2^{\pm}$ are the second invariants of $\bar{s}^{\pm}$, which is the deviatoric components of effective stresses. E_0 and v_0 are the Young's modulus and the Poisson's ratio of the initial undamaged material; and b_0 is also a material parameter. The parameter α is related to the biaxial strength increase as follows

$$\alpha = \frac{\dfrac{f_{bc}}{f_c} - 1}{2\dfrac{f_{bc}}{f_c} - 1} \tag{17}$$

where f_c and f_{bc} are the uniaxial compressive strength and the biaxial compressive strength, respectively. It is observed from Eq. (16) that the developed tensile damage release rate Y^+ is governed by the elastic energy and the developed compressive damage release rate Y^- is governed by the plastic energy.

Moreover, the damage evolutions are defined by the energy release rates as follows

$$D^{\pm} = G^{\pm}(r_Y^{\pm}), \quad r_Y^{\pm} = \max_{\tau \in [0, t]}\{Y^{\pm}(\tau)\} \tag{18}$$

where the monotonic functions $G^{\pm}(\cdot)$ are the damage evolution functions; and the damage thresholds $r_Y^{\pm}$ denote the maximum values of $Y^{\pm}$ throughout the whole loading process

$[0, t]$. Based on the damage release rate dependent damage evolution and the damage consistent condition, Li and Ren[23] further developed the damage evolution driven by the energy equivalent strain as follows

$$D^{\pm}=G^{\pm}(r_e^{\pm}),\quad r_e^{\pm}=\max_{\tau\in[0,t]}\{\bar{\epsilon}^{e\pm}(\tau)\} \tag{19}$$

where the energy equivalent strain

$$\begin{cases}\bar{\epsilon}^{e+}=\sqrt{\dfrac{2Y^+}{E_0}}\\[2ex] \bar{\epsilon}^{e-}=\dfrac{1}{E_0(1-\alpha)}\sqrt{\dfrac{Y^-}{b_0}}\end{cases} \tag{20}$$

Under uniaxial loading, the energy equivalent strains are reduced to the uniaxial strain while the damage evolutions are reduced to the uniaxial damage evolution as follows

$$D^{\pm}=G^{\pm}(r_e^{\pm}),\quad r_e^{\pm}=\max_{\tau\in[0,t]}(\epsilon_\tau^{e\pm}) \tag{21}$$

It is observed that the multiaxial damage evolution Eq.(19) is consistent with the uniaxial damage evolution Eq.(21). The damage evolution functions $G^{\pm}(\cdot)$ could be determined by the uniaxial testing data or the theoretical development in the sub-scale.
The evolutionary models for the damage and the plasticity are developed in Chapter 3 and Chapter 4, respectively.

3 Damage evolution

To consider the material rate-sensitivity, the rate dependent damage evolutions could be developed based on the inviscid damage evolutions shown in Eq.(19). By introducing the rate-dependent energy equivalent strain $\bar{\epsilon}_r^{e\pm}$, the rate-dependent damage evolutions are

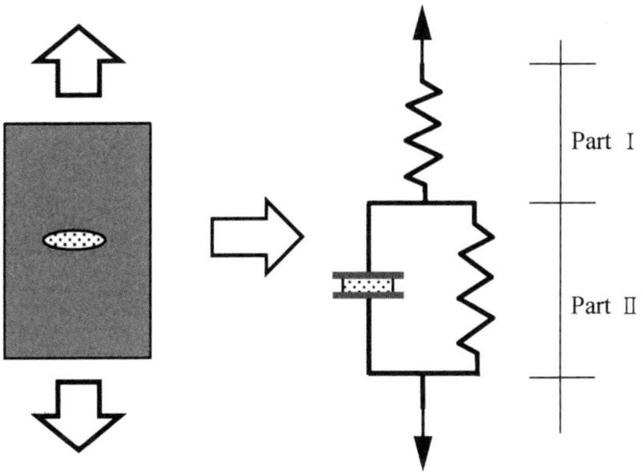

Fig. 2. Physical based rate-dependent model

$$\begin{cases} D^{\pm} = G^{\pm}(r_{\mathrm{d}}^{\pm}) \\ r_{\mathrm{d}}^{\pm} = \max_{\tau \in [0,\, t]} \{\bar{\epsilon}_{\mathrm{r}}^{\mathrm{e}\pm}(\tau)\} \\ \mathcal{R}^{\pm}(\bar{\epsilon}_{\mathrm{r}}^{\mathrm{e}\pm},\, \dot{\bar{\epsilon}}_{\mathrm{r}}^{\mathrm{e}\pm},\, \bar{\epsilon}^{\mathrm{e}\pm},\, \dot{\bar{\epsilon}}^{\mathrm{e}\pm}) = 0 \end{cases} \tag{22}$$

Based on the Perzyna's theory[30], Ren and Li[32] proposed the phenomenological expression of the differential operator $\mathcal{R}^{\pm}$. In the present work, an alternative model is proposed based on the concept of Stefan effect.

Stefan[42] and Reynolds[34] investigated the problem of two separating plates with a thin viscous film in-between. The solution of this problem indicated the linear dependencies between the applied forces and the separating velocity of the plates. Thereafter, the findings was defined as Stefan effects and widely used to describe the material rate-dependencies. The derivation of the Stefan effect could be found in Appendix A. In the proposed model, we consider that the material rate-sensitivity comes from the Stefan effect.

We assume that the rate-dependency is induced by the water (including the void water and the crystal water) within concrete. As shown in Fig. 2, two parts are considered in the model. Part Ⅰ represents the elastic part of concrete specimen and Part Ⅱ is the combination of the solid matrix and the viscous fluid. We obtain the following equation after analyzing the model shown in Fig. 2.

$$\begin{cases} \epsilon_1 = \dfrac{\sigma}{E_1} \\ \epsilon_2 = \dfrac{\sigma - \sigma_v}{E_2} \end{cases} \tag{23}$$

where σ is the uniaxial stress; ϵ_1 and ϵ_2 are the strains of Part Ⅰ and Part Ⅱ, respectively; and E_1 and E_2 are the elastic parameters of Part Ⅰ and Part Ⅱ, respectively. Based on the discussion of the Stefan effect in Appendix A, the viscous stress representing the rate effect of the model could be expressed as follows

$$\sigma_v = A\dot{\epsilon}_2 \tag{24}$$

The viscous coefficient A has been defined in Appendix A. Based on Eqs.(23) and (24), the overall strain of the model could be expressed as

$$\epsilon_{\mathrm{r}} = \epsilon_1 + \epsilon_2 = \sigma\left(\frac{1}{E_1} + \frac{1}{E_2}\right) - \frac{\sigma_v}{E_2} \tag{25}$$

For the quasi-static loading, we have $\sigma_v \to 0$. Eq.(25) reduces to

$$\epsilon = \epsilon_1 + \epsilon_2 = \frac{\sigma}{E_0} \tag{26}$$

where $\dfrac{1}{E_0} = \dfrac{1}{E_1} + \dfrac{1}{E_2}$ defines the equivalent uniaxial elastic parameters E_0. It is also observed that ϵ_r is the rate dependent strain of the system shown in Fig. 2 and ϵ is the quasi-static strain. And according to Eq.(25) we could further define $\epsilon_v = \frac{\sigma_v}{E_2}$ as the viscosity induced strain decrease. Manipulating Eqs. (23)—(26) yields the following first order differential system

$$\begin{cases} \gamma_a \dot{\epsilon}_2 + (\alpha_n + \beta_n)\epsilon_2 = \alpha_n \epsilon_r \\ \gamma_a \dot{\epsilon}_2 + \beta_n \epsilon_r = \beta_n \epsilon \end{cases} \tag{27}$$

where the coefficients

$$\gamma_a = \frac{A}{E_0}, \ \alpha_n = 1 + n, \ \beta_n = \frac{1+n}{n}, \ n = \frac{E_1}{E_2} \tag{28}$$

By giving the static strain ϵ, the rate-dependent strain ϵ_r and the strain of Part II ϵ_2 could be solved by the dual system shown in Eq.(27). It is indicated that ϵ_r is needed to further calculate the rate-dependent damage, and particularly ϵ_2 is an internal variable which memorizes the strain history.

The numerical algorithm to solve Eq. (27) under general loading conditions will be developed in Chapter 5. The closed form solution of Eq.(27) reads

$$\begin{cases} \epsilon_2 = \epsilon_r - \dfrac{\beta_n \epsilon}{\alpha_n + \beta_n} \\ \epsilon_r = \dfrac{\beta_n e^{-\frac{\beta_n t}{\gamma_a}}}{\gamma_a (\alpha_n + \beta_n)} \displaystyle\int e^{\frac{\beta_n t}{\gamma_a}} [\gamma_a \dot{\epsilon} + (\alpha_n + \beta_n)\epsilon] dt + C_1 e^{-\frac{\beta_n t}{\gamma_a}} \end{cases} \tag{29}$$

where C_1 is a constant related to the initial conditions. Considering the constant rate loading $\epsilon = \dot{\epsilon} t$ and the initial condition $[\epsilon(0) = 0, \epsilon_r(0) = 0]$, the solution Eq.(29) further gives

$$\epsilon_r = \epsilon - \epsilon_v \tag{30}$$

where the viscous strain

$$\epsilon_v = \frac{\alpha_n \gamma_a}{\beta_n (\alpha_n + \beta_n)} (1 - e^{-\frac{\beta_n t}{\gamma_a}}) \dot{\epsilon} \tag{31}$$

The closed form solution shown in Eqs.(30) and (31) could verify the simulating results of the numerical scheme.

The development of damage evolution function is refer to [17] and [23]. The stochastic damage evolution functions under static loading are expressed as follows

$$D = \int_0^1 H[\epsilon - \Delta(x)] dx \tag{32}$$

where $\Delta(x)$ is the 1-D micro-fracture strain random field; x is the spatial coordinate of the random field. And it is observed that ϵ is an constant subjected to spatial integration. By replacing the static strain ϵ with the dynamic strain ϵ_r, we have the rate-dependent stochastic damage evolution as follows

$$D = \int_0^1 H[\epsilon_r - \Delta(x)]\mathrm{d}x \hat{=} G(\epsilon_r) \tag{33}$$

Consider that $\Delta(x)$ is a homogeneous log-normal random field. Let λ and ζ be two distribution parameters which denote the mean value and the standard deviation of the corresponding normal random variable $Z(x) = \ln \Delta(x)$, respectively. And the auto-correlation coefficient function for $Z(x)$ is as follows

$$\rho_z(y) = \mathrm{e}^{-\xi|y|} \tag{34}$$

where ξ is the correlation parameter. Then the mean value and the standard deviation of damage evolutions expressed in Eq.(33) could be solved. The mean value reads

$$d = \mu_D = \Phi_i\left(\frac{\ln \epsilon_r - \lambda}{\zeta}\right) \hat{=} g(\epsilon_r) \tag{35}$$

and the standard deviation reads

$$V_D^2 = 2\int_0^1 (1-y)\Phi_{ii}\left(\frac{\ln \epsilon_r - \lambda}{\zeta}, \frac{\ln \epsilon_r - \lambda}{\zeta} \middle| \rho_z(y)\right)\mathrm{d}y - \mu_D^2 \tag{36}$$

where functions Φ_i and Φ_{ii} are the cumulative probability functions of the 1-D and the 2-D standard normal distributions, respectively. For the deterministic analysis, the mean value evolution Eq.(35) is adopted and the damage variable is denoted by the lower case variable d. Before implementing the developed uniaxial rate-dependent damage evolutionary model [Eqs.(27)—(36)] into the multiaxial framework, the following aspects should be emphasized:

(1) Due to the essential differences between the behaviors of quasi-brittle materials under tension and compression, two groups of equations with similar expressions but different parameters are developed to reproduce the tensile damage and the compressive damage, respectively. And the superscripts + and − are adopted to denote the quantities related the tensile and the compressive damages as appropriate.

(2) The strains mentioned in the proposed method are refereed to the elastic strain. And the additional plastic strains will be modeled in the next section. By substituting the energy equivalent strain $\bar{\epsilon}^{e\pm}$ into Eqs. (27) as the static strain ϵ, the rate-dependent energy equivalent strain $\bar{\epsilon}_r^{e\pm}$ could be solved as ϵ_r. Thereafter the dynamic damage thresholds $r_d^{\pm}$ could be determined by considering the maximum values of $\bar{\epsilon}_r^{e\pm}$ over the entire loading process. Further substituting $r_d^{\pm}$ into the damage evolution functions Eqs.(33) and (35), the stochastic as well as the deterministic damages could be obtained.

4 Multi-variable plasticity

As discussed in Chapter 2, the T-ESP exhibits numbers of deficiencies in application. Hence a phenomenological model to describe the ESP is developed in the present chapter. On one hand, the proposed phenomenological plastic model should be also defined in the effective stress space to take the advantage of the ESP. On the other hand, the plastic strain is usually defined as a function of strain in experimental studies. Thus we consider that the evolution of plastic strain is proportional to the evolution of elastic strain, which is closely related to the effective stress. One obtains

$$\dot{\boldsymbol{\epsilon}}^{\mathrm{p}} = f_{\mathrm{p}} \dot{\boldsymbol{\epsilon}}^{\mathrm{e}} \tag{37}$$

Besides the damage, the evolutions of plastic strain are also rather different under tension and compression. Thus the split of plastic strain rate is introduced as follows

$$\begin{cases} \dot{\boldsymbol{\epsilon}}^{\mathrm{p}} = \dot{\boldsymbol{\epsilon}}^{\mathrm{p}+} + \dot{\boldsymbol{\epsilon}}^{\mathrm{p}-} \\ \dot{\boldsymbol{\epsilon}}^{\mathrm{p}+} = f_{\mathrm{p}}^{+} \dot{\boldsymbol{\epsilon}}^{\mathrm{e}+} = f_{\mathrm{p}}^{+} \mathbb{E}_0^{-1} : \dot{\bar{\boldsymbol{\sigma}}}^{+} \\ \dot{\boldsymbol{\epsilon}}^{\mathrm{p}-} = f_{\mathrm{p}}^{-} \dot{\boldsymbol{\epsilon}}^{\mathrm{e}-} = f_{\mathrm{p}}^{-} \mathbb{E}_0^{-1} : \dot{\bar{\boldsymbol{\sigma}}}^{-} \end{cases} \tag{38}$$

The scalars $f_{\mathrm{p}}^{\pm}$ are considered as functions of the corresponding damage variables $D^{\pm}$. We define the plastic theory shown in Eq. (38) as multi-variable plasticity because more than one plastic strain tensors are considered. By systematic investigations of experimental data, we propose the following expressions

$$f_{\mathrm{p}}^{\pm} = f_{\mathrm{p}}^{\pm}(\dot{D}^{\pm}, D^{\pm}) = H(\dot{D}^{\pm}) \xi_{\mathrm{p}}^{\pm} (D^{\pm})^{n_{\mathrm{p}}^{\pm}} \tag{39}$$

where $\xi_{\mathrm{p}}^{\pm}$ and $n_{\mathrm{p}}^{\pm}$ are material parameters fitted by the experimental data; and the Heaviside function $H(\cdot)$ defines the associate evolutions between damage and plasticity. As discussed in Chapter 3, the inevitable randomness of damages could be described by a random field. Thus the plastic evolutions are also random and related to the stochastic damage evolutions. On the other hand, the coupling between the randomness of damage and the nonlinearity of plastic evolution yields extremely complicated solutions of the total stress strain relationship Eq. (1). Thus a reduced form of the plastic evolution Eq. (39) is proposed as follows by using the mean values of damage (denoted by the lower case character $d^{\pm}$).

$$f_{\mathrm{p}}^{\pm} = f_{\mathrm{p}}^{\pm}(\dot{D}^{\pm}, d^{\pm}) = H(\dot{D}^{\pm}) \xi_{\mathrm{p}}^{\pm} (d^{\pm})^{n_{\mathrm{p}}^{\pm}} \tag{40}$$

By combining Eqs. (38) and (40), we obtain the governing equation of plastic evolution as follows

$$\begin{aligned} \dot{\boldsymbol{\epsilon}}^{\mathrm{p}} &= H(\dot{D}^{+}) \xi_{\mathrm{p}}^{+} (d^{+})^{n_{\mathrm{p}}^{+}} \mathbb{E}_0^{-1} : \dot{\bar{\boldsymbol{\sigma}}}^{+} + H(\dot{D}^{-}) \xi_{\mathrm{p}}^{-} (d^{-})^{n_{\mathrm{p}}^{-}} \mathbb{E}_0^{-1} : \dot{\bar{\boldsymbol{\sigma}}}^{-} \\ &= H(\dot{D}^{+}) \xi_{\mathrm{p}}^{+} (d^{+})^{n_{\mathrm{p}}^{+}} \mathbb{E}_0^{-1} : \mathbb{Q}^{+} : \dot{\bar{\boldsymbol{\sigma}}} + H(\dot{D}^{-}) \xi_{\mathrm{p}}^{-} (d^{-})^{n_{\mathrm{p}}^{-}} \mathbb{E}_0^{-1} : \mathbb{Q}^{-} : \dot{\bar{\boldsymbol{\sigma}}} \end{aligned} \tag{41}$$

The definition of the effective stress rate projective tensors[11] is

$$\dot{\bar{\boldsymbol{\sigma}}}^{\pm} = \mathbb{Q}^{\pm} : \dot{\bar{\boldsymbol{\sigma}}} \tag{42}$$

It is clear that $\mathbb{Q}^{\pm}$ are rather different from the effective stress projective tensors $\mathbb{P}^{\pm}$. By referring to[47], the present paper gives the expression of them as follows

$$\begin{cases} \mathbb{Q}^{+} = \mathbb{P}^{+} + 2 \sum_{t=1}^{3} \sum_{s=0}^{t} \dfrac{\langle \hat{\bar{\sigma}}_t \rangle - \langle \hat{\bar{\sigma}}_s \rangle}{\hat{\bar{\sigma}}_t - \hat{\bar{\sigma}}_s} \boldsymbol{P}^{(ts)} \otimes \boldsymbol{P}^{(ts)} \\ \mathbb{Q}^{-} = \mathbb{I} - \mathbb{Q}^{+} \end{cases} \tag{43}$$

where the second order symmetric tensor

$$\boldsymbol{P}^{(ts)}=\frac{1}{2}(\boldsymbol{p}^{(t)}\otimes\boldsymbol{p}^{(s)}+\boldsymbol{p}^{(t)}\otimes\boldsymbol{p}^{(s)}) \tag{44}$$

where the superscripts t and s in brackets denote the orders of the eigenvalues or the eigenvectors of the effective stress. And the Macaulay brackets

$$\langle x\rangle=\frac{|x|+x}{2} \tag{45}$$

To determine the elasoplastic tangential stiffness tensor, the differentiation of Eq. (2) yields

$$\dot{\bar{\boldsymbol{\sigma}}}=\mathbb{E}_0:(\dot{\boldsymbol{\epsilon}}-\dot{\boldsymbol{\epsilon}}^{\mathrm{p}}) \tag{46}$$

Combining Eqs. (41) and (46) gives

$$[\mathbb{I}+H(\dot{D}^{+})\xi_p^{+}(d^{+})^{n_p^{+}}\mathbb{Q}^{+}+H(\dot{D}^{-})\xi_p^{-}(d^{-})^{n_p^{-}}\mathbb{Q}^{-}]:\dot{\bar{\boldsymbol{\sigma}}}=\mathbb{E}_0:\dot{\boldsymbol{\epsilon}} \tag{47}$$

Hence the elasoplastic tangential stiffness tensor reads

$$\mathbb{E}^{\mathrm{ep}}=[\mathbb{I}+H(\dot{D}^{+})\xi_p^{+}(d^{+})^{n_p^{+}}\mathbb{Q}^{+}+H(\dot{D}^{-})\xi_p^{-}(d^{-})^{n_p^{-}}\mathbb{Q}^{-}]^{-1}:\mathbb{E}_0 \tag{48}$$

The rate-dependency of plasticity is not explicitly included in the proposed plastic model. However, due to the rate dependencies of damage evolutions, the plastic evolution is technically related to the strain rates. Furthermore, although the linear rate-dependency is defined by Eq. (27), the proposed plastic damage model exhibits strong nonlinearities dependent to the strain rate. The reason could refer to the nonlinear dependency between the damage and the plasticity proposed in the present work.
Further define the plastic stress and plastic stress rates as follows

$$\begin{cases}\boldsymbol{\sigma}^{\mathrm{p}\pm}=\mathbb{E}_0:\boldsymbol{\epsilon}^{\mathrm{p}\pm}\\ \dot{\boldsymbol{\sigma}}^{\mathrm{p}\pm}=\mathbb{E}_0:\dot{\boldsymbol{\epsilon}}^{\mathrm{p}\pm}\end{cases} \tag{49}$$

Recall Eqs. (38) and (40), we have

$$\dot{\boldsymbol{\sigma}}^{\mathrm{p}\pm}=H(\dot{D}^{\pm})\xi_p^{\pm}(d^{\pm})^{n_p^{\pm}}\dot{\bar{\boldsymbol{\sigma}}}^{\pm} \tag{50}$$

Thus Eq. (46) could be recast into the following form

$$\dot{\bar{\boldsymbol{\sigma}}}=\mathbb{E}_0:\dot{\boldsymbol{\epsilon}}-\dot{\boldsymbol{\sigma}}^{\mathrm{p}+}-\dot{\boldsymbol{\sigma}}^{\mathrm{p}-} \tag{51}$$

Theoretically, the plastic stress $\boldsymbol{\sigma}^{\mathrm{p}\pm}$ are equivalent to the plastic strain $\boldsymbol{\epsilon}^{\mathrm{p}\pm}$. But Eq. (51) suggests that $\boldsymbol{\sigma}^{\mathrm{p}\pm}$ works better with the effective stress and yields very concise expressions. Thus the numerical scheme is developed in the form of plastic stress.

5 Numerical scheme

At each integration point of the structural finite element model, the unknown stress increments should be calculated by the constitutive model during each time-step. As shown in the previous chapters, the proposed constitutive relationship defines a strong nonlinear system made up of a set of coupled nonlinear equations. Thus the numerical implementation of the proposed model should be carefully considered to guarantee the accuracy and the stability of the structural simulation. The existing numerical schemes for the computation of the material inelasticity could be concluded as follows:

(1) The implicit methods are generally unconditionally stable. Hence a relatively long time step could be adopted so that less numbers of time step is needed. And another attractive aspect indicates that the results of the implicit method automatically satisfy the yield condition. On the other hand, these methods need iterations to solve the resulting implicit equations. The internal iterations may diverge for the strain softening problem or other ill-conditioned cases. And the second derivatives of the yield functions, which are rather complicated for the general forms of the yield functions, are required to calculate the algorithm consistent tangential tensor.

(2) The explicit methods are conditionally stable. Thus they require rather small time steps to avoid intolerable errors, so that a large number of time steps are needed. And the results of these methods may not satisfy the yield condition to the predefined tolerance. However, the explicit methods offer an extremely simple but robust computations in each time step. No iteration is needed and the order of derivatives remains to one. Moreover, the explicit methods fit the parallel code very well and is able to achieve excellent efficiency for the parallel computation.

In the present paper, an explicit scheme is developed to implement the proposed rate-dependent damage-plasticity model. Denote the subscript form x_k as the quantities at the k-th time step. And the finite difference is defined by $\Delta x_k = x_{k+1} - x_k$. By giving the values of the state variables and the internal variables in the k-th time step and also the strain in the (k+1)-th step, the proposed scheme approaches the values of the state variables and the internal variables in the (k+1)-th step.

We start with the elastic trial. By freezing the evolutions of plasticity and damage, the trial solution of the effective stress as follows

$$\bar{\boldsymbol{\sigma}}_{k+1}^{\text{trial}} = \mathbb{E}_0 : \boldsymbol{\epsilon}_{k+1} - \boldsymbol{\sigma}_k^{\text{p}} \tag{52}$$

Then the trial solutions of the damage energy release rates $(Y^{\pm})_{k+1}^{\text{trial}}$ and the energy equivalent strains $(\bar{\epsilon}^{\text{e}\pm})_{k+1}^{\text{trial}}$ could be explicitly determined. The rate dependent energy equivalent strains could be solved by the finite difference discretization of Eqs. (27) as follows

$$\begin{cases} \gamma_a^{\pm}[(\bar{\epsilon}_2^{\pm})_{k+1} - (\bar{\epsilon}_2^{\pm})_k] + (\alpha_n^{\pm} + \beta_n^{\pm})(\bar{\epsilon}_2^{\pm})_{k+1}\Delta t_k = \alpha_n^{\pm}(\bar{\epsilon}_r^{\text{e}\pm})_{k+1}\Delta t_k \\ \gamma_a^{\pm}[(\bar{\epsilon}_2^{\pm})_{k+1} - (\bar{\epsilon}_2^{\pm})_k] + \beta_n^{\pm}(\bar{\epsilon}_r^{\text{e}\pm})_{k+1}\Delta t_k = \beta_n^{\pm}(\bar{\epsilon}^{\text{e}\pm})_{k+1}\Delta t_k \end{cases} \tag{53}$$

Cast into matrix expression, we obtain

$$\begin{bmatrix} \gamma_a^{\pm} + (\alpha_n^{\pm} + \beta_n^{\pm})\Delta t_k & -\alpha_n^{\pm}\Delta t_k \\ \gamma_a^{\pm} & \beta_n^{\pm}\Delta t_\kappa \end{bmatrix} \begin{Bmatrix} (\bar{\epsilon}_2^{\pm})_{k+1} \\ \bar{\epsilon}_r^{\text{e}\pm})_{k+1} \end{Bmatrix} = \begin{Bmatrix} \gamma_a^{\pm}(\bar{\epsilon}_2^{\pm})_k \\ \gamma_a^{\pm}(\epsilon_2^{\pm})_k + \beta_n^{\pm}(\bar{\epsilon}^{\text{e}\pm})_{k+1}\Delta t_k \end{Bmatrix} \tag{54}$$

An inversion gives

$$\begin{Bmatrix}(\bar{\epsilon}_2^{\pm})_{k+1}\\ (\bar{\epsilon}_{\mathrm{r}}^{\mathrm{e}\pm})_{k+1}\end{Bmatrix}=\begin{bmatrix}\dfrac{\gamma_a^{\pm}}{\gamma_a^{\pm}+\beta_n^{\pm}\Delta t_k} & \dfrac{\alpha_n^{\pm}\beta_n^{\pm}\Delta t_k}{(\alpha_n^{\pm}+\beta_n^{\pm})(\gamma_a^{\pm}+\beta_n^{\pm}\Delta t_k)}\\ \dfrac{\gamma_a^{\pm}}{\gamma_a^{\pm}+\beta_n^{\pm}\Delta t_k} & \dfrac{\gamma_a^{\pm}\beta_n^{\pm}+(\alpha_n^{\pm}+\beta_n^{\pm})\Delta t_k\beta_n^{\pm}}{(\alpha_n^{\pm}+\beta_n^{\pm})(\gamma_a^{\pm}+\beta_n^{\pm}\Delta t_k)}\end{bmatrix}\begin{Bmatrix}(\bar{\epsilon}_2^{\pm})_k\\ (\bar{\epsilon}^{\mathrm{e}\pm})_{k+1}\end{Bmatrix} \tag{55}$$

Substituting the trial energy equivalent strains $(\bar{\epsilon}^{\mathrm{e}\pm})_{k+1}^{\mathrm{trial}}$ into the discrete system Eq. (54), the trial rate-dependent energy equivalent strains $(\bar{\epsilon}_r^{\mathrm{e}\pm})_{k+1}^{\mathrm{trial}}$ and the trial internal energy equivalent strains $(\bar{\epsilon}_2^{\pm})_{k+1}^{\mathrm{trial}}$ could be solved. Thus the trial damage criteria gives

$$(R^{\pm})_{k+1}^{\mathrm{trial}}=|(\bar{\epsilon}_r^{\mathrm{e}\pm})_{k+1}^{\mathrm{trial}}|-|(r_{\mathrm{d}}^{\pm})_k| \tag{56}$$

If

$$(R^{\pm})_{k+1}^{\mathrm{trial}}\leqslant 0 \tag{57}$$

the elastic state holds. We obtain

$$\begin{aligned}&D_{k+1}^{\pm}=D_k^{\pm},\ \boldsymbol{\sigma}_{k+1}^{\mathrm{p}}=\boldsymbol{\sigma}_k^{\mathrm{p}}\\ &(r_{\mathrm{d}}^{\pm})_{k+1}=(r_{\mathrm{d}}^{\pm})_k,\ (\bar{\epsilon}_2^{\pm})_{k+1}=(\bar{\epsilon}_2^{\pm})_k,\ \bar{\boldsymbol{\sigma}}_{k+1}=\bar{\boldsymbol{\sigma}}_{k+1}^{\mathrm{trial}}\\ &\boldsymbol{\sigma}_{k+1}=\bar{\boldsymbol{\sigma}}_{k+1}-D_{k+1}^{+}\bar{\boldsymbol{\sigma}}_{k+1}^{+}-D_{k+1}^{-}\bar{\boldsymbol{\sigma}}_{k+1}^{-}\end{aligned} \tag{58}$$

For $(R^{\pm})_{k+1}^{\mathrm{trial}}>0$, the damage and the plasticity evolve. The explicit methods usually require special attentions to the case that the effective stress point changes from the elastic stage to the nonlinear stage (Fig. 3).

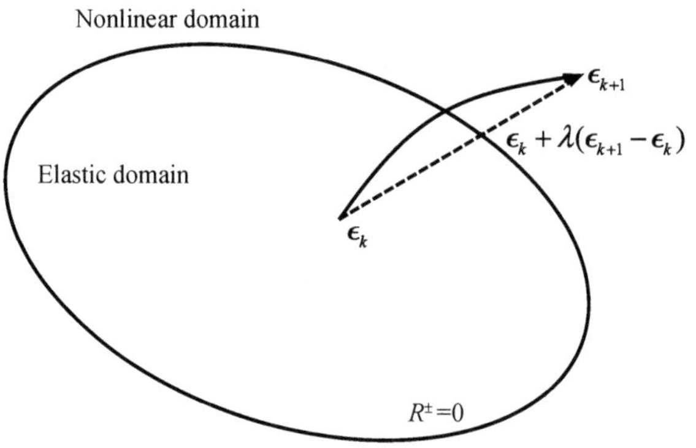

Fig. 3. Damage criterion intersection

The current time-step includes an elastic sub-step following by a nonlinear sub-step. And different governing equations hold for different sub-steps, respectively. If the state variables in the nonlinear domain are directly predicted by the starting point located in the elastic domain, the abnormal errors would happen during the calculation. Thus the transition point should be determined beforehand. Then the integration could be performed through two sub-steps with corresponding starting points and governing equations. As the time increments are usually very small for the explicit schemes to guarantee the overall numerical stability, we could linearize the load path within the time increment into a straight path without loss of much accuracy. Sloan et al.[39], Soan[40] systematically investigated the explicit methods for the plastic models and proposed the Pegasus algorithm to find the intersection point of a straight loading path. In the present work, the Pegasus algorithm developed for damage model is shown in Algorithm 1. Actually, the elastic sub-step could be also finished during the execution of the Pegasus algorithm with the damage and the plasticity remain unchanged.

By starting at the end of the elastic sub-step, we evaluate the evolutions of damage and plasticity based on the forward Euler scheme. By recalling Eqs. (47) and (48), the forward Euler scheme gives

$$\mathbb{T}^{\mathrm{ep}}_{k+\lambda} : (\bar{\boldsymbol{\sigma}}_{k+1} - \bar{\boldsymbol{\sigma}}_{k+\lambda}) = \mathbb{E}_0 : (\boldsymbol{\epsilon}_{k+1} - \boldsymbol{\epsilon}_{k+\lambda}) = \bar{\boldsymbol{\sigma}}^{\mathrm{trial}}_{k+1} - \bar{\boldsymbol{\sigma}}_{k+\lambda} \tag{59}$$

begin
- Initialization with
- $\lambda_0^{\pm} \leftarrow 0;\ \lambda_1^{\pm} \leftarrow 1;$
- $R_0^{\pm} \leftarrow |(\bar{\epsilon}_{\mathrm{r}}^{\mathrm{e}\pm})_k| - |(r_{\mathrm{d}}^{\pm})_k|;\ R_1^{\pm} \leftarrow |(\bar{\epsilon}_{\mathrm{r}}^{\mathrm{e}\pm})_{k+1}^{\mathrm{trial}}| - |(r_{\mathrm{d}}^{\pm})_k|;$
- /*interations to solve the intersection point */
- **for** $i \leftarrow 1$ ***to*** MAXITS **do**
  - $\lambda^{\pm} \leftarrow \lambda_1^{\pm} - \frac{R_1^{\pm}}{R_1^{\pm} - R_0^{\pm}}(\lambda_1^{\pm} - \lambda_0^{\pm});$
  - $\boldsymbol{\epsilon}_{k+\lambda^{\pm}} \leftarrow \boldsymbol{\epsilon}_k + \lambda^{\pm}(\boldsymbol{\epsilon}_{k+1} - \boldsymbol{\epsilon}_k);$
  - $t_{k+\lambda^{\pm}} \leftarrow t_k + \lambda^{\pm}(t_{k+1} - t_k);$
  - Calculate $(\bar{\epsilon}_{\mathrm{r}}^{\mathrm{e}\pm})_{k+\lambda^{\pm}}$ and $(\bar{\epsilon}_2^{\mathrm{e}\pm})_{k+\lambda^{\pm}}$ by Eqs. (54);
  - $R_{\lambda^{\pm}}^{\pm} \leftarrow |(\bar{\epsilon}_{\mathrm{r}}^{\mathrm{e}\pm})_{k+\lambda^{\pm}}| - |(r_d^{\pm})_k|;$
  - **if** $|R_{\lambda^{\pm}}^{\pm}| \leqslant$ RTOL **then**
    - **break**;
  - **end**
  - **if** $R_{\lambda^{\pm}}^{\pm} R_0^{\pm} < 0$ **then**
    - $\lambda_1^{\pm} \leftarrow \lambda^{\pm};\ R_1^{\pm} \leftarrow R_{\lambda^{\pm}}^{\pm};$
    - **else**
      - $\lambda_0^{\pm} \leftarrow \lambda^{\pm};\ R_0^{\pm} \leftarrow R_{\lambda^{\pm}}^{\pm};$
    - **end**
  - **end**
- **end**
- **if** $|R_{\lambda^{\pm}}^{\pm}| \leqslant$ RTOL **then**
  - $\bar{\boldsymbol{\sigma}}_{k+\lambda^{\pm}} \leftarrow \bar{\boldsymbol{\sigma}}_k + \mathbb{E}_0 : (\boldsymbol{\epsilon}_{k+\lambda^{\pm}} - \boldsymbol{\epsilon}_k);$
  - $\boldsymbol{\sigma}_{k+\lambda^{\pm}} \leftarrow \bar{\boldsymbol{\sigma}}_{k+\lambda^{\pm}} - D_k^{+}\bar{\boldsymbol{\sigma}}_{k+\lambda^{\pm}}^{+} - D_k^{-}\bar{\boldsymbol{\sigma}}_{k+\lambda^{\pm}}^{-};$
  - **else**
    - **exit** with error message;
  - **end**
- **end**

end

Algorithm 1: Pegasus algorithm

Further manipulation yields

$$\mathbb{T}^{\mathrm{ep}}_{k+\lambda} : \bar{\boldsymbol{\sigma}}_{k+1} = \bar{\boldsymbol{\sigma}}^{\mathrm{trial}}_{k+1} - (\mathbb{I} - \mathbb{T}^{\mathrm{ep}}_{k+\lambda}) : \bar{\boldsymbol{\sigma}}_{k+\lambda} \tag{60}$$

where the fourth order tensor $\mathbb{T}^{\mathrm{ep}}$ at the k-th step is

$$\mathbb{T}^{\mathrm{ep}}_{k+\lambda} = \mathbb{I} + \theta_k^{\mathrm{p}+}\mathbb{Q}^{+}_{k+\lambda} + \theta_k^{\mathrm{p}-}\mathbb{Q}^{-}_{k+\lambda} \tag{61}$$

and the plastic factors are

$$\begin{aligned}\theta_k^{p+} &= H[(D^+)_{k+1}^{\text{trial}} - D_k^+]\xi_p^+[d_k^+]^{n_p^+}\\ \theta_k^{p-} &= H[(D^-)_{k+1}^{\text{trial}} - D_k^-]\xi_p^-[d_k^-]^{n_p^-}\end{aligned} \tag{62}$$

Recall Eq.(38), the plastic stress could be determined by

$$\boldsymbol{\sigma}_{k+1}^{p} = \boldsymbol{\sigma}_k^{p} + \theta_k^{p+}(\bar{\boldsymbol{\sigma}}_{k+1}^{+} - \bar{\boldsymbol{\sigma}}_{k+\lambda^+}^{+}) + \theta_k^{p-}(\bar{\boldsymbol{\sigma}}_{k+1}^{-} - \bar{\boldsymbol{\sigma}}_{k+\lambda^-}^{-}) \tag{63}$$

Then the energy equivalent strains $(\bar{\epsilon}_2^{\pm})_{k+1}$, $(\bar{\epsilon}^{\pm})_{k+1}$ and $(\bar{\epsilon}_r^{\pm})_{k+1}$ could be explicitly calculated. The damage criteria

$$\begin{cases}(r_d^+)_{k+1} = \operatorname{sgn}\{(r_d^+)_k\}\max\{|(\bar{\epsilon}_r^+)_{k+1}|, |(r_d^+)_k|\}\\ (r_d^-)_{k+1} = \operatorname{sgn}\{(r_d^-)_k\}\max\{|(\bar{\epsilon}_r^-)_{k+1}|, |(r_d^-)_k|\}\end{cases} \tag{64}$$

and the damage variables

$$\begin{cases}(D^{\pm})_{k+1} = G^{\pm}[(r_d^{\pm})_{k+1}]\\ (d^{\pm})_{k+1} = g^{\pm}[(r_d^{\pm})_{k+1}]\end{cases} \tag{65}$$

Finally the stress

$$\begin{aligned}\boldsymbol{\sigma}_{k+1} &= (\mathbb{I} - \mathbb{D}_{k+1}) : (\mathbb{E}_0 : \boldsymbol{\epsilon}_{k+1} - \boldsymbol{\sigma}_{k+1}^{p})\\ &= (\mathbb{I} - \mathbb{D}_{k+1}) : (\bar{\boldsymbol{\sigma}}_{k+1}^{\text{trial}} + \boldsymbol{\sigma}_k^{p} - \boldsymbol{\sigma}_{k+1}^{p})\end{aligned} \tag{66}$$

The time steps without elastic sub-step could be considered as the particular case of the elastic-nonlinear step with $\lambda^{\pm}=0$.

The inconsistency induced by the extrapolation could be observed for the effective stress. Recalling Eq.(60), we find

$$\begin{aligned}\bar{\boldsymbol{\sigma}}_{k+1} &= (\mathbb{T}_{k+\lambda}^{ep})^{-1} : [\bar{\boldsymbol{\sigma}}_{k+1}^{\text{trial}} - (\mathbb{I} - \mathbb{T}_{k+\lambda}^{ep})\bar{\boldsymbol{\sigma}}_{k+\lambda}]\\ &\neq \mathbf{C}_0 : \boldsymbol{\epsilon}_{k+1} - \boldsymbol{\sigma}_{k+1}^{p} \mathrel{\hat{=}} \tilde{\bar{\boldsymbol{\sigma}}}_{k+1}\end{aligned} \tag{67}$$

Thus the drift of damage criteria is

$$\begin{aligned}\tilde{R}_{k+1}^{\pm} &= |\bar{\epsilon}_{\mathrm{r}}^{\pm}(\tilde{\bar{\boldsymbol{\sigma}}}_{k+1})| - |(r_{\mathrm{d}}^{\pm})_{k+1}|\\ &= |\bar{\epsilon}_{\mathrm{r}}^{\pm}(\tilde{\bar{\boldsymbol{\sigma}}}_{k+1})| - |r_{\mathrm{d}}^{\pm}(\bar{\boldsymbol{\sigma}}_{k+1})| \neq 0\end{aligned} \tag{68}$$

If the drift exceeds the tolerance, the global convergence of the explicit algorithm will be lost. Thus the criteria to control the global convergence of the numerical simulation should be put as follows

$$|\tilde{R}_{k+1}^{\pm}| \leqslant \text{GTOL} \tag{69}$$

According to Reference [39], the suitable values of the global tolerance GTOL are typically in the range 10^{-9}–10^{-6}.

The complete explicit algorithm for the proposed damage-plasticity model is summarized in Algorithm 2.

6 Stability analysis

6.1 Physical stability analysism

The regularization operator $\mathcal{R}^{\pm}$ in Eq. (22) represents the material rate-dependency by slowing down the evolutions of damages $D^{\pm}$ under dynamic loading. It has been extensively investigated that the damage induced softening may yield strain localization, which is usually referred as the damage induced instability. On the other hand, the strain rate in the strain localized domain is basically larger than that in the rest part of solid. Thus the damage evolution in the strain localized domain would be slowed down much more when the rate-dependency is taken into account. This effect could be considered as a localization limiter. Needleman first showed that the material rate-dependency can remove the instability[26]. In the present work, the regularization operator $\mathcal{R}^{\pm}$ also introduces a kind of non-locality for the spatial damage propagation, which may relieve the instability in the simulation of softening solids.

begin
Initialization with $\boldsymbol{\sigma}_0, \bar{\boldsymbol{\sigma}}_0, \boldsymbol{\epsilon}_0, \boldsymbol{\sigma}_0^{\mathrm{p}}, (\bar{\epsilon}_2^{\pm})_0, D_0^{\pm}, d_0^{\pm}, (r_d^{\pm})_0, \tilde{R}_0^{\pm}$;
for $k \leftarrow 0$ ***to*** Nk **do**
/*elastic trial */
$\bar{\boldsymbol{\sigma}}_{k+1}^{\mathrm{trial}} \leftarrow \mathbb{E}_0 : \boldsymbol{\epsilon}_{k+1} - \boldsymbol{\sigma}_k^{\mathrm{p}}$;
Calculate $(\bar{\epsilon}^{\mathrm{e}\pm})_{k+1}^{\mathrm{trial}}$;
Calculate $(\bar{\epsilon}_{\mathrm{r}}^{\mathrm{e}\pm})_{k+1}^{\mathrm{trial}}$ and $(\bar{\epsilon}_2^{\mathrm{e}\pm})_{k+1}^{\mathrm{trial}}$ by Eqs.(54);
$(R^{\pm})_{k+1}^{\mathrm{trial}} \leftarrow |(\bar{\epsilon}_{\mathrm{r}}^{\mathrm{e}\pm})_{k+1}^{\mathrm{trial}}| - |(r_{\mathrm{d}}^{\pm})_k|$;
if $(R^{\pm})_{k+1}^{\mathrm{trial}} \leqslant 0$ **then**
/*pure elastic step */
$D_{k+1}^{\pm} \leftarrow D_k^{\pm}; \boldsymbol{\sigma}_{k+1}^{\mathrm{p}} \leftarrow \boldsymbol{\sigma}_k^{\mathrm{p}}; (r_d^{\pm})_{k+1} \leftarrow (r_d^{\pm})_k$;
$(\bar{\epsilon}_2^{\pm})_{k+1} \leftarrow (\bar{\epsilon}_2^{\pm})_{k+1}^{\mathrm{trial}}; \bar{\boldsymbol{\sigma}}_{k+1} \leftarrow \bar{\boldsymbol{\sigma}}_{k+1}^{\mathrm{trial}}$;
$\boldsymbol{\sigma}_{k+1} \leftarrow \bar{\boldsymbol{\sigma}}_{k+1} - D_{k+1}^{+}\bar{\boldsymbol{\sigma}}_{k+1}^{+} - D_{k+1}^{-}\bar{\boldsymbol{\sigma}}_{k+1}^{-}$;
$\tilde{R}_{k+1}^{\pm} \leftarrow (R^{\pm})_{k+1}^{\mathrm{trial}}$;
return ;
end
/*Nonlinear correction */
if $\tilde{R}_k^{\pm} < -\mathrm{GTOL}$ **then**
Perform Algorithm 1 to determine $\lambda^{\pm}$;
else
$\lambda^{\pm} \leftarrow 0$;
end
end
$\bar{\boldsymbol{\sigma}}_{k+1} \leftarrow (\mathbb{T}_{k+\lambda}^{\mathrm{ep}})^{-1} : [\bar{\boldsymbol{\sigma}}_{k+1}^{\mathrm{trial}} - (\mathbb{I} - \mathbb{T}_{k+\lambda}^{\mathrm{ep}})\bar{\boldsymbol{\sigma}}_{k+\lambda}]$;
$\boldsymbol{\sigma}_{k+1}^{\mathrm{p}} \leftarrow \boldsymbol{\sigma}_k^{\mathrm{p}} + \theta_k^{\mathrm{p}+}\left(\bar{\boldsymbol{\sigma}}_{k+1}^{+} - \bar{\boldsymbol{\sigma}}_{k+\lambda^{+}}^{+}\right) + \theta_k^{\mathrm{p}-}\left(\bar{\boldsymbol{\sigma}}_{k+1}^{-} - \bar{\boldsymbol{\sigma}}_{k+\lambda^{-}}^{-}\right)$;
$(r_d^{+})_{k+1} \leftarrow \mathrm{sgn}\{(r_d^{+})_k\}\max\{|(\bar{\epsilon}_r^{+})_{k+1}|, |(r_d^{+})_k|\}$;
$(r_d^{-})_{k+1} \leftarrow \mathrm{sgn}\{(r_d^{-})_k\}\max\{|(\bar{\epsilon}_r^{-})_{k+1}|, |(r_d^{-})_k|\}$;
$(D^{\pm})_{k+1} \leftarrow G^{\pm}\left[(r_d^{\pm})_{k+1}\right]; (d^{\pm})_{k+1} \leftarrow g^{\pm}\left[(r_d^{\pm})_{k+1}\right]$;
$\boldsymbol{\sigma}_{k+1} \leftarrow (\mathbb{I} - \mathbb{D}_{k+1}) : (\mathbb{E}_0 : \boldsymbol{\epsilon}_{k+1} - \boldsymbol{\sigma}_{k+1}^{\mathrm{p}})$;
$\tilde{\bar{\boldsymbol{\sigma}}}_{k+1} \leftarrow \mathbb{E}_0 : \boldsymbol{\epsilon}_{k+1} - \boldsymbol{\sigma}_{k+1}^{\mathrm{p}}$;
$\tilde{R}_{k+1}^{\pm} \leftarrow |\bar{\epsilon}_r^{\pm}(\tilde{\bar{\boldsymbol{\sigma}}}_{k+1})| - |r_d^{\pm}(\bar{\boldsymbol{\sigma}}_{k+1})|$;
/*check global convergence */
if $|\tilde{R}_{k+1}^{\pm}| > \mathrm{GTOL}$ **then**
exit with error message;
end
end
end

Algorithm 2: Explicit algorithm

For simplicity, researchers[5, 26, 27, 46] usually investigated the stability of the dynamically damaged system based on the uniaxial bar problem in tension, as shown in Fig. 4.
The equation of motion for the uniaxial bar problem reads

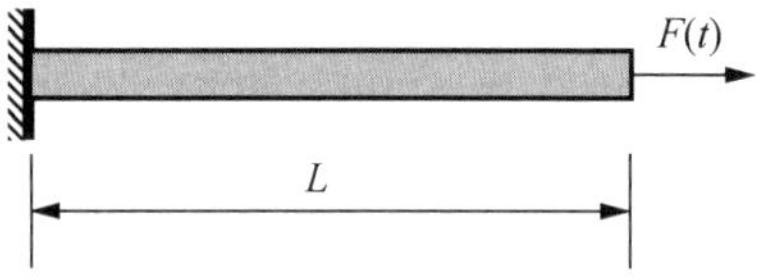

Fig. 4. Uniaxial bar problem in dynamic tensile

$$\frac{\partial\sigma}{\partial x}=\rho\frac{\partial^2 u}{\partial t^2} \tag{70}$$

with ρ the density of the material, σ the axial stress and u the axial displacement, x and t are spatial and temporal coordinates, respectively.

The equation of continuity reads

$$\epsilon=\frac{\partial u}{\partial x} \tag{71}$$

where ϵ denotes the uniaxial strain.

And the constitutive equation in 1-D is

$$\sigma=(1-D)E_0(\epsilon-\epsilon^{\mathrm{p}}) \tag{72}$$

where D is the uniaxial damage, E_0 is the undamaged Young's modulus, and ϵ^{p} is the plastic strain. For simplicity, we only consider the tensile loading bar and neglect the compressive damage in this discussion.

For the rate independent system, the damage are directly defined as the function of uniaxial elastic strain ϵ^{e}, we have

$$D=G(\epsilon^{\mathrm{e}}),\ \epsilon^{\mathrm{e}}=\epsilon-\epsilon^{\mathrm{p}} \tag{73}$$

Moreover, the differentiation of Eq.(72) gives

$$\frac{\partial\sigma}{\partial x}=(1-D-D'\epsilon^{\mathrm{e}})E_0\left(\frac{\partial\epsilon}{\partial x}-\frac{\partial\epsilon^{\mathrm{p}}}{\partial x}\right) \tag{74}$$

Recalling the plastic model proposed in the former part, we have the standard first order expression of the governing equations for the rate independent system as follows

$$\begin{cases}(1-D-D'\epsilon^{\mathrm{e}})E_0\left(\dfrac{\partial\epsilon}{\partial x}-\dfrac{\partial\epsilon^{\mathrm{p}}}{\partial x}\right)-\rho\dfrac{\partial v}{\partial t}=0\\ \dfrac{\partial\epsilon}{\partial t}-\dfrac{\partial v}{\partial x}=0\\ f_p(D)\dfrac{\partial\epsilon}{\partial t}-[1+f_p(D)]\dfrac{\partial\epsilon^{\mathrm{p}}}{\partial t}=0\end{cases} \tag{75}$$

where the velocity $v=\dfrac{\partial u}{\partial t}$.

Consider the strain rate dependent damage evolutions, we have the damage evolution

$$D = G(\epsilon_r) \tag{76}$$

The rate dependent strain ϵ_r is governed by Eq.(27) with the total strain ϵ replaced by the elastic strain ϵ^{e}. In this case, the differentiation of Eq.(72) gives

$$\frac{\partial \sigma}{\partial x} = (1-D)E_0\left(\frac{\partial \epsilon}{\partial x} - \frac{\partial \epsilon^{\mathrm{p}}}{\partial x}\right) - D'E_0\frac{\partial \epsilon_r}{\partial x} \tag{77}$$

Therefore, the standard first order expression of the governing equations for the rate dependent system is as follows

$$\begin{cases} (1-D)E_0\left(\dfrac{\partial \epsilon}{\partial x} - \dfrac{\partial \epsilon^{\mathrm{p}}}{\partial x}\right) - D'E_0\epsilon^{\mathrm{e}}\dfrac{\partial \epsilon_r}{\partial x} - \rho\dfrac{\partial v}{\partial t} = 0 \\ \dfrac{\partial \epsilon}{\partial t} - \dfrac{\partial v}{\partial x} = 0 \\ f_{\mathrm{p}}(D)\dfrac{\partial \epsilon}{\partial t} - [1 + f_{\mathrm{p}}(D)]\dfrac{\partial \epsilon^{\mathrm{p}}}{\partial t} = 0 \\ \gamma_a\dfrac{\partial e_2}{\partial t} - \alpha_n\dfrac{\partial \epsilon_r}{\partial t} = -(\alpha_n + \beta_n)e_2 \\ \gamma_a\dfrac{\partial e_2}{\partial t} + \beta_n\dfrac{\partial \epsilon_r}{\partial t} - \beta_n\left(\dfrac{\partial \epsilon}{\partial t} - \dfrac{\partial \epsilon^{\mathrm{p}}}{\partial t}\right) = 0 \end{cases} \tag{78}$$

where $e_2 = \dfrac{\partial \epsilon_2}{\partial t}$.

According to the theory of partial differential equation[29], the quasi-linear PDE could be expressed in the following form

$$\mathbf{P}(\mathbf{U})\frac{\partial \mathbf{U}}{\partial t} + \mathbf{Q}(\mathbf{U})\frac{\partial \mathbf{U}}{\partial x} = \mathbf{R}(\mathbf{U}) \tag{79}$$

Consider the generalized eigenproblem as follows

$$\mathbf{W}^{\mathrm{T}}(\mathbf{P} - \lambda\mathbf{Q}) = \mathbf{0}^{\mathrm{T}} \Rightarrow |\mathbf{P} - \lambda\mathbf{Q}| = 0 \tag{80}$$

Eq.(79) is hyperbolic if all the eigenvalues of Eq.(80) are real, which means the travelling wave occurs within the 1-D bar. On the other hand, if the complex eigenvalues are solved for Eq. (80), the wave could not travel within the bar and the deformation would concentrate in a local part[5, 46]. Then the system loses its stability and the strain localization occurs.

By solving the eigenvalue problem for the rate independent system expressed in Eq.(75), we have the following expression of eigenvalues:

$$\begin{cases} \lambda = \pm\sqrt{\dfrac{\rho}{\dfrac{\partial \sigma}{\partial \epsilon}}} \\ \dfrac{\partial \sigma}{\partial \epsilon} = \dfrac{(1 - D - D'\epsilon^{\mathrm{e}})E_0}{1 + f_{\mathrm{p}}(D)} \end{cases} \tag{81}$$

It is observed from Eq.(81) that the stability of the rate independent system Eq.(75) is governed by the tangential stiffness $\frac{\partial \sigma}{\partial \epsilon}$. If the material appears to be softening, the negative tangential stiffness leads to the loss of stability. This conclusion also agrees with the celebrated work of Bazant and Belytschko[2].

The eigenvalues of the the rate-dependent system expressed by Eq.(78) could solved as follows

$$\begin{cases} \lambda = \pm\sqrt{\dfrac{\rho}{\widetilde{E}}} \\ \widetilde{E} = \dfrac{\left(1 - D - \dfrac{\beta}{\alpha+\beta} D' \epsilon^{\mathrm{e}}\right) E_0}{[1 + f_{\mathrm{p}}(D)]\gamma} \end{cases} \tag{82}$$

A comparison between Eqs.(81) and (82) gives

$$1 - D - \frac{\beta}{\alpha+\beta} D' \epsilon^{\mathrm{e}} > 1 - D - D' \epsilon^{\mathrm{e}} \tag{83}$$

Thus we could conclude that the rate-dependent system expressed by Eq.(78) is more stable than the rate independent system expressed in Eq.(75). On the other hand, Eq.(78) is conditionally stable, which means the instability could be relieved but not be totally removed by the rate-dependence model introduced in the present work. The further improvements of the stability is undoubtedly an interesting topic, and the next step-forward in this direction will be reported in our forthcoming works.

6.2 Numerical stability analysis

Beside the physical stability, the numerical stability especially for the time integration scheme is worthwhile to know.

In the present work, the numerical stability is discussed in the structural level and the material level respectively.

Considering the structural system discretized by the finite element method, we have the governing ODE as follows

$$\begin{cases} \mathbf{Ma} + \mathbf{Cv} + \mathbf{f}^{\mathrm{int}} = \mathbf{f}^{\mathrm{ext}} \\ \mathbf{a} = \ddot{\mathbf{u}},\ \mathbf{v} = \dot{\mathbf{u}} \end{cases} \tag{84}$$

where $\mathbf{M}$ and $\mathbf{C}$ are the mass matrix and the damping matrix, respectively; $\mathbf{u}$, $\mathbf{v}$ and $\mathbf{a}$ are the displacement vector, the velocity vector and the acceleration vector, respectively; $\mathbf{f}^{\mathrm{int}}$ and $\mathbf{f}^{\mathrm{ext}}$ are the internal force vector and the external force vector, respectively.

To solve the ODE Eq.(84) in an explicit way, the central difference method is usually applied. We have the following scheme[3]

(1) First partial update of velocity

$$\mathbf{v}_{n+\frac{1}{2}} = \mathbf{v}_n + \frac{1}{2}(t_{n+1} - t_n)\mathbf{a}_n \tag{85}$$

(2) Update displacement

$$\mathbf{d}_{n+1}=\mathbf{d}_n+(t_{n+1}-t_n)\mathbf{v}_{n+\frac{1}{2}} \tag{86}$$

(3) Compute acceleration

$$\mathbf{a}_{n+1}=\mathbf{M}^{-1}(\mathbf{f}_{n+1}^{\text{ext}}-\mathbf{f}_{n+1}^{\text{int}}-\mathbf{C}\mathbf{v}_{n+\frac{1}{2}}) \tag{87}$$

(4) Second partial update of velocity

$$\mathbf{v}_{n+1}=\mathbf{v}_{n+\frac{1}{2}}+\frac{1}{2}(t_{n+1}-t_n)\mathbf{a}_{n+1} \tag{88}$$

The numerical stability in the structural level is governed by the following condition[3]

$$\Delta t=\alpha_{\text{NL}}\Delta t_{\text{crit}} \tag{89}$$

The critical time step Δt_{crit} could be expressed as

$$\Delta t_{\text{crit}}=\frac{2}{\omega_{\max}}\leqslant\min_e\frac{l_e}{c_e} \tag{90}$$

where $\omega_{\max}$ is the maximum natural frequency of the linearised system; l_e and c_e are the characteristic length and wave speed in element e. The phenomenological coefficient α_{NL} is introduced to account for the destabilizing effect of nonlinearities. The suggested values are $0.80\leqslant\alpha_{NL}\leqslant 0.98$. On the other hand, although the material damage, plasticity and rate dependency may yield slower wave speed than the linear elastic material, Belytschko et al.[3] suggested that the critical time step evaluated by Eq.(90) should not be increased. In the material level, the proposed numerical scheme in Section 5 could offer exact simulating results in the elastic loading stage and the unloading stage. Thus the pure nonlinear loading steps are considered in the discussion of numerical stability. For simplicity, we still consider the uniaxial tension. Recall Eq.(63), the plastic strain in the $k+1$ step reads

$$\epsilon_{k+1}^{\text{p}}=\epsilon_k^{\text{p}}+\theta_k^{\text{p}+}(\epsilon_{k+1}-\epsilon_k) \tag{91}$$

Perform Eq.(91) recursively and assume at the starting point $\epsilon_1=\epsilon_1^p=0$, we have

$$\epsilon_{k+1}^{\text{p}}=\sum_{i=1}^{k}\theta_i^{\text{p}+}(\epsilon_{i+1}-\epsilon_i)\leqslant\theta_k^{\text{p}+}\sum_{i=1}^{k}(\epsilon_{i+1}-\epsilon_i)=\theta_k^{\text{p}+}\epsilon_{k+1} \tag{92}$$

Expression Eq.(92) indicated that the numerically solved plastic strain is bounded. Thus the proposed scheme is unconditionally stable for the computation of plastic strain.

Then we should compute the $(\bar{\epsilon}_r^{\text{e}\pm})_{k+1}$ based on Eq.(55). The stability of Eq.(55) is governed by the transformation matrix on the right hand side. By solving the eigenvalue equation of the transformation matrix in Eq.(55), we have the eigenvalues

$$\begin{cases}\lambda_1=1\\ \lambda_2=\dfrac{\gamma_a^{\pm}\beta_n^{\pm}}{(\alpha_n^{\pm}+\beta_n^{\pm})(\gamma_a^{\pm}+\beta_n^{\pm}\Delta t_k)}\in(0,1)\end{cases} \tag{93}$$

Hence it is indicated that Eq.(55), with which $(\bar{\epsilon}_r^{\text{e}\pm})_{k+1}$ could be numerically solved, is unconditionally stable.

The last step is to solve the stress by Eq.(66). For the simple uniaxial tension case, we have

$$\begin{aligned}\sigma_{k+1} &= (1-d_{k+1})E_0(\epsilon_{k+1}-\epsilon^{\mathrm{p}}_{k+1}) \\ &= \{1-g^{\pm}[(r^{\pm}_{\mathrm{d}})_{k+1}]\}E_0(\epsilon_{k+1}-\epsilon^{\mathrm{p}}_{k+1})\end{aligned} \tag{94}$$

It is observed that all the terms on the RHS of Eq.(94) are bounded. Thus we find that the proposed numerical scheme in material level is unconditional stable. All that we need is to pay attention to the numerical convergence which is governed by Eq.(69).

In the end, we reach the conclusion that the numerical stability is governed by the structural level. Thus during the numerical simulation, the critical time step is only confirmed by Eq.(89).

7 Numerical tests

To learn the performance of the proposed model, we perform a series of numerical tests under various of loading conditions. Except particular statement, the time step is chosen to be 10^{-6} s in the following numerical simulations.

7.1 Uniaxial tests

Numbers of model parameters(Table 1) are introduced in the proposed model to describe the properties of concrete. The bracketed values in Table 1 are chosen for the numerical simulations of uniaxial tests.

Table 1. Model parameters for uniaxial tests

	Tension	Compression
Elasticity	E_0 (35 GPa), v(0.18)	
Biaxial strength increase	N/A	$\frac{f_{\mathrm{bc}}}{f_{\mathrm{c}}}$ (1.16)
Static random damage	λ^+(5.3)	λ^-(7.6)
	ζ^+(0.6)	ζ^-(0.6)
	ξ^+(20)	ξ^-(10)
Dynamic damage	γ_a^+(12)	γ_a^-(1.5)
	n^+(0.000 2)	n^-(0.000 5)
Plasticity	ξ_{p}^+(0.6)	ξ_{p}^-(0.4)
	n_{p}^+(0.1)	n_{p}^-(0.1)

7.1.1 Monotonic static tests

Zeng et al.[50] carried out systematic experiments for the uniaxial and confined stress-strain performances of concrete. The mean values and the standard deviations(STD) of stress strain curves are characterized based on the testing data. On the other hand, the stochastic damage evolutions adopted in the present work yield the model results of the mean value and the standard deviations of stress strain curves. As shown in Fig. 5, the simulated mean value curve agrees well with the experimental results. The results of the standard deviation deserve further improvements, although they also show certain trends of the experimental results.

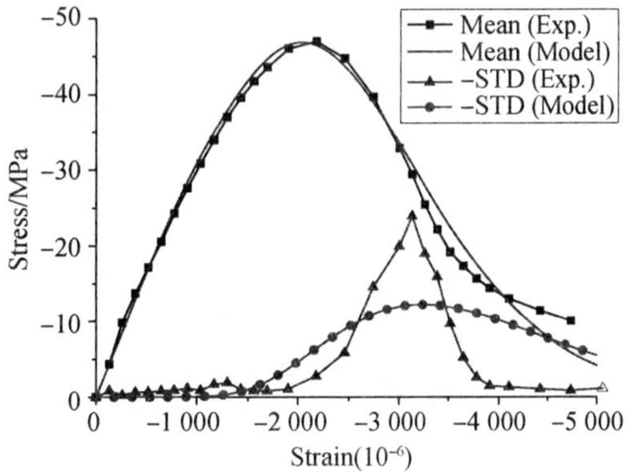

Fig. 5. Stress-strain curves under compression

7.1.2 Uniaxial dynamic behaviors

The numerical results of the dynamic increase factor, which is defined by the dynamic strength over static strength, are plotted in Figs. 6 and 7 against the collected existing experimental results[6, 8, 19, 25, 41, 45, 49]. The main trend of the increase of strength under fast loading is well reproduced.

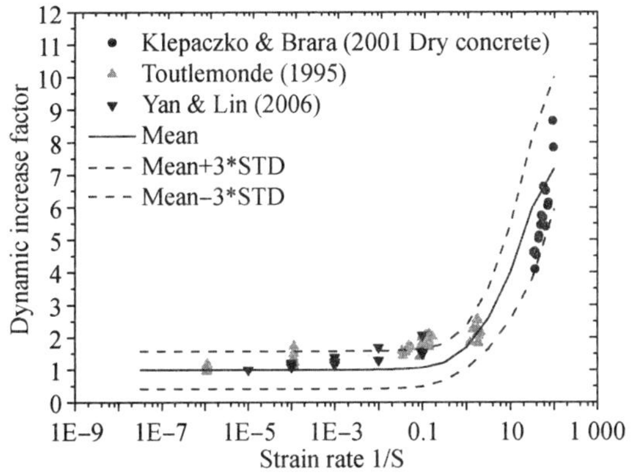

Fig. 6. Dynamic increase factor under tension

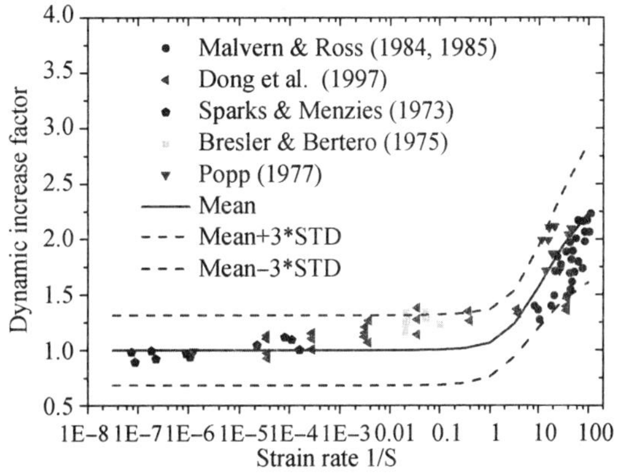

Fig. 7. Dynamic increase factor under compression

Meanwhile, the distributed domain of the testing data is also banded by the calculated Mean $\pm 3\times$STD curves.

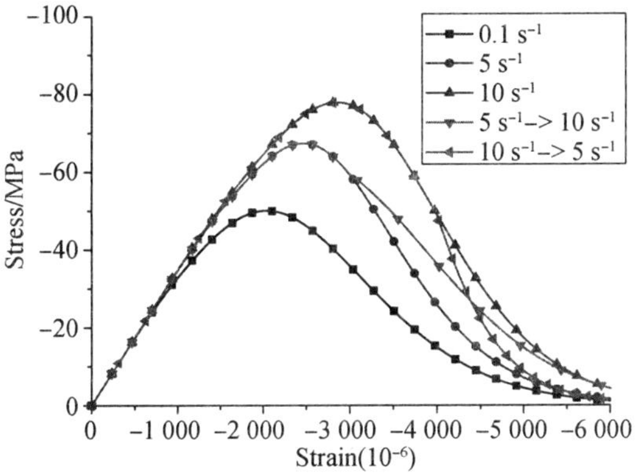

Fig. 8. Stress-strain curves subjected to different strain rates

The stress-strain curves under different loading rates are shown in Fig. 8. The stress strengthening induced by the material rate-dependency is adequately reproduced. And the transitions of the stress-strain curves across different loading rates are smooth and convergent.

7.1.3 Cyclic tests

When subjected to the cyclic or the repeated loading, concrete often experiences residual strain, which is described by the proposed plastic model. The simulated results are

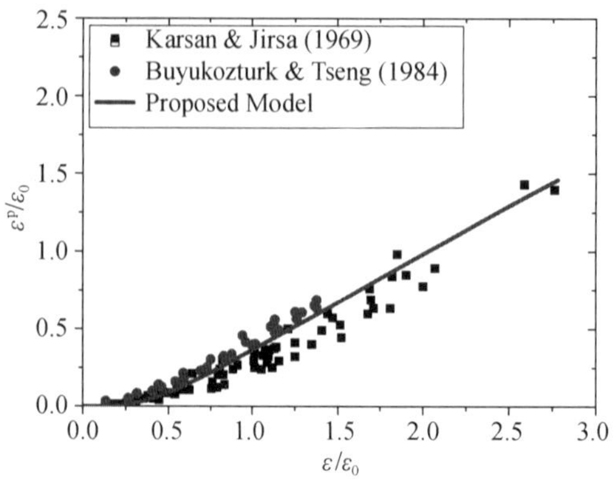

Fig. 9. Plastic strain of concrete under compression

plotted in Fig. 9 against the experimental data[7, 18]. The total strain ϵ and the plastic strain ϵ^p are normalized by the peak strain ϵ_0 of the total stress-strain curve. And it is indicated that the model results agree well with the experiential data. The proposed plastic model works well in reproducing the residual strain of plain concrete.

The numerical tests for the concrete specimen under uniaxial cyclic loading are performed. The results are plotted in Figs. 10 and 11. The typical nonlinear behaviors of concrete, e. g. the strength softening, the residual strain, the degradation of stiffness, and the stiffness change between compression and tension, could be easily located on the simulated curves. And the hysteretic cycles could be also found, which dissipate energy throughout the unloading-reloading loops. The other interesting point is shown in the dynamic curve (Fig. 11). We could find that the curve is smoothed at the transition points linking the loading curve and the unloading curve by the rate-dependency.

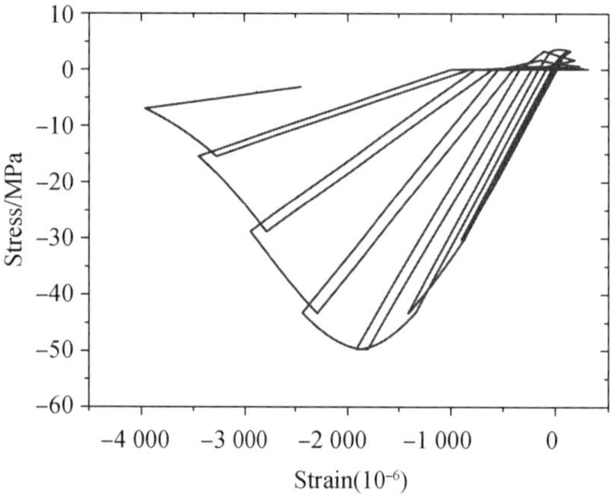

Fig. 10. Cyclic stress-strain curve under quasi-static loading

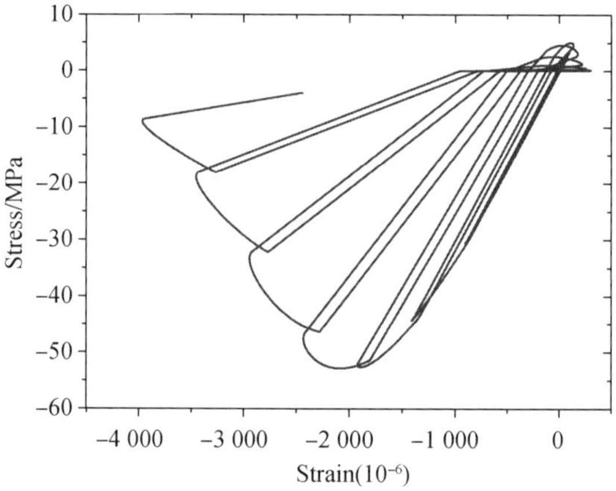

Fig. 11. Cyclic stress-strain curve under dynamic loading ($\dot{\epsilon}=1.0\ s^{-1}$)

7.2 Biaxial tests

The results of biaxial compressive experiments carried out by Ren and coworkers[33] are simulated to verify the multiaxial performance of the proposed model. The experimental set-up for the biaxial tests is shown in Fig. 12. The behaviors of high performance concrete specimens, with the dimension of 150 mm×150 mm×50 mm, subjected to uniaxial and biaxial loading were experimentally investigated. The loads are applied by the top and left loading platens. And the strain rates were set to be less than $10^{-5}\ s^{-1}$ to simulate the static loading condition. To avoid the friction between concrete and steel platen, two slice of

Teflon antifriction sheets are inserted into each contact surface. The uniaxial and biaxial compressive complete stress-strainstrain curves were obtained under strain control loading scheme. The tested stress is calculated by the applied force divided by the sectional area. And the tested strain the calculated by the relative displacement between correspond loading platens divided by the length of the specimen.

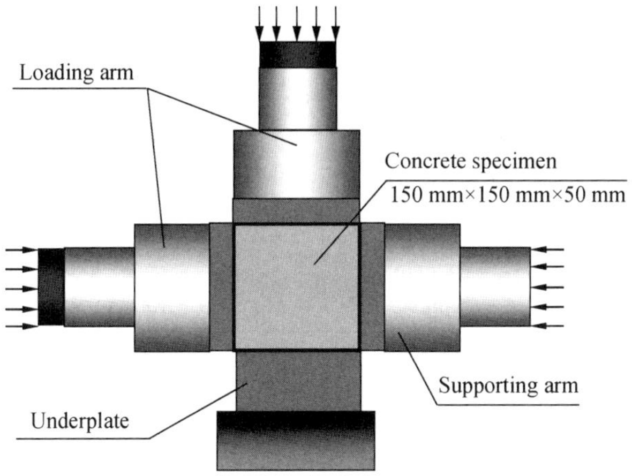

Fig. 12. Biaxial test set-up

Due to the lack of reliable biaxial dynamic tests of concrete, only the static testing results are simulated in the present work. The bracketed values in Table 2 are chosen for the numerical simulations of biaxial tests. We assume the specimen is uniformed stressed because of the Teflon antifriction sheets. Thus only a single element is adopted to simulate the specimen. The simulated average stress-strain curves for the biaxial tests are shown in Fig. 13. The results in Fig. 13 are acceptably accurate up to the peak and in the first stage of the softening branch, but they diverge from the experimental curve in the late stage of the test. It suggests that the tail of the adopted average damage evolution functions in Eq.(35) may deserve further improvement.

Table 2. Model parameters for biaxial tests

	Tension	Compression
Elasticity	E_0(37.6 GPa), v(0.20)	
Biaxial strength increase	N/A	$\frac{f_{bc}}{f_c}$(1.30)
Static random damage	λ^+(4.92)	λ^-(7.70)
	ζ^+(0.30)	ζ^-(0.30)
Plasticity	ξ_p^+(0.6)	ξ_p^-(0.4)
	n_p^+(0.1)	n_p^-(0.1)

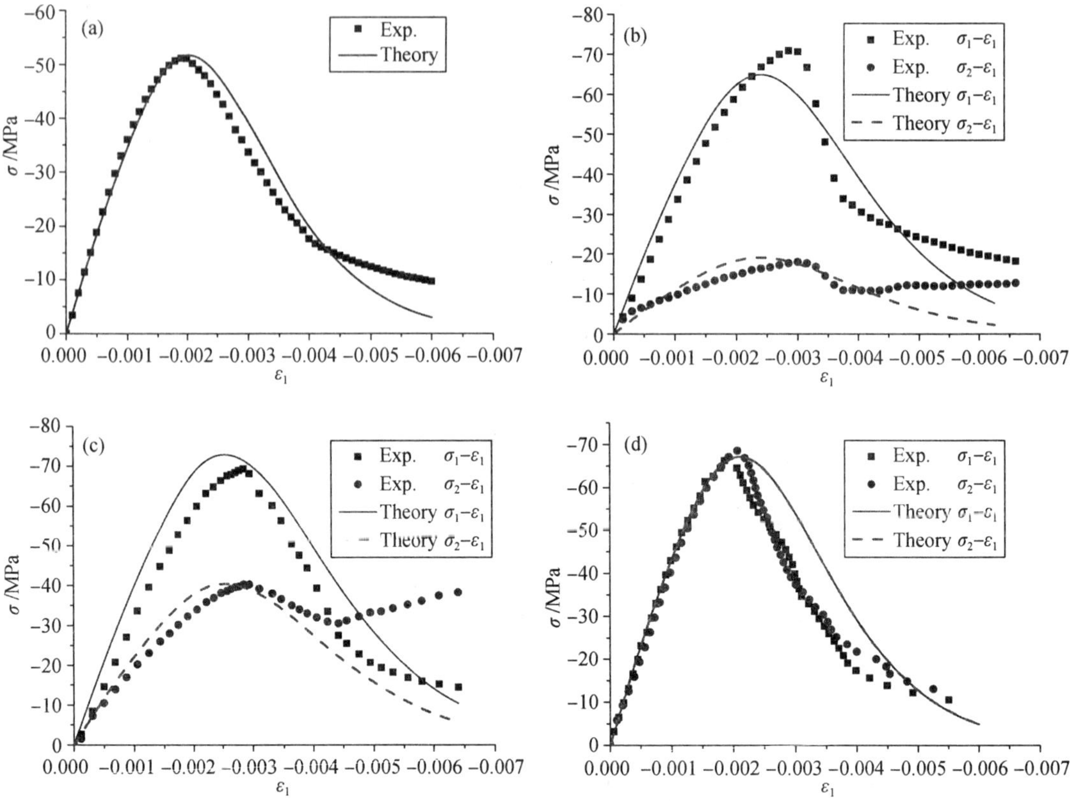

Fig. 13. Biaxial tests. (a) Uniaxial compression, (b) $\varepsilon_1 : \varepsilon_2 = 1.0 : 0.1$, (c) $\varepsilon_1 : \varepsilon_2 = 1.0 : 0.4$, (d) $\varepsilon_1 : \varepsilon_2 = 1.0 : 1.0$

7.3 Impact tests of RC beams

Saatci and Vecchio[36] performed experiments for the shear mechanisms of RC beams in University of Toronto Structural Laboratories. The experimental program consisted of two phases: static tests and impact tests. Thereafter, the experimental results were numerically simulated in the work of Ozbolt and Sharma[28]. The experimental set-up is shown in Fig. 14. The length of the beam is 4 880 mm, while the span is 3 000 mm. The overhangs were designed to amplify the inertia effect of the beam under impact loading. The upper supports were devised to prevent the uplift of the specimen. All specimens were doubly reinforced with two longitudinal reinforcing bars (nominal diameter=29.9 mm) on each side. And the closed stirrups (nominal diameter = 7.01 mm) were used as transverse reinforcement. A 38 mm clear cover was provided between the beam surface and the steel bars.

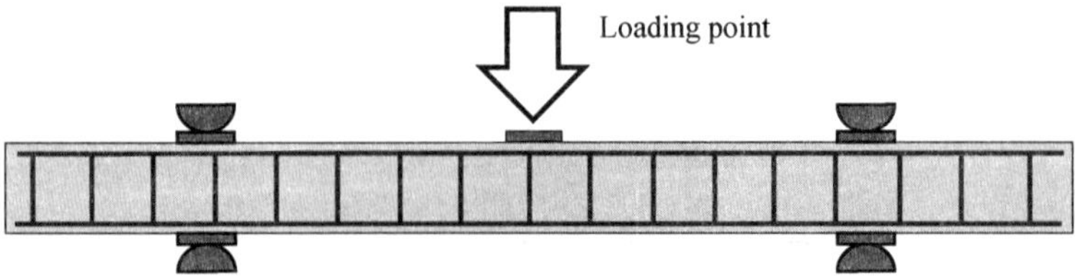

Fig. 14. Experimental set-up of RC beam

Four beams (Table 3) of the experiments are simulated based on the proposed model. The bracketed values in Table 4 are chosen for the numerical simulations of RC beams. The drop-weight for the specimens SS0b and SS1b was 211 kg and the impact velocity is 8.0 m/s. Our numerical simulations are performed based on ABAQUS platform with our proposed damage model implemented by VUMAT.

Table 3. Simulated specimens

Specimen	Transverse reinforcement ratio/%	loading pattern
SS0a	0	Static
SS0b	0	Impact
SS1a	0.1	Static
SS1b	0.1	Impact

Table 4. Model parameters

	Tension	Compression
Elasticity	E_0(35 GPa), v(0.20)	
Biaxial strength increase	N/A	$\frac{f_{bc}}{f_c}$(1.16)
Static random damage	λ^+(4.2)	λ^-(7.1)
	ζ^+(0.45)	ζ^-(0.6)
	ξ^+(20)	ξ^-(10)
Dynamic damage	γ_a^+(6)	γ_a^-(1.5)
	n^+(0.000 2)	n^-(0.000 5)
Plasticity	ξ_p^+(0.6)	ξ_p^-(0.4)
	n_p^+(0.1)	n_p^-(0.1)

The simulated damage contours, which actually indicate the crack patters of the beams, are shown in Fig. 15. The shear failure patterns could be clearly observed. The simulated maximum reactions of the beams are shown in Table 5. Agreements could be found among the experimental results of Saatci and Vecchio[36], the simulated results of Ozbolt and Sharma[28] and the simulated results of the proposed model.

Table 5. Simulated maximum reactions for the beams (kN)

Specimen	SS0a	SS0b	SS1a	SS1b
Experiment[36]	98	149	305	356
Ozbolt and Sharma[28]	112	158	346	372
Proposed model	101	145	267	313

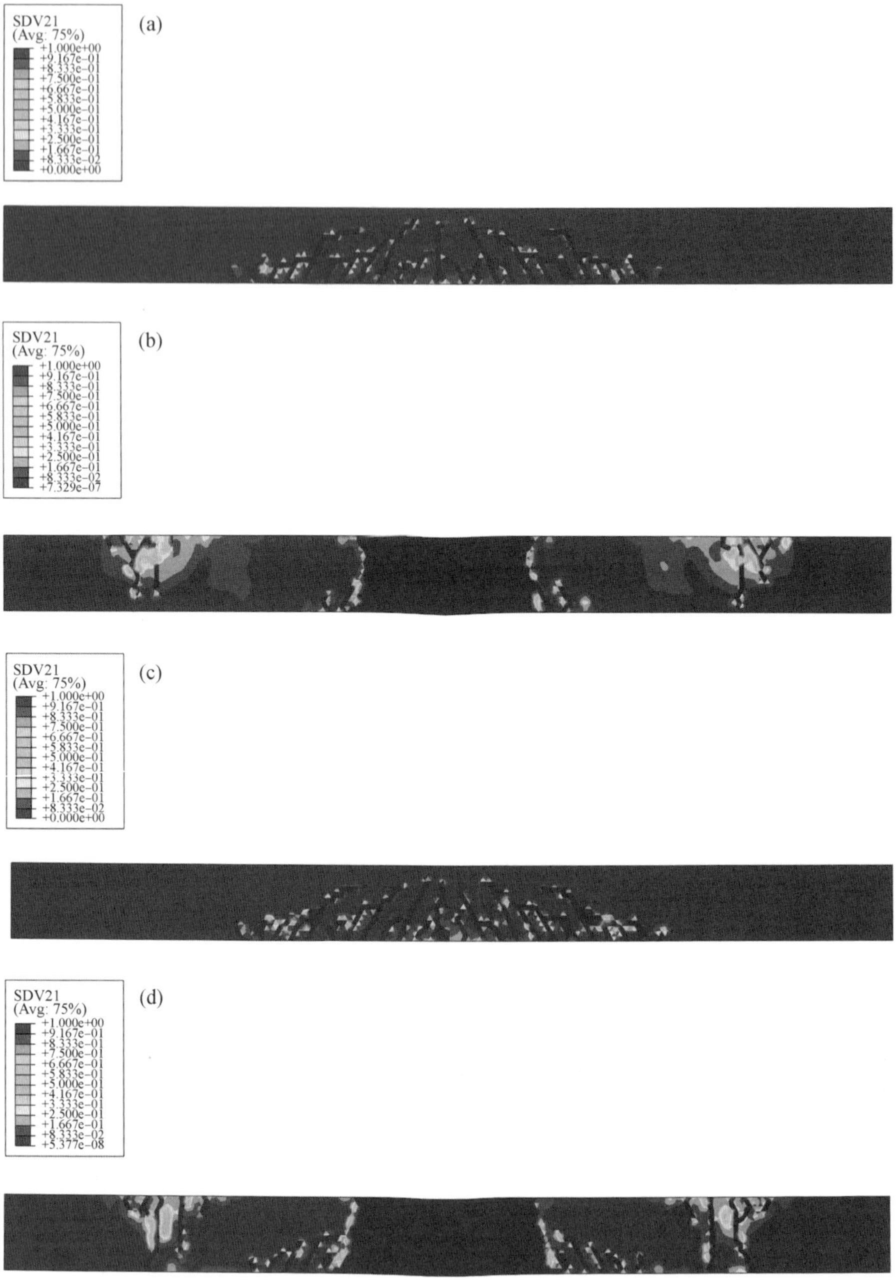

Fig. 15. Simulated contours of tensile damage D^+. a SS0a, b SS0b, c SS1a, d SS1b

8 Conclusions

Based on the theoretical developments and the numerical simulations in the present paper, numbers of conclusions could be given as follows:

(1) The typical nonlinear properties of concrete, including the strain softening, the residual strain, the unilateral effect and the hysteretic cycles, could be properly reproduced by the

proposed model. Thus the structural nonlinear simulation performed based on the proposed model may capture the salient features of structural performance.

(2) By giving the scattering region of the testing data, the randomness of concrete is also reproduced by the proposed model. This aspect still deserve further investigations, which may develop the appropriate descriptions of the structural randomness and offer possible methods for the nonlinear structural reliability analysis.

(3) An explicit algorithm is developed for the numerical implementation of the proposed model in structural analysis. The algorithm tolerates excessive localization of strain and damage due to its robustness, but its global convergence requires particular attentions.

(4) The description of the material rate-dependency in the present work agrees well with experimental results. Moreover, a rather interesting point is detected. Although a linear differential system is proposed to develop the rate-dependent damage evolution, the nonlinear dynamic increase of material strength has been achieved due to the nonlinear coupling between the damage and the plasticity, which still deserves special attention in the upcoming study.

(5) Last but not the least, the rate-dependency not only enhances the material strength but also improves the material behaviors in a series of details. As shown by the numerical results, the transition points of the cyclic stress-strain curves are polished by the rate-dependency. Moreover, the increase of smoothness for the stress-strain curves may enhance the convergence of simulation and relieve the localization of deformation.

Acknowledgments

Financial supports from the National Science Foundation of China are gratefully appreciated (GNs: 51261120374, 91315301 and 51208374). We would also like to express our sincere appreciations the anonymous referee for many valuable suggestions and corrections.

Appendix Derivations of Stefan effect

In a viscous fluid, the stress σ_{ij} is often decomposed into the pressure p and the shear stress τ_{ij} as follows

$$\sigma_{ij} = \tau_{ij} - p\delta_{ij} \tag{95}$$

where δ_{ij} is the Kronecker delta. The constitutive law for the incompressible Newtonian fluid reads

$$\tau_{ij} = \mu(v_{i,j} + v_{j,i}) \tag{96}$$

where μ is the viscosity and v_i is the velocity in the i-th direction. And the incompressibility reads

$$v_{k,k} = 0 \tag{97}$$

The equilibrium without body forces and inertia effects reads

$$\sigma_{ij,j} = 0 \tag{98}$$

Substituting Eq.(95) into Eq.(98) and using Eqs.(96) and (97), we have

$$\mu v_{i,jj} = p_{,i} \tag{99}$$

Combining Eqs. (97) and (99) and expanding the tensor expressions into regular expressions, we have the following governing equation for the incompressible Newtonian inertia free flow.

$$\begin{cases} \mu\left(\dfrac{\partial^2 v_1}{\partial x_1^2}+\dfrac{\partial^2 v_1}{\partial x_2^2}+\dfrac{\partial^2 v_1}{\partial x_3^2}\right)=\dfrac{\partial p}{\partial x_1} \\ \mu\left(\dfrac{\partial^2 v_2}{\partial x_1^2}+\dfrac{\partial^2 v_2}{\partial x_2^2}+\dfrac{\partial^2 v_2}{\partial x_3^2}\right)=\dfrac{\partial p}{\partial x_2} \\ \mu\left(\dfrac{\partial^2 v_3}{\partial x_1^2}+\dfrac{\partial^2 v_3}{\partial x_2^2}+\dfrac{\partial^2 v_3}{\partial x_3^2}\right)=\dfrac{\partial p}{\partial x_3} \\ \dfrac{\partial v_1}{\partial x_1}+\dfrac{\partial v_2}{\partial x_2}+\dfrac{\partial v_3}{\partial x_3}=0 \end{cases} \tag{100}$$

The model problem to solve the Stefan effect is shown in Fig. 16. Two parallel, circular plates with radius R that are separated by a Newtonian (incompressible) liquid with viscosity μ and thickness h. The next step is to obtain the expression of the force F_v applied on the plates by solving Eqs.(100) with the boundary conditions shown in Fig. 16.

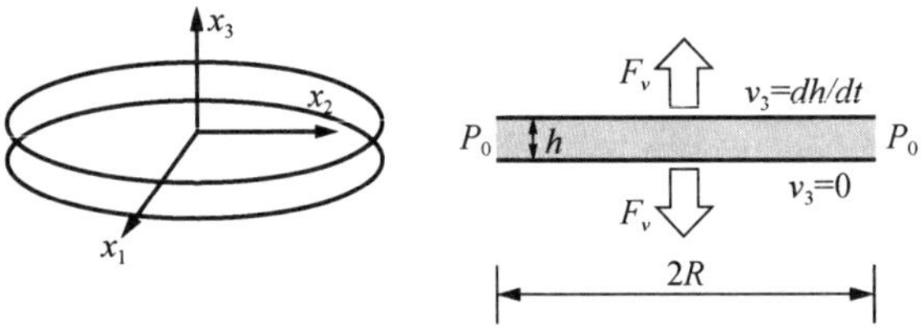

Fig. 16. Model problem of Stefan effect

As the model problem is axisymmetric, Eqs.(100) could be casted into the following form.

$$\begin{cases} \mu\left(\dfrac{\partial^2 v_r}{\partial r^2}+\dfrac{1}{r}\dfrac{\partial v_r}{\partial r}+\dfrac{\partial^2 v_r}{\partial z^2}-\dfrac{v_r}{r^2}\right)=\dfrac{\partial p}{\partial r} \\ \mu\left(\dfrac{\partial^2 v_z}{\partial r^2}+\dfrac{1}{r}\dfrac{\partial v_z}{\partial r}+\dfrac{\partial^2 v_z}{\partial z^2}\right)=\dfrac{\partial p}{\partial z} \\ \dfrac{\partial v_r}{\partial r}+\dfrac{v_r}{r}+\dfrac{\partial v_z}{\partial z}=0 \end{cases} \tag{101}$$

Reynolds [34] further assumes

$$\begin{cases} \dfrac{\partial^2 v_1}{\partial x_1^2} \ll \dfrac{\partial^2 v_1}{\partial x_3^2},\ \dfrac{\partial^2 v_1}{\partial x_2^2} \ll \dfrac{\partial^2 v_1}{\partial x_3^2} \\ \dfrac{\partial^2 v_2}{\partial x_1^2} \ll \dfrac{\partial^2 v_1}{\partial x_3^2},\ \dfrac{\partial^2 v_2}{\partial x_2^2} \ll \dfrac{\partial^2 v_1}{\partial x_3^2} \\ \dfrac{\partial v_3}{\partial x_3}=0,\ \dfrac{\partial p}{\partial x_3}=0 \end{cases} \tag{102}$$

Subtstuting Eq.(102) into Eq.(101) yields

$$\begin{cases} \mu \dfrac{\partial^2 v_r}{\partial z^2} = \dfrac{\mathrm{d}p}{\mathrm{d}r} \\ \dfrac{\partial p}{\partial z} = 0 \\ \dfrac{\partial v_r}{\partial r} + \dfrac{v_r}{r} + \dfrac{\mathrm{d}v_z}{\mathrm{d}z} = 0 \end{cases} \tag{103}$$

The first equation in Eqs.(103) suggests the solution of v_r in the following form

$$v_r = \frac{z(z-h)}{2\mu}\frac{\mathrm{d}p}{\mathrm{d}r} \tag{104}$$

Substituting Eq.(104) into the third equation of Eqs.(103), we obtain

$$\frac{\mathrm{d}v_z}{\mathrm{d}z} = -\frac{z(z-h)}{2\mu}\left(\frac{\mathrm{d}^2 p}{\mathrm{d}r^2} + \frac{1}{r}\frac{\mathrm{d}p}{\mathrm{d}r}\right) \tag{105}$$

An integration from 0 to h with respect to z gives

$$\frac{h^3}{12\mu}\left(\frac{\mathrm{d}^2 p}{\mathrm{d}r^2} + \frac{1}{r}\frac{\mathrm{d}p}{\mathrm{d}r}\right) = v_z\Big|_{z=h} = \dot{h} \tag{106}$$

The solution of Eq.(106) could be expressed as follows

$$p(r) = \frac{3\mu\dot{h}}{h^3}r^2 + C_1\ln r + C_2 \tag{107}$$

Considering the conditions that $p(R) = p_0$ and $p(0)$ is bounded, we further have

$$p(r) = p_0 - \frac{3\mu\dot{h}}{h^3}(R^2 - r^2) \tag{108}$$

By integrating the additional pressure over the plate, we have the force induced by the viscosity

$$F_v = \int_0^R 2\pi r[p_0 - p(r)]\mathrm{d}r = \frac{3\mu\pi R^4}{2h^3}\dot{h} \tag{109}$$

Considering the following expressions of stress and strain

$$\sigma_v = \frac{F_v}{\pi R^2},\ \dot{\epsilon} = \frac{\dot{h}}{h} \tag{110}$$

Eq.(109) further reads

$$\sigma_v = A\dot{\epsilon},\ A = \frac{\alpha_S\mu}{\gamma_h^2} \tag{111}$$

where the shape coefficient $\alpha_S = \dfrac{3}{2}$ and the aspect ratio $\gamma_h = \dfrac{h}{R}$ for circular plates. Although derived considering the circular plate, Eqs. (111) offers the general form of viscous coefficients A. And it is also observed that the aspect ratio γ_h plays a very important role for the rate-dependency between stress and strain rate besides the viscosity μ.

References

[1] ABRAMS D A. Effect of rate of application of load on the compressive strength of concrete[C]// Proceeding of ASTM. 1917, 17: 364-377.

[2] BAŽANT Z P, BELYTSCHKO T B. Wave propagation in a strain-softening bar: exact solution[J]. Journal of Engineering Mechanics, 1985, 111(3): 381-389.

[3] BELYTSCHKO T, LIU W K, MORAN B, et al. Nonlinear finite elements for continua and structures [M]. New York: John Wiley & Sons, 2000.

[4] BISCHOFF P H, PERRY S H. Compressive behaviour of concrete at high strain rates[J]. Materials and Structures, 1991, 24(144): 425-450.

[5] BORST R. Some recent issues in computational failure mechanics [J]. International Journal for Numerical Methods in Engineering, 2001, 52(1-2): 63-95.

[6] BRESLER B, BERTERO V. Influence of high strain rate and cyclic loading of unconfined and confined concrete in compression [C]//Proc. of 2nd Canadian Conference on Earthquake Engineering, Hamilton, Ontario. 1975: 1-13.

[7] BUYUKOZTURK O, TSENG T. Concrete in biaxial cyclic compression[J]. Journal of Structural Engineering, 1984, 110(3): 461-476.

[8] DONG Y L, XIE H P, ZHAO P. Experimental study and constitutive model on concrete under compression with different strain rate[J]. J hydraul Eng, 1997, 7: 72-77.

[9] DUBÉ J F, PIJAUDIER-CABOT G, BORDERIE C L A. Rate dependent damage model for concrete in dynamics[J]. Journal of Engineering Mechanics, 1996, 122(10): 939-947.

[10] DUVANT G, LIONS J L. Inequalities in mechanics and physics[M]. Springer Science & Business Media, 2012.

[11] FARIA R, OLIVER J, CERVERA M. A strain-based plastic viscous-damage model for massive concrete structures[J]. International Journal of Solids and Structures, 1998, 35(14): 1533-1558.

[12] FREUND L B. Crack propagation in an elastic solid subjected to general loading — Ⅲ. Stress wave loading[J]. Journal of the Mechanics and Physics of Solids, 1973, 21(2): 47-61.

[13] HATANO T, TSUTSUMI H. Dynamical compressive deformation and failure of concrete under earthquake load[J]. Transactions of the Japan Society of Civil Engineers, 1960, 1960(67): 19-26.

[14] HSU T T C, MO Y L. Unified theory of concrete structures[M]. John Wiley & Sons, 2010.

[15] JASON L, HUERTA A, PIJAUDIER-CABOT G, et al. An elastic plastic damage formulation for concrete: Application to elementary tests and comparison with an isotropic damage model [J]. Computer Methods in Applied Mechanics and Engineering, 2006, 195(52): 7077-7092.

[16] JU J W. On energy-based coupled elastoplastic damage theories: constitutive modeling and computational aspects[J]. International Journal of Solids and structures, 1989, 25(7): 803-833.

[17] KANDARPA S, KIRKNER D J, SPENCER JR B F. Stochastic damage model for brittle materials subjected to monotonic loading[J]. Journal of Engineering Mechanics, 1996, 122(8): 788-795.

[18] KARSAN I D, JIRSA J O. Behavior of concrete under compressive loadings [J]. Journal of the Structural Division, 1969, 95(12): 2543-2564.

[19] KLEPACZKO J R, BRARA A. An experimental method for dynamic tensile testing of concrete by spalling[J]. International Journal of Impact Engineering, 2001, 25(4): 387-409.

[20] KRAJCINOVIC D, FANELLA D. A micromechanical damage model for concrete[J]. Engineering Fracture Mechanics, 1986, 25(5-6): 585-596.

[21] LEE J, FENVES G L. Plastic-damage model for cyclic loading of concrete structures[J]. Journal of Engineering Mechanics, 1998, 124(8): 892-900.

[22] LEMAITRE J. Evaluation of dissipation and damage in metals submitted to dynamic loading (Constitutive equations for defining dissipation and damage in metals submitted to dynamic loading) [J]. ICAM-1, Japan, 1971.

[23] LI J, REN X D. Stochastic damage model for concrete based on energy equivalent strain [J]. International Journal of Solids and Structures, 2009, 46(11-12): 2407-2419.

[24] LOREFICE R, ETSE G, CAROL I. Viscoplastic approach for rate-dependent failure analysis of concrete joints and interfaces[J]. International Journal of Solids and Structures, 2008, 45(9): 2686-2705.

[25] MALVAR L J, ROSS C A. Review of strain rate effects for concrete in tension[J]. ACI Materials Journal, 1998, 95(6): 735-739.

[26] NEEDLEMAN A. Material rate dependence and mesh sensitivity in localization problems [J]. Computer Methods in Applied Mechanics and Engineering, 1988, 67(1): 69-85.

[27] NIAZI M S, WISSELINK H H, MEINDERS T. Viscoplastic regularization of local damage models: revisited[J]. Computational Mechanics, 2013, 51(2): 203-216.

[28] OŽBOLT J, SHARMA A. Numerical simulation of reinforced concrete beams with different shear reinforcements under dynamic impact loads[J]. International Journal of Impact Engineering, 2011, 38 (12): 940-950.

[29] PEERLINGS R H J, BORST R, BREKELMANS W A M, et al. Some observations on localization in non-local and gradient damage models[J]. European Journal of Mechanics. A, Solids, 1996, 15(6): 937-953.

[30] PERZYNA P. Fundamental problems in viscoplasticity in Advances in Applied Mechanics[M]// Advances in Applied Mechanics, 1966(9):243-277.

[31] REN X D, LI J. Dynamic fracture in irregularly structured systems[J]. Physical Review E, 2012, 85 (5): 055102.

[32] REN X D, LI J. A unified dynamic model for concrete considering viscoplasticity and rate-dependent damage[J]. International Journal of Damage Mechanics, 2013, 22(4): 530-555.

[33] REN X D, YANG W, ZHOU Y, et al. Behavior of high-performance concrete under uniaxial and biaxial loading[J]. ACI Materials Journal, 2008, 105(6): 548.

[34] REYNOLDS O. On the Theory of Lubrication and Its Application to Mr. Beauchamp Tower's Experiments, Including an Experimental Determination of the Viscosity of Olive Oil[J]. Phil. Trans. Roy. Soc., 1885, 1: 157.

[35] ROSS C A, TEDESCO J W, KUENNEN S T. Effects of strain rate on concrete strength[J]. Materials Journal, 1995, 92(1): 37-47.

[36] SAATCI S, VECCHIO F J. Effects of shear mechanisms on impact behavior of reinforced concrete beams[J]. ACI Structural Journal, 2009, 106(1): 78-86.

[37] SIMO J C, HUGHES T J R. Computational inelasticity[M]. Springer Science & Business Media, 2006.

[38] SIMO J C, JU J W. Strain-based and stress-based continuum damage models — Ⅰ. Formulation[J]. International Journal of Solids and Structures, 1987, 23(7): 821-840.

[39] SLOAN S W, ABBO A J, SHENG D. Refined explicit integration of elastoplastic models with automatic error control[J]. Engineering Computations, 2001, 18(1-2): 121-154.

[40] SLOAN S W. Substepping schemes for the numerical integration of elastoplastic stress-strain relations [J]. International Journal for Numerical Methods in Engineering, 1987, 24(5): 893-911.

[41] SPARKS P R, MENZIES J B. The effect of rate of loading upon the static and fatigue strengths of plain concrete in compression[J]. Magazine of Concrete Research, 1973, 25(83): 73-80.

[42] STEFAN M J. Versuche furdie scheinbare adhäsion[C]//Sitzungsberichte der Kaiserlichen Akademie der Wissenschaften Wiens, Mathematisch Naturwissenschaftliche Klasse, 1874: 713-735.

[43] TAKEDA J I, TACHIKAWA H. Deformation and fracture of concrete subjected to dynamic load [C]//Proceedings of the Conference on Mechanical Behavior of Materials. 1972 (Conf Paper).

[44] TEDESCO J W, ROSS C A. Strain-rate-dependent constitutive equations for concrete[J]. J Press Vessel Technol-Trans ASME, 1998, 120(4):398-405.

[45] TOUTLEMONDE F. Impact resistance of concrete structures[D]. Laboratory of Bridges and Roads (LCPC), Paris,1995.

[46] WANG W M, SLUYS L J, DE BORST R. Viscoplasticity for instabilities due to strain softening and strain-rate softening[J]. International Journal for Numerical Methods in Engineering, 1997, 40(20): 3839-3864.

[47] WU J Y, LI J, FARIA R. An energy release rate-based plastic-damage model for concrete[J]. International Journal of Solids and Structures, 2006, 43(3-4): 583-612.

[48] XU X P, Needleman A. Numerical simulations of fast crack growth in brittle solids[J]. Journal of the Mechanics and Physics of Solids, 1994, 42(9): 1397-1434.

[49] YAN D, LIN G. Dynamic properties of concrete in direct tension[J]. Cement and Concrete Research, 2006, 36(7): 1371-1378.

[50] ZENG SJ, REN XD, LI J. Hydrostatic behavior of concrete subjected to dynamic compression[J]. J Struct Eng ASCE, 2013, 139(9): 1582-1592.

A physically motivated model for fatigue damage of concrete①

Zhaodong Ding[1], **Jie Li**[2]

1 Department of Building Engineering, Hefei University of Technology, Hefei, China
2 Department of Building Engineering, Tongji University, Shanghai, China

Abstract The fatigue problem of concrete is still a challenging topic in the researches and applications of concrete engineering. This paper aims to develop a fatigue damage evolution law based model for concrete motivated by the analysis of physical mechanism. In this model, the fatigue energy dissipation process at microscale is investigated with rate process theory. The concept of self-similarity is employed to bridge the scale gap between microscale cracking and mesoscale dissipative element. With the stochastic fracture model, the crack avalanches and macro-crack nucleation processes from mesoscale to macroscale are simulated to obtain the behaviors of macroscope damage evolution of concrete. In conjunction with continuum damage mechanics framework, the fatigue damage constitutive model for concrete is then proposed. Numerical simulations are carried out to verify the model, revealing that the proposed model accommodates well with physical mechanism of fatigue damage evolution of concrete whereby the fatigue life of concrete structures under different stress ranges can be predicted.
Keywords Fatigue; Damage; Energy dissipation; Multiscale modelling; Concrete

1 Introduction

The fatigue of concrete is frequently encountered in the service period of civil structures. Traditional analysis of concrete fatigue in engineering generally follows the empirical methods associated with the statistics of experiment results. The conventional manner is to check the stress history of hot point with available *S-N* curve and some fatigue damage accumulation law such as Miner's law. However, this treatment can neither track the stress redistribution due to the fatigue damage nor explain the mechanism of fatigue damage evolution properly. Therefore, even the researches about the interesting problem have been carried out 60 years or more, the fatigue problem of concrete is still a challenging topic in the civil engineering.

In retrospect of the researches of concrete fatigue, there are roughly three categories of approaches from the historical clue, i.e. the empirical studies based on experiment results; the studies of fatigue cracks propagation based on fracture mechanics and the studies of fatigue damage based on damage mechanics. The main purpose of experimental studies was to derive the critical contents

① Originally published in *International Journal of Damage Mechanics*, 2017, 27(8): 1192–1212.

for engineering demands such as the S-N curve or ε-N curve[1-5]. More interest in physical arguments like the correlation between irreversible strain and fatigue life, the degradation of elastic modulus due to damage evolution and the scatter results of fatigue life, etc. were also concerned and investigated with advanced experimental methods[6-9]. These experimental findings provide an intuitional understanding as to the concrete fatigue, but the phenomenological treatment limits its application in engineering practices. In physical reality, the fatigue phenomenon in concrete, as one of quasi-brittle materials, hinges upon the initiation and propagation of micro-cracks and micro-voids. It is thus reasonable to quantify the fatigue in a function of crack growth rate in fracture mechanics.

Due to the nonlinearity of crack tips of concrete, the fracture mechanics serves as a basic tool to study the propagation of cracks in concrete. Among the classical crack models, the cohesive crack model has been well developed and widely accepted. For the case of fatigue cracks in concrete, the cohesive crack models underlies a better stress-crack opening displacement (COD) curve and a numerical simulation scheme[10-13]. However, most of these models deal with a single crack, which is actually not suitable for concrete fatigue. For this reason, damage mechanics was introduced, and the fatigue damage variable was used to model the material degradation induced by propagation of micro-crack group.

In the scope of damage mechanics, the evolution of damage variable plays a critical role. Marigo[14] suggested a classical way to modify the damage evolution law which makes "fatigue damage" distinct from "brittle damage" first. Most of researchers in this area followed his step[15-17]. Under this framework, the damage surface, which is analogous to the yielding surface in plastic mechanics, was introduced to determine damage evolution. Obviously, the concept of damage surface submits to a phenomenological expression. Thus in the continuum damage mechanics, the fatigue damage evolution is still based on empirical estimation and lacks of physical basis.

In order to better understand the fatigue damage evolution at macroscale, it is necessary to investigate the micro-crack group evolution, which forms the physical basis of damage phenomenon. In this way, the rate process theory may provide a reasonable explanation upon the single micro-crack propagation under external stress, which, meanwhile, provides a quantitative measure of the bonds breakage rate at the micro-crack tip[18-22]. However, although the microscope fracture process of various solids under monotonous loading is well studied with rate process theory, the fatigue process is seldom analyzed with convincing arguments. Also, the multiscale modeling of damage and fracture from microscope to macroscope is still far from mature.

This paper devotes to build a fatigue damage evolution law with physical mechanism. Integrating with the energy analysis of the breaking process on microscale, the energy dissipation expression of concrete under fatigue loading is derived through rate process theory. In order to bridge the gap from microscale to mesoscale, the scaling method in renormalization group theory is introduced, whereby the fatigue energy dissipation of micro-crack group at mesoscale is well explained based on the similarity of crack distribution at each scale. The fatigue damage evolution equation is then built with the application of stochastic fracture model, revealing how the failure of mesoscale elements would shape macro fracture and damage evolution. Finally, with the established fatigue damage evolution equation, a fatigue constitutive model is established based on continuum damage mechanics and is validated with experiment results. It is noted that the fatigue life of concrete shows great randomness in related experiments[6, 9] and its distribution is an important problem of which much progress is achieved by recent researches[19, 23]. This paper mainly focuses on the average aspect of fatigue damage evolution and the fatigue life distribution is to be explored in the future researches.

2 Formulation of concrete fatigue damage evolution law

2.1 Energy dissipation at microscale under fatigue loading

Various ingredients included in concrete result in a heterogeneous and multi-phase material. The complex physical and chemical reactions produce a lot of micro-cracks whose tips are constituted by disordered nanoparticles. Taking one micro-crack tip in concrete into consideration, the nanoparticles of calcium silicate hydrate (C-S-H) are connected by nanoscale adhesion force[24]. The action of adhesion force would form a barrier which is called activation energy barrier for arresting the breakage of nanoparticles at micro-crack tip. Such an activation energy barrier is connected with the competition between the surface energy and elastic energy when the micro-crack tip moves(Fig. 1).

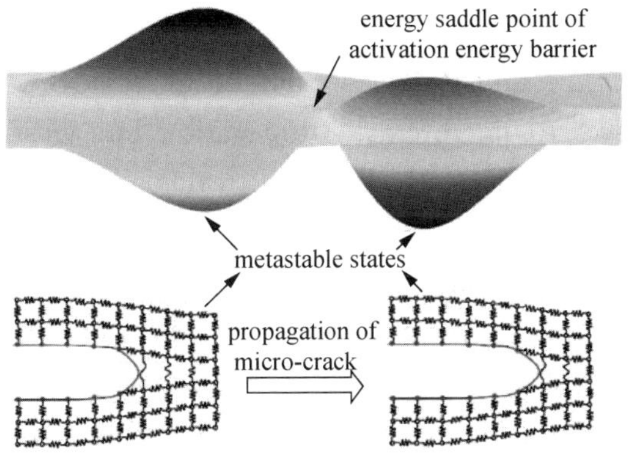

Fig. 1. The diagram of activation energy barrier in a simplified lattice model

The random thermal activated vibration of nanoparticles might drive the system cross the activation energy barrier, which behaves as bidirectional movements of breaking and healing of bonds between the nanoparticles at micro-crack tip. The probabilities of breaking and healing are the same in case that there is no external loading, indicating that the micro-crack tip locates in a statistical equilibrium state and the micro-crack is stationary. However, the external loading would change the energy balance between two adjacent metastable states caused by bond's breakage of adjacent nanoparticles, and promote a net effect of micro-crack propagation. The rate of propagation relates to the free energy change ΔQ, which could be derived by fracture analysis.

According to linear fracture mechanics, with defined microscale character length l_a (for example the scale of nanoscale), the stress intensity factor of a planar three dimensional microscale crack is expressed as[25]

$$K_a = \sigma_a \sqrt{l_a} k_a(\alpha) \tag{1}$$

where σ_a is the microscale stress, related to macroscale stress σ through a micro-macro stress coefficient c : $\sigma_a = c\sigma$; $\alpha = a/l_a$ is the relative crack length and a is the absolute half crack length; $k_a(\alpha)$ is the dimensionless stress intensity factor.

Accordingly, the energy released per unit length of crack tip advance is

$$G(\alpha) = \frac{K_a^2}{E_a} = \frac{k_a^2(\alpha) l_a c^2 \sigma^2}{E_a} \tag{2}$$

where E_a denotes the microscale elastic modulus.

Supposing η is a geometry parameter whereby $\eta\alpha$ is the width of crack, the energy released, in case of crack advance δ_a, is

$$\Delta Q_f = \delta_a (\eta\alpha l_a) G(\alpha) = V_a(\alpha) \frac{c^2\sigma^2}{E_a} \tag{3}$$

where $V_a(\alpha) = \delta_a \eta\alpha l_a^2 k_a^2(\alpha)$ is activation volume, which is connected with transformation of local activated nanoparticles at the energy saddle with respect to the initial metastable state[26].

Inspired by the works of Le and Bažant[19] and Le et al.[23], we also relate the energy released at nanoscale level to micro-crack propagation but with a modification of local energy dissipation process. Traditionally, in Griffith theory of fracture[27] the elastic energy released in the fracture process is equal to the surface energy increased, which implies the system is in a reversible equilibrium state. Rice[28] extends the Griffith concept to a more general framework of irreversible thermodynamics, which admits entropy production and relates the crack propagation to the net energy released. Noting γ_0 as the surface energy produced in the process of forming the micro-surfaces(we omit the conventional coefficient 2 representing two surfaces of a crack here for simplicity), the net energy released during the process of micro-crack advancing δ_a is

$$\Delta Q_{net} = \Delta Q_f - \gamma_0 \tag{4}$$

On the other hand, the transfer frequency, according to the rate process theory, between two metastable states can be expressed herein[18, 20]

$$f = v\{ e^{[-Q_0 + (\Delta Q_f - \gamma_0)/2]/kT} - e^{[-Q_0 - (\Delta Q_f - \gamma_0)/2]/kT} \} = 2v e^{-Q_0/kT} \sinh\left(\frac{\Delta Q_f - \gamma_0}{kT} \right) \tag{5}$$

where v is characteristic frequency $v = kT/h$; k, T and h are Boltzmann constant, absolute temperature and Planck constant respectively.
Since $\sinh(x) \approx x$ for small x and generally $kT \gg \Delta Q_f - \gamma_0$[20], Eq. (5) becomes

$$f = 2v e^{-Q_0/kT} \frac{\Delta Q_f - \gamma_0}{kT} \tag{6}$$

Thus, the crack propagation velocity can be expressed as the drift of nanoparticles at crack tip[20]

$$\dot{a} = \delta_a f \tag{7}$$

Eq. (7) is the basic expression of micro-crack propagation under static loading. While referring to the fatigue loading, Eq. (7) might underestimate the fracture velocity. A logical explanation to the fact is that there is a certain accumulative physical quantity in fatigue case promoting the effect of loading-unloading irreversibility[29-30]. For example, in the cohesive crack model, the accumulation of COD is such kind of physical quantity. However, the cohesive crack model still cannot explain how the accumulative physical quantity takes effect. Here we try to introduce another explanation: the fatigue disturbance effect would change the relative stable configuration in fracture process zone and raise the local energy that may influence the breakage process.
Actually, the C-S-H particles in cement hydrates consist of water molecules, inter- and intra-layer calcium ions and silica tetrahedral, etc.[31]. The initial undisturbed configurations of these basic particles at the existed micro-crack tip are well arranged and form a local minimum free energy state. When there exist external fatigue actions, the repeating fatigue disturbance would break the relative regular configurations and make a chaotic status at the local crack tip. Since the chaotic configurations have a higher free energy than the regular ones, it would make nanoparticles moving easier to cross the energy barriers. This physical mechanism is illustrated as Fig. 2.
Therefore, when a static sustained load is applied, the crack tip would be mainly in a stable state of lower energy. While when the fatigue load is applied, the crack tip would experience a cycle from lower

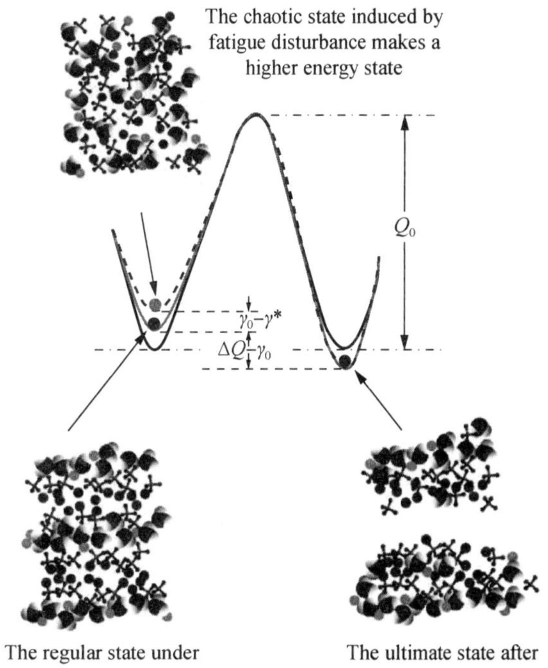

Fig. 2. Schematic diagram of fatigue disturbance effect

energy(energy valley) to higher energy(energy peak), indicating that the crack tip would not be trapped in the relative stable state in fatigue case. Obviously, in this case, the surface energy in Eq.(4) shall be modified. Under fatigue loading, the longer the opening and closure process lasts, the larger the frequency of crossing energy peaks and the quicker the propagation of the crack. In order to quantify the fatigue effect under consideration, the equivalent accumulation strain is firstly introduced as

$$\vartheta = \int_0^t |\dot{\varepsilon}_{eq}| \, dt \tag{8}$$

where $\varepsilon_{eq} = \sqrt{2Y/E}$ denotes the energy equivalent strain[32]; Y denotes the damage energy release rate conjugated with damage variable in function of Helmholtz free energy (HFE)[33].

Under fatigue loading the modified surface energy is supposed to be in an exponential decay form with the equivalent accumulation strain

$$\gamma^*(\vartheta) = \gamma_0 e^{-\beta\vartheta} \tag{9}$$

where β is a constant coefficient.

Then, in fatigue loading case, the propagation rate of micro-crack in the fracture process zone at microscale should be

$$\dot{a} = \delta_a f = 2\delta_a v e^{-Q_0/kT} \frac{\Delta Q_f - \gamma^*(\vartheta)}{kT} = 2\delta_a v e^{-Q_0/kT} \frac{\Delta Q_f - \gamma_0 e^{-\beta\vartheta}}{kT} \tag{10}$$

In fact, the number of micro-cracks at the bottom scale is very large and their sizes are in bound of same scale. Further, since the energy dissipation of micro-cracks depends upon the dimensions of cracks, the statistical average process is thus a reasonable objective to be illustrated for the sake of simplicity. In this case, it is appropriate to use the statistical average value as the representative value without considering the difference of micro-cracks dimensions in the same scale. Supposing the average initial and ultimate micro-crack lengths are a_0 and a_c, respectively, the average fracture energy released by micro-cracks in a same scale is

$$\Delta\widetilde{Q}_f = \frac{\int_{a_0}^{a_c} \Delta Q_f \, da}{a_c - a_0} = \widetilde{V}_a \frac{c^2 \sigma^2}{E_a} \tag{11}$$

where the symbol "~" denotes the statistical average quantity.

From the viewpoint of damage mechanics, the cracking process is actually a damage process of materials. The average fracture energy released is same as the damage energy released here

$$\Delta\widetilde{Q}_f = \Delta Q_d = V\int_{d_0}^{d_0+\delta_d} Y \, dd \tag{12}$$

where ΔQ_d denotes the change of damage energy; d_0 denotes the initial damage variable; δ_d

denotes the damage increment due to the average fracture energy released; V denotes the activated volume of damage which is related to representative volume element in classical damage mechanics.

The fracture process between the connections of two nanoparticles is very short and therefore the damage energy release rate could be considered as constant in the process. Eq.(12) can thus be simplified as

$$\Delta\widetilde{Q}_{\mathrm{f}} = \widetilde{V}_{\mathrm{a}} \frac{c^2\sigma^2}{E_{\mathrm{a}}} = \Delta Q_{\mathrm{d}} \approx Y\delta_{\mathrm{d}}V \tag{13}$$

Meanwhile, the average rate of micro-crack propagation can be expressed with damage energy release rate

$$\dot{a} = \delta_{\mathrm{a}}\widetilde{f} = 2\frac{\delta_{\mathrm{a}}\delta_{\mathrm{d}}V}{h}[Y - \gamma(\vartheta)]\mathrm{e}^{-Q_0/kT} \tag{14}$$

where $\gamma(\vartheta) = \gamma^*(\vartheta)/V\delta_{\mathrm{d}}$ is the volumetric homogenized surface energy, which is expressed in terms of volumetric energy.

The average energy released rate in the same scale can be written as

$$G = \frac{\Delta\widetilde{Q}_{\mathrm{f}}}{\delta_{\mathrm{a}}\eta a_{\mathrm{m}}} = \frac{Y\delta_{\mathrm{d}}V}{\delta_{\mathrm{a}}\eta a_{\mathrm{m}}} \tag{15}$$

where a_{m} is the average length of micro-cracks.

The average energy dissipation of a single micro-crack is then given by

$$\dot{E} = G\eta a_{\mathrm{m}}\dot{a} = C_0 Y[Y - \gamma(\vartheta)]\mathrm{e}^{-Q_0/kT} \tag{16}$$

where $C_0 = \dfrac{2\delta_{\mathrm{d}}^2V^2}{h}$ can be seen as material constant.

2.2 Fatigue energy dissipation at mesoscale

Since the fatigue energy dissipation at microscope is known, it is necessary to determine the energy dissipation at larger scale based on this primary scale's result. However, how to upscale from microscale to macroscale is a challenging issue in multiscale modeling. In the field of concrete fatigue, a simple scaling scheme states there exists a hierarchical structure of crack distribution in the fracture process zone at different scales[20] and the function of crack number at each scale should be self-similar which features a fractal characteristic and has the form of power laws[34]. It is indicated that the scaling form is same from microscope to macroscope.

While our approach herein is somewhat different from above approach, here we emphasize on the statistical aspect of fracture and damage. Actually, due to the similarity between fracture and phase transition process, the related methodology such as renormalization group theory and related concepts such as self-similarity, scaling and critical point are introduced to study fracture process[35-36]. For the fatigue problem, the influence of microscopic fatigue dissipation upon the macroscopic fatigue evolution in disordered media were also studied with fiber bundle model and random fuse model, respectively[37, 38]. Further researches based on these approaches and models reveal that the fracture failure modes of quasi-brittle material mainly rely on the size effect and disorder distribution of material mechanical property[39]. Small size and large disorder tend to provoke percolation-like failure where smeared small cracks dominate. As the increase of size and decrease of disorder, series of small cracks synthesized into series of larger ones, and further few dominating cracks would form and induce nucleation-like failure. The phenomenon of concrete failure shows that there are few dominant macro-cracks at macroscale while there are mount of smaller cracks at subscales, especially at the fracture process zone (Fig. 3). This hints that there is a transition of damage mechanism at different scales, i. e. the damage pattern at microscale is of percolation-like and the failure mode tend to avalanches-like and nucleation-like as the scales becomes larger. With this hypothesis, the methodology adopted herein can be stated as: from the microscale to mesoscale the fracture process is percolation-like, where cracks distribution exposes a self-similarity pattern; while from mesoscale to macroscale the crack avalanches dominate and the stochastic fracture model (it is similar to fiber bundle model and can catch avalanches process) is employed to study the process.

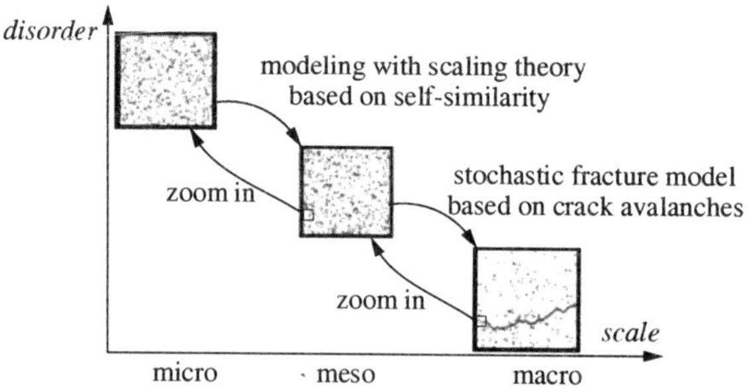

Fig. 3. Scaling diagram of cracks in concrete

It is well recognized that the mechanical drive force of cracks is the external loading of which the intensity can be measured by damage energy release rate Y from the viewpoint of damage mechanics. Correspondingly, the resistance force origins from surface energy of which the effect is usually called "R-curve" behavior in quasi-brittle materials[40]. Here the representative volumetric homogenized surface energy $\Gamma = K\gamma^*$ is defined to include the overall resistance effect of mesoscale due to the micro-cracks in this domain, where K is a scale coefficient. With the drive force and resistance, it is natural to conjecture that the total number of micro-cracks in unit volume is determined by the non-dimensional parameter $\alpha = Y/\Gamma$. Then the density of micro-cracks ρ_c at mesoscale could be expressed as the function of α

$$\rho_c = f(\alpha) = f\left(\frac{Y}{\Gamma(\vartheta)}\right) \tag{17}$$

Assuming that the activated process from microscale to mesoscale is similar to phase transition phenomenon, the function f features scale invariance that could reproduce itself on successive scaling process of parameter α. It is indicated that there exist rescaling factors c_1 and c_2 making $\rho_c = f(\alpha) = c_1 f(\alpha/c_2)$ to be invariant under arbitrary scale change. In order to reach this purpose, the renormalization scale is supposed to be $1 + p\delta$ if the parameter α is expanded with a small factor $c_2 = 1/(1-\delta)$, now

$$\rho_c = (1+p\delta) f[(1-\delta)\alpha] \tag{18}$$

It is readily seen that when δ tends to zero, the above equation yields to a simple differential equation: $pf(\alpha) = \alpha \cdot \mathrm{d}f/\mathrm{d}\alpha$ and further its solution has the form of power law

$$\rho_c = C_1 \alpha^{p} = C_1 \left(\frac{Y}{\Gamma(\vartheta)}\right)^{p} \tag{19}$$

where p is a critical exponent; C_1 is a constant coefficient.

The total energy dissipation at mesoscale should be the summation of all micro-cracks contained in the mesoscale volume, i.e.

$$\dot{E}_f = \rho_c \dot{E} = CY[Y-\gamma(\vartheta)]\left(\frac{Y}{\Gamma(\vartheta)}\right)^{p} \tag{20}$$

where $C = C_1 C_0 e^{-Q_0/kT}$.

In fact, the theoretical micro-crack density in Eq. (19) is a rough result, without considering the interaction between the micro-cracks. In the damage developing process, the interaction among micro-cracks in high density will reduce the number of activated micro-cracks. The situation is similar to that high density of dislocations will cause the locking effect, which reduces the activated dislocations in the plastic mechanics[41]. Taking the damage variable as the measure of effective micro-crack density, the decrease of activated micro-crack density could be expressed as

$$\frac{\partial \rho_c}{\partial d} = -\kappa \rho_c \tag{21}$$

The solution of equation above is

$$\rho_c^{\mathrm{eff}} = \rho_c e^{-\kappa d} \tag{22}$$

where ρ_c is the solution of Eq. (19), which denotes the theoretical micro-crack density; ρ_c^{eff} is effective activated micro-crack density; κ is a material constant.

With the representative average energy dissipation and the activated micro-crack density modification, the total energy dissipation at mesoscale then be expressed as

$$\dot{E}_f = Ce^{-\kappa d} Y[Y-\gamma(\vartheta)]\left[\frac{Y}{\Gamma(\vartheta)}\right]^{p} \tag{23}$$

2.3 Macroscope fatigue damage evolution

As stated in previous section, the visible cracks appear in larger scales and the self-similarity of small-crack-group distribution pattern breaks when the scale transfers from mesoscale to macroscale. The statistical analysis based on the fiber bundle model results shows that the model is able to reflect the avalanches process that would lead to macro-crack. Historically, this kind of model was proposed to describe the progressive damage propagation of a brittle rod subjected to uniaxial loading[42-43]. Li et al. modeled the fracture strain of model elements as a 1-D random field and developed the uniaxial stochastic damage model for concrete[44-46]. In this paper, an updated model is proposed accounting for the fatigue accumulation effect.

A macroscale element could be modeled as a series of discrete mesoscale elements in parallel(Fig. 4), in which each single element would damage gradually as the energy dissipating induced by the external loading. When the energy dissipation in an element reaches its energy threshold the element is supposed to break.

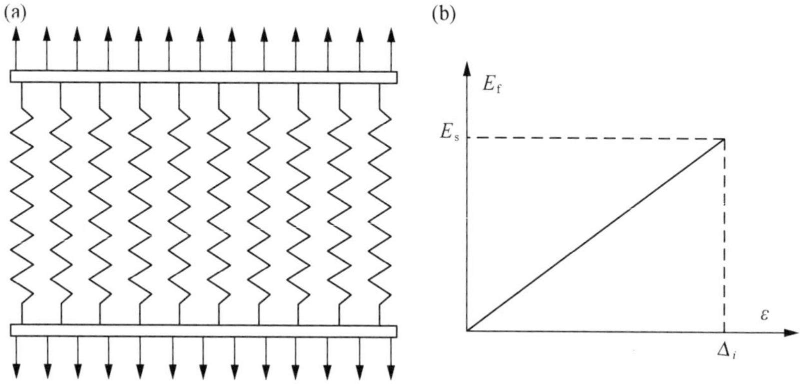

Fig. 4. Schematic diagram of stochastic fracture model: (a) mesoscale fiber bundle and (b) the fatigue fracture criteria

Assuming that the total number of elements in the model tends to infinite, the damage variable can be presented as a stochastic integration

$$\begin{cases} d = \int_0^1 H(E_f - E_s)\mathrm{d}x \\ E_s = \dfrac{1}{2}E_0\,\Delta_i^2(x) \end{cases} \tag{24}$$

where $H(\cdot)$ denotes a Heaviside function, indicating that when the total energy dissipated E_f by cracks at different scales exceeds the inherent energy of the representative volume element E_s, the element breaks. E_0 denotes the initial elastic modulus. $\Delta(x)$ denotes one-dimensional random fracture strain field that describes the characteristic distribution of mesoscale elements. x denotes the spatial coordinate of each mesoscale elements. Generally, the random fracture strain field of concrete is viewed to submit to the lognormal distribution

$$N(\ln\Delta(x);\lambda,\zeta) = \frac{1}{\zeta\sqrt{2\pi}}\exp\left(-\frac{[\ln\Delta(x)-\lambda]^2}{2\zeta^2}\right) \tag{25}$$

where λ and ζ are the mean value and the standard deviation of the corresponding normal distribution. The autocorrelation coefficient in exponential function for the random field is taken as

$$\varphi(\Delta x) = e^{-\xi |x_j - x_i|} \tag{26}$$

where ξ is correlation parameter; $|x_j - x_i|$ denotes the distance of two spatial coordinates. Actually, a meso-element could be broken either from brittle fracture when external strain exceeds the inherent strain threshold or from fatigue damage when fatigue energy dissipation exceeds the inherent energy threshold. One might realize that there exists a competitive mechanism between brittle damage and fatigue damage. The unified damage evolution expression, integrating with the competitive mechanism, can be expressed as

$$\begin{cases} d = \int_0^1 [1 - H(E_s - E_f) H(E_s - Y)] dx \\ E_f = \int_0^t C e^{-\kappa d} Y (Y - \gamma(\vartheta)) \left(\dfrac{Y}{\Gamma e^{-\beta\vartheta}} \right)^p dt \\ E_s = \dfrac{1}{2} E_0 \Delta^2 (x) \end{cases} \tag{27}$$

where $H(E_s - E_f)$ is the criterion for fatigue damage, and the $H(E_s - Y)$ is the criterion for brittle damage. The expression of damage means that the element will break when either of the two criteria is satisfied.

Considering that the volumetric homogenized surface energy of microscale $\gamma(\vartheta)$ is sufficiently small compared with actual damage energy release rate Y, it is then omitted and only the representative volumetric homogenized surface energy Γ is included. The damage evolution expression can be thus simplified as

$$\begin{cases} d = \int_0^1 [1 - H(E_s - E_f) H(E_s - Y)] dx \\ E_f = \int_0^t C e^{-\kappa d} Y^2 \left(\dfrac{Y}{\Gamma e^{-\beta\vartheta}} \right)^p dt \\ E_s = \dfrac{1}{2} E_0 \Delta^2 (x) \end{cases} \tag{28}$$

The derivation of above fatigue damage evolution is not concerned with particular loading condition and the compression damage is essentially induced by lateral tension and shear. So the form of Eq. (28) should be applicable to both condition of under tension and under compression. The only differences between tension and compression are the values of related parameters, which will be specified in next paragraph.

3 Macro-damage mechanics framework

3.1 Thermodynamics of damage evolution

When the scale of interest rises to the macro field, the thermodynamic based continuum damage mechanics is appropriate in practical use.

The general form of constitutive model with damage and plasticity can be written as[33]

$$\sigma=(\mathbf{I}-\mathbf{D}):\bar{\sigma}=(\mathbf{I}-\mathbf{D}):\mathbf{E}_0:(\boldsymbol{\varepsilon}-\boldsymbol{\varepsilon}^{\mathrm{p}}) \tag{29}$$

where σ, $\boldsymbol{\varepsilon}$, and $\boldsymbol{\varepsilon}^{\mathrm{p}}$ are the second-order stress, strain and plastic strain tensors, respectively; $\mathbf{E}_0$ is the fourth-order tensor denoting the initial undamaged elasticity tensor; $\mathbf{D}$ the fourth-order damage tensor; $\bar{\sigma}$ is the effective stress based on Lemaitre's strain equivalence hypothesis[47]. $\boldsymbol{\varepsilon}^{\mathrm{p}}$ can be determined by the traditional treatment with the yield condition and the evolution potential based on the effective stress, which is called as effective stress space plasticity.

The quasi-brittle materials such as concrete expose a significant different behavior under tension and compression, which makes it appropriate to decompose the effective stress as[48]

$$\begin{cases}\bar{\sigma}^{+}=\mathbf{P}^{+}:\bar{\sigma}=\sum_i H(\hat{\bar{\sigma}}_i)(\mathbf{p}_i\otimes\mathbf{p}_i)\\ \bar{\sigma}^{-}=\mathbf{P}^{-}:\bar{\sigma}=\bar{\sigma}-\bar{\sigma}^{+}\end{cases} \tag{30}$$

where $\hat{\bar{\sigma}}_i$ and $\mathbf{p}_i$ are the ith eigenvalue and eigenvector of the effective stress tensor, respectively. $H(\cdot)$ denotes the Heaviside function. $\mathbf{P}^{+}$ and $\mathbf{P}^{-}$ are the fourth-order projective tensors which can be expressed as

$$\begin{cases}\mathbf{P}^{+}=\sum_i H(\hat{\bar{\sigma}}_i)(\mathbf{p}_i\otimes\mathbf{p}_i\otimes\mathbf{p}_i\otimes\mathbf{p}_i)\\ \mathbf{P}^{-}=\mathbf{I}-\mathbf{P}^{+}\end{cases} \tag{31}$$

Also it is natural to simplify the second-order damage tensor using a bi-scalar damage representation, which connects the corresponding effective stress tensor and Cauchy stress

$$\mathbf{D}=d^{+}\mathbf{P}^{+}+d^{-}\mathbf{P}^{-} \tag{32}$$

where d^{+} and d^{-} denote the tensile and the compressive damage scalars, which can be expressed as Eq.(28) for tensile and the compressive cases, respectively.

Besides, the initial HFE potential can be decomposed as the elastic part and plastic part

$$\psi_0=\psi_0^{\mathrm{e}}+\psi_0^{\mathrm{p}} \tag{33}$$

Accordingly, the HFE potentials could be expressed as a reduced form from initial states due to the damage evolution under tensile and compression

$$\begin{cases}\psi^{+}(\boldsymbol{\varepsilon}^{\mathrm{e}},d^{+},d^{-})=(1+d^{+})\psi_0^{\mathrm{e}+}+(1+d^{-})\psi_0^{\mathrm{p}+}\\ \psi^{-}(\boldsymbol{\varepsilon}^{\mathrm{e}},d^{+},d^{-})=(1+d^{-})\psi_0^{\mathrm{e}-}+(1+d^{-})\psi_0^{\mathrm{p}-}\end{cases} \tag{34}$$

where the elastic parts is defined as the elastic strain energy, and the plastic part is defined as the plastic accumulation works with respect to the positive and negative components, separately[49]

$$\psi_0^{\mathrm{e}\pm}=\frac{1}{2}\bar{\sigma}^{\pm}:\boldsymbol{\varepsilon}^{\mathrm{e}} \tag{35}$$

$$\psi_0^{p\pm}=\int_0^{\varepsilon^p}\bar{\sigma}^{\pm}:\mathrm{d}\varepsilon^p \tag{36}$$

The energy release rate is a thermodynamic-relevant variable, which also has positive and negative components defined as

$$Y^{\pm}=-\frac{\partial\psi^{\pm}}{\partial d^{\pm}} \tag{37}$$

With neglecting the tensile plastic HFE ψ_0^{p+} and adopting the Drucker-Prager plastic model for the compressive HFE ψ_0^{p-}, Wu et al.[33] derived the approximate explicit expressions of the damage release rates as follows

$$Y^{+}\approx\frac{1}{2E_0}\left[\frac{2(1+v_0)}{3}3\bar{J}_2^{+}+\frac{1-2v_0}{3}(\bar{I}_1^{+})^2-v_0\bar{I}_1^{+}\bar{I}_1^{-}\right] \tag{38}$$

$$Y^{-}\approx b_0\left(\alpha\bar{I}_1^{-}+\sqrt{3\bar{J}_2^{-}}\right)^2 \tag{39}$$

where $\bar{I}_1^{\pm}$ are the first invariants of the tensile and the compressive effective stresses, respectively; $\bar{J}_2^{\pm}$ are the second invariants of the deviatoric components of effective stresses; E_0 and v_0 are the Young's modulus and the Poisson's ratio of the initial undamaged material, respectively; b_0 is also a material parameter. The parameter α is related to the biaxial strength increase usually taken as 0.121 2.

Under the circumstance of uniaxial loading, the expression of damage release rates in Eqs.(38) and (39) can be simply written as

$$Y^{+}=\frac{(\bar{\sigma}^{+})^2}{2E_0}=\frac{1}{2E_0}\left[\frac{\sigma^{+}}{(1-d^{+})}\right]^2 \tag{40}$$

$$Y^{-}=b_0[(\alpha-1)\bar{\sigma}^{-}]^2=b_0\left[\frac{(\alpha-1)\sigma^{-}}{(1-d^{-})}\right]^2 \tag{41}$$

The explicit expressions of damage energy release rate combining with the damage evolution equations given in Eq. (28) have formed a complete fatigue damage evolution expression.

3.2 The evolution of fatigue plasticity

The plasticity evolution usually can be theoretically determined by defining corresponding yield condition and evolution potential based on the effective stress with the classical plastic theory. But considering the solution of plastic strain in effective stress space using traditional return-mapping algorithm is of computational costs, a phenomenological model connecting the evolution of plasticity with the evolution of fatigue damage is presented herein. Since the plastic strain is usually defined as a function of elastic strain in experimental studies and the plasticity of concrete mainly comes from the slips at the crack surfaces, the plastic strain as a function of elastic strain and damage variable is defined by

$$\dot{\varepsilon}^p=f_p(d)\dot{\varepsilon}^e \tag{42}$$

where f_p is called the plastic evolution function. Its explicit expression can be established from regression analysis of relative experiment results[50]

$$f_p(d)=\xi_p(d)^{n_p} \tag{43}$$

where ξ_p and n_p are both plastic coefficients.

For the case of fatigue loading, the change of elastic strain is no longer monotonic but in a cyclic manner and a modification is made to the equation above in a holonomic form

$$\varepsilon^p=f_p(d)\varepsilon^e_{max} \tag{44}$$

where ε^e_{max} denotes the maximum elastic strain in fatigue loading history.

In the condition of concrete fatigue, the plasticity under tension is very small. There is, moreover, few data in the fatigue tests under tension. Therefore the plastic strain under tension is omitted in our model.

The involvement of the plastic strain will not affect the crack tip as much as elastic strain under the condition of compression, the equivalent accumulation strain is thus modified as

$$\vartheta=\int_0^t(|\dot{\varepsilon}_{eq}|-|\dot{\varepsilon}_p|)dt \tag{45}$$

With the fatigue damage evolution as well as the plasticity evolution, a complete fatigue damage constitutive relationship is built under the framework of continuum damage mechanics and it is worth mentioning that the fatigue damage evolution is established not just by empirically fitting but with physically based analysis. The numerical implementation of the proposed model is a conventional routine where the forward Euler scheme is adopted to evaluate the evolutions of damage and plasticity.

4 Validation of fatigue damage evolution

4.1 Fatigue damage evolution and fatigue life of concrete

A very important aspect of the fatigue model is the prediction of fatigue damage development, which can be expressed as a decline of modulus observed in fatigue experiments. Here the experimental results derived by Gao and Hsu[51] is taken to validate the model. In their experiment, the dimensions of test specimens are 100 mm×100 mm×300 mm. The compression strength is $f_c=34.6$ MPa and the upper bound of fatigue stress is $\sigma_{max}=20.8$ MPa. The fatigue damage in the tests is defined as $d=1-(E_{fa}/E_0)$, where E_{fa} is fatigue modulus after N cycles, and E_0 is initial fatigue modulus. Under the condition of normal temperature, $e^{-Q_0/kT}\sim 3.75\times 10^{-18}$ in concrete[18]. For comparative purpose with related literatures, the parameters of C30 level concrete are included in the model validation.

Table 1 presents the parameters that describe the random field of fracture strain in stochastic fracture model. These parameters are identified from static test results by Zeng[52], where λ, ζ, and ξ are the mean, variance, and correlation length of the random field, respectively. The other parameters in the model of fatigue damage evolution are listed in Table 2.

Table 1. Parameters of fracture strain of concrete in lognormal distribution

Type	Tensile strength/MPa Elastic modulus/MPa	λ	Tension Compression	ζ	Tension Compression	ξ	Tension Compression
C30	2.21	4.753 6		0.656 0		0.027 0	
	34 700	7.471 0		0.376 7		0.015 1	
C40	2.50	4.541 6		0.692 7		0.021 7	
	38 200	7.560 1		0.273 4		0.016 3	
C50	2.67	4.869 6		0.582 8		0.032 2	
	36 600	7.566 8		0.254 6		0.023 8	

Table 2. Parameters in the model of fatigue damage evolution

	κ	$\Gamma/(\mathrm{J\cdot m^{-3}})$	$C/(\mathrm{J^{-1}\cdot m^{3}\cdot s^{-1}})$	β	P	n_p	ξ_p
Tension	25	0.155	1.20×10^{-3}	0.006	10.5	—	—
Compression			1.52×10^{-14}	0.036	8.5	5.0	0.8

Fig. 5 shows the fatigue damage evolution under uniaxial compression using the model proposed in the paper. It is seen that the fatigue damage evolution exhibits three stages obviously, which is in accordance with the experimental results. The three stages corresponding to the initial stage of brittle damage coupled with fatigue damage, the stable stage of energy dissipation process by fatigue cracks and the instability stage of macroscale cracks propagation.

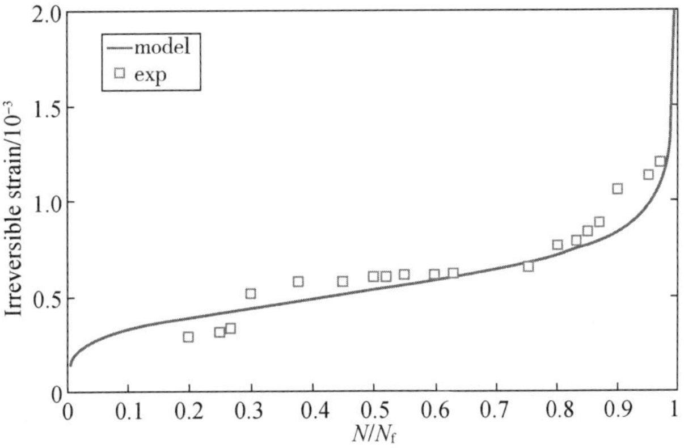

Fig. 5. Fatigue damage evolution under uniaxial compression

The irreversible strain measured in the experiments and the plastic strain predicted by the model are shown in Fig. 6, revealing that there is a good accordance between model prediction and test results.

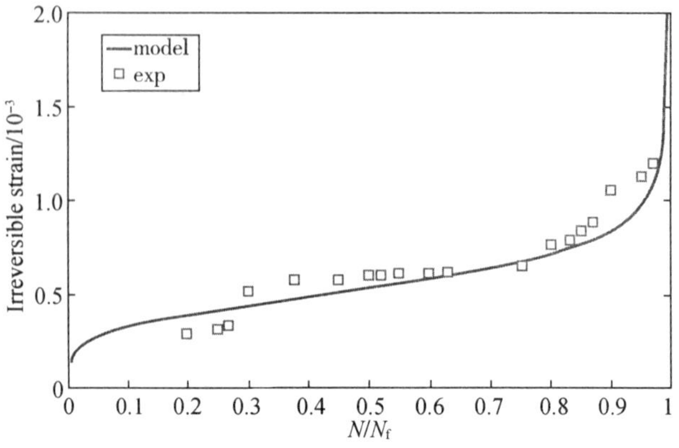

Fig. 6. Plastic strain evolution under uniaxial compression

One of the main applications of fatigue damage model is the prediction of fatigue period. Fig. 7 shows the results of fatigue period under different tensile stress levels ($r=\sigma_{min}/f_t$). For validating purpose, the Cornelissen's experimental results (1984) are exposed in the same figure. Similarly, Fig. 8 shows the results of fatigue period under different compressive stress levels. The comparison is proceeded against the Tepfers' regression results from experimental data[5]. Evidently, theoretical results by proposed model almost coincide with the average results of experiment at different stress level.

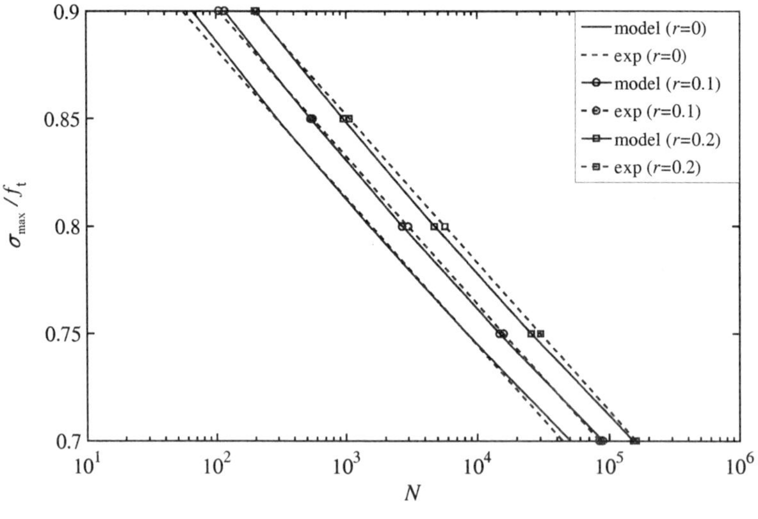

Fig. 7. *S-N* curve under different stress range in uniaxial tension

4.2 Fatigue of concrete notched beam

The fatigue experiment of concrete notched beam was carried out by Kolluru et al.[53], of which the experimental set-up and specimen size are shown in Fig. 9. In this experiment, the average value of compression strength of the specimens is $f_c=35$ MPa and the loading

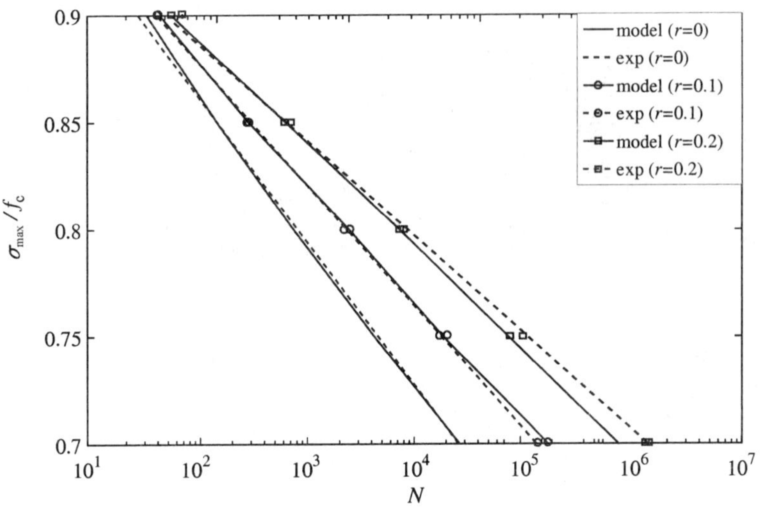

Fig. 8. ***S-N*** **curve under different stress range in uniaxial compression**

range of concentrated force is (0–0.74) F_{peak}, where F_{peak} denotes the quasi-static peak load. In the numerical simulation, the parameters of C40 level concrete are taken as showed in Table 1. All the simulation is performed on ABAQUS platform with the proposed fatigue damage model implemented by user subroutine.

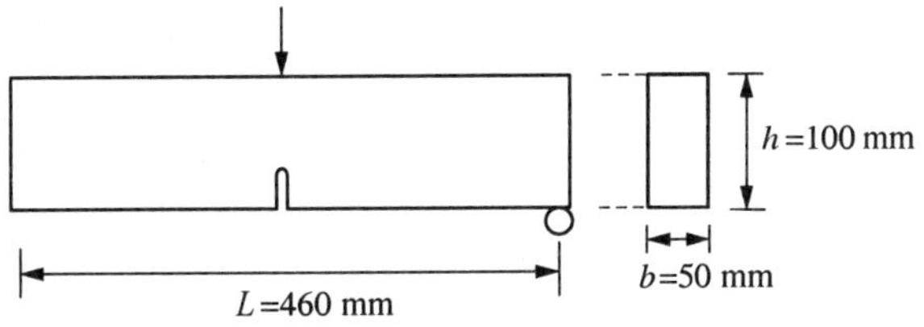

Fig. 9. Experimental set-up and specimen size of concrete notched beam[53]

The simulated damage contours and corresponding von Mises stress contours are shown in Fig. 10. One could recognize the fatigue crack growth in the beams in case of different fatigue cycles. Numerical results reveal that the fatigue damage is concentrated on the local scope along the notch direction.

Although the fatigue damage model cannot represent the fatigue crack growth directly, the propagation of damage zone with amplitude of damage variable near to 1.0 is equivalent to growth of crack in some sense. The comparison between crack length and damage zone length is shown in Fig. 11, indicating that the development of total damage is in good accordance with crack propagation.

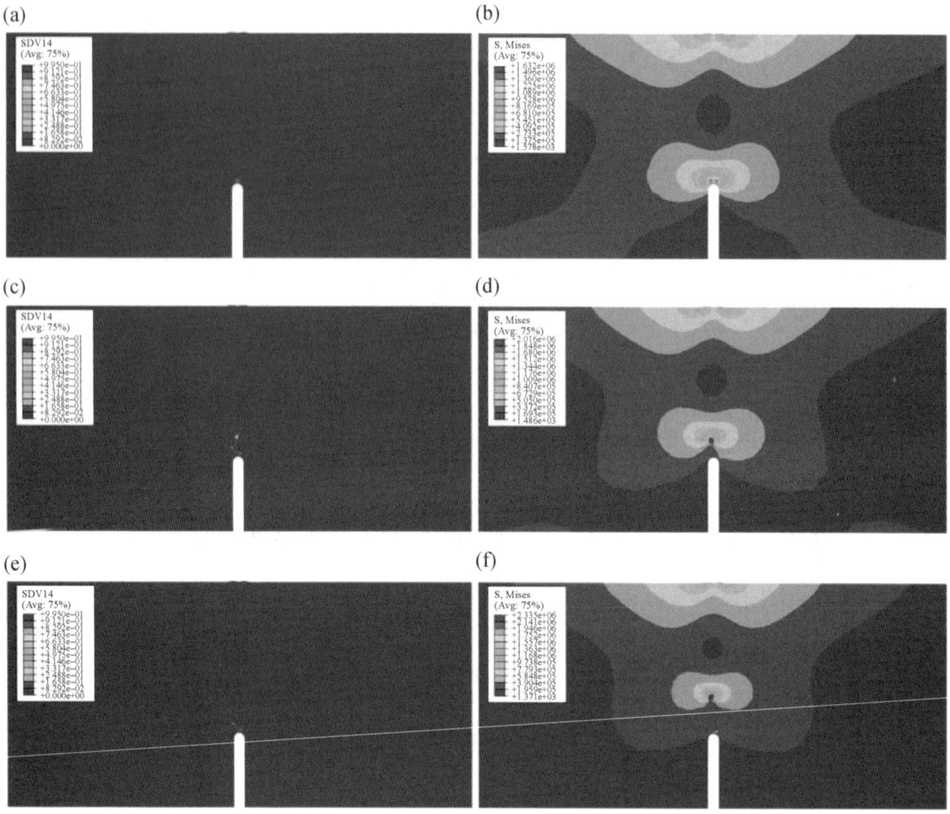

Fig. 10. Fatigue damage evolution and corresponding stress states near the crack tip (the damage threshold is set as 0.995 and corresponding von Mises stress near the crack tip is given descriptively): (a) damage distribution under cycle 1 557, (b) Von Mises stress contour under cycle 1 557, (c) damage distribution under cycle 15 066, (d) Von Mises stress contour under cycle 15,066, (e) damage distribution under cycle 25 220, and (f) Von Mises stress contour under cycle 25 220

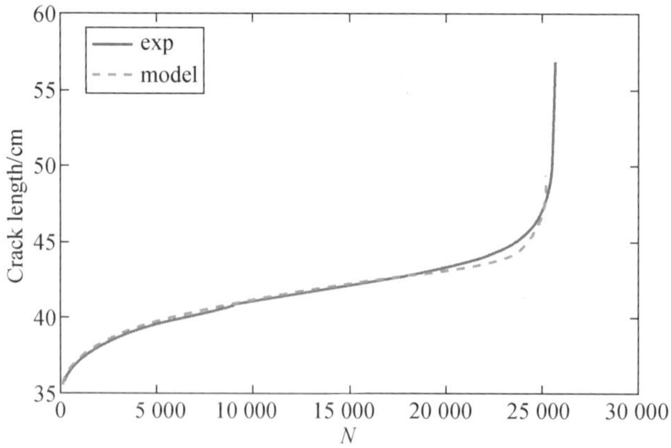

Fig. 11. The comparison between crack length and damage zone length

5 Conclusions

Current macro continuum mechanics lacks of the physical basis relevant to the fatigue damage evolution of quasi-brittle material such as concrete. This paper starts from the energy analysis in the propagation process of micro-cracks and derives the fatigue energy dissipation at mesoscale with heuristic modeling based on the self-similarity of cracks distribution. The macro-damage evolution is established in conjunction with the stochastic fracture model, in which the fracture of each single element correlates to the energy dissipation induced failure at mesoscale. With the physical mechanism of concrete fatigue damage evolution, an updated fatigue damage model is established accounting for the fatigue accumulation effect. The model is verified through comparison against the testing results and it is shown that the model secures the main character of concrete fatigue damage process. Some remarkable conclusions can be summarized as follows:

(1) The fatigue disturbance effect measured by the equivalent accumulation strain changes the local energy distribution and reduces the activation energy barrier. This is a critical factor resulting in the difference between static and fatigue loading conditions.

(2) The upscaling from the microscale to mesoscale could be implemented utilizing the physical essence of self-similarity of cracks distribution. The damage process from mesoscale to macroscale can be represented by the stochastic fracture model in which damage evolution is reflected by avalanches of elements.

(3) In conjunction with the fatigue damage evolution law obtained with physical basis, the fatigue damage degradation process and fatigue life of concrete structural member can be well represented and evaluated in the framework of macro continuum damage mechanics.

Declaration of Conflicting Interests

The author(s) declared no potential conflicts of interest with respect to the research, authorship, and/or publication of this article.

Funding

The author(s) disclosed receipt of the following financial support for the research, authorship, and/or publication of this article: The National Natural Science Foundation of China(Grant No. 51538010 and 51608158).

References

[1] AAS-JAKOBSEN K. Fatigue of Concrete Beams and Columns[M]. Trondheim: NTH Institute of Betonkonstruksjoner, 1970.

[2] CORNELISSEN H. Fatigue failure of concrete in tension[J]. Heron, 1984, 29: 1-68.

[3] HOLMEN J. Fatigue of concrete by constant and variable amplitude loading[J]. ACI Special Publication, 1982, 75: 71-110.

[4] HSU T. Fatigue of plain concrete[J]. ACI Journal, 1981, 78: 292-305.

[5] TEPFERS R, KUTTI T. Fatigue strength of plain, ordinary, and lightweight concrete[J]. ACI Journal, 1979, 76: 635-652.

[6] BREITENBÜCHER R, IBUK H. Experimentally based investigations on the degradation-process of concrete under cyclic load[J]. Materials and Structures, 2006, 39: 717-724.

[7] BREITENBÜCHER R, IBUK H, ALAWIEH H. Influence of cyclic loading on the degradation of mechanical concrete properties[M]//Advances in Construction Materials. Berlin: Springler Berlin Heidelberg, 2007: 317-324.

[8] HOHBERG R. Zum Ermüdungsverhalten von Beton[D]. PhD Thesis, TU Berlin, Berlin, 2004.

[9] OH B. Fatigue life distributions of concrete for various stress levels[J]. ACI Materials Journal, 1991, 88: 122-128.

[10] HORDIJK D. Local approach to fatigue of concrete[D]. PhD Thesis, Delft University of Technology, Delft, 1991.

[11] KESSLER-KRAMER C, MECHTCHERINE V, MÜLLER H. Testing and modeling the behavior of concrete under cyclic tensile loading[M]//Fracture Mechanics of Concrete Structures. Ia-FraMCos, 2004: 995-1004.

[12] NGUYEN O, REPETTO E, ORTIZ M, et al. A cohesive model of fatigue crack growth[J]. International Journal of Fracture, 2001, 110: 351-369.

[13] YANG B, MALL S, RAVI-CHANDAR K. A cohesive zone model for fatigue crack growth in quasibrittle materials[J]. International Journal of Solids and Structures, 2001, 38: 3927-3944.

[14] MARIGO J. Modelling of brittle and fatigue damage for elastic material by growth of microvoids[J]. Engineering Fracture Mechanics, 1985, 21: 861-874.

[15] ALLICHE A. Damage model for fatigue loading of concrete[J]. International Journal of Fatigue, 2004, 26: 915-921.

[16] MAI SH, LE-CORRE F, FORET G, et al. A continuum damage modeling of quasi-static fatigue strength of plain concrete[J]. International Journal of Fatigue, 2012, 37: 79-85.

[17] PAPA E, TALIERCIO A. Anisotropic damage model for the multiaxial static and fatigue behaviour of plain concrete[J]. Engineering Fracture Mechanics, 1996, 55: 163-179.

[18] KRAUSZ AS, KRAUSZ K. Fracture Kinetics of Crack Growth[M]. Dordrecht: Springer, 1988.

[19] LE JL, BAŽANT ZP. Unified nano-mechanics based probabilistic theory of quasibrittle and brittle structures: Ⅱ. Fatigue crack growth, lifetime and scaling[J]. Journal of the Mechanics and Physics of Solids, 2011, 59: 1322-1337.

[20] LE JL, BAŽANT ZP, BAZANT MZ. Unified nano-mechanics based probabilistic theory of quasibrittle and brittle structures: Ⅰ. Strength, static crack growth, lifetime and scaling[J]. Journal of the Mechanics and Physics of Solids, 2011, 59: 1291-1321.

[21] TOBOLSKY A, EYRING H. Mechanical properties of polymeric materials[J]. The Journal of Chemical Physics, 1943, 11: 125-134.

[22] ZHURKOV SN. Kinetic concept of the strength of solids[J]. International Journal of Fracture, 1965, 1: 311-323.

[23] LE J-L, BAŽANT ZP. Finite weakest-link model of lifetime distribution of quasibrittle structures under fatigue loading[J]. Mathematics and Mechanics of Solids, 2014, 19: 56-70.

[24] MURILLO JSR, MOHAMED A, HODO W, et al. Computational modeling of shear deformation and failure of nanoscale hydrated calcium silicate hydrate in cement paste: Calcium silicate hydrate Jennite [J]. International Journal of Damage Mechanics, 2016, 25: 98-114.

[25] GROSS D, SEELIG T. Fracture Mechanics: With an Introduction to Micromechanics[M]. Berlin: Springer Science & Business Media, 2011.
[26] LI J. The mechanics and physics of defect nucleation[J]. MRS Bulletin, 2007, 32: 151-159.
[27] GRIFFITH AA. The phenomena of rupture and flow in solids[J]. Philosophical Transactions of the Royal Society of London Series A, Containing Papers of a Mathematical or Physical Character, 1921, 221: 163-198.
[28] RICE JR. Thermodynamics of the quasi-static growth of Griffith cracks[J]. Journal of the Mechanics and Physics of Solids, 1978, 26: 61-78.
[29] ABDELMOULA R, MARIGO JJ, WELLER T. Construction and justification of Paris-like fatigue laws from Dugdale-type cohesive models[J]. Annals of Solid and Structural Mechanics, 2010, 1: 139-158.
[30] DESMORAT R, RAGUENEAU F, PHAM H. Continuum damage mechanics for hysteresis and fatigue of quasi-brittle materials and structures[J]. International Journal for Numerical and Analytical Methods in Geomechanics, 2007, 31: 307-329.
[31] PELLENQ RJ-M, KUSHIMA A, SHAHSAVARI R, et al. A realistic molecular model of cement hydrates[J]. Proceedings of the National Academy of Sciences, 2009, 106: 16102-16107.
[32] LI J, REN X D. Stochastic damage model for concrete based on energy equivalent strain[J]. International Journal of Solids and Structures, 2009, 46: 2407-2419.
[33] WU J Y, LI J, FARIA R. An energy release rate-based plastic-damage model for concrete[J]. International Journal of Solids and Structures, 2006, 43: 583-612.
[34] BARENBLATT GI. Scaling, Self-similarity, and Intermediate Asymptotics: Dimensional Analysis and Intermediate Asymptotics[M]. Cambridge, MA: Cambridge University Press, 1996.
[35] ALAVA MJ, NUKALA PK, ZAPPERI S. Statistical models of fracture[J]. Advances in Physics, 2006, 55: 349-476.
[36] BONAMY D, BOUCHAUD E. Failure of heterogeneous materials: A dynamic phase transition? [J]. Physics Reports, 2011, 498: 1-44.
[37] OLIVEIRA C, VIEIRA ADP, HERRMANN H, et al. Subcritical fatigue in fuse networks[J]. EPL (Europhysics Letters), 2012, 100: 36006.
[38] VIEIRA AP, ANDRADE JS, JR. HERRMANN HJ. Subcritical crack growth: The microscopic origin of Paris' law[J]. Physical Review Letters, 2008, 100: 195503.
[39] SHEKHAWAT A, ZAPPERI S, SETHNA JP. From damage percolation to crack nucleation through finite size criticality[J]. Physical Review Letters, 2013, 110: 185505.
[40] SURESH S. Fatigue of Materials[M]. Cambridge, MA: Cambridge University Press, 1998.
[41] JOHNSTON WG, GILMAN JJ. Dislocation velocities, dislocation densities, and plastic flow in lithium fluoride crystals[J]. Journal of Applied Physics, 1959, 30: 129-144.
[42] DANIELS H. The statistical theory of the strength of bundles of threads: I [J]. Proceedings of the Royal Society of London Series A Mathematical and Physical Sciences, 1945, 183: 405-435.
[43] KRAJCINOVIC D. Damage Mechanics[M]. Amsterdam: Elsevier Science B.V, 1996.
[44] LI J. Research on the stochastic damage mechanics for concrete materials and structures[J]. Journal of Tongji University(Natural Science Edition), 2004, 32: 1270-1277.
[45] LI J, LU Z H, ZHANG Q Y. Study on stochastic damage constitutive law for concrete material subjected to uniaxial compressive stress[J]. Journal of Tongji University(Natural Science Edition), 2003, 31: 505-509.

[46] LI J, ZHANG Q Y. Study of stochastic damage constitutive relationship for concrete material[J]. Journal of Tongji University(Natural Science Edition), 2001, 29: 1135-1141.

[47] LEMAITRE J. Evaluation of dissipation and damage in metals[C]//Proceedings of international conference on the mechanical behavior of materials 1(ICM 1), Kyoto, Japan, 15-20 August 1971.

[48] FARIA R, OLIVER J, CERVERA M. A strain-based plastic viscous-damage model for massive concrete structures[J]. International Journal of Solids and Structures, 1998, 35: 1533-1558.

[49] JU JW. On energy-based coupled elastoplastic damage theories: Constitutive modeling and computational aspects[J]. International Journal of Solids and Structures, 1989, 25: 803-833.

[50] REN X D, ZENG S J, LI J. A rate-dependent stochastic damage-plasticity model for quasi-brittle materials[J]. Computational Mechanics, 2014, 55: 267-285.

[51] GAO L, HSU CTT. Fatigue of concrete under uniaxial compression cyclic loading[J]. ACI Materials Journal, 1998, 95: 575-581.

[52] ZENG S J. Dynamic experimental research and stochastic damage constitutive model for concrete[D]. PhD Thesis, Tongji University, Shanghai, 2012.

[53] KOLLURU S, O'NEIL E, POPOVICS J, et al. Crack propagation in flexural fatigue of concrete[J]. Journal of Engineering Mechanics, 2000, 126: 891-898.

A probabilistic analyzed method for concrete fatigue life①

Junsong Liang[1], **Zhaodong Ding**[2], **Jie Li**[1,3]

1 *Department of Structural Engineering, College of Civil Engineering, Tongji University, 1239 Siping Road, Shanghai 200092, China*

2 *School of Civil Engineering, Hefei University of Technology, 193 Tunxi Road, Hefei, 230009, China*

3 *State Key Laboratory of Disaster Reduction in Civil Engineering, Tongji University, 1239 Siping Road, Shanghai, 200092, China*

Abstract In this manuscript, a novel analyzed method is proposed for stochastic analysis of concrete fatigue life. Starting from the material randomness and the typical fatigue damage accumulation behavior, a newly developed stochastic fatigue damage model (SFDM) is introduced to calculate the fatigue life. Then, a sensitivity analysis towards SFDM is carried out, based on which, the random model parameters representing the concrete fatigue mechanisms are verified. Based on the collected test data of different material strength levels, the probabilistic distributions of the random model parameters are identified, and these are used for the subsequent probabilistic analysis of concrete fatigue life. To implement the probabilistic analysis, a probability density evolution equation is developed by employing the probability density evolution method (PDEM). Through solving this equation, the probability density functions (PDF) of random concrete fatigue life and the corresponding mean and variance as well as their evolution with different loading levels are obtained.

Keywords Concrete; Fatigue life; Damage; SFDM; Probabilistic analysis; PDEM

1 Introduction

Concrete material is one of the widest used materials in civil engineering area, and many concrete structures are subjected to cyclic live loads during their service life, such as wind power tower bases, bridge decks, pavements of high way and airport etc. In many circumstances, fatigue or time-delayed damage will occur in these kinds of structures. Thus, it is important to study the fatigue degradation for their residual life evaluation. It has long been recognized that the degradation of concrete material under external loads usually shows two essential characteristics[1-2]: ① the response (such as displacement, strain, fatigue life etc.) is subjected to enormous statistical scatter; and ② the material experiences a typical non-linear damage accumulation process before its final fracture. Consequently, it is important to apply the stochastic approach on concrete fatigue analysis,

① Originally published in *Probabilistic Engineering Mechanics*, 2017, 49: 13-21.

in which the inherent fatigue damage accumulation behavior should be captured as well.
A review of the previous studies towards concrete fatigue problem suggests that, a lot of methods could be applied to model the random fatigue life or the fatigue damage accumulation behavior. However, few of the methods could consider both of the basic characteristics properly. To analyze the random fatigue life, extensive studies[3-8] used statistical and data fitting methods based on concrete fatigue experiments. In some of the representative studies, Holmen[3] suggested the lognormal distribution for the concrete fatigue life, whereas Oh[5] recommended the Weibull distribution. Based on these statistical results, the so-called probabilistic S-N(S-N-P) curve was proposed. Although this kind of method gives a direct means for probabilistic study on concrete fatigue life, the following imperfectness might be encountered in the analysis: ① the damage accumulation process which reflects the material mechanical property under external loads is totally unknown; ② the S-N-P models are pure empirical and lack universality. That is to say, different S-N-P curves are fitted in different tests, despite the usage of concrete in the same strength level; ③ the mechanisms of the stochastic phenomena remain unclear and 4 a unified and direct calculation method for the probability density function (PDF) of concrete fatigue life, and the corresponding mean and variance as well as their evolution with different loading levels is still lacking. To model the fatigue damage accumulation process, the continuum damage mechanics (CDM) theory has often been used in recent years[9-13]. By introducing the framework of thermodynamics and the energy based representations, the general material constitutive relationship under uniaxial loads could be established as follows

$$\sigma=(1-d)E_0\varepsilon, \tag{1}$$

where σ represents the Cauchy stress; ε represents the macroscopic strain; E_0 denotes the initial Young's Modulus of concrete; and d is damage variable. The material is in a perfect undamaged state when d equals 0 and a totally invalid state when d equals 1. Then, the fatigue damage behavior could be well described by introducing a specific damage evolution law. However, most CDM-Models could only be used for deterministic analysis, and the fatigue damage evolution laws are always phenomenological. Hence, neither the probabilistic information nor the essence of the fatigue randomness could be properly described by the CDM-based analysis.
In order to eliminate the above insufficient points, a novel analyzed method for concrete fatigue life estimation is still needed, in which the fatigue life analysis and its damage accumulation process should be combined to gain a comprehensive understanding of the concrete fatigue mechanisms. Fortunately, the developments of the stochastic damage mechanics (SDM)[14-19] and the probability density evolution method (PDEM)[20-25] during the last several decades offer a feasible way to handle this problem. A brief introduction to the SDM and PDEM is given as follows. Similar to the CDM, the constitutive equation based on SDM is deduced from the thermodynamics framework as well[18]. However, the

key progress is that the concrete material is idealized as a parallel fiber bundle, through which the inherent mechanisms of concrete variability could be well described. According to the SDM, the stochastic damage accumulation under external loads could be calculated by representing the fracture strains of each fiber as a 1-D random field. The SDM is originally used for the monotonous loading cases. Recently, through a detailed study towards the physical mechanisms of concrete fatigue, Ding and Li[19] extended the SDM to fatigue cases, and created the so-called stochastic fatigue damage model (SFDM). On the other hand, by integrating the principle of preservation of probability with the uncoupled physical equation, a new probabilistic analyzed method, namely, the probability density evolution method (PDEM) has been developed[23]. Through this method, a completely uncoupled 1-D governing partial differential equation is derived, which is capable of capturing the instantaneous probability density function (PDF) and its evolution of any response of the non-linear stochastic system.

With inspirations from the previous work, this paper aims at developing a novel and universal probabilistic analyzed method for the evaluation of concrete fatigue life. Based on the existed studies of SDM and PDEM, the main research work of this paper is summarized as follows: starting from the idea that concrete fatigue life is dependent on the corresponding damage accumulation process, the SFDM is introduced to calculate the fatigue life. Then, in order to study the inherent material randomness, which causes the stochastic fatigue phenomena, a sensitivity analysis of the model parameters is carried out, based on which the random parameters of SFDM are verified. Subsequently, the probabilistic distributions of the random model parameters are systematically identified based on the collected test data[3, 26-27] of different concrete strength levels. Finally, the PDEM is applied to fulfill the probabilistic assessment of concrete fatigue life, through which a universal probability density evolution equation for concrete fatigue life is developed. By solving this equation, the PDF of concrete fatigue life as well as the mean and variance and their evolutions due to different fatigue loading levels could be obtained directly. In addition, the analysis of concrete fatigue life and its damage accumulation process are well unified based on the present analyzed method, which could give out a preliminarily understanding of the mechanisms of the random fatigue life.

2 Stochastic fatigue damage model (SFDM)

2.1 Framework of SDM

According to the previous studies towards the SDM[16, 18] the representative volume element (RVE) of concrete can be idealized as a fiber bundle system (Fig. 1). In this system, the fibers are represented by a series of individual micro-springs jointed in parallel. The springs are linked with a rigid bar on the ends so that each of them bears uniform deformation during the loading process. Through this idealization, the individual spring represents the micro-properties of concrete, and the bundle is used to describe the response of the RVE[28].

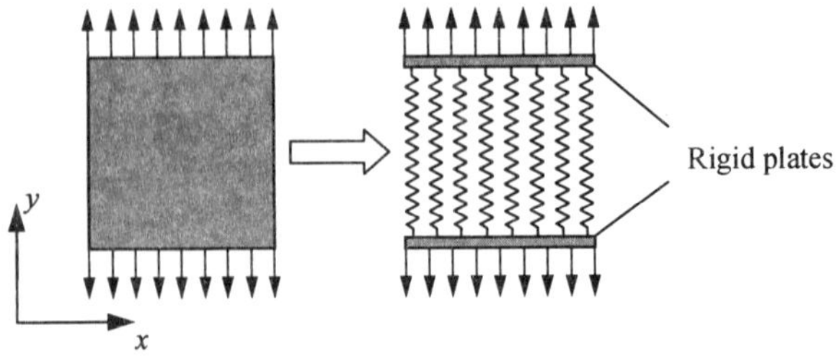

Fig. 1. Fiber bundle system

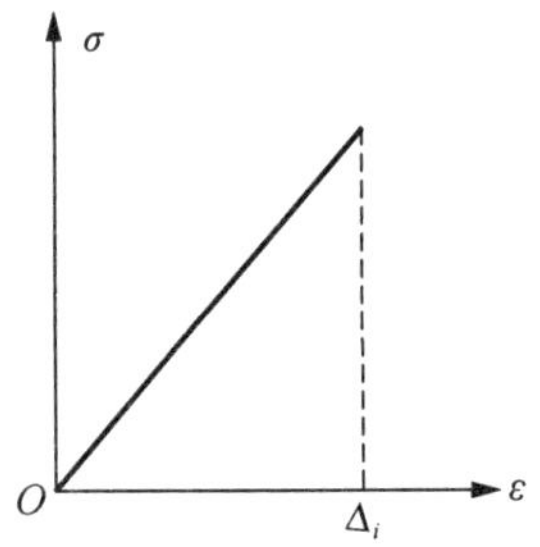

Fig. 2. Stress-strain relationship of micro-spring

Generally, the stress-strain relationship of each micro-spring is considered as the perfect elasto-brittle type with random fracture strain Δ_i (Fig. 2).

By introducing the random distribution of the fracture strain, the damage variable for the fiber bundle system can be defined as follows[18, 28-29]

$$d=\int_0^1 H[\varepsilon-\Delta(x)]\mathrm{d}x, \tag{2}$$

where $\Delta(x)$ is the 1-D micro fracture strain random field; x denotes the spatial coordinate of the micro-spring; and $H(\cdot)$ is the Heaviside function which is defined as

$$H(x)=\begin{cases}0 & x\leqslant 0\\ 1 & x>0\end{cases}. \tag{3}$$

Usually, $\Delta(x)$ is assumed to obey lognormal distribution and has been widely studied[29-30]. According to Eq. (2), concrete damage will develop when the material strain induced by external loads is larger than the fracture strain of the spring element. In this case, the damage mechanisms could be depicted as that the material area represented by the micro-springs causes dynamic instability and is cut through by micro cracks in a certain instant. In this case, the material energy dissipates transiently and the fracture behavior obeys the brittle Griffith criterion[31]. However, in the fatigue case, the loading levels are relatively low. Hence, the external loads are not sufficient to stimulate an instant fracture of the micro-spring. In this circumstance, the micro cracks are in a metastable state and propagate gradually in the fracture process zone (FPZ) of an upper-scale crack[19, 32-33]. Hence, the energy of the micro-spring input from external loads cumulates little by little, until it is larger than the elastic energy of the spring and a fracture occurs. Taking this physical mechanism into consideration, the SDM-model could be extended to the fatigue case by introducing the rate process theory[33-35]. Thus, Ding and Li[19, 30] developed the following fatigue damage accumulation criterion

$$d=\int_0^1 H(E_\mathrm{f}-E_\mathrm{s})\mathrm{d}x \tag{4}$$

in which

$$E_s = \frac{1}{2} \cdot E_0 \cdot [\Delta(x)]^2 \tag{5}$$

with E_f denoting the cumulated energy induced by the fatigue loads, which will be subsequently described.

2.2 Fatigue energy accumulation

According to Le et al.[33], the cumulated fatigue energy, E_j, is related to the thermal motion of micro particles in nano scale. Based on this mechanism, the homogenized cracking velocity of a single micro concrete crack can be expressed as[30, 36]:

$$\dot{\bar{a}}_m = 2\delta e^{-Q_0/kT} \frac{\Delta\bar{Q} - \Gamma e^{-\beta|\Delta\varepsilon|} \cdot V}{h} \tag{6}$$

where $\bar{a}_m$ is the average half micro crack length; the superposed dot denotes the total time derivative of a variable; δ is the micro particle gap; Q_0 is the energy barrier for the fracture of the particle bonds in the micro crack tip; k, T, h are Boltzmann constant, absolute temperature and Planck constant, respectively; Γ is the homogenized surface energy density of the micro crack; V is the activated volume in the micro crack tip area; $\Delta\bar{Q}$ is the average released energy caused by the propagation of the micro crack tip to a unit length; and item $e^{-\beta \cdot |\Delta\varepsilon|}$ represents the oscillation effect of the homogenized surface energy due to repeated loading with β denoting the material constant. For more details of the parameters and the physical mechanisms described in this section, one can refer to the early literatures[30, 32-33, 36]. In order to consider the micro cracking mechanism under the damage theory framework, Ding and Li[19, 32, 36] proposed the following equation to unify the material damage and the fracture energy dissipation process

$$\Delta\bar{Q} \approx Y\delta_d V \tag{7}$$

in which δ_d is the damage accumulation caused by unit advance of the micro crack; and Y represents the macro damage energy release rate(DERR)[18, 37].

The fatigue energy accumulation, which results from the micro cracking behavior as is described above, could be written as follows

$$\dot{E}_f = \sum_{i=1}^{N_a} \bar{G} \cdot \dot{\bar{a}}_m = N_a \cdot \bar{G} \cdot \dot{\bar{a}}_m, \tag{8}$$

where $\bar{G}$ denotes the average energy release rate; and N_a is the activated crack density of a upper scale crack tip[32, 36]. By referring to the representations of $\bar{G}$ and N_a from Le et al.[33], and by considering the micro cracking behavior and its energy dissipation process as Eqs. (6) and (7) describe, Ding[30, 36] further developed the following equation to calculate the fatigue energy accumulation.

$$\dot{E}_f = C_0 e^{-Q_0/kT} e^{-\kappa d} Y^2 \left(\frac{Y}{\Gamma \cdot e^{-\beta \cdot |\Delta\varepsilon|}} \right)^p, \tag{9}$$

where C_0 represents the material constant; κ is the material parameter, which is used to consider the shielding effect of the multi cracking behavior[19]; and p is the spatial scale transfer parameter.

In the present paper, the SFDM under uniaxial tensile or compressive loading cases is summarized as follows[30], which is used for the subsequent fatigue life analysis.

$$d^{\pm}=\int_0^1 H[E_f^{\pm}-E_s^{\pm}]\mathrm{d}x \tag{10a}$$

$$E_s^{\pm}=\frac{1}{2}\cdot E_0\cdot[\Delta^{\pm}(x)]^2 \tag{10b}$$

$$E_f^{\pm}=\int_0^t C_0 e^{-Q_0/kT}e^{-\kappa d}(Y^{\pm})^2\left(\frac{Y^{\pm}}{\Gamma\cdot e^{-\beta\cdot|\Delta\varepsilon|}}\right)^p \mathrm{d}t \tag{10c}$$

$$Y^{+}=\frac{1}{2E_0}\left[\frac{\sigma}{(1-d^{+})}\right]^2;\ Y^{-}=\frac{(\alpha-1)}{2E_0}\left[\frac{\sigma}{(1-d^{-})}\right]^2 \tag{10d}$$

in which symbol "±" means that the material is under tensile("+") or compressive("−") loading cases, respectively; and α is the material constant reflecting the shear effect of concrete under compression, which usually takes the value of 0.121 2[37].

3 Probabilistic distribution identification of random model parameters

According to the experimental studies[2, 38], a three-stage nonlinear damage accumulation process is always observed from the specimens(Fig. 3). However, the damage behavior of different specimens is subjected to large scatter.

Since concrete material contains large numbers of initial micro cracks, it may be suggested that the stochastic fatigue damage behavior comes from the random propagation of these micro cracks. Consequently, the origination of fatigue randomness should be described by the parameters which could reflect the corresponding mechanism in the damage accumulation process. In this chapter, these random parameters are identified and their probabilistic distributions are recognized.

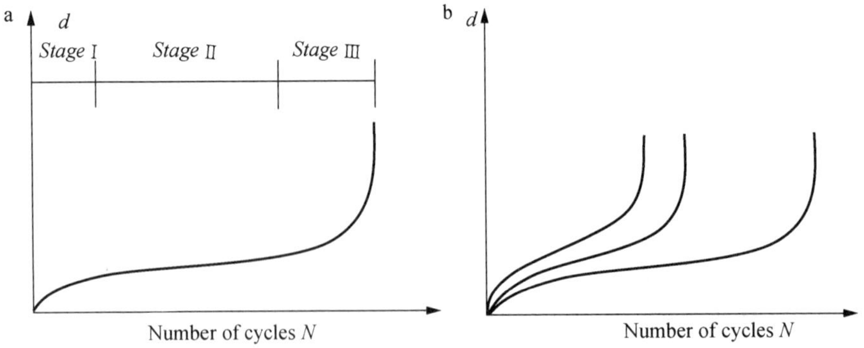

Fig. 3. Fatigue damage accumulation process of concrete: (a) Three-stage non-linear behavior; (b) Scatter of damage evolution

3.1 Sensitivity analysis of model parameters

In general, by setting an objective function L, the sensitivity of the model parameters can be acquired through perturbation analysis. By considering the variation of the objective function, which is caused by the small changes of the model parameters, the local sensitivity is expressed as follows[30, 39]

$$\frac{\delta L}{L}=\sum_{i}\left|S_{a_i}^{L}\right|\cdot\frac{\delta a_i}{a_i},\tag{11}$$

where $|S_{a_i}^{L}|$ represents the sensitivity coefficient of variable a_i; δa_i denotes the small change of the ith parameter a_i. For convenience, one could take the logarithm of Eq.(9) and gets:

$$\ln(\dot{E}_{\mathrm{f}})=\ln C_0-Q_0/kT-\kappa d+2\ln Y+p\ln(Y)-p\ln(\Gamma)+p\beta\mid\Delta\varepsilon\mid.\tag{12}$$

In Eq.(12), d and Y are state variables rather than model parameters; Q_0/kT is micro constant; $\Delta\varepsilon$ is the strain range; and p is the spatial transfer factor which shows no physical meaning and should be deterministic[30]. Therefore, parameters C_0, κ, Γ and β should be mainly investigated.

By taking partial derivative of each model parameter in Eq.(12), sensitivity coefficients could be obtained

$$\begin{cases}S_{C_0}^{\dot{E}_{\mathrm{f}}}=1\\ S_{\kappa}^{\dot{E}_{\mathrm{f}}}=\kappa d\\ S_{\Gamma}^{\dot{E}_{\mathrm{f}}}=p\\ S_{\beta}^{\dot{E}_{\mathrm{f}}}=p\beta\mid\Delta\varepsilon\mid\end{cases}\tag{13}$$

In the calculations, the initial values of the model parameters can usually be set as $p=8.5$, $\beta=0.006$, $\kappa=25$, $\Gamma=0.155\ \mathrm{J/m^3}$[30]. Hence, the sensitivity of the model parameters in the fatigue damage accumulation process could be calculated, which is shown in Fig. 4.

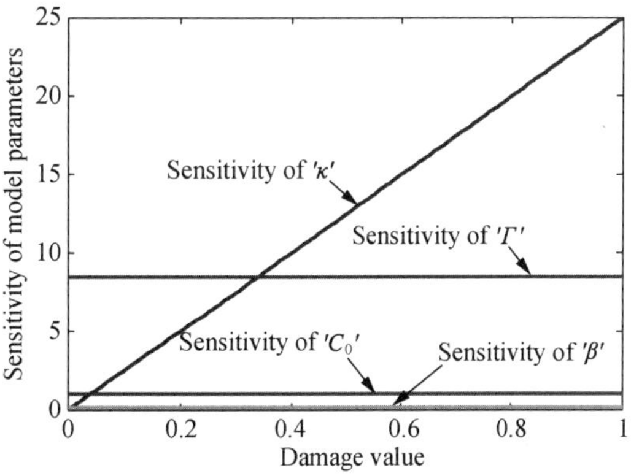

Fig. 4. Sensitivity analysis of model parameters

It can be realized that, κ and Γ are the most sensitive parameters in the fatigue damage accumulation process. Thus, this paper considers these two parameters as random variables. As is mentioned above, concrete material contains large numbers of micro cracks. These micro cracks are initially formed and distributed in a random way. Therefore, they will randomly propagate when external loads apply, which causes the corresponding stochastic material responses. κ and Γ are parameters representing the inherent micro cracking behavior[19, 36]. Hence, by randomizing these parameters, the random cracking behavior, which leads to the random fatigue damage evolution, can be properly reflected. On the other hand, concrete fatigue life is dependent on its damage accumulation process; as a result, one can notice that it is the randomness in the damage evolution process that leads to the random concrete fatigue life.

3.2 Distribution identification

In this section, test data from the previous studies[3, 26-27] are adopted to identify the distributions of the random model parameters. These data are shown in Fig. 5, in which $\sigma_{\max}$, f_c and "$\sigma_{\max}/f_c$" represent the maximum fatigue loading stress, average compressive strength of concrete and fatigue loading ratio, respectively. It is noticed that, the statistic information of each experiment is quite different. The reason is quite likely that the concrete used in these tests are in different strength levels. As a matter of fact, the material properties of concrete (e.g. strength, initial Young's modulus, etc.) vary with different concrete strength levels. Therefore, the distributions of κ and Γ should also be identified on the basis of different concrete strength levels. Hence, in the present paper the collected data are classified into three levels based on the design code of concrete structures[40].

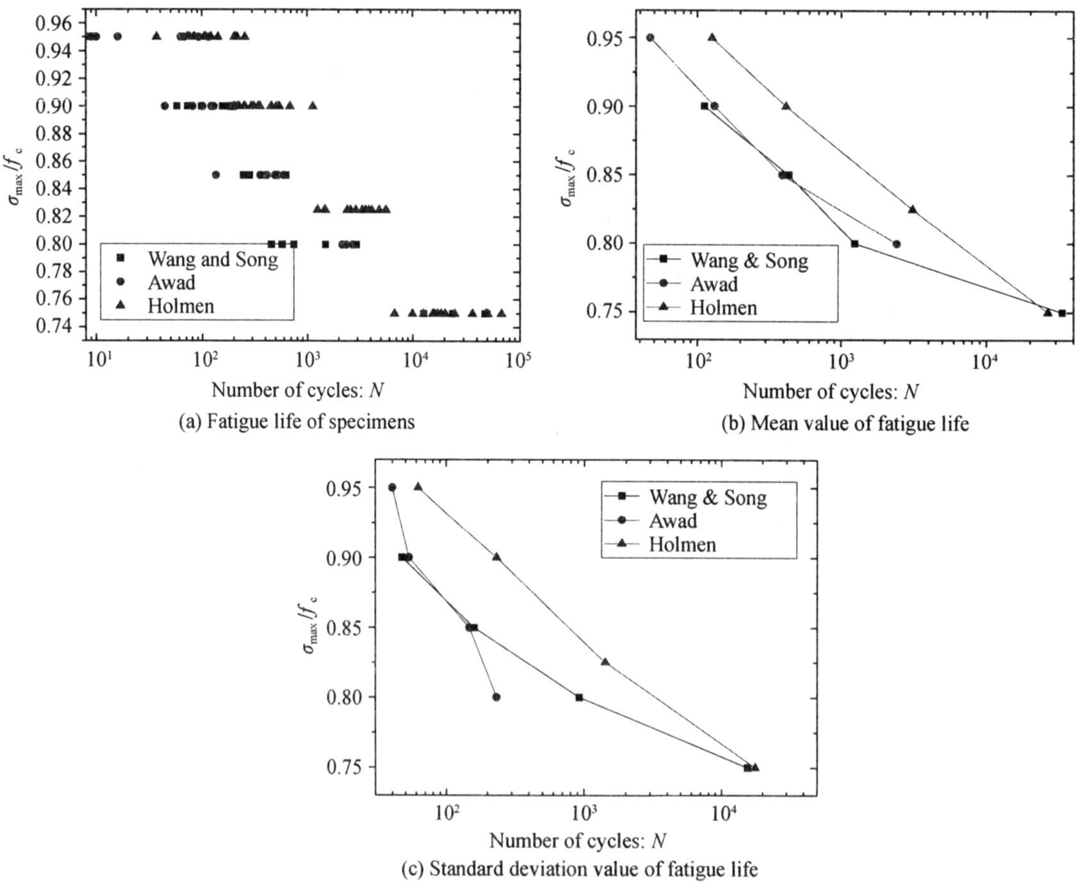

(a) Fatigue life of specimens

(b) Mean value of fatigue life

(c) Standard deviation value of fatigue life

Fig. 5. Compressive fatigue test data

According to SFDM, concrete fatigue life can be assigned as the cycle in which damage variable equals one. Therefore, the identification criterion could be defined as follows

$$\Delta N = | N_{\text{model}} - N_{\text{test}} | \rightarrow 0 \tag{14}$$

in which N_{model} and N_{test} represent the fatigue life from SFDM calculation and test, respectively. In the identification calculation, Eq.(14) could be satisfied by adjusting the values of κ and Γ, until N_{model} is infinitely close to N_{test}. Hence, it is a typical optimization problem. In the present paper, this problem is solved by adopting the non-gradient simplex method. Based on this method, an object function, such as Eq.(14), is established as a first step. Then, through setting a set of initial solutions, the optimum solutions of parameters κ and Γ could be obtained after a series of optimization computations, which include reflection, shrink, expansion and contraction[41-42].

Based on the above knowledge, the identification results are shown in Figs. 6-8, and the distributions of the parameters of different concrete strength levels are listed in Tables 1 and 2. It can be noticed that, the mean value of the parameters increases with concrete strength, whereas the coefficient of variation shows the opposite trend. This phenomenon is quite similar to other inherent mechanical properties of concrete, such as strength, initial Young's modulus etc. For the sake of simplification, it is assumed that κ and Γ are independently distributed in the identification.

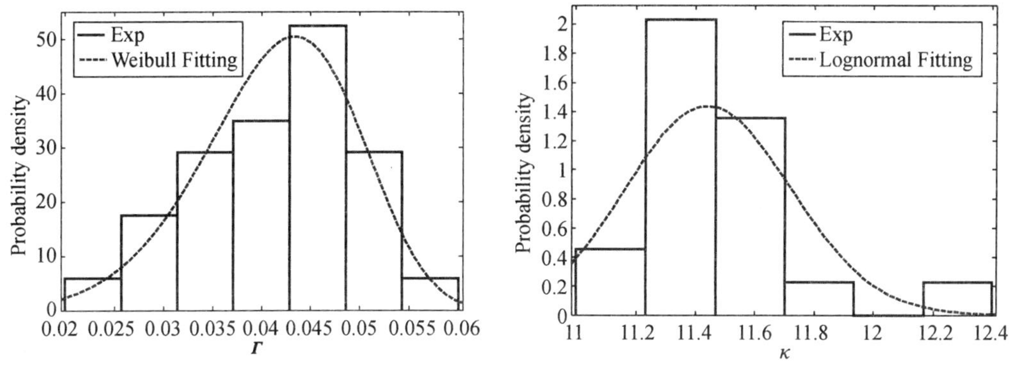

Fig. 6. Identification result based on Wang and Song[27]

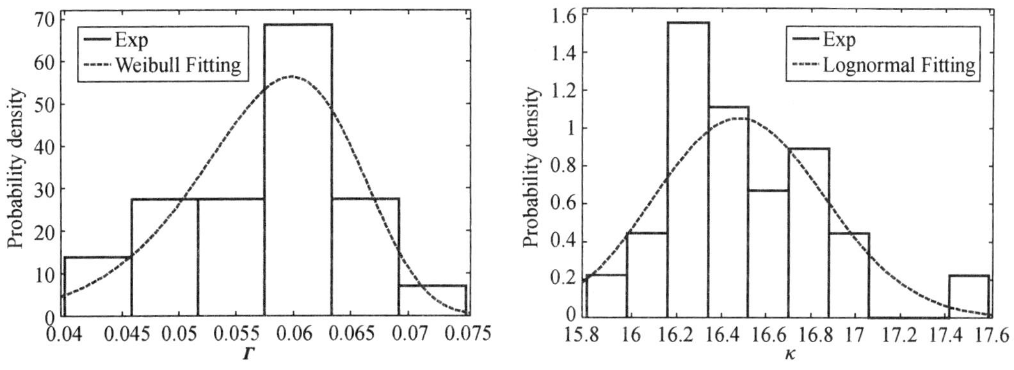

Fig. 7. Identification result based on Awad[26]

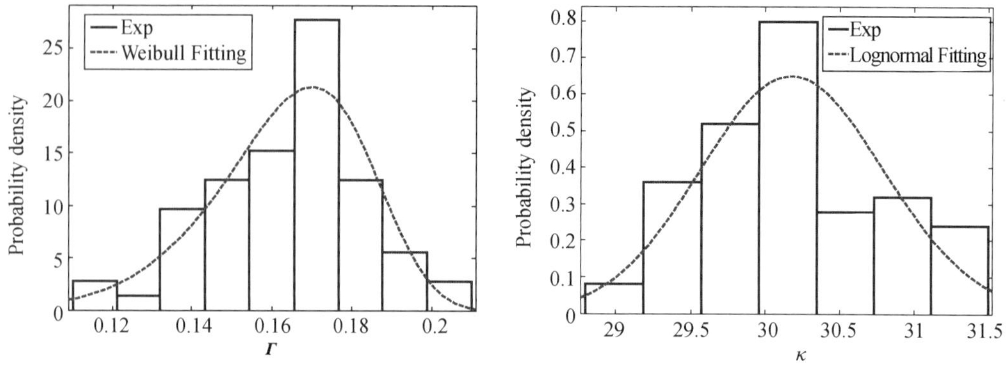

Fig. 8. Identification result based on Holmen[3]

Table 1 Identification results of parameter Γ

Concrete strength levels	Distribution type	Mean value/($J \cdot m^{-3}$)	Variance	Coefficient of variation/%
C20	Weibull	0.044 274 8	5.901 1e-05	17.35
C25	Weibull	0.057 443 4	5.565 0e-05	12.98
C30	Weibull	0.163 441	0.000 393 64	12.14

C20 is used in Wang and Song's test[34] in which $f_c = 24.64$ MPa;
C25 is used in Awad's test[33] in which $f_c = 29.63$ MPa;
And C30 is used in Holmen's test[8] in which $f_c = 40$ MPa.

Table 2 Identification results of parameter κ

Concrete grade	Distribution type	Mean value	Variance	Coefficient of variation/%
C20	Lognormal	11.449 5	0.077 136	2.43
C25	Lognormal	16.489 0	0.143 708	2.29
C30	Lognormal	30.190 1	0.377 852	2.04

4 Probabilistic analysis method for concrete fatigue life

4.1 The generalized probability density evolution method(PDEM)

In general, a random system could be expressed as the following function

$$\mathbf{X}(t) = \mathbf{G}(\mathbf{\Theta}, t) \text{ or } X_l(t) = G_l(\mathbf{\Theta}, t),\ l = 1, 2, \cdots, m \tag{15}$$

where $X(t) = [X_1(t), X_2(t), \cdots, X_l(t)]$ represents the state vector; $\mathbf{\Theta} = (\Theta_1, \Theta_2, \cdots, \Theta_s)$ denotes the random vector, which characterizes the system's randomness; t is the generalized time; and m is the dimension of the system. Further, the generalized velocity of the system could be expressed as

$$\dot{\mathbf{X}}(t) = \dot{\mathbf{H}}(\mathbf{\Theta}, t) \text{ or } \dot{X}_l(t) = \dot{H}_l(\mathbf{\Theta}, t),\ l = 1, 2, \cdots, m \tag{16}$$

in which $\dot{H}(\mathbf{\Theta}, t) = \partial G / \partial t$.

Suppose that $\mathbf{Z}(t) = [Z_1(t), Z_2(t), \cdots, Z_m(t)]$ denotes the quantities of interests in the system(e.g., the displacement, strain, fatigue life etc.). These quantities can generally be determined by the state vector $\mathbf{X}$, i.e.

$$\dot{\mathbf{Z}}(t) = \mathrm{T} \mid \dot{\mathbf{X}}(t) \mid \text{ or } \dot{\mathbf{Z}}_l(t) = \mathbf{T} \mid \dot{\mathbf{X}}_l(t) \mid, \tag{17}$$

where $\mathrm{T}(\cdot)$ is the transfer operator; and $\dot{\mathbf{Z}}(t)$ is the so-called generalized velocity. Combining Eqs.(16) and (17) yields

$$\dot{\mathbf{Z}}(t) = \mathrm{T} \mid \dot{\mathbf{X}}(t) \mid = \mathrm{T}[\dot{\mathbf{H}}(\mathbf{\Theta}, t)] = \mathbf{h}(\mathbf{\Theta}, \mathbf{t}). \tag{18}$$

For convenience, consider a stochastic system $[\mathbf{Z}(t), \mathbf{\Theta}]$ and its joint probability density function(PDF) is defined as $p_{\mathbf{Z\Theta}}(\mathbf{z}, \boldsymbol{\theta}, t)$. The random event upon the system could be define by $\{[\mathbf{Z}(t), \mathbf{\Theta}] \in \Omega_t \times \Omega_\theta\}$, where Ω_t denotes the distribution domain of $\mathbf{Z}$ at time t, and Ω_θ represents an arbitrary subdomain of the distribution domain of $\mathbf{\Theta}$. Taking a small time increment $\mathrm{d}t$ into account, the random event will evolve to $\{[\mathbf{Z}(t+\mathrm{d}t), \mathbf{\Theta}] \in \Omega_{t+\mathrm{d}t} \times \Omega_\theta\}$. According to the principle of preservation of probability[22], the following equations could be obtained

$$\Pr\{[\mathbf{Z}(t), \mathbf{\Theta}] \in \Omega_t \times \Omega_\theta\} = \Pr\{[\mathbf{Z}(t+\mathrm{d}t), \mathbf{\Theta}] \in \Omega_{t+\mathrm{d}t} \times \Omega_\theta\} \tag{19}$$

$$\int_{\Omega_t \times \Omega_\theta} p_{\mathbf{Z\Theta}}(\mathbf{z}, \boldsymbol{\theta}, t)\mathrm{d}\mathbf{z}\mathrm{d}\boldsymbol{\theta} = \int_{\Omega_{t+\mathrm{d}t} \times \Omega_\theta} p_{\mathbf{Z\Theta}}(\mathbf{z}, \boldsymbol{\theta}, t+\mathrm{d}t)\mathrm{d}\mathbf{z}\mathrm{d}\boldsymbol{\theta}. \tag{20}$$

Based on these equations and the divergence theorem, Li and Chen[22, 23] further deduced the following so-called generalized probability density evolution equation

$$\frac{\partial p_{\mathbf{Z\Theta}}(\mathbf{z}, \boldsymbol{\theta}, t)}{\partial t} + \sum_{l=1}^{m} h_l(\theta, t) \frac{\partial p_{\mathbf{Z\Theta}}(\mathbf{z}, \boldsymbol{\theta}, t)}{\partial z_l} = 0 \tag{21}$$

or in an alternative form

$$\frac{\partial p_{\mathbf{Z\Theta}}(\mathbf{z}, \boldsymbol{\theta}, t)}{\partial t} + \sum_{l=1}^{m} \dot{\mathbf{Z}}_l(\boldsymbol{\theta}, t) \frac{\partial p_{\mathbf{Z\Theta}}(\mathbf{z}, \boldsymbol{\theta}, t)}{\partial z_l} = 0, \tag{22}$$

where $h_l(\boldsymbol{\theta}, t)$ denotes the components of $\mathbf{h}(\mathbf{\Theta}, \mathbf{t})$.

By introducing the initial and boundary conditions of Eqs.(21) and (22), which are expressed as follows

$$p_{\mathbf{Z\Theta}}(\mathbf{z}, \boldsymbol{\theta}, t)\Big|_{t=t_0} = \delta(\mathbf{z} - \mathbf{z}_0)p_{\mathbf{\Theta}}(\boldsymbol{\theta}) \tag{23}$$

$$p_{\mathbf{Z\Theta}}(\mathbf{z}, \boldsymbol{\theta}, t)\Big|_{z_l \to \pm\infty} = 0, \ l = 1, 2, \cdots, m. \tag{24}$$

The joint PDF, $p_{\mathbf{Z\Theta}}(\mathbf{z}, \boldsymbol{\theta}, t)$, can be calculated, and the PDF of $Z(t)$ could be further obtained by the following integration

$$p_{\mathbf{Z}}(\mathbf{z}, t) = \int_{\Omega_\theta} p_{\mathbf{Z\Theta}}(\mathbf{z}, \boldsymbol{\theta}, t)\mathrm{d}\boldsymbol{\theta}, \tag{25}$$

where $\mathbf{z}_0$ denotes the initial value of $\boldsymbol{Z}$; and $\delta(\cdot)$ represents the Dirac function.

4.2 Probability density evolution equation for concrete fatigue life

As mentioned above, the generalized probability density evolution method provides a feasible way to obtain the PDF of the quantities and their evolution process of a random system. In this paper, fatigue life is mainly concerned. Hence, the corresponding probability density evolution equation can be written in 1-D domain as follows

$$\frac{\partial p_{Z\Theta}(z, \boldsymbol{\theta}, t)}{\partial t}+\frac{\partial Z(\boldsymbol{\theta}, t)}{\partial t}\frac{\partial p_{Z\Theta}(z, \boldsymbol{\theta}, t)}{\partial z}=0. \tag{26}$$

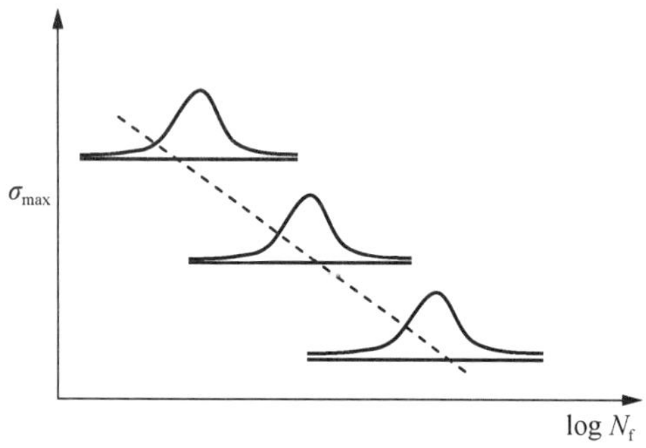

Fig. 9. Example of PDF evolution figure for concrete fatigue life

For S-N curve, the maximum fatigue stress, $\sigma_{\max}$, is usually considered as the independent variable, and the fatigue life, N_f, is taken as the dependent variable. Thus, in the probabilistic analysis, the generalized time axis is assigned to $\sigma_{\max}$. In other words, the fatigue life distribution will be given according to different loading levels (Fig. 9).

Based on the above knowledge, the random fatigue life, N_f, can be written as

$$N_f=Z(\boldsymbol{\Theta}, \sigma_{\max})=N_f(\boldsymbol{\Theta}, \sigma_{\max}) \tag{27}$$

in which the random vector $\boldsymbol{\Theta}$ takes the following form

$$\boldsymbol{\Theta}=\boldsymbol{\Theta}(\kappa, \Gamma). \tag{28}$$

Combining Eqs.(26)—(28), the probability density evolution function for fatigue life can be expressed as follows

$$\frac{\partial p_{N\Theta}(N_{\mathrm{f}}, \boldsymbol{\Theta}(\kappa, \Gamma), \sigma_{\max})}{\partial \sigma_{\max}}+\frac{\partial N_{\mathrm{f}}(\boldsymbol{\Theta}(\kappa, \Gamma), \sigma_{\max})}{\partial \sigma_{\max}}\cdot\frac{\partial p_{N\Theta}(N_{\mathrm{f}}, \boldsymbol{\Theta}(\kappa, \Gamma), \sigma_{\max})}{\partial N_{\mathrm{f}}}=0. \tag{29}$$

4.3 Solution procedures for probability density evolution equations

The procedures to solve the probability density evolution equations can generally be performed as the following steps[28, 43]:

Step 1: The distribution domain of $\boldsymbol{\Theta}$ is divided into a series of subdomains, through which a set of discrete representative points are selected and defined as

$$\boldsymbol{\theta}_q=(\theta_{q,1}, \theta_{q,2}, \cdots, \theta_{q,s});\ q=1, 2, \cdots, n_{\mathrm{sel}} \tag{30}$$

where n_{sel} is the total number of the selected points and s is the number of random variables. The assigned probability of each point in the point set is calculated through the following equation

$$p_q=\int_{V_q} p_{\Theta}(\boldsymbol{\theta})\mathrm{d}\boldsymbol{\theta}, \tag{31}$$

where V_q represents the representative volume of each point. In the present, the points are selected based on the point selection method proposed by Chen[44].

Step 2: For a given point set $\{\boldsymbol{\theta}_i, i=q\}$, the deterministic analysis of the fatigue life is conducted by solving Eqs.(10a)—(10d). Then the generalized velocity, $\frac{\partial Z(\boldsymbol{\theta}_q, t)}{\partial t}$, can be obtained by adopting other point sets $\{\boldsymbol{\theta}_j, j=q \text{ and } j \neq i\}$ in the calculations.

Step 3: With the representative point set and the assigned probability from Step 1, as well as the generalized velocity obtained from Step 2, the generalized probability density evolution equation, expressed by Eq.(26), can be written as the following discrete form

$$\frac{\partial p_{Z\Theta}(z, \boldsymbol{\theta}_q, t)}{\partial t}+\sum_{l=1}^{m} \frac{\partial Z(\boldsymbol{\theta}_q, t)}{\partial t} \frac{\partial p_{Z\Theta}(z, \boldsymbol{\theta}_q, t)}{\partial z}=0. \tag{32}$$

Eq.(32) can be solved by using the finite-difference method based on the initial conditions given as follows[23, 28, 43]

$$p_{Z\Theta}(z, \boldsymbol{\theta}_q, t)\Big|_{t=t_0}=\delta(t-t_0) p_q. \tag{33}$$

Step 4: Taking the numerical integration with respect to $\boldsymbol{\theta}_q$ yields the PDF as follows

$$P_Z(z, t)=\sum_{q=1}^{n_{\mathrm{sel}}} p_{Z\Theta}(z, \boldsymbol{\theta}_q, t). \tag{34}$$

5 Calculation of concrete damage accumulation and fatigue life distribution

In this chapter, damage accumulation under compressive fatigue loads and the corresponding fatigue life distribution are analyzed using the present method. In the calculation, concrete strength level is C30 and the test data which is used for comparison are given by Holmen[3].

The 1-D random field $\Delta(x)$ obeys lognormal distribution, of which the mean value, standard deviation and correlation length are represented by λ, ζ^2, ξ, respectively. For concrete of C30, these parameters can take the following values: $\lambda=4.753\,6$, $\zeta=0.656\,0$, $\xi=0.027\,0$[19, 29]. Based on the distribution types verified in Chapter 3, 200 sample points are selected for κ and Γ. Other model parameters are $e^{-Q_0/kT}=3.75\times10^{-18}$ in room temperature, and $C_0=1.52\times10^{-5}\ \mathrm{J^{-1}\ Nm/s}$[19].

The analyzed results are shown in Figs. 10-13. In Fig. 10, the calculated S-N curve is compared to the empirical S-N model proposed by Tepfers[45]. The agreement verifies the validity of the present method for fatigue life calculation. In Fig. 11 the stochastic damage accumulation of same selected samples are plotted. Obviously, the inherent randomness of concrete, which is represented by κ and Γ, has great effect on fatigue damage evolution and the corresponding fatigue life. Figs. 12 and 13 demonstrate the probabilistic analyzed results. In Fig. 12, the analyzed mean and standard deviation values(Std.V.) coincide with the experimental results very well; and all the test data are almost covered by the mean value plus/minus 1.5 times of Std. V. of the analyzed results. In Fig. 13, the calculated probability density evolution contours, and the PDF slices at several typical fatigue loading levels are shown. As mentioned before, the fatigue loading levels represent the generalized time axis of the random system; thus, it is obvious that PDF of concrete fatigue life evolves with the generalized time due to the inherent material randomness.

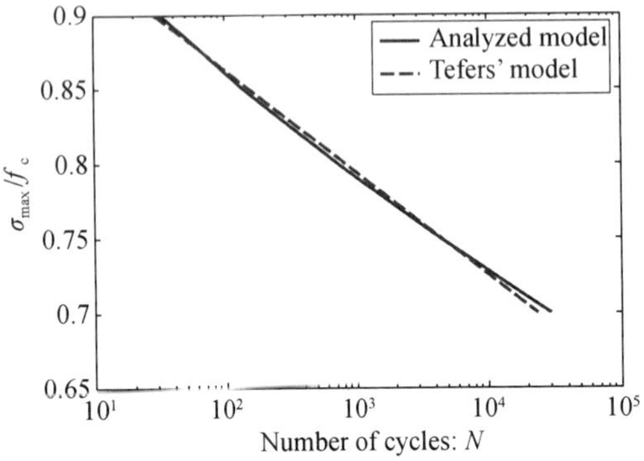

Fig. 10. Comparison of analyzed S-N curve with empirical model

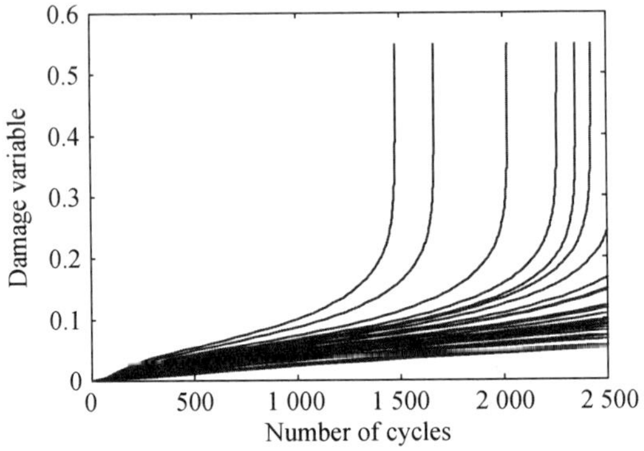

Fig. 11. Stochastic damage accumulation

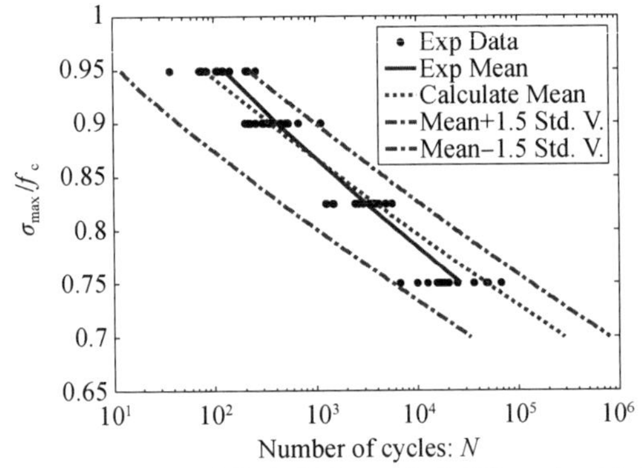

(a) Probabilistic S-N curve by PDEM and test data

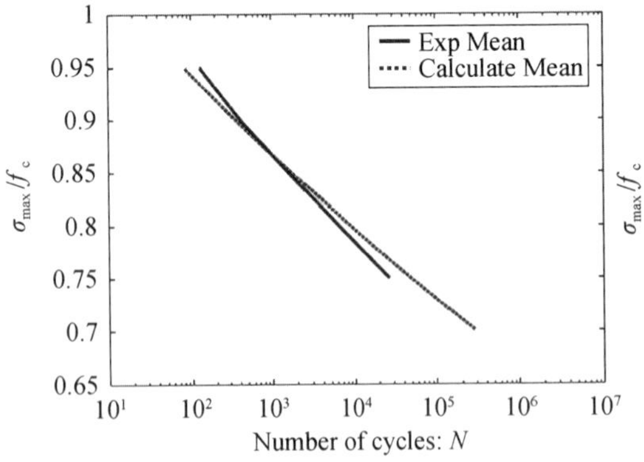

(b) Mean value

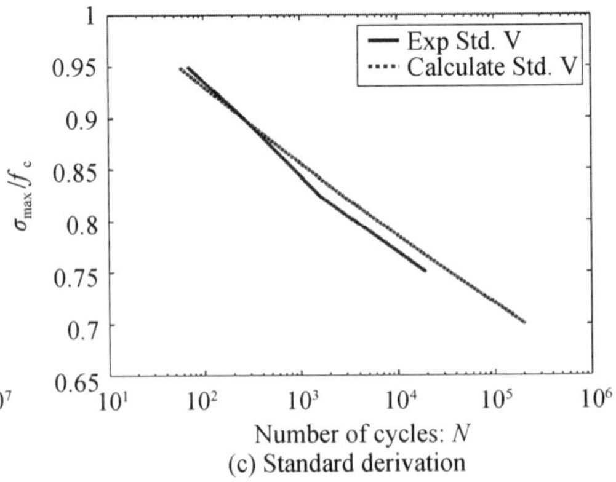

(c) Standard derivation

Fig. 12. Comparison of analyzed results and test data

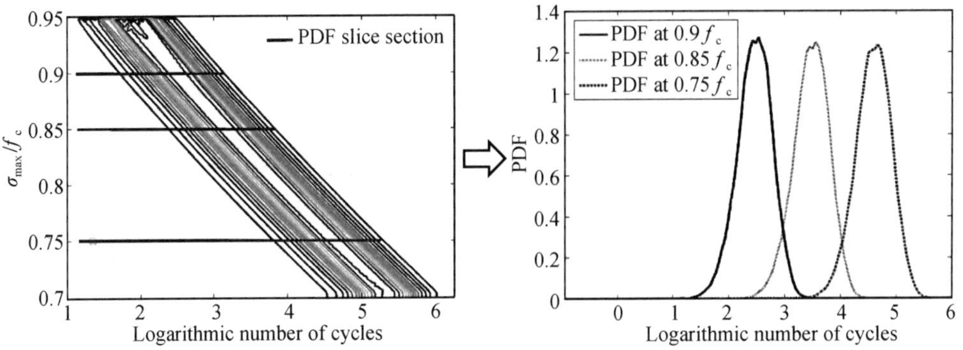

Fig. 13. PDF contours and the corresponding PDF slices

6 Conclusion

An innovative probabilistic analyzed method for concrete fatigue life evaluation is provided by this paper. According to this method, the random fatigue life is dependent on the stochastic damage accumulation process, which is caused by the inherent material randomness. Based on the present work, the following developments for the probabilistic analysis towards concrete fatigue life may be concluded:

(1) the analysis of random fatigue life and the corresponding damage accumulation process are unified by combining SFDM and PDEM;

(2) the mechanisms of the random fatigue phenomena are preliminarily clarified by applying the SFDM on fatigue life calculation and by performing the sensitivity analysis;

(3) the random model parameters and their probabilistic distributions are systematically identified, which provides the necessary guarantees for the analysis of concrete of different strength levels;

(4) the probability density evolution equation for concrete fatigue life is developed based on PDEM, which offers a general way to obtain the mean and standard variation values, PDF and their evolutions in the random system directly.

Acknowledgment

Financial supports from the National Natural Science Foundation of China(Grant Numbers 51261120374 and 51538010) are gratefully appreciated.

References

[1] LI J. Research on the stochastic damage mechanics for concrete material and structures[J]. Journal-Tongji University, 2004, 32: 1270-1277.

[2] BREITENBÜCHER R, IBUK H. Experimentally based investigations on the degradation process of concrete under cyclic load[J]. Materials and Structures, 2006, 39(7): 717-724.

[3] HOLMEN J. Fatigue of concrete by constant and variable amplitude loading[J]. Special Publication, 1982, 75: 71-110.

[4] PETRYNA Y S, PFANNER D, STANGENBERG F, et al. Reliability of reinforced concrete structures under fatigue[J]. Reliability Engineering & System Safety, 2002, 77(3): 253-261.
[5] OH B H. Fatigue life distributions of concrete for various stress levels[J]. Materials Journal, 1991, 88(2): 122-128.
[6] YAO J T P, KOZIN F, WEN Y K, et al. Stochastic fatigue, fracture and damage analysis[J]. Structural Safety, 1986, 3(3-4): 231-267.
[7] SAIN T, KISHEN J M C. Probabilistic assessment of fatigue crack growth in concrete[J]. International Journal of Fatigue, 2008, 30(12): 2156-2164.
[8] MENG X. Experimental and theoretical research on residual strength of concrete under fatigue loading[D]. Dalian: Dalian University of Technology, 2006.
[9] MARIGO J J. Modelling of brittle and fatigue damage for elastic material by growth of microvoids[J]. Engineering Fracture Mechanics, 1985, 21(4): 861-874.
[10] CHOW C L, WEI Y. A model of continuum damage mechanics for fatigue failure[J]. International Journal of Fracture, 1991, 50(4): 301-316.
[11] ALLICHE A. Damage model for fatigue loading of concrete[J]. International Journal of Fatigue, 2004, 26(9): 915-921.
[12] MAI S H, LE-CORRE F, FORÊT G, et al. A continuum damage modeling of quasi-static fatigue strength of plain concrete[J]. International Journal of Fatigue, 2012, 37: 79-85.
[13] LIANG J, REN X, LI J. A competitive mechanism driven damage-plasticity model for fatigue behavior of concrete[J]. International Journal of Damage Mechanics, 2015, 25(3): 377-399.
[14] KRAJCINOVIC D, FANELLA D. A micromechanical damage model for concrete[J]. Engineering Fracture Mechanics, 1986, 25(5-6): 585-596.
[15] KANDARPA S, KIRKNER D J, SPENCER JR B F. Stochastic damage model for brittle materials subjected to monotonic loading[J]. Journal of Engineering Mechanics, 1996, 122(8): 788-795.
[16] LI J, ZHANG Q. Study of stochastic damage constitutive relationship for concrete material[J]. Journal-Tongji University, 2001, 29(10): 1135-1141.
[17] LI J, REN X. Stochastic damage model for concrete based on energy equivalent strain[J]. International Journal of Solids and Structures, 2009, 46(11-12): 2407-2419.
[18] J. LI, J. WU, J. CHEN. Stochastic Damage Mechanics of Concrete Structures[M]. Beijing: Science Press, 2014.
[19] Z. DING, J. LI. Modeling of fatigue damage of concrete with stochastic character[C]//Proceedings of IALCCE (2014) Tokyo, Japan, 2014.
[20] LI J, CHEN J B. Probability density evolution method for dynamic response analysis of structures with uncertain parameters[J]. Computational Mechanics, 2004, 34(5): 400-409.
[21] LI J, CHEN J. The number theoretical method in response analysis of nonlinear stochastic structures[J]. Computational Mechanics, 2007, 39(6): 693-708.
[22] LI J, CHEN J. The principle of preservation of probability and the generalized density evolution equation[J]. Structural Safety, 2008, 30(1): 65-77.
[23] LI J, CHEN J. Stochastic dynamics of structures[M]. John Wiley & Sons, 2009.
[24] CHEN J B, LI J. A note on the principle of preservation of probability and probability density evolution equation[J]. Probabilistic Engineering Mechanics, 2009, 24(1): 51-59.
[25] LI J, CHEN J, SUN W, et al. Advances of the probability density evolution method for nonlinear stochastic systems[J]. Probabilistic Engineering Mechanics, 2012, 28: 132-142.

[26] AWAD M E L M. Strength and deformation characteristics of plain concrete subjected to high repeated and sustained loads[M]. University of Illinois at Urbana-Champaign, 1971.

[27] WANG H L, SONG Y P. Fatigue capacity of plain concrete under fatigue loading with constant confined stress[J]. Materials and Structures, 2011, 44(1): 253-262.

[28] ZHOU H, LI J, REN X. Multiscale stochastic structural analysis toward reliability assessment for large complex reinforced concrete structures[J]. International Journal for Multiscale Computational Engineering, 2016, 14(3).

[29] ZENG S J. Dynamic experimental research and stochastic damage constitutive model for concrete[D]. Ph. D. dissertation, Tongji Univ., Shanghai, China (in Chinese), 2012.

[30] DING Z D. The fatigue constitutive relation of concrete and stochastic fatigue response of concrete structure[D]. Doctoral dissertation. Tongji University, Shanghai, China, 2015.

[31] GURTIN M E. Thermodynamics and the Griffith criterion for brittle fracture[J]. International Journal of Solids and Structures, 1979, 15(7): 553-560.

[32] LI J, DING Z. A fatigue constitutive model for concrete[C]//Proceeding of the 4th CanCNSM, Montreal, Canada, 2013.

[33] LE J L, BAŽANT Z P, BAZANT M Z. Unified nano-mechanics based probabilistic theory of quasibrittle and brittle structures: I. Strength, static crack growth, lifetime and scaling[J]. Journal of the Mechanics and Physics of Solids, 2011, 59(7): 1291-1321.

[34] ZHURKOV S N. Kinetic concept of the strength of solids[J]. International Journal of Fracture Mechanics, 1965, 1(4): 311-323.

[35] KRAUSZ A S. Fracture kinetics of crack growth[M]. Springer Science & Business Media, 1988.

[36] DING Z, LI J. The fatigue constitutive model of concrete based on micro-mesomechanics (in Chinese) [J]. Chinese Journal of Theoretical and Applied Mechanics, 2012,46 (6): 911-919.

[37] WU J Y, LI J, FARIA R. An energy release rate-based plastic-damage model for concrete[J]. International Journal of Solids and Structures, 2006, 43(3-4): 583-612.

[38] GAO L, HSU C T T. Fatigue of concrete under uniaxial compression cyclic loading[J]. Materials Journal, 1998, 95(5): 575-581.

[39] AMAR G, DUFAILLY J. Identification and validation of viscoplastic and damage constitutive equations[J]. European journal of mechanics. A. Solids, 1993, 12(2): 197-218.

[40] Code for design of concrete structures: GB 50010—2010[S]. Beijing: China Architecture and Building Press, 2010.

[41] SPENDLEY W, HEXT G R, HIMSWORTH F R. Sequential application of simplex designs in optimisation and evolutionary operation[J]. Technometrics, 1962, 4(4): 441-461.

[42] NELDER J A, MEAD R. A simplex method for function minimization[J]. The Computer Journal, 1965, 7(4): 308-313.

[43] LI J, CHEN J, SUN W, et al. Advances of the probability density evolution method for nonlinear stochastic systems[J]. Probabilistic Engineering Mechanics, 2012, 28: 132-142.

[44] CHEN J, ZHANG S. Improving point selection in cubature by a new discrepancy[J]. SIAM Journal on Scientific Computing, 2013, 35(5): A2121-A2149.

[45] TEPFERS R, KUTTI T. Fatigue strength of plain, ordinary, and lightweight concrete[C]//Journal Proceedings. 1979, 76(5): 635-652.

Two-scale random field model for quasi-brittle materials①

Jie Li, Lu Hai, Taozhi Xu

School of Civil Engineering, Tongji University, 200092 Shanghai, China

Abstract A new idea of two-scale random field is presented by taking the mechanical behaviors of quasi-brittle materials as an example. In this way, the random fluctuation of material damage at the micro-meso-scale and the spatial correlation of mechanical properties at the macro-scale are both well described by the random distribution of fracture strain. Moreover, it is proved that by introducing a transform matrix, the mesoscopic random fields can be described by a two-scale random field. Based on this, a two-step numerical implementation of the two-scale random field is proposed, and the mechanical behaviors of quasi-brittle materials can be analyzed by incorporating the probability density evolution theory. Finally, an experimental example is presented to demonstrate the capability of the proposed model. Ideal agreements are achieved against the experimental results. Particularly, the randomness of material properties at both scales can be well reproduced.

Keywords Two-scale random field; Spatial correlation; Transform matrix; Mesoscopic stochastic fracture model; Concrete

1 Introduction

The material uncertainties have considerable influences on the behaviors of engineering structures and the randomness of material properties in engineering structures and its quantification have been the subject of intensive studies for many years[1-2]. In this way, from the pioneering works of Vanmarcke[3-4], random field has become a rational and efficient mathematical model for characterizing the randomness of material properties. However, in the most studies, only single-scale random field is considered. The idea of two-scale or even multi-scale random field has not been fully investigated. In fact, for most practical problems, the single-scale random field may not be suitable, especially for the problems in which multiple-scale random fluctuations are encountered. For example, apart from the material uncertainties at one spatial location of structures due to mesoscopic heterogeneity, the spatial correlations of mechanical properties of materials at different locations should also be incorporated within the framework of random field. This is rather important for predicting the dynamic responses and failure modes of engineering systems under extreme loads[5-6].

---

① Originally published in *Probabilistic Engineering Mechanics*, 2021, 66: 103154.

The fracture and damage of quasi-brittle materials may be a suitable ground for studying multi-scale random field, of which concrete is a typical research object for the target. In fact, due to the intrinsic heterogeneity of material meso-structures, concrete usually exhibits highly complicated mechanical responses when subjected to loadings. Therefore, the constitutive modeling of material behaviors and nonlinear analysis of concrete structures are still quite challenging problems in solid mechanics and engineering practice. Over the last few decades, the formulation of continuum damage mechanics (CDM) has been introduced to the nonlinear constitutive modeling of concrete. Due to its solid foundation of the irreversible thermodynamic and relevant physical mechanisms, CDM provides a powerful framework for the derivation of nonlinear constitutive model for concrete[7-10]. Using these models, the typical mechanical behaviors of concrete bearing the loading, i.e. the unilateral effect, the stress softening, the stiffness degradation and the residual deformation after unloading can be well reproduced. However, the inherent randomness of materials cannot be reflected by the deterministic analysis model. Actually, as a main theoretical foundation for damage mechanics, the principal of irreversible thermodynamic just specifies the necessary condition of damage evolution law, while the damage evolution functions for material models have to be determined empirically. Considering this background, stochastic damage mechanics was developed in the last twenty years[11]. In this way, a mesoscopic stochastic fracture model (MSFM) was proposed and developed systematically[12-15]. Being able to reflect the nonlinearity and randomness simultaneously, such kind of model has been applied for many engineering practices[6, 16].

In addition, though not focus of this paper, the work of stochastic multi-scale homogenization should be also mentioned, since it constitutes another important method of characterizing the material randomness. Within stochastic computational homogenization, the determination of representative volume element (RVE) size is mainly dealt with by the concept of stochastic volume element(SVE), in which spatially averaged mechanical response quantities over the domain of SVE converge to deterministic quantities[17-18]. However, due to damage evolution and stress softening of concrete-like quasi-brittle solids, the commonly strategies to determine RVE size would become limited; and thus the application of computational homogenization in quasi-brittle materials needs further investigations[19-21].

Within the implement of mesoscopic stochastic fracture model(MSFM), one RVE is idealized as parallel micro-springs model and the damage evolution is defined in terms of the stochastic fracture of micro-springs, by which the stochastic damage and fluctuate behavior of materials at the mesoscopic scale can be well represented. However, when the model is used for the stochastic analysis of structures, the damage evolution functions at different spatial locations have to be assumed identical, which results that the macroscopic spatial variation of material properties of engineering structures, e.g. the compressive strength of concrete of a beam, cannot be effectively reflected. This may lead to a risk in safety and reliability evaluation of complex concrete structures. Detailed investigation shows that there exists two-scale random field, i.e. the fracture strain random field at the mesoscopic level of materials as well as the strength random field at the macroscopic level of structures. Only a single-scale random field is inapplicable to reflect complex probabilistic mechanical behavior of concrete structures. To address this issue, we introduce a novel idea of two-scale random field in this paper and present a detail investigation cooperating with quasi-brittle materials.

The remainder of this paper is organized as follows: Section 2 revisits the framework of MSFM. Section 3 is devoted to the two-scale random field by cooperating with MSFM. Section 4 proves that there is a relationship between two-scale random field and mesoscopic fracture strain random field. Section 5 provides the numerical generation method of two-scale random field. Finally, taking concrete as an example, Section 6 presents a series of comparisons between analytical results and experimental results to demonstrate the ability of the proposed model. The most relevant conclusions are drawn in Section 7.

2 Stochastic damage model

Within the framework of MSFM[11-13], a structural RVE can be modeled as a series of micro-springs jointed in parallel, as shown in Fig. 1(a). The springs are linked with a rigid bar on the ends so that each of them bears uniform deformation during the loading process. By this idealization, the micro-properties of the materials can be represented by the individual micro-spring while the macro-response of the RVE can be described by the parallel system. The stress-strain behavior of each micro-spring is assumed to be perfect elastic-brittle type with random fracture strain Δ [Fig. 1(b)] For simplicity, the elastic stiffness of each spring is assumed to be a material constant. Therefore, given the probability distribution of the fracture strain of the micro-springs, the stochastic sequential fractures of the springs throughout the entire loading process could illustrate the macroscopic stochastic nonlinear behavior of the RVE. The macroscopic damage variable of the parallel micro-springs system can be defined as[11-13]

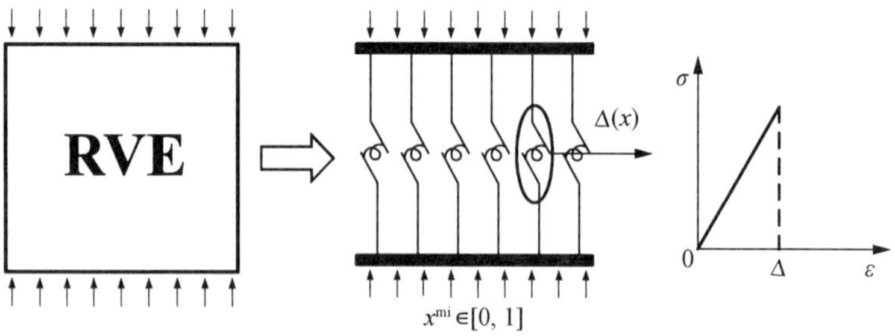

Fig. 1. Stochastic damage model: (a) parallel micro-springs system; (b) stress-strain relationship of micro-spring

$$D(\varepsilon^e)=\int_0^1 H[\varepsilon^e-\Delta(x^{mi})]\mathrm{d}x^{mi} \tag{1}$$

where ε^{e} is the elastic strain; $\Delta(x^{mi})$ is one-dimensional random field of the fracture strain; $x^{mi} \in [0, 1]$ denotes the mesoscopic spatial coordinate of the micro-spring; $H(\cdot)$ is the Heaviside function, which is expressed as

$$H(y)=\begin{cases}1 & y>0 \\ 0 & y \leqslant 0\end{cases} \tag{2}$$

Then, upon the framework of CDM, the energy equivalent strain was developed to extent Eq.(1) to the multiaxial damage evolution function, which was determined as[13]

$$\varepsilon_{eq}^{e+}=\sqrt{\frac{2\bar{Y}^{+}}{E_0}}, \quad \varepsilon_{eq}^{e-}=\frac{1}{(\alpha-1)E_0}\sqrt{\frac{\bar{Y}^{-}}{b_0}} \tag{3}$$

where Y^{+} and Y^{-} are the tensile and shear damage energy release rates, respectively; E_0 is Young's modulus of intact material; b_0 is a material constant and α is the strengthening coefficient under biaxial compression. In Eq. (3), the superscript "+" denotes a tensile parameter and the superscript "−" denotes a shear parameter. Thus, the multiaxial damage evolution laws read

$$D^{\pm}(\varepsilon_{eq}^{e\pm})=\int_0^1 H[\varepsilon_{eq}^{e\pm}-\Delta^{\pm}(x^{mi})]\mathrm{d}x^{mi} \tag{4}$$

Accordingly, the constitutive relationship can be given by[11]

$$\boldsymbol{\sigma}=(1-D^{+})\bar{\boldsymbol{\sigma}}^{+}+(1-D^{-})\bar{\boldsymbol{\sigma}}^{-}=(\mathbf{I}-\mathbf{D}):\bar{\boldsymbol{\sigma}}=(\mathbf{I}-\mathbf{D}):\mathbf{E}_0:(\varepsilon-\varepsilon^{p}) \tag{5}$$

where $\boldsymbol{\sigma}$ represents the Cauchy stress; $\bar{\boldsymbol{\sigma}}^{+}$ and $\bar{\boldsymbol{\sigma}}^{-}$ are the positive and negative components of the effective stress $\bar{\boldsymbol{\sigma}}$; $\mathbf{E}_0$ denotes the fourth-order isotropic linear-elastic stiffness tensor; ε is the total strain; ε^{p} is the plastic strain; $\mathbf{I}$ is the fourth-order identity tensor; $\mathbf{D}$ is the damage tensor.

Apparently, due to the stochastic characteristic of the fracture strain of the micro-springs, the damage variable defined by Eq. (4) is a random variable. By introducing the random distribution of the micro fracture strain, the probability characteristics of the damage variable can be obtained[11].

Suppose $\Delta(x^{mi})$ is a homogeneous single-scale random field with the first-order and second-order probability density functions expressed as

$$f(\delta, x^{mi})=f(\delta) \tag{6}$$

$$f(\delta_1, \delta_2; x_1^{mi}, x_2^{mi})=f(\delta_1, \delta_2; |x_1^{mi}-x_2^{mi}|)=f(\delta_1, \delta_2; \eta^{mi}) \tag{7}$$

where δ is a sample realization of Δ and $\eta^{mi}=|x_1^{mi}-x_2^{mi}|$. Eq. (7) shows that the joint probability density function of the random variables Δ_1 and Δ_2 depends only on the relative distance η^{mi}.

As an approximation to the real physical background, it is usually assumed that the homogeneous random field $\Delta(x^{mi})$ satisfies the lognormal distribution with expectation μ_Δ and standard deviation σ_Δ [11, 13, 22]. Then $Z(x^{mi})=\ln\Delta(x^{mi})$ is a homogeneous, normal random field with the mean λ and variance ζ^2 which are defined as

$$\lambda = E[\ln \Delta(x^{mi})] = \ln\left(\frac{\mu_\Delta}{\sqrt{1+\sigma_\Delta^2/\mu_\Delta^2}}\right) \tag{8}$$

$$\zeta^2 = \mathrm{var}[\ln \Delta(x^{mi})] = \ln(1+\sigma_\Delta^2/\mu_\Delta^2) \tag{9}$$

The covariance function of $Z(x^{mi})$ is[11]

$$\begin{aligned} C_Z(\eta^{mi}) &= E[Z(x^{mi})Z(x^{mi}+\eta^{mi})] - E[Z(x^{mi})]E[Z(x^{mi}+\eta^{mi})] \\ &= \zeta^2 \rho_Z(\eta^{mi}) = \zeta^2 \exp(-\omega_Z^{mi}\eta^{mi}) \end{aligned} \tag{10}$$

in which $\omega_Z^{mi} > 0$ is mesoscopic correlation parameter.

Accordingly, the autocorrelation function for $Z(x^{mi})$ is given by

$$\rho_Z(\eta^{mi}) = \frac{C_Z(\eta^{mi})}{\zeta^2} = \exp(-\omega_Z^{mi}\eta^{mi}) \tag{11}$$

Based on the definition of the correlation length of the homogeneous random field[4], which is called the fluctuation scale as well, the mesoscopic correlation length L_Z^{mi} for $Z(x^{mi})$ is given by

$$L_Z^{mi} = 2\int_0^\infty \rho_Z(\eta^{mi})\mathrm{d}\eta^{mi} = 2\int_0^\infty \exp(-\omega_Z^{mi} \cdot \eta^{mi})\mathrm{d}\eta^{mi} = \frac{2}{\omega_Z^{mi}} \tag{12}$$

Moreover, the autocorrelation function $\rho_\Delta^{mi}(\eta^{mi})$ for $\Delta(x^{mi})$ is determined as[23]

$$\rho_\Delta(\eta^{mi}) = \frac{\mu_\Delta^2\{\exp[\zeta^2\rho_Z^{mi}(\eta^{mi})]-1\}}{\sigma_\Delta^2} \tag{13}$$

Correspondingly, the mesoscopic correlation length L_Δ^{mi} for $\Delta(x^{mi})$ is given by

$$L_\Delta^{mi} = 2\int_0^\infty \rho_\Delta(\eta^{mi})\mathrm{d}\eta^{mi} \tag{14}$$

3 Two scale random field

According to above stochastic damage model, the probabilistic mechanical behaviors of one RVE can be demonstrated. However, a question arises that apart from the correlation of micro-meso-scale material properties, the macro-scale correlation among RVEs at different spatial locations of structures should be also considered. If the random field of the microscopic fracture strain are the same for the different spatial locations of a structure, the strength of the RVEs at different spatial locations would be the same. This is not true for real engineering structures. Actually, the macro-strength of concrete at structural scale could be regarded as another random field. Regarding this, a two-scale random field is introduced.

In order to consider the stochastic mechanical behavior of concrete at both scales, the random field of the fracture strain can be generalized from the mesoscopic to the macroscopic scale. As Fig. 2 depicts, the two-scale fracture strain random field is denoted as $\Delta(x^{mi}, x)$, where $x^{mi} \in [0, 1]$ denotes the mesoscopic spatial coordinate and $x \in [0, +\infty]$ indicates the macroscopic spatial coordinate. By introducing of two-scale random field $\Delta(x^{mi}, x)$, the correlation between RVEs at different spatial locations of structures can be established. This makes it possible to model the spatial variability of strength of materials.

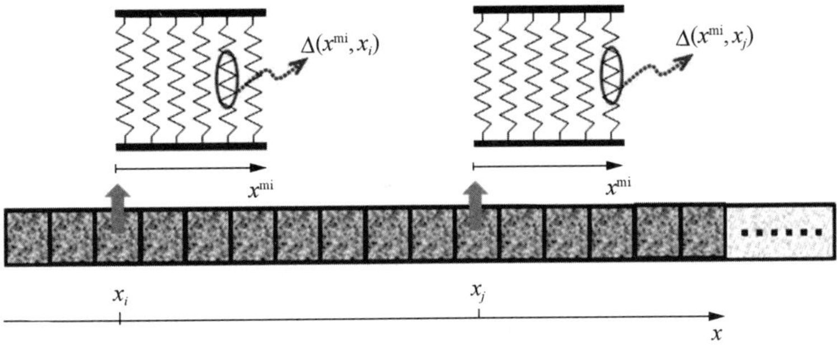

Fig. 2. Two scale random field-based stochastic damage model

Without loss of generality, it is supposed that $\Delta(x^{mi}, x)$ is a homogeneous two-scale random field with the one-dimensional and two-dimensional probability density functions expressed as following

$$f[\delta;\ (x^{mi}, x)] = f(\delta) \tag{15}$$

$$\begin{aligned} f[\delta_1, \delta_2; (x_1^{mi}, x), (x_2^{mi}, x)] &= f(\delta_1, \delta_2;\ |x_1^{mi} - x_2^{mi}|, x) \\ &= f(\delta_1, \delta_2;\ \eta^{mi}) \end{aligned} \tag{16}$$

$$\begin{aligned} f[\delta_1, \delta_2;\ (x_1^{mi}, x_1), (x_2^{mi}, x_2)] &= f(\delta_1, \delta_2;\ |x_1^{mi} - x_2^{mi}|, |x_1 - x_2|) \\ &= f(\delta_1, \delta_2;\ \eta^{mi}, \eta^{ma}) \end{aligned} \tag{17}$$

where $\eta^{mi} = |x_1^{mi} - x_2^{mi}|$; $\eta^{ma} = |x_1 - x_2|$; $f(\delta)$ is one-dimensional probability density function for $\Delta(x^{mi}, x)$; $f(\delta_1, \delta_2;\ \eta^{mi})$ denotes two-dimensional probability density function within one RVE, which depends only on the mesoscopic relative distance η^{mi} between x_1^{mi} and x_2^{mi}; $f(\delta_1, \delta_2;\ \eta^{mi}, \eta^{ma})$ indicates two-dimensional probability density function for $\Delta(x^{mi}, x)$, which depends on the mesoscopic relative distance η^{mi} between x_1^{mi} and x_2^{mi} as well as the macroscopic relative distance η^{ma} between x_1 and x_2.

Actually, Eq.(16) is a special case of Eq.(17) when η^{ma} is equal to be zero, as follows

$$f(\delta_1, \delta_2;\ \eta^{mi}, 0) = f(\delta_1, \delta_2;\ \eta^{mi}) \tag{18}$$

Obviously, the constitutive relationship of materials can be still given by Eq.(5) in which the damage variable of the RVE is defined as

$$D(\varepsilon_{eq}^{e}, x) = \int_0^1 H[\varepsilon_{eq}^{e} - \Delta(x^{mi}, x)] dx^{mi} \tag{19}$$

Let

$$\varphi(x^{mi}, x) = H[\varepsilon_{eq}^{e} - \Delta(x^{mi}, x)] \tag{20}$$

such that $\varphi(x^{mi}, x)$ obeys (0, 1) distribution.

Performing expectation operator on Eq. (19), the expectation of the damage variable $D(\varepsilon_{eq}^{e}, x)$ is obtained as

$$\mu_D(\varepsilon_{eq}^{e}, x) = E\left[\int_0^1 \varphi(x^{mi}, x)\mathrm{d}x^{mi}\right] = \int_0^1 E[\varphi(x^{mi}, x)]\mathrm{d}x^{mi}$$
$$= \int_0^1 F(\varepsilon_{eq}^{e})\mathrm{d}x^{mi} = F(\varepsilon_{eq}^{e}) = \mu_D(\varepsilon_{eq}^{e}) \tag{21}$$

where $F(\varepsilon_{eq}^{e})$ represents the one-dimensional marginal distribution function of $\Delta(x^{mi}, x^{ma})$. Similarly, the variance of $D(\varepsilon_{eq}^{e}, x)$ can be given as

$$V_D^2(\varepsilon_{eq}^{e}, x) = E\{[D(\varepsilon_{eq}^{e}, x)]^2\} - \{E[D(\varepsilon_{eq}^{e}, x)]\}^2$$
$$= E\left[\int_0^1 \varphi(x_1^{mi}, x)\mathrm{d}x_1^{mi}\int_0^1 \varphi(x_2^{mi}, x)\mathrm{d}x_2^{mi}\right] - [\mu_D(\varepsilon_{eq}^{e}, x)]^2 \tag{22}$$

Applying the properties of mean square stochastic integral, the first item of Eq. (22) is determined by

$$E\left[\int_0^1 \varphi(x_1^{mi}, x)\mathrm{d}x_1^{mi}\int_0^1 \varphi(x_2^{mi}, x)\mathrm{d}x_2^{mi}\right] = \int_0^1\int_0^1 R_\varphi[(x_1^{mi}, x), (x_2^{mi}, x)]\mathrm{d}x_1^{mi}\mathrm{d}x_2^{mi}$$
$$= \int_0^1\int_0^1 F(\varepsilon_{eq}^{e}, \varepsilon_{eq}^{e}; \eta^{mi}, 0)\mathrm{d}x_1^{mi}\mathrm{d}x_2^{mi}$$
$$= 2\int_0^1 (1-\eta^{mi})F(\varepsilon_{eq}^{e}, \varepsilon_{eq}^{e}; \eta^{mi})\mathrm{d}\eta^{mi} \tag{23}$$

where $F(\varepsilon_{eq}^{e}, \varepsilon_{eq}^{e}; \eta^{mi})$ is two-dimensional probability distribution function at the mesoscopic scale of $\Delta(x^{mi}, x)$.

Therefore, Eq. (22) could be rewritten as

$$V_D^2(\varepsilon_{eq}^{e}, x) = 2\int_0^1 (1-\eta^{mi})F(\varepsilon_{eq}^{e}, \varepsilon_{eq}^{e}; \eta^{mi})\mathrm{d}\eta^{mi} - [F(\varepsilon_{eq}^{e})]^2 = V_D^2(\varepsilon_{eq}^{e}) \tag{24}$$

Eq. (21) and Eq. (24) indicate that the expectation and variance of the macroscopic damage of the RVE are independent of the macroscopic spatial locations. That is, the introduction of two-scale random field does not change the probabilistic characteristics of damage evolution within one RVE.

As illustrated before, by introducing the two-scale fracture strain random field, the spatial correlation of the macroscopic strength at the structural or macroscopic scale can be reflected. In fact, the covariance function of the damage can be given as

$$C_D(\varepsilon_{eq}^{e}, \eta^{ma}) = E[D(\varepsilon_{eq}^{e}, x_1)D(\varepsilon_{eq}^{e}, x_2)] - \{E[D(\varepsilon_{eq}^{e}, x_1)]E[D(\varepsilon_{eq}^{e}, x_2)]\}$$
$$= E\left[\int_0^1 \varphi(x_1^{mi}, x_1)\mathrm{d}x_1^{mi}\int_0^1 \varphi(x_2^{mi}, x_2)\mathrm{d}x_2^{mi}\right] - [\mu_D(\varepsilon_{eq}^{e})]^2 \tag{25}$$

By taking a similar deduction as Eq. (23), Eq. (25) can be rewritten as

$$C_D(\varepsilon_{eq}^{e}, \eta^{ma}) = 2\int_0^1 (1-\eta^{mi})F(\varepsilon_{eq}^{e}, \varepsilon_{eq}^{e}; \eta^{mi}, \eta^{ma})\mathrm{d}\eta^{mi} - [\mu_D(\varepsilon_{eq}^{e})]^2 \tag{26}$$

Obviously, the covariance function is related with the macroscopic variable η^{ma}. When η^{ma} equals to zero, the covariance of the damage can be simplified to be the variance as follows

$$C_D(\varepsilon_{eq}^e, 0) = V_D^2(\varepsilon_{eq}^e) \tag{27}$$

Furthermore, the autocorrelation function of the damage can be obtained as

$$\rho_D(\varepsilon_{eq}^e, \eta^{ma}) = \frac{C_D(\varepsilon_{eq}^e, \eta^{ma})}{V_D^2(\varepsilon_{eq}^e)} \tag{28}$$

Apparently, the autocorrelation function equals to one when η^{ma} is zero.

On the basis of above formula, the mean and the variance of the stress can be obtained from Eq.(5). For example, for one-dimensional stress state of compression, the mean and the variance of the stress are as follows

$$\mu_\sigma(\varepsilon, x) = \mu_\sigma(\varepsilon) = (1 - \mu_D)E_0(\varepsilon - \varepsilon^p) \tag{29}$$

$$V_\sigma^2(\varepsilon, x) = V_\sigma^2(\varepsilon) = V_D^2[E_0(\varepsilon - \varepsilon^p)]^2 \tag{30}$$

The covariance function $C_\sigma(\varepsilon, \eta^{ma})$ for the $\sigma(\varepsilon)$ at the macroscopic scale is given by

$$C_\sigma(\varepsilon, \eta^{ma}) = C_D(\varepsilon - \varepsilon^p, \eta^{ma})[E_0(\varepsilon - \varepsilon^p)]^2 \tag{31}$$

Obviously, the maximum value of $\sigma - \varepsilon$ curve gives the macro strength σ_{max} which can be regarded as a three-dimensional random field at the structural scale. According to specific problems, it can be simplified as two-dimensional or one-dimensional random field.

4 Relationship between two-scale random field and meso-scale random field

Assume $\Delta(x^{mi}, x)$ satisfying the lognormal distribution with expectation μ_Δ and standard deviation σ_Δ, $Z(x^{mi}, x) = \ln \Delta(x^{mi}, x)$ is a homogeneous two scale random field with normal distribution and the cross-covariance function for $Z(x^{mi}, x)$ is

$$\begin{aligned} &C_Z(\eta^{mi}, \eta^{ma}) \\ &= E[Z(x^{mi}, x)Z(x^{mi} + \eta^{mi}, x + \eta^{ma})] - E[Z(x^{mi}, x)]E[Z(x^{mi} + \eta^{mi}, x + \eta^{ma})] \\ &= \zeta^2 \rho_Z(\eta^{mi}, \eta^{ma}) \end{aligned} \tag{32}$$

where $\zeta^2 = \ln(1 + \sigma_\Delta^2/\mu_\Delta^2)$; $\rho_Z(\eta^{mi}, \eta^{ma})$ denotes the cross-correlation function for $Z(x^{mi}, x)$, which can be expressed as

$$\rho_Z(\eta^{mi}, \eta^{ma}) = \frac{C_Z(\eta^{mi}, \eta^{ma})}{\zeta^2} \tag{33}$$

Obviously,

$$\rho_Z(\eta^{mi}, 0) = \rho_Z(\eta^{mi}) \tag{34}$$

$$\rho_Z(0, \eta^{ma}) = \rho_Z(\eta^{ma}) \tag{35}$$

$$\rho_Z(0, 0) = 1 \tag{36}$$

where $\rho_Z(\eta^{mi})$ denotes the autocorrelation function in terms of mesoscopic scale for $Z(x^{mi}, x^{ma})$; $\rho_Z(\eta^{ma})$ denotes the autocorrelation function in terms of the macroscopic scale.

It is noticed that $\rho_Z(\eta^{mi})$ actually reflects the micro-mesoscopic spatial correlation of the fracture strain(see Eq.(11)) and thus the mesoscopic correlation length L_Z^{mi} (see Eq.(12)) is much less than the size of the RVE L_{RVE}. That is

$$L_Z^{mi} \ll L_{RVE} \tag{37}$$

Moreover, $\rho_Z(\eta^{ma})$ in Eq.(35) reflects the macroscopic spatial correlation of fracture strain random field among different RVEs. Therefore, the macroscopic correlation length L_Z^{ma} should be larger than the size of the RVE. That is

$$L_{RVE} < L_Z^{ma} \tag{38}$$

By combining Eqs.(37) and (38), we have

$$L_Z^{mi} \ll L_{RVE} < L_Z^{ma} \tag{39}$$

Therefore, the cross-correlation function $\rho_Z(\eta^{mi}, \eta^{ma})$ for $Z(x^{mi}, x)$ can be assumed to be a product of the mesoscopic and macroscopic autocorrelation functions, which is expressed as

$$\rho_Z(\eta^{mi}, \eta^{ma}) = \rho_Z(\eta^{mi})\rho_Z(\eta^{ma}) \tag{40}$$

Then the cross-covariance function for $Z(x^{mi}, x)$ can be rewritten as

$$C_Z(\eta^{mi}, \eta^{ma}) = \zeta^2 \rho_Z^{mi}(\eta^{mi})\rho_Z^{ma}(\eta^{ma}) \tag{41}$$

In addition, the exponential decaying form can be adopted for $\rho_Z(\eta^{mi})$ and $\rho_Z(\eta^{ma})$ [24], as follows

$$\rho_Z^{mi}(\eta^{mi}) = \exp(-\omega_Z^{mi}\eta^{mi}) \tag{42}$$

$$\rho_Z^{ma}(\eta^{ma}) = \exp(-\omega_Z^{ma}\eta^{ma}) \tag{43}$$

where $\omega_Z^{mi} > 0$ is mesoscopic correlation parameter; $\omega_Z^{ma} > 0$ is macroscopic correlation parameter. The mesoscopic correlation length L_Z^{mi} can be given by $L_Z^{mi} = 2/\omega_Z^{mi}$. Similarly, the macroscopic correlation length L_Z^{ma} is determined as $L_Z^{ma} = 2/\omega_Z^{ma}$.

On the basis of above background, it can be proved that there is a deterministic relationship between the two-scale random field $Z(x^{mi}, x)$ and meso-scale random field $Z(x^{mi})$. In fact, the two-scale random field of the fracture strain can be discretized as $\Delta(x^{mi}, x_1)$, $\Delta(x^{mi}, x_2)$, …, $\Delta(x^{mi}, x_N)$. Then the corresponding Gaussian random fields are denoted as $Z(x^{mi}, x_1)$, $Z(x^{mi}, x_2)$, …, $Z(x^{mi}, x_N)$. That is

$$\mathbf{Z}(x^{mi}, x) = [Z(x^{mi}, x_1), Z(x^{mi}, x_2), \cdots, Z(x^{mi}, x_N)]^T \tag{44}$$

On the other hand, the mesoscopic random field $\Delta(x^{mi})$ at different positions x_i can be expressed as $\Delta_1(x^{mi})$, $\Delta_2(x^{mi})$, …, $\Delta_N(x^{mi})$. The corresponding Gaussian random fields are denoted as $Z_1(x^{mi})$, $Z_2(x^{mi})$, …, $Z_N(x^{mi})$. That is

$$\mathbf{Z}(x^{mi}) = | Z_1(x^{mi}), Z_2(x^{mi}), \cdots, Z_N(x^{mi}) |^T \tag{45}$$

By introducing a transform matrix λ, the following formula exists

$$\mathbf{Z}(x^{mi}, x) = \lambda \mathbf{Z}(x^{mi}) \tag{46}$$

The above formula can be proved as following.

Consider two RVEs at x_i and x_j, as shown in Fig. 2. The cross-covariance function $C_{Z_{ij}}(\eta^{mi}, \eta^{ma})$ of the two-scale Gaussian random fields $Z(x^{mi}, x)$ can be written as

$$\begin{aligned} & C_{Z_{ij}}(\eta^{mi}, \eta^{ma}) \\ & = E[Z(x^{mi}, x_i) \cdot Z(x^{mi}+\eta^{mi}, x_j)] - E[Z(x^{mi}, x_i)]E[Z(x^{mi}+\eta^{mi}, x_j)] \\ & = \zeta^2 \rho_{Z_{ij}}(\eta^{mi}, \eta^{ma}) \end{aligned} \tag{47}$$

where $\rho_{Z_{ij}}(\eta^{mi}, \eta^{ma})$ is cross-correlation function of the two-scale Gaussian random fields. Then the cross-covariance function matrix for the discretized Gaussian random fields $Z(x^{mi}, x)$ will be

$$\boldsymbol{C}_Z(\eta^{mi}, \eta^{ma}) = \begin{bmatrix} C_{Z_{11}}(\eta^{mi}, \eta^{ma}) & C_{Z_{12}}(\eta^{mi}, \eta^{ma}) & \cdots & C_{Z_{1N}}(\eta^{mi}, \eta^{ma}) \\ C_{Z_{21}}(\eta^{mi}, \eta^{ma}) & C_{Z_{22}}(\eta^{mi}, \eta^{ma}) & \cdots & C_{Z_{2N}}(\eta^{mi}, \eta^{ma}) \\ \vdots & \vdots & \ddots & \vdots \\ C_{Z_{N1}}(\eta^{mi}, \eta^{ma}) & C_{Z_{N1}}(\eta^{mi}, \eta^{ma}) & \cdots & C_{Z_{NN}}(\eta^{mi}, \eta^{ma}) \end{bmatrix} \tag{48}$$

Obviously, the elements of the cross-covariance matrix satisfy the following conditions

$$C_{Z_{ij}}(\eta^{mi}, \eta^{ma}) = C_{Z_{ji}}(\eta^{mi}, \eta^{ma}) i, j = 1, 2, \cdots, N \tag{49}$$

When $i = j$, the elements of the cross-covariance matrix will be transformed into the mesoscopic auto-covariance function due to $\eta^{ma} = 0$. That is

$$C_{Z_{ii}}(\eta^{mi}, \eta^{ma}) = C_Z^{mi}(\eta^{mi}) = \zeta^2 \rho_Z^{mi}(\eta^{mi}), \ i = 1, 2, \cdots, N \tag{50}$$

When $i \neq j$, if Eq.(40) holds, the elements of the cross-covariance matrix can be expressed as

$$C_{Z_{ij}}(\eta^{mi}, \eta^{ma}) = \zeta^2 \rho_Z(\eta^{mi})(\rho_Z^{ma})_{ij}, \ i, j = 1, 2, \cdots, N \tag{51}$$

where $(\rho_Z^{ma})_{ij}$ is the macroscopic correlation function between the two mesoscopic Gaussian random fields $Z(x^{mi}, x_i)$ and $Z(x^{mi}, x_j)$, satisfying

$$(\rho_Z^{ma})_{ij} = \rho_Z(\eta_{ij}^{ma}) \tag{52}$$

where $\eta_{ij}^{ma} = | x_i - x_j |$.

Consequently, the cross-covariance matrix can be re-expressed as

$$\begin{aligned} \boldsymbol{C}_Z(\eta^{mi}, \eta^{ma}) & = \zeta^2 \rho_Z(\eta^{mi}) \begin{bmatrix} 1 & (\rho_Z^{ma})_{12} & \cdots & (\rho_Z^{ma})_{1N} \\ (\rho_Z^{ma})_{21} & 1 & \cdots & (\rho_Z^{ma})_{2N} \\ \vdots & \vdots & \ddots & \vdots \\ (\rho_Z^{ma})_{N1} & (\rho_Z^{ma})_{N2} & \cdots & 1 \end{bmatrix} \\ & = \zeta^2 \rho_Z(\eta^{mi}) \cdot \boldsymbol{r} \end{aligned} \tag{53}$$

where

$$\boldsymbol{r} = \begin{bmatrix} 1 & (\rho_Z^{ma})_{12} & \cdots & (\rho_Z^{ma})_{NN} \\ (\rho_Z^{ma})_{21} & 1 & \cdots & (\rho_Z^{ma})_{N1} \\ \vdots & \vdots & \ddots & \vdots \\ (\rho_Z^{ma})_{N1} & (\rho_Z^{ma})_{N2} & \cdots & 1 \end{bmatrix} \tag{54}$$

Generally, $\boldsymbol{r}$ is a real symmetric positive definite matrix, which allows the Cholesky decomposition performing as

$$\boldsymbol{r} = \boldsymbol{h}\boldsymbol{h}^{\mathrm{T}} \tag{55}$$

where $(\bullet)^{\mathrm{T}}$ indicates the transpose of a matrix; $\boldsymbol{h}$ is lower triangular matrix, which can be expressed as

$$\boldsymbol{h} = \begin{bmatrix} 1 & 0 & \cdots & 0 \\ h_{21} & h_{22} & \cdots & 0 \\ \vdots & \vdots & \ddots & \vdots \\ h_{N_l 1} & h_{N_l 2} & \cdots & h_{N_l N_l} \end{bmatrix} \tag{56}$$

Noting Eq.(50), there exists

$$C_{Z_{ij}}(\eta^{\mathrm{mi}}, \eta^{\mathrm{ma}}) = C_Z(\eta^{\mathrm{mi}})\boldsymbol{h}\boldsymbol{h}^{\mathrm{T}} \tag{57}$$

On the other hand, based on Eq.(46), the cross-covariance function $C_{Z_{ij}}(\eta^{\mathrm{mi}}, \eta^{\mathrm{ma}})$ can be given as

$$C_{Z_{ij}}(\eta^{\mathrm{mi}}, \eta^{\mathrm{ma}}) = C_Z(\eta^{\mathrm{mi}})\lambda\lambda^{\mathrm{T}} \tag{58}$$

Obviously, we can get

$$\lambda = \boldsymbol{h} \tag{59}$$

The above theorem indicates that the mesoscopic random fields can be transformed to macroscopic spatial correlative random fields by the transform matrix. It has been proved by experiments that the macroscopic correlation function ρ_Z^{ma} is actually the correlation function $\rho_{\sigma_{\max}}^{\mathrm{ma}}$ of macro strength random field $\sigma_{\max}$ ([25]).

5 Numerical implementation

For the stochastic response analysis of structures, a series of sample realizations for two-scale random field $Z(x^{\mathrm{mi}}, x)$ should be obtained. Regarding this, a numerical implementation method is introduced here.

(1) Given the probability property of mesoscopic Gaussian random field $Z(x^{\mathrm{mi}})$, a series of samples can be obtained by using the stochastic harmonic function representation[26-27]. That is

$$Z(x^{\mathrm{mi}}) = \lambda + \sqrt{2}\sum_{i=1}^{N} A_i \cos(w_i^{\mathrm{mi}} x^{\mathrm{mi}} + \phi_i),\ i = 1, 2, \cdots, N \tag{60}$$

where ϕ_i is independent random phase angle of the ith cosine component with uniformly distribution in the range $[0, 2\pi]$; N is the number of the cosine components; w_i^{mi} is the random frequency of the ith cosine component with the probability density function satisfying

$$p_{w_i^{mi}}=\begin{cases}\dfrac{1}{w_{i-1}^{mi(p)}-w_i^{mi(p)}} & w_i^{mi}\in[w_{i-1}^{mi(p)},\ w_i^{mi(p)}]\\ 0 & \text{otherwise}\end{cases}\tag{61}$$

A_i represents the amplitude of the ith cosine component, which is given by

$$A_i=\sqrt{2S_Z(w_i^{mi})\Delta w^{mi}}\tag{62}$$

where $S_Z(w^{mi})$ is the power spectral density function(PSD) function for $Z(x^{mi})$, which can be obtained based on the Wiener-Khinchin transform of the auto-covariance function $C_Z(\eta^{mi})$

$$\begin{aligned}S_Z(w^{mi})&=\frac{1}{2\pi}\int_{-\infty}^{+\infty}C_Z(\eta^{mi})\exp(-iw^{mi}\eta^{mi})\mathrm{d}\eta^{mi}\\&=\frac{1}{2\pi}\int_{-\infty}^{+\infty}\zeta^2\exp(-\xi_Z^{mi}\eta^{mi})\exp(-iw^{mi}\eta^{mi})\mathrm{d}\eta^{mi}\\&=\frac{1}{2\pi}\frac{2\xi_Z^{mi}\zeta^2}{(\xi_Z^{mi})^2+(w^{mi})^2}\end{aligned}\tag{63}$$

Then the obtained sample of Gaussian random fields for N_l RVEs can be expressed as

$$\boldsymbol{Z}(x^{mi})=[Z_1(x^{mi}),\ Z_2(x^{mi}),\ \cdots,\ Z_{N_l}(x^{mi})]^{\mathrm{T}}\tag{64}$$

(2) Based on the macroscopic spatial correlation function $\rho_Z(\eta^{ma})=\exp(-\omega_Z^{ma}\eta^{ma})$, the transform matrix is determined by Eqs.(55) and (56). Then the sample of the two-scale random field $\boldsymbol{Z}(x^{mi},x)=[Z(x^{mi},x_1),\ Z(x^{mi},x_2),\ \cdots,\ Z(x^{mi},x_{N_l})]^{\mathrm{T}}$ is given as

$$\boldsymbol{Z}(x^{mi},x)=\lambda\boldsymbol{Z}_g(x^{mi})\tag{65}$$

(3) $\Delta(x^{mi},x)=\exp[Z(x^{mi},x)]$ is employed to obtain the numerical realization of the fracture strain field.

6 Illustrative example

To demonstrate the ability of the proposed model, a series of tests has been carried out[24]. In the experiment, there were ten concrete beams and ten cylindrical core specimens were obtained from each concrete beam, as Fig. 3 demonstrates. The diameter and the length of the core specimen. were respectively 100 mm and 200 mm with the length-diameter ratio (L/D) of 2.0. The distance between two adjacent core specimens was 160 mm. For simplicity, the center of the first test area along the length of the beam was taken as the origin ($x=0$), as Fig. 3 depicts. Then, one hundred complete compressive stress-strain curves of the cylindrical core specimens were obtained from the compression tests, which indicates that ten complete compressive stress-strain curves were given at the same location of concrete beam. The typical characteristics of nonlinear and randomness of concrete material under the uniaxial compressive loading can be apparently observed from the test results[24].

To determine the model parameters (E_0, λ, ζ, ω_Z^{mi}, ω_Z^{ma}), a series of parameters identifications

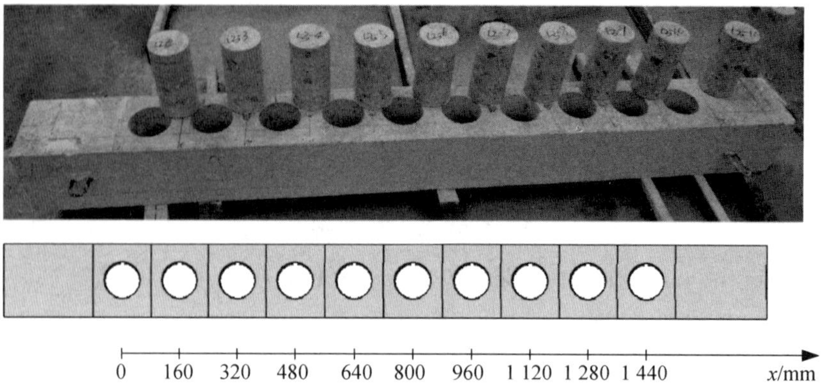

Fig. 3. Core specimens drilled from the plain concrete beams

are conducted based on the statistical characteristics of the test results, which are set to be: $E_0 = 30.8$ GPa, $\lambda = 7.58$, $\zeta = 0.39$, $\omega_Z^{\mathrm{mi}} = 188.0$ and $\omega_Z^{\mathrm{ma}} = 4.44/\mathrm{m}$, respectively.

Based on the proposed two-step implementation method, the two-scale Gaussian random field can be numerically simulated. Fig. 4 presents one sample realization of the random field $Z(x^{\mathrm{mi}}, x)$ including ten RVEs. Then by using Eq.(5), the damage evolution curves and stress-strain curves can be derived for the different RVEs. Figs. 5 and 6 gives a set of numerical results for one sample realization of the random field, which indicate that the mechanical behaviors are different among the RVEs.

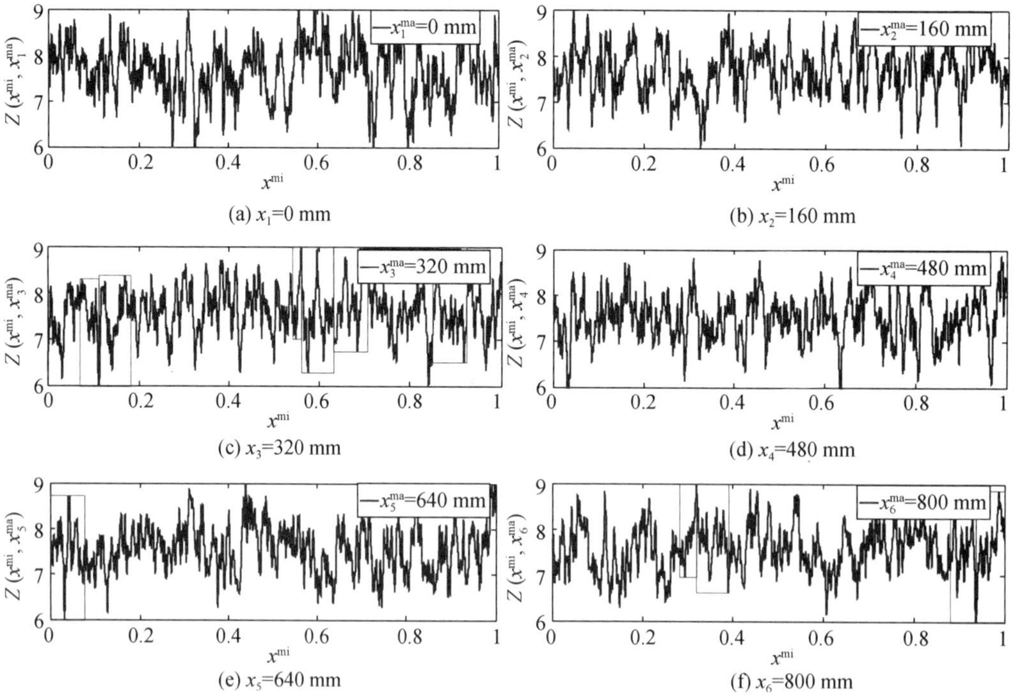

(a) x_1=0 mm

(b) x_2=160 mm

(c) x_3=320 mm

(d) x_4=480 mm

(e) x_5=640 mm

(f) x_6=800 mm

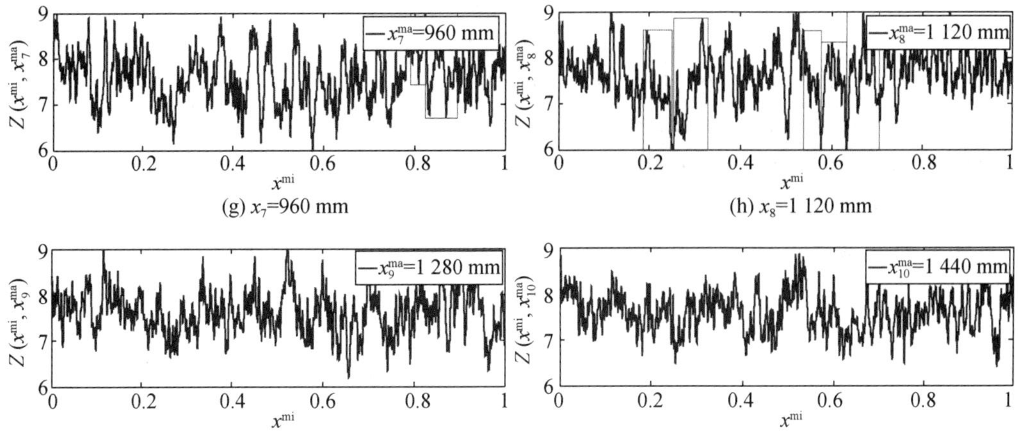

(g) x_7=960 mm (h) x_8=1 120 mm

(i) x_9=1 280 mm (j) x_{10}=1 440 mm

Fig. 4. Mesoscopic Gaussian random field samples for ten RVEs, respectively

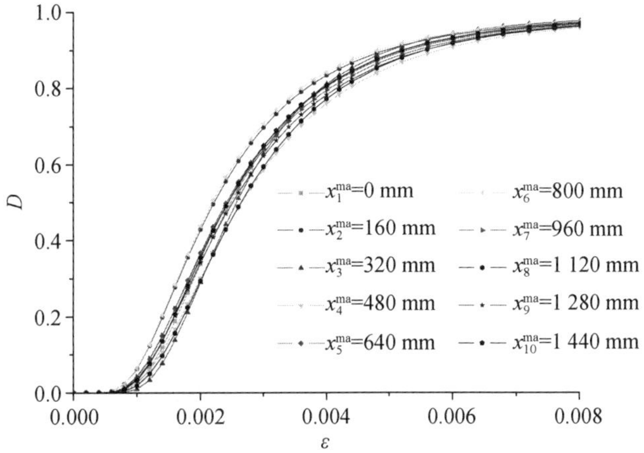

Fig. 5. Damage evolution curves for ten RVEs of one sample

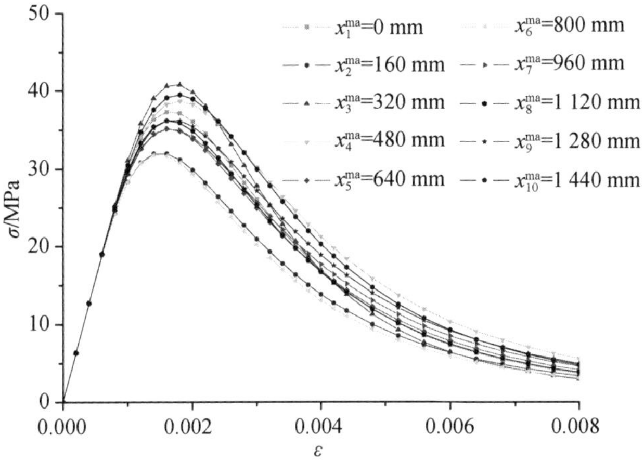

Fig. 6. Stress-strain relation for ten RVEs of one sample

With the help of probability density evolution theory[28], the stochastic stress-strain relationship can be analytically predicted and its statistical characteristics such as the mean and the standard deviation can be obtained and compared with the experimental results. In this way, for simulating the mesoscopic random field $\boldsymbol{Z}(x^{mi})$, 200 represent points are selected in the probability space specified by the stochastic harmonic function as Eq. (60). For each represent point, the two-scale random field are numerically generated by using above numerical implementation method. Then the stochastic analysis can be carried out by probability density evolution method taking Eq. (5) as a basic physical equation. The comparisons between theoretical results and experimental results are shown in Fig. 7. Obviously, these figures demonstrate that good agreement is achieved for the mean and the standard deviation of $\sigma-\varepsilon$ curves. Moreover, Fig. 8 presents the comparison of the autocorrelation function for strength of concrete between the numerical results and experimental results.

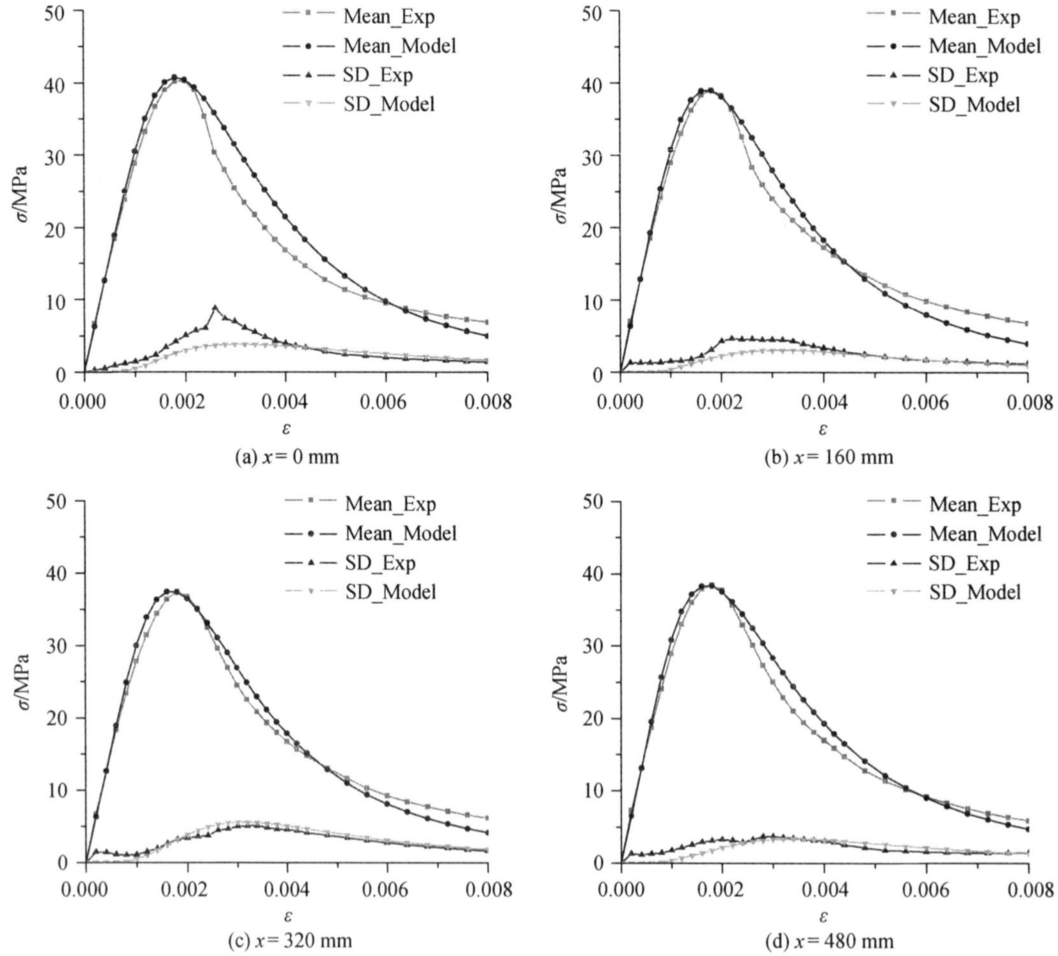

(a) x= 0 mm

(b) x= 160 mm

(c) x= 320 mm

(d) x= 480 mm

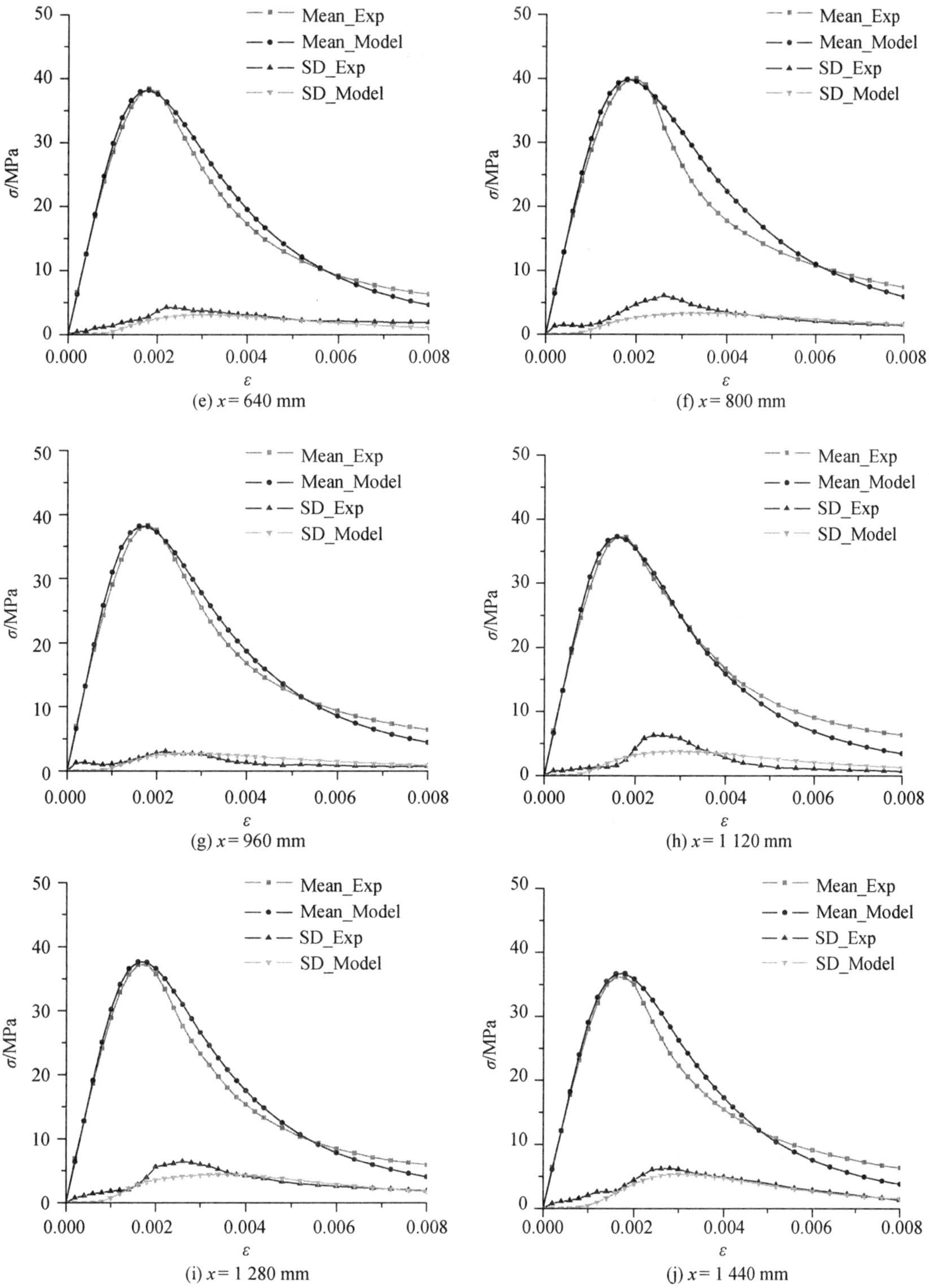

Fig. 7. Analytically predicted variability of stress-strain relations at the different locations of concrete beams

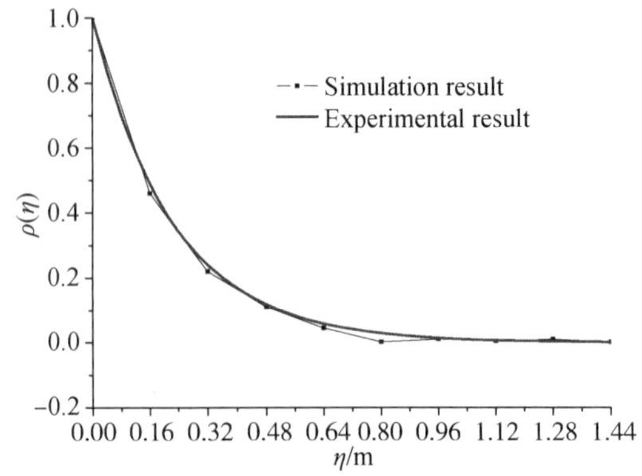

Fig. 8. Autocorrelation coefficient function for peak strength of concrete

The previous analyses manifest that the mesoscopic stochastic fracture model based on the proposed two-scale random field and its numerical implementation can well reproduce the stochastic damage evolution and nonlinear behavior of one RVE as well as the spatial variability of damage evolution and mechanical characteristics among different RVEs at the structural scale.

7 Conclusions

In this paper, a novel two-scale random field model is proposed for the stochastic nonlinear analysis of quasi-brittle materials. To incorporate the random fluctuation at the micro-meso-levels and the spatial correlation of nonlinear mechanical behaviors of materials at the macro-scale within a unified framework, a two-scale homogeneous random field considering the probability characteristics at both mesoscopic and macroscopic scales is investigated. It is found that with the propagation of randomness across scales, the damage evolution and stress-strain curve of materials both present the characteristics of random fluctuation and spatial variability. Furthermore, it is proved that the mesoscopic random fields can be transformed to macroscopic spatial correlative random fields by a transform matrix that indeed describes the spatial correlations of macroscopic strength of materials. Based on this, a two-step implementation method is developed for the numerical simulations. Comparisons are made with the experimental data of a systematical tests, which demonstrate the validity of the proposed model and numerical method.

It is worth to point out that, the idea of two-scale random field or its extended edition, multi-scale random field, is not only can be applied for mechanical analysis of structures, but also can be extended to other area, such as earthquake engineering, wind engineering, and materials science, etc.

Declaration of competing interest

The authors declare that they have no known competing financial interests or personal relationships that could have appeared to influence the work reported in this paper.

Acknowledgments

Financial support from the National Natural Science Foundation of China (Grant No. 51538010) and the Education Commission of Shanghai China (Grant No. 2017-01-07-00-07-E00006) are gratefully appreciated.

References

[1] GHANEM R G, SPANOS P D. Stochastic finite elements: a spectral approach[M]. New York: Springer, 1991.

[2] STEFANOU G. The stochastic finite element method: past, present and future[J]. Computer Methods in Applied Mechanics and Engineering, 2009, 198(9-12): 1031-1051.

[3] VANMARCKE E. Random Fields: Analysis and Synthesis[M]. Massachusetts: The MIT Press, Cambridge, 1983.

[4] VANMARCKE E. Random fields: analysis and synthesis[M]. World Scientific, 2010.

[5] SANTOSO A M, PHOON K K, QUEK S T. Effects of soil spatial variability on rainfall-induced landslides[J]. Computers & Structures, 2011, 89(11-12): 893-900.

[6] LI J, ZHOU H, DING Y. Stochastic seismic collapse and reliability assessment of high-rise reinforced concrete structures[J]. The Structural Design of Tall and Special Buildings, 2018, 27(2): e1417.

[7] MAZARS J. A description of micro-and macroscale damage of concrete structures[J]. Engineering Fracture Mechanics, 1986, 25(5-6): 729-737.

[8] JU J W. On energy-based coupled elastoplastic damage theories: constitutive modeling and computational aspects[J]. International Journal of Solids and Structures, 1989, 25(7): 803-833.

[9] FARIA R, OLIVER J, CERVERA M. A strain-based plastic viscous-damage model for massive concrete structures[J]. International Journal of Solids and Structures, 1998, 35(14): 1533-1558.

[10] WU J Y, LI J, FARIA R. An energy release rate-based plastic-damage model for concrete[J]. International Journal of Solids and Structures, 2006, 43(3-4): 583-612.

[11] LI J, WU J, CHEN J. Stochastic Damage Mechanics of Concrete Structures[M]. Beijing: Science Press, 2014 (in Chinese).

[12] LI J, ZHANG Q. Study of stochastic damage constitutive relationship for concrete material[J]. Journal-Tongji University, 2001, 29(10): 1135-1141.

[13] LI J, REN X. Stochastic damage model for concrete based on energy equivalent strain[J]. International Journal of Solids and Structures, 2009, 46(11-12): 2407-2419.

[14] REN X, ZENG S, LI J. A rate-dependent stochastic damage-plasticity model for quasi-brittle materials [J]. Computational Mechanics, 2015, 55(2): 267-285.

[15] DING Z, LI J. A physically motivated model for fatigue damage of concrete[J]. International Journal of Damage Mechanics, 2018, 27(8): 1192-1212.

[16] LI J, GAO R. Fatigue reliability analysis of concrete structures based on physical synthesis method[J]. Probabilistic Engineering Mechanics, 2019, 56: 14-26.

[17] OSTOJA-STARZEWSKI M. Material spatial randomness: From statistical to representative volume element[J]. Probabilistic Engineering Mechanics, 2006, 21(2): 112-132.

[18] TEFERRA K, GRAHAM-BRADY L. A random field-based method to estimate convergence of apparent properties in computational homogenization[J]. Computer Methods in Applied Mechanics and Engineering, 2018, 330: 253-270.

[19] GITMAN I M, GITMAN M B, ASKES H. Quantification of stochastically stable representative volumes for random heterogeneous materials[J]. Archive of Applied Mechanics, 2006, 75(2): 79-92.

[20] GITMAN I M, ASKES H, SLUYS L J. Coupled-volume multi-scale modelling of quasi-brittle material [J]. European Journal of Mechanics-A/Solids, 2008, 27(3): 302-327.

[21] NGUYEN V P, LLOBERAS-VALLS O, STROEVEN M, et al. On the existence of representative volumes for softening quasi-brittle materials-a failure zone averaging scheme[J]. Computer Methods in Applied Mechanics and Engineering, 2010, 199(45-48): 3028-3038.

[22] KANDARPA S, KIRKNER D J, SPENCER JR B F. Stochastic damage model for brittle materials subjected to monotonic loading[J]. Journal of Engineering Mechanics, 1996, 122(8): 788-795.

[23] GRIGORIU M. Crossings of non-Gaussian translation processes[J]. Journal of Engineering Mechanics, 1984, 110(4): 610-620.

[24] XU T, LI J. Assessing the spatial variability of the concrete by the rebound hammer test and compression test of drilled cores[J]. Construction and Building Materials, 2018, 188: 820-832.

[25] XU T. Study on Random Field of the Mechanical Properties of Concrete andStochastic Nonlinear Analysis of Structure[D]. Shanghai: Tongji University, Shanghai, 2018.

[26] CHEN J, SUN W, LI J, et al. Stochastic harmonic function representation of stochastic processes[J]. Journal of Applied Mechanics, 2013, 80(1).

[27] CHEN J, HE J, REN X, et al. Stochastic harmonic function representation of random fields for material properties of structures[J]. Journal of Engineering Mechanics, 2018, 144(7): 04018049.

[28] LI J, CHEN J. Stochastic dynamics of structures[M]. John Wiley & Sons, 2009.

A random medium model for simulation of concrete failure①

Shixue Liang[1], **Xiaodan Ren**[1], **Jie Li**[1, 2]

1 School of Civil Engineering, Tongji University, Shanghai 200092, China

2 The State Key Laboratory on Disaster Reduction in Civil Engineering, Tongji University, Shanghai 200092, China

Abstract A random medium model is developed to describe damage and failure of concrete. In the first place, to simulate the evolving cracks in a mesoscale, the concrete is randomly discretized as irregular finite elements. Moreover, the cohesive elements are inserted into the adjacency of finite elements as the possible cracking paths. The spatial variation of the material properties is considered using a 2-D random field, and the stochastic harmonic function method is adopted to simulate the sample of the fracture energy random field in the analysis. Then, the simulations of concrete specimens are given to describe the different failure modes of concrete under tension. Finally, based on the simulating results, the probability density distributions of the stress-strain curves are solved by the probability density evolution methods. Thus, the accuracy and efficiency of the proposed model are verified in both the sample level and collection level.

Keywords Failure simulation; Cohesive elements; Random field; Probability density evolution

1 Introduction

Concrete, as the most widely used construction material for infrastructures, is featured by its stochastic nonlinearities due to the random distribution of the multiple phases and defects[1]. After years of investigations, two groups of models, i.e., the continuum models based on the continuum damage models and the fracture simulation based discontinuous models, have been proposed to characterize the nonlinear behaviors of concrete.

Within the framework of the continuum damage mechanics, the continuum models[2-5] incline to consider the degeneration of mechanical behaviors by using the continuum damage variable. Based on the definition of the generalized damage variable (avoiding the explicit simulation of cracks and voids), the continuum damage mechanics has its intrinsic advantage to drive the physical mechanisms into the irreversible thermodynamics principles to model a wide range of softening materials, especially the concrete. However, the irreversible thermodynamics only provides a framework for the damage evolution. That is to say, the thermo dynamical inequalities could not define the specified forms of the damage evolution functions, which also play a very important role in continuum damage

① Originally published in *Science China Technological Sciences*, 2013, 56(05): 1273-1281.

mechanics. Hence, several empirical expressions of damage evolution were proposed based on the curve fitting of experimental results. In this case, the forms of damage evolution and the corresponding parameters are often lack of physical meanings. What's more, the generalized damage variable cannot describe the detailed crack propagation and interaction, which are critical to the strength and the failure modes of concrete.

Meanwhile, some researchers turned to the discontinuous models, which tend to simulate the propagation and coalescence of micro-voids and cracks in a direct way. As for a single crack, lying within the infinite homogenous solid, the closed-form solution[6] can be derived based on the classical linear elastic fracture mechanics (LEFM) for several stress conditions. Referring to the interaction of the cracks, many analytical techniques were also developed, such as differential method, Mori-Tanaka method and so on[7]. However, most of these methods are restricted to the weak interaction of cracks. When it comes to the complex coalescence, bifurcation even the intersection of cracks, it is still extraordinarily challenging to obtain the analytical solution nowadays. Some other researchers tried the numerical approaches that explicitly model the cracking process, e.g., the extended finite element method[8] (X-FEM) and the interface finite element method. In the X-FEM, the additional discontinuous enrichments are introduced to model the presence of cracks, voids or heterogeneities within the finite element, which is the geometric domain. In the interface finite element method, the variational equations are built with the terms representing the matrix and the interfaces separately. Then, a solid matrix is modeled as elastic media and the interface is described by the cohesive law. The cohesive law was firstly proposed by Dugdale[9] to describe the crack growth in metal. Later, Barenblatt[10] gave the mathematical explanation of the cohesive models. Hillerborg[11] investigated the applied formulations of several materials, such as the metal and concrete. Xu and Needleman[12] attained the convincible simulations of the fracture in solid, which was based on the combination of the finite element and cohesive element method.

When taking a close observation at the concrete, it is naturally to find out that its mechanical properties, ranging from the strength to the Young's modulus, are endowed with stochastic features due to the random distribution of the aggregates and the micro-defects(cracks and voids). Thereby, to deem the concrete as a kind of random medium will be more reasonable, as compared to the homogeneous description[1]. The random nature within the domain of concrete can be defined by the random medium. As for the simulation of concrete fracture and failure process, the random medium is modeled by a 2-D random field in this paper.

In the present paper, the randomly distributed finite elements and cohesive elements are developed in the first step to introduce the possible cracking paths. Then, the concrete is considered as a random medium. In this case, we use the 2-D random field, which is generated by the newly developed stochastic harmonic function method to model the spatial variation of fracture energy. It turns out that the numerical results of the tensile tests not only agree well with the experimental results in the mean and standard deviation curves, but also offer the probability density distributions of the stress-strain curves against the histograms of the experimental data.

2 Cohesive model based simulation

2.1 Cohesive crack method

Consider a solid Ω, which contains a series of zero-thickness cracks or shear bands. And let S be the interface within the solid[Fig. 1(a)]. As it depicts, S cuts Ω into two parts. Herein, we denote the two sides of interface S by S^+ and S^-, and the two parts of solid Ω by Ω^+ and Ω^-.

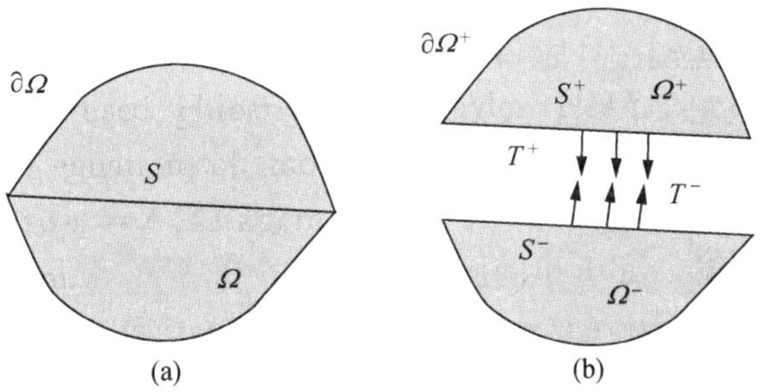

Fig. 1. Solid analysis. (a) Solid with an interface; (b) split the solid along the interface

To establish the weak form equation of the solid in both parts, we have

$$\int_{\Omega^+} \boldsymbol{\sigma}(\boldsymbol{u}) : \boldsymbol{\varepsilon}(\boldsymbol{v})\mathrm{d}\Omega = \int_{\partial\Omega^+} \boldsymbol{t} \cdot \boldsymbol{v}\mathrm{d}\Gamma + \int_{S^+} \boldsymbol{T}^+ \cdot \boldsymbol{v}^+ \mathrm{d}\Gamma, \tag{1}$$

$$\int_{\Omega^-} \boldsymbol{\sigma}(\boldsymbol{u}) : \boldsymbol{\varepsilon}(\boldsymbol{v})\mathrm{d}\Omega = \int_{\partial\Omega^-} \boldsymbol{t} \cdot \boldsymbol{v}\mathrm{d}\Gamma + \int_{S^+} \boldsymbol{T}^- \cdot \boldsymbol{v}^- \mathrm{d}\Gamma, \tag{2}$$

where T^+, T^- indicate the stress at the interfaces of the solid and $T = T^+ = -T^-$. Combining Eq.(1) and Eq.(2), we have

$$\int_{\Omega} \boldsymbol{\sigma}(\boldsymbol{u}) : \boldsymbol{\varepsilon}(\boldsymbol{v})\mathrm{d}\Omega + \int_{S} \boldsymbol{T} \cdot (\boldsymbol{v}^- - \boldsymbol{v}^+)\mathrm{d}\Gamma = \int_{\partial\Omega} \boldsymbol{t} \cdot \boldsymbol{v}\mathrm{d}\Gamma. \tag{3}$$

To solve the interfacial stress $\boldsymbol{T}$ in Eq.(3), a nonlinear zone is introduced between the real crack and the noncracking elastic area. Thanks to the nonlinear zone, the stress at the tip of the crack is reasonable, rather than the analytical infinity by the LEFM. Generally, the stress at the crack-tip is named as cohesive stress. Hillerborg proposed that the crack would propagate when the cohesive stress reached the strength of concrete, and the cohesive stress reduced along the opening of the crack. To apply the cohesive stress as the external tractions applying on the crack

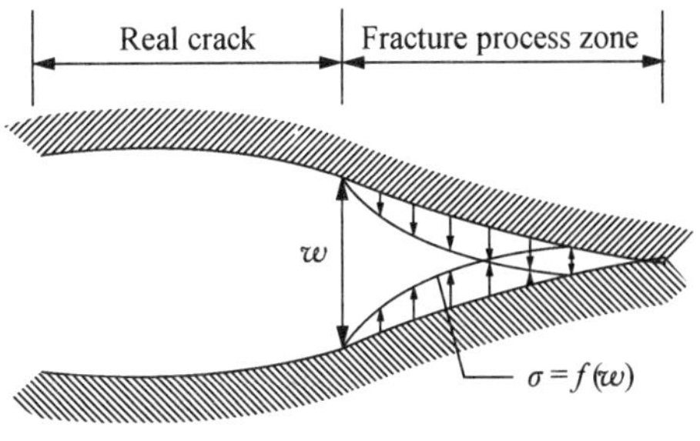

Fig. 2. Cohesive crack model

surfaces, the cohesive crack model is established(Fig. 2)[10-11].

As for the tensile fracture(Model-Ⅰ fracture), the cohesive stress can be deemed as the function of COD(crack opening displacement) as follows

$$\boldsymbol{T}=\boldsymbol{T}[\boldsymbol{w}(\boldsymbol{u})]. \tag{4}$$

In this paper, the following stress-COD relationship distribution[Fig. 3(a)] of the cohesive model is adopted, among various expressions[11]

$$f=f_t-kw, \tag{5}$$

Where $f=\boldsymbol{T}\cdot\boldsymbol{n}$ denotes the normal cohesive stress, $w=\boldsymbol{w}\cdot\boldsymbol{n}$ denotes the crack width, $\boldsymbol{n}$ is the normal unit vector, and k is the strength intensity factor.

Introduce the fracture energy as an intrinsic property of concrete, and specify its formulation as

$$G_c=\int_0^{w_1} f\mathrm{d}w, \tag{6}$$

where w_1 is the maximum width of cracks.

Apparently, for the f-w diagram shown in Fig. 3(a), G_c can be expressed as

$$\int_0^{w_1} f\mathrm{d}w=f_t\cdot w_1/2. \tag{7}$$

Hillerborg suggests that the stress-strain relationship of the cohesive element should be linear before the stress reaches the tensile strength. The ascending line is also linear which is determined by G_c after the tensile peak stress[Fig. 3(b)].

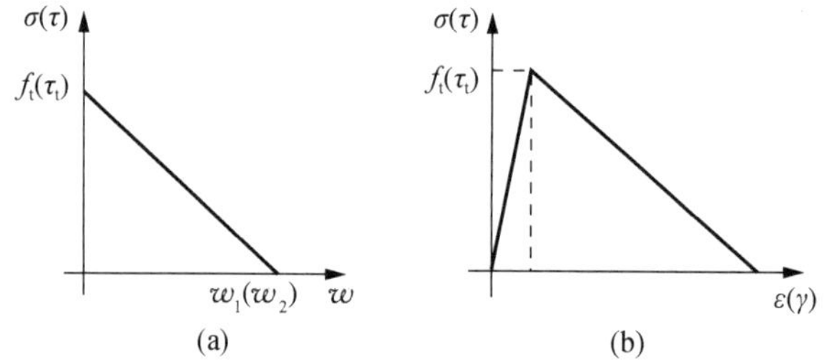

Fig. 3. The stress diagram of cohesive model. (a) Stress-crack width relationship; (b) stress-strain relationship of cohesive model

As for the shear fracture(Model-II fracture), the cohesive stress can be expressed as a function of CSD(crack shear displacement). Then the shear fracture energy G_s is introduced for Model-II fracture and the cohesive stress can be obtained as the Model-I fracture[13]. Although in this paper, the tensile fracture is critical to the failure of the concrete, the shear fracture is considered for the integrity of the model.

Substituting Eq.(4) into Eq.(3), we obtain the functional equation as

$$\int_\Omega \boldsymbol{\sigma}(\boldsymbol{u}):\boldsymbol{\varepsilon}(\boldsymbol{v})\mathrm{d}\Omega+\int_S \boldsymbol{T}[\boldsymbol{w}(\boldsymbol{u})]\cdot\boldsymbol{w}(\boldsymbol{v})\mathrm{d}\Gamma=\int_{\partial\Omega}\boldsymbol{t}\cdot\boldsymbol{v}\mathrm{d}\Gamma. \tag{8}$$

It could be noted in Eq.(8) that the presence of a cohesive surface results in the addition of a new term to the functional equation of the finite element[12]. Thus, the finite elements and the cohesive elements are compatible in a unified model.

2.2 Random cohesive model

As discussed in section 2.1, the cohesive elements may give the potential crack paths when connecting the finite elements. As we know, the cracks in concrete are highly irregular and randomly oriented. Herein, the irregular elements are developed to model the random initial micro cracks in the concrete. The irregular cohesive model[14] is established as follows.

(1) Define the boundary of the 2-D domain.

(2) Generate the set of random points in the domain and boundary.

(3) Decompose the domain by the Delaunay triangulation scheme.

(4) Contract the triangles to gain the cohesive elements that connect the adjacent finite elements.

Fig. 4. Generation of the irregular cohesive elements. (a) Random point set; (b) delauney triangulation; (c) finite elements and cohesive elements

3 Stochastic harmonic function and random field

When it comes to the classic fracture mechanics, the fracture energy is considered as the intrinsic property of the material. Admittedly, the deterministic description, which relies on the mean value of the fracture energy, simplifies the model as well as the simulation. However, the heterogeneity and the randomness that are induced by the randomly distributed micro-cracks cannot be taken into account in the deterministic simulation. What's more, the various cracking branches and the unpredictable failure modes cannot be well tackled either. Therefore, to define the concrete as a kind of random media would represent the essential randomness of the concrete[1]. In order to capture the random propagation of cracks and the failure modes, the fracture energy is described as a 2-D random field in the present work.

Chen and Li[15] originally developed the stochastic harmonic function for the simulation of random process, such as the time histories of strong ground motion and the wind. Based on the development of the stochastic harmonic function for spectrum representation, Liang et al.[16] extended it into the multi-dimensional random fields. The stochastic harmonic function method is introduced to characterizing the random field of fracture energy in the present paper. Without loss of generality, the 2-D, homogeneous Gaussian random field

$f_0(x, y)$ with $\mu=0$ and $\sigma=1$ can be represented as follows

$$f(x, y)=\sqrt{2}\sum_{n_1=1}^{N_1}\sum_{n_2=1}^{N_2}[A_{n_1n_2}\cos(K_{1n_1}x+K_{2n_2}y+\Phi_{n_1n_2}^{(1)})+\tilde{A}_{n_1n_2}\cos(K_{1n_1}x-K_{2n_2}y+\Phi_{n_1n_2}^{(2)})], \tag{9}$$

where $f(x, y)$ is the random field generated by the stochastic harmonic function; $A_{n_1n_2}$ and $\tilde{A}_{n_1n_2}$ denote the amplitudes of the n_1-th and n_2-th harmonic components; K_{1n_1} and K_{2n_2} indicate respectively the wave numbers. $\Phi_{n_1n_2}^{(1)}$, $\Phi_{n_1n_2}^{(2)}$, $n_1=1, 2, \cdots, N_1$, and $n_2=1, 2, \cdots, N_2$ are the independent random phases which uniformly distribute in the range $[0, 2\pi]$. K_{1n_1} and K_{2n_2} are the independent random wave numbers, which obey the uniform distribution in the ranges $(K_{1n_1-1}^{\mathrm{p}}, K_{1n_1}^{\mathrm{p}}]$ and $(K_{2n_2-1}^{\mathrm{p}}, K_{2n_2}^{\mathrm{p}}]$, respectively, i.e.,

$$p_{K_{jn_j}}(K_j)=\begin{cases}\dfrac{1}{K_{jn_j}^{\mathrm{p}}-K_{jn_j-1}^{\mathrm{p}}}=\dfrac{1}{\Delta K_{jn_j}}, & K\in(K_{jn_j-1}^{\mathrm{p}}, K_{jn_j}^{\mathrm{p}}], \quad j=1, 2,\\ 0, & \text{other},\end{cases} \tag{10}$$

$A_{n_1n_2}$ and $\tilde{A}_{n_1n_2}$ are determined by the following equations

$$A_{n_1n_2}=\sqrt{2S_{f_0f_0}(K_{1n_1}, K_{2n_2})\Delta K_{1n_1}\Delta K_{2n_2}}, \tag{11}$$

$$\tilde{A}_{n_1n_2}=\sqrt{2S_{f_0f_0}(K_{1n_1}, -K_{2n_2})\Delta K_{1n_1}\Delta K_{2n_2}}, \tag{12}$$

where $S_{f_0f_0}$ is the target spectrum density function. From Eqs. (11) and (12), it can be clearly observed that the random field could be fully determined by the target spectrum density function. It can be also proved that the power spectral density function of the stochastic harmonic function based random filed is equal to the target one. What's more, it has been proved that the stochastic harmonic random field asymptotically approaches the normal distribution.

According to Eqs. (5) and (8), the fracture energy will not only affect the tensile strength of the concrete but also influence the descending branch of the stress-strain relationship of the concrete. Therefore, the mean value, fluctuation and the correlation of fracture energy should be determined.

The correlation function[17] is chosen as

$$R(\tau_1, \tau_2)=\exp\left[-\left(\frac{\xi_1}{b_1}\right)^2-\left(\frac{\xi_2}{b_2}\right)^2\right], \tag{13}$$

where $\xi_1=x_2-x_1$ indicates the distance of x direction in cartesian coordinates; $\xi_2=y_2-y_1$ indicates the distance of y direction; b_1 and b_2 are the correlation length of x and y directions, respectively.

Using the Fourier transform, the corresponding power spectrum density function can be obtained as follows

$$S_{f_0f_0}(K_1, K_2)=\frac{b_1b_2}{4\pi}\exp\left[-\left(\frac{b_1K_1}{2}\right)^2-\left(\frac{b_2K_2}{2}\right)^2\right],$$

$$-K_{1u} \leqslant K_1 \leqslant K_{1u},\ -K_{2u} \leqslant K_2 \leqslant K_{2u}. \tag{14}$$

Substituting Eq.(14) into Eqs.(11) and (12), the random field of the fracture energy can be established by Eq.(9).

The random field $f_1(x, y)$ with μ_{G_c} and σ_{G_c} could be expressed as follows

$$f_1(x, y) = \mu_{G_c} + \sigma_{G_c}^2 f(x, y). \tag{15}$$

4 Tensile failure simulation of concrete

4.1 Numerical model

The simulations for the tensile failure of concrete were carried out based on the proposed methods in the former sections. The geometric dimensions of the numerical specimens were $b = 150$ mm and $h = 150$ mm. According to the test results obtained by Ren et al.[18], the material parameters were taken to be $E = 37\ 559$ MPa and $\nu = 0.2$. The mean value of the tensile strength of the concrete applied to both the finite elements and the cohesive elements was $f_t = 3.28$ MPa.

The random field of fracture energy was introduced based on the methods proposed in section 3. Due to the strong nonlinearities induced by the cracking process, we chose the explicit solution to get the temporal integration of the crack process. The numerical specimen was developed by the finite element package ABAQUS Explicit. Based on the previous irregular discretization method, more than 20 000 finite elements and 30 000 cohesive elements were generated. The numerical specimen and its boundary conditions are given in Fig. 5.

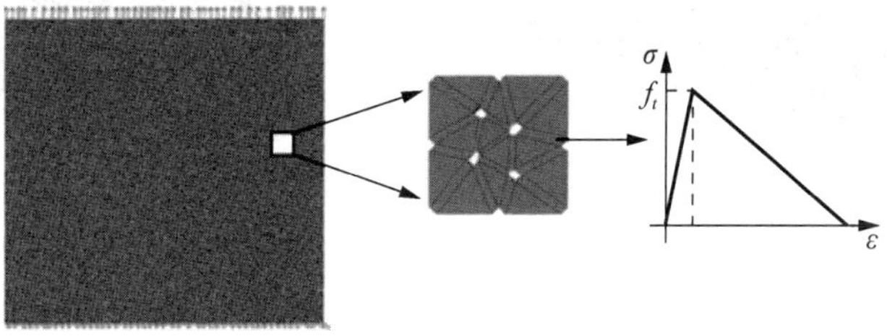

Fig. 5. The numerical model and the stress-strain relationship of cohesive elements

4.2 Random field model

Bazant and Pfeiffer[19] compared the notched and nonnotched beams to localize the heterogeneities of concrete. With the test results, the correlation length could be considered as the fracture influenced length governed by the heterogeneous of material. Thus in the present work, the correlation lengths in Eq.(13) were specified as $b_1 = b_2 = 3d_{max} = 24$ mm, where d_{max} is the maximum aggregate size.

Carpinteri and Chiaia[20-21] introduced the multifractal scaling law for fracture energy based on a series of compact tests and three-point bending tests. As for the cohesive crack model, the fracture energy is

$$G_c = G_c^{\infty}\left(1+\frac{l_{ch}}{b}\right)^{-1/2}, \tag{16}$$

$$l_{ch} = \alpha d_{max}, \tag{17}$$

where G_c^{∞} is the nominal asymptotic fracture energy valid within the limit of infinite structure size ($b \to \infty$), b is the size of the structure and is 150 mm in the simulation, l_{ch} is the characteristic length in determining the size dependent G_c along with the direction of the fracture process, and α is the non-dimensional parameter. The material parameters adopted in the present paper are $G_c^{\infty} = 160$ N/m and $\alpha = 30$.

Based on Eqs. (16) and (17) as well as the parameters, the mean value of the fracture energy was calculated as $\mu_{G_c} = 100$ N/m, and the standard deviation of the fracture energy was $\sigma_{G_c} = 0.1\,\mu_{G_c} = 10$ N/m. Thus, the fracture energy random field could be modeled by Eq. (15).

The target power spectrum density function and the power spectrum density function based on the stochastic harmonic function are plotted in Figs. 6 and 7, respectively. In addition, the comparison between these two PSDs is depicted in Fig. 8 at given wave numbers. The samples of the random field are shown in Fig. 9.

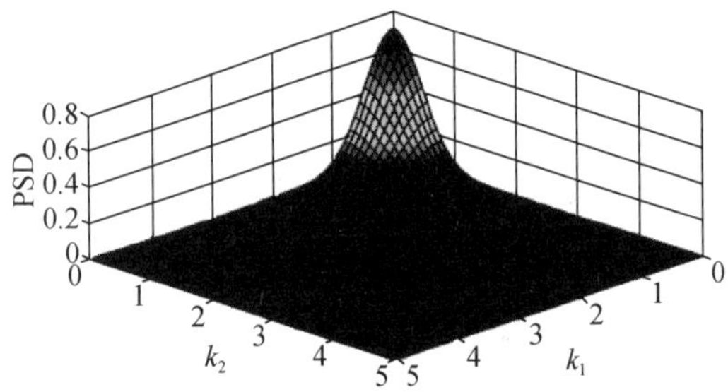

Fig. 6. Target power spectral density function

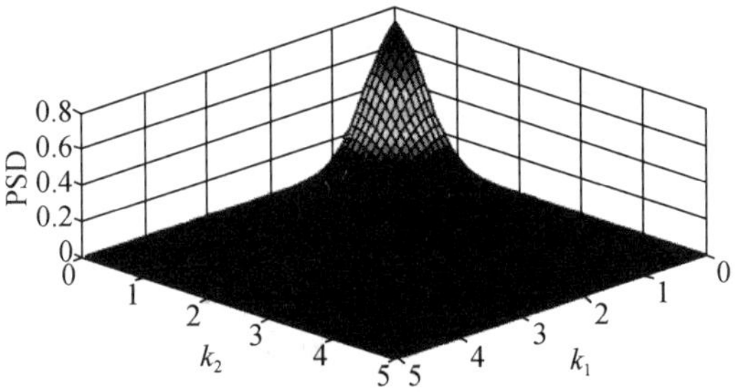

Fig. 7. Power spectral density function by stochastic harmonic function

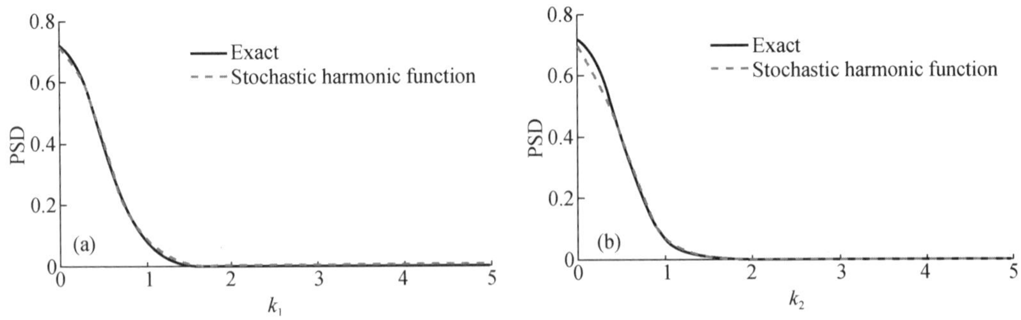

Fig. 8. Power spectral density function. (a) $K_2 = 0$; (b) $K_1 = 0$

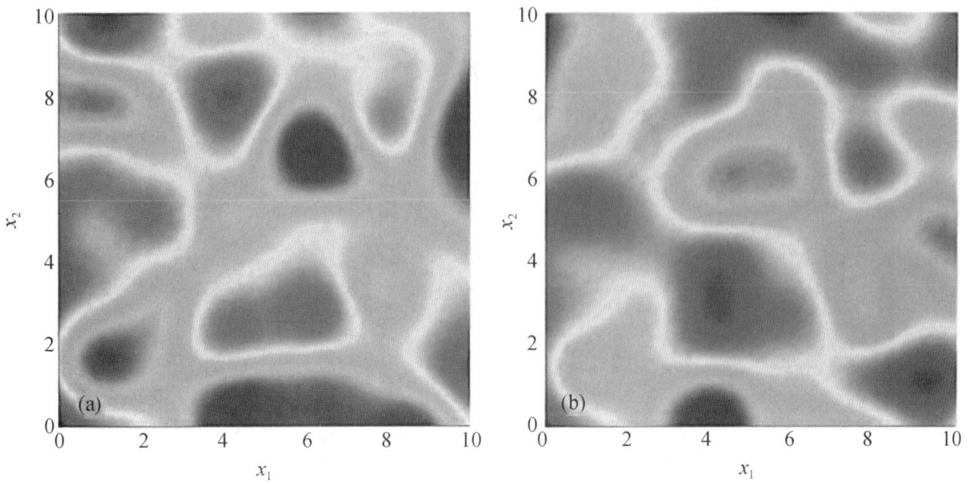

Fig. 9. Samples by stochastic harmonic function for fracture energy. (a) Sample 1; (b) sample 2

4.3 Failure simulation

As is shown in Figs. 10 and 11, the overall procedure of the nonlinearity and the cracks can be observed as follows: At the very beginning, the stress and strain both increase linearly;

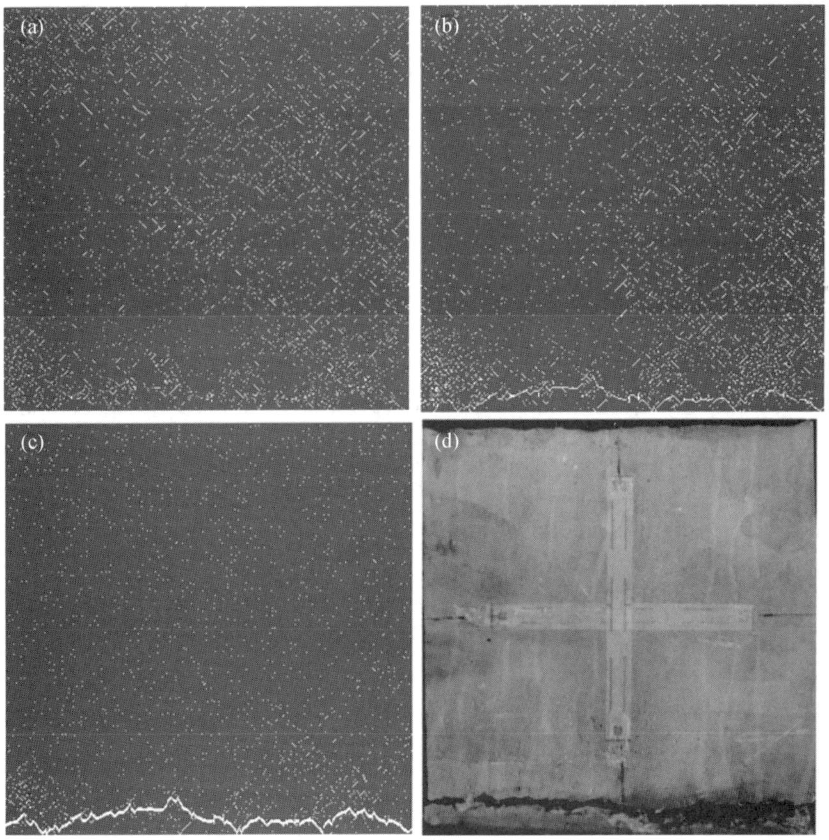

Fig. 10. Simulation and test result of end-crack failure (ε_u ultimate tensile strain). (a) $\varepsilon = 0.2\varepsilon_u$; (b) $\varepsilon = 0.7\varepsilon_u$; (c) $\varepsilon = \varepsilon_u$; (d) test result

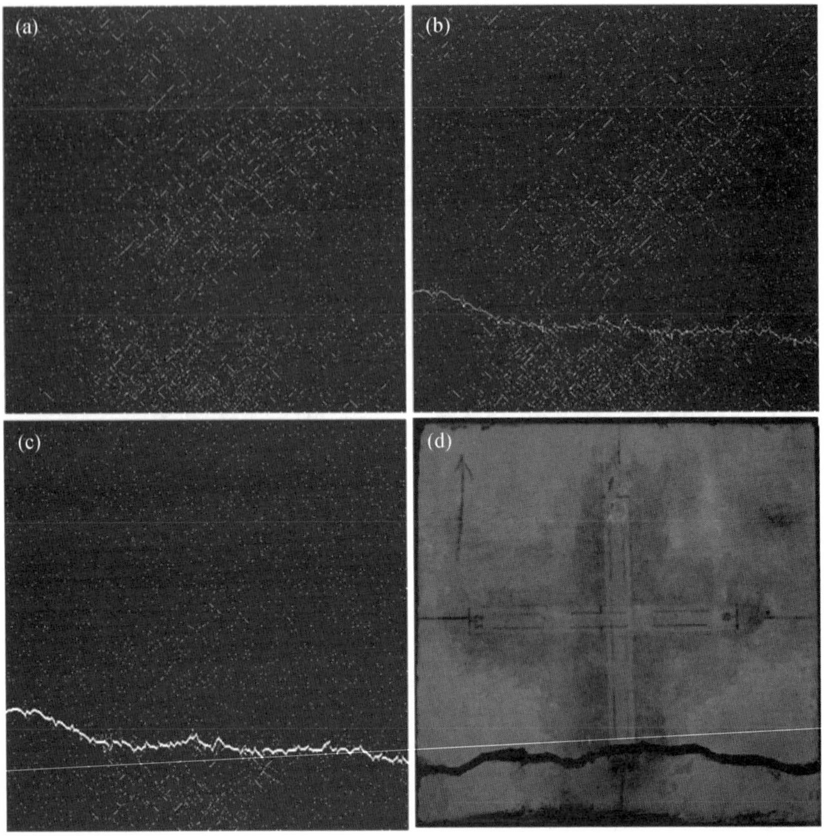

Fig. 11. Simulation and test result of non-end-crack failure (ε_u ultimate tensile strain). (a) $\varepsilon=0.2\varepsilon_u$; (b) $\varepsilon=0.7\varepsilon_u$; (c) $\varepsilon=\varepsilon_u$; (d) test result

when the loading increases, there are micro-cracks uniformly distributed in the whole specimen; then, with further increasing of the loading, the micro cracks concentrate on a certain area and the stress concentration happens on the tips of the cracks; at the final stage, a severe strain concentration occurs on the cracking zones and a main crack which is perpendicular to the tensile loading cuts through the specimen of the concrete. Fig. 10 indicates the comparison between the numerical experiment and the test result of the end-crack failure of the concrete under tensile test, and Figure 11 indicates the non-end-crack failure as well.

5 Probability density analysis of the stress-strain relationship

When taking a close look at the concrete based on the random medium model, the most accurate way to describe the random stress-strain relationship is the probability density function. Thus, the probability density evolution method[22] can be applied. Without loss of generality, the monotonic stress-strain relationship is governed by the following equation

$$\sigma=f(\Theta, \varepsilon), \tag{18}$$

where $\Theta=(\theta_1, \theta_2, \cdots, \theta_n)$ is the random parameter vector, whose joint probability density function is $p_\Theta(\theta)$. When it comes to the fracture energy random field by Eq. (9), Θ is composed of the random wave numbers K_{1n_1} and K_{2n_2} and the random phase angles $\Phi^{(1)}_{n_1n_2}$ and $\Phi^{(2)}_{n_1n_2}$.

To replace the generalized time parameter t by ε, the probability density evolution equation[23] is

$$\frac{\partial p_{\sigma\Theta}(\sigma, \theta, \varepsilon)}{\partial \varepsilon}+\dot{\sigma}(\theta, \varepsilon)\frac{\partial p_{\sigma\Theta}(\sigma, \theta, \varepsilon)}{\partial \sigma}=0, \tag{19}$$

Where $\dot{\sigma}$ denotes the first-order derivative of the stress.

Eq.(19) can be solved by combining the initial conditions, the simulations and the finite-difference method. The solving procedures are as follows.

(1) Select the representative points of the random wave numbers K_{1n_1}, K_{2n_2} and the random phase angles, $\Phi^{(1)}_{n_1n_2}$, $\Phi^{(2)}_{n_1n_2}$. The selected points are $\theta_q (q=1, 2, \cdots, N_{sel})$, where N_{sel} is the total number of the selected points.

(2) Simulation based on the proposed cohesive model is done for each representative point.

(3) The finite-difference computation method[24] is used to get the joint probability density function.

(4) Take numerical integration with respect to θ_q to acquire the numerical PDF.

In this paper, the quasi-symmetric method[25] was chosen in selecting the representative points and the number of random variables was 32. The total number of the representative points was 300. As is illustrated in Figure 12, the comparison between the simulating results and the test results suggests the validation of the model. Fig.13 depicts the probability density function of the stress-strain relationship of the concrete. Fig. 14 illustrates the comparison between the experimental results and the theoretical results of the stress probability density function at certain strain.

It can be noted from Eq. (19) that the probability density evolution of the stress will transfer in the whole loading process and will be affected by the random parameters and the

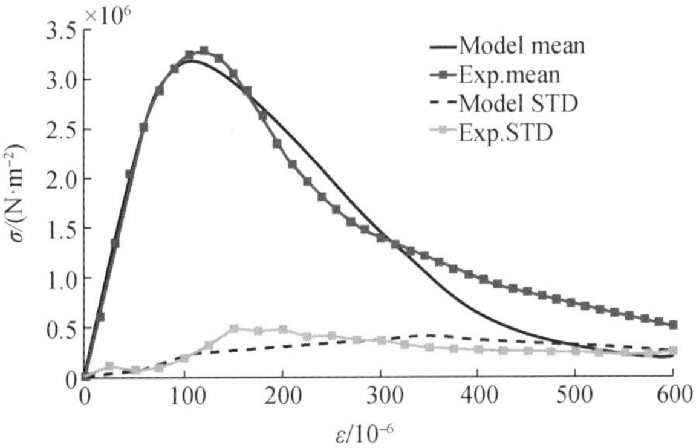

Fig. 12. Mean and standard deviation curves for stress

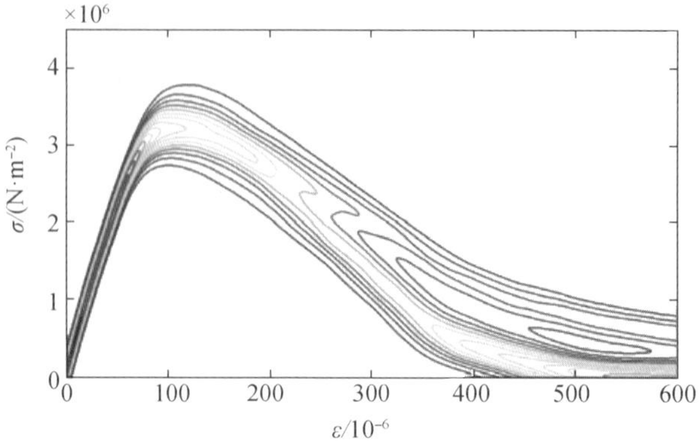

Fig. 13. Probability density functions of stress-strain curve

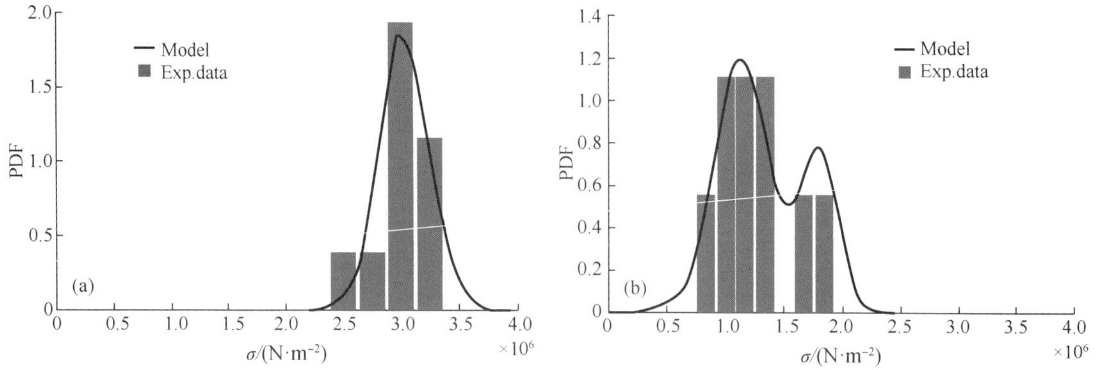

Fig. 14. Comparison of the probability density functions for the experimental and theoretical results. (a) $\varepsilon=0.000\,1$; (b) $\varepsilon=0.000\,3$

complexity of the model. To the best knowledge of the authors, the probability flow of the stress in the numerical model will lead to the further random cracks and the failure modes. As for the simulation in this paper, there were 187 end-crack failures and 113 non-end-crack failures of the numerical samples. It can be observed in Figs. 13 and 14 that the probability density of the stress changed from one peak to two peaks with the increasing of strain due to the random failure modes. Thus, the probability density of the stress-strain relationship can represent the stochastic features of concrete and reveal the essential transfer of probability density.

6 Conclusion

In this paper, the failure simulation of the concrete based on the random medium model is presented. The randomness of concrete is considered in two aspects. On the one hand, the random micro cracks are introduced by the irregular finite elements and cohesive elements. On the other hand, the stochastic harmonic function is adopted to model the 2-D random field of fracture energy. Then, simulations of the cracking process and the failure modes are given to testify the proposed model in the sample level. In addition, the probability distribution evolution method is introduced to clarify the probability density distributions of the stress-strain relationship in the collection level. With the efforts of the present work, a new random medium based model for the accurate simulation of concrete is developed.

This work was supported by the National Natural Science Foundation of China (Grant Nos. 90715033, 51261120374, 51208374).

References

[1] LI J. Research on the stochastic damage mechanics for concrete materials and structures[J]. Journal of Tongji University, 2004, 32: 1270-1277.

[2] MAZARS J. A description of micro-and macroscale damage of concrete structures[J]. Engineering Fracture Mechanics, 1986, 25(5-6): 729-737.

[3] SIMO J C, JU J W. Strain-and stress-based continuum damage models— I. Formulation[J]. Int J Solids Struct, 1987, 23(7): 821-840.

[4] LUBLINER J, OLIVER J, OLLER S, et al. A plastic-damage model for concrete[J]. International Journal of Solids and Structures, 1989, 25(3): 299-326.

[5] WU J Y, LI J, FARIA R. An energy release rate-based plastic-damage model for concrete[J]. International Journal of Solids and Structures, 2006, 43(3-4): 583-612.

[6] NEMAT-NASSER S, HORI M. Micromechanics: Overall Properties of Heterogeneous Materials[M]. North-Holland: Elsevier Amsterdam, 1999.

[7] GROSS D, SEELIG T. Fracture mechanics: with an introduction to micromechanics[M]. Heidelberg: Springer, 2011.

[8] MOËS N, BELYTSCHKO T. Extended finite element method for cohesive crack growth[J]. Eng Fract Mech, 2002, 69(7): 813-833.

[9] DUGDALE D S. Yielding of steel sheets containing slits[J]. J Mech Phys Solids, 1960, 8(2): 100-104.

[10] BARENBLATT G I. The mathematical theory of equilibrium cracks in brittle fracture[J]. Adv Appl Mech, 1962, 7: 55-129.

[11] HILLERBORG A, MODEER M, PETERSSON P E. Analysis of crack formation and crack growth in concrete by means of fracture mechanics and finite elements[J]. Cement Concr Res, 1976, 6(6): 773-781.

[12] XU X P, NEEDLEMAN A. Numerical simulations of fast crack growth in brittle solids[J]. J Mech Phys Solids, 1994, 42(9): 1397-1434.

[13] ROTS J G, DE BORST R. Analysis of mixed-mode fracture in concrete[J]. J Eng Mech, 1987, 113 (11): 1739-1758.

[14] REN X, LI J. Dynamic fracture in irregularly structured systems[J]. Phys Rev E, 2012, 85(5): 55102.

[15] CHEN J B, LI J. Stochastic harmonic function and spectrum representations[J]. Chinese J Theo Appl Mech, 2011, 43(3): 505-513.

[16] LIANG S, SUN W, LI J. Simulation of multi-dimensional random fields by stochastic harmonic functions[J]. Journal of Tongji University. Natural Science, 2012, 40(7):965-970.

[17] XU X F, GRAHAM-BRADY L. A stochastic computational method for evaluation of global and local behavior of random elastic media[J]. Comput Method Appl M, 2005, 194(42): 4362-4385.

[18] REN X D, YANG W Z, ZHOU Y, et al. Behavior of high-performance concrete under uniaxial and biaxial loading[J]. Aci Mater J, 2009, 105(6): 548-557.

[19] BAZANT Z P, PFEIFFER P A. Determination of fracture energy from size effect and brittleness number[J]. Aci Mater J, 1987, 84(6): 755-767.

[20] CARPINTERI A, CHIAIA B. Size effects on concrete fracture energy: Dimensional transition from order to disorder[J]. Mater Struct, 1996, 29(5): 259-266.

[21] CARPINTERI A, CHIAIA B, Ferro G. Size effects on nominal tensile strength of concrete structures: Multifractality of material ligaments and dimensional transition from order to disorder[J]. Mater Struct, 1995, 28(6): 311-317.

[22] LI J, CHEN J B. Probability density evolution method for analysis of stochastic structural dynamic response[J]. Chinese J Theo Appl Mech, 2003, 35(4): 437-442.

[23] ZENG S J, LI J. Analysis on constitutive law of plain concrete subjected to uniaxial compressive stress based on generalized probability density evolution method[J]. J Tongji Univ (Nat Sci), 2010, 38(6): 798-804.

[24] CHEN J B, LI J. Difference method for probability density evolution equation of stochastic structural response[J]. Chinese Quart Mech, 2005, 25(1): 22-28.

[25] LI J, XU J, CHEN J B. The use of quasi-symmetric point method in probability density evolution theory[J]. J Wuhan Tech Univ, 2010, 32(9): 1-5.

Part 2

Continuum Damage Mechanics

An energy release rate-based plastic-damage model for concrete①

Jianying Wu[1], **Jie Li**[2], **Rui Faria**[3]

1 *Department of Civil Engineering, College of Architecture & Civil Engineering, South China University of Technology, Guangzhou 200092, China*

2 *Department of Building Engineering, Tongji University, Shanghai 200092, China*

3 *Faculdade de Engenharia da Universidade do Porto, Civil Engineering Department, Rua Dr. Roberto Frias s/n, 4200-465 Porto, Portugal*

Abstract A new plastic-damage constitutive model for concrete is proposed in this paper. A tensile and a shear damage variable are adopted to describe the degradation of the macromechanical properties of concrete. Within the framework of continuum damage mechanics, the elastic Helmholtz free energy is defined to establish the plastic-damage constitutive relation with the internal variables. Regarding the specific format for the effective stress space plasticity, the evolution law for the plastic strains and the explicit expression for the elastoplastic Helmholtz free energy are determined and the damage energy release rates that are conjugated to the damage variables are derived. Thus, damage energy release rate-based damage criteria can be established in conformity to thermodynamical principles. In accordance with the normality rule, evolution laws for the damage variables are obtained to complete the proposed plastic-damage model. Some computational aspects concerning the numerical algorithm implementation are discussed as well. Several numerical simulations are presented at the end of the paper, whose results allow for validating the capability of the proposed model for reproducing the typical nonlinear performances of concrete structures under different monotonic and cyclic load conditions.

Keywords Continuum damage mechanics; Damage criteria; Plastic deformations; Concrete; Thermodynamics

1 Introduction

Despite the noteworthy recent research efforts and contributions, the task of nonlinear finite element analysis (NFEA) of concrete structures is still quite challenging. The accurate modeling of concrete performances, the key issue to NFEA, remains a somewhat controversial matter.

The most commonly used theories for the modeling of concrete are plasticity, fracture-based approaches and continuum damage mechanics (CDM). Plasticity, which has been successfully applied to metals, is nowadays theoretically consolidated and world-widely

① Originally published in *International Journal of Solids and Structures*, 2006, 43(3-4): 583-612.

recognized as computationally efficient. Examples of its application to concrete can be found in Ohtani and Chen[1], Etse and Willam[2] and Feenstra and de Borst[3], as well as in Chen[4] reviews and references therein. Though the plasticity models are far superior to elastic approaches in representing hardening and softening characteristics, they fail to address the process of damage due to microcracks growth, such as the stiffness degradation, the unilateral effect, etc.

Intermingled with classical continuum mechanics in uncoupled manner, fracture mechanics suggests an approach to describe localized damage as to be represented by the ideal or regular discrete cracks with definite geometries and locations, and it has been extensively used in engineering practice[5]. However, the associated questions whether the J integrals and stress intensity factors are material parameters or not are far from being settled[6]. Besides, before the appearance of macrocracks in concrete there exist a lot of smeared microcracks whose geometries and locations could not be determined precisely, so it appears difficult to apply fracture mechanics for modeling concrete.

Based on the thermodynamics of irreversible processes, the internal state variable theory and relevant physical considerations[7], CDM provides a powerful and general framework for the derivation of consistent constitutive models suitable for many engineering materials, including concrete. In the earlier literature[8-10], CDM was restricted to linear "elastic-damage" mechanics for brittle materials, i.e., linear elastic solids with distributed microcracks, affording different ways for handling observed phenomena like the stiffness degradation, the tensile softening and the unilateral effect due to the development of microcracks and microvoids. More examples of elastic CDM models for concrete can be found in Lubarda et al.[11], Cervera et al.[12], Halm and Dragon[13], Comi and Perego[14], etc. Coupled with plasticity or by means of empirical definitions, the irreversible strains due to plastic flow can also be accounted for in elastoplastic damage theories, such as those proposed by Ortiz[15], Lemaitre[16], Dragon[17], Resende[18], Simo and Ju[19], Yazdani and Schreyer[6], Carol et al.[20], di Prisco and Mazars[21], Lee and Fenves[22], Faria et al.[23], Hatzigeorgioiu et al.[24], Hansen et al.[25] among others.

One of the critical issues associated with damage model of concrete is selection of damage criteria. Several different criteria, such as the equivalent strain-based[8], and the stress-based approaches[15, 26] as well as the damage energy release rate-based (DERR-based) proposals[7, 9, 19], are generally adopted in the current practice. Among them, it has been pointed out that[7] the equivalent strain-based damage criteria are only appropriate for the elastic state, and the stress-based damage criteria are inherently inadequate for predicting perfect plastic damage growth. As is well known, in classical plasticity the stress tensor is thermodynamically conjugated to the plastic strain tensor, so the yield criteria are defined as functions of the stress tensor invariants. Analogously thermodynamically consistent damage criteria might be based on DERRs that are the conjugated forces to the damage variables.

However, most of the above referred damage models, except in Ju[7], are based on the

Helmholtz free energy(HFE) and the DERR proposed by Lemaitre[16], where contribution of the plastic strains to the microcrack growth process is not considered, and the driving force of damage growth is the elastic DERR alone, which might not be physically appropriate. Therefore, if the damage criteria are based on the elastic DERRs, the strength enhancement observed in the compression-compression domain[27] cannot be predicted(see the unilateral damage model of Mazars[9]). To deal with above problem, Ju[7] suggested an elastoplastic one-scalar damage model in which both the plastic strains and the plastic HFE were defined based on the effective stress space plasticity to cope with the coupled effects between the damage evolutions and the plastic flow. However, only under very simple situations such as Von Mises plasticity with linear isotropic hardening, the plastic HFE was provided explicitly in the model, and for other plasticity models more appropriate for concrete, it could only be computed incrementally with numerical integration, which might be time-consuming and unsteady. Moreover, though the model is thermodynamically more reasonable, but only under uniaxial compression its predicted results might agree with experimental observations; and no examples for concrete under more complex stress states, e.g., biaxial compression, were provided yet.

To improve the fitting of the model predictions to the concrete experimental data, other researchers[14, 23] abandoned the DERR-based damage criteria, and turned to the empirically defined ones. In spite of the satisfactory results obtained under pure tension and pure compression stress states, in the tension-compression domain the model of Faria et al.[23] neglects the influence on the compressive strength in one direction due to the tensile stress or strain acting perpendicularly and vice versa, which is obviously conflicting with the experimentally observed strength decays[27-28]. In Comi and Perego[14] a pure elastic-damage model is proposed, where the damage criteria were described by a hyperbola for the tensile stress domains and an ellipse for the pure compressive stress quadrants. Many parameters with not very clear physical meaning are required to be determined in this model, and only under monotonic loadings the numerical results fit the experimental test data adequately.

With inspiration from all the previous works and understandings mentioned above, this paper aims to present a novel rate-independent plastic-damage model for concrete that should be consistent with thermodynamics and fit well with the concrete experimental observations, including the stiffness degradation, the enhancement of strength and ductility under compressive confinement, the strength decay induced by orthogonal tensile cracking and the unilateral effect under cyclic loading. The paper is organized as follows.

In Section 2 the peculiarities of the experimentally observed behaviour of concrete are reviewed. A tensile damage variable and a shear damage variable leading to a fourth-order damage tensor, are chosen to describe the degradation of the macromechanical properties of concrete. A decomposition of the effective stress tensor is presented thereafter to define an elastic HFE, after which the plastic-damage constitutive relation with internal variables is derived. Section 3 is devoted to the evolution laws for the internal variables. The evolution law for plastic strains is obtained and then the plastic HFE of concrete is determined. The elastic and plastic components of HFE are added up to the total elastoplastic one, from which the elastoplastic DERRs conjugated to the damage variables are derived to establish the damage criteria. Subsequently, in accordance with the normality rule, the evolution laws for the damage variables are obtained. Pertinent computational aspects concerning the numerical implementation and the algorithmic consistent tangent modulus for the constitutive model are presented in Section 4. In Section 5 some numerical applications of the model to experimental tests of concrete specimens and structures under different loadings are provided to validate and demonstrate the capability of the proposed model. Section 6 closes this paper with the most relevant conclusions.

2 Elastic-damage model

2.1 Damage mechanisms

To identify the basic mechanisms that lead to concrete damage growth, some primary features of concrete behaviour experimentally observed in tests are presented first.

When loaded in compression concrete experiences progressive loss of stiffness in the deviatoric space along with observed volumetric changes when approaching collapse putting into evidence strong dilatancy. The available experimental results in tension indicate a deviatoric behaviour similar to that observed in compression, but the volumetric performance, unlike in compression, is of a softening nature. Finally, under hydrostatic compression concrete exhibits somewhat softening behaviour at low stress levels, but as stress increases a predominantly stiffening behaviour is observed; upon unloading, concrete behaves almost elastically at first, losing stiffness progressively afterwards[18].

Consequently, three damage mechanisms can be consensually identified from the experimental observations, which are: ① a tensile one, which represents the separation of material particles leading to cracks developed predominantly in fracture mode I; ② a shear one, inherent to mode II, associated to the breaking of internal bonds during the loading of concrete in shear; and finally ③ a compressive consolidation one, due to collapse of the microporous structure of cement matrix under triaxial compression.

From these damage mechanisms it is possible to conclude that:

(1) Loaded in tension concrete damage is activated as the result of both the tensile and shear mechanisms in the deviatoric space. In general the former one develops much quicker than the latter one[18], and thus the shear damage mechanism can be ignored for convenience. In the volumetric space the damage of concrete is activated by the tensile damage mechanism.

(2) Loaded in compression concrete damage is activated by the shear damage mechanism. The compressive consolidation mechanism of concrete in the volumetric space will be partially taken into consideration via minor modification of the shear damage criteria, as it

will be further discussed in Section 3 of the paper.

Thus, two basic but distinct mechanisms, namely the tensile and shear damage ones, respectively, are consistently referred to in the derivation of the constitutive model to be presented herein.

2.2 Types of damage models

The effectiveness and performance of a damage model depends heavily on its particular choice of a set of damage variables, which actually serves as a macroscopic approximation for describing the underlying micromechanical process of microcracking. In current literature, there are many ways to define appropriate damage variables which are phenomenologically defined or micromechanically derived. In this paper the attention is restricted to the phenomenological approach. According to the damage variables adopted, the CDM models can be classified essentially in two categories: ① the scalar damage ones, where one or several scalars are adopted to characterize the isotropic damage processes[8, 22-24]; ② the tensor damage ones, where second-order, fourth-order or even eighth-order damage tensor are necessary to account for anisotropic damage effects[6, 15, 17, 29].

The one-scalar damage models[7-8, 19] are somewhat limited to describe the unilateral effect inherent to concrete behaviour. Moreover, Poisson's ratio is inferred to be a constant in this kind of damage model, which is inconsistent with the phenomena that Poisson's ratio decrease under tensile load and increase under compressive load due to microcracking. In fact, even for isotropic damage, not a one-scalar damage variable but at least a fourth-order damage tensor should be employed to characterize the state of damage in materials[30]. However, the inherent complexities of numerical algorithms required by most of the anisotropic tensor damage models restrict their applicability to practical engineering.

Based on the considerations about the damage mechanisms discussed in Section 2.1, a tensile damage scalar d^+ and a shear damage scalar d^- will be adopted here to describe the degradation of the concrete macromechanical properties under tension and compression, respectively. As it will be shown, these two damage scalars will lead to a fourth-order damage tensor, which agrees with the conclusion drawn in Ju[30].

2.3 Decomposition of the effective stress tensor

In CDM the effective stress $\bar{\boldsymbol{\sigma}}$ in damaged material may be assumed to follow the classical elastoplastic behaviour[7, 23], i.e.

$$\bar{\boldsymbol{\sigma}} = \mathbf{C}_0 : \boldsymbol{\varepsilon}^e = \mathbf{C}_0 : (\boldsymbol{\varepsilon} - \boldsymbol{\varepsilon}^p) \tag{1}$$

or equivalently

$$\boldsymbol{\varepsilon}^e = \boldsymbol{\Lambda}_0 : \bar{\boldsymbol{\sigma}} \tag{2}$$

where $\mathbf{C}_0$ and $\boldsymbol{\Lambda}_0 = \mathbf{C}_0^{-1}$ denote the usual fourth-order isotropic linear-elastic stiffness and compliance tensors, respectively; $\boldsymbol{\varepsilon}$, $\boldsymbol{\varepsilon}^e$ and $\boldsymbol{\varepsilon}^p$ are rank-two tensors, denoting the strain tensor, its elastic and plastic tensor components.

To account for the different nonlinear performances of concrete under tension and compression, and to explicitly reproduce the dissimilar effects of the tensile and shear damage mechanisms, a decomposition of the effective stress tensor $\bar{\boldsymbol{\sigma}}$ into positive and negative components ($\bar{\boldsymbol{\sigma}}^{+}$, $\bar{\boldsymbol{\sigma}}^{-}$) is performed as[7, 15, 23]

$$\bar{\boldsymbol{\sigma}}^{+}=\mathbf{P}^{+}:\bar{\boldsymbol{\sigma}} \tag{3a}$$

$$\bar{\boldsymbol{\sigma}}^{-}=\bar{\boldsymbol{\sigma}}-\bar{\boldsymbol{\sigma}}^{+}=\mathbf{P}^{-}:\bar{\boldsymbol{\sigma}} \tag{3b}$$

with the fourth-order projection tensors $\mathbf{P}^{+}$ and $\mathbf{P}^{-}$ expressed as[31]

$$\mathbf{P}^{+}=\sum_{i}H(\bar{\sigma}_{i})(\mathbf{p}_{ii}\otimes\mathbf{p}_{ii}) \tag{4a}$$

$$\mathbf{P}^{-}=\mathbf{I}-\mathbf{P}^{+} \tag{4b}$$

where $\mathbf{I}$ is the fourth-order identity tensor; $H(\bar{\sigma}_i)$ denotes the Heaviside function computed for the ith eigenvalue $\bar{\sigma}_i$, of $\bar{\boldsymbol{\sigma}}$; and the second-order tensor $\mathbf{p}_{ii}$ will be defined in Eq.(7). In order to express the rate format of the proposed model and to facilitate the derivation of the algorithmic consistent modulus, it is also necessary to split $\dot{\bar{\boldsymbol{\sigma}}}$ into its positive and negative components ($\dot{\bar{\boldsymbol{\sigma}}}^{+}$, $\dot{\bar{\boldsymbol{\sigma}}}^{-}$) as

$$\dot{\bar{\boldsymbol{\sigma}}}^{+}=\frac{\mathrm{d}}{\mathrm{d}t}(\mathbf{P}^{+}:\bar{\boldsymbol{\sigma}})=\mathbf{Q}^{+}:\dot{\bar{\boldsymbol{\sigma}}} \tag{5a}$$

$$\dot{\bar{\boldsymbol{\sigma}}}^{-}=\frac{\mathrm{d}}{\mathrm{d}t}(\mathbf{P}^{-}:\bar{\boldsymbol{\sigma}})=\mathbf{Q}^{-}:\dot{\bar{\boldsymbol{\sigma}}} \tag{5b}$$

where the fourth-order tensors $\mathbf{Q}^{+}$ and $\mathbf{Q}^{-}$ are the corresponding projection tensors of $\dot{\bar{\boldsymbol{\sigma}}}$, and due to the nonlinear nature of $\mathbf{P}^{+}$ and $\mathbf{P}^{-}$, a somewhat more complicated procedure (refer to Faria et al.[31] and see Appendix I for more discussions) is necessary to derive the expressions for them, leading to

$$\mathbf{Q}^{+}=\mathbf{P}^{+}+2\sum_{i=1,\,j>i}^{3}\frac{\langle\bar{\sigma}_i\rangle-\langle\bar{\sigma}_j\rangle}{\bar{\sigma}_i-\bar{\sigma}_j}\mathbf{p}_{ij}\otimes\mathbf{p}_{ij} \tag{6a}$$

$$\mathbf{Q}^{-}=\mathbf{I}-\mathbf{Q}^{+} \tag{6b}$$

To be used in Eqs.(4a) and (6), the second-order symmetric tensor $\mathbf{p}_{ij}$ should be defined as (see Faria et al. for details[31])

$$\mathbf{p}_{ij}=\mathbf{p}_{ji}=\frac{1}{2}(\mathbf{n}_i\otimes\mathbf{n}_j+\mathbf{n}_j\otimes\mathbf{n}_i) \tag{7}$$

where $\mathbf{n}_i$ is the ith normalized eigenvector corresponding to $\bar{\sigma}_i$; symbols $\langle\cdot\rangle$ are the McAuley brackets (ramp function) defined as

$$\langle x\rangle=(x+|x|)/2 \tag{8}$$

2.4 Elastic HFE potential

To establish the intended constitutive law, an elastic HFE potential should be introduced as function of the free and internal variables. The initial elastic HFE potential ψ_0^e is here defined as the elastic strain energy. It can be written as the addition of its positive and negative components (ψ_0^{e+}, ψ_0^{e-}) with the decomposition of the effective stress tensor given in Eq.(3)

$$\psi_0^e(\boldsymbol{\varepsilon}^e)=\frac{1}{2}\bar{\sigma}:\boldsymbol{\varepsilon}^e=\frac{1}{2}\bar{\boldsymbol{\sigma}}^+:\boldsymbol{\varepsilon}^e+\frac{1}{2}\bar{\boldsymbol{\sigma}}^-:\boldsymbol{\varepsilon}^e=\psi_0^{e+}(\boldsymbol{\varepsilon}^e)+\psi_0^{e-}(\varepsilon^e) \tag{9}$$

where superscript "e" refers to "elastic" and subscript "0" refers to "initial" states. ψ_0^{e+} and ψ_0^{e-} are further expressed as

$$\psi_0^{e\pm}(\boldsymbol{\varepsilon}^e)=\frac{1}{2}\bar{\boldsymbol{\sigma}}^\pm:\boldsymbol{\varepsilon}^e=\frac{1}{2}\bar{\boldsymbol{\sigma}}:(\mathbf{P}^\pm:\boldsymbol{\Lambda}_0):\bar{\boldsymbol{\sigma}}=\frac{1}{2}\boldsymbol{\varepsilon}^e:(\mathbf{C}_0:\mathbf{P}^\pm):\boldsymbol{\varepsilon}^e \tag{10}$$

with symbol "±" denoting "+" or "−", as appropriate.

Considering the tensile and shear damage mechanisms mentioned in Section 2.1, an elastic HFE potential with the form

$$\psi^e(\boldsymbol{\varepsilon}^e, d^+, d^-)=\psi^{e+}(\boldsymbol{\varepsilon}^e, d^+)+\psi^{e-}(\boldsymbol{\varepsilon}^e, d^-) \tag{11}$$

can be postulated, with ψ^{e+} and ψ^{e-} defined as

$$\psi^{e\pm}(\boldsymbol{\varepsilon}^e, d^\pm)=(1-d^\pm)\psi_0^{e\pm}(\boldsymbol{\varepsilon}^e) \tag{12}$$

2.5 Constitutive law and dissipation

In general, the total elastoplastic HFE potential can be defined as the sum of the elastic component ψ^e defined in Eqs.(10)—(12), and plastic component ψ^p to be defined in Section 3.2, assumed uncoupled in the following manner

$$\psi(\boldsymbol{\varepsilon}^e, \boldsymbol{\kappa}, d^+, d^-)=\psi^e(\boldsymbol{\varepsilon}^e, d^+, d^-)+\psi^p(\boldsymbol{\kappa}, d^-) \tag{13}$$

where $\boldsymbol{\kappa}$ denotes a suitable set of plastic variables to be discussed in the next section.

On a purely isothermal mechanical process, the second principle of thermodynamics[32] states that any irreversible process should satisfy the Clausius-Duheim inequality, whose reduced form is

$$\dot{\gamma}=-\dot{\psi}+\boldsymbol{\sigma}:\dot{\boldsymbol{\varepsilon}}\geqslant 0 \tag{14}$$

By differentiating Eq.(13) with respect to time, one gets

$$\dot{\psi}=\frac{\partial\psi^e}{\partial\boldsymbol{\varepsilon}^e}:\dot{\boldsymbol{\varepsilon}}^e+\frac{\partial\psi}{\partial d^+}\dot{d}^++\frac{\partial\psi}{\partial d^-}\dot{d}^-+\frac{\partial\psi^p}{\partial\boldsymbol{\kappa}}\cdot\dot{\boldsymbol{\kappa}} \tag{15}$$

Referring to the standard thermodynamics arguments[33], along with the assumption that damage and plastic unloading are elastic processes, for any admissible process the following conditions have to be fulfilled

$$\boldsymbol{\sigma}=\frac{\partial\psi^e}{\partial\boldsymbol{\varepsilon}^e} \tag{16}$$

$$\dot{\gamma}^d = -\left(\frac{\partial \psi}{\partial d^+}\dot{d}^+ + \frac{\partial \psi}{\partial d^-}\dot{d}^-\right) \geqslant 0 \tag{17}$$

$$\dot{\gamma}^p = \boldsymbol{\sigma} : \dot{\boldsymbol{\varepsilon}}^p - \frac{\partial \psi^p}{\partial \boldsymbol{\kappa}} \cdot \dot{\boldsymbol{\kappa}} \geqslant 0 \tag{18}$$

From Eq.(16) it can be clearly seen that the Cauchy stress tensor is only dependent on the elastic HFE potential, which is a variant to Faria et al.[23] model where the total one are considered.

Considering the definition for the elastic HFE potential expressed in Eqs.(10)—(12), Eq.(16) leads to

$$\boldsymbol{\sigma} = (1-d^+)\mathbf{P}^+ : \bar{\boldsymbol{\sigma}} + (1-d^-)\mathbf{P}^- : \bar{\boldsymbol{\sigma}} = \bar{\boldsymbol{\sigma}} - (d^+\mathbf{P}^+ + d^-\mathbf{P}^-) : \bar{\boldsymbol{\sigma}} \tag{19}$$

It is then possible to obtain a final form for the constitutive law, which is a rather intuitive expression for the Cauchy stress tensor $\boldsymbol{\sigma}$ [34]

$$\boldsymbol{\sigma} = (\mathbf{I} - \mathbf{D}) : \bar{\boldsymbol{\sigma}} = (\mathbf{I} - \mathbf{D}) : \mathbf{C}_0 : (\boldsymbol{\varepsilon} - \boldsymbol{\varepsilon}^p) \tag{20}$$

where the fourth-order damage tensor $\mathbf{D}$ is expressed as

$$\mathbf{D} = d^+\mathbf{P}^+ + d^-\mathbf{P}^- \tag{21}$$

It should be noted that Eq.(20) is the standard relation between the Cauchy stress tensor $\boldsymbol{\sigma}$ and the effective stress tensor $\bar{\boldsymbol{\sigma}}$ in CDM, according to the "equivalence strain" assumption[7, 16], and Eq.(21) is also coincident with Ju[30].

From the observation of Eq.(17) the tensile and the shear DERRs Y^+ and Y^-, conjugated to the corresponding damage variables, can be expressed as

$$Y^{\pm} = -\frac{\partial \psi}{\partial d^{\pm}} \tag{22}$$

Eq.(22) demonstrates that the DERRs depend on the total elastoplastic HFE potential, and not just on the elastic one as in the classical damage model of Lemaitre[16]. Accordingly, it is physically incorrect to consider the damage criteria based on the elastic DERRs alone, since disregarding the contribution of plastic strains would prevent the model from predicting the enhancement of the concrete strength under the biaxial compression[9].

3 Plastic-damage model

3.1 Plastic strains

To determine the required effective stress tensor, the evolution law for the irreversible plastic strains tensor $\boldsymbol{\varepsilon}^p$ has to be established first. Owing to the coupling between the damage evolutions and the plastic flows, the so-called "effective stress space plasticity"[7] should be resorted to establish the evolution laws for the plastic strains[34]

$$\dot{\boldsymbol{\varepsilon}}^p = \dot{\lambda}^p \partial_{\bar{\boldsymbol{\sigma}}} F^p \tag{23a}$$

$$\dot{\boldsymbol{\kappa}} = \dot{\lambda}^{\mathrm{p}} \mathbf{H} \tag{23b}$$

$$F(\bar{\boldsymbol{\sigma}}, \boldsymbol{\kappa}) \leqslant 0,\ \dot{\lambda}^{\mathrm{p}} \geqslant 0,\ \dot{\lambda}^{\mathrm{p}} F(\bar{\boldsymbol{\sigma}}, \boldsymbol{\kappa}) \leqslant 0 \tag{23c}$$

where F and F^{p} are the plastic yield function and the plastic potential, respectively; $\dot{\lambda}^{\mathrm{p}}$ is the plastic flow parameter; $\mathbf{H}$ denotes the vectorial hardening function; and operator $\partial_x y = \partial y / \partial x$ is the partial differentiation operator.

Plastic potential F^{p} adopted in the present model is the Drucker-Prager function expressed as[15, 22]

$$F^{\mathrm{p}} = \alpha^{\mathrm{p}} \bar{I}_1 + \sqrt{2\bar{J}_2} \tag{24}$$

where $\bar{I}_1$ is the first invariant of $\bar{\boldsymbol{\sigma}}$; $\bar{J}_2$ is the second invariant of $\bar{\mathbf{s}}$, the deviatoric component of $\bar{\boldsymbol{\sigma}}$; and $\alpha^{\mathrm{p}} \geqslant 0$ is a parameter chosen to provide proper dilatancy with common range between 0.2 and 0.3 for concrete.

Combining Eqs.(23a) and (24), the evolution law for the plastic strains is obtained

$$\dot{\boldsymbol{\varepsilon}}^{\mathrm{p}} = \dot{\lambda}^{\mathrm{p}} \left(\frac{\bar{\mathbf{s}}}{\|\bar{\mathbf{s}}\|} + \alpha^{\mathrm{p}} \mathbf{1} \right) \tag{25}$$

where $\|\bar{\mathbf{s}}\| = \sqrt{\bar{\mathbf{s}} : \bar{\mathbf{s}}} = \sqrt{2\bar{J}_2}$ is the norm of $\bar{\mathbf{s}}$; $\mathbf{1}$ is the second-order identity tensor.

If one introduces the hardening parameters κ^{+} and κ^{-} as the equivalent plastic strains under uniaxial tension and compression, respectively, with the definition

$$\kappa^{+} = \int \sqrt{\dot{\varepsilon}^{\mathrm{p}}_{\max} \dot{\varepsilon}^{\mathrm{p}}_{\max}}\, \mathrm{d}t;\ \kappa^{-} = \int \sqrt{\dot{\varepsilon}^{\mathrm{p}}_{\min} \dot{\varepsilon}^{\mathrm{p}}_{\min}}\, \mathrm{d}t \tag{26a, b}$$

where $\dot{\varepsilon}^{\mathrm{p}}_{\max}$ and $\dot{\varepsilon}^{\mathrm{p}}_{\min}$ are the maximum and minimum eigenvalues of the plastic strain rate tensor $\dot{\boldsymbol{\varepsilon}}^{\mathrm{p}}$, the evolution law for vector $\boldsymbol{\kappa}$ may be postulated

$$\dot{\boldsymbol{\kappa}} = \{w\dot{\kappa}^{+}, (1-w)\dot{\kappa}^{-}\}^{\mathrm{T}} = \{w\dot{\varepsilon}^{\mathrm{p}}_{\max}, -(1-w)\dot{\varepsilon}^{\mathrm{p}}_{\min}\}^{\mathrm{T}} \tag{27}$$

with w being a weight factor expressed as

$$w = \sum_{i=1}^{3} \langle \bar{\sigma}_i \rangle \Big/ \sum_{i=1}^{3} |\bar{\sigma}_i| \tag{28}$$

that is equal to one if all the eigenstresses $\bar{\sigma}_i$ are positive and equal to zero if they are all negative[22]. Calling for Eq.(23b), the vectorial hardening function $\mathbf{H}$ is thus derived[34]

$$\mathbf{H} = \begin{Bmatrix} H^{+} \\ H^{-} \end{Bmatrix} = \begin{Bmatrix} w \partial_{\bar{\sigma}_{i,\max}} F^{\mathrm{p}} \\ -(1-w) \partial_{\bar{\sigma}_{i,\min}} F^{\mathrm{p}} \end{Bmatrix} = \operatorname{diag}[w, -(1-w)]\{\partial_{\hat{\bar{\boldsymbol{\sigma}}}} F^{\mathrm{p}}\} \tag{29}$$

where diag[•] is a diagonal matrix; vector $\hat{\bar{\boldsymbol{\sigma}}} = \{\bar{\sigma}_{i,\max}, \bar{\sigma}_{i,\min}\}^{\mathrm{T}}$, with $\bar{\sigma}_{i,\max}$ and $\bar{\sigma}_{i,\min}$ being the maximum and minimum values of $\bar{\sigma}_i$.

Any yield function F appropriate for concrete can be adopted in Eq.(23c). Here a modified function proposed in Lubliner et al.[35] and Lee and Fenves[22] is adopted

$$F(\bar{\boldsymbol{\sigma}}, \boldsymbol{\kappa}) = \left(\alpha \bar{I}_1 + \sqrt{3\bar{J}_2} + \beta \langle \bar{\sigma}_{i,\max} \rangle\right) - (1-\alpha)c \tag{30}$$

with following expressions for α , β and c

$$\alpha = (\vartheta - 1)/(2\vartheta - 1) \tag{31a}$$

$$\beta(\boldsymbol{\kappa}) = (1-\alpha)\bar{f}^-(\boldsymbol{\kappa})/\bar{f}^+(\boldsymbol{\kappa}) - (1+\alpha) \tag{31b}$$

$$c(\boldsymbol{\kappa}) = \bar{f}^-(\boldsymbol{\kappa}) \tag{31c}$$

where ϑ is the ratio between the yield strengths under equibiaxial and uniaxial compression, usually taking a value in the interval 1.10~1.20; and $\bar{f}^+(\boldsymbol{\kappa})$ and $\bar{f}^-(\boldsymbol{\kappa})$ denote evolution stresses (positive values are used here) in the effective stress space due to plastic hardening/softening under uniaxial tension and compression, respectively.

Though many hardening rules can be adopted to describe the expansion of the yield surface in the effective stress space, the following linear isotropic hardening expressions are used here for simplification

$$\bar{f}^{\pm}(\boldsymbol{\kappa}) = \bar{f}_y^{\pm} + E^{p\pm}\kappa^{\pm} \tag{32}$$

where $\bar{f}_y^+$ and $\bar{f}_y^-$ are the effective yield strengths under uniaxial tension and compression, approximated as $\bar{f}_y^+ = f_t$ and $\bar{f}_y^- = f_c$, with f_t and f_c being the uniaxial tensile and compressive strengths. In Eq.(32) $E^{p\pm}$ are the effective plastic hardening modulus under uniaxial tension and compression relating to the elastoplastic tangent modulus $E^{ep\pm}$ as follows[36]

$$E^{ep\pm} = \frac{E_0 E^{p\pm}}{E_0 + E^{p\pm}} = \left(1 - \frac{1}{1+R_E^{\pm}}\right)E_0 \tag{33}$$

where $R_E^{\pm} = E^{p\pm}/E_0$ denote the ratios between $E^{p\pm}$ and the initial Young's modulus E_0.

Then, in rate form the relation between the effective stress and the strain tensors can be deduced by the standard procedures in classical plasticity as follows[7]

$$\dot{\bar{\boldsymbol{\sigma}}} = \mathbf{C}^{ep} : \dot{\boldsymbol{\varepsilon}} \tag{34}$$

where $\mathbf{C}^{ep}$ is the continuum effective elastoplastic tangent tensor

$$\mathbf{C}^{ep} = \begin{cases} \mathbf{C}_0 & \text{if } \dot{\lambda}^p = 0 \\ \mathbf{C}_0 - \dfrac{(\mathbf{C}_0 : \partial_{\bar{\boldsymbol{\sigma}}} F^p) \otimes (\mathbf{C}_0 : \partial_{\bar{\boldsymbol{\sigma}}} F)}{\partial_{\bar{\boldsymbol{\sigma}}} F : \mathbf{C}_0 : \partial_{\bar{\boldsymbol{\sigma}}} F^p - \partial_{\boldsymbol{\kappa}} F \cdot \mathbf{H}} & \text{if } \dot{\lambda}^p > 0 \end{cases} \tag{35}$$

3.2 Plastic and elastoplastic HFE potentials

In order to clearly define concepts such as "loading", "unloading", or "reloading", damage criteria analogous to the yield criteria in plasticity should be introduced.

Since elastoplastic DERRs are the conjugated forces to the damage variables, thermodynamically consistent damage criteria should be based on $Y^{\pm}$ defined in Eq.(22). To this end, one has to determine the plastic HFE potential first.

Comparing contribution to the plastic HFE potential from plastic strains of concrete in compression, the one from tension is so much smaller that $\psi_0^{p+} = 0$ can be assumed.

Accordingly only the negative component ψ_0^{p-} is taken into consideration, in the present model through the following expression(Wu[37] and see also Appendix Ⅱ)

$$\psi_0^p(\boldsymbol{\kappa})=\psi_0^{p-}(\boldsymbol{\kappa})=\frac{b}{2E_0}\left(3\bar{J}_2^-+\eta^p\bar{I}_1^-\sqrt{3\bar{J}_2}-\frac{1}{2}\bar{I}_1^+\bar{I}_1^-\right) \tag{36}$$

where $\bar{I}_1^{\pm}$ are the first invariants of $\bar{\boldsymbol{\sigma}}^{\pm}$; $\bar{J}_2^-$ is the second invariant of $\bar{\mathbf{s}}^-$, the deviatoric tensorial components of $\bar{\boldsymbol{\sigma}}^-$; $\eta^p=\sqrt{3/2}\alpha^p$ is an alternative parameter to describe the dilatancy; and parameter b is a material property(see Appendix Ⅱ for further details), devised so that the ratio between the equibiaxial and the uniaxial compressive strengths could match the usual 1.10~1.20 values[27].

According to the analysis of concrete micromechanics[16], resulting mainly from internal slips on the interface between the matrix and the aggregates, plastic flow is generally controlled by the shear damage mechanism. Hence, plastic HFE potential is assumed to be a function with the form

$$\psi^p(\boldsymbol{\kappa},d^-)=(1-d^-)\psi_0^p(\boldsymbol{\kappa}) \tag{37}$$

Substituting Eqs.(10)—(12) and (37) into Eq.(13), the elastoplastic HFE potential becomes

$$\psi(\boldsymbol{\varepsilon}^e,\boldsymbol{\kappa},d^+,d^-)=\psi^+(\boldsymbol{\varepsilon}^e,d^+)+\psi^-(\boldsymbol{\varepsilon}^e,\boldsymbol{\kappa},d^-) \tag{38}$$

where

$$\psi^+(\boldsymbol{\varepsilon}^e,d^+)=(1-d^+)\psi_0^+(\boldsymbol{\varepsilon}^e) \tag{39a}$$

$$\psi^-(\boldsymbol{\varepsilon}^e,\boldsymbol{\kappa},d^-)=(1-d^-)\psi_0^-(\boldsymbol{\varepsilon}^e,\boldsymbol{\kappa}) \tag{39b}$$

Calling for Eqs.(10) and (36), the positive and negative component of the initial elastoplastic HFE ψ_0^+ and ψ_0^- can be expressed as

$$\psi_0^+=\psi_0^{e+}=(\bar{\boldsymbol{\sigma}}^+:\boldsymbol{\Lambda}_0:\bar{\boldsymbol{\sigma}})/2 \tag{40a}$$

$$\psi_0^-=\psi_0^{e-}+\psi_0^{p-}=b_0(\alpha\bar{I}_1+\sqrt{3\bar{J}_2})^2 \tag{40b}$$

where the introduced approximations and the expression for parameters b_0 are detailed in Appendix II.

3.3 Damage criteria and domain of linear behaviour

According to the definitions in Eq.(22), and taking Eq.(39) into consideration, the tensile and shear DERRs are expressed as

$$Y^{\pm}=-\frac{\partial\psi^{\pm}}{\partial d^{\pm}}=\psi_0^{\pm} \tag{41}$$

With these definitions for the DERRs, the state of damage in concrete can then be characterized by means of damage criteria with the following two equivalent functional forms

$$G^{\pm}(Y^{\pm},r^{\pm})=Y^{\pm}-r^{\pm}\leqslant 0\Leftrightarrow\bar{G}^{\pm}(Y^{\pm},r^{\pm})=g^{\pm}(Y^{\pm})-g^{\pm}(r^{\pm})\leqslant 0 \tag{42}$$

where $r^{\pm}$ are the current damage thresholds(energy barriers), which control the size of the expanding damage surfaces. Denoting by $r_0^{\pm}$ the initial damage thresholds before any loading applied, Eq.(42) implies $r^{\pm} \geqslant r_0^{\pm}$, which states that damages are initiated when the DERRs $Y^{\pm}$ exceed the corresponding initial damage thresholds $r_0^{\pm}$.

In Eq. (42) $g^{\pm}(\cdot)$ can be any monotonically increasing scalar functions, and in the proposed model functions $\sqrt{2E_0(\cdot)}$ and $\sqrt{(\cdot)/b_0}$ are postulated for convenience, therefore the tensile and shear DERRs given in Eq.(41) can now be redefined as

$$Y^{+} = \sqrt{2E_0 \psi_0^{+}} = \sqrt{E_0 (\bar{\boldsymbol{\sigma}}^{+} : \boldsymbol{\Lambda}_0 : \bar{\boldsymbol{\sigma}})} \tag{43a}$$

$$Y^{-} = \sqrt{\psi_0^{-}/b_0} = \alpha \bar{I}_1 + \sqrt{3\bar{J}_2} \tag{43b}$$

Correspondingly, the initial tensile damage and shear damage thresholds are calculated as

$$r_0^{+} = f_0^{+}; \; r_0^{-} = (1-\alpha) f_0^{-} \tag{44}$$

where f_0^{+} and f_0^{-} are the stresses(positive values) beyond which nonlinearity becomes visible under uniaxial tension and compression, respectively.

It should be noted that Eq. (43a) is unconditionally correct under all stress states, and Eq.(43b) holds strictly under plane stress states. And it does exist stress states under which Y^{-} is less than zero(e.g., under hydrostatic compression), demonstrating that the compressive consolidation mechanism could play a significant role.

Due to the above fact and the inherent shortcomings of the Drucker-Prager function, the predicted results of the present model would be somewhat conservative in triaxial compression. Under a not too high confinement compressive pressure, the shear DERR defined in Eq.(43b) can be modified to include the influence of the third invariant of $\bar{\boldsymbol{\sigma}}$, namely through equation[38]

$$Y^{-} = \alpha \bar{I}_1 + \sqrt{3\bar{J}_2} - \gamma \langle -\bar{\sigma}_{i,\max} \rangle \tag{45}$$

with

$$\gamma = 3(1-K_c)/(2K_c - 1) \tag{46}$$

where K_c, assumed as a constant[35], denotes the ratio of corresponding values of $\sqrt{\bar{J}_2}$ under tensile meridian and compressive meridian stress states for any given value of hydrostatic pressure $\bar{I}_1$. A value of $K_c = 2/3$ typical for concrete, gives $\gamma = 3.0$, which is adopted here.

Note that in Eq. (45) the influence of coefficient γ disappears in stress states other than triaxial compression, namely when $\bar{\sigma}_{i,\max} \geqslant 0$. Therefore, Eq.(45) can be viewed as a minor modification of Eq.(43b) to improve the predictive capability of the proposed model under stress state when the later does not hold.

Under biaxial stress states the damage criteria defined through Eqs.(42)—(45) lead to the domain of linear behaviour shown in Fig. 1, which agrees rather well with the experimental results from Kupfer et al.[27].

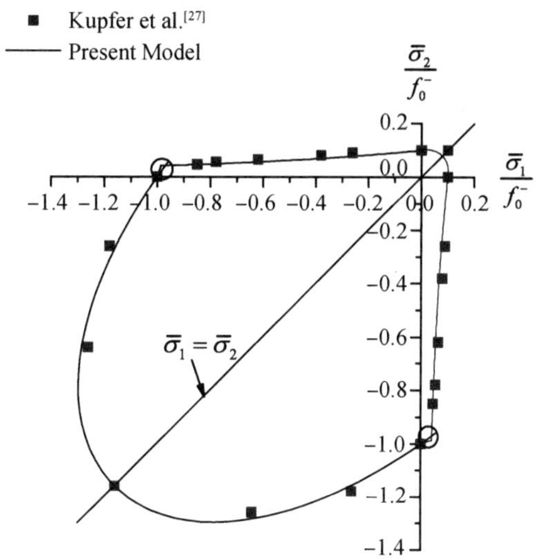

Fig. 1. Domain of linear behaviour of concrete under biaxial stresses

3.4 Characterization of damage

To describe the growth of microcracks and the evolution of the damage surfaces, it is necessary to specify the evolution laws for the damage variables $d^{\pm}$ and the corresponding damage thresholds $r^{\pm}$. Analogous to the normality rule in classical plasticity, the evolution laws for the damage variables can be defined by applying it to the damage criteria prescribed in Eq.(42)

$$\dot{d}^{\pm}=\dot{\lambda}^{d\pm}\frac{\partial g^{\pm}(Y^{\pm})}{\partial Y^{\pm}};\ \dot{\lambda}^{d\pm}=\dot{r}^{\pm} \tag{47a, b}$$

where $\dot{\lambda}^{d\pm}$ are damage consistency parameters.

In compacted form loading or unloading can be expressed through the Kuhn-Tucker relations

$$\dot{\lambda}^{d\pm}\geqslant 0;\ \bar{G}^{\pm}(Y^{\pm},\ r^{\pm})\leqslant 0;\ \dot{\lambda}^{d\pm}\bar{G}^{\pm}(Y^{\pm},\ r^{\pm})=0 \tag{48a, b, c}$$

Calling for the damage consistency condition, one obtains

$$\bar{G}^{\pm}(Y^{\pm},\ r^{\pm})=\dot{\bar{G}}^{\pm}(Y^{\pm},\ r^{\pm})=0 \tag{49}$$

which yields

$$r^{\pm}=Y^{\pm};\ \dot{\lambda}^{d\pm}=\dot{r}^{\pm}=\dot{Y}^{\pm}\geqslant 0 \tag{50}$$

So for a generic instant n thresholds $r^{\pm}$ are given by[7, 23]

$$r^{\pm}=\max\left\{r_0^{\pm},\ \max_{\tau\in[0,\ n]} Y_{\tau}^{\pm}\right\} \tag{51}$$

Introducing Eq.(50) into Eq.(47a), the evolution laws for the damage variables during loading are expressed as

$$\dot{d}^{\pm}=\dot{r}^{\pm}\frac{\partial g^{\pm}(r^{\pm})}{\partial r^{\pm}}=\dot{g}^{\pm}(r^{\pm}) \tag{52}$$

Performing a trivial integration and accounting for the initial conditions $d^{\pm}(r_0^{\pm})=0$, one obtains

$$d^{\pm}=g^{\pm}(r^{\pm}) \tag{53}$$

Usually concrete under one-dimensional tension is assumed to behave elastically until the stress reaches the tensile strength $f_t=f_0^{+}$, hardening is disregarded and the following function is adopted for damage variable d^{+}

$$d^{+}=g^{+}=1-\frac{r_0^{+}}{r^{+}}\left\{(1-A^{+})+A^{+}\exp\left[B^{+}\left(1-\frac{r^{+}}{r_0^{+}}\right)\right]\right\} \tag{54}$$

where parameters A^{+} and B^{+} are related to the percentage of steel ρ_s in reinforced concrete, which in accordance with experimental tests can be expressed as following[39]

$$A^{+}=1-c_s\rho_s/d_b \tag{55a}$$

$$B^{+}=(270/\sqrt{A^{+}})f_0^{+}/E_0\leqslant 1\,000f_0^{+}/E_0 \tag{55b}$$

in which d_b is the rebar diameter (in mm); and c_s has the dimension of length, and generally takes the value of 75 mm.

Eq. (54) actually describes the overall tension-stiffening effect due to the interaction between steel rebars and concrete. In plain concrete $A^{+}=1.0$ is obtained from Eq. (55a), thus Eq. (54) reduces to the function first proposed by Oliver et al.[40]

$$d^{+}=g^{+}=1-\frac{r_0^{+}}{r^{+}}\exp\left[B^{+}\left(1-\frac{r^{+}}{r_0^{+}}\right)\right] \tag{56}$$

To reduce sensitivity in the analysis of plain concrete as regards to the refinement of the finite element meshes, parameter B^{+} involved in Eq. (56) should be computed so as to satisfy the requirement of mesh objectivity[40]

$$B^{+}=\left[\frac{G_fE_0}{l_{ch}(f_0^{+})^2}-\frac{1}{2}\right]^{-1}\geqslant 0 \tag{57}$$

where l_{ch} is the "geometrical" characteristic length of finite element mesh.

To allow reproducing the hardening of concrete in compression, as well as the softening which characterizes the post-peak behaviour, the following evolution for damage variable d^{-} is adopted[9, 23]

$$d^{-}=g^{-}=1-\left\{\frac{r_0^{-}}{r^{-}}(1-A^{-})+A^{-}\exp\left[B^{-}\left(1-\frac{r^{-}}{r_0^{-}}\right)\right]\right\} \tag{58}$$

where parameters A^{-} and B^{-} may be determined by imposing the simulation curve to fit the one obtained from a one-dimensional experimental test.

Correspondingly, from Eqs.(52),(54) and (58) the evolution laws for $d^{\pm}$ are expressed as

$$\dot{d}^{\pm}=\dot{r}^{\pm}h^{\pm}\geqslant 0 \tag{59}$$

where $h^{\pm}$ are the following hardening/softening functions

$$h^{+}=\frac{\partial g^{+}(r^{+})}{\partial r^{+}}=(1-A^{+})\frac{r_0^{+}}{(r^{+})^2}+A^{+}\frac{B^{+}r^{+}+r_0^{+}}{(r^{+})^2}\exp\left[B^{+}\left(1-\frac{r^{+}}{r_0^{+}}\right)\right] \tag{60a}$$

$$h^{-}=\frac{\partial g^{-}(r^{-})}{\partial r^{-}}=(1-A^{-})\frac{r_0^{-}}{(r^{-})^2}+\frac{A^{-}B^{-}}{r_0^{-}}\exp\left[B-\left(1-\frac{r^{-}}{r_0^{-}}\right)\right] \tag{60b}$$

4 Computational aspects

4.1 Numerical algorithm

To explore the numerical implementation algorithm for the proposed plastic-damage model, the constitutive law in Eq.(20) has to be differentiated with respect to time, leading to

$$\dot{\boldsymbol{\sigma}}=(\mathbf{I}-\boldsymbol{\omega}):\mathbf{C}_0:(\dot{\boldsymbol{\varepsilon}}-\dot{\boldsymbol{\varepsilon}}^{\mathrm{p}})-(\bar{\boldsymbol{\sigma}}^{+}\dot{d}^{+}+\bar{\boldsymbol{\sigma}}^{-}\dot{d}^{-}) \tag{61}$$

with the symmetric fourth-order tensors $\boldsymbol{\omega}$ being

$$\boldsymbol{\omega}=d^{+}\mathbf{Q}^{+}+d^{-}\mathbf{Q}^{-} \tag{62}$$

According to the concept of operator split[7, 36], Eq.(61) can be decomposed into elastic, plastic and damage parts, leading to the corresponding numerical algorithm including elastic-predictor, plastic-corrector and damage-corrector steps, as established in Table 1 with Eq.(63).

Table 1. Numerical algorithm(operator split)

Elastic-predictor(63a)	Plastic-corrector(63b)	Damage-corrector(63c)
$\dot{\boldsymbol{\varepsilon}}=\nabla^s\dot{\mathbf{u}}(t)$	$\dot{\boldsymbol{\varepsilon}}=0$	$\dot{\boldsymbol{\varepsilon}}=0$
$\dot{\boldsymbol{\varepsilon}}^{\mathrm{p}}=0$	$\dot{\boldsymbol{\varepsilon}}^{\mathrm{p}}=\begin{cases}\dot{\lambda}^{\mathrm{p}}\partial_{\bar{\boldsymbol{\sigma}}}F^{\mathrm{p}} & \text{if } F=\dot{F}=0\\ 0 & \text{otherwise}\end{cases}$	$\dot{\boldsymbol{\varepsilon}}^{\mathrm{p}}=0$
$\dot{\boldsymbol{\kappa}}=0$	$\dot{\boldsymbol{\kappa}}=\dot{\lambda}^{\mathrm{p}}\mathbf{H}$	$\dot{\boldsymbol{\kappa}}=0$
$\dot{\bar{\boldsymbol{\sigma}}}=\mathbf{C}_0:\dot{\boldsymbol{\varepsilon}}$	$\dot{\bar{\boldsymbol{\sigma}}}=-\mathbf{C}_0:\dot{\boldsymbol{\varepsilon}}^{\mathrm{p}}$	$\dot{\bar{\boldsymbol{\sigma}}}=0$
$\dot{d}^{\pm}=0$	$\dot{d}^{\pm}=0$	$\dot{d}^{\pm}=\begin{cases}\dot{r}^{\pm}h^{\pm} & \text{if } \bar{G}^{\pm}=\dot{\bar{G}}^{\pm}=0\\ 0 & \text{otherwise}\end{cases}$
$\dot{r}^{\pm}=0$	$\dot{r}^{\pm}=0$	$\dot{r}^{\pm}=\begin{cases}\dot{Y}^{\pm} & \text{if } G^{\pm}=\dot{\bar{G}}^{\pm}=0\\ 0 & \text{otherwise}\end{cases}$
$\dot{\boldsymbol{\sigma}}=(\mathbf{I}-\boldsymbol{\omega}):\mathbf{C}_0:\dot{\boldsymbol{\varepsilon}}$	$\dot{\boldsymbol{\sigma}}=-(\mathbf{I}-\boldsymbol{\omega})\mathbf{C}_0:\dot{\boldsymbol{\varepsilon}}^{\mathrm{p}}$	$\dot{\boldsymbol{\sigma}}=-(\bar{\boldsymbol{\sigma}}^{+}\dot{d}^{+}+\bar{\boldsymbol{\sigma}}^{-}\dot{d}^{-})$

It is noted that during the elastic-predictor and the plastic-corrector steps the damage variables are fixed, so Eqs.(63a) and (63b) are decoupled with the damage part(63c), constituting a standard elastoplastic problem in the effective stress space. Regarding to the plastic yield function Eq. (30) adopted, the spectral decomposition form[41] of return mapping algorithm[36] is modified and improved to update the effective stress tensor $\bar{\boldsymbol{\sigma}}$. Once $\bar{\boldsymbol{\sigma}}$ is updated in the elastic-predictor and plastic-corrector steps, the damage variables $d^{\pm}$ and the Cauchy stress $\boldsymbol{\sigma}$ can thus be updated correspondingly in the damage-corrector step.

4.2 Algorithmic consistent tangent modulus

In the nonlinear finite element analysis, especially when post-peak behaviour is expected, the full Newton-Raphson method is usually adopted in the global iteration; hence, algorithmic consistent tangent modulus is required. According to Eq. (61), the total differential of Eq. (20) leads to

$$\mathrm{d}\boldsymbol{\sigma} = (\mathbf{I} - \boldsymbol{\omega}) : \mathrm{d}\bar{\boldsymbol{\sigma}} - [\bar{\boldsymbol{\sigma}}^{+}\mathrm{d}(d^{+}) + \bar{\boldsymbol{\sigma}}^{-}\mathrm{d}(d^{-})] \tag{64}$$

Calling for Eqs. (43a), (45) and (59); and taking Eq. (50b) into consideration, the differentials of damage variables in Eq. (64) are

$$\mathrm{d}(d^{+}) = h^{+}\mathrm{d}(r^{+}) = h^{+}\frac{E_0}{2Y^{+}}(\bar{\boldsymbol{\sigma}} : \boldsymbol{\Lambda}_0 : \mathbf{Q}^{+} + \bar{\boldsymbol{\sigma}}^{+} : \boldsymbol{\Lambda}_0) : \mathrm{d}\bar{\boldsymbol{\sigma}} \tag{65a}$$

$$\mathrm{d}(d^{-}) = h^{-}\mathrm{d}(r^{-}) = h^{-}\left(\alpha\mathbf{1} + \frac{3}{2\sqrt{3\bar{J}_2}}\bar{\mathbf{s}} + \gamma H(-\bar{\sigma}_{i,\max})\mathbf{p}_{ii,\max}\right) : \mathrm{d}\bar{\boldsymbol{\sigma}} \tag{65b}$$

with $\mathbf{p}_{ii,\max}$ being the tensor $\mathbf{p}_{ij}$ associated to $\bar{\sigma}_{i,\max}$. Then it can be easily concluded that

$$\bar{\boldsymbol{\sigma}}^{+}\mathrm{d}(d^{+}) = \mathbf{R}^{+} : \mathrm{d}\bar{\boldsymbol{\sigma}};\ \bar{\boldsymbol{\sigma}}^{-}\mathrm{d}(d^{-}) = \mathbf{R}^{-} : \mathrm{d}\bar{\boldsymbol{\sigma}} \tag{66}$$

where

$$\mathbf{R}^{+} = h^{+}\frac{E_0}{2Y^{+}}[(\bar{\boldsymbol{\sigma}}^{+} \otimes \bar{\boldsymbol{\sigma}}) : \boldsymbol{\Lambda}_0 : \mathbf{Q}^{+} + (\bar{\boldsymbol{\sigma}}^{+} \otimes \bar{\boldsymbol{\sigma}}^{+}) : \boldsymbol{\Lambda}_0] \tag{67a}$$

$$\mathbf{R}^{-} = h^{-}\left[\bar{\boldsymbol{\sigma}}^{-} \otimes \left(\alpha\mathbf{1} + \frac{3}{2\sqrt{3\bar{J}_2}}\bar{\mathbf{s}} + \gamma H(-\bar{\sigma}_{i,\max})\mathbf{p}_{ii,\max}\right)\right] \tag{67b}$$

By introducing Eqs. (66) into Eq. (64), one gets

$$\mathrm{d}\boldsymbol{\sigma} = (\mathbf{I} - \boldsymbol{\omega} - \mathbf{R}) : \mathrm{d}\bar{\boldsymbol{\sigma}} = [(\mathbf{I} - \boldsymbol{\omega} - \mathbf{R}) : \bar{\mathbf{C}}^{\mathrm{alg}}] : \mathrm{d}\boldsymbol{\varepsilon} = \mathbf{C}^{\mathrm{alg}} : \mathrm{d}\boldsymbol{\varepsilon} \tag{68}$$

with the algorithmic consistent tangent modulus $\mathbf{C}^{\mathrm{alg}}$ expressed as

$$\mathbf{C}^{\mathrm{alg}} = (\mathbf{I} - \boldsymbol{\omega} - \mathbf{R}) : \bar{\mathbf{C}}^{\mathrm{alg}} \tag{69}$$

where $\mathbf{R} = \mathbf{R}^{+} + \mathbf{R}^{-}$; and $\bar{\mathbf{C}}^{\mathrm{alg}}$ denotes the usual effective elastoplastic tangent modulus consistent with the algorithm for updating the effective stress in the previous mentioned elastic-predictor and plastic-corrector steps, satisfying

$$\mathrm{d}\bar{\boldsymbol{\sigma}}_{n+1} = \bar{\mathbf{C}}^{\mathrm{alg}} : \mathrm{d}\boldsymbol{\varepsilon}_{n+1} \tag{70}$$

with the following expression[36]

$$\bar{\mathbf{C}}^{alg} = \left(\Lambda_0 + \partial_{\bar{\boldsymbol{\sigma}}} F^p \otimes \frac{d\Delta\lambda^p}{d\bar{\boldsymbol{\sigma}}_{n+1}} + \Delta\lambda^p \partial^2_{\bar{\boldsymbol{\sigma}}} F^p\right)^{-1} \tag{71}$$

where

$$\frac{d\Delta\lambda^p}{d\bar{\boldsymbol{\sigma}}_{n+1}} = -\frac{\partial_{\bar{\boldsymbol{\sigma}}} F}{\partial_{\boldsymbol{\kappa}} F \cdot (\mathbf{H} + \Delta\lambda^p \partial_{\hat{\bar{\boldsymbol{\sigma}}}} \mathbf{H} \cdot \partial_{\Delta\lambda^p} \hat{\bar{\boldsymbol{\sigma}}})} \tag{72}$$

5 Applications

To illustrate the applicability and effectiveness of the proposed model, several numerical examples of various loading conditions of concrete are presented in this section. The results obtained by the suggested model are compared with corresponding experimental results to learn its performance. For all cases, unless otherwise specified, Poisson's ratio is 0.20; the equibiaxial to uniaxial compressive strength ratio ϑ is 1.16, and the dilatancy parameter α^p is chosen as 0.20.

5.1 Monotonic uniaxial tensile tests

The experimental results from two typical monotonic uniaxial tensile tests[42-43] are employed for the comparison purpose. The material properties used in the two test were: ① for Geopalaeratnam and Shah's test, $E_0 = 3.1 \times 10^4$ MPa, $f_0^+ = 3.48$ MPa, $G_f = 100$ N/m; ② for Zhang's test, $E_0 = 3.8 \times 10^4$ MPa, $f_0^+ = 3.40$ MPa, $G_f = 70$ N/m. Fig. 2 compares the predicted stress-strain curves from the proposed model with those obtained from the experimental tests. For both tests, as it can be observed from Fig. 2, predictions from the numerical model agree well with the experimental data, especially for the post-peak nonlinear softening branches.

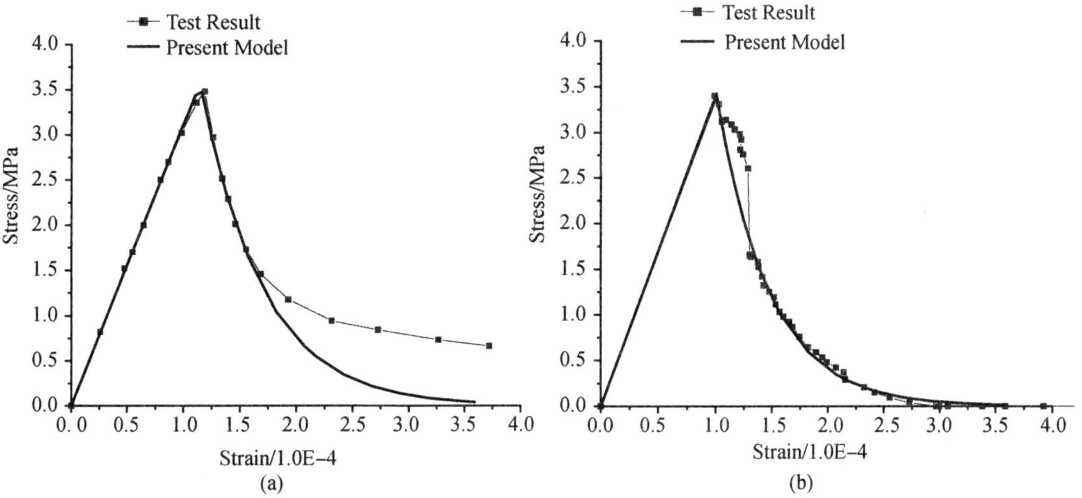

Fig. 2. Monotonic uniaxial tensile tests: (a) Geopalaeratnam and Shah[42]; (b) Zhang[43]

5.2 Monotonic uniaxial compressive tests

The model ability to reproduce the concrete behaviour under monotonic uniaxial compression can be checked in Fig. 3, where two types of experimental results taken from Karson and Jirsa[44] and Zhang[43] are plotted against the numerical predictions. The material properties adopted in the simulations were: for the former one, $E_0 = 3.17 \times 10^4$ MPa, $f_0^+ = 3.0$ MPa and $f_0^- = 10.2$ MPa; and for the latter one, $E_0 = 3.8 \times 10^4$ MPa, $f_0^+ = 3.40$ MPa and $f_0^- = 43.0$ MPa. As shown in Fig. 3, either in the hardening or in the softening regimes, the overall nonlinear performances numerically predicted and the experimental obtained stress-strain curves are rather close.

5.3 Monotonic biaxial stress test

The proposed model is also validated with the results under biaxial compression ($\sigma_3 = 0$) reported in Kupfer et al.[27]. The material properties adopted in the simulation were: $E_0 = 3.1 \times 10^4$ MPa, $f_0^+ = 3.0$ MPa, $f_0^- = 15.0$ MPa, $G_f = 75$ N/m. For specimens under load conditions $\sigma_2/\sigma_1 = -1/0$, $\sigma_2/\sigma_1 = -1/-1$ and $\sigma_2/\sigma_1 = -1/0.52$, the predicted stress-strain curves illustrated in Fig. 4(a)—(c) agree well with the experimental ones, capturing the overall experimental behaviour.

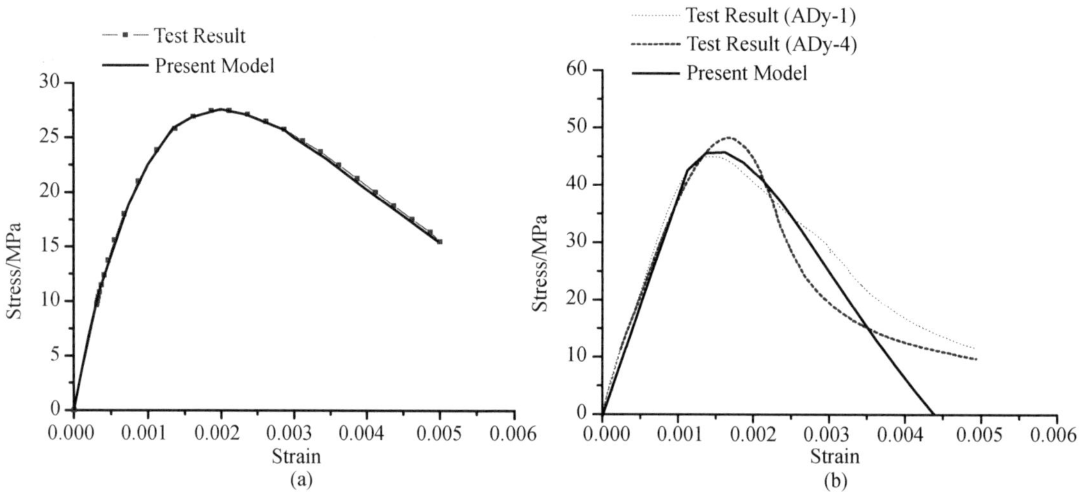

Fig. 3. Monotonic uniaxial compressive tests: (a) Karson and Jirsa[44]; (b) Zhang[43]

To illustrate the capability of the proposed model for predicting the nonlinear behaviour of concrete under other biaxial stress states, using the same material properties as above, the numerical biaxial strength envelope is reproduced in Fig. 4(d) as well, which is found to be almost coincident with the experimental one from Kupfer et al.[27]. The predicted envelope without considering the contribution of the plastic HFE potential is also provided in Fig. 4(d), which reproduces the same conservative results as those of Mazars[9]. As clearly perceptible in Fig. 4(d), another important attribute of the present model is its ability to predict not only the enhancement of concrete strength under biaxial compression, but also the reduction on the compressive strength induced by orthogonal tensile cracking under tension-compression stress states.

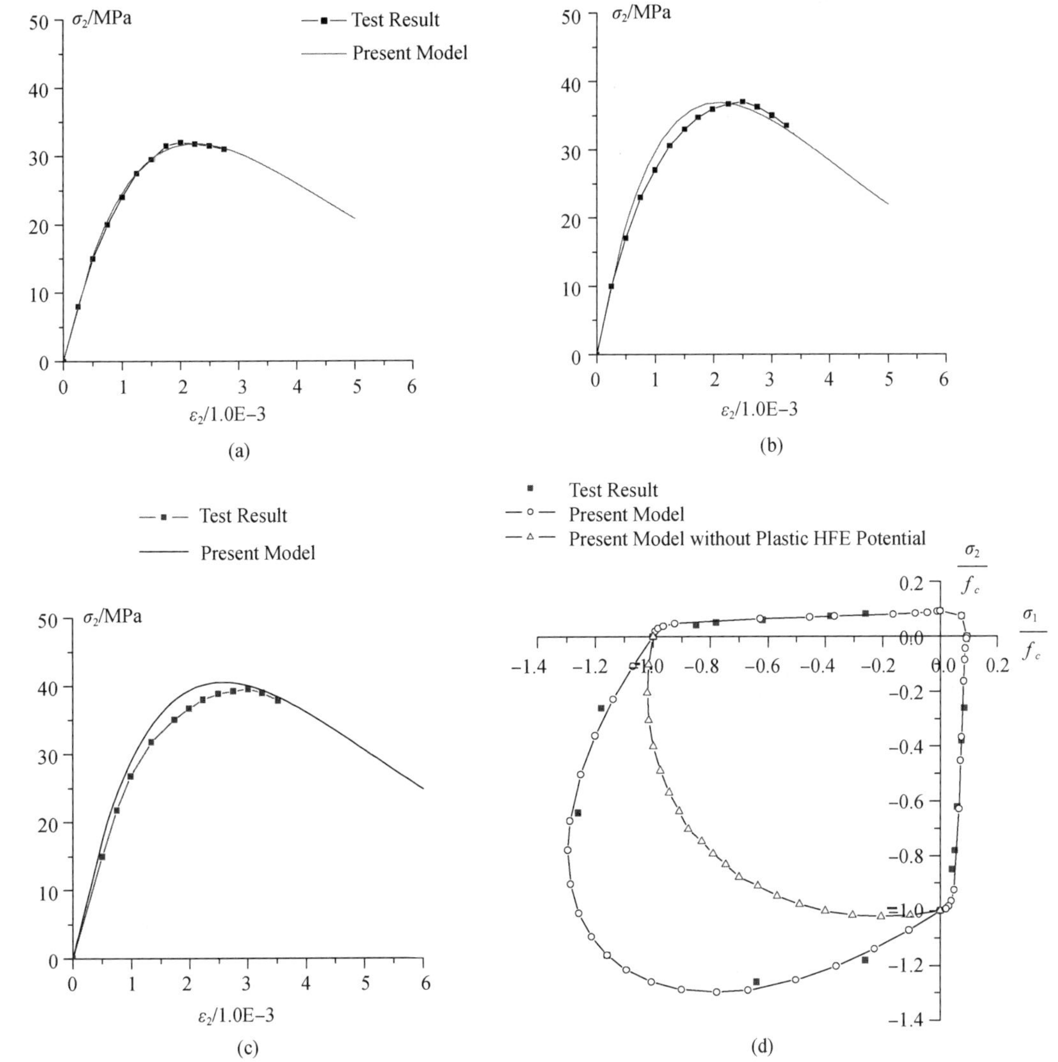

Fig. 4. Monotonic biaxial compressive tests[27]: (a) $\sigma_2/\sigma_1=-1/0$; (b) $\sigma_2/\sigma_1=-1/-1$; (c) $\sigma_2/\sigma_1=-1/0.52$; (d) strength envelope under biaxial stresses

5.4 Concrete under 3D compression

To check the application of the proposed model to concrete in compression under confinement, Fig. 5 compares the numerical predictions with the experimental results obtained by Green and Swanson[45].

To fit the uniaxial compressive stress-strain curve also reported in Fig. 5, the material properties adopted in the simulation were: $E_0 = 3.7 \times 10^4$ MPa, $f_0^+ = 4.0$ MPa, $f_0^- = 15.0$ MPa. The numerical predictions of specimens under three sets of confining stresses, namely, $\sigma_1 = \sigma_2 = 0.0$ MPa, $\sigma_1 = \sigma_2 = -6.895$ MPa and $\sigma_1 = \sigma_2 = -13.79$ MPa, are

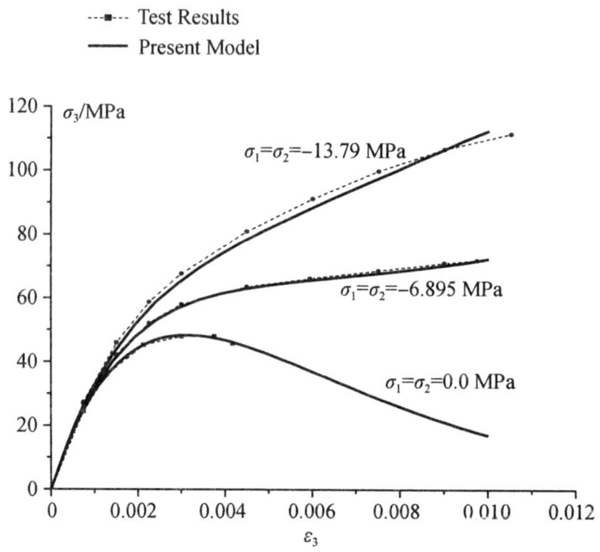

Fig. 5. 3D compression test[45]

reproduced in Fig. 5. It can be clearly seen that the enhancement of strength and ductility due to the compressive confinement, as well as the overall stress-strain experimental curves, are satisfactorily predicted by the proposed model.

5.5 Cyclic uniaxial tests

In Fig. 6 the cyclic uniaxial tensile test of Taylor[46] and the cyclic compressive test of Karson and Jirsa[44] are reproduced numerically to demonstrate the capability of the proposed model under cyclic load conditions. The following properties were adopted: for Taylor's simulation, $E_0 = 3.17 \times 10^4$ MPa, $f_0^+ = 3.50$ MPa, $G_f = 24$ N/m; and for Karsan and Jirsa's one, $E_0 = 3.0 \times 10^4$ MPa, $f_0^- = 15.0$ MPa. As shown in Fig. 6, under both tension and compression, the experimentally observed strength softening and stiffness degrading, as well as the irreversible strains upon unloading, are well reproduced by the proposed model.

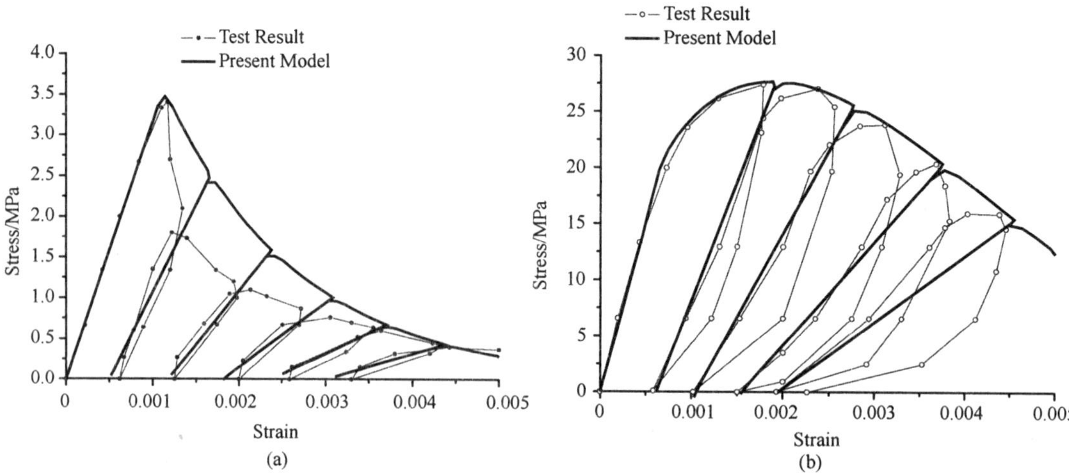

Fig. 6. Cyclic uniaxial tests: (a) cyclic uniaxial tension test[46]; (b) cyclic uniaxial compressive test[44]

5.6 Concrete dam under earthquake motions

To further illustrate the capability of the proposed model, the Koyna dam subjected to the recorded transverse and vertical components of the ground accelerations during the earthquake motions in 1967, extensively studied by other investigators[22, 47-50], is analyzed here.

Following the work of above investigators, the dam-foundation interactions was ignored assuming a rigid foundation, and a finite element mesh consisting of 760 4-node plane stress elements with 2×2 Gauss integration was adopted to model the dam. The dam-reservoir dynamic interactions were modeled using a 2-node element by the added mass technique of Westergaard[51]. A Rayleigh stiffness-proportional damping factor was assumed to provide a 3% fraction of the critical damping for the first mode of vibration of the dam. The generally used Hilber-Hughes-Taylor HHT-α method was adopted to integrate the dynamic equation of motion. The material properties adopted in the simulation were: density $\rho_0 = 2\ 643\ \mathrm{kg/m^3}$, $E_0 = 31\ 027$ MPa, $f_t = f_0^+ = 2.9$ MPa, $f_c = 24.1$ MPa, $G_f = 200$ N/m.

The predicted results of the horizontal displacements at the left corner of the dam crest are shown in Fig. 7(a) and (b) (the positive values represent the displacement towards downstream), which agree well with those by Lee and Fenves[22]. The evolution of tensile damage and the damage patterns predicted by the proposed model[37] are also in fair agreement with the observed damages reported by other investigators[22, 48].

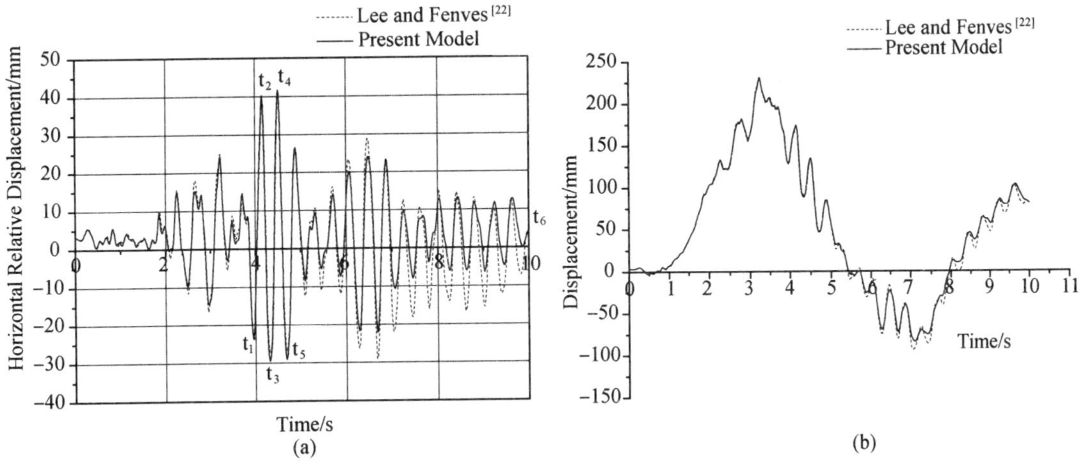

Fig. 7. Koyna dam under earthquake motions: (a) horizontal crest displacement relative to ground; (b) spatial horizontal crest displacement

5.7 Reinforced concrete slab

A square slab experimentally tested by McNeice[52], supported at the four corners and loaded by a point load at its center, is to be simulated herein. The slab was reinforced in the two directions at 75% of its depth, and the reinforcement ratio was 0.85% in each direction. The material properties for the concrete were: $E_0 = 28.6$ GPa; $v_0 = 0.15$; $f_0^+ = 3.17$ MPa, $f_0^- = 10.68$ MPa. For steel rebars, the material properties were: Young's modulus equal to 200 GPa and yield strength equal to 345 MPa(an elastic-perfectly plastic behaviour was assumed). One quarter of the slab is modeled regarding to the geometry symmetries. A 3×3 mesh of 8-node shell elements is used for modeling the concrete where the proposed model was enforced, and the two-way reinforcement was modeled using smeared rebar layers, each one with an appropriate thickness of the exact cross-sectional area of reinforcement.

The comparison between the experimental result and the numerical simulation is presented in Fig. 8 in terms of a curve representing the central load versus the central deflection of the slab. It can be clearly seen that the predicted results of the proposed model, namely the cracking and the ultimate loads, as well as the entire load-deflection curve, fit rather well the experimental ones.

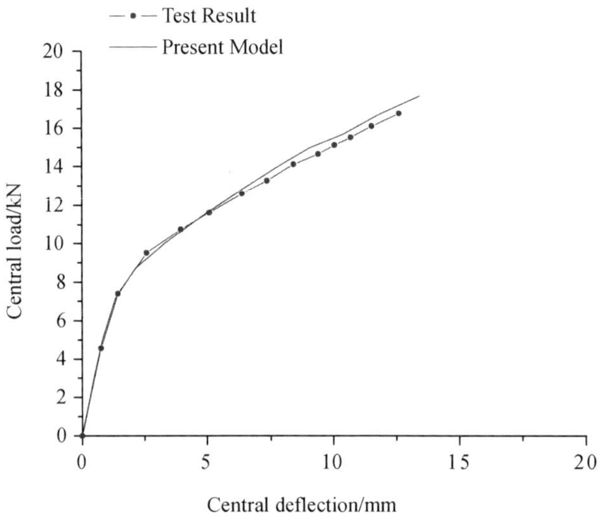

Fig. 8. RC slab test[52]

5.8 Reinforced concrete shearwalls

The proposed model was also used to simulate the nonlinear performances of 13 large-scale reinforced concrete shearwalls, tested by Lefas et al.[53] under various axial and monotonically increasing horizontal forces. The experiment consisted of two types of geometries: type 1 with a height/width ratio equal to 1.0, and type 2, where such ratio is 2.0. The steel configurations, the vertical loadings and other details of the experimental tests are referred in Lefas et al.[53] and Vecchio[54]. The material properties of concrete adopted in the simulation were: $E_0 = 30.7$ GPa, $f_0^+ = 3.0$ MPa, $f_0^- = 10.0$ MPa. For both types of shearwalls, 8-node plane stress elements and 2-node truss element are used for modeling the concrete and the steel reinforcement, respectively.

Table 2. Peak horizontal forces and corresponding displacements for the shearwalls

Wall	$F_{u,\text{test}}$/kN	$F_{u,\text{model}}$/kN	$F_{u,\text{model}}/F_{u,\text{test}}$	$U_{u,\text{test}}$/mm	$U_{u,\text{model}}$/mm	$U_{u,\text{model}}/U_{u,\text{test}}$
SW11	260	259.6	0.998	9.8	10.9	1.107
SW12	340	328.4	0.966	9.8	10.0	1.019
SW13	330	352.0	1.067	8.9	11.4	1.284
SW14	265	259.6	0.980	11.2	10.9	0.969
SW15	320	321.5	1.005	8.8	10.0	1.129
SW16	355	354.9	1.000	5.8	6.9	1.197
SW17	247	241.8	0.979	10.8	10.5	0.972
SW21	127	121.7	0.958	20.6	20.0	0.970
SW22	150	148.9	0.993	15.3	14.6	0.954
SW23	180	170.2	0.946	13.2	14.6	1.107
SW24	120	121.7	1.014	18.1	20.0	1.103
SW25	150	164.3	1.095	9.5	13.1	1.383
SW26	123	120.1	0.976	20.9	21.7	1.036
Mean	/	/	0.998	/	/	1.095
C.O.V.	/	/	0.042	/	/	0.131

The numerically predicted load-displacement responses agree well with the experimental results, one of which is shown in Fig. 9, reproducing the horizontal force and displacement at the top of the shearwalls SW11 and SW14. The distributions of the tensile damage of SW11 under the horizontal loadings of 122.0 kN and 153.1 kN are shown in Fig. 10, providing useful information about the flexural-shear induced cracking.

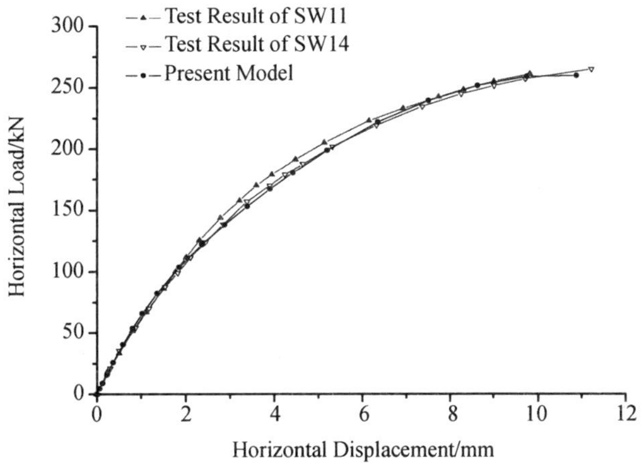

Fig. 9. RC shearwall tests

For the 13 shearwalls the peak loads and the corresponding displacements computed with the proposed model are compared with the experimental ones in Table 1. It can be clearly seen that the peak forces F and the corresponding displacements U of all the shearwalls are reproduced rather well by the proposed model: ① the ratio of the numerically predicted peak loads to the experimental ones has a mean value of 0.998 and a coefficient of variation

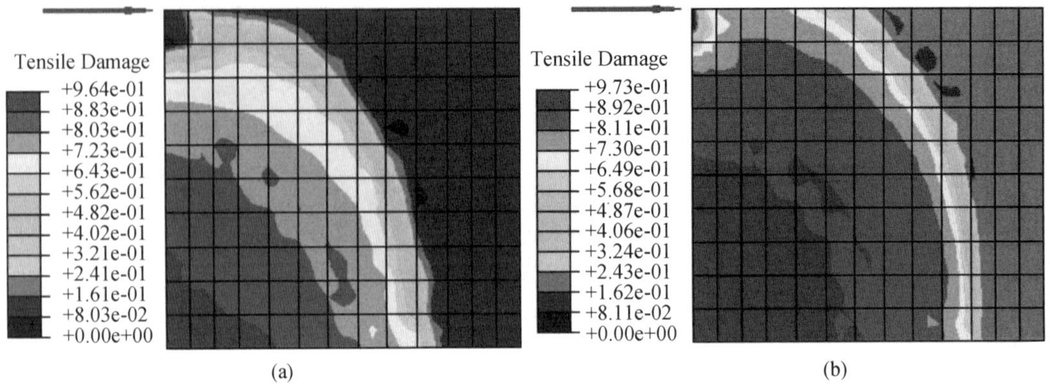

Fig. 10. Distribution of tensile damage in shearwall SW11—(a) horizontal force: 122.0 kN; (b) horizontal force: 153.1 kN

of 0.042, and ② the corresponding values for the displacement ratios are 1.095 and 0.131, respectively.

6 Conclusions

This paper presents a damage energy release rate-based plastic-damage constitutive model, mainly intended for the nonlinear analysis of plain and reinforced concrete structures. Within the framework of continuum damage mechanics, two damage scalars that lead to a fourth-order damage tensor are adopted to describe the degradation of the macromechanical properties of concrete. An elegant constitutive law is formulated on the basis of a decomposition of effective stress tensor. The plastic Helmholtz free energy is taken into accounted for the damage growth, and the damage criteria are based on the elastoplastic damage energy release rates.

The numerical predictions from the proposed model when applied to plain concrete and reinforced concrete specimens and structures were checked in several experimental examples, demonstrating the following capabilities:

(1) It is able to reproduce most of the typical nonlinear performances of concrete under monotonic and cyclic load conditions, as well as under 2D or 3D stress states;

(2) It is capable of providing not only the experimentally observed load-deflection response curves, but also the real distributions of structural damages.

Acknowledgements

The project under which this paper was prepared was supported by the National Natural Science Foundation for Outstanding Youth in China (No. 59825105) and the Natural Science Foundations of China for Innovative Research Groups (No. 50321803), which are gratefully acknowledged.

Appendix Ⅰ. Derivation of the projection tensors of $\bar{\boldsymbol{\sigma}}$ and $\dot{\bar{\boldsymbol{\sigma}}}$

The spectral decomposition of the effective stress tensor proposed in Ortiz[15] is considered here

$$\bar{\boldsymbol{\sigma}} = \sum_i \bar{\sigma}_i \mathbf{n}_i \otimes \mathbf{n}_i \quad \text{(A-Ⅰ.1)}$$

where $\bar{\sigma}_i$ and $\mathbf{n}_i$ ($i=1, 2, 3$) are the extracted effective eigenstresses and the corresponding normalized eigenvectors, which the following properties hold

$$\mathbf{n}_i \cdot \bar{\boldsymbol{\sigma}} \cdot \mathbf{n}_i = \boldsymbol{\sigma}_i ; \ \mathbf{n}_i \cdot \mathbf{n}_j = \delta_{ij} \quad \text{(A-Ⅰ.2a, b)}$$

The positive component of the effective stress tensor $\bar{\boldsymbol{\sigma}}^+$ can be expressed as

$$\bar{\boldsymbol{\sigma}}^+ = \sum_i \langle \bar{\sigma}_i \rangle \mathbf{n}_i \otimes \mathbf{n}_i = \sum_i H(\bar{\sigma}_i) \mathbf{n}_i \otimes \boldsymbol{n}_i \bar{\sigma}_i \quad \text{(A-Ⅰ.3)}$$

Substituting Eq.(A-Ⅰ.2a) into Eq.(A-Ⅰ.3), one obtains

$$\bar{\boldsymbol{\sigma}}^+ = \sum_i H(\bar{\sigma}_i) \mathbf{n}_i \otimes \mathbf{n}_i [(\mathbf{n}_i \otimes \mathbf{n}_j) : \bar{\boldsymbol{\sigma}}] = \mathbf{P}^+ : \bar{\boldsymbol{\sigma}} \quad \text{(A-Ⅰ.4)}$$

with the positive projection tensor $\mathbf{P}^+$ being[31]

$$\mathbf{P}^+ = \sum_i H(\bar{\sigma}_i)(\mathbf{n}_i \otimes \mathbf{n}_i) \otimes (\mathbf{n}_i \otimes \mathbf{n}_i) \quad \text{(A-Ⅰ.5)}$$

which were already defined in Eqs.(3a) and (4a), accordingly, $\bar{\boldsymbol{\sigma}}^-$ and $\mathbf{P}^-$ defined in Eqs.(3b) and (4b) can thus be easily obtained.

The total differential of Eq.(A-Ⅰ.3) gives

$$\mathrm{d}\bar{\boldsymbol{\sigma}}^+ = \sum_i H(\bar{\sigma}_i)(\mathbf{n}_i \otimes \mathbf{n}_i)\mathrm{d}\bar{\sigma}_i + \sum_i \langle \bar{\sigma}_i \rangle \mathrm{d}(\mathbf{n}_i \otimes \mathbf{n}_i) \quad \text{(A-Ⅰ.6)}$$

Differentiating of Eq.(A-Ⅰ.2a), one obtains

$$\mathrm{d}\bar{\sigma}_i = \mathbf{n}_i \cdot \mathrm{d}\bar{\boldsymbol{\sigma}} \cdot \mathbf{n}_i + 2\mathrm{d}\mathbf{n}_i \cdot \bar{\boldsymbol{\sigma}} \cdot \mathbf{n}_i = (\mathbf{n}_i \otimes \mathbf{n}_i) : \mathrm{d}\bar{\boldsymbol{\sigma}} \quad \text{(A-Ⅰ.7)}$$

where Eq.(A-Ⅰ.2b) is taken into consideration to conclude that the above second right term is equal to zero.

In Faria et al.[31] the expression for $\mathrm{d}(\mathbf{n}_i \otimes \mathbf{n}_i)$ was derived as

$$\mathrm{d}(\mathbf{n}_i \otimes \mathbf{n}_i) = \mathrm{d}\mathbf{n}_i \otimes \boldsymbol{n}_i + \mathbf{n}_i \otimes \mathrm{d}\mathbf{n}_i = \left[2\sum_{j\neq i} \frac{1}{\bar{\sigma}_i - \bar{\sigma}_j}(\mathbf{p}_{ij} \otimes \mathbf{p}_{ij})\right] : \mathrm{d}\bar{\boldsymbol{\sigma}} \quad \text{(A-Ⅰ.8)}$$

where $\mathbf{p}_{ij}$ is the second-order symmetry tensor defined in Eq.(7).

Substituting Eqs.(A-Ⅰ.7) and (A-Ⅰ.8) into Eq.(A-Ⅰ.6), and after some simple simplifications, $\mathbf{Q}^+$ and $\mathbf{Q}^-$ which are the fourth-order symmetry projection tensor of $\dot{\bar{\boldsymbol{\sigma}}}$, can thus be derived as those expressions defined in Eqs.(6a) and (6b).

Appendix Ⅱ. Derivation of the plastic and elastoplastic HFE potential

Since the contribution to the HFE potential of plastic strains in tension is neglected in the proposed mode, i.e, $\psi_0^{p+}=0$ is assumed, the plastic HFE potential is defined

$$\psi_0^{\mathrm{p}}(\boldsymbol{\kappa})=\psi_0^{\mathrm{p}-}(\boldsymbol{\kappa})=\int_0^{\boldsymbol{\varepsilon}^{\mathrm{p}}}\bar{\boldsymbol{\sigma}}^{-}:\mathrm{d}\boldsymbol{\varepsilon}^{\mathrm{p}} \tag{A-Ⅱ.1}$$

Substituting Eq. (25) into (A-Ⅱ.1), and after some simplifications[34], the above initial plastic HFE potential becomes Eq.(36) as

$$\psi_0^{\mathrm{p}}=\frac{b}{2E_0}\left(3\bar{J}_2^{-}+\eta^{\mathrm{p}}\bar{I}_1^{-}\sqrt{3\bar{J}_2}-\frac{1}{2}\bar{I}_1^{+}\bar{I}_1^{-}\right) \tag{A-Ⅱ.2}$$

where parameter $b\geqslant 0$ will be determined in the following procedure. If $b=0$ is assumed, the contribution of plastic HFE would not be taken into consideration.

Calling for Eq.(2) and with ν_0 denoting the initial Poisson's ratio, the negative component of the elastic HFE potential defined in Eq.(10b) is rewritten as

$$\psi_0^{\mathrm{e}-}=\frac{1}{2E_0}\left[\frac{2(1+\nu_0)}{3}3\bar{J}_2^{-}+\frac{1-2\nu_0}{3}(\bar{I}_1^{-})^2-\nu_0\bar{I}_1^{+}\bar{I}_1^{-}\right] \tag{A-Ⅱ.3}$$

Then ψ_0^{-} defined in Eq.(39b) becomes

$$\psi_0^{-}=\psi_0^{\mathrm{e}-}+\psi_0^{\mathrm{p}-}=b_0[3\bar{J}_2^{-}+b_1\bar{I}_1^{-}\sqrt{3\bar{J}_2}+b_2(\bar{I}_1^{-})^2+b_3\bar{I}_1^{+}\bar{I}_1^{-}] \tag{A-Ⅱ.4}$$

with parameters b_0, b_1, b_2 and b_3 being

$$b_0=a/(6E_0);\ b_1=3b\eta^{\mathrm{p}}/a;\ b_2=(1-2\nu_0)/a;\ b_3=-1.5(b-2\nu_0)/a \tag{A-Ⅱ.5}$$

and $a=3b+2(1+\nu_0)$.

Calling for Eqs.(41) and (A-Ⅱ.4), under pure compression Y^{-} can then be reduced to the following expression:

$$Y^{-}=\psi_0^{-}=b_0\left[3\bar{J}_2^{-}+b_1\bar{I}_1^{-}\sqrt{3\bar{J}_2^{-}}+b_2(\bar{I}_1^{-})^2\right] \tag{A-Ⅱ.6}$$

Denoting by f_{b0}^{-} the stresses (positive values) beyond which nonlinearity becomes visible under equibiaxial compression, the initial shear damage threshold r_0^{-} is thus established as[34]

$$r_0^{-}=\frac{1}{2E_0}[1+b(1-\eta^{\mathrm{p}})](f_0^{-})^2=\frac{1}{2E_0}[2(1-\nu_0)+b(1-2\eta^{\mathrm{p}})](f_{b0}^{-})^2 \tag{A-Ⅱ.7}$$

leading to the following expression for parameter b introduced in Eq.(A-Ⅱ.2) as

$$b=\frac{1-2(1-\nu_0)\vartheta_0^2}{(1-2\eta^{\mathrm{p}})\vartheta_0^2-(1-\eta^{\mathrm{p}})} \tag{A-Ⅱ.8}$$

with $\vartheta_0=f_{b0}^{-}/f_0^{-}$, assumed to take the same value as ϑ defined in Eq.(31a), i.e, $\vartheta_0=\vartheta$[35]. For concrete, the following inequality is inherently satisfied

$$1-2(1-\nu_0)\vartheta^2<0 \tag{A-Ⅱ.9}$$

Owing to the non-negative nature of parameter b, Eqs. (A-Ⅱ.8) and (A-Ⅱ.9) lead to condition that dilatancy parameter α^{p} should fulfill

$$\alpha^{\mathrm{p}} = \sqrt{2/3}\,\eta^{\mathrm{p}} \geqslant \sqrt{2/3}\,(\vartheta^2 - 1)/(2\vartheta^2 - 1) \qquad \text{(A-II.10)}$$

Since ratio ϑ usually lies in the interval 1.10–1.20, from Eq. (A-Ⅱ.10) one gets $\alpha^{\mathrm{p}} \geqslant 0.12-0.19$, which can be strictly satisfied since the value of 0.20 – 0.30 is usually adopted for concrete[22].

Taking into consideration Eqs. (A-Ⅱ.6) and (A-Ⅱ.7), for different dilatancy parameters η^{p}, the corresponding domains of linear behaviour of concrete under biaxial compression, using the typical parameters $\nu_0 = 0.2$ and $\vartheta = 1.16$, are shown in Fig. 11; the plastic potential function introduced in Eq. (24), corresponding to different values of η^{p} are also reproduced. As the dilatancy parameter η^{p} (or α^{p}) has substantial influence on the plastic potential, but almost no effect on the domains of linear behaviour under biaxial compression, it allows great simplifications to be introduced on the shear DERR expressed in Eq. (A-Ⅱ.6). Let η^{p} take the value of 0.257 56, i.e., $\alpha^{\mathrm{p}} = 0.21$, then $b_2 = (b_1/2)^2$ is obtained, and consequently Eq. (A-Ⅱ.6) can be approximated by the following form of a perfect square polynomial:

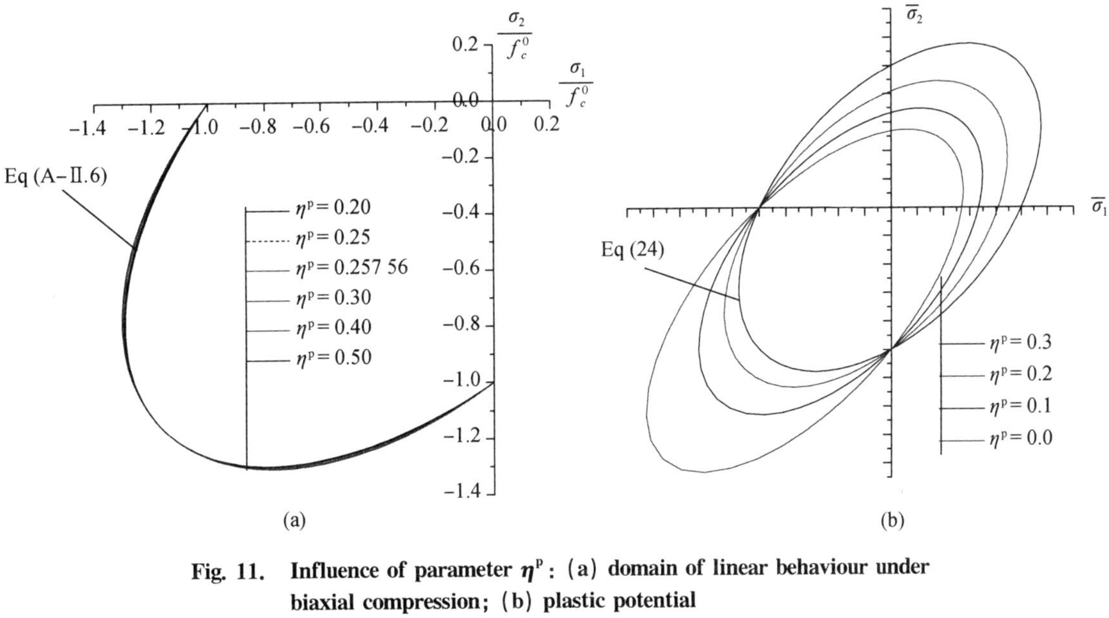

Fig. 11. Influence of parameter η^{p}: (a) domain of linear behaviour under biaxial compression; (b) plastic potential

$$Y^- = \psi_0^- = b_0\left(\alpha \bar{I}_1^- + \sqrt{3\bar{J}_2^-}\right)^2 \qquad \text{(A-II.11)}$$

In a tension-included stress states (e.g. tensile-compressive quadrants), according to Eq. (A-Ⅱ.4) the tensile stresses do have some influence on the domain of linear behaviour, localized in the circle areas close to the tension-compression corners depicted in Fig. 1, whose effect is so subtle that can be ignored. Consequently, the negative component of initial elastoplastic HFE potential ψ_0^- and the corresponding shear DERR Y^- defined in Eq. (A-Ⅱ.11), can be further approximated as the expression defined in Eq. (40b).

It should be noted that the above approximations lead to great simplicity without significant loss of accuracy, and at the same time they ensure the convexity of the damage surfaces, which not only can greatly facilitate its numerical implementation, but also may enhance the numerical stability of the proposed model.

References

[1] OHTANI Y, CHEN W. Multiple hardening plasticity for concrete materials[J]. Journal of Engineering Mechanics, 1988, 114(11): 1890-1910.

[2] ETSE G, WILLAM K. Fracture energy formulation for inelastic behavior of plain concrete[J]. Journal of Engineering Mechanics, 1994, 120(9): 1983-2011.

[3] FEENSTRA P H, DE BORST R. A composite plasticity model for concrete[J]. International Journal of Solids and Structures, 1996, 33(5): 707-730.

[4] CHEN W F, SALEEB A F. Constitutive equations for engineering materials: Elasticity and modeling [M]. Elsevier, 1994.

[5] KRAJCINOVIC D. Constitutive theories for solids with defective microstructure[C]//Coastal Zone'89. ASCE, 1985: 39-56.

[6] YAZDANI S, SCHREYER H L. Combined plasticity and damage mechanics model for plain concrete [J]. Journal of Engineering Mechanics, 1990, 116(7): 1435-1450.

[7] JU J W. On energy-based coupled elastoplastic damage theories: constitutive modeling and computational aspects[J]. International Journal of Solids and Structures, 1989, 25(7): 803-833.

[8] MAZARS J. Application de la mécanique de l'endommagement au comportement non linéaire et à la rupture du béton de structure[J]. These De Docteur Es Sciences Presentee A L'universite Pierre Et Marie Curie-Paris 6, 1984.

[9] MAZARS J. A model of a unilateral elastic damageable material and its application to concrete[J]. Fracture Toughness and Fracture Energy of Concrete, 1986: 61-71.

[10] MAZARS J, Pijaudier-Cabot G. Continuum damage theory—application to concrete[J]. Journal of Engineering Mechanics, 1989, 115(2): 345-365.

[11] LUBARDA V A, KRAJCINOVIC D, MASTILOVIC S. Damage model for brittle elastic solids with unequal tensile and compressive strengths[J]. Engineering Fracture Mechanics, 1994, 49(5): 681-697.

[12] CERVERA M, OLIVER J, FARIA R. Seismic evaluation of concrete dams via continuum damage models[J]. Earthquake Engineering & Structural Dynamics, 1995, 24(9): 1225-1245.

[13] HALM D, DRAGON A. A model of anisotropic damage by mesocrack growth; unilateral effect[J]. International Journal of Damage Mechanics, 1996, 5(4): 384-402.

[14] COMI C, PEREGO U. Fracture energy based bi-dissipative damage model for concrete [J]. International Journal of Solids and Structures, 2001, 38(36-37): 6427-6454.

[15] ORTIZ M. A constitutive theory for the inelastic behavior of concrete[J]. Mechanics of Materials, 1985, 4(1): 67-93.

[16] LEMAITRE J. A continuous damage mechanics model for ductile fracture[J]. Journal of Engineering Materials Technology, 1985,107:83-89.

[17] DRAGON A. Plasticity and ductile fracture damage: study of void growth in metals[J]. Engineering Fracture Mechanics, 1985, 21(4): 875-885.

[18] RESENDE L. A damage mechanics constitutive theory for the inelastic behaviour of concrete[J].

Computer Methods in Applied Mechanics and Engineering, 1987, 60(1): 57-93.

[19] SIMO J C, JU J W. Strain- and stress-based continuum damage models-Ⅰ: formulation [J]. International Journal of Solids andStructures, 1987, 23 (7): 821-840.

[20] CAROL I, RIZZI E, WILLAM K. A unified theory of elastic degradation and damage based on a loading surface[J]. International Journal of Solids and Structures, 1994, 31(20): 2835-2865.

[21] DI PRISCO M, MAZARS J. Crush-crack': a non-local damage model for concrete[J]. Mechanics of Cohesive-frictional Materials: An International Journal on Experiments, Modelling and Computation of Materials and Structures, 1996, 1(4): 321-347.

[22] LEE J, FENVES G L. Plastic-damage model for cyclic loading of concrete structures[J]. Journal of Engineering Mechanics, 1998, 124(8): 892-900.

[23] FARIA R, OLIVER J, CERVERA M. A strain-based plastic viscous-damage model for massive concrete structures[J]. International Journal of Solids and Structures, 1998, 35(14): 1533-1558.

[24] HATZIGEORGIOU G, BESKOS D, THEODORAKOPOULOS D, et al. A simple concrete damage model for dynamic FEM applications[J]. International Journal of Computational Engineering Science, 2001, 2(02): 267-286.

[25] HANSEN E, WILLAM K, CAROL I. A two-surface anisotropic damage/plasticity model for plain concrete[C]//Fourth International Conference on Fracture Mechanics of Concrete and Concrete Structures. 2001: 549-556.

[26] CHOW C L, WANG J. An anisotropic theory of continuum damage mechanics for ductile fracture[J]. Engineering fracture mechanics, 1987, 27(5): 547-558.

[27] KUPFER H, HILSDORF H K, RUSCH H. Behavior of concrete under biaxial stresses[C]//Journal proceedings. 1969, 66(8): 656-666.

[28] VECCHIO F J, COLLINS M P. The modified compression-field theory for reinforced concrete elements subjected to shear[J]. ACI Structure Journal, 1986, 83(2): 219-231.

[29] CHABOCHE J L, LESNE P M, MAIRE J F. Continuum damage mechanics, anisotropy and damage deactivation for brittle materials like concrete and ceramic composites[J]. International Journal of Damage Mechanics, 1995, 4(1): 5-22.

[30] JU J W. Isotropic and anisotropic damage variables in continuum damage mechanics[J]. Journal of Engineering Mechanics, 1990, 116(12): 2764-2770.

[31] FARIA R, OLIVER J O O, Ruiz M C. On isotropic scalar damage models for the numerical analysis of concrete structures [M]. Barcelona, Spain: Centro Internacional de Métodos Numéricos en Ingeniería, 2000.

[32] LUBLINER J. On the thermodynamic foundations of non-linear solid mechanics[J]. International Journal of Non-Linear Mechanics, 1972, 7(3): 237-254.

[33] COLEMAN B D, GURTIN M E. Thermodynamics with internal state variables[J]. The Journal of Chemical Physics, 1967, 47(2): 597-613.

[34] WU J Y, LI J. A new energy-based elastoplastic damage model for concrete[C]//Proc. of XXI Int. Conf. of Theoretical and Applied Mechanics (ICTAM), Warsaw, Poland. 2004.

[35] LUBLINER J, OLIVER J, OLLER S, et al. A plastic-damage model for concrete[J]. International Journal of Solids and Structures, 1989, 25(3): 299-326.

[36] SIMO J C, HUGHES T J R. Computational Inelasticity[M]. Springer-Verlag. New York, 1998: 143-149.

[37] WU J Y. Damage energy release rate-based elastoplastic damage constitutive model for concrete and its

application to nonlinear analysis of structures [D]. Shanghai, China: Department of Building Engineering, Tongji University, 2004.

[38] LI J, WU J Y. Energy-based CDM model for nonlinear analysis of confined concrete[C]//ACI SP-238, In: Proc. of the Symposium on Confined Concrete, Changsha, China. 2004.

[39] STEVENS N J, UZUMERI S M, WILL G T. Constitutive model for reinforced concrete finite element analysis[J]. Structural Journal, 1991, 88(1): 49-59.

[40] OLIVER J, CERVERA M, OLLER S, et al. Isotropic damage models and smeared crack analysis of concrete, SCI-C 1990, Second Int[C]//Conf. on Computer Aided Design of Concrete Structure, Zell am See, Austria. 1990: 945-957.

[41] LEE J, FENVES G L. A return-mapping algorithm for plastic-damage models: 3-D and plane stress formulation[J]. International Journal for Numerical Methods in Engineering, 2001, 50(2): 487-506.

[42] GOPALARATNAM V S, SHAH S P. Softening response of plain concrete in direct tension[C]// Journal Proceedings. 1985, 82(3): 310-323.

[43] ZHANG Q. Research on the stochastic damage constitutive of concrete material [D]. Ph. D. Dissertation, Tongji University, Shanghai, China, 2001.

[44] KARSAN I D, JIRSA J O. Behavior of concrete under compressive loadings [J]. Journal of the Structural Division, 1969, 95(12): 2543-2564.

[45] GREEN S J, SWANSON S R. Static constitutive relations for concrete[R]. Terra Tek Inc Salt Lake City Ut, 1973.

[46] TAYLOR R L. FEAP: a finite element analysis program for engineering workstation. Rep. No[R]. UCB/SEMM-92 (Draft Version), Department of Civil Engineering, University of California, Berkeley, 1992.

[47] CHOPRA A K, CHAKRABARTI P. The Koyna earthquake and the damage to Koyna dam [J]. Bulletin of the Seismological Society of America, 1973, 63(2): 381-397.

[48] BHATTACHARJEE S S, LEGER P. Seismic cracking and energy dissipation in concrete gravity dams [J]. Earthquake Engineering & Structural Dynamics, 1993, 22(11): 991-1007.

[49] GHRIB F, TINAWI R. An application of damage mechanics for seismic analysis of concrete gravity dams[J]. Earthquake Engineering & Structural Dynamics, 1995, 24(2): 157-173.

[50] CERVERA M, OLIVER J, MANZOLI O. A rate-dependent isotropic damage model for the seismic analysis of concrete dams [J]. Earthquake Engineering & Structural Dynamics, 1996, 25(9): 987-1010.

[51] WESTERGAARD H M. Water pressures on dams during earthquakes [J]. Transactions of the American Society of Civil Engineers, 1933, 98(2): 418-433.

[52] MCNEICE G M. Elastic-plastic bending analysis of plates and slabs by the finite element method[D]. University of London, 1967.

[53] LEFAS I D, KOTSOVOS M D, AMBRASEYS N N. Behavior of reinforced concrete structural walls: strength, deformation characteristics, and failure mechanism [J]. Structural Journal, 1990, 87(1): 23-31.

[54] VECCHIO F J. Finite element modeling of concrete expansion and confinement [J]. Journal of Structural Engineering, 1992, 118(9): 2390-2406.

Elastoplastic damage model for concrete based on consistent free energy potential①

Ji Zhang, Jie Li

State Key Laboratory for Disaster Reduction in Civil Engineering, Tongji University, Shanghai 200092, China

Abstract The aim of this study is to formulate an appropriate free energy potential for inelastic behavior of concrete and construct an elastoplastic damage model on a more rational basis. The concept of effective plastic energy storage rates is proposed, which are conjugate forces of hardening variables in an undamaged configuration. Then an analogy between the evolution of hardening variables and that of a plastic strain is used to postulate the formulation of plastic free energy. This formulation reflects the specific characteristics of a certain plasticity model, so it can serve well as a thermodynamic link between plasticity and damage. By combination of the general formulation of free energy with the double hardening plasticity theory and two-parameter damage expression, a thermodynamically well-founded elastoplastic damage model for concrete is constructed. The operator split algorithm is employed, and the numerical simulations agree well with a series of material tests.

Keywords Free energy; Plastic energy storage; Damage energy release; Constitutive model; Concrete

1 Introduction

Nonlinear analysis of concrete structures has drawn wide attention and extensive efforts in the civil engineering community for half a century, in which constitutive modeling remains a critical issue owing to the complex mechanical behavior of concrete materials under external actions.

Research on constitutive relations is of everlasting interest in solid mechanics. As for concrete, various kinds of theories have been proposed, among which plasticity models[1-7] and damage models[8-17] are the most fully developed. Nowadays continuum damage mechanics along with micromechanical and multi-scale damage theories is progressing rapidly and has become a quite promising approach to characterize concrete behavior.

Continuum damage mechanics is based on the solid foundation of thermodynamics, and one of its central concepts, a damage energy release rate, is directly derived from a Helmholtz free energy potential. How to determine and formulate free energy for inelastic solids has always been a difficult and controversial issue while they always need to be prescribed in

① Originally published in *Science China Technological Sciences*, 2014, 57(11): 2278-2286.

the first place for the construction of damage models.

Free energy equal to the elastic strain energy has served as the basis of many elastic damage models[8, 13] and some elastoplastic damage models as well[5, 18]. Although this simple formulation seems natural, it is not applicable when plastic deformation is taken into account. Justification of this argument has been discussed in depth in refs.[19-20].

By introducing a plastic part of free energy, Ju[11] takes a step from elastic damage to elastoplastic damage in a thermodynamic sense. Wu et al.[14] propose an approximation for the integration of the plastic free energy of Ju[11]. This approximate expression of the plastic free energy has been used in a series of subsequent further studies[16, 21-23]. Another form of inelastic free energy for concrete[15] is transplanted from certain elastoplastic damage models developed for ductile materials[24] and has been inherited in some following research work[25-26].

The above two kinds of formulations of free energy for concrete by Wu et al.[14] and Cicekli et al.[15], which are in active use in current investigation, fail to include the specific characteristics of a plasticity model. Consequently, the damage sub-model derived from the free energy is unrelated to the plasticity sub-model employed. This is against the inherent coupling between damage and plastic evolutions.

In the field of crystal plasticity, micromechanical and physical analysis has been employed to determine free energy[20, 27]. As regards concrete and similar brittle materials, although some success has been achieved in the description of (stochastic) damage evolution by micromechanical and multi-scale theories[16-17, 21, 26, 28], very little is known about the micro-mechanisms of plastic evolution. Zhang and Li[29] give a microelement formulation of elastoplastic free energy, which, however, is not yet directly applicable to multidimensional constitutive modeling.

The present investigation aims to find an appropriate free energy potential for inelastic behavior of concrete and then construct an elastoplastic damage model on a more rational basis. We intend that this free energy reflects the specific characteristics of a plasticity model so that the damage sub-model derived from it is in consistency with the plasticity sub-model employed, and damage and plasticity are properly combined in one model.

2 Free energy

2.1 Mapping of free energy

For infinitesimal elastoplastic deformation, the strain $\boldsymbol{\varepsilon}$ is split additively into an elastic part $\boldsymbol{\varepsilon}^{\mathrm{e}}$ and a plastic part $\boldsymbol{\varepsilon}^{\mathrm{p}}$ as

$$\boldsymbol{\varepsilon} = \boldsymbol{\varepsilon}^{\mathrm{e}} + \boldsymbol{\varepsilon}^{\mathrm{p}}. \tag{1}$$

According to Lubliner[30], to be consistent with Eq.(1), the Helmholtz free energy (per unit volume) ψ must can be decomposed as the sum of its elastic part ψ^{e} and plastic part ψ^{p}. Within the context of an elastoplastic isotropic damage theory for isothermal processes, this is usually expressed as

$$\psi(\boldsymbol{\varepsilon}^{\mathrm{e}}, d, \kappa_1, \kappa_2, \cdots, \kappa_n) = \psi^{\mathrm{e}}(\boldsymbol{\varepsilon}^{\mathrm{e}}, d) + \psi^{\mathrm{p}}(d, \kappa_1, \kappa_2, \cdots, \kappa_n), \tag{2}$$

where d is the damage variable, and κ_i the hardening variables. Such a decomposition of ψ is also supported by some experimental results[19].

The elastic free energy is calculated as

$$\psi^{\mathrm{e}} = \frac{1}{2}\boldsymbol{\sigma} : \boldsymbol{\varepsilon}^{\mathrm{e}} = \frac{1}{2}\boldsymbol{\varepsilon}^{\mathrm{e}} : \mathbb{C} : \boldsymbol{\varepsilon}^{\mathrm{e}}, \tag{3}$$

where $\boldsymbol{\sigma}$ is the nominal (Cauchy) stress, and $\mathbb{C}$ the elastic stiffness. With the mapping between $\boldsymbol{\sigma}$ and the effective stress $\bar{\boldsymbol{\sigma}}$,

$$\boldsymbol{\sigma} = (1-d)\bar{\boldsymbol{\sigma}}, \tag{4}$$

and the hypothesis of strain equivalence,

$$\boldsymbol{\sigma} = \mathbb{C} : \boldsymbol{\varepsilon}^{\mathrm{e}}, \ \bar{\boldsymbol{\sigma}} = \bar{\mathbb{C}} : \boldsymbol{\varepsilon}^{\mathrm{e}}, \tag{5}$$

where $\bar{\mathbb{C}}$ is the effective elastic stiffness, a mapping between ψ^{e} and its counterpart in the undamaged configuration, the effective elastic free energy

$$\bar{\psi}^{\mathrm{e}} = \frac{1}{2}\bar{\boldsymbol{\sigma}} : \boldsymbol{\varepsilon}^{\mathrm{e}} = \frac{1}{2}\boldsymbol{\varepsilon}^{\mathrm{e}} : \bar{\mathbb{C}} : \boldsymbol{\varepsilon}^{\mathrm{e}}, \tag{6}$$

can be obtained as

$$\psi^{\mathrm{e}} = (1-d)\bar{\psi}^{\mathrm{e}}. \tag{7}$$

How to determine ψ^{p} for damaged concrete is still unknown. For the sake of simplicity, if it is assumed that a mapping similar to Eq. (7) also exists for ψ^{p}[11], that is

$$\psi^{\mathrm{p}} = (1-d)\bar{\psi}^{\mathrm{p}}, \tag{8}$$

where the effective plastic free energy $\bar{\psi}^{\mathrm{p}}$ is independent of d, one can get

$$\psi = (1-d)\bar{\psi}, \tag{9}$$

where the effective free energy

$$\bar{\psi}(\boldsymbol{\varepsilon}^{\mathrm{e}}, \kappa_1, \kappa_2, \cdots, \kappa_n) = \bar{\psi}^{\mathrm{e}}(\boldsymbol{\varepsilon}^{\mathrm{e}}) + \bar{\psi}^{\mathrm{p}}(\kappa_1, \kappa_2, \cdots, \kappa_n). \tag{10}$$

A comparison between Eqs. (2) and (10) shows that the assumption in Eq. (8) makes $\bar{\psi}$ independent of d. Consequently, with Eq. (9), the damage energy release rate takes a simple form

$$Y = -\frac{\partial \psi}{\partial d} = \bar{\psi}. \tag{11}$$

2.2 Clausius-Duhem inequality

According to Coleman and Gurtin[31], for isothermal (pure mechanical) processes, the Clausius-Duhem inequality takes the form

$$T\dot{S}_{\mathrm{i}} = \boldsymbol{\sigma} : \dot{\boldsymbol{\varepsilon}} - \dot{\psi} \geqslant 0, \tag{12}$$

where T is the absolute temperature, and $\dot{S}_i$ the intrinsic entropy production rate. Substitution of Eqs. (1) and (2) into Eq. (12) gives

$$\left(\boldsymbol{\sigma}-\frac{\partial \psi^{e}}{\partial \boldsymbol{\varepsilon}^{e}}\right): \dot{\boldsymbol{\varepsilon}}^{e}+\boldsymbol{\sigma}: \dot{\boldsymbol{\varepsilon}}^{p}-\sum_{i} \frac{\partial \psi^{p}}{\partial \kappa_{i}} \dot{\kappa}_{i}-\frac{\partial \psi}{\partial d} \dot{d} \geqslant 0. \tag{13}$$

For Eq. (13) to be valid for any admissible processes, one gets

$$\boldsymbol{\sigma}-\frac{\partial \psi^{e}}{\partial \boldsymbol{\varepsilon}^{e}}=0, \tag{14}$$

and

$$\boldsymbol{\sigma}: \dot{\boldsymbol{\varepsilon}}^{p}-\sum_{i} \frac{\partial \psi^{p}}{\partial \kappa_{i}} \dot{\kappa}_{i}-\frac{\partial \psi}{\partial d} \dot{d} \geqslant 0. \tag{15}$$

It is often assumed that plastic and damage dissipations are decoupled from each other, and then Eq. (15) is replaced by

$$\boldsymbol{\sigma}: \dot{\boldsymbol{\varepsilon}}^{p}-\sum_{i} \frac{\partial \psi^{p}}{\partial \kappa_{i}} \dot{\kappa}_{i} \geqslant 0, \tag{16}$$

and

$$-\frac{\partial \psi}{\partial d} \dot{d} \geqslant 0. \tag{17}$$

Substitution of Eqs. (4) and (8) into Eq. (16) leads to

$$\bar{\boldsymbol{\sigma}}: \dot{\boldsymbol{\varepsilon}}^{p}-\sum_{i} \frac{\partial \bar{\psi}^{p}}{\partial \kappa_{i}} \dot{\kappa}_{i} \geqslant 0, \tag{18}$$

which shows that the mapping of free energy brings a plastic dissipation inequality in the undamaged configuration.

2.3 Effective plastic energy storage rates

With the definition of the concept of damage energy release rate Y in Eq. (11), the damage dissipation inequality Eq. (17) can be rephrased as

$$Y\dot{d} \geqslant 0. \tag{19}$$

This conjugate pair in Eq. (19) consists of an important internal variable d and an important thermodynamic force Y. For an elastoplastic damage model, another important set of internal variables are κ_i, whose conjugate forces are introduced here as

$$Z_{i}=\frac{\partial \psi^{p}}{\partial \kappa_{i}}=\frac{\partial \psi}{\partial \kappa_{i}}, \tag{20}$$

and tentatively referred to as the plastic energy storage rates. The concept of Z_i has been introduced in literature[32]; in the present study, we try to use them to find free energy for elastoplastic damage models.

Further, we define the counterparts of Z_i in the undamaged configuration, the effective plastic energy storage rates, as

$$\bar{Z}_i = \frac{\partial \bar{\psi}^{\mathrm{p}}}{\partial \kappa_i} = \frac{\partial \bar{\psi}}{\partial \kappa_i}. \tag{21}$$

With the definition in Eq. (21), the plastic dissipation inequality in the undamaged configuration Eq.(18) can be rephrased as

$$\bar{\boldsymbol{\sigma}} : \dot{\boldsymbol{\varepsilon}}^{\mathrm{p}} - \sum_i \bar{Z}_i \dot{\kappa}_i \geqslant 0. \tag{22}$$

And since $\bar{\psi}^{\mathrm{p}}$ is dependent only on κ_i, it can be calculated from Eq.(21) as

$$\bar{\psi}^{\mathrm{p}} = \sum_i \int_0^{\kappa_i} \bar{Z}_i \mathrm{d}\kappa_i, \tag{23}$$

as long as $\bar{Z}_i$ are known.

Remark: Some single-crystal plasticity theories[20, 27] use plastic free energy functions of similar forms to Eq. (23), where κ_i are dislocation densities, and Z_i are related to the interaction forces between dislocations (Peach-Koehler forces).

For concrete and similar brittle materials, however, appropriate definitions of κ_i with clear physical mechanisms have not been found, let alone those of Z_i. Plastic free energy might come from the microscopic residual strain energy locked by microcracks.

In order to find $\bar{Z}_i$, an idea in Ju[11] is resorted to, which is referred to here as flow rule analogy. For the generalized force-displacement pairs $\bar{\boldsymbol{\sigma}} - \dot{\boldsymbol{\varepsilon}}^{\mathrm{p}}$ and $\bar{Z}_i - \dot{\kappa}_i$ in Eq.(22), it is postulated that κ_i evolve in an analogous way as $\boldsymbol{\varepsilon}^{\mathrm{p}}$. The evolution of $\boldsymbol{\varepsilon}^{\mathrm{p}}$ is prescribed by the flow rule

$$\dot{\boldsymbol{\varepsilon}}^{\mathrm{p}} = \dot{\lambda} \frac{\partial g^{\mathrm{p}}}{\partial \bar{\boldsymbol{\sigma}}}, \tag{24}$$

where $\dot{\lambda}$ is the consistency parameter, and g^{p} the plastic potential. Then the evolution of κ_i may be prescribed by

$$\dot{\kappa}_i = \dot{\lambda} \frac{\partial g^{\mathrm{p}}}{\partial \bar{Z}_i}, \tag{25}$$

which shares the same $\dot{\lambda}$ and g^{p} as Eq.(24).

For a specific definition of κ_i, their evolution law is already known as

$$\dot{\kappa}_i = \dot{\lambda} h_i, \tag{26}$$

where h_i are the hardening functions. A comparison between Eqs.(25) and (26) gives

$$h_i = \frac{\partial g^{\mathrm{p}}}{\partial \bar{Z}_i}, \tag{27}$$

from which $\bar{Z}_i$ can be obtained.

Remark: Ju[11] assumes $\bar{\psi}^{\mathrm{p}}$ to be

$$\bar{\psi}^{\mathrm{p}} = \int_0^{\boldsymbol{\varepsilon}^{\mathrm{p}}} \bar{\boldsymbol{\sigma}} : \mathrm{d}\boldsymbol{\varepsilon}^{\mathrm{p}}, \tag{28}$$

and so do some subsequent elastoplastic damage models[14, 16]. This formulation ignores the effect of the specific definitions of κ_i and Z_i. Substitution of Eq.(28) into Eq.(18) leads to zero plastic dissipation. In other words, the entropy production contributed by plastic evolution in Eq.(12) is neglected.

In the present formulation of $\bar{\psi}^{\mathrm{p}}$ [Eq.(23)], we take into account the internal variables κ_i and their conjugate thermodynamic forces Z_i. By doing so and developing the flow rule analogy, the characteristics of the hardening law [Eq.(26)] and flow rule [Eq.(24)] are embodied in our free energy.

2.4 Special form for Lubliner-Lee plasticity

The above general formulation of free energy is specialized now for a concrete plasticity model widely used in elastoplastic damage modeling. It is originally developed by Lubliner et al.[2] and then extended by Lee and Fenves[3] to a tensile-compressive double hardening model.

The tensile and compressive hardening variables are defined in a rate form as

$$\dot{\kappa}^{+}=w\dot{\varepsilon}_{\max}^{\mathrm{p}},\ \dot{\kappa}^{-}=-(1-w)\dot{\varepsilon}_{\min}^{\mathrm{p}}, \tag{29}$$

where $\dot{\varepsilon}_{\max}^{\mathrm{p}}$ and $\dot{\varepsilon}_{\min}^{\mathrm{p}}$ are the maximum and minimum principal plastic strain rates, respectively, and the weight factor

$$w=\frac{\sum_i \langle\bar{\sigma}_i\rangle}{\sum_i |\bar{\sigma}_i|}, \tag{30}$$

in which $\bar{\sigma}_i$ are the principal effective stresses, and the Mcauley brackets $\langle\ \rangle$ are defined as $\langle x\rangle=\frac{1}{2}(|x|+x)$.

From the flow rule Eq.(24), $\dot{\varepsilon}_{\max}^{\mathrm{p}}$ and $\dot{\varepsilon}_{\min}^{\mathrm{p}}$ are calculated respectively as

$$\dot{\varepsilon}_{\max}^{\mathrm{p}}=\dot{\lambda}\frac{\partial g^{\mathrm{p}}}{\partial\bar{\sigma}_{\max}},\ \dot{\varepsilon}_{\min}^{\mathrm{p}}=\dot{\lambda}\frac{\partial g^{\mathrm{p}}}{\partial\bar{\sigma}_{\min}}, \tag{31}$$

where $\bar{\sigma}_{\max}$ and $\bar{\sigma}_{\min}$ are the maximum and minimum principal effective stresses, respectively. Substituting Eq.(31) into Eq.(29) and comparing it to Eq.(26), one gets respectively the tensile and compressive hardening functions,

$$h^{+}=w\frac{\partial g^{\mathrm{p}}}{\partial\bar{\sigma}_{\max}},\ h^{-}=-(1-w)\frac{\partial g^{\mathrm{p}}}{\partial\bar{\sigma}_{\min}}. \tag{32}$$

Comparing Eq.(32) to Eq.(27), we get respectively the tensile effective plastic energy storage rate

$$\bar{Z}^{+}=\begin{cases}\dfrac{\bar{\sigma}_{\max}}{w}, & 0<w\leqslant 1\\ 0, & w=0\end{cases} \tag{33}$$

and the compressive effective plastic energy storage rate

$$\bar{Z}^{-}=\begin{cases}-\dfrac{\bar{\sigma}_{\min}}{1-w}, & 0\leqslant w<1,\\ 0, & w=1.\end{cases} \tag{34}$$

Substituting Eqs.(33) and (34) into Eq.(23), we obtain

$$\begin{aligned}\bar{\psi}^{\mathrm{p}}&=\int_{0}^{\kappa^{+}}\bar{Z}^{+}\,\mathrm{d}\kappa^{+}+\int_{0}^{\kappa^{-}}\bar{Z}^{-}\,\mathrm{d}\kappa^{-}\\&=\int_{0}^{\varepsilon_{\max}^{\mathrm{p}}}\langle\bar{\sigma}_{\max}\rangle\mathrm{d}\varepsilon_{\max}^{\mathrm{p}}+\int_{0}^{\varepsilon_{\min}^{\mathrm{p}}}\langle-\bar{\sigma}_{\min}\rangle\mathrm{d}(-\varepsilon_{\min}^{\mathrm{p}}),\end{aligned} \tag{35}$$

with which we completely determine the free energy potential for our construction of the elastoplastic damage model.

Remark: Only the maximum and minimum principal values of $\bar{\boldsymbol{\sigma}}$ and $\dot{\boldsymbol{\varepsilon}}^{\mathrm{p}}$ appear in this expression of $\bar{\psi}^{\mathrm{p}}$ [Eq.(35)]. This is because only $\dot{\varepsilon}_{\max}^{\mathrm{p}}$ and $\dot{\varepsilon}_{\min}^{\mathrm{p}}$ are used to define κ^{+} and κ^{-}, which features in the Lubliner-Lee plasticity. In our theory, different plasticity models result in different forms of free energy.

3 Constitutive model

3.1 Two-parameter damage

In order to describe the distinct damage behavior of concrete in tension and compression, $\boldsymbol{\sigma}(\bar{\boldsymbol{\sigma}})$ is split into its positive part $\boldsymbol{\sigma}^{+}$ $(\bar{\boldsymbol{\sigma}}^{+})$ and negative part $\boldsymbol{\sigma}^{-}$ $(\bar{\boldsymbol{\sigma}}^{-})$,

$$\boldsymbol{\sigma}=\boldsymbol{\sigma}^{+}+\boldsymbol{\sigma}^{-},\ \bar{\boldsymbol{\sigma}}=\bar{\boldsymbol{\sigma}}^{+}+\bar{\boldsymbol{\sigma}}^{-}, \tag{36}$$

as first advocated by Ladeveze[8]. The meaning of this positive-negative split can be clearly shown from the dyadic forms of these stress tensors in the principal coordinate system,

$$\begin{aligned}&\bar{\boldsymbol{\sigma}}=\sum_{i}\bar{\sigma}_{i}\,\vec{n}_{i}\otimes\vec{n}_{i},\\&\bar{\boldsymbol{\sigma}}^{+}=\sum_{i}\langle\bar{\sigma}_{i}\rangle\,\vec{n}_{i}\otimes\vec{n}_{i},\ \bar{\boldsymbol{\sigma}}^{-}=-\sum_{i}\langle-\bar{\sigma}_{i}\rangle\,\vec{n}_{i}\otimes\vec{n}_{i},\end{aligned} \tag{37}$$

where $\vec{n}_{i}$ are the orthonormal basis along the principal directions. $\bar{\boldsymbol{\sigma}}^{+}$ and $\bar{\boldsymbol{\sigma}}^{-}$ can be calculated from $\bar{\boldsymbol{\sigma}}$ as

$$\bar{\boldsymbol{\sigma}}^{+}=\mathbb{P}^{+}:\bar{\boldsymbol{\sigma}},\ \bar{\boldsymbol{\sigma}}^{-}=\mathbb{P}^{-}:\bar{\boldsymbol{\sigma}}, \tag{38}$$

with the fourth-order projection operators[33]

$$\mathbb{P}^{+}=\sum_{i}H(\bar{\sigma}_{i})\,\vec{n}_{i}\otimes\vec{n}_{i}\otimes\vec{n}_{i}\otimes\vec{n}_{i},\ \mathbb{P}^{-}=\mathbb{I}-\mathbb{P}^{-}, \tag{39}$$

where $\mathbb{I}$ is the fourth-order symmetric identity tensor.

Two independent damage variables, d^{+} and d^{-}, are introduced to describe tensile and compressive damage respectively as

$$\boldsymbol{\sigma}^{+}=(1-d^{+})\bar{\boldsymbol{\sigma}}^{+},\ \boldsymbol{\sigma}^{-}=(1-d^{-})\bar{\boldsymbol{\sigma}}^{-}. \tag{40}$$

Combination of Eqs.(36) and (40) gives the two-parameter damage expression[12]

$$\boldsymbol{\sigma} = (1-d^+)\bar{\boldsymbol{\sigma}}^+ + (1-d^-)\bar{\boldsymbol{\sigma}}^-. \tag{41}$$

Remark: Comparing Eq.(41) to the general damage expression

$$\boldsymbol{\sigma} = (\mathbb{I} - \mathbb{D}) : \bar{\boldsymbol{\sigma}}, \tag{42}$$

where $\mathbb{D}$ is the fourth-order damage tensor, and using Eqs.(36) and (38), one can get

$$\mathbb{D} = d^+\mathbb{P}^+ + d^-\mathbb{P}^-. \tag{43}$$

According to Eq.(43), it can be regarded that $\mathbb{P}^+$ and $\mathbb{P}^-$ represent tensile and compressive damage mechanisms, respectively, whose linear combination constitutes the overall damage $\mathbb{D}$, d^+ and d^- being the combination coefficients.

3.2 Damage energy release rates

In accordance with the positive-negative split of $\bar{\boldsymbol{\sigma}}$ in Eq.(36), the positivc and negative parts of $\bar{\psi}^{\mathrm{e}}$ are introduced from Eq.(6) as

$$\bar{\psi}^{\mathrm{e}+} = \frac{1}{2}\bar{\boldsymbol{\sigma}}^+ : \boldsymbol{\varepsilon}^{\mathrm{e}},\ \bar{\psi}^{\mathrm{e}-} = \frac{1}{2}\bar{\boldsymbol{\sigma}}^- : \boldsymbol{\varepsilon}^{\mathrm{e}}. \tag{44}$$

The definitions in Eqs.(33) and (34) are consistent with the positive-negative split of $\bar{\boldsymbol{\sigma}}$ in Eq.(36), so the positive and negative parts of $\bar{\psi}^{\mathrm{p}}$ can be introduced from Eq.(35) as

$$\begin{aligned} \bar{\psi}^{p+} &= \int_0^{\kappa^+} \bar{Z}^+ \,\mathrm{d}\kappa^+ = \int_0^{\varepsilon^{\mathrm{p}}_{\max}} \langle \bar{\sigma}_{\max} \rangle \mathrm{d}\varepsilon^{\mathrm{p}}_{\max}, \\ \bar{\psi}^{\mathrm{p}-} &= \int_0^{\kappa^-} \bar{Z}^- \,\mathrm{d}\kappa^- = \int_0^{\varepsilon^{\mathrm{p}}_{\min}} \langle -\bar{\sigma}_{\min} \rangle \mathrm{d}(-\varepsilon^{\mathrm{p}}_{\min}). \end{aligned} \tag{45}$$

In accordance with the scalar based damage expressions in Eq.(40), the mapping of the free energy in Eq.(9) also takes the scalar forms,

$$\psi^+ = (1-d^+)\bar{\psi}^+,\ \psi^- = (1-d^-)\bar{\psi}^-, \tag{46}$$

where

$$\bar{\psi}^+ = \bar{\psi}^{\mathrm{e}+} + \bar{\psi}^{\mathrm{p}+},\ \bar{\psi}^{\ } = \bar{\psi}^{\mathrm{e}-} + \bar{\psi}^{\mathrm{p}-}. \tag{47}$$

Two scalar damage energy release rates, Y^+ and Y^-, are defined from Eq.(46) as

$$Y^+ = -\frac{\partial \psi^+}{\partial d^+} = \bar{\psi}^+,\ Y^- = -\frac{\partial \psi^-}{\partial d^-} = \bar{\psi}^-. \tag{48}$$

Substituting Eqs.(44) and (45) into Eq.(47) and then into Eq.(48), we obtain respectively the positive and negative damage energy release rates

$$\begin{aligned} Y^+ &= \frac{1}{2}\bar{\boldsymbol{\sigma}}^+ : \boldsymbol{\varepsilon}^{\mathrm{e}} + \int_0^{\hat{\varepsilon}^{\mathrm{p}}_{\max}} \langle \bar{\sigma}_{\max} \rangle \mathrm{d}\varepsilon^{\mathrm{p}}_{\max}, \\ Y^- &= \frac{1}{2}\bar{\boldsymbol{\sigma}}^- : \boldsymbol{\varepsilon}^{\mathrm{e}} + \int_0^{\hat{\varepsilon}^{\mathrm{p}}_{\min}} \langle -\bar{\sigma}_{\min} \rangle \mathrm{d}(-\varepsilon^{\mathrm{p}}_{\min}). \end{aligned} \tag{49}$$

3.3 Damage evolution

Here, we formulate damage evolution on the basis of the function for tensile damage evolution of Oliver et al.[34],

$$d^+=g^+(r^+)=1-\frac{r_0^+}{r^+}\exp\left[A^+\left(1-\frac{r^+}{r_0^+}\right)\right], \tag{50}$$

and that for compressive damage evolution developed by Faria et al.[12],

$$d^-=g^-(r^-)=1-\frac{r_0^-}{r^-}(1-A^-)-A^-\exp\left[B^-\left(1-\frac{r^-}{r_0^-}\right)\right], \tag{51}$$

where $r^\pm$ are the tensile/compressive damage thresholds, $r_0^\pm$ the initial thresholds (linear limits), and $A^\pm$ and B^- dimensionless constants.

In Eqs. (50) and (51), the damage thresholds $r^\pm$ and $r_0^\pm$ are effective stresses. For the present elastoplastic damage model, the introduced damage thresholds $R^\pm$ and $R_0^\pm$ are effective free energy. So we extract the square roots of $R^\pm/R_0^\pm$ and modify Eqs. (50) and (51) as

$$d^+=G^+(R_0^+, R^+)=1-\sqrt{\frac{R_0^+}{R^+}}\exp\left[A^+\left(1-\sqrt{\frac{R^+}{R_0^+}}\right)\right], \tag{52}$$

and

$$\begin{aligned} d^- &= G^-(R_0^-, R^-) \\ &= 1-\sqrt{\frac{R_0^-}{R^-}}(1-A^-)-A^-\exp\left[B^-\left(1-\sqrt{\frac{R^-}{R_0^-}}\right)\right]. \end{aligned} \tag{53}$$

The tensile and compressive damage thresholds are defined from Y^+ and Y^- respectively as

$$R^+=\max\{R_0^+, \max_{\tau\in[0, t]} Y_\tau^+\}, \quad R^-=\max\{R_0^-, \max_{\tau\in[0, t]} Y_\tau^-\}, \tag{54}$$

where t is the instant denoting present state, and τ any instant throughout the loading history. And the corresponding initial thresholds are defined respectively as

$$R_0^+=\bar{\psi}^{e+}\Big|_{\text{initial damage}}, \quad R_0^-=\bar{\psi}^{e-}\Big|_{\text{initial damage}}, \tag{55}$$

since it is believed that for concrete plasticity occurs after damage.

To get specific expressions for Eq. (55), we resort to strength theory. From Eqs. (44) and (55), we calculate R_0^+ and R_0^- respectively as

$$\begin{aligned} R_0^+ &= \frac{1}{2}(k\bar{\boldsymbol{\sigma}}^+) : \bar{\mathbb{C}}^{-1} : (k\bar{\boldsymbol{\sigma}})\Big|_{f_0^d(k\bar{\boldsymbol{\sigma}})=0}, \\ R_0^- &= \frac{1}{2}(k\bar{\boldsymbol{\sigma}}^-) : \bar{\mathbb{C}}^{-1} : (k\bar{\boldsymbol{\sigma}})\Big|_{f_0^d(k\bar{\boldsymbol{\sigma}})=0}, \end{aligned} \tag{56}$$

where $f_0^d(\ )$ is a function describing the initial damage surface in the stress space, and k a coefficient projecting the present effective stress state $\bar{\boldsymbol{\sigma}}$ to the initial damage surface proportionally. The Hsieh-Ting-Chen criterion

$$f_0^{\mathrm{d}}(\boldsymbol{\sigma})=a\frac{J_2}{f_0^{-2}}+b\frac{\sqrt{J_2}}{f_0^-}+c\frac{\hat{\sigma}_{\max}}{f_0^-}+d\frac{I_1}{f_0^-}-1, \tag{57}$$

is selected here for initial damage, where J_2 is the second invariant of deviatoric stress, $\hat{\sigma}_{\max}$ the maximum principal stress, I_1 the first invariant of stress, and a, b, c, and d four material parameters to be calibrated from the initial damage stresses in uniaxial tension (f_0^+), uniaxial compression (f_0^-), equibiaxial compression (f_{b0}^-), and a certain triaxial compression (f_{tr0}^-).

Remark: Wu et al.[14] approximate $\bar{\psi}^-$ to be

$$\bar{\psi}^-=\bar{\psi}^{e-}+\bar{\psi}^{p-}\approx b_0(\alpha\bar{I}_1+\sqrt{3\bar{J}_2})^2, \tag{58}$$

in biaxial compression and then get r_0^- from Eq.(58). The approximation in Eq.(58) has been derived for the sum of $\bar{\psi}^{e-}$ and $\bar{\psi}^{p-}$ while there is only $\bar{\psi}^{e-}$ and no $\bar{\psi}^{p-}$ at initial damage. Hence, calculating r_0^- by Eq.(58) is unjustified and unnecessary.

In the present formulation of R_0^- [Eqs.(55) and (56)], we calculate the elastic free energy directly, and no approximation is introduced.

3.4 Effective stress space elastoplasticity

From Eqs.(1) and (5), one gets the elastoplastic relation in the effective stress space,

$$\bar{\boldsymbol{\sigma}}=\bar{\mathbb{C}}:(\boldsymbol{\varepsilon}-\boldsymbol{\varepsilon}^{\mathrm{p}}). \tag{59}$$

$\bar{\mathbb{C}}$ is all we need for the linear elastic relation while the plastic relation comprises several intertwined components. The Lubliner-Lee plasticity, which has been used to calculate the free energy potential in the previous section, is completed here with some simplified assumptions.

The yield function assumes the form

$$\begin{aligned}f^{\mathrm{p}}(\bar{\boldsymbol{\sigma}},\kappa^+,\kappa^-)=&\alpha\bar{I}_1+\sqrt{3\bar{J}_2}+\beta(\kappa^+,\kappa^-)\langle\bar{\sigma}_{\max}\rangle-\\&\gamma\langle-\bar{\sigma}_{\max}\rangle-(1-\alpha)c(\kappa^-),\end{aligned} \tag{60}$$

where the parameter

$$\alpha=\frac{f_{bc}/f_c-1}{2f_{bc}/f_c-1}, \tag{61}$$

in which f_c is the uniaxial compressive strength, and f_{bc} the equibiaxial compressive strength, the parameter

$$\beta(\kappa^+,\kappa^-)=(1-\alpha)\frac{\bar{f}^-(\kappa^-)}{\bar{f}^+(\kappa^+)}-(1+\alpha), \tag{62}$$

and the cohesion

$$c(\kappa^-) = \bar{f}^-(\kappa^-), \tag{63}$$

in which $\bar{f}^+$ and $\bar{f}^-$ are the subsequent effective yield stresses in uniaxial tension and compression respectively. The parameter γ can be calibrated according to strength increase in triaxial compression[35-36]. The fact that the parameters α and γ still remain constants during hardening and the simple evolution function for β make the backward Euler algorithm applicable[37].

Simple linear hardening is assumed as

$$\bar{f}^+(\kappa^+) = \bar{f}_y^+ + \bar{H}^+\kappa^+, \ \bar{f}^-(\kappa^-) = \bar{f}_y^- + \bar{H}^-\kappa^-, \tag{64}$$

where $\bar{f}_y^+$ and $\bar{f}_y^-$ are the initial effective yield stresses in uniaxial tension and compression, respectively, and $\bar{H}^+$ and $\bar{H}^-$ the corresponding effective hardening moduli. The definitions for the hardening variables κ^+ and κ^- are as Eq.(29).

The flow rule is expressed in Eq.(24), where the plastic potential used here is the Drucker-Prager type function

$$g^p(\bar{\boldsymbol{\sigma}}) = \alpha^p \bar{I}_1 + \sqrt{2\bar{J}_2}, \tag{65}$$

in which the parameter α^p controls plastic dilation of concrete.

Finally, with the Karush-Kuhn-Tucker loading/unloading conditions,

$$f^p \leqslant 0, \ \dot{\lambda} \geqslant 0, \ \dot{\lambda} f^p = 0, \tag{66}$$

and the consistency condition

$$\dot{\lambda}\dot{f}^p = 0, \tag{67}$$

the plastic relation is complete.

4 Validation

The method of operator split[11, 38] is used to numerically implement the elastoplastic damage relation. For the iteration process of the plastic part, the spectral version[3] of backward Euler return mapping algorithms[38] is employed.

To validate the present constitutive theory, several typical mechanical tests of concrete specimens are simulated with the commercial finite element program Abaqus and its user material subroutine UMAT. Since the stress states throughout a specimen are uniform, only one element is used to build each finite element model.

In the following simulations of uniaxial cyclic loading, material parameters are just calibrated to reproduce experimental results. As for biaxial loading, first uniaxial mechanical properties are calibrated to fit the experimental uniaxial stress-strain curves, and then they are used to predict biaxial behavior. The strength ratio $f_{b0}^-/f_0^- = 1.15$ is calibrated from the biaxial tests by Kupfer et al.[39], and the common values for the Poisson's ratio $\nu = 0.2$ and dilation parameter $\alpha^p = 0.2$ are adopted.

4.1 Cyclic loading

For cyclic uniaxial loading, the tests by Geopalaeratnam and Shah[40] and Karson and Jirsa[41] are simulated. In the uniaxial tension simulation, the effective Young's modulus $\bar{E}$ =29.8 × 10^3 N/mm^2, f_0^+ =3.48 N/mm^2 and $\bar{f}_y^+$ =3.48 N/mm^2 are set. The calculated cyclic stress-strain curve is shown along with the test data in Fig. 1.

In the uniaxial compression simulation, $\bar{E}$ =32.3 × 10^3 N/mm^2, f_0^- =10.8 N/mm^2 and $\bar{f}_y^-$ =27.6 N/mm^2 are set. The calculated cyclic stress-strain curve is shown along with the test data in Fig. 2.

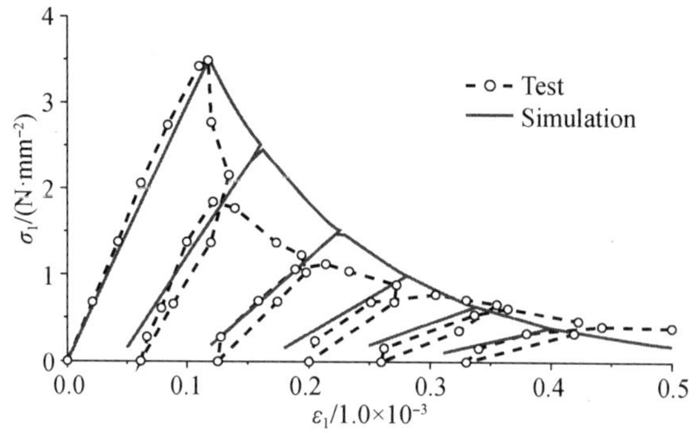

Fig. 1. Cyclic tension[40]

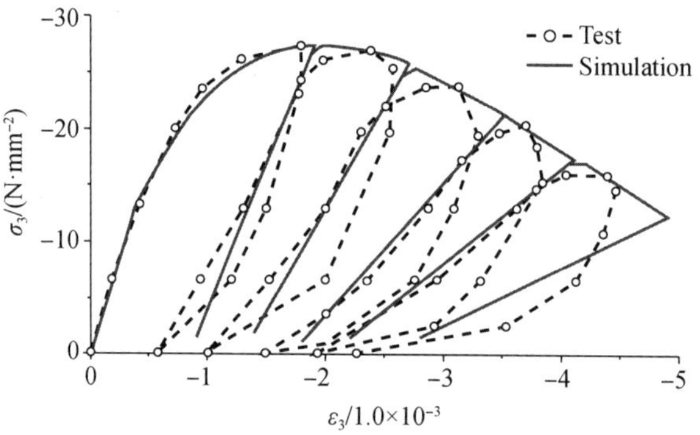

Fig. 2. Cyclic compression[41]

4.2 Biaxial loading

For biaxial loading, the tests by Kupfer et al.[39] are simulated. In the simulations, $\bar{E}$ = 28.2 × 10^3 N/mm^2, f_0^+ = 3.2 N/mm^2, $\bar{f}_y^+$ = 3.2 N/mm^2, f_0^- = 19.3 N/mm^2 and $\bar{f}_y^-$ = 32.2 N/mm^2 are set. The calculated stress-strain curves of uniaxial compression [Fig. 3(a)], biaxial tension-compression[Fig. 3(b)], equibiaxial compression[Fig. 3(c)], and non-equibiaxial compression[Fig. 3(d)] are shown respectively along with the test data.

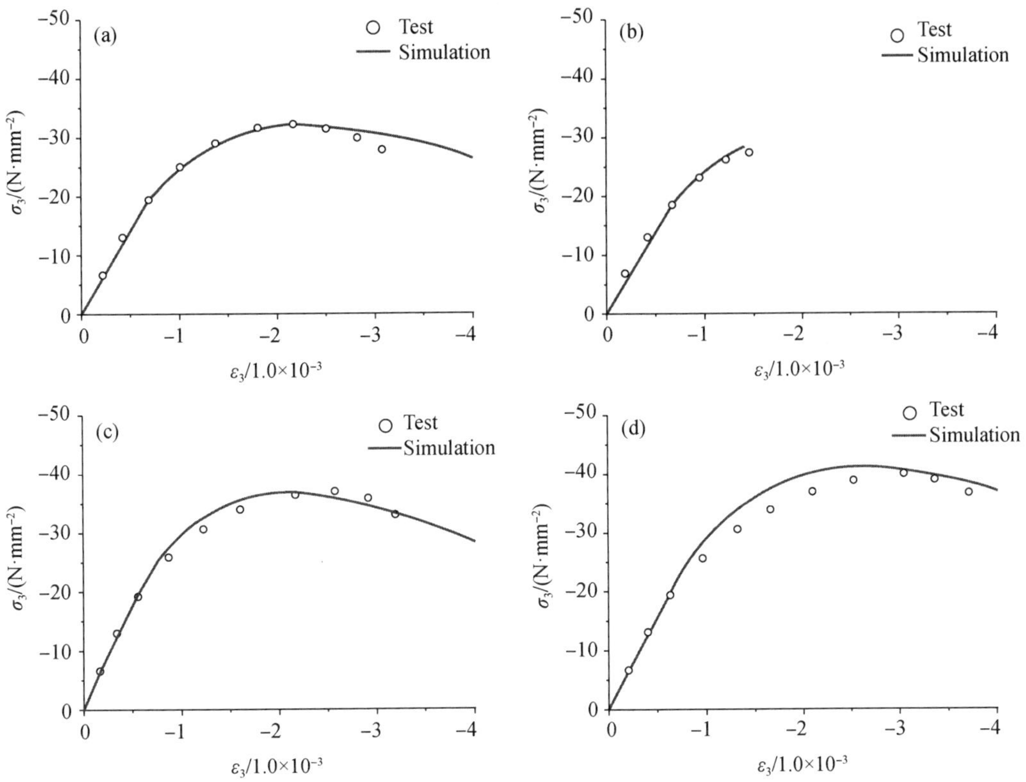

Fig. 3. **Biaxial loading[39]. (a) Compression ($\sigma_1 : \sigma_2 : \sigma_3 = 0 : 0 : -1$); (b) tension-compression ($\sigma_1 : \sigma_2 : \sigma_3 = 0.052 : 0 : -1$); (c) compression-compression ($\sigma_1 : \sigma_2 : \sigma_3 = 0 : -1 : -1$); (d) compression-compression ($\sigma_1 : \sigma_2 : \sigma_3 = 0 : -0.52 : -1$)**

5 Conclusion

The general formulation of free energy proposed in this study possesses the basic feature we desire. It inherently embodies the hardening law and flow rule of a certain plasticity model. Its special form for Lubliner-Lee plasticity is a typical example to demonstrate this feature. For concrete materials, plasticity is still a phenomenological theory, any yield criterion, hardening law, or flow rule is a priori. Thus, an elastoplastic damage model for concrete built upon the present formulation is consistent with the mathematical assumptions about plastic evolution.

With this feature, the proposed free energy potential can serve as a thermodynamic link between plastic and damage evolutions. Plastic evolution is embodied in the free energy, from which damage evolution is derived. Entropy production comes from both plastic and damage dissipations.

By combining this general formulation of free energy with the double hardening plasticity theory and two-parameter damage expression, we have constructed a thermodynamically well-founded elastoplastic damage model for concrete, which is validated by a series of tests in plane stress condition.

Compared to similar elastoplastic damage models in literature, the new model proposed in this study not only represents a progress in the thermodynamic aspect but also provides flexibility and opportunities to extend to a wider scope of application through the introduction of a third hardening mechanism and a third damage mechanism in triaxial compression.

This work was supported by the National Natural Science Foundation of China(Grant Nos. 51261120374, 51108336 and 51378377).

References

[1] CHEN A C T, CHEN W F. Constitutive relations for concretc[J]. Journal of the Engineering Mechanics Division, 1975, 101(4): 465-481.

[2] LUBLINER J, OLIVER J, OLLER S, et al. A plastic-damage model for concrete[J]. International Journal of Solids and Structures, 1989, 25(3): 299-326.

[3] LEE J, FENVES G L. Plastic-damage model for cyclic loading of concrete structures[J]. Journal of Engineering Mechanics, 1998, 124(8): 892-900.

[4] IMRAN I, PANTAZOPOULOU S J. Plasticity model for concrete under triaxial compression[J]. Journal of Engineering Mechanics, 2001, 127(3): 281-290.

[5] GRASSL P, JIRASEK M. Damage-plastic model for concrete failure[J]. Int J Solids Struct, 2006, 43: 7166-7196.

[6] PAPANIKOLAOU V K, KAPPOS A J. Confinement-sensitive plasticity constitutive model for concrete in triaxial compression[J]. International Journal of Solids and Structures, 2007, 44(21): 7021-7048.

[7] ZHENG F G, WU Z R, GU C S, et al. A plastic damage model for concrete structure cracks with two damage variables[J]. Science China Technological Sciences, 2012, 55(11): 2971-2980.

[8] LADEVEZE P. Sur une theorie de l'endommagement anisotrope[M]. Dissertation for the Doctoral Degree. Cachan, France: Laboratoire de Mecanique et Technologie, 1983.

[9] MAZARS J. A description of micro-and macroscale damage of concrete structures[J]. Engineering Fracture Mechanics, 1986, 25(5-6): 729-737.

[10] SIMO J C, JU J W. Strain-and stress-based continuum damage models—Ⅰ. Formulation[J]. International Journal of Solids and Structures, 1987, 23(7): 821-840.

[11] JU J W. On energy-based coupled elastoplastic damage theories: constitutive modeling and computational aspects[J]. International Journal of Solids and Structures, 1989, 25(7): 803-833.

[12] FARIA R, OLIVER J, CERVERA M. A strain-based plastic viscous-damage model for massive concrete structures[J]. International Journal of Solids and Structures, 1998, 35(14): 1533-1558.

[13] COMI C, PEREGO U. Fracture energy based bi-dissipative damage model for concrete[J]. International Journal of Solids and Structures, 2001, 38(36-37): 6427-6454.

[14] WU J Y, LI J, FARIA R. An energy release rate-based plastic-damage model for concrete[J]. International Journal of Solids and Structures, 2006, 43(3-4): 583-612.

[15] CICEKLI U, VOYIADJIS G Z, AL-RUB R K A. A plasticity and anisotropic damage model for plain

concrete[J]. International Journal of Plasticity, 2007, 23(10-11): 1874-1900.

[16] LI J, REN X D. Stochastic damage model for concrete based on energy equivalent strain[J]. International Journal of Solids and Structures, 2009, 46(11-12): 2407-2419.

[17] ZHU Q Z, SHAO J F, KONDO D. A micromechanics-based thermodynamic formulation of isotropic damage with unilateral and friction effects[J]. European Journal of Mechanics-A/Solids, 2011, 30(3): 316-325.

[18] LEMAITRE J. A continuous damage mechanics model for ductile fracture[J]. Journal of Engineering Materials and Technology, 1985, 107: 83-89.

[19] LEMAITRE J, DESMORAT R. Engineering damage mechanics: ductile, creep, fatigue and brittle failures[M]. Springer Science & Business Media, 2006.

[20] GURTIN M E, FRIED E, ANAND L. The mechanics and thermodynamics of continua[M]. New York: Cambridge University Press, 2010.

[21] LI J, REN X D. Homogenization-based multi-scale damage theory[J]. Science China Physics, Mechanics and Astronomy, 2010, 53: 690-698.

[22] REN X D, CHEN J S, LI J, et al. Micro-cracks informed damage models for brittle solids[J]. International Journal of Solids and Structures, 2011, 48: 1560-1571.

[23] REN X D, LI J. A unified dynamic model for concrete considering viscoplasticity and rate-dependent damage[J]. International Journal of Damage Mechanics, 2012, 22: 530-555.

[24] VOYIADJIS G Z, AL-RUB R K A, PALAZOTTO A N. Thermodynamic framework for coupling of non-local viscoplasticity and non-local anisotropic viscodamage for dynamic localization problems using gradient theory[J]. International Journal of Plasticity, 2004, 20(6): 981-1038.

[25] AL-RUB R K A, KIM S M. Computational applications of a coupled plasticity-damage constitutive model for simulating plain concrete fracture[J]. Engineering Fracture Mechanics, 2010, 77(10): 1577-1603.

[26] KIM S M, AL-RUB R K A. Meso-scale computational modeling of the plastic-damage response of cementitious composites[J]. Cement and Concrete Research, 2011, 41(3): 339-358.

[27] GURTIN M E, ANAND L, LELE S P. Gradient single-crystal plasticity with free energy dependent on dislocation densities[J]. Journal of the Mechanics and Physics of Solids, 2007, 55(9): 1853-1878.

[28] SUMARAC D, KRAJCINOVIC D, MALLICK K. Elastic parameters of brittle, elastic solids containing slits—mean field theory[J]. International Journal of Damage Mechanics, 1992, 1(3): 320-346.

[29] ZHANG J, LI J. Microelement formulation of free energy for quasi-brittle materials[J]. Journal of Engineering Mechanics, 2014, 140(8): 06014008.

[30] LUBLINER J. On the thermodynamic foundations of nonlinear solid mechanics[J]. International Journal of Non-Linear Mechanics, 1972, 7: 237-254.

[31] COLEMAN B D, GURTIN M. Thermodynamics with internal state variables[J]. The Journal of Chemical Physics, 1967, 47: 597-613.

[32] CHABOCHE J L. Constitutive equations for cyclic plasticity and cyclic viscoplasticity[J]. International Journal of Plasticity, 1989, 5: 247-302.

[33] ORTIZ M A. Constitutive theory for inelastic behavior of concrete[J]. Mechanics of Materials, 1985, 4: 67-93.

[34] OLIVER J, CERVERA M, OLLER S, et al. Isotropic damage models and smeared crack analysis of concrete[C]//Proceedings of 2nd International Conference on Computer Aided Analysis and Design of Concrete Structures, Zell am See, Austria, 1990.

[35] ZHANG J, ZHANG Z X, CHEN C Y. Yield criterion in plastic-damage models for concrete[J]. Acta Mechanica Solida Sinica, 2010, 23: 220-230.

[36] ZHANG J, LI J. Investigation into Lubliner yield criterion of concrete for 3D simulation[J]. Engineering Structure, 2012, 44: 122-127.

[37] ZHANG J, LI J. Construction of homogeneous loading functions for elastoplastic damage models for concrete[J]. Science China Physics, Mechanics and Astronomy, 2014, 57: 490-500.

[38] SIMO J C, HUGHES T J R. Computational Inelasticity[M]. New York: Springer-Verlag, 1998.

[39] KUPFER H, HILSDORF H K, RUSCH H. Behavior of concrete under biaxial stresses[J]. ACI Journal Proceedings, 1969, 66: 656-666.

[40] GEOPALAERATNAM V S, SHAH S P. Softening response of plain concrete in direct tension[J]. ACI Journal Proceedings, 1985, 82: 310-323.

[41] KARSON I D, JIRSA J O. Behaviour of concrete under compressive loadings[J]. Journal of the Structural Division, 1969, 95: 2535-2563.

3D elastoplastic damage model for concrete based on novel decomposition of stress①

Ji Zhang[1, 2], Jie Li[2], J. Woody Ju[2, 3]

1 *School of Civil Engineering and Mechanics, Huazhong University of Science and Technology, Wuhan 430074, China*

2 *School of Civil Engineering, Tongji University, Shanghai 200092, China*

3 *Department of Civil and Environmental Engineering, University of California, Los Angeles, CA 90095, USA*

Abstract This study aims to develop a versatile constitutive framework for concrete that can be used in various stress states and under reversed cyclic loading. A multi-axial compression stress state is decomposed into a hydrostatic confining part and a biaxial net part to distinguish concrete behavior in triaxial and biaxial compression while preserving their similarity. This novel decomposition is combined with the positive-negative decomposition of stress, and on its basis is developed a 3D elastoplastic damage model. The suppression on damage of the net part of stress from the presence of the confining part is introduced through an increase in microscopic fracture strains, and confining effects in ductility and lateral deformation are introduced, respectively, by hardening slowdown and dilation reduction. In the meantime, independent damage/plasticity evolution in tension and compression is incorporated in the proposed 3D model with two damage variables and two hardening variables. An elastic-plastic-damage operator split integration algorithm with the backward Euler return mapping is derived, and a series of numerical simulations are carried out in comparison with tests under uniaxial, biaxial, pseudo-triaxial, and true triaxial loading with satisfactory agreement.

Keywords Stress decomposition; Confining effect; Unilateral effect; Damage; Plasticity; Concrete

1 Introduction

Nonlinear analysis of concrete structures has drawn broad attention and extensive efforts in the civil engineering community for half a century, in which constitutive modeling remains a critical issue owing to the diverse and complex mechanical behavior of concrete materials under various loading scenarios.

Research on constitutive relations is of everlasting interest in solid mechanics. As for concrete, different types of theories have been proposed, among which plasticity models[1-5] and damage models[6-13] are the most fully developed.

While plasticity/damage models have achieved more and more success in their

① Originally published in *International Journal of Solids and Structures*, 2016, 94: 125-137.

characterization of concrete behavior under biaxial loading (and triaxial loading with tensile stress components), the situations are much more difficult for nonlinear analysis involving triaxial compression. There exist plasticity and/or damage models developed with special attention to triaxial compression: some are pure plasticity and fail to describe stiffness degradation[3-5, 14-15], which are also dedicated to multiaxial compression and are not intended for use in tension; some are pure damage mechanics and cannot represent irrecoverable deformation[12]; some employ damage mechanics for tension and plasticity for compression but not simultaneously[16-17]; and some really combine plasticity and damage simultaneously and take tension into account, but do not reflect the independent hardening[11, 18-22] or damage[18-20, 22] in tension and compression (under multiaxial tension-compression or reversed cyclic loading), among which the powerful microplane models[21] are formulated in terms of stress and strain vectors rather than tensors, leading to much greater computational cost.

In particular, Caner and Bažant[21] and Grassl et al.[11] possess a wide scope of applicability since not only are triaxial confining effects successfully characterized, but the stiffness recovery during transition between tension and compression can be captured with independent formulation for tensile and compressive damage. In these theories, however, the independency between tensile and compressive hardening is not included, which, we believe, is also necessary for a realistic characterization of unilateral effects and situations of mixed tensile-compressive stress states.

It is the purpose of the present study to develop (within the classical tensor-based context) a more versatile constitutive framework that is able to meet the manifold demands in nonlinear analysis of concrete structures. It is intended that the distinct behavior of concrete in tension, uni-/bi-axial compression, and triaxial compression (except for hydrostatic compression) is treated appropriately, which is required for various stress states, and at the same time the independent plasticity/damage evolution in tension and compression is allowed, which is important for multiaxial tension-compression (shear and torsion) as well as reversed cyclic loading (seismic analysis).

2 Stress decomposition

The positive-negative decomposition of stress[23] has proved effective in distinguishing tension and compression behavior[8-9, 11]. Traditionally, the spherical-deviatoric decomposition is performed upon a symmetric second-order tensor, and frequently used for stress and/or strain in constitutive modeling. However, a spherical part of stress is present also in stress states other than triaxial compression and, therefore, is not a good indicator of triaxial confining effects. This study attempts to decompose stress in a new way and, by doing so, distinguish triaxial and biaxial compression behavior.

2.1 Positive-negative decomposition

To describe the distinct damage behavior of concrete in tension and compression, the nominal stress $\boldsymbol{\sigma}$ / effective stress $\bar{\boldsymbol{\sigma}}$ is decomposed into its positive part ${}^{+}\boldsymbol{\sigma}/{}^{+}\bar{\boldsymbol{\sigma}}$ and negative part ${}^{-}\boldsymbol{\sigma}/{}^{-}\bar{\boldsymbol{\sigma}}$,

$$\boldsymbol{\sigma} = {}^{+}\boldsymbol{\sigma} + {}^{-}\boldsymbol{\sigma},\ \bar{\boldsymbol{\sigma}} = {}^{+}\bar{\boldsymbol{\sigma}} + {}^{-}\bar{\boldsymbol{\sigma}}, \tag{1}$$

as first advocated by Ladeveze[23] and Ortiz[24], and nowadays inherited in a lot of successful damage models for concrete[8-11, 13].

The meaning of this popular positive-negative decomposition can be clearly shown from the dyadic forms of these stress tensors in the principal coordinate system,

$$\bar{\boldsymbol{\sigma}} = \sum_i \bar{\sigma}_i \underset{\rightarrow i}{n} \otimes \underset{\rightarrow i}{n},\ {}^{+}\bar{\boldsymbol{\sigma}} = \sum_i \langle \bar{\sigma}_i \rangle \underset{\rightarrow i}{n} \otimes \underset{\rightarrow i}{n},\ {}^{-}\bar{\boldsymbol{\sigma}} = -\sum_i \langle -\bar{\sigma}_i \rangle \underset{\rightarrow i}{n} \otimes \underset{\rightarrow i}{n}, \tag{2}$$

where $\bar{\sigma}_i$ are the principal effective stresses, $\underset{\rightarrow i}{n}$ the orthonormal basis along the principal directions, and the Mcauley brackets $\langle\rangle$ are defined as

$$\langle x \rangle = \frac{1}{2}(|x| + x). \tag{3}$$

This positive-negative decomposition is illustrated in Fig. 1(a) with the principal stresses, where the dotted area represents the positive part, and the blank area the negative part.

2.2 Confining-net decomposition

In order to further describe the distinct behavior in triaxial compression, the authors have been seeking a proper way to make further decomposition, and are inspired by an interesting similarity between concrete behavior in biaxial and triaxial compression: with the magnitude of the intermediate principal stress $|\sigma_2|$ increasing, the multiaxial strength $|\sigma_3|$ first increases, reaching its peak value around $\sigma_2 : \sigma_3 = -0.5 : -1$, then decreases, and remains still higher than the starting point ($\sigma_1 = \sigma_2$) until $\sigma_2 : \sigma_3 = -1 : -1$. This characteristic of concrete in triaxial compression can be found directly from some targeted experimental study[25], and indirectly from the geometric properties of failure surfaces based on early test data.

Based on this similarity, the multi-axial compression stress ${}^{-}\bar{\boldsymbol{\sigma}}$ is further decomposed into a hydrostatic confining part $\bar{\boldsymbol{\sigma}}^{\equiv}$ and a biaxial net part $\bar{\boldsymbol{\sigma}}^{-}$ as

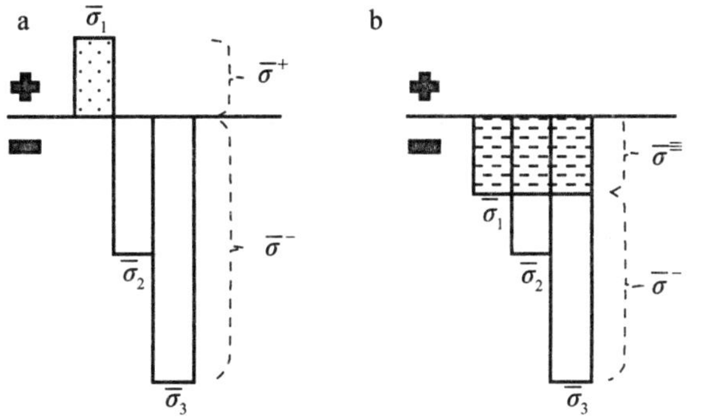

Fig. 1. Stress decomposition in terms of principal values. (a) $\bar{\boldsymbol{\sigma}}_1 \geq 0$[23]. (b) $\bar{\boldsymbol{\sigma}}_1 < 0$

$$^{-}\bar{\boldsymbol{\sigma}} = \bar{\boldsymbol{\sigma}}^{\equiv} + \bar{\boldsymbol{\sigma}}^{-}, \tag{4}$$

where

$$\bar{\boldsymbol{\sigma}}^{\equiv} = -\langle -\bar{\sigma}_1 \rangle \mathbf{I} = -\langle -\bar{\sigma}_1 \rangle \sum_i \underset{\rightarrow i}{n} \otimes \underset{\rightarrow i}{n} \tag{5}$$

with $\mathbf{I}$ being the second-order identity tensor, and

$$\bar{\boldsymbol{\sigma}}^{-} = -\sum_i (\langle -\bar{\sigma}_i \rangle - \langle -\bar{\sigma}_1 \rangle) \underset{\rightarrow i}{n} \otimes \underset{\rightarrow i}{n}. \tag{6}$$

This confining-net decomposition is illustrated in Fig. 1(b) with the principal stresses, where the dashed area represents the confining part, and the blank area the net part. Combining Eqs.(1) and (4) gives complete decomposition of stress

$$\bar{\boldsymbol{\sigma}} = \begin{cases} ^{+}\bar{\boldsymbol{\sigma}} + \bar{\boldsymbol{\sigma}}^{-}, & \bar{\sigma}_1 > 0 \\ \bar{\boldsymbol{\sigma}}^{-}, & \bar{\sigma}_1 = 0 \\ \bar{\boldsymbol{\sigma}}^{\equiv} + \bar{\boldsymbol{\sigma}}^{-}, & \bar{\sigma}_1 < 0 \end{cases}. \tag{7}$$

With the definitions in Eqs.(2)b,(5), and (6) and $^{+}\bar{\boldsymbol{\sigma}} = \bar{\boldsymbol{\sigma}}^{+}$, Eq.(7) can be rephrased in a unified form

$$\bar{\boldsymbol{\sigma}} = \bar{\boldsymbol{\sigma}}^{+} + \bar{\boldsymbol{\sigma}}^{-} + \bar{\boldsymbol{\sigma}}^{\equiv}. \tag{8}$$

In the present decomposition, $\bar{\boldsymbol{\sigma}}^{+}$ stands for a uni-/bi-/tri-axial tension state, $\bar{\boldsymbol{\sigma}}^{-}$ a uni-/bi-axial compression state, and $\bar{\boldsymbol{\sigma}}^{\equiv}$ an equi-triaxial compression state.

3 Damage framework

3.1 Damage variables

On the basis of the positive-negative decomposition in Eq.(1), the two-parameter damage expression by[7]

$$\boldsymbol{\sigma} = (1-d^{+})^{+}\bar{\boldsymbol{\sigma}} + (1-d^{-})^{-}\bar{\boldsymbol{\sigma}}, \tag{9}$$

where $d^{\pm}$ is the tensile/compressive damage variable, provides a simple and effective platform for the construction of damage models[8, 11, 13].

With the present decomposition in Eq.(8), it is natural to formulate a three-parameter damage expression as

$$\boldsymbol{\sigma} = (1-d^{+})\bar{\boldsymbol{\sigma}}^{+} + (1-d^{-})\bar{\boldsymbol{\sigma}}^{-} + (1-d^{\equiv})\bar{\boldsymbol{\sigma}}^{\equiv}, \tag{10}$$

where $d^{\equiv}$ is the damage variable in equi-triaxial compression. Since hydrostatic compression tests of concrete[26-27] indicate that stiffness degradation hardly occurs (despite considerable plastic deformation) until the stress level reaches a very high value, $d^{\equiv}$ is neglected (i.e. $\boldsymbol{\sigma}^{\equiv} = \bar{\boldsymbol{\sigma}}^{\equiv}$) here and Eq.(10) is simplified to

$$\boldsymbol{\sigma} = (1-d^{+})\bar{\boldsymbol{\sigma}}^{+} + (1-d^{-})\bar{\boldsymbol{\sigma}}^{-} + \bar{\boldsymbol{\sigma}}^{\equiv}. \tag{11}$$

3.2 Damage energy release rates

Within the traditional thermodynamic framework, the tensile/compressive damage energy release rate $Y^{\pm}$ is introduced as the conjugate force of $d^{\pm}$ with respect to the Helmholz free energy ψ

$$Y^{\pm}=-\frac{\partial\psi}{\partial d^{\pm}}. \tag{12}$$

The free energy ψ of concrete consists of an elastic part ψ^{e} and a plastic part ψ^{p}[6, 28]: ψ^{e} is the macroscopic (elastic) strain energy, and ψ^{p} can be interpreted as the microscopic additional strain energy locked by frictional microcracks.

Because tensile plastic deformation of concrete comes mainly from the microscopic stress-free opening rather than the frictional sliding and roughness-induced opening of microcracks, the plastic free energy ψ^{p} can be neglected in spite of considerable irrecoverable deformation in tension. Hence, the tensile damage energy release rate is simply

$$Y^{+}=\frac{1}{2}\bar{\boldsymbol{\sigma}}^{+}:\boldsymbol{\varepsilon}^{e}, \tag{13}$$

where ε^{e} is the elastic strain.

In the compression case, it is still a challenge to formulate elastoplastic free energy from micromechanics. Here the macroscopic approximation of the compressive damage energy release rate by Wu et al.[8]

$$Y^{-}\approx\frac{C}{\bar{E}}\left(\alpha\bar{I}_1^{-}+\sqrt{3\bar{J}_2^{-}}\right)^2 \tag{14}$$

is adopted, where C is a dimensionless material constant, $\bar{E}$ the effective (initial) Young's modulus, $\bar{I}_1^{-}$ the first invariant of $\bar{\boldsymbol{\sigma}}^{-}$, $\bar{J}_2^{-}$ the second invariant of the deviatoric part of $\bar{\boldsymbol{\sigma}}^{-}$, and

$$\alpha=\frac{f_{bc}/f_c-1}{2f_{bc}/f_c-1} \tag{15}$$

with f_c being the uniaxial compressive strength and f_{bc} being the equi-biaxial compressive strength.

3.3 Damage criteria

The maximum of $Y^{\pm}$ throughout the loading history τ is defined as the tensile/compressive damage threshold

$$r^{\pm}=\max_{\tau\in[0,\,t]}Y^{\pm}(\tau), \tag{16}$$

where t is the current instant. $Y^{\pm}$ and $r^{\pm}$ build up the tensile/compressive damage criterion

$$g^{\pm}(Y^{\pm},\,r^{\pm})=Y^{\pm}-r^{\pm}=0. \tag{17}$$

3.4 Damage evolution laws

Damage is driven by the thresholds as reflected in the tensile/compressive damage evolution function

$$d^{\pm}=G^{\pm}(r^{\pm}), \tag{18}$$

which is expected to meet the requirements for initial/final value and monotonicity

$$G^{\pm}(0)=0,\ \lim_{r^{\pm}\to\infty} G^{\pm}(r^{\pm})=1,\ \frac{\mathrm{d}G^{\pm}(r^{\pm})}{\mathrm{d}r^{\pm}}\geqslant 0. \tag{19}$$

3.5 Loading/unloading conditions

Loading or unloading is expressed by the Karush-Kuhn-Tucker conditions

$$g^{\pm}\leqslant 0,\ \dot{d}^{\pm}\geqslant 0,\ \dot{d}^{\pm}g^{\pm}=0. \tag{20}$$

Moreover, the consistency condition

$$\dot{d}^{\pm}\dot{g}^{\pm}=0 \tag{21}$$

is always required.

4 Damage evolution

While the expression of a damage state [Eq. (11)] treats a damaged material as a continuum, the evolution laws of damage [Eq. (18)] are rooted in the behavior of individual microcracks. Because the formulation or simulation of a discontinuum is extremely complex and difficult, in this section a simplified micromechanical formulation of damage evolution by Li and Ren[29] is used with an extension to triaxial compression in accordance with the present decomposition of stress [Eq.(8)].

4.1 Microscopic parallel system

Three kinds of basic stress states are selected to represent the damage evolution of the three parts from the stress decomposition[Eq.(8)], respectively: the uniaxial tension σ^{+} for the tension part $\bar{\boldsymbol{\sigma}}^{+}$, the uniaxial compression σ^{-} for the net compression part $\bar{\boldsymbol{\sigma}}^{-}$, and the equi-triaxial compression $\sigma^{\equiv}$ for the confining part $\bar{\boldsymbol{\sigma}}^{\equiv}$.

For each basic stress state, a parallel system (bundle) of microsprings (fibers) is built up to formulate damage evolution as exhibited in Fig. 2, where v is the Poisson's ratio and $\varepsilon^{+,e}$, $\varepsilon^{-,e}$, $\varepsilon^{\equiv,e}$ the elastic strains in the directions of σ^{+}, σ^{-}, $\sigma^{\equiv}$, respectively. It is assumed that all the fibers in a bundle have the same modulus of elasticity but random fracture strains Δ_l as displayed in Fig. 3, where σ_l and A_l are the stress and cross-sectional area of the lth fiber, respectively. Since the damage variable $d^{\equiv}$ for $\bar{\boldsymbol{\sigma}}^{\equiv}$ in Eq.(10) has been neglected in the present study, no fracture occurs in Fig. 2(c), i.e. $\Delta_l^{\equiv}=\infty$.

Remark 4.1. The rupture of fibers in Fig. 2(a) and (c) represents the mode Ⅰ fracture, mode Ⅱ fracture, and no fracture of microcracks, respectively, in σ^{+}, σ^{-}, and $\sigma^{\equiv}$.

4.2 Basic stress states

According to the classical definition of damage in terms of cross-sectional area, the tensile/compressive damage variable [Fig. 2(a)/(b)] can be calculated as

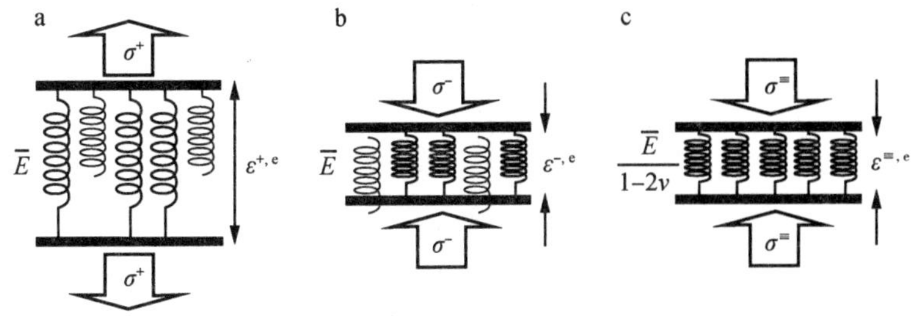

Fig. 2. Microspring bundles. (a) Uniaxial tension. (b) Uniaxial compression. (c) Equi-triaxial compression

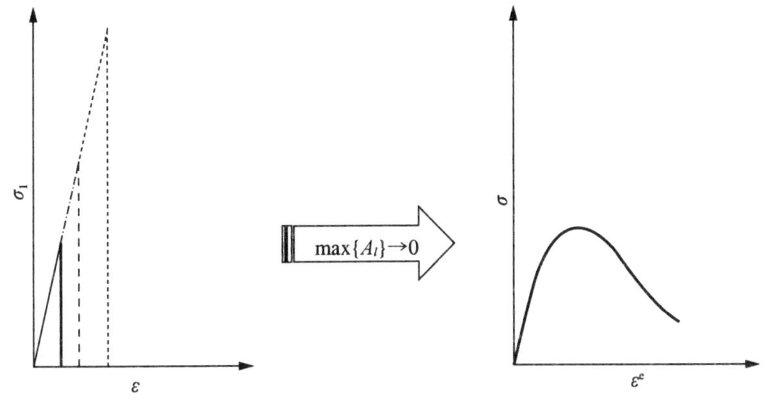

Fig. 3. Microscopic and macroscopic stress-strain curves

$$d^{\pm}=\frac{\sum_{l}A_{l}H(\varepsilon^{\pm,\mathrm{e}}-\Delta_{l}^{\pm})}{\sum_{l}A_{l}},\tag{22}$$

where $H(\ )$ is the Heaviside step function. As the fibers become infinitely thin, i. e. $\max\{A_l\}\to 0$, the limit of Eq.(22) is

$$d^{\pm}=\int_{0}^{1}H[\varepsilon^{\pm,\mathrm{e}}-\Delta^{\pm}(x)]\mathrm{d}x,\tag{23}$$

where x is the normalized spatial coordinate of a fiber within a bundle.

$\Delta^{\pm}(x)$ constitute a microscopic random field, consequently $d^{\pm}$ determined from Eq.(23) is a random variable, whose mean value and variance can be derived[29] under the assumption of homogeneity (stationarity) to be

$$\varepsilon[d^{\pm}]=F^{\pm}(\varepsilon^{\pm,\mathrm{e}})\tag{24}$$

and

$$\mathcal{D}[d^{\pm}]=2\int_{0}^{1}(1-r)F^{\pm}(\varepsilon^{\pm,\mathrm{e}},\varepsilon^{\pm,\mathrm{e}};r)\mathrm{d}r-[F^{\pm}(\varepsilon^{\pm,\mathrm{e}})]^{2}\tag{25}$$

respectively, where $F^{\pm}()=F^{\pm}(,x)$ is the 1D cumulative distribution function of $\Delta^{\pm}(x)$, $r=|x_1-x_2|$, and $F^{\pm}(,;r)=F^{\pm}(,x_1;,x_2)$ the 2D cumulative distribution function of $\Delta^{\pm}(x)$. Assume that the microscopic fracture strains $\Delta^{\pm}(x)$ are lognormally distributed similar to the fracture strength of concrete[30], and then the general expressions in Eqs.(24) and (25) can be specifically calculated as

$$\varepsilon[d^{\pm}]=\Phi\left(\frac{\ln \varepsilon^{\pm,e}-\mu^{\pm}}{V^{\pm}}\right) \tag{26}$$

and

$$\mathcal{D}[d^{\pm}]=2\int_0^1(1-r)\Phi\left(\frac{\ln \varepsilon^{\pm,e}-\mu^{\pm}}{V^{\pm}},\frac{\ln \varepsilon^{\pm,e}-\mu^{\pm}}{V^{\pm}};r\right)\mathrm{d}r-\Phi^2\left(\frac{\ln \varepsilon^{\pm,e}-\mu^{\pm}}{V^{\pm}}\right), \tag{27}$$

where $\Phi()$ and $\Phi(,;r)$ are the 1D and 2D cumulative distribution functions of the standard normal distribution, respectively,

$$\mu^{\pm}=\varepsilon[\ln \Delta^{\pm}], \tag{28}$$

and

$$V^{\pm}=\sqrt{\mathcal{D}[\ln \Delta^{\pm}]}. \tag{29}$$

The stochastic damage evolution is formulated above, nevertheless only the mean values [Eq.(26)] are to be used in the following sections as available data from triaxial tests are very limited.

Remark 4.2. The formulation of uniaxial damage evolution can be connected to the fracture energy G_f by

$$\int_0^{+\infty}\sigma^{+}\,\mathrm{d}\varepsilon^{+}=\frac{G_f}{l_{ch}}, \tag{30}$$

where l_{ch} is the characteristic length of an element, which is introduced to achieve mesh objectivity[31].

4.3 Superimposed basic stress states

The damage evolution in uniaxial tension [Fig. 2(a)] and that in uniaxial compression [Fig. 2(b)] are tentatively taken to be independent from each other for convenience as treated in many other theories[8-9, 11, 21].

The presence of the third kind of basic stress state, equi-triaxial compression [Fig. 2(c)], suppresses the damage evolution in uniaxial compression [Fig. 2(b)]. To account for this interaction, the superposition of uniaxial compression and equi-triaxial compression is considered, i.e. the pseudo-triaxial compression in the compressive meridional plane ($\bar{\sigma}_1=\bar{\sigma}_2$).

In the present bundle model, this suppression on damage is reflected through an increase in the microscopic fracture strains under confinement

$$\Delta_{\equiv}^{-}(x)=\Delta^{-}(x)\exp\left(c_{\Delta}\frac{\langle-\bar{\sigma}_1\rangle}{f_c}\right), \tag{31}$$

where $\Delta^{-}_{\equiv}(x)$ are the fracture strains of fibers in the superimposed σ^{-} [Fig. 2(b)] and $\sigma^{\equiv}$ [Fig. 2(c)], and c_{Δ} the coefficient of fracture enhancement.

Replacing $\Delta^{-}(x)$ in Eqs.(23), (28), and (29) with $\Delta^{-}_{\equiv}(x)$ leads to

$$d^{-}=\int_{0}^{1} H[\varepsilon^{-,e}-\Delta^{-}_{\equiv}(x)]dx \tag{32}$$

and

$$\varepsilon[d^{-}]=\Phi\left(\frac{\ln \varepsilon^{-,e}-\mu^{-}_{\equiv}}{V^{-}_{\equiv}}\right), \tag{33}$$

where

$$\mu^{-}_{\equiv}=\varepsilon[\ln \Delta^{-}_{\equiv}]=\mu^{-}+c_{\Delta}\frac{\langle-\bar{\sigma}_1\rangle}{f_c}, \tag{34}$$

and

$$V^{-}_{\equiv}=\sqrt{\mathcal{D}[\ln \Delta^{-}_{\equiv}]}=V^{-}. \tag{35}$$

Remark 4.3. From a micromechanical view, this interaction between the net part $\bar{\boldsymbol{\sigma}}^{-}$ and confining part $\bar{\boldsymbol{\sigma}}^{\equiv}$ comes from the suppression on shear sliding between crack surfaces (damage) by the normal pressure on them (confinement).

4.4 General stress states

With the concepts of energy-driven thresholds [Eq.(16)] and thresholds-driven damage [Eq.(18)], one can extend the above damage evolution functions to general stress states.

In uniaxial tension, Eq.(13) is calculated to be

$$Y^{+}=\frac{1}{2}\bar{E}(\varepsilon^{+,e})^2. \tag{36}$$

From Eqs.(16)a and (35), an equivalent tensile elastic strain can be determined as

$$\tilde{\varepsilon}^{+,e}(r^{+})=\sqrt{\frac{2r^{+}}{\bar{E}}}, \tag{37}$$

which takes the place of $\varepsilon^{+,e}$ [Fig. 2(a)] in multiaxial tension.

Replacing $\varepsilon^{+,e}$ in Eqs.(23) and (26) with $\tilde{\varepsilon}^{+,e}(r^{+})$ leads to

$$d^{+}=G^{+}(r^{+})=\int_{0}^{1} H[\tilde{\varepsilon}^{+,e}(r^{+})-\Delta^{+}(x)]dx \tag{38}$$

and

$$\varepsilon[d^{+}]=\Phi\left[\frac{\ln \tilde{\varepsilon}^{+,e}(r^{+})-\mu^{+}}{V^{+}}\right]. \tag{39}$$

In uniaxial compression, Eq.(14) is calculated to be

$$Y^{-}=C(1-\alpha)^2\bar{E}(\varepsilon^{-,e})^2. \tag{40}$$

From Eqs.(16) and (39), an equivalent compressive elastic strain can be determined as

$$\tilde{\varepsilon}^{-,e}(r^-)=\frac{1}{1-\alpha}\sqrt{\frac{r^-}{\bar{C}\bar{E}}}, \tag{41}$$

which takes the place of $\varepsilon^{-,e}$ [Fig. 2(b)] in general multiaxial compression.
Replacing $\varepsilon^{-,e}$ in Eqs.(32) and (33) with $\tilde{\varepsilon}^{-,e}(r^-)$ leads to

$$d^-=G^-(r^-)=\int_0^1 H[\tilde{\varepsilon}^{-,e}(r^-)-\Delta_{\equiv}^-(x)]\mathrm{d}x \tag{42}$$

and

$$\varepsilon[d^-]=\Phi\left[\frac{\ln\tilde{\varepsilon}^{-,e}(r^-)-\mu_{\equiv}^-}{V_{\equiv}^-}\right]. \tag{43}$$

5 Plasticity

For infinitesimal elastoplastic deformation, the strain $\boldsymbol{\varepsilon}$ is split additively into an elastic part $\boldsymbol{\varepsilon}^e$ and a plastic part $\boldsymbol{\varepsilon}^p$ as

$$\boldsymbol{\varepsilon}=\boldsymbol{\varepsilon}^e+\boldsymbol{\varepsilon}^p. \tag{44}$$

The hypothesis of elastic strain equivalence is taken here:

$$\boldsymbol{\sigma}=\mathbb{C}:\boldsymbol{\varepsilon}^e,\ \bar{\boldsymbol{\sigma}}=\bar{\mathbb{C}}:\boldsymbol{\varepsilon}^e, \tag{45}$$

where $\mathbb{C}$ and $\bar{\mathbb{C}}$ are the nominal and effective elastic stiffness, respectively. From Eqs.(44) and (45), one gets the elastoplastic relation in the effective stress space

$$\bar{\boldsymbol{\sigma}}=\bar{\mathbb{C}}:(\boldsymbol{\varepsilon}-\boldsymbol{\varepsilon}^p). \tag{46}$$

In isotropic elasticity, $\bar{E}$ and v are all that are needed to determine $\bar{\mathbb{C}}$ and construct a linear elastic relation. A plastic relation comprises several intertwined components and becomes rather complex when multiple hardening is introduced. Here the relatively simple Lubliner-Lee plasticity[1-2] with tension-compression double hardening is adopted with some enrichment for triaxial compression.

5.1 Yield criterion

The yield function assumes the form

$$f^p(\bar{\boldsymbol{\sigma}},\kappa^+,\kappa^-)=\sqrt{3\bar{J}_2}+\alpha\bar{I}_1+\beta(\kappa^+,\kappa^-)\langle\bar{\sigma}_1\rangle-(1-\alpha)\bar{f}^-(\kappa^-), \tag{47}$$

where $\kappa^{\pm}$ is the tensile/compressive hardening variable, $\bar{I}_1$ the first invariant of $\bar{\boldsymbol{\sigma}}$, $\bar{J}_2$ the second invariant of the deviatoric part of $\bar{\boldsymbol{\sigma}}$, and

$$\beta(\kappa^+,\kappa^-)=(1-\alpha)\frac{\bar{f}^-(\kappa^-)}{\bar{f}^+(\kappa^+)}-(1+\alpha) \tag{48}$$

with $\bar{f}^{\pm}(\kappa^{\pm})$ being the subsequent effective yield stress (cohesion) in uniaxial tension/

compression.

Remark 5. 1. Since the yield surface described by Eq. (47), which consists of two cones[32-33], is open in the negative direction of the hydrostatic axis, the present 3D model cannot characterize concrete behavior in hydrostatic compression.

5.2 Flow rule

The non-associated flow rule is adopted to characterize the plastic dilation of concrete

$$\dot{\boldsymbol{\varepsilon}}^{\mathrm{p}} = \dot{\lambda} \frac{\partial g^{\mathrm{p}}}{\partial \bar{\boldsymbol{\sigma}}}, \tag{49}$$

where $\dot{\lambda}$ is the plastic consistency parameter, and g^{p} the plastic potential function. The simple Drucker-Prager type function

$$g^{\mathrm{p}}(\bar{\boldsymbol{\sigma}}) = \sqrt{2\bar{J}_2} + \eta_0 \bar{I}_1, \tag{50}$$

where η_0 is the parameter controlling dilation, is sufficient for biaxial loading. To reflect the decrease in plastic dilation of concrete in triaxial compression, η_0 in Eq.(49) is reduced in the present formulation, leading to

$$g^{\mathrm{p}}(\bar{\boldsymbol{\sigma}}) = \sqrt{2\bar{J}_2} + \eta \bar{I}_1, \tag{51}$$

where

$$\eta = \eta_0 \exp\left(-c_\eta \frac{\langle -\bar{\sigma}_1 \rangle}{f_{\mathrm{c}}} \right) \tag{52}$$

with c_η being the coefficient of dilation reduction.

5.3 Hardening laws

The Ludwik power law is employed to formulate the strain hardening as follows

$$\bar{f}^{\pm}(\kappa^{\pm}) = \bar{f}_{\mathrm{y}}^{\pm} + \bar{K}^{\pm}(\kappa^{\pm})^{n^{\pm}}, \tag{53}$$

where $\bar{f}_{\mathrm{y}}^{\pm}$ is the initial effective yield stress in uniaxial tension/compression, $\bar{K}^{\pm}$ the effective tensile/compressive strength index, and $n^{\pm}$ the tensile/compressive hardening exponent.

5.4 Hardening variables

The rate evolution equations for the hardening variables are originally

$$\dot{\kappa}^{+} = w\dot{\varepsilon}_1^{\mathrm{p}}, \ \dot{\kappa}^{-} = -(1-w)\dot{\varepsilon}_3^{\mathrm{p}}, \tag{54}$$

where $\varepsilon_1^{\mathrm{p}}$ and $\varepsilon_3^{\mathrm{p}}$ are the major and minor principal plastic strains, respectively, and the weight factor

$$w = \frac{\sum_i \langle \bar{\sigma}_i \rangle}{\sum_i |\bar{\sigma}_i|}. \tag{55}$$

In the present model, the increase in ductility of concrete in triaxial compression is reflected by slowing down the evolution in Eq.(54)b as

$$\dot{\kappa}^{+}=w\dot{\varepsilon}_{1}^{p},\ \dot{\kappa}^{-}=-(1-w)\dot{\varepsilon}_{3}^{p}\exp\left(-c_{\kappa}\frac{\langle-\bar{\sigma}_{1}\rangle}{f_{c}}\right),\tag{56}$$

where c_{κ} is the coefficient of hardening slowdown.

With the flow rule Eq.(49), Eq.(56) can be rephrased with $\dot{\lambda}$ as

$$\dot{\kappa}^{\pm}=\dot{\lambda}h^{\pm},\tag{57}$$

where the tensile and compressive hardening functions are, respectively,

$$h^{+}=w\frac{\partial g^{p}}{\partial\bar{\sigma}_{1}},\ h^{-}=-(1-w)\frac{\partial g^{p}}{\partial\bar{\sigma}_{3}}\exp\left(-c_{\kappa}\frac{\langle-\bar{\sigma}_{1}\rangle}{f_{c}}\right).\tag{58}$$

5.5 Loading/unloading conditions

Finally with the Karush-Kuhn-Tucker loading/unloading conditions

$$f^{p}\leqslant 0,\ \dot{\lambda}\geqslant 0,\ \dot{\lambda}f^{p}=0\tag{59}$$

and the consistency condition

$$\dot{\lambda}\dot{f}^{p}=0,\tag{60}$$

the plastic relation is complete.

Remark 5.2. It is indicated by Eqs.(47),(48) and (53) that the size and shape of the yield surface is controlled by the independent evolution of the two cohesions $\bar{f}^{+}(\kappa^{+})$ and $\bar{f}^{-}(\kappa^{-})$. The expansion and distortion of the yield surface during tensile/compressive hardening are illustrated in Fig. 4(a)/(b), where $\bar{f}^{\pm}(\kappa^{\pm})$ increases while $\bar{f}^{\mp}(\kappa^{+})=\bar{f}_{y}^{\mp}$ remains unchanged.

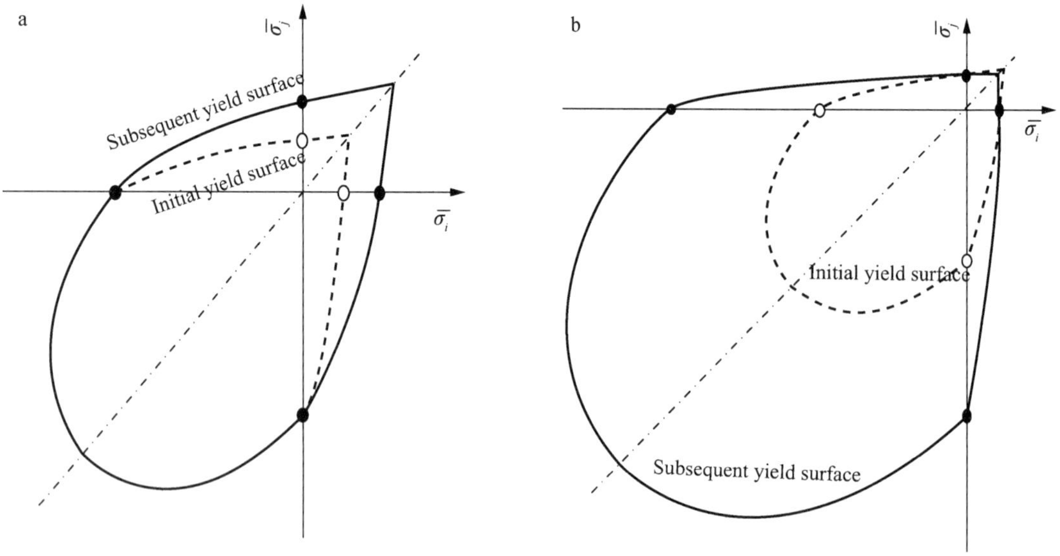

Fig. 4. Evolution of yield surface. (a) Tensile hardening. (b) Compressive hardening

It is noticed from Fig. 4(b) that the biaxial tensile strength decreases with compressive hardening. Detailed discussion on this issue can be found in Zhang and Li[33]. For multiaxial tension, tensile hardening is not independent in a strict sense from compressive hardening.

Remark 5.3. From Eqs. (5), (31), (52), and (56), it can be seen that the present decomposition of stress uses $\langle -\sigma_1 \rangle$, the minimum principal compressive stress, to evaluate triaxial confinement, in contrast with the traditional spherical-deviatoric decomposition, which employs I_1. In the critical case of biaxial compression ($\sigma_1 = 0$), there is no triaxial confinement at all according to the present theory.

6 Algorithm

6.1 Operator split

In accordance with the concept and methods of product formulas and operator split[34-35], the constitutive equations presented in the previous sections are replaced here with the elastic, plastic, and damage parts in Table 1, where $\underset{\rightarrow}{u}$ is the displacement. It can serve as a mathematical foundation for developing first-order accurate algorithms.

These three parts are taken as split problems in a series, which add up to the original elastoplastic damage problem. The elastic predictor problem and damage corrector problem are straightforward and involve no iteration. For the plastic corrector problem, the spectral version[6, 36] of the closest-point projection return mapping algorithm[35] is employed.

Table 1. Split of rate constitutive equations

Total	=Elastic predictor	+Plastic corrector	+Damage corrector
$\dot{\boldsymbol{\varepsilon}} = (\underset{\rightarrow}{u} \otimes \underset{\rightarrow}{\nabla} + \underset{\rightarrow}{\nabla} \otimes \underset{\rightarrow}{u})/2$	$\dot{\boldsymbol{\varepsilon}} = (\underset{\rightarrow}{u} \otimes \underset{\rightarrow}{\nabla} + \underset{\rightarrow}{\nabla} \otimes \underset{\rightarrow}{u})/2$		
$\dot{\boldsymbol{\varepsilon}}^p = \dot{\lambda} \partial g^p / \partial \bar{\boldsymbol{\sigma}}$		$\dot{\boldsymbol{\varepsilon}}^p = \dot{\lambda} \partial g^p / \partial \bar{\boldsymbol{\sigma}}$	
$\dot{\boldsymbol{\kappa}}^{\pm} = \dot{\lambda} h^{\pm}$		$\dot{\boldsymbol{\kappa}}^{\pm} = \dot{\lambda} h^{\pm}$	
$\dot{\bar{\boldsymbol{\sigma}}} = \bar{\mathbb{C}} : (\dot{\boldsymbol{\varepsilon}} - \dot{\boldsymbol{\varepsilon}}^p)$	$\dot{\bar{\boldsymbol{\sigma}}} = \bar{\mathbb{C}} : \dot{\boldsymbol{\varepsilon}}$	$\dot{\bar{\boldsymbol{\sigma}}} = -\bar{\mathbb{C}} : \dot{\boldsymbol{\varepsilon}}^p$	
$\dot{r}^{\pm} = \begin{cases} \dot{Y}^{\pm}, & Y^{\pm} = r^{\pm} \text{ and } \dot{Y}^{\pm} > 0 \\ 0, & Y^{\pm} < r^{\pm} \text{ or } \dot{Y}^{\pm} \leqslant 0 \end{cases}$			$\dot{r}^{\pm} = \begin{cases} \dot{Y}^{\pm}, & Y^{\pm} = r^{\pm} \text{ and } \dot{Y}^{\pm} > 0 \\ 0, & Y^{\pm} < r^{\pm} \text{ or } \dot{Y}^{\pm} \leqslant 0 \end{cases}$
$\dot{d}^{\pm} = \dot{r}^{\pm} \partial G^{\pm} / \partial r^{\pm}$			$\dot{d}^{\pm} = \dot{r}^{\pm} \partial G^{\pm} / \partial r^{\pm}$
$\dot{\boldsymbol{\sigma}} = (1-d^+)\dot{\bar{\boldsymbol{\sigma}}}^+ + (1-d^-)\dot{\bar{\boldsymbol{\sigma}}}^- + \dot{\bar{\boldsymbol{\sigma}}}^= - \dot{d}^+ \bar{\boldsymbol{\sigma}}^+ - \dot{d}^- \bar{\boldsymbol{\sigma}}^-$	$\dot{\boldsymbol{\sigma}} = (1-d^+)\dot{\bar{\boldsymbol{\sigma}}}^+ + (1-d^-)\dot{\bar{\boldsymbol{\sigma}}}^- + \dot{\bar{\boldsymbol{\sigma}}}^=$	$\dot{\boldsymbol{\sigma}} = (1-d^+)\dot{\bar{\boldsymbol{\sigma}}}^+ + (1-d^-)\dot{\bar{\boldsymbol{\sigma}}}^- + \dot{\bar{\boldsymbol{\sigma}}}^=$	$\dot{\boldsymbol{\sigma}} = -\dot{d}^+ \bar{\boldsymbol{\sigma}}^+ - \dot{d}^- \bar{\boldsymbol{\sigma}}^-$

6.2 Consistent tangent stiffness

The differentiation with respect to the strain increment $\Delta\boldsymbol{\varepsilon}_{n+1}$ of the resulting $\bar{\boldsymbol{\sigma}}_{n+1}$ and $\boldsymbol{\sigma}_{n+1}$ give, respectively, the tangent stiffness consistent with the updating algorithms in elastic-plastic steps and elastic-plastic-damage steps,

$$\bar{\mathbb{C}}_{n+1} = \frac{\partial \bar{\boldsymbol{\sigma}}_{n+1}}{\partial \boldsymbol{\varepsilon}_{n+1}}, \ \mathbb{C}_{n+1} = \frac{\partial \boldsymbol{\sigma}_{n+1}}{\partial \boldsymbol{\varepsilon}_{n+1}}, \tag{61}$$

which leasd to

$$\mathbb{C}_{n+1}=\frac{\partial \boldsymbol{\sigma}_{n+1}}{\partial \bar{\boldsymbol{\sigma}}_{n+1}}:\bar{\mathbb{C}}_{n+1}. \tag{62}$$

In Eq.(62) $\bar{\mathbb{C}}_{n+1}$ is already given as[36]

$$\bar{\mathbb{C}}_{n+1}=\left[\bar{\mathbb{C}}^{-1}+\frac{\partial g^{\mathrm{p}}}{\partial \bar{\boldsymbol{\sigma}}_{n+1}}\otimes\frac{\partial(\Delta\lambda)}{\partial \bar{\boldsymbol{\sigma}}_{n+1}}+\Delta\lambda\frac{\partial^2 g^{\mathrm{p}}}{\partial \bar{\boldsymbol{\sigma}}_{n+1}^2}\right]^{-1} \tag{63}$$

from conventional elastic-plastic algorithm, and $\partial\boldsymbol{\sigma}_{n+1}/\partial\bar{\boldsymbol{\sigma}}_{n+1}$ needs to be derived in accordance with the present damage formulation.

From Eq.(11) one can obtain

$$\frac{\partial \boldsymbol{\sigma}}{\partial \bar{\boldsymbol{\sigma}}}=(1-d^+)\frac{\partial \bar{\boldsymbol{\sigma}}^+}{\partial \bar{\boldsymbol{\sigma}}}+(1-d^-)\frac{\partial \bar{\boldsymbol{\sigma}}^-}{\partial \bar{\boldsymbol{\sigma}}}+\frac{\partial \bar{\boldsymbol{\sigma}}^=}{\partial \bar{\boldsymbol{\sigma}}}-\bar{\boldsymbol{\sigma}}^+\otimes\frac{\partial d^+}{\partial \bar{\boldsymbol{\sigma}}}-\bar{\boldsymbol{\sigma}}^-\otimes\frac{\partial d^-}{\partial \bar{\boldsymbol{\sigma}}}. \tag{64}$$

In Eq.(64) $\partial d^{\pm}/\partial\bar{\boldsymbol{\sigma}}$ is broken down into

$$\frac{\partial d^{\pm}}{\partial \bar{\boldsymbol{\sigma}}}=\frac{\partial d^{\pm}}{\partial r^{\pm}}\frac{\partial r^{\pm}}{\partial \bar{\boldsymbol{\sigma}}}, \tag{65}$$

where $\partial d^{\pm}/\partial r^{\pm}$ can be acquired from Eqs.(39)/(43) and (37)/(40) as

$$\frac{\partial d^+}{\partial r^+}=\frac{1}{2\sqrt{2\pi}V^+ r^+}\exp\left[-\frac{1}{2}\left(\frac{\ln\bar{\varepsilon}^{+\mathrm{e}}-\mu^+}{V^+}\right)^2\right] \tag{66}$$

and

$$\frac{\partial d^-}{\partial r^-}=\frac{1}{2\sqrt{2\pi}V^- r^-}\exp\left\{-\frac{1}{2}\left[\frac{\ln\bar{\varepsilon}^{-\mathrm{e}}-\left(\mu^-+c_\Delta\dfrac{\langle-\bar{\sigma}_1\rangle}{f_{\mathrm{c}}}\right)}{V^-}\right]^2\right\}, \tag{67}$$

and for $\partial r^{\pm}/\partial\bar{\boldsymbol{\sigma}}$ one can use Eqs.(13)/(14) and (16) to get

$$\frac{\partial r^+}{\partial \bar{\boldsymbol{\sigma}}}=\frac{1}{2}(\bar{\boldsymbol{\sigma}}^+:\bar{\mathbb{C}}^{-1}+\mathbb{P}^+:\bar{\mathbb{C}}^{-1}:\bar{\boldsymbol{\sigma}}) \tag{68}$$

and

$$\frac{\partial r^-}{\partial \bar{\boldsymbol{\sigma}}}=\frac{2C}{E}\left(\alpha\bar{I}_1^-+\sqrt{3\bar{J}_2^-}\right)\left(\alpha\mathbf{I}+\sqrt{\frac{3}{2}}\frac{\bar{\mathbf{s}}^-}{\|\bar{\mathbf{s}}^-\|}\right). \tag{69}$$

The differentiation of the three parts from stress decomposition in Eq.(64) generates three fourth-order tensors: $\partial\bar{\boldsymbol{\sigma}}^=/\partial\bar{\boldsymbol{\sigma}}$ can be obtained from Eq.(5) to be

$$\mathbb{P}^==\frac{\partial \bar{\boldsymbol{\sigma}}^=}{\partial \bar{\boldsymbol{\sigma}}}=H(-\bar{\sigma}_1)\mathbf{I}\otimes\mathbf{N}_{11}; \tag{70}$$

$\partial\bar{\boldsymbol{\sigma}}^+/\partial\bar{\boldsymbol{\sigma}}$ is derived in Faria et al.[37] as

$$\mathbb{P}^+=\frac{\partial \bar{\boldsymbol{\sigma}}^+}{\partial \bar{\boldsymbol{\sigma}}}=\sum_i H(\bar{\sigma}_i)\mathbf{N}_{ii}\otimes\mathbf{N}_{ii}+2\sum_{i<j}\frac{\langle\bar{\sigma}_i\rangle-\langle\bar{\sigma}_j\rangle}{\bar{\sigma}_i-\bar{\sigma}_j}\mathbf{N}_{ij}\otimes\mathbf{N}_{ij}; \tag{71}$$

and the remaining $\partial \bar{\boldsymbol{\sigma}}^{-}/\partial \bar{\boldsymbol{\sigma}}$ is

$$\begin{aligned}\mathbb{P}^{-}=&\frac{\partial \bar{\boldsymbol{\sigma}}^{-}}{\partial \bar{\boldsymbol{\sigma}}}\\ =&\underset{[}{H}(\bar{\sigma}_1)\left[\sum_i \underset{[}{H}(-\bar{\sigma}_i)\boldsymbol{N}_{ii}\otimes\boldsymbol{N}_{ii}+2\sum_{i<j}\frac{\langle-\bar{\sigma}_j\rangle-\langle-\bar{\sigma}_i\rangle}{\bar{\sigma}_i-\bar{\sigma}_j}\boldsymbol{N}_{ij}\otimes\boldsymbol{N}_{ij}\right]+\\ &\underset{]}{H}(-\bar{\sigma}_1)(\mathbb{I}-\mathbf{I}\otimes\boldsymbol{N}_{11}).\end{aligned}\tag{72}$$

where $N_{ij}=\operatorname{sym}(\underset{\rightarrow}{n}_i\otimes\underset{\rightarrow}{n}_j)$, $\mathbb{I}$ is the fourth order symmetric identity tensor, and the left-/right-continuous Heaviside step function is defined as

$$\underset{]}{H}(x)=\begin{cases}1, & x>0\\ 0, & x\leqslant 0\end{cases},\quad \underset{[}{H}(x)=\begin{cases}1, & x\geqslant 0\\ 0, & x<0\end{cases}.\tag{73}$$

Further, they constitute a partition of unity:

$$\mathbb{I}=\mathbb{P}^{+}+\mathbb{P}^{-}+\mathbb{P}^{=}.\tag{74}$$

Remark 6.1. Wu and Xu[38] demonstrate that the derivative expressed by Eq.(70) also serves as the thermodynamically consistent projection operator such that

$$\bar{\boldsymbol{\sigma}}^{+}=\mathbb{P}^{+}:\bar{\boldsymbol{\sigma}}.\tag{75}$$

Apparently, the derivative expressed by Eq.(69) satisfies

$$\bar{\boldsymbol{\sigma}}^{=}=\mathbb{P}^{=}:\bar{\boldsymbol{\sigma}}.\tag{76}$$

And with Eq.(73), one has

$$\bar{\boldsymbol{\sigma}}^{-}=\mathbb{P}^{-}:\bar{\boldsymbol{\sigma}}.\tag{77}$$

Substituting Eqs.(74)—(76) into Eq.(10), comparing it to the common representation of $\bar{\sigma}$ with the damage effect tensor $\mathbb{M}$

$$\bar{\boldsymbol{\sigma}}=\mathbb{M}:\boldsymbol{\sigma},\tag{78}$$

and using Eq.(73), one gets

$$\mathbb{M}=(\mathbb{I}-d^{+}\mathbb{P}^{+}-d^{-}\mathbb{P}^{-}-d^{=}\mathbb{P}^{=})^{-1}.\tag{79}$$

Since $d^{=}$ is neglected in the present study, Eq.(78) is simplified to

$$\mathbb{M}=(\mathbb{I}-d^{+}\mathbb{P}^{+}-d^{-}\mathbb{P}^{-})^{-1}.\tag{80}$$

7 Validations

To validate the present constitutive theory, a series of typical mechanical tests of concrete specimens are simulated. For all these simulations, some of the material parameters are set fixed as listed in Table 2, where f_t is the uniaxial tensile strength.

In the following simulations of uniaxial loading, the uniaxial material parameters are just calibrated to reproduce the experimental results. As for multiaxial loading, part of the multiaxial test data are set aside to calibrate the multiaxial material parameters, which, together with the already calibrated uniaxial material parameters, are then used to blindly predict the other multiaxial test data.

Table 2. Parameters for all simulations

Poisson's ratio	v	0.2
Dilation parameter	η_0	0.3
Tensile hardening exponent	n^+	0.5
Compressive hardening exponent	n^-	0.5
Initial effective yield stress in uniaxial tension	$\bar{f}_y^+$	f_t
Initial effective yield stress in uniaxial compression	$\bar{f}_y^-$	$0.5f_c$

7.1 Uniaxial loading

7.1.1 Cyclic tension

For cyclic uniaxial tension, the test by Geopalaeratnam and shah[39] is simulated. The material parameters are calibrated as listed in Table 3. The calculated stress-strain curve is shown along with the test data in Fig. 5.

Table 3. Parameters for cyclic tension simulation

Effective Young's modulus	$\bar{E}$	31.9×10^3 N/mm^2
Uniaxial tensile strength	f_t	3.48 N/mm^2
Mean value of natural logarithm of tensile fracture strain	μ^+	4.85
Standard deviation of natural logarithm of tensile fracture strain	V^+	0.07
Effective tensile strength index	K^+	$0.002\bar{E}$

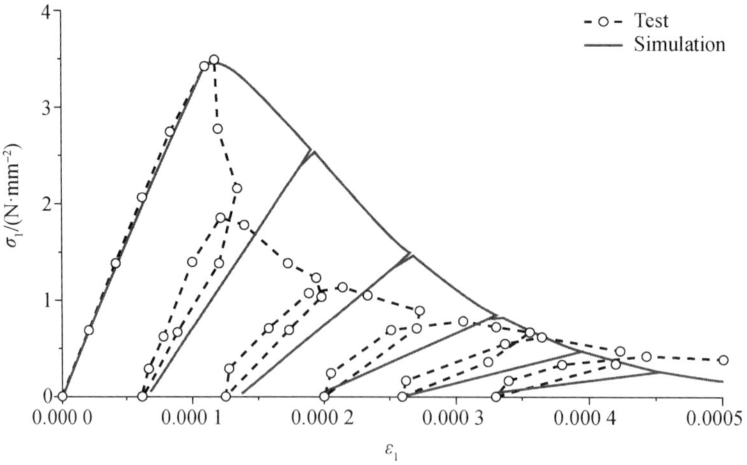

Fig. 5. Cyclic tension[39]

Remark 7.1. Although plastic deformation of concrete in tension is much smaller than that in compression, it is still quite considerable compared to elastic deformation in tension. The reason to neglect tensile plastic free energy is that tensile plastic deformation does not involve much frictional sliding and energy locking as discussed before Eq. (13). Plastic flow in tension generated by the present model is attributed to microcrack opening without prominent frictional effects.

7.1.2 Cyclic compression

For cyclic uniaxial compression, the test by Karsan and Jirsa[40] is simulated. The material parameters are calibrated as listed in Table 4. The calculated stress-strain curve is shown along with the test data in Fig. 6.

Table 4. Parameters for cyclic compression simulation

Effective Young's modulus	$\bar{E}$	34.7×10^3 N/mm^2
Uniaxial compressive strength	f_c	27.4 N/mm^2
Mean value of natural logarithm of compressive fracture strain	μ^-	7.33
Standard deviation of natural logarithm of compressive fracture strain	V^-	0.53
Effective compressive strength index	K^-	$0.045\bar{E}$

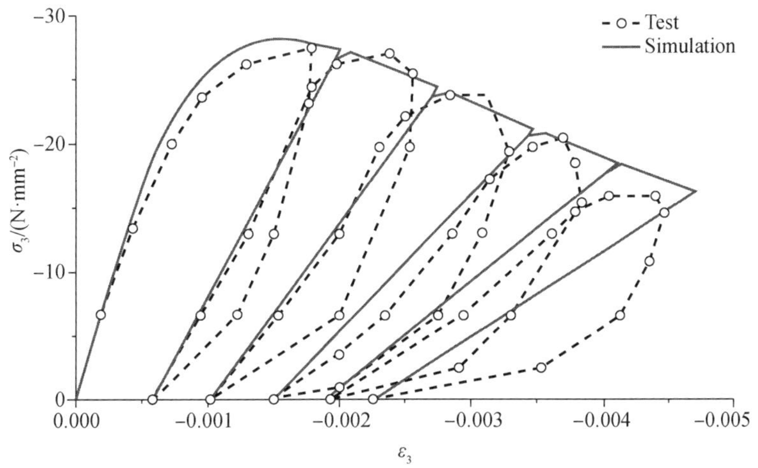

Fig. 6. Cyclic compression[40]

7.1.3 Reversed cyclic loading

For reversed cyclic uniaxial loading, appropriate test results are not available. The material parameters listed in Table 5 in the next section are used in this simulation, which are calibrated from the tests by Kupfer et al.[41]. The calculated stress-strain curve is shown in Fig. 7.

Table 5. Parameters for biaxial loading simulations

Effective Young's modulus	$\bar{E}$	31.0×10^3 N/mm^2
Uniaxial tensile strength	f_t	2.89 N/mm^2
Uniaxial compressive strength	f_c	32.1 N/mm^2
Mean value of natural logarithm of tensile fracture strain	μ^+	4.75
Mean value of natural logarithm of compressive fracture strain	μ^-	7.56
Standard deviation of natural logarithm of tensile fracture strain	V^+	0.1
Standard deviation of natural logarithm of compressive fracture strain	V^-	0.5
Effective tensile strength index	K^+	$0.003\bar{E}$
Effective compressive strength index	K^-	$0.045\bar{E}$
Equi-biaxial compressive strength	f_{bc}	$1.16f_c$

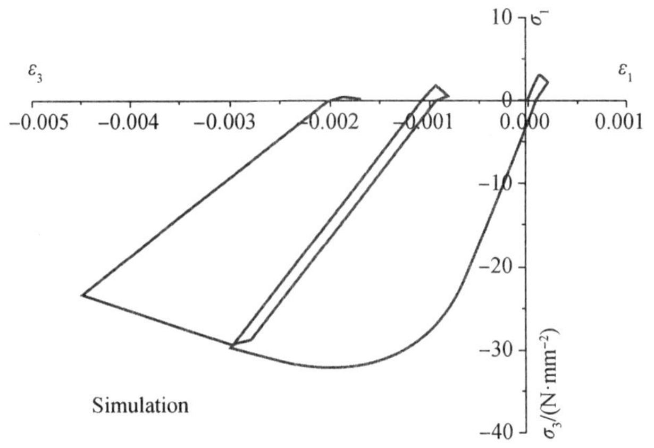

Fig. 7. Reversed cyclic loading

In comparison with those successful 3D models[42, 11], which employ single hardening plasticity, the present model (with tension-compression double hardening) appears to produce more realistic behavior for concrete under reversed cyclic loading. The performance of double hardening will be further examined later in Fig. 10 for biaxial loading.

It is seen that most of the basic characteristics of concrete under uniaxial loading, such as hardening, softening, stiffness degradation, irrecoverable deformation, and unilateral effects, can be represented appropriately with the proposed model. It is also noted that the hysteresis between unloading and reloading fails to be characterized, which can be included through introduction of microsliders into the microspring systems in Fig. 2, leading to much more complex formulation.

7.2 Biaxial loading

For proportional biaxial loading, the tests by Kupfer et al.[41] are simulated. First the uniaxial material parameters $\overline{E}$, f_t, f_c, μ^+, μ^-, V^+, V^-, K^+, and K^- are calibrated to fit the experimental uniaxial stress-strain curves; and then they are used to predict the biaxial behavior with f_{bc} calibrated from the peak stresses. These parameters are collected in Table 5.

7.2.1 Tension-compression

The calculated stress-strain curves in biaxial tension-compression are displayed along with the test data in Fig. 8.

7.2.2 Compression-compression

The calculated stress-strain curves in biaxial compression are displayed along with the test data in Fig. 9.

7.2.3 Successive tension and compression

For compression that follows preceding tensile failure in a perpendicular direction, a special loading scenario is simulated: first a specimen is loaded in tension to almost complete failure along the x axis; and then the displacement in this direction is kept constant ($\varepsilon_x \equiv 0.4\times10^{-3}$), and the specimen is subjected to compression along the y axis. The material parameters listed in Table 5 are used. The calculated stress-strain curves are displayed in Fig. 10. A simulation of free uniaxial

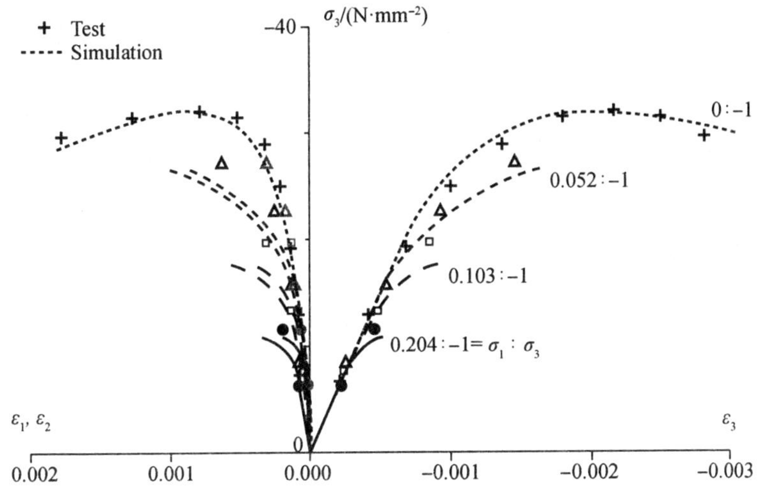

Fig. 8. Biaxial tension-compression[40]

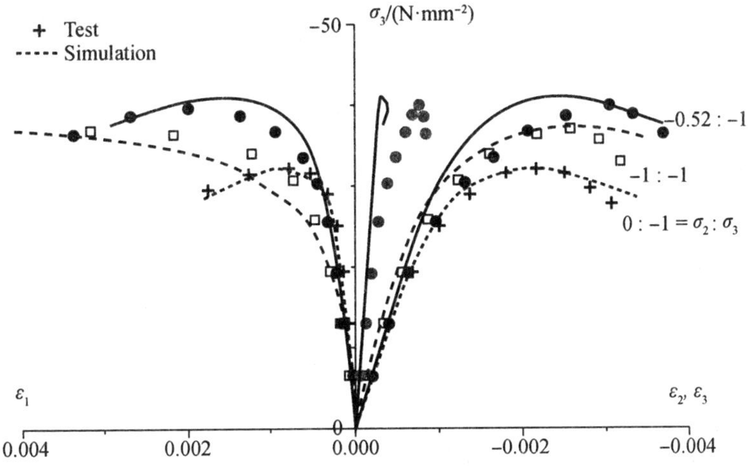

Fig. 9. Biaxial compression[41]

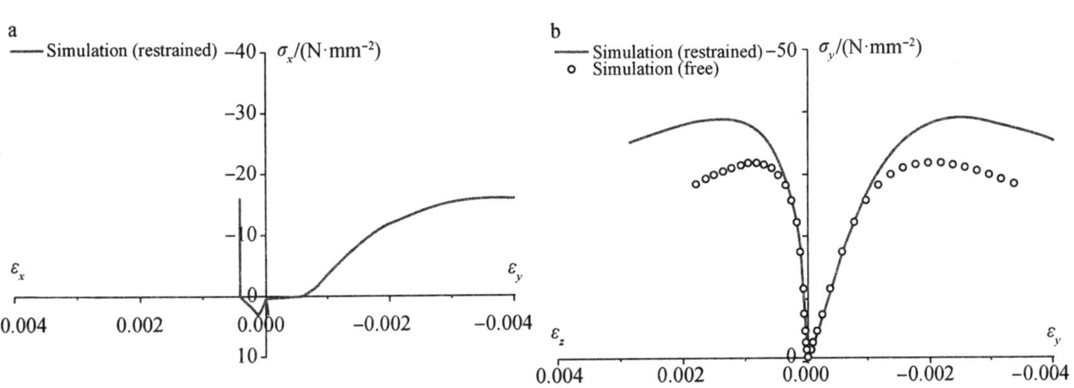

Fig. 10. Longitudinal compression after transversal tension. (a) Transversal stress. (b) Longitudinal stress

compression (without lateral restraint) is also presented in Fig. 10(b) as a reference. Figs. 7 and 10 demonstrate the ability of the present model to formulate independent plasticity/damage evolution in tension and compression.

It is observed that the variation in strength and ductility, the lateral deformation, and the independent tensile and compressive plasticity/damage evolution of concrete under biaxial loading can be represented appropriately with the proposed model.

7.3 Triaxial loading

7.3.1 Pseudo-triaxial compression

For pseudo-triaxial compression, the tests by Candappa et al.[43] are simulated. First the uniaxial material parameters $\bar{E}$, f_c, μ^-, V^-, and K^- are calibrated to fit the experimental uniaxial stress-strain curves, the triaxial material parameters c_Δ and c_κ to fit the longitudinal stress-strain ($\sigma_3 - \varepsilon_3$) curve, and c_η to fit the longitudinal stress-lateral strain ($\sigma_3 - \varepsilon_1$) curve under the confining pressure $\sigma_1 = \sigma_2 = -4$ N/mm^2; and then all these calibrated parameters are used to predict the triaxial behavior under $\sigma_1 = \sigma_2 = -8$ and -12 N/mm^2. They are collected in Table 6. The calculated stress-strain curves are exhibited along with the test data in Fig. 11.

Table 6. Parameters for pseudo-triaxial compression simulations

Effective Young's modulus	$\bar{E}$	50.9×10^3 N/mm^2
Uniaxial compressive strength	f_c	60.6 N/mm^2
Mean value of natural logarithm of compressive fracture strain	μ^-	7.65
Standard deviation of natural logarithm of compressive fracture strain	V^-	0.4
Effective compressive strength index	K^-	$0.035\bar{E}$
Coefficient of fracture enhancement	c_Δ	3
Coefficient of hardening slowdown	c_κ	5
Coefficient of dilation reduction	c_η	10.5

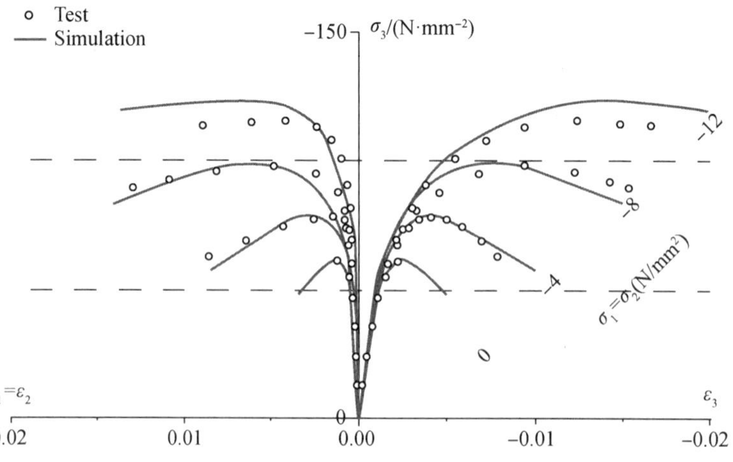

Fig. 11. Pseudo-triaxial compression[43]

This set of numerical examples demonstrate the ability of the proposed model to reflect three basic confining effects: increase in strength, increase in ductility, and decrease in dilation.

7.3.2 True triaxial compression

For true triaxial compression, the tests by van Mier[44] are simulated. First the uniaxial material parameters $\overline{E}$, f_c, μ^-, V^-, and K^- are set to produce the uniaxial stress-strain curve, and the triaxial material parameters c_Δ, c_κ, and c_η are calibrated to fit the stress-strain curves in the loading path $\sigma_1 : \sigma_2 : \sigma_3 = -0.05 : -0.1 : -1$; and then all these calibrated parameters are used to predict the triaxial behavior in $\sigma_1 : \sigma_2 : \sigma_3 = -0.05 : -0.33 : -1$(the cylindrical strength of the specimen is a bit lower than that of the one for calibration). They are collected in Table 7. The calculated stress-strain curves are exhibited along with the test data in Fig. 12.

Table 7. Parameters for true triaxial compression simulations

Effective Young's modulus	$\overline{E}$	38.0×10^3 N/mm^2
Uniaxial compressive strength	f_c	39.9 N/mm^2
Mean value of natural logarithm of compressive fracture strain	μ^-	7.47
Standard deviation of natural logarithm of compressive fracture strain	V^-	0.35
Effective compressive strength index	K^-	$0.035\overline{E}$
Coefficient of fracture enhancement	c_Δ	5
Coefficient of hardening slowdown	c_κ	12
Coefficient of dilation reduction	c_η	36

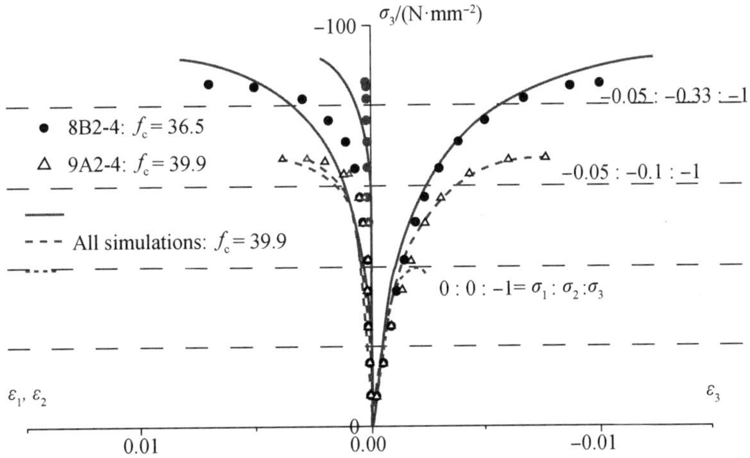

Fig. 12. True triaxial compression[44]

This set of numerical examples preliminarily demonstrate the ability of the proposed model to reflect the effects of the intermediate principal stress.

Remark 7.2. It can be found that the ability of the present model to calculate lateral deformation in true triaxial compression(Fig. 12) as well as in asymmetric biaxial loading

(Figs. 8 and 9) is limited by the simple circular deviatoric trace of the plastic potential surface. More delicate plastic potential function is required to improve this aspect, which, however, would make the implicit return-mapping algorithm very difficult especially when distortion of the yield surface(Fig. 4) resulting from tension-compression double hardening [Eq.(47)] is present at the same time.

Remark 7.3. It is expected that with adequate cyclic and multiaxial experimental stress-strain curves, many of the material parameters in the present model can be determined in a statistical sense from the strength and type of a certain concrete material.

8 Conclusions

The novel confining-net decomposition of stress in the presented work is effective in distinguishing concrete behavior in triaxial compression from that in biaxial compression and at the same time preserving their similarity. In contrast with the traditional spherical-deviatoric decomposition, the confining part out of the proposed decomposition, which is also a spherical tensor, is present only in triaxial compression, thus serving as an indicator of triaxial confining effects in both damage and plasticity formulation.

This confining-net decomposition can be combined naturally with the positive-negative decomposition, which deals with unilateral effects successfully. Consequently, the 3D damage model developed on the basis of the novel decomposition of stress, in conjunction with the tension-compression double hardening plasticity, is also applicable under multiaxial tension-compression or reversed cyclic loading.

More reliable experimental results in true triaxial compression are warranted to further examine and improve the proposed theory and to calibrate stochastic damage evolution.

Finally, with the neglected third damage variable restored and a closed yield surface employed, the presented framework can be extended to a more complex form for use under very high confinement and hydrostatic compression.

Acknowledgments

The financial support from the National Science Foundation of China (NSFC) Projects (Grant nos. 51378377 and 51108336) and Fundamental Research Funds for the Central Universities (Grant no. 2016YXMS094) is greatly appreciated. The third co-author (J. W. Ju) acknowledges the financial support from the 1 000 Talents Program of Tongji University and the Central Organization Department.

References

[1] LUBLINER J, OLIVER J, OLLER S, et al. A plastic-damage model for concrete[J]. International Journal of Solids and Structures, 1989, 25(3): 299-326.

[2] LEE J, FENVES G L. Plastic-damage model for cyclic loading of concrete structures[J]. Journal of

Engineering Mechanics, 1998, 124(8): 892-900.
[3] GRASSL P, LUNDGREN K, GYLLTOFT K. Concrete in compression: a plasticity theory with a novel hardening law[J]. International Journal of Solids and Structures, 2002, 39(20): 5205-5223.
[4] PAPANIKOLAOU V K, KAPPOS A J. Confinement-sensitive plasticity constitutive model for concrete in triaxial compression[J]. International Journal of Solids and Structures, 2007, 44(21): 7021-7048.
[5] CARRAZEDO R, MIRMIRAN A, DE HANAI J B. Plasticity based stress-strain model for concrete confinement[J]. Engineering Structures, 2013, 48: 645-657.
[6] JU J W. On energy-based coupled elastoplastic damage theories: constitutive modeling and computational aspects[J]. International Journal of Solids and Structures, 1989, 25(7): 803-833.
[7] FARIA R, OLIVER J, CERVERA M. A strain-based plastic viscous-damage model for massive concrete structures[J]. International Journal of Solids and Structures, 1998, 35(14): 1533-1558.
[8] WU J Y, LI J, FARIA R. An energy release rate-based plastic-damage model for concrete[J]. International Journal of Solids and Structures, 2006, 43(3-4): 583-612.
[9] CICEKLI U, VOYIADJIS G Z, AL-RUB R K A. A plasticity and anisotropic damage model for plain concrete[J]. International Journal of Plasticity, 2007, 23(10-11): 1874-1900.
[10] AL-RUB R K A, KIM S M. Computational applications of a coupled plasticity-damage constitutive model for simulating plain concrete fracture[J]. Engineering Fracture Mechanics, 2010, 77(10): 1577-1603.
[11] GRASSL P, XENOS D, NYSTRÖM U, et al. CDPM2: A damage-plasticity approach to modelling the failure of concrete[J]. International Journal of Solids and Structures, 2013, 50(24): 3805-3816.
[12] MAZARS J, HAMON F, GRANGE S. A new 3D damage model for concrete undermonotonic, cyclic and dynamic loadings[J]. Materials and Structures, 2014: 1-15.
[13] ZHANG J, LI J. Elastoplastic damage model for concrete based on consistent free energy potential[J]. Science China Technological Sciences, 2014, 57(11): 2278-2286.
[14] IMRAN I, PANTAZOPOULOU S J. Plasticity model for concrete under triaxial compression[J]. Journal of Engineering Mechanics, 2001, 127(3): 281-290.
[15] CHI Y, XU L, YU H. Constitutive modeling of steel-polypropylene hybrid fiber reinforced concrete using a non-associated plasticity and its numerical implementation[J]. Composite Structures, 2014, 111: 497-509.
[16] ČERVENKA J, PAPANIKOLAOU V K. Three dimensional combined fracture-plastic material model for concrete[J]. International Journal of Plasticity, 2008, 24(12): 2192-2220.
[17] SÁNCHEZ P J, HUESPE A E, OLIVER J, et al. A macroscopic damage-plastic constitutive law for modeling quasi-brittle fracture and ductile behavior of concrete[J]. International Journal for Numerical and Analytical Methods in Geomechanics, 2012, 36(5): 546-573.
[18] LUCCIONI B M, ROUGIER V C. A plastic damage approach for confined concrete[J]. Computers & Structures, 2005, 83(27): 2238-2256.
[19] GRASSL P, JIRÁSEK M. Damage-plastic model for concrete failure[J]. International Journal of Solids and Structures, 2006, 43(22-23): 7166-7196.
[20] JASON L, HUERTA A, PIJAUDIER-CABOT G, et al. An elastic plastic damage formulation for concrete: Application to elementary tests and comparison with an isotropic damage model[J]. Computer Methods in Applied Mechanics and Engineering, 2006, 195(52): 7077-7092.
[21] CANER F C, BAŽANT Z P. Microplane model M7 for plain concrete. Ⅰ: Formulation[J]. Journal of

Engineering Mechanics, 2013, 139(12): 1714-1723.

[22] HOFSTETTER B V G. Review and enhancement of 3D concrete models for large-scale numerical simulations of concrete structures[J]. International Journal for Numerical and Analytical Methods in Geomechanics, 2013, 37(3): 221-246.

[23] LADEVEZE P. Sur une theorie de l'endommagement anisotrope (Internal Report no. 34)[J]. 1983.

[24] ORTIZ M. A constitutive theory for the inelastic behavior of concrete[J]. Mechanics of Materials, 1985, 4(1): 67-93.

[25] SHANG H, JI G. Mechanical behaviour of different types of concrete under multiaxial compression[J]. Magazine of Concrete Research, 2014, 66(17): 870-876.

[26] GREEN S J, SWANSON S R. Static constitutive relations for concrete[R]. Air Force Weapons Laboratory, Kirtland Air Force Base, Albuq uerque,NM, 1973.

[27] GABET T, MALÉCOT Y, DAUDEVILLE L. Triaxial behaviour of concrete under high stresses: Influence of the loading path on compaction and limit states[J]. Cement and Concrete Research, 2008, 38(3): 403-412.

[28] ZHANG J, LI J. Microelement formulation of free energy for quasi-brittle materials[J]. Journal of Engineering Mechanics, 2014, 140(8): 06014008.

[29] LI J, REN X D. Stochastic damage model for concrete based on energy equivalent strain[J]. International Journal of Solids and Structures, 2009, 46(11-12): 2407-2419.

[30] BAŽANT Z P, BECQ-GIRAUDON E. Statistical prediction of fracture parameters of concrete and implications for choice of testing standard[J]. Cement and Concrete Research, 2002, 32(4): 529-556.

[31] OLIVER J, CERVERA M, OLLER S, et al. Isotropic damage models and smeared crack analysis of concrete[C]//Proc. SCI-C Computer Aided Analysis and Design of Concrete Structures. Swansea: Pineridge Press, 1990, 945-958.

[32] ZHANG J, ZHANG Z, CHEN C. Yield criterion in plastic-damage models for concrete[J]. Acta Mechanica Solida Sinica, 2010, 23(3): 220-230.

[33] ZHANG J, LI J. Investigation into Lubliner yield criterion of concrete for 3D simulation[J]. Engineering Structures, 2012, 44: 122-127.

[34] CHORIN A J, HUGHES T J R, MCCRACKEN M F, et al. Product formulas and numerical algorithms[J]. Communications on Pure and Applied Mathematics, 1978, 31(2): 205-256.

[35] LEE J, FENVES G L. A return-mapping algorithm for plastic-damage models: 3-D and plane stress formulation[J]. International Journal for Numerical Methods in Engineering, 2001, 50(2): 487-506.

[35] SIMO J C, HUGHES T J R. Computational Inelasticity Springer-Verlag[M]. New York, 1998: 143-149.

[36] WU J Y. Damage energy release rate-based elastoplastic damage constitutive model for concrete and its application to nonlinear analysis of structures[D]. Shanghai, China: Department of Building Engineering, Tongji University, 2004.

[37] FARIA R, OLIVER J O O, RUIZ M C. On isotropic scalar damage models for the numerical analysis of concrete structures[M]. Barcelona, Spain: Centro Internacional de Métodos Numéricos en Ingeniería, 2000.

[38] WU J Y, XU S L. Reconsideration on the elastic damage/degradation theory for the modeling of microcrack closure-reopening (MCR) effects[J]. International Journal of Solids and Structures, 2013, 50(5): 795-805.

[39] GOPALARATNAM V S, SHAH S P. Softening response of plain concrete in direct tension[C]// Journal Proceedings. 1985, 82(3): 310-323.

[40] KARSAN I D, JIRSA J O. Behavior of concrete under compressive loadings[J]. Journal of the Structural Division, 1969, 95(12): 2543-2564.

[41] KUPFER H, HILSDORF H K, RUSCH H. Behavior of concrete under biaxial stresses[C]//Journal Proceedings. 1969, 66(8): 656-666.

[42] CANER F C, BAŽANT Z P. Microplane model M7 for plain concrete. Ⅱ: Calibration and verification [J]. Journal of Engineering Mechanics, 2013, 139(12): 1724-1735.

[43] CANDAPPA D C, SANJAYAN J G, SETUNGE S. Complete triaxial stress-strain curves of high-strength concrete[J]. Journal of Materials in Civil Engineering, 2001, 13(3): 209-215.

[44] VAN MIER J G M. Strain-softening of concrete under multiaxial loading conditions[M]. Eindhoven: Technische Hogeschool Eindhoven, 1984.

A unified dynamic model for concrete considering viscoplasticity and rate-dependent damage①

Xiaodan Ren[1], **Jie Li**[1,2]

1 *School of Civil Engineering, Tongji University, Shanghai, China*

2 *The State Key Laboratory on Disaster Reduction in Civil Engineering, Tongji University, Shanghai, China*

Abstract Precise numerical simulation of structures under dynamic loading is often hampered by the lack of the elaborate rate-dependent nonlinear model of materials. Thus in this article, a unified damage model is proposed to represent the nonlinear behaviors and the rate-dependency of plain concrete in an energy consistent manner. The coupled inviscid plastic damage theory with two damage scalars is first adopted as the multi-dimensional framework of this model. Then based on the strain equivalence hypothesis, the "effective stress space plasticity" is introduced and its rate-dependent extension is proposed by analogy of the Perzyna-type viscoplasticity. To consider the dynamic damage evolution, the rate-dependent damage evolution is introduced according to the simplified Perzyna-type flow rule. Finally, the unified model is developed by introducing both the "effective stress space viscoplasticity" and the rate-dependent damage evolution into the elastoplastic damage framework. Under uniaxial tension, the closed form solution of the dynamic increase factor is analytically derived with two material parameters which have been thoroughly investigated in the literatures. The model results also show well agreements with the experimental data in both material and structural levels.

Keywords Damage; Rate-dependency; Plasticity; Perzyna model; Concrete

1 Introduction

Concrete is the most widely used material in civil engineering. A tremendous amount of concrete is utilized for construction throughout the world each year. However, the development of concrete science especially concrete mechanics has not been able to meet the demand of design and analysis of significant concrete structures. Some key scientific and technological problems still remain somewhat controversial, such as the constitutive modeling and the nonlinear analysis of concrete structures. Even though numbers of breakthroughs in solid mechanics have been introduced to improve the nonlinear modeling of concrete, no widely accepted nonlinear model for concrete has been proposed up till now. Therefore, it is quite a challenging task for us to develop the elaborate nonlinear material model of concrete which is able to reproduce its salient features consistently. Also, two of

① Originally published in *International Journal of Damage Mechanics*, 2013, 22(4): 530–555.

the essential properties of concrete, which should be taken into account are the nonlinearity and the rate-dependency.

Continuum damage mechanics (CDM) has been introduced to the nonlinear modeling of concrete for more than 20 years[1-5]. Based on the irreversible thermodynamics, the CDM provides a powerful and general framework for the representation of essential nonlinear performances of concrete including the strain softening, the stiffness degradation, and the unilateral effect. The early damage models of concrete are mainly restricted in the "elastic damage" mechanics for brittle materials, which could not predict some of the typical nonlinearities of concrete such as volumetric dilation and strength increase under multi-dimensional compression. Hence, the research of the coupled plastic damage model (CPDM) of concrete has been increasingly stressed in recent 20 years[6-13]. Due to its comprehensive grasp of the essential nonlinear properties of concrete, the CPDM has been adopted as the standard framework for the nonlinear simulation of concrete structures.

The strain rate effect is activated by dynamic loading such as earthquake, impact, and explosion that act on the existing concrete structures. The classic experimental researches[14-15] indicate that the properties of concrete are significantly deviated by the strain rate. The rate-dependent tensile and compressive strengths of concrete are typically represented by means of the dynamic increase factor (DIF) curves that are defined as the ratio between the dynamic and the static uniaxial strengths. The various rate-dependent performances of materials can hardly be well reproduced by the viscoelastic model, because the most important effects lie in the nonlinear portion of the material response once cracking or micro-cracking has started[16]. Hence, the viscoplastic theory was introduced to the dynamic constitutive model of concrete[17-20]. The Perzyna-type viscoplasticity[21], which defines the evolution of viscoplastic strain based on "overstress" is widely adopted. On the other side, the rate-dependent damage evolution was also proposed in constitutive modeling. Simo and Ju[11-22], Ju[22], and Dube et al.[16] proposed the Perzyna-type rate-dependent isotropic and anisotropic damage models. Cervera et al.[23] and Faria et al.[6] simplified the Perzyna-type rate-dependent evolution equation into the explicit form. By taking a careful inspection, we find that the viscoplastic model cannot describe the stress degradation correctly while the rate-dependent damage model intends to overestimate the strain at peak stress. Thus, in this article we combine the viscoplasticity and rate-dependent damage together in a consistent way to achieve the reasonable dynamic increases in the senses of stress as well as strain.

Wu et al.[24] proposed the biscalar damage framework based on the elastoplastic damage release rate. Then, by defining the damage consistent condition, Li and Ren[25] proposed the energy equivalent strain (EES), based on which the multi-axial damage evolution could be easily defined using the uniaxial damage evolution function. And the theoretical model of the uniaxial damage evolution is also proposed in Li and Ren[25]. On the basis of these previous works, the purpose of this article is to represent the nonlinearity and the rate-dependency of concrete in a unified constitutive model. By understanding that the nonlinearity is comprised of plasticity and damage, the coupled plastic damage framework is adopted to represent the nonlinear features of concrete. Combining the "effective stress space viscoplasticity" with the simplified Perzyna-type rate-dependent damage evolution, the dynamic increases of the strength and the peak strain are well reproduced. Under uniaxial loading, the analytical solution of the DIF is derived and compared with the existing experimental data. Finally, the proposed model is verified by the numerical simulations of a series of reinforced concrete beams under static as well as dynamic loading.

2 Coupled plastic damage framework

According to the strain equivalence hypothesis[22, 26], the effective stress tensor $\bar{\boldsymbol{\sigma}}$ in damaged material is defined as

$$\bar{\boldsymbol{\sigma}} = \mathbf{C}_0 : \boldsymbol{\varepsilon}^{\mathrm{e}} = \mathbf{C}_0 : (\boldsymbol{\varepsilon} - \boldsymbol{\varepsilon}^{\mathrm{p}}) \tag{1}$$

where $\mathbf{C}_0$ is the fourth-order linear-elastic stiffness and $\boldsymbol{\varepsilon}$, $\boldsymbol{\varepsilon}^{\mathrm{e}}$, and $\boldsymbol{\varepsilon}^{\mathrm{p}}$ the second-order tensors, denoting the strain tensor and its elastic and plastic components, respectively. For the brittle materials with the unilateral effect, the effective stress tensor is usually decomposed as follows

$$\bar{\boldsymbol{\sigma}} = \bar{\boldsymbol{\sigma}}^{+} + \bar{\boldsymbol{\sigma}}^{-} \tag{2}$$

where the superscripts + and − denote the tensile component as well as the shear component of the corresponding quantities, respectively. According to Wu et al.[24], we have the following expressions of each component of the effective stress

$$\bar{\boldsymbol{\sigma}}^{+} = \mathbf{P}^{+} : \bar{\boldsymbol{\sigma}} \tag{3}$$

$$\bar{\boldsymbol{\sigma}}^{-} = \bar{\boldsymbol{\sigma}} - \bar{\boldsymbol{\sigma}}^{+} = \mathbf{P}^{-} : \bar{\boldsymbol{\sigma}} \tag{4}$$

where the fourth-order projection tensors $\mathbf{P}^{+}$ and $\mathbf{P}^{-}$ are expressed as[6]

$$\mathbf{P}^{+} = \sum_i H(\hat{\bar{\sigma}}_i)(\mathbf{p}_i \otimes \mathbf{p}_i \otimes \mathbf{p}_i \otimes \mathbf{p}_i) \tag{5}$$

$$\mathbf{P}^{-} = \mathbf{I} - \mathbf{P}^{+} \tag{6}$$

in which $\mathbf{I}$ is the fourth-order identity tensor and $\hat{\bar{\sigma}}_i$ and $\mathbf{p}_i$ the ith eigenvalue and normalized eigenvector of the effective stress tensor $\bar{\boldsymbol{\sigma}}$, respectively. Actually i is neither dummy index nor free index of a tensor. It is an index for the sequence of the eigenvalues with the range from 1 to rank ($\bar{\boldsymbol{\sigma}}$). $H(\cdot)$ is the Heaviside function, which is defined as

$$H(x) = \begin{cases} 0 & x < 0 \\ x & x \geqslant 0 \end{cases} \tag{7}$$

To establish the constitutive law, the Helmholtz free energy (HFE) potential should be introduced as function of the state variables and the internal variables. The initial elastic

HFE potential ψ_0^e is defined as the elastic strain energy. It can be expressed as the summation of its positive and negative components (ψ_0^{e+}, ψ_0^{e-}) considering the decomposition of the effective stress tensor in Eqs.(2)— (4). We have

$$\psi_0^e = \frac{1}{2}\bar{\boldsymbol{\sigma}} : \varepsilon^e = \frac{1}{2}\bar{\boldsymbol{\sigma}}^+ : \boldsymbol{\varepsilon}^e + \frac{1}{2}\bar{\boldsymbol{\sigma}}^- : \boldsymbol{\varepsilon}^e = \psi_0^{e+} + \psi_0^{e-} \tag{8}$$

where the superscript "e" refers to "elastic" whereas the subscript "0" refers to the "initial" undamaged states.

Considering the tensile failure as well as the shear failure mechanisms, two damage scalars, e.g. d^+ and d^-, are adopted here and, therefore, the elastic HFE with the form

$$\psi^e(\boldsymbol{\varepsilon}^e, d^+, d^-) = \psi^{e+}(\boldsymbol{\varepsilon}^e, d^+) + \psi^{e-}(\boldsymbol{\varepsilon}^e, d^-) \tag{9}$$

can be postulated, where ψ^{e+} and ψ^{e-} are defined as

$$\psi^{e\pm}(\boldsymbol{\varepsilon}^e, d^\pm) = (1-d^\pm)\psi_0^{e\pm} \tag{10}$$

with symbol "$\pm$" denoting "$+$" or "$-$" as appropriate.

The total elastoplastic HFE potential could be defined as the summation of the elastic component ψ^e and plastic component ψ^p [24, 27], that is

$$\psi(\boldsymbol{\varepsilon}^e, \boldsymbol{\kappa}, d^+, d^-) = \psi^e(\boldsymbol{\varepsilon}^e, d^+, d^-) + \psi^p(\boldsymbol{\varepsilon}^e, \boldsymbol{\kappa}, d^+, d^-) \tag{11}$$

where $\boldsymbol{\kappa}$ denotes a suitable set of plastic variables; the plastic HFE potential ψ^p is defined as

$$\psi^p(\boldsymbol{\varepsilon}^e, \boldsymbol{\kappa}, d^+, d^-) = \psi^p(\boldsymbol{\varepsilon}^e, \boldsymbol{\kappa}, d^-) = (1-d^-)\psi_0^p = (1-d^-)\int_0^{\boldsymbol{\varepsilon}^p} \bar{\boldsymbol{\sigma}}^- : \mathrm{d}\boldsymbol{\varepsilon}^p \tag{12}$$

According to the second principle of thermodynamics, any arbitrary irreversible process satisfies the Clausius-Duhem inequality as follows

$$\dot{\gamma} = -\dot{\psi} + \boldsymbol{\sigma} : \dot{\boldsymbol{\varepsilon}} \geqslant 0 \tag{13}$$

Differentiating both sides of Eq.(11) with respect to time yields

$$\dot{\psi} = \frac{\partial \psi^e}{\partial \boldsymbol{\varepsilon}^e} : \dot{\boldsymbol{\varepsilon}}^e + \frac{\partial \psi}{\partial d^+}\dot{d}^+ + \frac{\partial \psi}{\partial d^-}\dot{d}^- + \frac{\partial \psi^p}{\partial \boldsymbol{\kappa}}\dot{\boldsymbol{\kappa}} \tag{14}$$

Taking the additional assumption that the damage and plastic unloading are elastic processes, we obtain the constitutive equality

$$\boldsymbol{\sigma} = \frac{\partial \psi^e}{\partial \boldsymbol{\varepsilon}^e} \tag{15}$$

And the dissipative inequalities

$$\dot{\gamma}^d = -\left(\frac{\partial \psi}{\partial d^+}\dot{d}^+ + \frac{\partial \psi}{\partial d^-}\dot{d}^-\right) \geqslant 0 \tag{16}$$

$$\dot{\gamma}^p = \boldsymbol{\sigma} : \dot{\boldsymbol{\varepsilon}}^p - \frac{\partial \psi^p}{\partial \boldsymbol{\kappa}}\dot{\boldsymbol{\kappa}} \geqslant 0 \tag{17}$$

From Eq.(15), it can be clearly indicated that the Cauchy stress tensor $\bar{\boldsymbol{\sigma}}$ is only dependent on the elastic HFE potential ψ^{e}.

Substituting the elastic HFE potential of Eqs.(8)—(11) in Eq.(15) leads to

$$\boldsymbol{\sigma}=(1-d^{+})\mathbf{P}^{+}:\bar{\boldsymbol{\sigma}}+(1-d^{-})\mathbf{P}^{-}:\bar{\boldsymbol{\sigma}}=(\mathbf{I}-d^{+}\mathbf{P}^{+}-d^{-}\mathbf{P}^{-}):\bar{\boldsymbol{\sigma}} \tag{18}$$

Then we obtain the final form for the constitutive law, which is a very visual expression for the Cauchy stress tensor $\boldsymbol{\sigma}$ [27]

$$\boldsymbol{\sigma}=(\mathbf{I}-\mathbf{D}):\bar{\boldsymbol{\sigma}}=(\mathbf{I}-\mathbf{D}):\mathbf{C}_0:\boldsymbol{\varepsilon}^{e}=(\mathbf{I}-\mathbf{D}):\mathbf{C}_0:(\boldsymbol{\varepsilon}-\boldsymbol{\varepsilon}^{p}) \tag{19}$$

where the fourth-order damage tensor $\mathbf{D}$ is given by

$$\mathbf{D}=d^{+}\mathbf{P}^{+}+d^{-}\mathbf{P}^{-} \tag{20}$$

From Eq.(16), the tensile and the shear damage energy release rate(DERR) Y^{+} and Y^{-}, conjugated to the corresponding damage variables, can be expressed as

$$Y^{\pm}=-\frac{\partial\psi}{\partial d^{\pm}} \tag{21}$$

Eq.(21) indicates that the DERRs depend on the total elastoplastic HFE potential. Further, substituting Eqs.(10)—(12) in Eq.(21), one obtains

$$Y^{+}=\psi_0^{e+} \tag{22}$$

$$Y^{-}=\psi_0^{e-}+\psi_0^{p} \tag{23}$$

Furthermore, Li and Ren[25] proposed the damage consistent condition for the modeling of the damage evolution. Accordingly, the EESs for the damage evolution were proposed as follows

$$\bar{\varepsilon}^{e+}=\sqrt{\frac{2Y^{+}}{E_0}} \tag{24}$$

$$\bar{\varepsilon}^{e-}=\frac{1}{(\alpha-1)}\sqrt{\frac{Y^{-}}{b_0}} \tag{25}$$

The material parameters α and b_0 which are related to the plastic behaviors being discussed in the next section. The CPDM can be proposed based on the definition of either damage energy release rate (DERR) or energy equivalent strains (EESs). In this article, we intend to adopt the EES as the damage driven force.

3 Plastic evolution

3.1 Mechanisms of plasticity

When loaded in compression concrete experiences progressive loss of stiffness in the deviatoric space along with observed volumetric changes when approaching collapse putting into evidence strong dilatancy[10]. The compressive plastic strain is composed of the sliding between both sides of cracks and the volumetric dilation of concrete. Both of them perform critical roles in the nonlinear performance of concrete in compression. On the other hand, the available experimental results under tension indicate a brittle behavior. The failure is sudden with comparatively low strength and deformation. As we know, the opening tensile crack could not be completely closed during unloading due to the roughness of the cracking surfaces. That leads to the tensile remnant strain, which is usually considered as plastic strain in material modeling. Hence the plastic strain just propagates in the direction of the dominant tensile stress, which is perpendicular to the section of crack. Also, its absolute value is much less than the compression induced plastic strain. Furthermore, this kind of plastic strain could not influence the nonlinear performance under the monotonic tensile loading unless both sides of the tensile crack contact each other. Therefore, we neglect the tensile induced plasticity and develop the plastic framework for compression in the following two sections.

3.2 Static plastic evolution

Different types of plasticity models combined with damage expressions have been proposed in the literatures. Among them, one family of models rely on the stress-based plasticity formulated in the Cauchy stress space, see Ortiz[9], Lubliner et al.[28], and Carol et al.[29]. However, the Cauchy stress may descend due to the softening property of concrete, which will result in local concave and global shrinkage of the plastic potential function. Thus, it is difficult to conduct reliable numerical simulation under this situation. Another family of models is based on the plasticity formulated in the effective stress space, see Ju[22], Lee and Fenves[8], Faria et al.[6], Li and Wu[27], and Jason et al.[30]. The effective stress represents the micro-stress on the undamaged material. Hence, the plastic potential function will keep expanding throughout the loading process and thus it is convenient to develop reliable numerical algorithms for this model. According to the "effective stress space plasticity", the evolution law of plastic strain is expressed as follows

$$\dot{\boldsymbol{\varepsilon}}^{\mathrm{p}}=\dot{\lambda}^{\mathrm{p}}\frac{\partial F^{\mathrm{p}}(\bar{\boldsymbol{\sigma}},\boldsymbol{\kappa})}{\partial\bar{\boldsymbol{\sigma}}} \tag{26}$$

$$\dot{\boldsymbol{\kappa}}=\dot{\lambda}^{\mathrm{p}}\mathbf{H} \tag{27}$$

$$F^{\mathrm{p}}\leqslant 0,\ \dot{\lambda}^{\mathrm{p}}\geqslant 0,\ \dot{\lambda}^{\mathrm{p}}F^{\mathrm{p}}\leqslant 0 \tag{28}$$

where F^{p} is the plastic potential, which is the yield function in the associated flow rule; λ^{p} the plastic flow parameter; and $\mathbf{H}$ the hardening function.

Differentiating Eq.(1) with respect to time yields

$$\dot{\bar{\boldsymbol{\sigma}}}=\mathbf{C}_0:\boldsymbol{\varepsilon}^{\mathrm{e}}=\mathbf{C}_0:(\dot{\boldsymbol{\varepsilon}}-\dot{\boldsymbol{\varepsilon}}^{\mathrm{p}})=\mathbf{C}_0:\left[\dot{\boldsymbol{\varepsilon}}-\dot{\lambda}^{\mathrm{p}}\frac{\partial F^{\mathrm{p}}(\dot{\boldsymbol{\sigma}},\boldsymbol{\kappa})}{\partial\bar{\boldsymbol{\sigma}}}\right] \tag{29}$$

The plastic consistent condition could be expressed as follows

$$\dot{F}^{\mathrm{p}}(\dot{\boldsymbol{\sigma}},\boldsymbol{\kappa})=\frac{\partial F^{\mathrm{p}}}{\partial\bar{\boldsymbol{\sigma}}}:\dot{\bar{\boldsymbol{\sigma}}}+\frac{\partial F^{\mathrm{p}}}{\partial\boldsymbol{\kappa}}\dot{\boldsymbol{\kappa}}=0 \tag{30}$$

Then by substituting Eqs. (27) and (29) into Eq. (30), the plastic flow parameter can be solved and expressed as

$$\dot{\lambda}^{\mathrm{p}}=\frac{\dfrac{\partial F^{\mathrm{p}}}{\partial\bar{\boldsymbol{\sigma}}}:\mathbf{C}_0:\dot{\boldsymbol{\varepsilon}}}{\dfrac{\partial F^{\mathrm{p}}}{\partial\bar{\boldsymbol{\sigma}}}:\mathbf{C}_0:\dfrac{\partial F^{\mathrm{p}}}{\partial\bar{\boldsymbol{\sigma}}}-\dfrac{\partial F^{\mathrm{p}}}{\partial\boldsymbol{\kappa}}\mathbf{H}} \tag{31}$$

The Drucker-Prager function is adopted as the plastic potential function as follows

$$F^{\mathrm{p}}(\bar{\boldsymbol{\sigma}},\ \boldsymbol{\kappa})=F_0^{\mathrm{p}}(\bar{\boldsymbol{\sigma}})-C(\boldsymbol{\kappa}) \tag{32}$$

$$F_0^{\mathrm{p}}(\bar{\boldsymbol{\sigma}})=\alpha\bar{I}_1^-+\sqrt{3\bar{J}_2^-} \tag{33}$$

$$C(\boldsymbol{\kappa})=(1-\alpha)c(\boldsymbol{\kappa}) \tag{34}$$

where $\bar{I}_1^-$ is the first invariants of $\bar{\boldsymbol{\sigma}}^-$; $\bar{J}_2^-$ the second invariants of $\bar{\mathbf{s}}^-$, the deviatoric components of $\bar{\boldsymbol{\sigma}}^-$; $c(\cdot)$ the effective cohesive stress function; and α the material parameter. It is observed that we only consider the compressive components of effective stress for the plastic potential functions.

Considering Eq. (26), the initial plastic HFE potential could be expressed as

$$\psi_0^{\mathrm{p}}=\int_0^{\boldsymbol{\varepsilon}^{\mathrm{p}}}\bar{\boldsymbol{\sigma}}^-:\mathrm{d}\boldsymbol{\varepsilon}^{\mathrm{p}}=\int_0^{\lambda^{\mathrm{p}}}\bar{\boldsymbol{\sigma}}^-:\frac{\partial F^{\mathrm{p}}}{\partial\bar{\boldsymbol{\sigma}}}\mathrm{d}\lambda^{\mathrm{p}} \tag{35}$$

Then, substituting Eq. (32) in Eq. (35) yields the initial plastic HFE as follows[24]

$$\begin{aligned}\psi_0^{\mathrm{p}}&=\int_0^{\lambda^{\mathrm{p}}}\bar{\boldsymbol{\sigma}}^-:\frac{\partial\left[\alpha\bar{I}_1^-+\sqrt{3\bar{J}_2^-}-(1-\alpha)c(\boldsymbol{\kappa})\right]}{\partial\bar{\boldsymbol{\sigma}}}\mathrm{d}\lambda^{\mathrm{p}}\\&=\int_0^{\lambda^{\mathrm{p}}}\frac{1}{\|\bar{\mathbf{s}}\|}\left(\sqrt{\frac{3}{2}}\bar{\mathbf{s}}^-:\bar{\mathbf{s}}^-+\alpha\bar{I}_1^-\|\bar{\mathbf{s}}\|\right)\mathrm{d}\lambda^{\mathrm{p}}\approx\frac{b}{2E_0}\left(3\bar{J}_2^-+\alpha\bar{I}_1^-\cdot\sqrt{3\bar{J}_2}\right)\geqslant 0\end{aligned} \tag{36}$$

where $b_0=\sqrt{\dfrac{8}{3}}\dfrac{E_0\lambda^{\mathrm{p}}}{\sqrt{\bar{\mathbf{s}}:\bar{\mathbf{s}}}}\geqslant 0$ could be defined as the material parameter.

By adopting the equivalent plastic strain as the plastic variable, we have

$$\kappa=\int\sqrt{\frac{2}{3}\dot{\boldsymbol{\varepsilon}}_{\mathrm{p}}:\dot{\boldsymbol{\varepsilon}}_{\mathrm{p}}}\,\mathrm{d}t \tag{37}$$

Then, the equivalent plastic strain rate is

$$\dot{\kappa}=\sqrt{\frac{2}{3}\dot{\boldsymbol{\varepsilon}}_{\mathrm{p}}:\dot{\boldsymbol{\varepsilon}}_{\mathrm{p}}} \tag{38}$$

Substituting Eq. (26) in Eq. (38), we obtain

$$\dot{\kappa}=\dot{\lambda}^{\mathrm{p}}\sqrt{\frac{2}{3}\frac{\partial F^{\mathrm{p}}}{\partial\bar{\boldsymbol{\sigma}}}:\frac{\partial F^{\mathrm{p}}}{\partial\bar{\boldsymbol{\sigma}}}} \tag{39}$$

Combining Eq.(27) with Eq.(39), we obtain the hardening function as follows

$$\mathbf{H}=\sqrt{\frac{2}{3}\frac{\partial F^{\mathrm{p}}}{\partial \bar{\boldsymbol{\sigma}}}:\frac{\partial F^{\mathrm{p}}}{\partial \bar{\boldsymbol{\sigma}}}} \tag{40}$$

Under uniaxial compression ($\bar{\sigma}_3=0$, $\bar{\sigma}_2=0$, $\bar{\sigma}_1=\bar{\sigma}<0$), we obtain

$$\kappa=\int\sqrt{\frac{2}{3}\dot{\boldsymbol{\varepsilon}}_{\mathrm{p}}:\dot{\boldsymbol{\varepsilon}}_{\mathrm{p}}}\,\mathrm{d}t=|\varepsilon_{\mathrm{p}}| \tag{41}$$

$$\bar{I}_1=\bar{\sigma}\quad\sqrt{3\bar{J}_2}=-\bar{\sigma} \tag{42}$$

Substituting Eqs.(41) and (42) in Eq.(32) and considering plastic consistent condition, we get

$$-E_0\varepsilon^{\mathrm{e}}-c(\varepsilon^{\mathrm{p}})=0 \tag{43}$$

where E_0 is the initial elastic modulus of concrete. If the functional relation between the uniaxial elastic strain ε^{e} and the plastic strain ε^{p} is determined, we are able to obtain the effective cohesive stress function $c(\cdot)$.

On the other hand, Karson and Jirsa[31] proposed the empirical expression for concrete under uniaxial compression as follows

$$\left(\frac{\varepsilon^{\mathrm{p}}}{\varepsilon_{\mathrm{c}}}\right)=a\left(\frac{\varepsilon}{\varepsilon_{\mathrm{c}}}\right)^2+b\left(\frac{\varepsilon}{\varepsilon_{\mathrm{c}}}\right) \tag{44}$$

where ε_{c} is the total strain corresponding to the peak stress and a and b the material parameters identified based on the repeated loading test. Their proposed values are

$$\begin{cases}a=0.145\\ b=0.130\end{cases} \tag{45}$$

It is clear that the uniaxial total strain in Eq.(44) could be decomposed as follows

$$\varepsilon=\varepsilon^{\mathrm{e}}+\varepsilon^{\mathrm{p}} \tag{46}$$

Then, substituting Eq.(46) in Eq.(44) and combining with Eq.(43), we obtain

$$c(\varepsilon^{\mathrm{p}})=\frac{\sqrt{b^2+4a(\varepsilon^{\mathrm{p}}/\varepsilon_{\mathrm{c}})}-[2a(\varepsilon^{\mathrm{p}}/\varepsilon_{\mathrm{c}})+b]}{2a}E_0\varepsilon_{\mathrm{c}} \tag{47}$$

Replacing ε^{p} with $\boldsymbol{\kappa}$ in Eq.(47), we get the explicit expression of function $c(\boldsymbol{\kappa})$ for Eq.(32).

3.3 Dynamic plastic evolution

Based on the viscoplastic theory proposed by Perzyna[21], the "effective stress space viscoplasticity" may be written as

$$\dot{\bar{\boldsymbol{\sigma}}}=\mathbf{C}_0:(\dot{\boldsymbol{\varepsilon}}-\dot{\boldsymbol{\varepsilon}}_{\mathrm{vp}}) \tag{48}$$

$$\dot{\boldsymbol{\varepsilon}}_{\mathrm{vp}}=\gamma_{\mathrm{p}}\phi(F^{\mathrm{p}})\frac{\partial F^{\mathrm{p}}}{\partial \bar{\boldsymbol{\sigma}}} \tag{49}$$

$$\psi(F^{\mathrm{p}}) = \left\langle \frac{F^{\mathrm{p}}(\bar{\boldsymbol{\sigma}}, \boldsymbol{\kappa})}{C(\boldsymbol{\kappa})} \right\rangle^{n_{\mathrm{p}}} = \left\langle \frac{F_0^{\mathrm{p}}(\bar{\boldsymbol{\sigma}})}{C(\boldsymbol{\kappa})} - 1 \right\rangle^{n_{\mathrm{p}}} \tag{50}$$

$$\dot{\boldsymbol{\kappa}} = \gamma_{\mathrm{p}} \psi(F^{\mathrm{p}}) \mathbf{H} \tag{51}$$

where $\dot{\boldsymbol{\varepsilon}}_{\mathrm{vp}}$ denotes the viscoplastic component of total strain rate $\dot{\boldsymbol{\varepsilon}}$; F^{p} the inviscid plastic potential which is adopted as Eq.(32) in this article; $\mathbf{H}$ the hardening function expressed in Eq.(40); γ_{p} the fluidity parameter, and n_{p} the order of Perzyna's viscoplasticity.

Comparing Eqs.(26) and (49), the flow parameter could be expressed as follows under dynamic loading

$$\dot{\lambda}_{\mathrm{vp}} = \gamma_{\mathrm{p}} \psi(F^{\mathrm{p}}) \tag{52}$$

Considering $\dfrac{F^{\mathrm{p}}(\bar{\boldsymbol{\sigma}}, \boldsymbol{\kappa})}{C(\boldsymbol{\kappa})} > 0$ and applying inverse function in the both sides, one obtains

$$F^{\mathrm{p}} = \psi^{-1}\left(\frac{\dot{\lambda}_{\mathrm{vp}}}{\gamma_{\mathrm{p}}}\right) \tag{53}$$

where $\psi^{-1}(\cdot)$ is the inverse function of $\psi(\cdot)$. Define

$$F^{\mathrm{vp}} = F^{\mathrm{p}} - \psi^{-1}\left(\frac{\dot{\lambda}_{\mathrm{vp}}}{\gamma_{\mathrm{p}}}\right) = 0 \tag{54}$$

The continuum form of "effective stress space viscoplasticity" could be easily derived by replacing the plastic potential F^{p} with F^{vp} in "effective stress space plasticity." By differentiating both sides of Eq. (54), the generalized consistent condition could be expressed as

$$\dot{F}^{\mathrm{vp}} = \frac{\partial F^{\mathrm{p}}}{\partial \bar{\boldsymbol{\sigma}}} : \dot{\bar{\boldsymbol{\sigma}}} + \left(\frac{\partial F^{\mathrm{p}}}{\partial \boldsymbol{\kappa}} \mathbf{H}\right) \dot{\lambda}_{\mathrm{vp}} - \frac{\mathrm{d}}{\mathrm{d}\dot{\lambda}_{\mathrm{vp}}}\left[\psi^{-1}\left(\frac{\dot{\lambda}_{\mathrm{vp}}}{\gamma_{\mathrm{p}}}\right)\right] \ddot{\lambda}_{\mathrm{vp}} = 0 \tag{55}$$

Eq.(55) plays an important role in the numerical integration of continuum Perzyna model.

4 Damage evolution

On the basis of the normality structure in continuum mechanics, the evolution law for damage variables can be defined as[32]

$$\dot{d}^{\pm} = \dot{\lambda}^{\pm} \frac{\partial g^{\pm}(q^{\pm})}{\partial q^{\pm}} \tag{56}$$

where $g^{\pm}$ is the Lemaitre potential of dissipation related to damage and q the damage driven force. The energy consistent damage driven force is the energy conjugated force $Y^{\pm}$ of damage variable $d^{\pm}$.

Calling for the associative flow rule and damage consistency condition[6, 22, 24] for static loading, one obtains

$$\dot{\lambda}^{\pm} = \dot{q}^{\pm} \tag{57}$$

Introducing Eq.(57) into Eq.(56), the evolution laws for damage variables during loading are expressed as

$$\dot{d}^{\pm}=\dot{q}^{\pm}\frac{\partial g^{\pm}(q^{\pm})}{\partial q^{\pm}} \tag{58}$$

Performing a trivial integration and accounting with the initial conditions $d^{\pm}(q_0^{\pm})=0$, one obtains

$$d^{\pm}=g^{\pm}(q^{\pm}) \tag{59}$$

Eq.(59) is the static damage evolution law. In this article, we chose EESs expressed in Eqs.(24) and (25) as the damage driven forces $q^{\pm}$.

4.1 Rate-independent damage evolution model

Static uniaxial damage evolution function based on the parallel buddle model is introduced in this section. According to Kandarpa et al.[33], Li and Zhang[34], Li and Ren[25], etc., a structural element in uniaxial loading condition can be idealized as a series of micro-elements jointed in parallel(Fig. 1). The elements are linked with rigid bar on the ends so that each of them bears uniform deformation during the loading process. The individual element represents the micro-properties of the material and the element system describes the macro- or structural response. Therefore, the complicated material behaviors can be represented based on the parallel system.

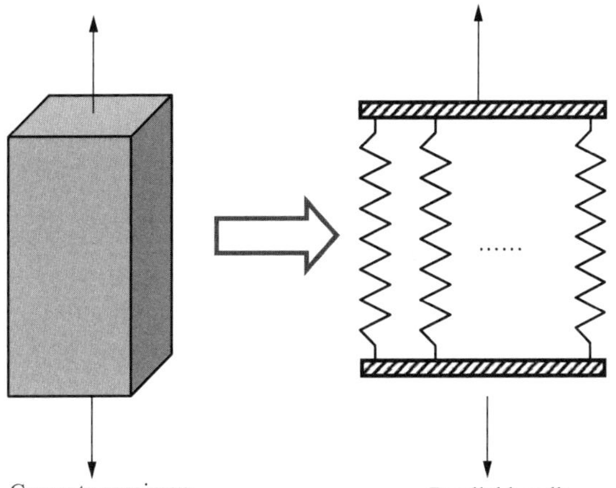

Fig. 1. The parallel element model

On the basis of the spatial averaging for the random field, the damage evolution could be established as follows

$$d(\varepsilon^{e})=F(\varepsilon^{e};\lambda,\zeta)=\int_0^{\varepsilon^{e}}\left\{\frac{1}{x\zeta\sqrt{2\pi}}e^{-\frac{(\ln x-\lambda)^2}{2\zeta^2}}\right\}dx \tag{60}$$

where $F(\cdot)$ is the cumulative function of the lognormal distribution and λ and ζ the parameters which denote the mean value and the SD of the variable's natural logarithm, respectively.

Let μ_Δ and σ_Δ be the first-order expectation and SD of the lognormal distribution, respectively. Then, the relations among parameters can be given as follows

$$\lambda = E[\ln \Delta(x)] = \ln\left(\frac{\mu_\Delta}{\sqrt{1+\sigma_\Delta^2/\mu_\Delta^2}}\right) \tag{61}$$

$$\zeta^2 = \mathrm{Var}[\ln \Delta(x)] = \ln(1+\sigma_\Delta^2/\mu_\Delta^2) \tag{62}$$

According to the properties of the lognormal distribution, the cumulative function could be also calculated using the following expression

$$F(\varepsilon^e; \lambda, \zeta) = \Phi\left[\frac{\ln \varepsilon^e - \lambda}{\zeta}\right] \tag{63}$$

where $\Phi(\cdot)$ is the cumulative distribution function of a standard normal distribution. The parameters (λ, ζ) are introduced in the parallel element model to represent the rate-independent damage evolution under uniaxial loading. As we known, the behavior of concrete shows strong dependency to the loading conditions. Thus, two groups of material parameters are introduced to reproduce the damage evolutions under tension and compression, respectively. In this article, (λ^+, ζ^+) denote the tensile material parameters related to the evolution of d^+, whereas (λ^-, ζ^-) the compressive material parameters related to the evolution of d^-. Furthermore, the damage and failure under uniaxial compression are mainly due to the degradation induced by the shear stress, hence (λ^-, ζ^-) are also referred as the shear material parameters.

By substituting the EES Eqs. (24) and (25) in the damage evolution function (60), we obtain the multi-dimensional static damage evolution as follows

$$d^\pm(\bar{\varepsilon}^{e\pm}) = F(\bar{\varepsilon}^{e\pm}; \lambda^\pm, \zeta^\pm) = F^\pm(\bar{\varepsilon}^{e\pm}) \tag{64}$$

4.2 Dynamic damage evolution model

Simo and Ju[11-12] introduced the rate-dependent expression of damage flow parameter as follows

$$\dot{\lambda}^\pm = \gamma_d^\pm \left\langle \frac{q^\pm}{\lambda^\pm} - 1 \right\rangle^{n_d^\pm} \tag{65}$$

Combining Eq. (64) with Eq. (65), the rate-dependent damage evolution could be obtained. However, the implicit differential equations are usually computational expensive. Hence, Cervera et al.[23] proposed the reduced close form rate-dependent damage evolution function as follows

$$\begin{cases} d^\pm = g^\pm(q_r^\pm) \\ \dot{q}_r^\pm = \gamma_d^\pm \left\langle \dfrac{q^\pm}{q_r^\pm} - 1 \right\rangle^{n_d^\pm} \end{cases} \tag{66}$$

where $q_r^\pm$ is the dynamic damage driven force and $q^\pm$ the static damage driven force; $\gamma_d^\pm$ the fluidity parameter and $n_d^{\pm}$ the order of rate-dependent damage evolution. Under the static loading $\dot{q}_r^{\pm} \to 0$, we obtain

$$\dot{q}_r^{\pm} = \gamma_d^{\pm} \left\langle \frac{q^{\pm}}{q_r^{\pm}} - 1 \right\rangle^{n_d^{\pm}} = 0 \tag{67}$$

Then, the following equation could be easily derived

$$q^{\pm} = q_r^{\pm} \tag{68}$$

It is concluded that the dynamic evolution model adopted here is consistent with the static damage evolution model. Since the EES is adopted as damage driven force in this article, the dynamic damage evolution law could be obtained by substituting it into Eq.(66). We have

$$\begin{cases} d^{\pm} = g^{\pm}(q_r^{\pm}) \\ \dot{q}_r^{\pm} = \gamma_d^{\pm} \left\langle \dfrac{\bar{\varepsilon}^{e\pm}}{q_r^{\pm}} - 1 \right\rangle^{n_d^{\pm}} \end{cases} \tag{69}$$

5 Results and verification

5.1 Uniaxial tension ($\bar{\sigma}_1 = \bar{\sigma} > 0$, $\bar{\sigma}_2 = 0$, $\bar{\sigma}_3 = 0$)

1. Static model

Based on the aforementioned discussions, the plastic strain induced by the tensile loading is neglected in this model. Hence under the static uniaxial loading condition, we obtain

$$\varepsilon = \varepsilon^e + \varepsilon^p = \varepsilon^e \tag{70}$$

Then based on the static damage evolution function proposed in 'Dynamic damage evolution model', the rate-independent uniaxial constitutive relationship is expressed as

$$\begin{cases} \sigma = [1 - d^+(\varepsilon)]E_0\varepsilon \\ d^+(\varepsilon) = F^+(\varepsilon) \end{cases} \tag{71}$$

The simulated results of Eq.(71) versus experimental data[35] are illustrated in Fig. 2. The material parameters identified based on test results are: $E_0 = 45\,000$ MPa, $\lambda^+ = 4.857$, and $\zeta^+ = 0.748$. The agreement between the simulated and the experimental results shows that this model is able to represent the tensile behaviors of concrete.

2. Dynamic model

Based on Eqs.(69) and (71), the rate-dependent constitutive relationship of concrete under uniaxial tension could be expressed as

$$\sigma = [1 - d^+(q_r^+)]E_0\varepsilon \tag{72a}$$

$$d^+ = F^+(q_r^+) \tag{72b}$$

The evolution of the damage driven force q_r^+ is scaled by the following Perzyna-type equation.

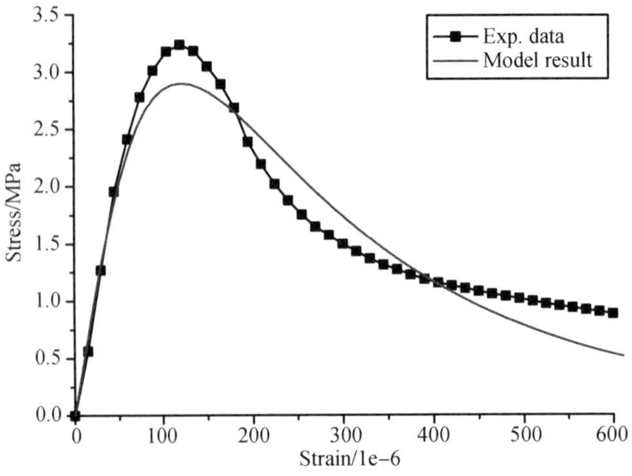

Fig. 2. Static uniaxial tensile tests

$$\dot{q}_{\mathrm{r}}^{+}=\gamma_{\mathrm{d}}^{+}\left\langle\frac{\varepsilon}{q_{\mathrm{r}}^{+}}-1\right\rangle^{n_d^+} \tag{73}$$

The results could be easily calculated by the numerical integration of Eq. (73). It is observed that Eq.(72) could be rewritten as

$$\sigma=[1-F^{+}(q_{\mathrm{r}}^{+})]E_0\varepsilon \tag{74}$$

with the function $F(\cdot)$ calculated by Eq.(63).

The simulated results of Eq.(73) are shown against the experimental data[36-37] in a range of velocity between 1×10^{-6} and 1 (Fig. 3). The material parameters are set as E_0 = 35 400 MPa, $\lambda^{+}=5.684$, $\zeta^{+}=0.541$, $\gamma_d^{+}=0.12$, and $n_d^{+}=4.4$. Unfortunately, the softening branch of concrete is not available, so the model predictions after peak point cannot be verified. It is observed that this model could predict the strength increase under dynamic loading precisely. But the corresponding strains are overestimated by this model due to the omission of plasticity.

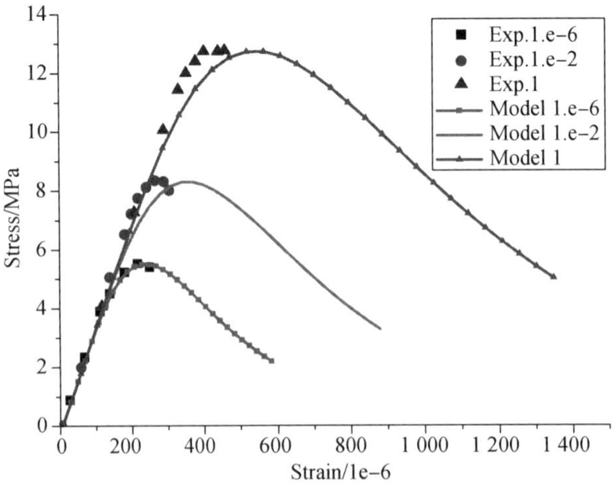

Fig. 3. Dynamic uniaxial tensile tests

3. The dynamic increase factor

According to the experimental results in the literatures, the strength of concrete under dynamic loading is higher than that under static loading. The ratio between dynamic and static strengths is defined as DIF. Thus, numbers of empirical expressions of DIF were proposed based on experimental data. In this section, the analytical expression of DIF is theoretically derived based on the aforementioned dynamic damage model.

It is assumed that the strain rate remains constant during the entire loading process. Then, we obtain

$$\varepsilon(t)=\dot{\varepsilon}t \tag{75}$$

where $\dot{\varepsilon}$ is the strain rate. By substituting Eq.(75) into Eq.(73), one obtains

$$\dot{q}_r^+=\gamma_d^+\left\langle\frac{\dot{\varepsilon}}{q_r^+}t-1\right\rangle^{n_d^+}=\gamma_d^+\left(\frac{\dot{\varepsilon}}{q_r^+}t-1\right)^{n_d^+} \tag{76}$$

Eq.(76) is the first-order differential equation of q_r^+ with respect to time t. Let the initial condition be

$$q_r^+(0)=0 \tag{77}$$

We adopt the trial solution of Eq.(77) as follows

$$q_r^+=kt \tag{78}$$

where k remains constant during the entire loading process. Substituting Eq. (78) into Eq.(76), we obtain

$$k=\gamma_d^+\left(\frac{\dot{\varepsilon}}{k}-1\right)^{n_d^+} \tag{79}$$

Eq.(79) is an algebraic equation of k. It is investigated that the trial solution (78) satisfies Eq.(76) with k determined from Eq.(79). On the other side, Eq.(78) satisfies the initial condition expressed in Eq.(77). Thus, it is derived that the trail solution (78) is the real solution of Eq.(76) with initial condition (77) based on the uniqueness theorem of ordinary differential equation. By substitutin $\dot{q}_r^+=k$ into Eq.(79), one obtains

$$\frac{\dot{\varepsilon}}{\dot{q}_r^+}\left(\frac{\dot{\varepsilon}}{\dot{q}_r^+}-1\right)^{n_d^+}=\frac{\dot{\varepsilon}}{\gamma_d^+} \tag{80}$$

Combining Eqs.(75) and (78) yields

$$q_r^+=\frac{\dot{q}_r^+}{\dot{\varepsilon}}\varepsilon \tag{81}$$

Then by substituting Eq.(81) into Eq.(72a), one obtains

$$\sigma=\frac{\dot{\varepsilon}}{\dot{q}_r^+}\left\{\left[1-d^+\left(\frac{\dot{q}_r^+}{\dot{\varepsilon}}\varepsilon\right)\right]E_0\left(\frac{\dot{q}_r^+}{\dot{\varepsilon}}\varepsilon\right)\right\} \tag{82}$$

Comparing Eq.(82) and Eq.(71), it is clear that the maximum value in the brace of Eq.(82) is the static tensile strength σ_s. On the other hand, $\dot{\varepsilon}$ and $\dot{q}_r^+$ keep constant under constant rate loading. Then the dynamic strength, which is the maximum value of Eq.(82), could be expressed as

$$\sigma_d = \frac{\dot{\varepsilon}}{\dot{q}_r^+}\sigma_s \tag{83}$$

where σ_d denotes the dynamic strength of concrete. Substituting Eq.(83) in Eq.(80), we obtain the analytical expression of DIF as follows

$$\frac{\sigma_d}{\sigma_s}\left(\frac{\sigma_d}{\sigma_s}-1\right)^{n_d^+} = \frac{\dot{\varepsilon}}{\gamma_d^+} \tag{84}$$

Eq.(84) is the implicit equation of DIF σ_d/σ_s, which could be numerically solved by the Newton iterative procedure. It is observed that the parameters γ_d^+ and n_d^+ in Eq.(84) govern the dynamic damage evolution of materials. Thus, we can further conclude that the dynamic increase of the strength is mainly controlled by the dynamic damage evolution rather than the viscoplasticity.

The predicted results of Eq.(84) in terms of DIF in tension are plotted in Fig. 4 against the corresponding experimental data in former literatures. The dynamic material parameters are adopted as $\gamma_d^+=0.05$, $n_d^+=2.5$. It can be noticed that the closed form solution of the DIF analytically derived from this model provides good predictions against the testing data.

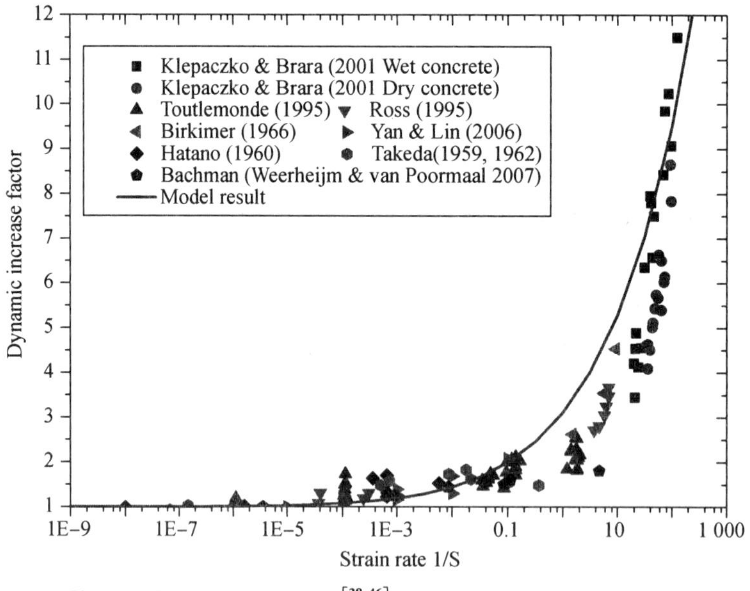

Fig. 4. The DIF in tension.[38-46] **DIF: dynamic increase factor**

5.2 Uniaxial compression($\bar{\sigma}_3=0$, $\bar{\sigma}_2=0$, $\bar{\sigma}_1=\bar{\sigma}<0$)

1. Static model

Under uniaxial compression, the former tensor framework is reduced to the following scalar expressions

$$\begin{cases}\sigma=[1-d^{-}(\varepsilon^{e})]E_0\varepsilon^{e}\\ \varepsilon=\varepsilon^{e}+\varepsilon^{p}\\ d^{-}(\varepsilon^{e})=F^{-}(\varepsilon^{e})\\ (\varepsilon^{p}/\varepsilon_c)=a(\varepsilon/\varepsilon_c)^2+b(\varepsilon/\varepsilon_c)\end{cases} \tag{85}$$

Fig. 5 shows the model results versus the experimental results[35]. The material parameters are set as $E_0=45\ 100$ MPa, $\lambda^{-}=7.500\ 5$, and $\zeta=0.295\ 5$. The predicted results agree well with experimental data.

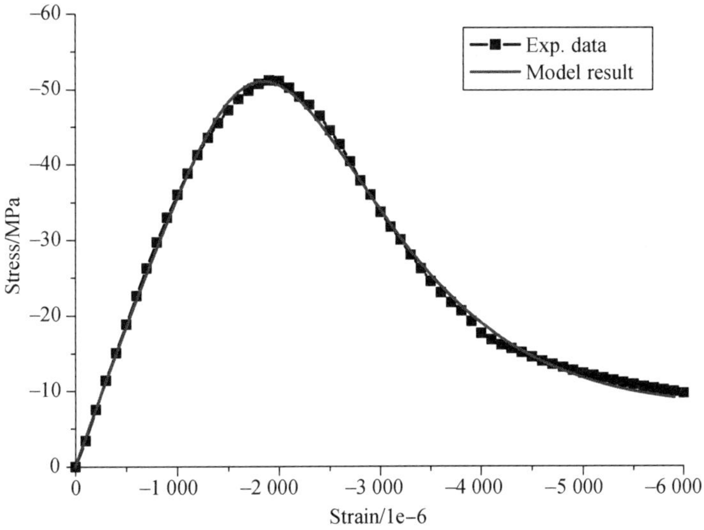

Fig. 5. Static uniaxial compressive tests

2. Dynamic model

Actually, under uniaxial loading condition, the former tensor framework of rate-dependent plastic damage model could be reduced into a rather simple expression as follows

$$\begin{cases}\sigma=[1-d^{-}(q_r^{-})]E_0\varepsilon^{e}\\ \dot{\varepsilon}=\dot{\varepsilon}^{e}+\dot{\varepsilon}^{p}\end{cases} \tag{86}$$

The rate-dependencies of damage and plasticity should be carefully considered. According to the discussions in the former section, the viscoplastic strain evolution could be derived as

$$\left\{\dot{\varepsilon}^{p}=\gamma_p(1-\alpha)\left\langle\frac{E_0\varepsilon^{e-}}{c(\varepsilon^{vp})}-1\right\rangle^{n_p}\triangleq\bar{\gamma}_p\left\langle\frac{E_0\varepsilon^{e-}}{c(\varepsilon^{vp})}-1\right\rangle^{n_p}\right. \tag{87a}$$

$$\left\{c(\varepsilon^{vp})=\frac{\sqrt{b^2+4a(\varepsilon^{vp}/\varepsilon_c)}-[2a(\varepsilon^{vp}/\varepsilon_c)+b]}{2a}E_0\varepsilon_c\right. \tag{87b}$$

where $\bar{\gamma}_p$ is defined as generalized flow parameter. It is clear that the strain rate approaches

0 under static loading. Hence, we get

$$\dot{\varepsilon}^{\mathrm{vp}} \to 0 \tag{88}$$

Substituting Eq.(88) in Eq.(87a), one gets

$$\bar{\gamma}_{\mathrm{p}} \left\langle \frac{E_0 \varepsilon^{\mathrm{e}}}{c(\varepsilon^{\mathrm{vp}})} - 1 \right\rangle^{n_p} = 0 \tag{89}$$

Combining Eqs.(89) and (87b), static plastic evolution function (44) can be derived. It is concluded that the dynamic plastic model proposed in the present is consistent with corresponding static plastic evolution model.

Based on the aforementioned discussions, dynamic evolution of compressive damage variable is

$$\begin{cases} d^- = F^-(q_{\mathrm{r}}^-) \\ \dot{q}_{\mathrm{r}}^- = \gamma_{\mathrm{d}}^- \left\langle \dfrac{\varepsilon^{\mathrm{e}}}{q_{\mathrm{r}}^-} - 1 \right\rangle^{n_{\mathrm{d}}^-} \end{cases} \tag{90}$$

Fig. 6 shows the model results versus the experimental data[16] for the dynamic loading of concrete cylinders under uniaxial compression in three different strain rates. Corresponding parameters of this model are identified as $E_0 = 31\ 700$ MPa, $\lambda^- = 7.42$, $\zeta^- = 0.318$, $\bar{\gamma}_{\mathrm{p}} = 100$, $n_{\mathrm{p}} = 6$, $\gamma_{\mathrm{d}}^- = 30$, and $n_{\mathrm{d}}^- = 4$. This model is able to represent the dynamic increase of the peak stress and the corresponding strain accurately. The simulated result agrees well with the experimental data. It is observed that the stress-strain curve shows a faster increase in its ascending part under higher loading rate and in the mean while drops faster in its descending part.

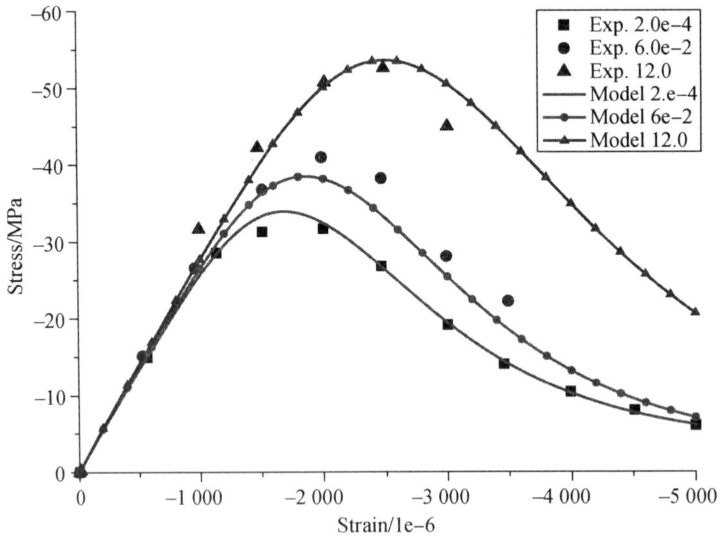

Fig. 6. Dynamic uniaxial compressive tests

3. The dynamic increased factor

Due to the coupling between rate-dependent evolutions of damage and plastic strain, the closed form expression of the DIF could not be analytically derived. Hence, the uniaxial compressive stress-strain curves of concrete are numerically simulated and then the DIF of peak stress and strain could be identified. The results are illustrated in Figs. 7 and 8 versus experimental data. The material properties adopted in numerical simulations are: $\bar{\gamma}_p = 5$, $n_p = 6$, $\gamma_d^- = 30$, and $n_d^- = 4$.

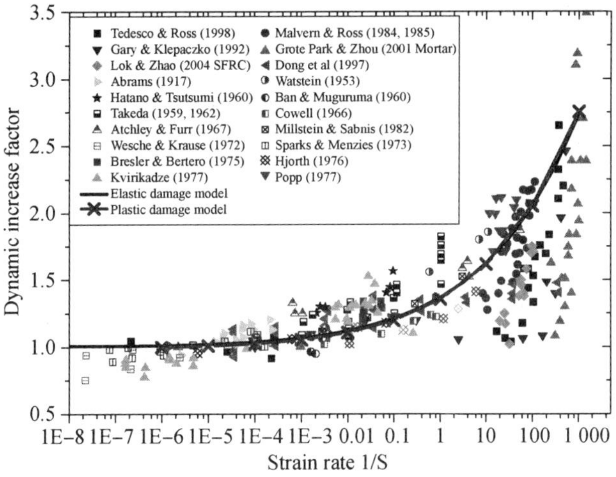

Fig. 7. The DIF of strength in compression[43-45, 47-64]

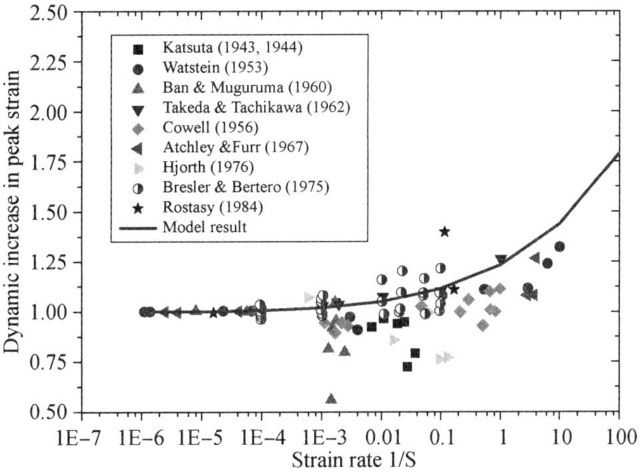

Fig. 8. The DIF of peak strain in compression.[45, 51, 53, 58-60, 63, 65-66] **DIF: dynamic increase factor**

The curve of DIF directly calculated using Eq.(84) is also shown in Fig. 7. The material parameters are adopted as $\gamma_d^- = 40$, $n_d^- = 4$. Although the evolution of plastic strain is neglected in the derivation of Eq.(84), the predicted results of dynamic increase of strength are essentially coincident with numerical results of the plastic damage formulations made up of Eqs.(86), (87), and (90). Hence, Eq.(84) is able to predict the dynamic increase of strength under tension and compression. It is concluded that the dynamic increase of the strength mainly relies on the rate-dependency of the damage evolution. On the other hand, it is shown in Figs. 7 and 8 that the dynamic increase of peak strain is visibly lower than that of strength. In other words, the rate-sensitivity of the plastic strain restrains the evolution of total strain corresponding to the certain stress. Then, it is concluded that the rate-dependency of damage and plasticity should be considered simultaneously in dynamic compressive model of concrete. On the other hand, we neglect the tensile plasticity to reduce the numerical costs while the pay is that the results for tensile tests are not as good as compressive ones.

5.3 Reinforced concrete beam

For the multi-dimensional cases, the constitutive model could not be solved directly. Hence, numerical implementations for the proposed model are developed based on the framework of return-mapping algorithm[67]. The numerical algorithm of the proposed model could be divided into three parts. First is the algorithm for the inviscid plastic damage model based on effective stress decomposition. This part could be found in Faria et al.[6] and Wu et al.[24]. The second part is the algorithm for the viscoplastic extension, which is thoroughly investigated in Ju[68] and Alfano et al.[69]. Third is the numerical integration for the rate-dependent damage evolution[22].

Kulkarni and Shah[70] reported a series of experimental data for reinforced concrete beams subjected to static and dynamic loadings. The experimental setups are indicated in Fig. 9. All the specimens were of the same section shown in Fig. 9 and no shear reinforcement was set. We simulated four of the testing specimens with test program presented in Table 1. to verify the proposed constitutive model, finite element models are developed in this study. A mesh of four node plane stress elements is adopted for modeling of the concrete where the proposed material model is implemented. The reinforcements are imposed by rebar layers. The material properties for concrete are:

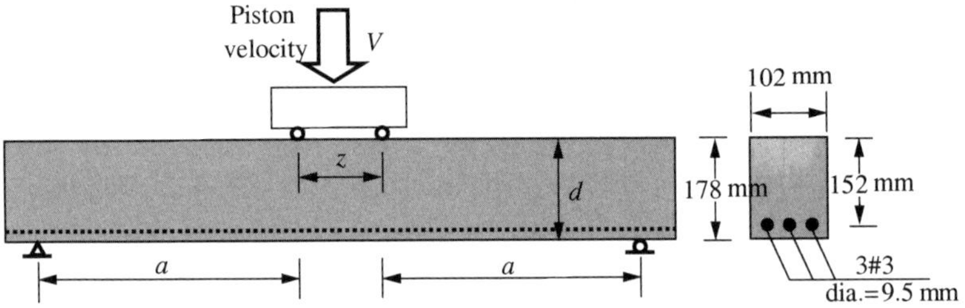

Fig. 9. Experimental setup

Table 1. Test program

Specimen designation	a/d	z/mm	f_c/MPa	V/(mm · s^{-1})	Failure mode
B3OC25-S	5.0	152	46.2	0.007 1	Flexure
B3OC25-D	5.0	152	46.2	380	Flexure
B3NO15-S	4.0	152	43.0	0.007 1	Shear
B3NO15-D	4.0	152	43.0	380	Flexure

$$E_0 = 37\,000 \text{ MPa},\ \nu_0 = 0.2,\ \rho_0 = 2\,450 \text{ kg/m}^3,\ \lambda^+ = 4.6,\ \zeta^+ = 0.4,\ n_d^+ = 1.5,\ \gamma_d^+ = 10,$$
$$\lambda^- = 7.3,\ \zeta^- = 0.5,\ n_d^- = 4.5,\ \gamma_d^- = 1.3 \times 10^4,\ n_p = 4, \text{ and } \bar{\gamma}_p = 100.$$

The steel reinforcing bars are modeled as the inviscid linear hardening plastic materials with material parameters: $E_s = 210$ GPa, $f_y = 580$ MPa, and hardening modulus $E_p = 0.01E_s$. The simulated results against the experimental results are plotted in Fig. 10. The specimens B3OC25-S and B3OC25-D were designed to show the strength enhancement of reinforced concrete beams. The numerical results well represent the enhancements and agrees well with the test data. The failure mode change is shown by the results of the specimens B3NO15-S and B3NO15-D. We employ Fig. 11, which is often mentioned in the technique books of reinforced concrete structures[71], to explain the dynamics strength enhancement induced failure mode change. The red arrows in Fig. 11 indicate the possible failure mode change induced by the increase of the concrete strength. And it is observed that the failure mode change is well reproduced by the proposed model.

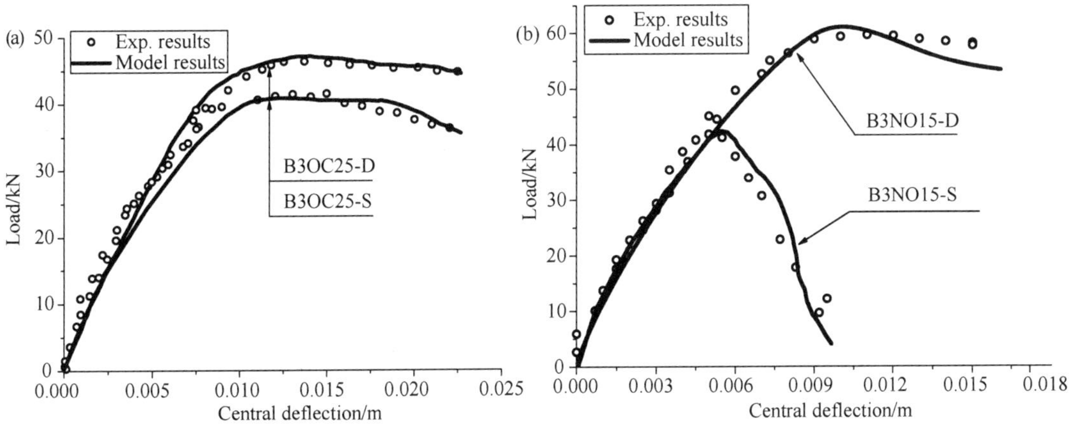

Fig. 10. Load-deflection diagrams: (a) specimens B3OC25-S and B3OC25-D and (b) specimens B3NO15-S and B3NO15-D

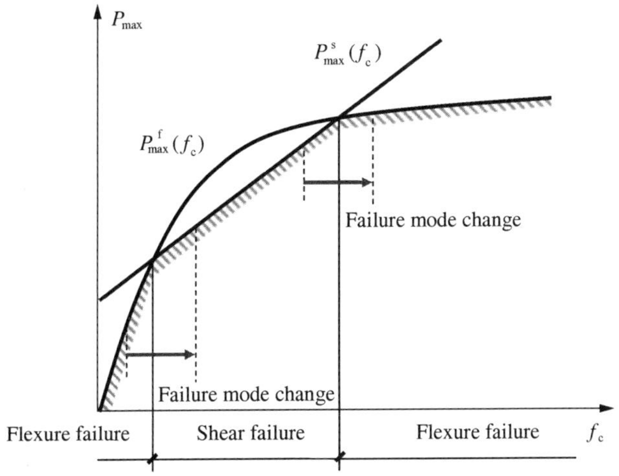

Fig. 11. Strength enhancement induced failure mode change. f_c is the compressive strength of concrete; P^s_{max} and P^f_{max} the maximum loading subjected to shear and flexure failure modes, respectively; and $P_{max} = \min(P^s_{max}, P^f_{max})$ the lower limit which represents the real failure load

6 Conclusions

A constitutive model aiming to describe the nonlinear behavior of concrete in statics and dynamics has been developed in this article. Based on the theoretical development and computations, several conclusions have been drawn:

(1) Energy release rate based continuum damage framework is adopted to represent the multi-dimensional nonlinear behaviors of concrete. And the EESs are developed to bridge the gap between one- and multi-dimensional damage evolution.

(2) Two kinds of rate-dependencies have been considered in this model. The rate-dependent damage model is adopted to represent the dynamic increase of stress and strength under high loading conditions. And the viscoplasticity is introduced to represent the dynamic evolution of plastic stain. The comparisons between the simulated and the experimental results reveal that the dual rate-dependent model could correctly predict the dynamic increase of concrete in the sense of strength and peak strain.

(3) The expression of the DIF has been analytically derived based on the proposed model. The agreement between the analytical results and the experimental data indicates that the proposed expression of DIF could predict the tensile as well as the compressive dynamic increase of strength with two material parameters.

(4) Numerical results suggest that not only the strength enhancement but also the failure mode change could be well described by the proposed model.

Funding

This research is supported by the National Science Foundation of China(grant no. 90715033).

References

[1] COMI C, BERTHAUD Y, BILLARDON R. On localization in ductile-brittle materials under compressive loadings[J]. European Journal of Mechanics, A/Solids, 1995, 14(1): 19-43.

[2] LUBARDA V A, KRAJCINOVIC D, MASTILOVIC S. Damage model for brittle elastic solids with unequal tensile and compressive strengths[J]. Engineering Fracture Mechanics, 1994, 49(5): 681-697.

[3] MAZARS J. Application de la mécanique de l'endommagement au comportement non linéaire et à la rupture du béton de structure[M]. These De Docteur Es Sciences Presentee A L'universite Pierre Et Marie Curie-Paris 6, 1984.

[4] MAZARS J. A description of micro-and macroscale damage of concrete structures[J]. Engineering Fracture Mechanics, 1986, 25(5-6): 729-737.

[5] MAZARS J, PIJAUDIER-CABOT G. Continuum damage theory—application to concrete[J]. Journal of Engineering Mechanics, 1989, 115(2): 345-365.

[6] FARIA R, OLIVER J, CERVERA M. A strain-based plastic viscous-damage model for massive concrete structures[J]. International Journal of Solids and Structures, 1998, 35(14): 1533-1558.

[7] HANSEN E, WILLAM K, CAROL I. A two-surface anisotropic damage/plasticity model for plain concrete[C]//Fourth International Conference on Fracture Mechanics of Concrete and Concrete Structures. 2001: 549-556.

[8] LEE J, FENVES G L. Plastic-damage model for cyclic loading of concrete structures[J]. Journal of Engineering Mechanics, 1998, 124(8): 892-900.

[9] ORTIZ M. A constitutive theory for the inelastic behavior of concrete[J]. Mechanics of Materials, 1985, 4(1): 67-93.

[10] RESENDE L. A damage mechanics constitutive theory for the inelastic behaviour of concrete[J]. Computer Methods in Applied Mechanics and Engineering, 1987, 60(1): 57-93.

[11] SIMO J C, JU J W. Strain-and stress-based continuum damage models—Ⅰ. Formulation[J]. International Journal of Solids and Structures, 1987, 23(7): 821-840.

[12] SIMO J C, JU J W. Strain-and stress-based continuum damage models—Ⅱ. Computational aspects[J]. International Journal of Solids and Structures, 1987, 23(7): 841-869.

[13] YAZDANI S, SCHREYER H L. Combined plasticity and damage mechanics model for plain concrete [J]. Journal of Engineering Mechanics, 1990, 116(7): 1435-1450.

[14] BISCHOFF P H, PERRY S H. Compressive behaviour of concrete at high strain rates[J]. Materials and Structures, 1991, 24(6): 425-450.

[15] MALVAR L J, ROSS C A. Review of strain rate effects for concrete in tension[J]. ACI Materials Journal, 1998, 95: 735-739.

[16] DUBÉ J F, PIJAUDIER-CABOT G, BORDERIE C L. Rate dependent damage model for concrete in dynamics[J]. Journal of Engineering Mechanics, 1996, 122(10): 939-947.

[17] ETSE J G, CAROSIO A. Diffuse and localized failure predictions of perzyna viscoplastic models for cohesive-frictional materials[J]. Latin American Applied Research 2002, 32: 21-31.

[18] LOREFICE R, ETSE G, CAROL I. Viscoplastic approach for rate-dependent failure analysis of concrete joints and interfaces[J]. International Journal of Solids and Structures, 2008, 45(9): 2686-2705.

[19] MEYER C, DE BORST R, BICANIC N, et al. Dynamic loading[M]//Finite Element Analysis of Reinforced Concrete Structures II. Chap. 6. New York, NY: ASCE, 1993: 314-366.

[20] WINNICKI A, PEARCE C J, BIĆANIĆ N. Viscoplastic Hoffman consistency model for concrete[J]. Computers & Structures, 2001, 79(1): 7-19.

[21] PERZYNA P. Fundamental problems in viscoplasticity[M]//Advances in applied mechanics. Elsevier, 1966, 9: 243-377.

[22] JU J W. On energy-based coupled elastoplastic damage theories: constitutive modeling and computational aspects[J]. International Journal of Solids and Structures, 1989, 25(7): 803-833.

[23] CERVERA M, OLIVER J, MANZOLI O. A rate-dependent isotropic damage model for the seismic analysis of concrete dams[J]. Earthquake Engineering & Structural Dynamics, 1996, 25(9): 987-1010.

[24] WU J Y, LI J, FARIA R. An energy release rate-based plastic-damage model for concrete[J]. International Journal of Solids and Structures, 2006, 43(3-4): 583-612.

[25] LI J, REN X D. Stochastic damage model for concrete based on energy equivalent strain[J]. International Journal of Solids and Structures, 2009, 46(11-12): 2407-2419.

[26] LEMAITRE J. Evaluation of dissipation and damage in metals[C]//Proc. ICM Kyoto. 1971, 1.

[27] LI J, WU J Y. Energy-based CDM model for nonlinear analysis of confined concrete[C]//ACI SP-238, In: Proc. of the Symposium on Confined Concrete, Changsha, China. 2004.

[28] LUBLINER J, OLIVER J, OLLER S, et al. A plastic-damage model for concrete[J]. International Journal of Solids and Structures, 1989, 25(3): 299-326.

[29] CAROL I, RIZZI E, WILLAM K. On the formulation of anisotropic elastic degradation. Ⅰ. Theory

based on a pseudo-logarithmic damage tensor rate[J]. International Journal of Solids and Structures, 2001, 38(4): 491-518.

[30] JASON L, HUERTA A, PIJAUDIER-CABOT G, et al. An elastic plastic damage formulation for concrete: Application to elementary tests and comparison with an isotropic damage model [J]. Computer Methods in Applied Mechanics and Engineering, 2006, 195(52): 7077-7092.

[31] KARSAN I D, JIRSA J O. Behavior of concrete under compressive loadings [J]. Journal of the Structural Division, 1969, 95(12): 2543-2564.

[32] LEMAITRE J. A course on damage mechanics[M]. Berlin: Spring-Verlag, 1992.

[33] KANDARPA S, KIRKNER D J, SPENCER JR B F. Stochastic damage model for brittle materials subjected to monotonic loading[J]. Journal of Engineering Mechanics, 1996, 122(8): 788-795.

[34] LI J AND ZHANG Q Y. Stochastic damage constitutive relationship for concrete material[J]. Journal of Tongji University, 2001, 29: 1135-1141.

[35] REN X D, YANG W, ZHOU Y, et al. Behavior of high-performance concrete under uniaxial and biaxial loading[J]. ACI Materials Journal, 2008, 105(6): 548.

[36] SUARIS W, SHAH S P. Rate sensitive damage theory for brittle solids[J]. Journal of Engineering Mechanics, 1984, 110(6): 985-997.

[37] SUARIS W, SHAH S P. Constitutive model for dynamic loading of concrete[J]. Journal of Structural Engineering, 1985, 111(3): 563-576.

[38] KLEPACZKO J R, BRARA A. An experimental method for dynamic tensile testing of concrete by spalling[J]. International journal of impact engineering, 2001, 25(4): 387-409.

[39] TOUTLEMONDE F. Impact resistance of concrete structures (in French)[D]. Laboratory of Bridgesand Roads (LCPC), Paris, 1995.

[40] ROSS C A, TEDESCO J W, KUENNEN S T. Effects of strain rate on concrete strength[J]. Materials Journal, 1995, 92(1): 37-47.

[41] BIRKIMER DL. Critical normal fracture strain of cement portland concrete[D]. University of Cincinnati, Ohio, 1966.

[42] YAN D, LIN G. Dynamic properties of concrete in direct tension[J]. Cement and Concrete Research, 2006, 36(7): 1371-1378.

[43] HATANO T AND TSUTSUMI H. Dynamic compressive deformation and failure of concrete under earthquakeload[C]//Proceedings of the 2nd WCEE. Japan: Tokyo, 8-10 July, 1960: 1963-1978.

[44] TAKEDA J. A Loading Apparatus for High Speed Testing of Building Materials and Structiutes[C]// Proc. of the 2nd Japan Congress on Testing Materials, 1959.

[45] TAKEDA J. The mechanical properties of several kinds of concretes at Compressive, tensile and flexural tests in high rates of loading[J]. Transactions of the Architectural Institute of Japan, 1962 (77): 1-6.

[46] WEERHEIJM J, VAN DOORMAAL J. Tensile failure of concrete at high loading rates: new test data on strength and fracture energy from instrumented spalling tests[J]. International Journal of Impact Engineering, 2007, 34(3): 609-626.

[47] TEDESCO J W, POWELL J C, ROSS C A, et al. A strain-rate-dependent concrete material model for ADINA[J]. Computers & Structures, 1997, 64(5-6): 1053-1067.

[48] GARY G, KLEPACZKO J R, ZHAO H. Corrections for wave dispersion and analysis of small strains with Split Hopkinson Bar[C]// Proceedings of the International Symposium of Impact Engineering, Sendai, Japan,2-4 November 1992: 73-78.

[49] LOK T S, ZHAO P J. Impact response of steel fiber-reinforced concrete using a split Hopkinson pressure bar[J]. Journal of Materials in Civil Engineering, 2004, 16(1): 54-59.

[50] ABRAMS D A. Effect of rate of application of load on the compressive strength of concrete[C]//Proceeding of ASTM. 1917, 17: 364-377.

[51] ATCHLEY B L, FURR H L. Strength and energy absorption capablities of plain concrete under dynamic and static loadings[C]//Journal Proceedings. 1967, 64(11): 745-756.

[52] WESCHE K, KRAUSE K. The effect of loading rate on compressive strength and modulus of elasticity of concrete[J]. Materialpruefung, 1972, 14(7): 212-218.

[53] BRESLER B. Influence of high strain rate and cyclic loading of unconfined and confined concrete in compression[C]//Proc. of 2nd Canadian Conference on Earthquake Engineering, Hamilton, Ontario. 1975: 1-13.

[54] KVIRIKADZE O P. Determination of the ultimate strength and modulus of deformation of concrete at different rates of loading[C]//International Symposium, Testing In-Situ of Concrete Structures, Budapest. 1977: 109-117.

[55] MALVERN L E, ROSS C A. Dynamic Response of Concrete and Concrete Structures[R]. Florida Univ Gainesville Dept of Engineering Sciences, 1986.

[56] GROTE D L, PARK S W, ZHOU M. Dynamic behavior of concrete at high strain rates and pressures: Ⅰ. experimental characterization[J]. International Journal of Impact Engineering, 2001, 25(9): 869-886.

[57] DONG Y L, XIE H P, ZHAO P. Compressive constitutive relationship of concrete under difference loading rate[J]. Shui Li Xue Bao, 1997, 7: 72-77.

[58] WATSTEIN D. Effect of straining rate on the compressive strength and elastic properties of concrete [C]//Journal Proceedings. 1953, 49(4): 729-744.

[59] BAN S. Behaviour of plain concrete under dynamic loading with straining rate comparable to earthquake loading[C]//Proc. of 2nd World Conference on Earthquake Engineering. 1960: 1979-1993.

[60] COWELL W L. Dynamic properties of plain portland cement concrete[R]. Naval Civil Engineering Lab Port Hueneme Calif, 1966.

[61] MILLSTEIN L, SABNIS GM. Concrete strength under impact loading[M]//Concrete Structures under Impact and Impulsive Loading. Berlin: BAM, 1982: 101-111.

[62] SPARKS P R, MENZIES J B. The effect of rate of loading upon the static and fatigue strengths of plain concrete in compression[J]. Magazine of Concrete Research, 1973, 25(83): 73-80.

[63] HJORTH O. Ein Beitrag zur Frage der Festigkeiten und des Verbundverhaltens von Stahl und Beton bei hohen Beanspruchungsgeschwindigkeiten[M]. Institut fur Baustoffkunde und Stahlbetonbau der Technischen Universitat Braunschweig, 1976.

[64] POPP C. Unterschungen fiber das Verhalten von Beton bei schlagartiger Beanspruchung (A study of the behaviour of concrete under impact loading)[M]. Deutsche Ausschuss fur Stahlbeton, Report no. 281. Washington, DC: Transportation Research Board, 1977.

[65] KATSUTA T. On the elastic and plastic properties of concrete in compression tests with high deformation velocity: Part 1[J]. Transactions of the Institute of Japanese Architects, 1943, 29: 268-274.

[66] ROSTASY F S, SCHEUERMANN J, SPRENGER K H. Mechanical behaviour of some construction materials subjected to rapid loading and low temperature[J]. Betonwerk+Fertigteil-Technik, 1984, 50(6): 393-401.

[67] SIMO JC, HUGHES TJR. Computational Inelasticity[M]. New York: Springer-Verlag, 1998.

[68] JU J W. Consistent tangent moduli for a class of viscoplasticity[J]. Journal of Engineering Mechanics, 1990, 116(8): 1764-1779.

[69] ALFANO G, DE ANGELIS F, ROSATI L. General solution procedures in elasto/viscoplasticity[J]. Computer Methods in Applied Mechanics and engineering, 2001, 190(39): 5123-5147.

[70] KULKARNI S M, SHAH S P. Response of reinforced concrete beams at high strain rates[J]. Structural Journal, 1998, 95(6): 705-715.

[71] PARK R, PAULAY T. Reinforced concrete structures[M]. New York: John Wiley & Sons, 1991.

A competitive mechanism driven damage-plasticity model for fatigue behavior of concrete①

Junsong Liang, Xiaodan Ren, Jie Li

Department of Building Engineering, Tongji University, Shanghai, China

Abstract In this paper, an innovative three-dimensional plastic-damage model is proposed to describe the behavior of concrete material under fatigue loading. In the proposed model, two damage variables are defined to describe the fatigue degradation of the mechanical properties of concrete under tension and shear, respectively. Meanwhile, by adopting the elastoplastic damage energy release rates, which are derived from the elastoplastic Helmholtz free energy potentials, a novel fatigue damage evolution law is developed. Specifically, the damage evolution process is governed by the competition between the damage healing effect and the damage driving effect induced by the propagation and closure of micro cracks under fatigue loading. The proposed model also allows the accumulation of fatigue damage within the classical damage surface by adopting the loading/unloading irreversibility concept. Moreover, the evolution of plastic strains is considered within the framework of the effective stress space plasticity. In the end, a set of numerical tests is presented, and the validity and effectiveness of the proposed model are illustrated according to the simulated results and the comparisons with the experimental data in literatures.

Keywords Continuum damage mechanics; Concrete; Fatigue; Damage evolution law; Competitive mechanism

1 Introduction

In recent years, the developments of structural design and material technique increase the stress level of concrete, which may also increase the risk of fatigue failure for concrete structures. As a matter of fact, the fatigue of concrete has been investigated for more than a hundred years[1-2], but the related research fields still remain open up till now. Much of the work has been done relying on the fitting of experimental data corresponding to the relatively simple loading conditions. Thus most of the existing models are not able to describe the physical mechanisms of the concrete fatigue process and the concrete fatigue behavior under complex loading conditions.

In general, the research methods for concrete fatigue can be mainly classified into three progressive categories: the experimental study, the study based on fracture mechanics and the study based on damage mechanics. To fulfill the engineering demands, concrete fatigue was first studied by experimental approach[3-6]. In this way, the fatigue properties are described by the empirical S-N and ε-N curves, which indicate the fatigue life of concrete

① Originally published in *International Journal of Damage Mechanics*, 2016, 25(3): 377-399.

under different stress or strain levels. Recent experimental studies mainly concentrate on new materials and new test contents[7-10]. With no doubt the experimental studies and the corresponding developed empirical models have provided a solid foundation for the study of concrete fatigue problem and meet the demands of the practical engineering. However, the phenomenological nature limits their further developments in science and engineering. From the physical perspective, fatigue comes from the initiation and propagation of micro cracks. Staring from this idea, the Paris's law, which defines the fatigue induced fracture, is introduced for metal in the beginning[11] and extended to concrete later on[12-14]. To consider the nonlinearity of crack tips, the cohesive crack model for concrete was introduced by Hillerborg et al.[15]. Then a lot of jobs has been done to give a suitable stress-crack opening displacement (COD) relation which is dependent on the fatigue loading and a better numerical scheme[16-18]. An inspection to the fracture mechanics based models suggests that these models are mainly focused on single macro crack analysis, thus it is not perfectly suitable to describe the fatigue behavior of concrete which is due to the propagation of multiple cracks with serious interactions and intersections.

In order to analyze the effects of multiple crack propagation subjected to fatigue loading, damage mechanics provides a more reliable way. Based on the irreversible thermodynamics, damage variables are introduced as the internal variables to describe the material degradation induced by the initiation and propagation of micro cracks and micro voids. Therefore, the evolution of damage variables is a core problem. For metal, the fatigue damage evolution has been studied in-depth[19-21]. As for concrete, Marigo[22] suggested a feasible way to extend the static damage evolution function into fatigue damage evolution equation in his celebrated work, which makes "fatigue damage" distinct from "brittle damage". Then many researchers followed up his way and made further significant contributions[23-24]. However, most of the existing concrete damage models are either too complicated by adopting an anisotropic fourth-order damage tensor, or too simple through adopting a single damage variable which cannot reflect the fatigue nonlinear process of concrete under complex loading cases such as alternative loading or multi-axial loading. Meanwhile, the damage mechanisms and physical background of concrete fatigue damage evolution process remain quite unclear. In addition, the residual strain and its dependence on the fatigue loading have not been fully considered. Thus the simulated results cannot always meet the engineering demands.

With the inspirations from the previous work, this paper aims to present a new bi-scalar elastoplastic fatigue damage constitutive model for concrete. Based on the elastoplastic damage energy release rates (DERRs), a novel fatigue damage evolution law is proposed. Meanwhile, the competitive mechanism between the damage healing effect and damage driving effect is proposed as the physical background for the irregular development of fatigue damage. Also, the plastic strains are considered within the effective stress space[25-26]. Systematic case studies show that the simulated results of the proposed model agree well with the experimental records, including the three-stage damage evolution process, fatigue stiffness degradation, cumulative irreversible deformation, and the fatigue behaviors under the complex loading cases.

2 Theoretical background

2.1 Damage mechanisms for fatigue loading

Generally, the continuum measuring of material degradation such as stiffness or strength degradation reflects the damage evolution process induced by the increase of cracks and defects in the subscales. Since the cracks and defects are initiated and driven by the applied stress, damage should also be governed by the corresponding stress states. Actually the damage in concrete is anisotropic, however, some researchers have adopted two damage scalars, a tensile and a shear ones (d^+, d^-), in the classical damage framework for simplification and approximation in engineering analysis[26-27], which correspond to mainly two types of damage mechanisms of concrete under tension and compression, respectively[28-29]. In sum, the tensile damage pattern represents the separations of material particles under tensile loading, leading to cracks of fracture mode I, while the shear damage pattern represents the breaking of internal bonds which is associated to the shear and slip effects under compressive or shear loading, and leads to cracks of fracture modes II and III(Fig. 1). One can easily recognize that concrete damage under complex loading cases such as alternating loading or multi-axial loading should be activated by the joint effects of these two basic mechanisms, and each part may dominate in different circumstances.

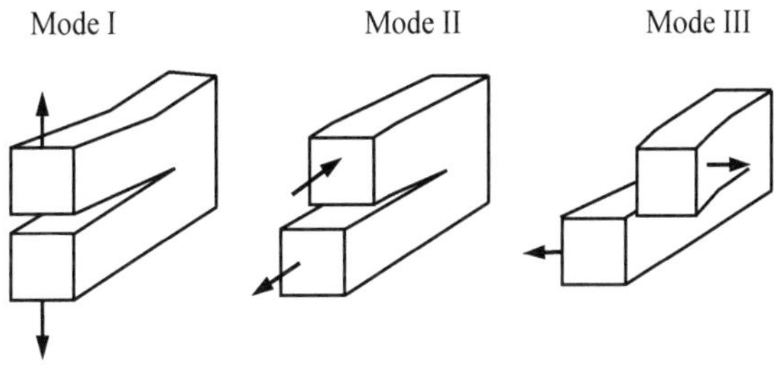

Fig. 1. Fracture modes and the corresponding damage patterns

Although the concrete damage evolution process under fatigue loading is different from the classical one, the damage mechanisms should be the same. Therefore, these two primary damage variables are adopted herein in the derivation of the fatigue damage model.

2.2 Split of effective stress tensor

As a starting point, the strain equivalence hypothesis[30] is introduced herein: *the stain associated with a damaged state under the applied stress is equivalent to the stain associated with its undamaged state under the effective stress.*

Let $\boldsymbol{\varepsilon}$ be the second-order total strain tensor in the real state, the effective stress $\bar{\boldsymbol{\sigma}}$ in the ideal undamaged state can be defined as follows[25, 31-32]

$$\bar{\boldsymbol{\sigma}} = \mathbb{S}_0 : (\boldsymbol{\varepsilon} - \boldsymbol{\varepsilon}^{\mathrm{p}}) = \mathbb{S}_0 : \boldsymbol{\varepsilon}^{\mathrm{e}} \tag{1}$$

or equivalently

$$\boldsymbol{\varepsilon}^{\mathrm{e}} = \mathbb{C}_0 : \bar{\boldsymbol{\sigma}} \tag{2}$$

where $\mathbb{S}_0$ and $\mathbb{C}_0 = \mathbb{S}_0^{-1}$ denote the rank-four initial undamaged elastic stiffness and compliance tensors, respectively, $\boldsymbol{\varepsilon}^{\mathrm{e}}$ and $\boldsymbol{\varepsilon}^{\mathrm{p}}$ denote the elastic and plastic components of the total strain tensor, respectively.

To inspect the different effects of the tensile and shear damage mechanisms, a decomposition of the effective stress tensor into its positive (tensile) and negative (compressive) components ($\bar{\boldsymbol{\sigma}}^{+}$, $\bar{\boldsymbol{\sigma}}^{-}$) is performed[25-26, 31, 33]

$$\bar{\boldsymbol{\sigma}}^{+} = \mathbb{P}^{+} : \bar{\boldsymbol{\sigma}} \tag{3a}$$

$$\bar{\boldsymbol{\sigma}}^{-} = \mathbb{P}^{-} : \bar{\boldsymbol{\sigma}} \tag{3b}$$

$$\bar{\boldsymbol{\sigma}} = \bar{\boldsymbol{\sigma}}^{-} + \bar{\boldsymbol{\sigma}}^{+} \tag{3c}$$

where $\mathbb{P}^{+}$ and $\mathbb{P}^{-}$ are rank-four stress projective tensors expressed as

$$\mathbb{P}^{+} = \sum_i H(\bar{\sigma}_i)\mathbf{n}^{(i)} \otimes \mathbf{n}^{(i)} \otimes \mathbf{n}^{(i)} \otimes \mathbf{n}^{(i)} \tag{4a}$$

$$\mathbb{P}^{-} = \mathbb{I} - \mathbb{P}^{+} \tag{4b}$$

with $\mathbb{I}$ expressing the fourth-order identity tensor; $\bar{\sigma}_i$ is the ith eigenvalue of the effective stress tensor $\bar{\boldsymbol{\sigma}}$, $\mathbf{n}^{(i)}$ is the ith normalized eigenvector corresponding to $\bar{\sigma}_i$, and the Heaviside function reads

$$H(x) = \begin{cases} 0 & \text{if} \quad x \leqslant 0 \\ 1 & \text{if} \quad x > 0 \end{cases} \tag{5}$$

It is also necessary to split the effective stress rate tensor $\dot{\bar{\boldsymbol{\sigma}}}$ into its positive and negative parts($\dot{\bar{\boldsymbol{\sigma}}}^{+}$, $\dot{\bar{\boldsymbol{\sigma}}}^{-}$) for the purpose of expressing the rate form of the proposed model and facilitating the derivation of the continuum tangent modulus

$$\dot{\bar{\boldsymbol{\sigma}}}^{+} = \frac{\mathrm{d}}{\mathrm{d}t}(\mathbb{P}^{+} : \bar{\boldsymbol{\sigma}}) = \mathbb{Q}^{+} : \dot{\bar{\boldsymbol{\sigma}}} \tag{6a}$$

$$\dot{\bar{\boldsymbol{\sigma}}}^{-} = \frac{\mathrm{d}}{\mathrm{d}t}(\mathbb{P}^{-} : \bar{\boldsymbol{\sigma}}) = \mathbb{Q}^{-} : \dot{\bar{\sigma}} \tag{6b}$$

$$\dot{\bar{\boldsymbol{\sigma}}} = \dot{\bar{\boldsymbol{\sigma}}}^{+} + \dot{\bar{\boldsymbol{\sigma}}}^{-} \tag{6c}$$

where the fourth-order tensors $\mathbb{Q}^{+}$ and $\mathbb{Q}^{-}$ are the corresponding stress rate projective tensors derived by Faria et al.[34] and Wu et al.[26] as

$$\mathbb{Q}^{+} = \mathbb{P}^{+} + 2\sum_{i=1,\, j>i}^{3} \frac{\langle \bar{\sigma}_i \rangle - \langle \bar{\sigma}_j \rangle}{\bar{\sigma}_i - \bar{\sigma}_j} \mathbf{p}_{ij} \otimes \mathbf{p}_{ij} \tag{7a}$$

$$\mathbb{Q}^{-} = \mathbb{I} - \mathbb{Q}^{+} \tag{7b}$$

in which the second-order symmetric tensor $\mathbf{p}_{ij}$ is formulated as

$$\mathbf{p}_{ij} = \mathbf{p}_{ji} = \frac{1}{2}(\mathbf{n}^{(i)} \otimes \mathbf{n}^{(j)} + \mathbf{n}^{(j)} \otimes \mathbf{n}^{(i)}) \tag{8}$$

and the McAuley bracket reads $\langle x \rangle = x \cdot H(x)$.

It should be pointed out that, the projective tensors of the stress tensor and the stress rate tensor are different. An alternative way to operate the decomposition work is proposed by Wu and Xu[35], in which these two kinds of projective tensors are unified.

3 Framework of double-scalar elastoplastic damage constitutive model

3.1 Elastoplastic damage constitutive relation

Based on the framework of continuum damage mechanics, the material damage and plasticity could be unified in the following form as the elastoplastic damage constitutive relation[26, 32].

$$\boldsymbol{\sigma} = (\mathbb{I} - \mathbb{D}) : \mathbb{S}_0 : (\boldsymbol{\varepsilon} - \boldsymbol{\varepsilon}^{\mathrm{p}}) \tag{9}$$

where $\boldsymbol{\sigma}$ is the second-order Cauchy stress tensor; and $\mathbb{D}$ is the fourth-order damage tensor. Substituting Eq.(1) into Eq.(9) yields the relationship between the stress tensor and the effective stress tensor

$$\boldsymbol{\sigma} = (\mathbb{I} - \mathbb{D}) : \bar{\boldsymbol{\sigma}} \tag{10}$$

According to Eq.(10) one can easily find that the real stress in the damaged state is actually the damage reduction of the effective stress in the undamaged state. However, the general fourth-order damage tensor is too complex and unnecessary in the application, so that the structure of $\mathbb{D}$ should be simplified.

Despite the anisotropic nature of $\mathbb{D}$, the bi-scalar damage representation[26, 32] is adopted for the sake of simplification

$$\mathbb{D} = d^+ \mathbb{P}^+ + d^- \mathbb{P}^- \tag{11}$$

Then, by substituting Eq.(11) into Eq.(10) and taking the decomposition of effective stress tensor expressed by Eq.(3) into consideration, we further have

$$\boldsymbol{\sigma} = (1 - d^+)\bar{\boldsymbol{\sigma}}^+ + (1 - d^-)\bar{\boldsymbol{\sigma}}^- \tag{12}$$

It is noted that after splitting the damage tensor $\mathbb{D}$ into the tensile and compressive components, the evolution of damage in concrete should be governed by the evolution of the corresponding eigen-damages d^+ and d^-, which are in accordance with the corresponding tensile and compressive stress states. Moreover, the effective stress can be obtained when the plastic strains have been solved, vice versa as indicated in Eq.(1). Thus, Eq.(12) defines the basic variables in the model calculation, that is, the plastic strains (or effective stress) and damage variables. The evolutionary models for these internal variables will be developed in the next two sections.

3.2 Plastic strains

On the basis of the strain equivalence hypothesis mentioned in section "Split of effective stress tensor", the plastic strains in the undamaged state should be equivalent to those of the damaged state. Hence, the evolution of the plastic strains can be governed by the stress of the undamaged state, that is, the effective stress. As a result, the constitutive model shown in Eq. (9) or Eq. (10) can be decoupled into the plasticity part and damage part, which is convenient for the calculation of the basic internal variables as can be seen in later chapters.

As a matter of fact, Eq. (1) defines the classical plastic constitutive relationship in the effect stress space. Thus, by defining the proper yield condition and the evolution potential based on effective stress, the so-called "effective stress space plasticity"[25-26, 28] can be obtained.

$$\dot{\boldsymbol{\varepsilon}}^{\mathrm{p}} = \dot{\lambda}^{\mathrm{p}} \frac{\partial F^{\mathrm{p}}(\bar{\boldsymbol{\sigma}}, \boldsymbol{\kappa})}{\partial \bar{\boldsymbol{\sigma}}} \tag{13a}$$

$$\dot{\boldsymbol{\kappa}} = \dot{\lambda}^{\mathrm{p}} \mathbf{H} \tag{13b}$$

$$F(\bar{\boldsymbol{\sigma}}, \boldsymbol{\kappa}) \leqslant 0, \quad \dot{\lambda}^{\mathrm{p}} \geqslant 0, \quad \dot{\lambda}^{\mathrm{p}} F(\bar{\boldsymbol{\sigma}}, \boldsymbol{\kappa}) = 0 \tag{13c}$$

where F and F^{p} are the yield function and the evolutionary potential function, respectively, $\dot{\lambda}^{\mathrm{p}}$ and $\boldsymbol{\kappa}$ are the plastic flow parameter and the hardening parameter, respectively, and $\mathbf{H}$ denotes the vectorial hardening function.

After some trivial derivations based on the standard procedures of classical plastic theory[36], we could further have the following rate form of the plastic constitutive law

$$\dot{\bar{\boldsymbol{\sigma}}} = \mathbb{S}^{\mathrm{ep}} : \dot{\boldsymbol{\varepsilon}} \tag{14}$$

with $\mathbb{S}^{\mathrm{ep}}$ denoting the continuum effective elastoplastic tangent tensor

$$\mathbb{S}^{\mathrm{ep}} = \begin{cases} \mathbb{S}_0 & \text{if } \dot{\lambda}^{\mathrm{p}} = 0 \\ \mathbb{S}_0 - \dfrac{(\mathbb{S}_0 : \partial_{\bar{\boldsymbol{\sigma}}} F^{\mathrm{p}}) \otimes (\mathbb{S}_0 : \partial_{\bar{\boldsymbol{\sigma}}} F)}{\partial_{\bar{\boldsymbol{\sigma}}} F : \mathbb{S}_0 : \partial_{\bar{\boldsymbol{\sigma}}} F^{\mathrm{p}} - \partial_{\boldsymbol{\kappa}} F \cdot \mathbf{H}} & \text{if } \dot{\lambda}^{\mathrm{p}} > 0 \end{cases} \tag{15}$$

in which operator $\partial_x f$ denotes the partial derivative $\partial f / \partial x$.

According to the framework of the classical plastic theory, Eqs. (1), (13) to (15) have provided a complete and rigorous way to describe the plastic evolution, hence it is often adopted in the theoretical development. However, numbers of deficiencies are also detected in its applications[32]. The yield function F and the evolutionary potential function F^{p} defined by the effective stress are difficult to measure experimentally; meanwhile the physical background of these two functions is quite unclear. Moreover, to obtain the plastic strains from the above nonlinear system of equations, an intricate numerical implementation is needed[37], which will bring large quantities of iterations at each integration points and will lead to low efficiency in the simulation of large scale structures.

Therefore, the empirical plastic models with the sound experimental support, but in a simple expression, are also proposed[31, 36]. According to Wu[36], the evolution of plastic strains is proportional to the effective stress

$$\dot{\boldsymbol{\varepsilon}}^{\mathrm{p}}=b^{\mathrm{p}}\bar{\boldsymbol{\sigma}} \tag{16}$$

where b^{p} can be regarded as the empirical plastic flow parameter.

The expression of b^{p} in the present paper is a modified version of Faria et al.[31], which is proposed as the following expression by Wu[36]

$$b^{\mathrm{p}}=\xi^{\mathrm{p}}E_0H(\dot{d}^-)\frac{\langle\boldsymbol{\varepsilon}^{\mathrm{e}}:\dot{\boldsymbol{\varepsilon}}\rangle}{\bar{\boldsymbol{\sigma}}:\bar{\boldsymbol{\sigma}}}\geqslant 0 \tag{17}$$

where $\xi^{\mathrm{p}}\geqslant 0$ is a plastic constant controlling the plastic strain rate, and E_0 is the initial Young's modulus.

By differentiating Eq. (1) with respect to time, and taking Eqs. (16) and (17) into consideration, the corresponding continuum effective elastoplastic tangent tensor in Eq.(14) can be deduced as follows

$$\mathbb{S}^{\mathrm{ep}}=\mathbb{S}_0-\xi^{\mathrm{p}}E_0H(\dot{d}^-)H(\mathbf{I}_{\bar{\sigma}}:\dot{\boldsymbol{\varepsilon}})(\mathbf{I}_{\bar{\sigma}}\otimes\mathbf{I}_{\bar{\sigma}}) \tag{18}$$

with $\mathbf{I}_{\bar{\sigma}}$ denoting the unit effective stress tensor taking the form

$$\mathbf{I}_{\bar{\sigma}}=\frac{\bar{\boldsymbol{\sigma}}}{\sqrt{\bar{\boldsymbol{\sigma}}:\bar{\boldsymbol{\sigma}}}} \tag{19}$$

It should be pointed out that the tensile plasticity is neglected in the present model since it is much smaller than the compressive one.

3.3 Damage criterion and damage evolution for static loading

As mentioned before, the tensile damage and the shear damage are related to the corresponding stress states. Hence, we could choose to characterize the damage evolution based on stress. However, the Cauchy stress is not a suitable choice since stress softening often happens, which will lead to complex calculation procedures. Thus, the effective stress should be appropriately adopted[32]. On the other hand, the DERRs are usually suggested as the driving forces of the internal damage variables according to the irreversible thermodynamics[28, 36]. Hence, by postulating the expression of Helmholtz free energy (HFE) ψ as the following form

$$\psi=(1-d^+)\psi_0^+ +(1-d^-)\psi_0^- \tag{20}$$

the tensile and compressive DERRs(Y^+, Y^-), conjugated to the corresponding damage variables, can be deduced as

$$Y^+=-\frac{\partial\psi}{\partial d^+}=\psi_0^+ \tag{21a}$$

$$Y^-=-\frac{\partial\psi}{\partial d^-}=\psi_0^- \tag{21b}$$

The initial tensile and compressive HFEs(ψ_0^+, ψ_0^-) can be further split into the elastic parts and plastic parts[26, 32]

$$\psi_0^+ = \psi_0^{e+} + \psi_0^{p+} \tag{22a}$$

$$\psi_0^- = \psi_0^{e-} + \psi_0^{p-} \tag{22b}$$

in which the elastic parts are

$$\psi_0^{e\pm} = \frac{1}{2}\bar{\boldsymbol{\sigma}}^{\pm} : \boldsymbol{\varepsilon}^e = \frac{1}{2}(\bar{\boldsymbol{\sigma}}^{\pm} : \mathbb{C}_0 : \bar{\boldsymbol{\sigma}}) \tag{23}$$

where symbol "$\pm$" denotes "$+$" or "$-$", as appropriate.

As for the plastic parts, Ju[25] first suggested the following expression

$$\psi_0^{p\pm} \approx \int_0^{\boldsymbol{\varepsilon}^p} \bar{\boldsymbol{\sigma}}^{\pm} : d\boldsymbol{\varepsilon}^p \tag{24}$$

Later, Wu et al.[26] neglected the tensile plastic HFE ψ_0^{p+} and introduced the Drucker-Prager plastic model for the compressive plastic HFE ψ_0^{p-}. They derived the approximating explicit expressions of the DERRs based on the effective stress which are adopted here

$$Y^+ = \psi_0^+ \approx \frac{1}{2}(\bar{\boldsymbol{\sigma}}^+ : \mathbb{C}_0 : \bar{\boldsymbol{\sigma}}) \tag{25a}$$

$$Y^- = \psi_0^- \approx b_0(\alpha \bar{I}_1^- + \sqrt{3\bar{J}_2^-})^2 \tag{25b}$$

where $\bar{I}_1^-$ is the first invariant of compressive effective stress $\bar{\boldsymbol{\sigma}}^-$; $\bar{J}_2^-$ is the second invariant of $\bar{\mathbf{s}}^-$, the deviator of compressive effective stress $\bar{\boldsymbol{\sigma}}^-$; b_0 is a material parameter; and α is the shear damage parameter usually taking the value of 0.121 2[36].

For convenience, the DERRs can be replaced by their equivalent forms which will not change the damage state and characteristics of damage evolution. Hence the following equivalent damage energy release rates (EDERRs) are often adopted in the model calculation[26]

$$\tilde{Y}^+ = \sqrt{2E_0\psi_0^+} = \sqrt{E_0(\bar{\boldsymbol{\sigma}}^+ : \mathbb{C}_0 : \bar{\boldsymbol{\sigma}})} \tag{26a}$$

$$\tilde{Y}^- = \sqrt{\psi_0^-/b_0} = \alpha \bar{I}_1^- + \sqrt{3\bar{J}_2^-} \tag{26b}$$

Then the damage state of concrete can be characterized by means of the damage criteria, formulated in the effective stress space, with the following functional form

$$g^{\pm}(\tilde{Y}^{\pm}, r^{\pm}) = \tilde{Y}^{\pm} - r^{\pm} \leqslant 0 \tag{27}$$

where $r^{\pm}$ are the current damage thresholds, which control the size of the expanding damage surfaces. Denoting by $r_0^{\pm}$ the initial damage thresholds before any loadings, the condition expressed by Eq.(27) implies $r^{\pm} \geqslant r_0^{\pm}$, which indicates the initiation of damages when the EDERRs $\tilde{Y}^{\pm}$ exceed the corresponding initial damage thresholds $r_0^{\pm}$. Here $r_0^{\pm}$ can be considered as the characteristic material properties.

We then define the evolution laws for damage variables and the corresponding damage thresholds respectively by the rate forms which are analogous to Ju et al.[38]

$$\dot{d}^{\pm} = \dot{\lambda}^{d^{\pm}} H^{\pm}(d^{\pm}, \tilde{Y}^{\pm}) \tag{28}$$

$$\dot{r}^{\pm} = \dot{\lambda}^{d^{\pm}} \tag{29}$$

where $\dot{\lambda}^{d^{\pm}} \geqslant 0$ are damage consistency parameters, which define the damage evolution condition through the Kuhn-Tucker relations

$$\dot{\lambda}^{d^{\pm}} \geqslant 0,\ g^{\pm}(\tilde{Y}^{\pm}, r^{\pm}) \leqslant 0,\ \dot{\lambda}^{d^{\pm}} \cdot g^{\pm}(\tilde{Y}^{\pm}, r^{\pm}) = 0 \tag{30}$$

and $H^{\pm}(d^{\pm}, \tilde{Y}^{\pm})$ are the damage hardening functions controlling the damage evolution process. Eq. (30) actually defines a standard unilateral constraint problem for damage evolutions. If $g^{\pm}(\tilde{Y}^{\pm}, r^{\pm}) < 0$, then $\dot{\lambda}^{d^{\pm}} = 0$, which indicates that the damages will not take place. However, if $\dot{\lambda}^{d^{\pm}} > 0$, we have $g^{\pm}(\tilde{Y}^{\pm}, r^{\pm}) = 0$, that is, the damage criteria are satisfied and damages will evolve. In this event, by the damage consistency condition we could obtain

$$g^{\pm}(\tilde{Y}^{\pm}, r^{\pm}) = \dot{g}^{\pm}(\tilde{Y}^{\pm}, r^{\pm}) = 0 \Rightarrow r^{\pm} = \tilde{Y}^{\pm},\ \dot{\lambda}^{d^{\pm}} = \dot{r}^{\pm} = \dot{\tilde{Y}}^{\pm} \tag{31}$$

and the current damage thresholds $r^{\pm}$ can be defined as the following expressions, where $t \in R^{+}$ refers to the current time[25-26, 38]

$$r^{\pm} = \max\left[r_0^{\pm}, \max_{\tau \in [0, t]} \tilde{Y}_{\tau}^{\pm}\right] \tag{32}$$

4 Damage model extended to fatigue loading

4.1 Fatigue damage evolution

In the case of fatigue loading, two distinctions in the damage evolution process must be emphasized. Firstly, since the stress levels are relatively low, the applied maximum loading states may be inside the damage surface and induce nevertheless the damage developments. This behavior cannot be predicted using the static damage model developed in last chapter because damages will not take place when loading levels are inside the damage surfaces. Moreover, the damage features are quite different from those of the static loading case, which, we consider, is due to the different physical mechanisms of the damage evolution process. Thus the static damage evolution law should be appropriately modified to fatigue.

In order to obtain the damage evolution laws under the low stress levels, the concept of damage surface should be replaced by the loading/unloading irreversibility concept as proposed by Marigo[22]. In other words, the damages will take place when the material is in the loading phase while the ones will not evolve if the material is in the neutral or unloading phase[23-24]. In the present work, the loading/unloading states can be identified through the EDERRs. That is,

if $\dot{\tilde{Y}}^{\pm} > 0$, the material is in the loading state, hence $\dot{d}^{\pm} > 0$;
if $\dot{\tilde{Y}}^{\pm} \leqslant 0$, the material is in the neutral or unloading state, hence $\dot{d}^{\pm} = 0$.
Moreover, to characterize the fatigue damage evolution process which is quite different from that of the static loading case, the fatigue damage hardening functions $H_f^{\pm}(d^{\pm}, \tilde{Y}^{\pm})$ are proposed to replace the static damage hardening functions in Eq.(28).
According to the above analysis, the fatigue damage evolution laws can be expressed as

$$\dot{d}^{\pm} = \begin{cases} \dot{\tilde{Y}}^{\pm} \cdot H_f^{\pm}(d^{\pm}, \tilde{Y}^{\pm}) & \text{if} \quad \dot{\tilde{Y}}^{\pm} > 0 \\ 0 & \text{if} \quad \dot{\tilde{Y}}^{\pm} \leqslant 0 \end{cases} \tag{33}$$

Based on the experimental observations and the inspections of the physical background of fatigue damage evolution process of concrete, this paper proposes the expression of the fatigue hardening functions as follows

$$H_f^{\pm}(d^{\pm}, \tilde{Y}^{\pm}) = \frac{1}{\kappa \cdot \tilde{Y}^{\pm}} \cdot \left(\frac{\tilde{Y}^{\pm}}{C^{\pm}} \right)^n \cdot h^{\pm}(d^{\pm}) \tag{34}$$

where κ and n are material constants related to the fatigue damage process; $C^{\pm}$ are material constants related to the dissipated energy density due to the micro cracks' propagation; and $h^{\pm}(d^{\pm})$ are the competition factors formulated as

$$h^{\pm}(d^{\pm}) = \exp(-\beta_1 \cdot d^{\pm}) + \exp[-\beta_2 \cdot (1 - d^{\pm})] \tag{35}$$

with β_1 and β_2 denoting the material parameters reflecting the competitive mechanism during the fatigue damage evolution process.
The physical background and the competitive mechanism mentioned above will be investigated in the next section.

4.2 Physical mechanisms of concrete fatigue damage process and modeling

The typical "three-stage" experimental result which reflects the fatigue damage process of concrete is shown in Fig. 2. Stage I is a rapid damage developing phase which lasts about 10% of the fatigue life. Stage II is a steady damage growing phase right after the damage growth rate decreasing to some extent in stage I, and it lasts about 80% of the fatigue life. Stage III is an abrupt damage evolving phase which lasts about the remaining 10% of the fatigue life; in this phase the damage growth rate is steep and the value of damage rises sharply; then the specimen ruptures in a sudden way.
Careful inspections to the "three-stage" damage evolution feature suggest the competitive mechanism of damage healing effect and damage driving effect during the propagation process of micro cracks of concrete. The driving effect here represents the damage accelerating behavior due to the crack initiation and growth from the stable state in the micro-scale to the unstable state in the macro-scale under sustaining repeated loading; the healing effect in the present paper might be different from the conventional damage healing pattern which is reflected by the decrease of damage variables. Since numbers of sub-scale cracks may not stop propagating under fatigue loading, the degradation of damage variables

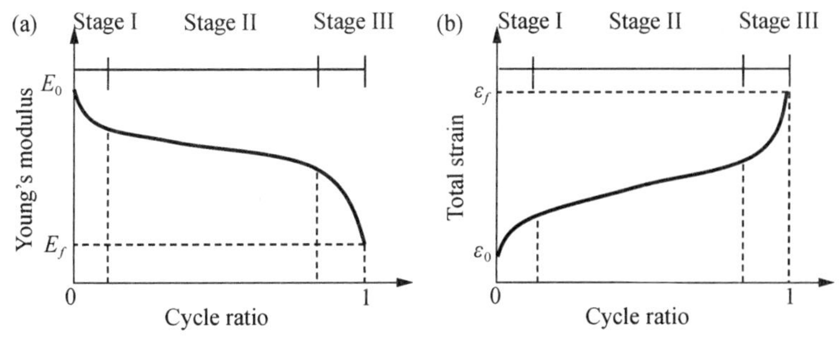

Fig. 2. Typical fatigue damage process of concrete: (a) degradation of Young's modulus and (b) evolution of total strain

in macro-scale should be covered. Thus in the fatigue loading case, the healing effect is considered to be the crack closing effect and healing of micro cracks due to the hydration of cement gel in concrete and these behaviors may be reflected in the degradation of damage growth rate in macro-scale.

Fig. 3 indicates the competitive mechanism due to the crack propagation process in concrete. In the beginning of a few loading-unloading cycles, large amount of micro cracks initiate; at the same time, the pre-existing micro cracks also extend stably. Since the crack density is low, the interaction between these cracks can be properly ignored. As a result, the propagation of micro cracks is dispersed in this period. Hence damage state is activated, which means that the stable micro cracks can propagate freely, leading to a fast growth of damage. If this cracking behavior remains unchanged, or we can consider as that the physical mechanism of damage evolution remains unchanged, the damage growth rate should be kept constant as the initial damage growth rate and leads to a shorter fatigue life than the real situation as shown in Fig. 4.

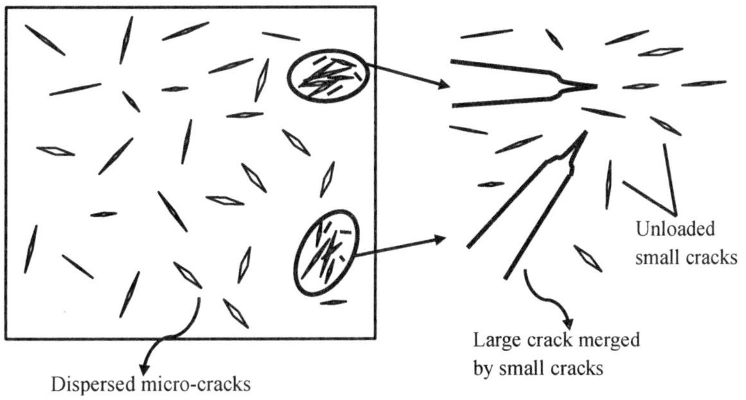

Fig. 3. Crack propagation process in concrete

However, the development of the dispersed micro cracks will not always be unconfined. Because when the micro cracks propagate to some extent, certain cracks will merge to form larger scale cracks and these large cracks may be checked by other large cracks as shown in

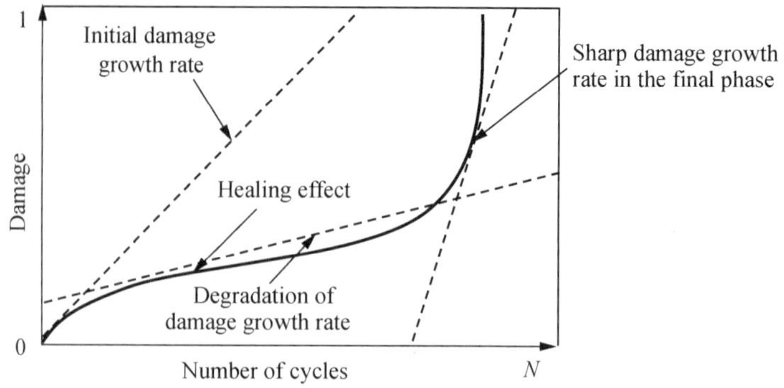

Fig. 4. Change of fatigue damage growth rate of concrete

Fig. 3. In this procedure, the original micro cracks far from the large crack tip will unload due to stress redistribution which will lead to the crack closing effect. Then due to the hydration of cement gel during the long term fatigue process, these closed cracks will heal with time, which may reduce the number of activated micro cracks. Since the material is still under loading, the merging and healing behaviors both exist, therefore a competitive mechanism is formed in this period. As a result, the damage growth rate decreases and approaches to be steady as shown in Fig.4. However, due to the sustaining fatigue loading, more and more large cracks are formed, these large cracks may also join together to form larger cracks until finally several macro unstable cracks emerge and dominate; meanwhile, the creep of concrete also contributes to the damage evolution from a stable stage to an unstable stage. In this process, the healing effect of stable micro cracks gradually loses its power in the competition, hence damage bursts in the final phase and a sharp damage growth rate can be observed as shown in Fig. 4.

The accurate descriptions for the above crack propagation process and the chemical reactions such as hydration of the cement gel during fatigue damage process are hard to acquire since the lack of observations in the micro-scale. Hence, from the perspective of phenomenology and for the sake of engineering practicability, an exponent damage evolution rate decaying function $\alpha_1(d)$ reflecting the damage healing effect is proposed as follows

$$\alpha_1(d)=\exp(-\beta_1 \cdot d) \tag{36}$$

Actually, the same kind of function has been used to describe the blocking effect in plastic mechanism of metal[39]. We also propose a function $\alpha_2(d)$ to reflect the damage driving effect formulated as

$$\alpha_2(d)=\exp[-\beta_2 \cdot (1-d)] \tag{37}$$

where β_1 and β_2 are material parameters appeared in Eq. (35). Then we suggest characterizing the competitive mechanism by superimposing the healing effect and driving effect as follows

$$h(d)=\alpha_1(d)+\alpha_2(d)=\exp(-\beta_1 \cdot d)+\exp[-\beta_2 \cdot (1-d)] \tag{38}$$

where $h(d)$ is the so-called competition factor.

The features of $\alpha_1(d)$, $\alpha_2(d)$ and the competition factor $h(d)$ are shown in Fig. 5. As analyzed above, the competitive mechanism characterizes the fatigue damage evolution process: in the beginning of repeated loading (Stage Ⅰ), large amount of micro cracks initiate and propagate freely, which gives rise to the fast damage development. Since the value of damage is small in this phase, the healing effect plays a leading role, which results in a degradation of damage growth rate. With the development of damage, the influence of the healing effect reduces gradually; meanwhile the impact of the driving effect becomes larger and larger as Fig. 5 indicates. Thus, a competitive mechanism between damage healing effect and damage driving effect is formed. As a result, the competition factor approaches to be steady, leading to a stable damage evolution phase in Stage Ⅱ. Finally, due to the progressive growth of damage, certain macro unstable cracks occur. Hence the influence of damage driving effect becomes even larger and dominates in the final phase. As a result, it leads to the invalidation of healing effect and the burst of damage in Stage Ⅲ.

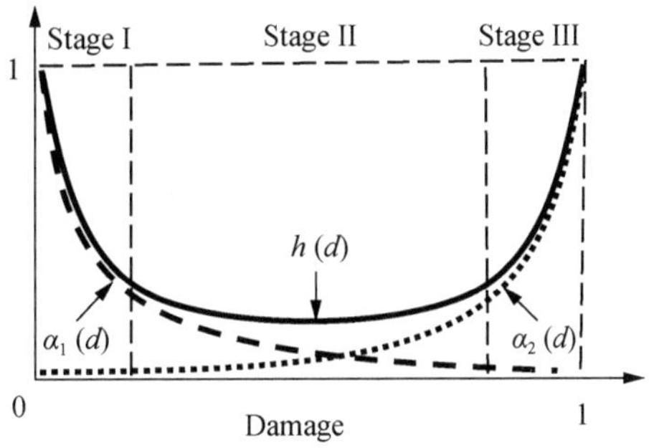

Fig. 5. Feature of competition factor

According to the knowledge in section "Damage mechanisms for fatigue loading", the tensile and compressive competition factors are recommended as expressed in Eq.(35).

5 Numerical algorithm

A numerical implementation algorithm for the proposed model is developed in this section. In a finite element context, the nodal displacements are usually updated iteratively through the global calculation of the structural finite element model, so that the strain increments are known within the local calculation during each time step. Thus the unknown stress should be updated by the constitutive relationship based on a strain-driven numerical procedure during the corresponding time step. Considering an arbitrary time interval $[t_k, t_{k+1}]$, this paper denotes the subscript form x_k as the value of variable x at time t_k. And the finite difference is defined by $\Delta x_k = x_{k+1} - x_k$. Then by giving the values of variables $\{\bar{\boldsymbol{\sigma}}_k$ or $\boldsymbol{\varepsilon}_k^{\mathrm{p}}$, $d_k^{\pm}\}$ and the strain at time t_{k+1}, the proposed scheme advances the solution to time t_{k+1} and update the values of variables to $\{\bar{\boldsymbol{\sigma}}_{k+1}$ or $\boldsymbol{\varepsilon}_{k+1}^{\mathrm{p}}$, $d_{k+1}^{\pm}\}$.

5.1 Numerical scheme based on the operator split method

According to the concept of operator split, the calculation of the elastoplastic damage constitutive model can be decomposed into the elastic, plastic, and damage parts[25-26, 37] leading to the corresponding numerical algorithm including elastic trial, plastic-corrector, and damage-corrector steps. By differentiating Eq.(1) with respect to time, we have

$$\dot{\bar{\boldsymbol{\sigma}}} = \mathbb{S}_0 : \dot{\boldsymbol{\varepsilon}}^{\mathrm{e}} = \mathbb{S}_0 : (\dot{\boldsymbol{\varepsilon}} - \dot{\boldsymbol{\varepsilon}}^{\mathrm{p}}) \tag{39}$$

Then, we start with the elastic trial through freezing the evolutions of plasticity and damage. That is, we let $\dot{d}^{\pm} = 0$ and $\dot{\boldsymbol{\varepsilon}}^{\mathrm{p}} = 0$ in the trial step. Then by approximating Eq.(39) to its finite difference scheme, the trial effective stress at time t_{k+1} can be obtained as follows

$$\bar{\boldsymbol{\sigma}}_{k+1}^{\mathrm{trial}} = \bar{\boldsymbol{\sigma}}_k + \mathbb{S}_0 : (\boldsymbol{\varepsilon}_{k+1} - \boldsymbol{\varepsilon}_k) \tag{40}$$

So the trial solutions of the EDERRs $\widetilde{Y}_{k+1}^{\mathrm{trial}\pm}$ can be explicitly determined according to Eq.(26) after splitting the trial effective stress tensor. If we have $\widetilde{Y}_{k+1}^{\mathrm{trial}\pm} \leqslant \widetilde{Y}_k^{\pm}$, the material is in the neutral or unloading phase and the elastic state holds. Hence, the elastic trial leads to the following results

$$\begin{aligned} &\bar{\boldsymbol{\sigma}}_{k+1} = \bar{\boldsymbol{\sigma}}_{k+1}^{\mathrm{trial}}, \quad \boldsymbol{\varepsilon}_{k+1}^{\mathrm{p}} = \boldsymbol{\varepsilon}_k^{\mathrm{p}}, \quad d_{k+1}^{\pm} = d_k^{\pm} \\ &\boldsymbol{\sigma}_{k+1} = (1 - d_{k+1}^{+})\bar{\boldsymbol{\sigma}}_{k+1}^{+} + (1 - d_{k+1}^{-})\bar{\boldsymbol{\sigma}}_{k+1}^{-} \end{aligned} \tag{41}$$

For $\widetilde{Y}_{k+1}^{\mathrm{trial}\pm} > \widetilde{Y}_k^{\pm}$, the plasticity and damage evolve. In this case, the trial solutions of effective stress and EDERRs are not equivalent to the real values so that the corresponding plastic-corrector and damage-corrector steps should be carried out[25-26, 37]. It is noted that during the elastic trial and plastic-corrector steps, the damage variables are fixed, leading to a standard elastoplastic problem expressed by Eqs.(13) and (39) in the effective stress space. Thus the return mapping algorithm[37] can be adopted to update the effective stress tensor and the corresponding EDERRs at time t_{k+1}.

A more convenient way for correcting the trial solutions is deduced by Wu[36] based on the empirical plastic model, and the real solutions of the basic variables at time t_{k+1} can be expressed as follows

$$\bar{\boldsymbol{\sigma}}_{k+1}^{\pm} = \xi_{k+1}\bar{\boldsymbol{\sigma}}_{k+1}^{\mathrm{trial}\pm}, \quad \widetilde{Y}_{k+1}^{\pm} = \xi_{k+1}\widetilde{Y}_{k+1}^{\mathrm{trial}\pm} \tag{42}$$

where ξ_{k+1} is the plastic modification parameter formulated as

$$\xi_{k+1} = 1 - \frac{\xi^{\mathrm{p}}}{\|\bar{\boldsymbol{\sigma}}_{k+1}^{\mathrm{trial}}\|} E_0 \left\langle \mathbf{I}_{\bar{\boldsymbol{\sigma}}_{k+1}^{\mathrm{trial}}} : (\boldsymbol{\varepsilon}_{k+1} - \boldsymbol{\varepsilon}_k) \right\rangle \tag{43}$$

in which $\|\mathbf{x}\|$ denotes the norm of tensor $\mathbf{x}$, and $\mathbf{I}_{\bar{\boldsymbol{\sigma}}_{k+1}^{\mathrm{trial}}}$ is the trial unit effective stress tensor.

Once the effective stress and EDERRs are updated in the above elastic trial and plastic-corrector steps, the damage variables can be correspondingly updated by approximating the evolution Eq.(33) to its implicit-like finite difference scheme

$$d_{k+1}^{\pm} = d_k^{\pm} + \langle \widetilde{Y}_{k+1}^{\pm} - \widetilde{Y}_k^{\pm} \rangle \cdot H_f^{\pm}(d_{k+1}^{\pm}, \widetilde{Y}_{k+1}^{\pm}) \tag{44}$$

The Newton-Raphson iterative method is adopted to solve the nonlinear Eq.(44). The procedures of this iterative method are omitted herein since it is simple.

Finally, the Cauchy stress $\boldsymbol{\sigma}$ can be updated in the damage-corrector step through Eq.(12).

5.2 The continuum tangent modulus

In the nonlinear finite element analysis, the full Newton-Raphson method is usually adopted in the global iteration. Hence, the continuum tangent modulus is often required to determine the rate constitutive relation due to the change of internal variables. For the double-scalar constitutive model, the rate equation can be expressed as the following form

$$\dot{\boldsymbol{\sigma}} = \mathbb{S}^{\tan} : \dot{\boldsymbol{\varepsilon}} \tag{45}$$

in which the continuum tangent modulus $\mathbb{S}^{\tan}$ can be expressed by[26]

$$\mathbb{S}^{\tan} = [\mathbb{I} - \mathbb{W} - (\mathbb{R}^{+} + \mathbb{R}^{-})] : \mathbb{S}^{ep} \tag{46}$$

where $\mathbb{S}^{ep}$ denotes the continuum effective elastoplastic tangent tensor consistent with the corresponding plastic model mentioned in section "Plastic strains", and the fourth-order tensor $\mathbb{W}$ can be expressed as the following equation

$$\mathbb{W} = d^{+} \mathbb{Q}^{+} + d^{-} \mathbb{Q}^{-} \tag{47}$$

The fourth-order tensors $\mathbb{R}^{+}$ and $\mathbb{R}^{-}$ were given by Wu et al.[26] as follows

$$\mathbb{R}^{+} = h_d^{+} \cdot \frac{E_0}{2\bar{Y}^{+}} [(\bar{\boldsymbol{\sigma}}^{+} \otimes \bar{\boldsymbol{\sigma}}) : \mathbb{C}_0 : \mathbb{Q}^{+} + (\bar{\boldsymbol{\sigma}}^{+} \otimes \bar{\boldsymbol{\sigma}}^{+}) : \mathbb{C}_0] \tag{48a}$$

$$\mathbb{R}^{-} = h_d^{-} \cdot \left[\bar{\boldsymbol{\sigma}}^{-} \otimes \left(\alpha \mathbf{1} + \frac{3}{2\sqrt{3\bar{J}_2}} \bar{\mathbf{s}} \right) \right] \tag{48b}$$

in which $\mathbf{1}$ denotes the second-order unit tensor, and $\bar{J}_2$ is the second invariant of $\bar{\mathbf{s}}$, the deviator of effective stress $\bar{\boldsymbol{\sigma}}$, and $h_d^{\pm}$ are the damage hardening functions which can take the forms of Eq.(34) in the present work.

6 Applications

In order to obtain the model properties, a numerical test is performed based on a concrete specimen of compressive strength $f_c = 34$ MPa, tensile strength $f_t = 3.4$ MPa, and initial Young's modulus $E_0 = 3.15 \times 10^4$ MPa. Then the experiments found in literatures are simulated and compared to illustrate the applicability and effectiveness of the proposed model. In the following sections, the axial direction represents the fatigue loading direction, and the total strain means the maximum axial tensile or compressive strain; N and N_f denote the number of cycles and maximum cycle number(fatigue life), respectively. For all cases, unless otherwise specified, Poisson's ratio is 0.20; and the shear damage parameter α is 0.121 2.

6.1 Uniaxial compressive fatigue

Numerical test is under the stress range of $0.8f_c - 0.2f_c$ and the results are shown in Fig. 6. As can be seen, simulations by the present model can capture the accumulation of permanent deformation, rigidity degeneration, and the "three-stage" damage evolution accurately. To check the validity of the model, comparisons are made to the experimental results of Gao and Hsu[40] and Kim and Kim[41]. In Gao and Hsu's research, the concrete parameters were $f_c = 34.6$ MPa, $E_0 = 3.5 \times 10^4$ MPa and the upper fatigue stress was 20.8 MPa; the fatigue damage was recorded. And in Kim and Kim's research, the compressive strength was $f_c = 84$ MPa and the stress range was $0.85f_c - 0.25f_c$; the total strain was observed and the average fatigue life recorded was $N_f = 1\ 045$. A good agreement of the experimental result and simulated result can be observed in Fig. 7. The simulated fatigue life for Fig. 7(a) and (b) is $N_f = 82\ 045$ and $N_f = 1\ 052$, respectively, which is close to the test records. Some parameters identified in the calculation are given in Table 1.

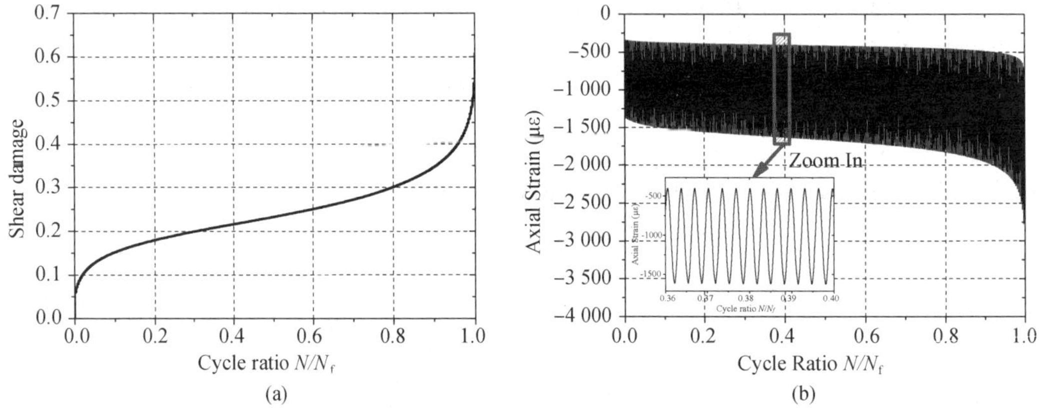

Fig. 6. Model properties under uniaxial compressive fatigue loading: (a) shear damage and (b) axial strain

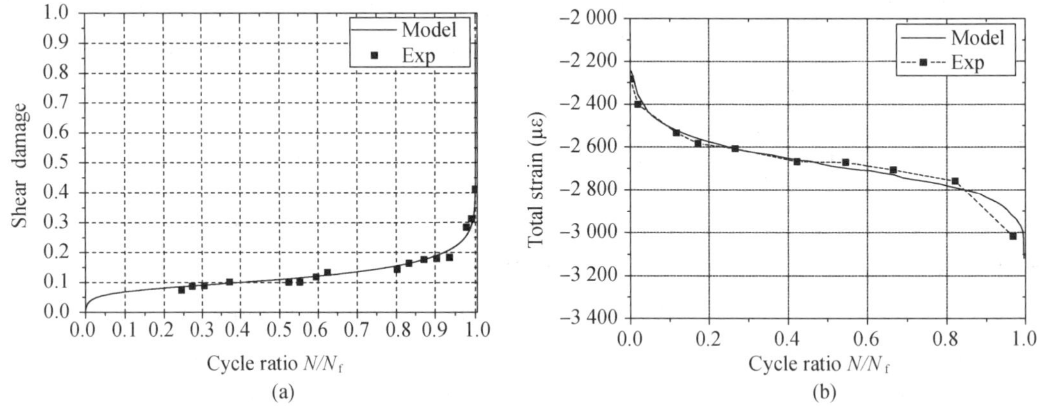

Fig. 7. Comparisons between experimental results and model simulations of uniaxial compressive fatigue: (a) shear damage and (b) maximum axial strain

Table 1. Parameters for uniaxial compressive fatigue loading

C^-/MPa	n	κ	β_1	β_2
32	8	8	64	48.6

6.2 Uniaxial tensile fatigue

Model test results under maximum stress $0.6f_t$ and minimum stress $0.02f_t$ are shown in

Figure 8, which indicate that the tensile fatigue damage and deformation process are well captured by the present model. Then comparison is made to the experiments performed by Cornelissen and Reinhardt[42]. In the experiment, the average values of tensile strength and the initial Young's modulus of the specimens were $f_t = 2.46$ MPa and $E_0 = 3.61 \times 10^4$ MPa; and the fatigue life of the specimen was $N_f = 10\ 154$. The total strain was observed under the stress range of $0.2f_t - 0.7f_t$. As shown in Fig. 9, the simulated result coincides well with the experimental result. The fatigue life in this simulation is $N_f = 10\ 188$, which is also close to the test record. The values of model parameters in the calculation are given in Table 2.

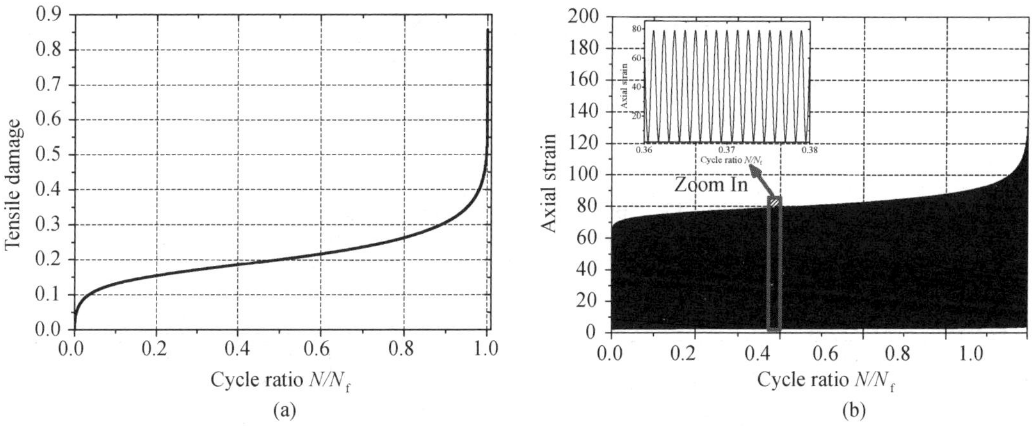

Fig. 8. Model properties under uniaxial tensile fatigue loading: (a) tensile damage and (b) axial strain

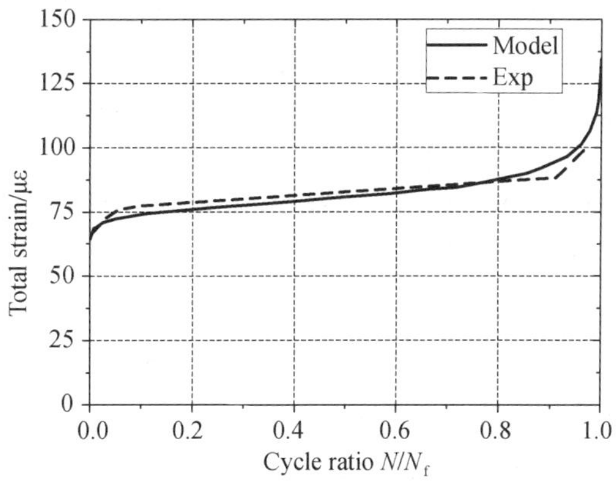

Fig. 9. Comparison between experimental result and model simulation of uniaxial tensile fatigue

Table 2. Parameters for uniaxial tensile fatigue loading

C^+/MPa	n	κ	β_1	β_2
2.8	6	8	48.0	48.0

6.3 Alternating fatigue

Some structures may endure alternating fatigue loading in practice and this circumstance is simulated as shown in Fig. 10.

The test results are shown in Fig. 11 in which the stress range is $0.8f_c - 0.4f_t$. Since the alternating fatigue loading brings both tensile and shear damages, the overall material degradation is larger than the uniaxial fatigue loading case. The contrast is made to the experimental results recorded by Cornelissen and Reinhardt[42]. The average material parameters were $f_c = 47.3$ MPa, $f_t = 2.46$ MPa, and $E_0 = 3.61 \times 10^4$ MPa; and the stress range was $0.7f_t - 0.15f_c$. The axial strain was observed and the fatigue life of the experimental specimen is $N_f = 5\ 733$.

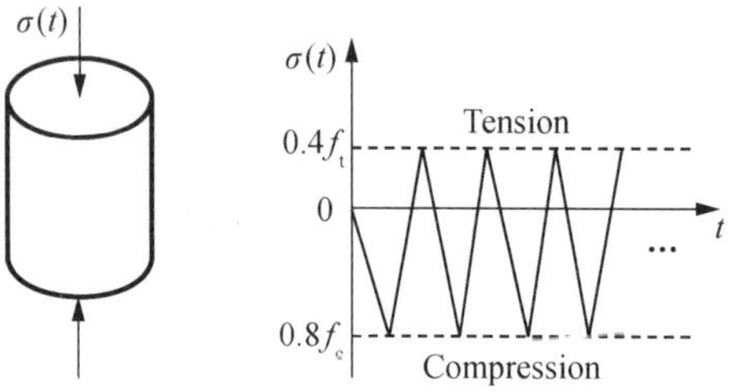

Fig. 10. Alternating fatigue loads

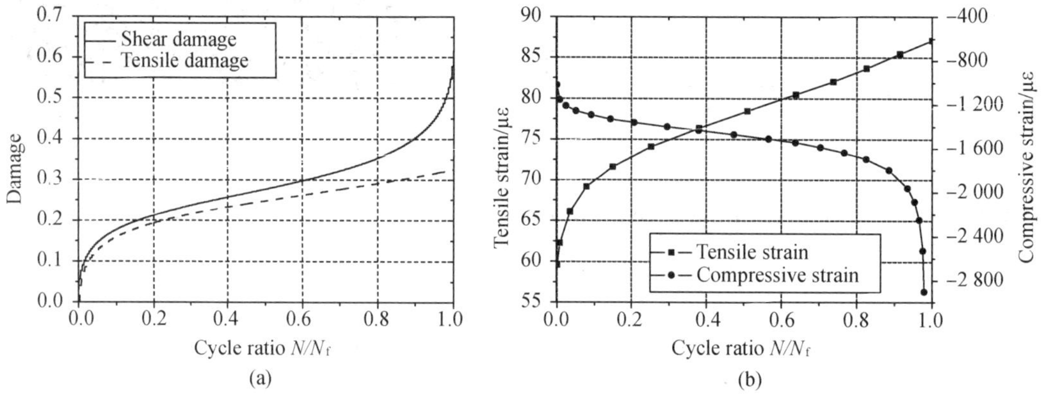

Fig. 11. Model properties under alternating fatigue loading: (a) tensile and shear damages and (b) total strain

As shown in Fig. 12, a good agreement between experimental result and model simulation can be seen, and the simulated fatigue life is $N_f = 5\ 728$. Some parameters identified for this simulation are given in Table 3.

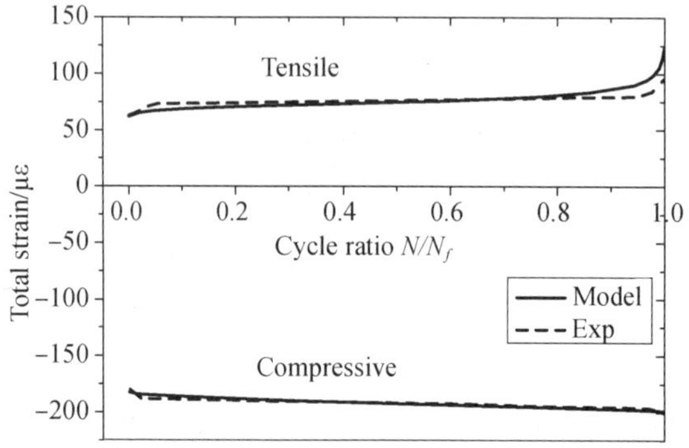

Fig. 12. Comparison between experimental results and model simulations of alternating fatigue

Table 3. Parameters for compressive-tensile fatigue loading

C^{-}/MPa	C^{+}/MPa	n	κ	β_1	β_2
32	2.8	4	8	32.0	40.0

6.4 Bi-axial compressive fatigue

The proposed model can also be applied to the multi-axial fatigue loading case and this circumstance is simulated as shown in Fig. 13. The constant lateral pressure is applied first, and then comes the axial fatigue stress.

The model test results including damage and axial strains are shown in Fig. 14. The stress range is $0.8f_c - 0.2f_c$ while the lateral pressures are $0f_c$, $0.1f_c$, $0.2f_c$, and $0.6f_c$. It is noticed that the axial strain is decreasing while the fatigue life is increasing due to the

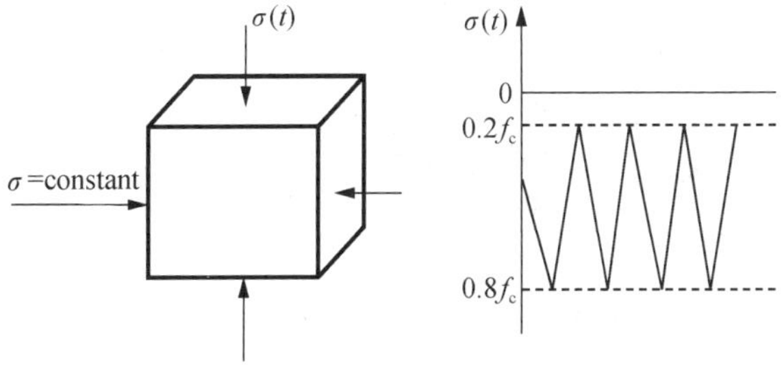

Fig. 13. Multi-axial fatigue loading process

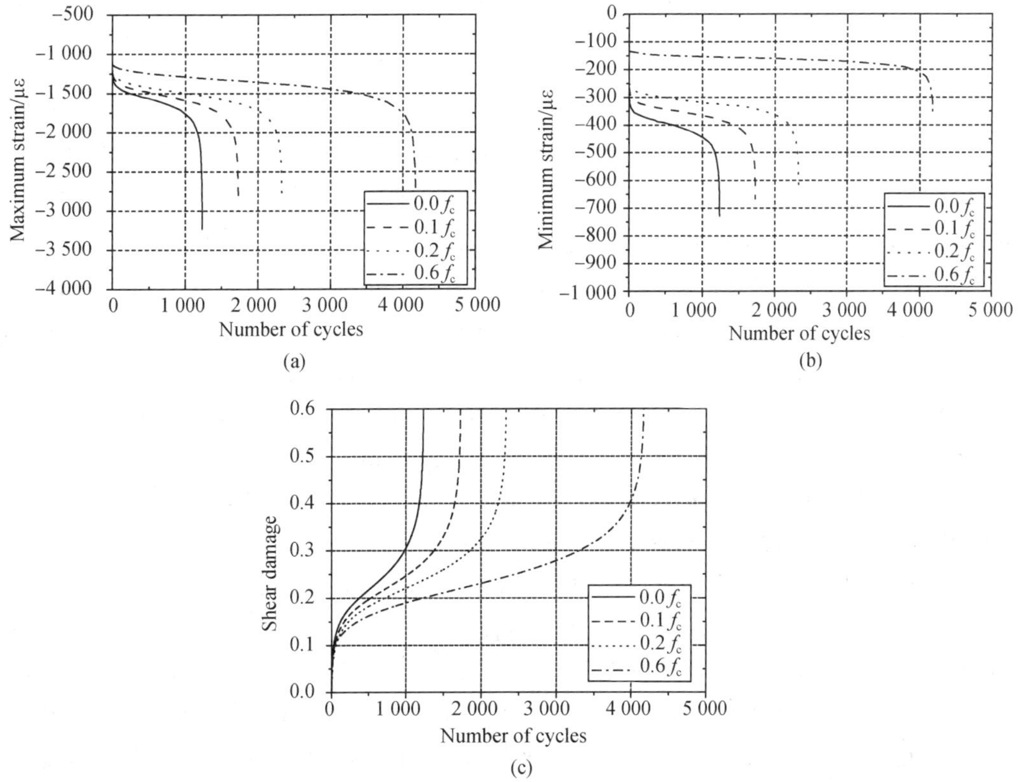

Fig. 14. Model properties under bi-axial compressive fatigue loading: (a) maximum axial strain, (b) minimum axial strain and (c) shear damage

increasing lateral pressures which is in accordance with the conclusion by Lu et al.[43]. The contrast is made to the experimental results of Lu et al.[43]. In the experiment, the average concrete compressive strength was f_c=20.47 MPa, while the lateral pressure was $0.5f_c$ and the axial stress range was $1.1f_c-0.22f_c$; the average fatigue life was $N_f=2\ 025$. The simulated results capture the experimental records well as shown in Fig. 15, and the simulated fatigue life is $N_f=2\ 028$. The values of model parameters are given in Table 4.

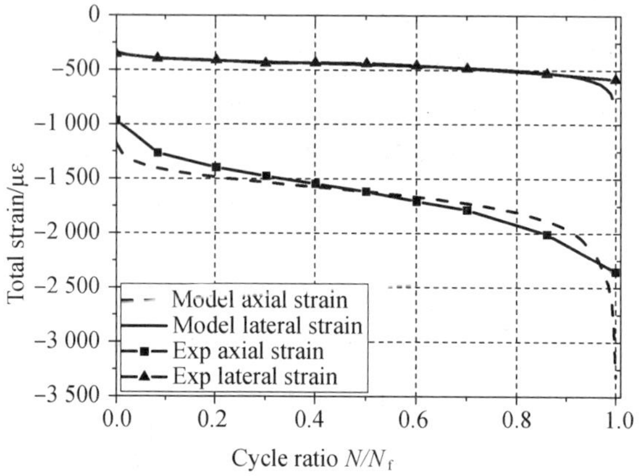

Fig. 15. Comparison between experimental results and model simulations of bi-axial compressive fatigue

Table 4. Parameters for bi-axial compressive fatigue loading

C^-/MPa	n	κ	β_1	β_2
16.4	3	8	24.0	22.4

7 Conclusions

This paper presents a new three-dimensional elastoplastic fatigue damage constitutive model for concrete. Based on the damage mechanisms and decomposition of effective stress tensor, two damage scalars are adopted to simplify and approximate the anisotropic fourth-order damage tensor and to describe the degradation of mechanical properties of concrete during fatigue loading. The elastoplastic DERRs, formulated in the effective stress space, are taken as the damage driving forces for the fatigue damage evolution. Meanwhile, the competitive mechanism of damage healing and damage driving effects is introduced to explain the typical concrete fatigue process, upon which an innovative fatigue damage evolution law is developed. Moreover, the plastic strains are solved in the effective stress space according to the strain equivalence concept.

The numerical simulations performed through the proposed model show good agreements with the experimental results in literatures, which demonstrate the following capabilities:
(1) it is able to predict the typical damage evolution of concrete under fatigue loading;
(2) it can be applied to analyze the concrete fatigue behavior under complex fatigue loads such as

multi-axial or alternating fatigue loads;

(3) the accumulation of permanent deformation during fatigue loading can be well predicted.

The macroscopic healing concept in this paper needs to be further studied in the micro-scale, and the specific model for the micro damage healing mechanism will be created in the future work.

Acknowledgement

We would like to express our sincere appreciations to the anonymous referee for many valuable suggestions and corrections.

Declaration of Conflicting Interests

The author(s) declared no potential conflicts of interest with respect to the research, authorship, and/or publication of this article.

Funding

The author(s) disclosed receipt of the following financial support for the research, authorship, and/or publication of this article: This study was financially supported by the National Science Foundation of China(Grant Nos. 51261120374, 91315301, 51208374).

References

[1] JOLY D. La résistance et l'élasticité des ciments porland[J]. Annales Des Ponts et Chausses 1898, 16(7): 198-224.

[2] NORDBY G M. Fatigue of concrete-a review of research[C]//Journal Proceedings. 1958, 55(8): 191-219.

[3] BENNETT E W, MUIR S E S J. Some fatigue tests of high-strength concrete in axial compression[J]. Magazine of Concrete Research, 1967, 19(59): 113-117.

[4] CORNELISSEN H A W. Fatigue failure of concrete in tension[J]. Heron, 1984, 29(4): 1-68.

[5] HOLMEN J O. Fatigue of concrete by constant and variable amplitude loading[J]. Special Publication, 1982, 75: 71-110.

[6] TEPFERS R, KUTTI T. Fatigue strength of plain, ordinary, and lightweight concrete[C]//Journal Proceedings. 1979, 76(5): 635-652.

[7] BREITENBÜCHER R, IBUK H. Experimentally based investigations on the degradation-process of concrete under cyclic load[J]. Materials and Structures, 2006, 39(7): 717-724.

[8] REESE R. Properties of aged asphalt binder related to asphalt concrete fatigue life[J]. Journal of the Association of Asphalt Paving Technologists, 1997, 66: 604-632.

[9] XIAO J, LI H, YANG Z. Fatigue behavior of recycled aggregate concrete under compression and bending cyclic loadings[J]. Construction and Building Materials, 2013, 38: 681-688.

[10] ZHANG J, STANG H, LI V C. Fatigue life prediction of fiber reinforced concrete under flexural load[J]. International Journal of Fatigue, 1999, 21(10): 1033-1049.

[11] PARIS P C. A rational analytic theory of fatigue[J]. The Trend in Engineering, 1961, 13: 9-14.

[12] BAZANT Z P, XU K. Size effect in fatigue fracture of concrete[J]. ACI Materials Journal, 1991, 88(4): 390-399.

[13] SHAH S P, CHANDRA S. Fracture of concrete subjected to cyclic and sustained loading[C]//Journal Proceedings. 1970, 67(10): 816-827.

[14] WU Z, ZHAO G, HUANG C. Investigation of fatigue fracture properties of concrete. China Civil Engineering Journal, 1995, 28(3): 59-65.

[15] HILLERBORG A, MODÉER M, PETERSSON P E. Analysis of crack formation and crack growth in concrete by means of fracture mechanics and finite elements[J]. Cement and Concrete Research, 1976, 6(6): 773-781.

[16] HORDIJK DA. Local approach to fatigue of concrete[D]. PhD Dissertation. Delft University of Technology, Delft, 1991.

[17] NGUYEN O, REPETTO E A, ORTIZ M, et al. A cohesive model of fatigue crack growth[J]. International Journal of Fracture, 2001, 110(4): 351-369.

[18] YANG B, MALL S, RAVI-CHANDAR K. A cohesive zone model for fatigue crack growth in quasibrittle materials[J]. International Journal of Solids and Structures, 2001, 38(22-23): 3927-3944.

[19] CHABOCHE J L, LESNE P M. A non-linear continuous fatigue damage model[J]. Fatigue & Fracture of Engineering Materials & Structures, 1988, 11(1): 1-17.

[20] CHOW C L, WEI Y. A model of continuum damage mechanics for fatigue failure[J]. International Journal of Fracture, 1991, 50(4): 301-316.

[21] CHOW CL AND WEI Y. A fatigue damage model for crack propagation[J]. Advances in Fatigue Lifetime Predictive Techniques, 1996, 1291(3): 86-99.

[22] MARIGO J J. Modelling of brittle and fatigue damage for elastic material by growth of microvoids[J]. Engineering Fracture Mechanics, 1985, 21(4): 861-874.

[23] ALLICHE A. Damage model for fatigue loading of concrete[J]. International Journal of Fatigue, 2004, 26(9): 915-921.

[24] MAI S H, LE-CORRE F, FORÊT G, et al. A continuum damage modeling of quasi-static fatigue strength of plain concrete[J]. International Journal of Fatigue, 2012, 37: 79-85.

[25] JU J W. On energy-based coupled elastoplastic damage theories: constitutive modeling and computational aspects[J]. International Journal of Solids and Structures, 1989, 25(7): 803-833.

[26] WU J Y, LI J, FARIA R. An energy release rate-based plastic-damage model for concrete[J]. International Journal of Solids and Structures, 2006, 43(3-4): 583-612.

[27] JU J W. Isotropic and anisotropic damage variables in continuum damage mechanics[J]. Journal of Engineering Mechanics, 1990, 116(12): 2764-2770.

[28] LI J, REN X D. A review on the constitutive model for static and dynamic damage of concrete[J]. Advances in Mechanics, 2010, 40(3): 284-297.

[29] RESENDE L. A damage mechanics constitutive theory for the inelastic behaviour of concrete[J]. Computer Methods in Applied Mechanics and Engineering, 1987, 60(1): 57-93.

[30] LEMAITRE J. Evaluation of dissipation and damage in metals[C]//Proc. ICM Kyoto. 1971, 1.

[31] FARIA R, OLIVER J, CERVERA M. A strain-based plastic viscous-damage model for massive concrete structures[J]. International Journal of Solids and Structures, 1998, 35(14): 1533-1558.

[32] REN X D, ZENG S, LI J. A rate-dependent stochastic damage-plasticity model for quasi-brittle materials[J]. Computational Mechanics, 2015, 55(2): 267-285.

[33] ORTIZ M. A constitutive theory for the inelastic behavior of concrete[J]. Mechanics of Materials, 1985, 4(1): 67-93.

[34] FARIA R, OLIVER J O O, RUIZ M C. On isotropic scalar damage models for the numerical analysis

of concrete structures [M]. Barcelona, Spain: Centro Internacional de Métodos Numéricos en Ingeniería, 2000.

[35] WU J Y, XU S L. Reconsideration on the elastic damage/degradation theory for the modeling of microcrack closure-reopening (MCR) effects[J]. International Journal of Solids and Structures, 2013, 50(5): 795-805.

[36] WU J Y. Damage energy release rate-based elastoplastic damage constitutive model for concrete and itsapplication to nonlinear analysis of structures[M]. Shanghai: Tongji University, 2004.

[37] SIMÓ JC, HUGHES TJR. Computational Inelasticity[M]. New York, NY: Springer-Verlag, 1998.

[38] JU J W, YUAN K Y, KUO A W. Novel strain energy based coupled elastoplastic damage and healing models for geomaterials — Part Ⅰ: Formulations[J]. International Journal of Damage Mechanics, 2012, 21(4): 525-549.

[39] JOHNSTON W G, GILMAN J J. Dislocation velocities, dislocation densities, and plastic flow in lithium fluoride crystals[J]. Journal of Applied Physics, 1959, 30(2): 129-144.

[40] GAO L, HSU C T T. Fatigue of concrete under uniaxial compression cyclic loading[J]. Materials Journal, 1998, 95(5): 575-581.

[41] KIM J K, KIM Y Y. Experimental study of the fatigue behavior of high strength concrete[J]. Cement and Concrete Research, 1996, 26(10): 1513-1523.

[42] CORNELISSEN H A W, REINHARDT H W. Uniaxial tensile fatigue failure of concrete under constant-amplitude and programme loading[J]. Magazine of Concrete Research, 1984, 36(129): 216-226.

[43] LU P, SONG Y, LI Q. Behavior of concrete under compressive fatigue loading with constant lateralstress[J]. Engineering Mechanics, 2004, 21(2): 173-177.

Softened damage-plasticity model for analysis of cracked reinforced concrete structures①

De-Cheng Feng[1], **Xiaodan Ren**[2], **Jie Li**[3]

1 *Key Laboratory of Concrete and Prestressed Concrete Structures of the Ministry of Education, Southeast Univ., 2 Sipailou, Nanjing 210096, China.*

2 *School of Civil Engineering, Tongji Univ., 1239 Siping Rd., Shanghai 200092, China.*

3 *School of Civil Engineering and State Key Laboratory on Disaster Reduction in Civil Engineering, Tongji Univ., 1239 Siping Rd., Shanghai 200092, China (corresponding author). E-mail: lijie@tongji.edu.cn*

Abstract This paper develops a new damage-plasticity model considering the compression-softening effect of RC. The model is based on the framework of a two-scalar damage-plasticity model, and adopts the elastoplastic damage energy release rates as the driven force of damage. To account for the compression-softening effect caused by transverse cracks of RC under shear, a softening coefficient is introduced in the model. With the modification, the new damage model can be used to simulate the typical shear behavior of cracked RC structures, which may not be captured by those models developed for plain concrete. Some computational aspects are also discussed. Finally, the model is validated through material tests and a series of RC member tests, and the results indicate that the proposed model has good performance in nonlinear analysis of RC structures.

Keywords Damage-plasticity model; Compression-softening; Shear; Reinforced concrete

1 Introduction

Shear behavior modeling of RC structures is a key problem in structural engineering. Although several celebrated works[1-7] have studied this, it is still quite challenging to accurately simulate the typical shear behavior of RC structures.

Traditional models for analysis of concrete structures and RC structures can be mainly grouped into two categories, orthotropic smeared-crack models[1, 3, 4, 6, 8-10] and damage models[11-15]. Smeared-crack models regard concrete as an orthotropic material and assume that cracks are smeared over the entire element. According to the assumption of the cracking direction, the model can be further divided into fixed-crack models[16-17] and rotating-crack models[18-19]. The major advantage of these kinds of models is their simplicity, but there also exists a serious objection against these models, i.e., lack of objectivity. The tangent stiffness matrix derived from the empirical stress-strain

① Originally published in *Journal of Structural Engineering*, 2018, 144(6):04018044.

relationships does not always give stable solutions for implicit time-integration methods[20]. On the other hand, damage mechanics offers an alternative way to construct the concrete constitutive model. The degradation of material is represented by a generalized internal variable, i.e., damage, and the typical characteristics of concrete material can be reflected by setting up corresponding damage evolution laws. Meanwhile, the residual deformations of concrete can be also explained by introducing the plastic strain in the model. Because of the solid theoretical foundation of damage models, damage and plasticity of concrete are organized in a unified framework[14-15].

However, most of the aforementioned models were proposed for plain concrete. The interaction effects between reinforcement steel and concrete in RC structures are usually neglected. In fact, when subjected to shear, RC will crack orthogonally to the principal tensile stress, and because of the presence of reinforcement bars, concrete between cracks can bear significant tensile stresses even at a large tensile strain, which is also known as *tension-stiffening*. Meanwhile, the large transverse tensile strain will cause a decrease of the compressive strength at the orthogonal direction, which is also known as *compression-softening*. These two effects make the properties of RC very different from those of plain concrete and are the key factors that should be considered in nonlinear analysis of RC structures, especially when modeling shear behavior(in a tension-compression stress state).

To represent the tension-stiffening effect, modifications have been introduced to both smeared-crack models[21] and damage models[14]. However, considerations for compression-softening effect of RC are relatively few. Vecchio and Collins[1] first conducted systematic tests of RC panels, and developed the modified compression field theory (MCFT) to quantify the effect of compression-softening in RC. Hsu[4] and Hsu and Zhu[6] also performed series tests of RC panels and derived the softened truss model(STM), including the stress-strain relationships of both the cracked concrete and reinforcement bars embedded in the concrete. These two theories are the pioneering works in determining the realistic shear behavior of cracked RC and have been implemented in finite-element codes[2, 7]. However, these models are orthotropic smeared-crack models, and although they provide a simple but detailed description of the shear mechanism of cracked RC, their phenomenological nature limits their future developments in engineering application from the computational point of view, especially for large-scale structures[15].

Based on the aforementioned background, this paper constructs a new damage-plasticity model for RC. The goal is to introduce the compression-softening effect of RC(like those in MCFT and STM) to the framework of a kind of damage-plasticity model by Wu et al.[14], herein called the Wu-Li model. The paper is structured as follows. The two-scalar damage model is adopted as the fundamental framework in section "Plasticity-Damage Model." Section "Modeling of Compression-Softening Effect" discusses the compression-softening effect and modifies the compressive damage evolution function to reflect the effect. The implementation of the proposed model and the nonlinear solution scheme are given in section "Computational Aspects." Section "Model Validation and Applications" demonstrates a series of numerical examples, showing the performance of the model in simulation of different kinds of problems. Finally, the conclusions are drawn in section "Conclusion."

2 Plasticity-Damage Model

2.1 Continuum Damage Framework

Starting from the hypothesis of strain equivalence[22], the effective stress of the material can be defined as follows[11-12, 14]

$$\bar{\boldsymbol{\sigma}} = \mathbb{E}_0 : (\boldsymbol{\epsilon} - \boldsymbol{\epsilon}^{\mathrm{p}}) \tag{1}$$

where $\bar{\boldsymbol{\sigma}}$ =second-order effective stress tensor; $\mathbb{E}_0$=fourth-order elastic stiffness tensor; and $\boldsymbol{\epsilon}$ and $\boldsymbol{\epsilon}^{\mathrm{p}}$ = second-order total strain tensor and its corresponding plastic component, respectively.

Because concrete behaviors in tensile part and compressive part are totally different, the effective stress tensor can be split as

$$\bar{\boldsymbol{\sigma}} = \bar{\boldsymbol{\sigma}}^+ + \bar{\boldsymbol{\sigma}}^- \tag{2}$$

where superscripts $+/-$ denote the positive (tensile) part and negative (compressive) part, respectively. Meanwhile, the detailed expressions of the positive and negative parts are given by[14]

$$\begin{aligned} \bar{\boldsymbol{\sigma}}^+ &= \mathbb{P}^+ : \bar{\boldsymbol{\sigma}} \\ \bar{\boldsymbol{\sigma}}^- &= \mathbb{P}^- : \bar{\boldsymbol{\sigma}} \end{aligned} \tag{3}$$

where $\mathbb{P}^{\pm}$ =projection tensors

$$\begin{aligned} \mathbb{P}^+ &= \sum_i H(\hat{\bar{\sigma}}_i) \boldsymbol{p}^{(i)} \otimes \boldsymbol{p}^{(i)} \otimes \boldsymbol{p}^{(i)} \otimes \boldsymbol{p}^{(i)} \\ \mathbb{P}^- &= \mathbb{I} - \mathbb{P}^+ \end{aligned} \tag{4}$$

where $\hat{\bar{\sigma}}_i$ = ith eigenvalue; $\boldsymbol{p}^{(i)}$ = corresponding normalized eigenvector; $\mathbb{I}$ = fourth-order identity tensor; and $H(\cdot)$=Heaviside function

$$H(x) = \begin{cases} 0 & x \leqslant 0 \\ 1 & x > 0 \end{cases} \tag{5}$$

Effective stress decomposition [Eq. (2)] makes it convenient to construct the tensile and compressive damage separately. Introducing two damage variables d^+ and d^-, the damage evolution can be defined according to the irreversible thermodynamics. The Helmholtz free energy (HFE) ψ can be written as

$$\psi = (1 - d^+)\psi_0^+ + (1 - d^-)\psi_0^- \tag{6}$$

where ψ_0 =initial HFE potential, and its corresponding positive and negative components (ψ_0^+, ψ_0^-) can be further split into two parts[14]

$$\psi_0^{\pm}=\psi_0^{e\pm}+\psi_0^{p\pm} \tag{7}$$

where $\psi_0^{e\pm}$ = elastic part; $\psi_0^{p\pm}$ = plastic part; and subscript $\pm$ indicates $+$ or $-$ as appropriate.

The expressions for the elastic part and the plastic part are[14]

$$\psi_0^{e\pm}=\frac{1}{2}\bar{\boldsymbol{\sigma}}^{\pm}:\boldsymbol{\epsilon}^{e}=\frac{1}{2}(\bar{\boldsymbol{\sigma}}^{\pm}:\mathbb{E}_0^{-1}:\bar{\boldsymbol{\sigma}}) \tag{8}$$

$$\psi_0^{p\pm}=\int\bar{\boldsymbol{\sigma}}^{\pm}:d\,\boldsymbol{\epsilon}^{p} \tag{9}$$

Using the Clausius-Duhem inequality in thermodynamics

$$\dot{\gamma}=-\dot{\psi}+\boldsymbol{\sigma}:\dot{\boldsymbol{\varepsilon}}\geqslant 0 \tag{10}$$

and substituting Eqs.(6)—(9) into Eq.(10), the final form of the constitutive law can be obtained[14]

$$\boldsymbol{\sigma}=(\mathbb{I}-\mathbb{D}):\mathbb{E}_0:(\boldsymbol{\epsilon}-\boldsymbol{\epsilon}^{p}) \tag{11}$$

where $\boldsymbol{\sigma}$ = second-order Cauchy stress tensor; and $\mathbb{D}$ = fourth-order damage tensor, which is determined by

$$\mathbb{D}=d^{+}\,\mathbb{P}^{+}+d^{-}\,\mathbb{P}^{-} \tag{12}$$

Meanwhile, the driven force of damage, i.e., the damage release rate(Y^{+}, Y^{-}), is also obtained according to Eq.(6) as

$$Y^{+}=-\frac{\partial\psi}{\partial d^{+}}=\psi_0^{+}$$

$$Y^{-}=-\frac{\partial\psi}{\partial d^{-}}=\psi_0^{-} \tag{13}$$

Because the tensile plastic strain is relatively small and has little influence on the overall behavior, Wu et al.[14] neglected the plastic part of the tensile HFE and proposed a new expression for compressive HFE. Based on Eq.(13), the damage release rates(Y^{+}, Y^{-}) are derived as

$$Y^{+}=\psi_0^{+}\approx\frac{1}{2}(\bar{\boldsymbol{\sigma}}^{+}:\mathbb{E}_0^{-1}:\bar{\boldsymbol{\sigma}})$$

$$Y^{-}=\psi_0^{-}\approx b_0(\alpha\bar{I}_1^{-}+\sqrt{3\bar{J}_2^{-}})^2 \tag{14}$$

where $\bar{I}_1^{-}$ = first invariant of the compressive effective stress $\bar{\sigma}^{-}$; $\bar{J}_2^{-}$ = second invariant of the deviator $\bar{\boldsymbol{s}}^{-}$ of the compressive effective stress $\bar{\boldsymbol{\sigma}}^{-}$; and b_0 and α = material parameters[14].

With the driven force $Y^{\pm}$, the damage state of concrete can be determined by some specific criteria

$$d^{\pm}=g^{\pm}(Y^{\pm},\,r^{\pm}) \tag{15}$$

where $g^{\pm}(\cdot)$ = monotonic damage evolution functions; and $r^{\pm}$ = damage thresholds expressed as

$$r^{\pm} = \max_{\tau \in [0, t]} \{Y^{\pm}(\tau)\} \tag{16}$$

which are maximum values of $Y^{\pm}$ in the entire loading history. The damage functions $g^{\pm}(\cdot)$ can be determined either by empirical formula form tests[14-15] or by theoretical derivation from the subscale[23-24].

2.2 Plastic Evolution

The evolution of plastic strains can be developed in either the Cauchy stress space or the effective stress space. In the first approach, softening of the Cauchy stress of concrete may cause shrinkage and concavity of the corresponding plastic potential functions, which will lead to certain numerical difficulty in simulation. On the other hand, in the second approach, the effective stress represents the undamaged part of concrete, and hence the plastic potential functions will keep expanding in the whole history, enabling development of robust schemes for plastic evolution.

There are two typical ways to develop the plastic model in the effective stress space, i.e., theoretical and empirical. The theoretical method[11, 14] relies on the standard plasticity theory and adopts the yield surface and flow rule to describe the plasticity evolution in a complete and rigorous way. However, it also requires a complex iterative process to calculate the plastic strain at each step, which increases the computational cost, especially for large-scale structural analysis[12, 15, 25]. To improve the numerical efficiency of the damage-type models, Faria et al.[12] introduced several simplifications and derived the empirical plastic model. This kind of model regards the plastic strain as an overall effect and uses an explicit expression to describe the plastic strain, and thus the yield surface and flow rule are not visible and the numerical efficiency is greatly enhanced for large, time-consuming analyses.

Considering the aforementioned aspects, this paper adopts the empirical plastic model proposed by Faria et al.[12] and then modified by Wu[26]

$$\dot{\boldsymbol{\epsilon}}^{\mathrm{p}} = b^{\mathrm{p}} \bar{\boldsymbol{\sigma}} \tag{17}$$

where b^{p} = empirical plastic flow parameter

$$b^{\mathrm{p}} = \xi^{\mathrm{p}} E_0 H(\dot{d}^-) \frac{\langle \boldsymbol{\epsilon}^{\mathrm{e}} : \dot{\boldsymbol{\epsilon}} \rangle}{\bar{\boldsymbol{\sigma}} : \bar{\boldsymbol{\sigma}}} \geqslant 0 \tag{18}$$

where $\xi^{\mathrm{p}} \geqslant 0$ is a plastic coefficient that controls the plastic strain rate; and E_0 = initial Young's modulus. Faria et al.[12], Wu[26], Tesser et al.[15], and Liang et al.[25] gave more details about the model parameters.

This paper neglects the tensile plastic strain because it is relatively small compared with the compressive plastic strain.

2.3 Damage Evolution

Continuum damage mechanics provides a rational and powerful framework for constitutive modeling of concrete, but the damage evolution cannot be directly given[27]. Generally, the literature adopts empirical damage expressions. Li and Ren[23] derived the energy equivalent strains by proposing the damage-consistent condition, making it possible to determine the multiaxial damage evolution from a uniaxial damage evolution function. The energy equivalent strains are expressed as

$$\bar{\epsilon}^{e+}=\sqrt{\frac{2Y^{+}}{E_0}}$$

$$\bar{\epsilon}^{e-}=\frac{1}{E_0(1-\alpha)}\sqrt{\frac{Y^{-}}{b_0}} \tag{19}$$

which can replace the damage release rates to be the driven force of damage evolution. Consequently, the damage variables can be attained by the uniaxial experiential functions[28-29]

$$d^{\pm}=g^{\pm}(\bar{\epsilon}^{e\pm})=\begin{cases}1-\dfrac{\rho^{\pm}n^{\pm}}{n^{\pm}-1+(x^{\pm})^{n^{\pm}}} & x^{\pm}\leqslant 1\\ 1-\dfrac{\rho^{\pm}}{\alpha^{\pm}(x^{\pm}-1)^2+x^{\pm}} & x^{\pm}\geqslant 1\end{cases} \tag{20}$$

where

$$x^{\pm}=\frac{\bar{\epsilon}^{e\pm}}{\epsilon_c^{\pm}},\quad \rho^{\pm}=\frac{f_c^{\pm}}{E_c\epsilon_c^{\pm}},\quad n^{\pm}=\frac{E_c\epsilon_c^{\pm}}{E_c\epsilon_c^{\pm}-f_c^{\pm}} \tag{21}$$

where E_c=elastic modulus of concrete; $f_c^{\pm}$ and $\epsilon_c^{\pm}$ = tensile/compressive strength and the corresponding strain, respectively; and $\alpha^{\pm}$=descending parameters that control the shape of the descending part of stress-strain curves.

Furthermore, to avoid mesh dependency problems, the determination of descending parameters $\alpha^{\pm}$ are based on tensile and compressive fracture energies[13-14, 29]. Because the damage evolution function [Eq.(20)] is rather complex, the explicit expression between the material parameters and fracture energies cannot be directly derived. To select the appropriate descending parameters $\alpha^{\pm}$ to ensure constant energy dissipations under different types of mesh dimension, an approach similar to that of Berto et al.[30] is adopted, i.e.

$$\frac{G_f^{+}}{l_{ch}}=\int\sigma^{+}d\epsilon^{+}$$

$$\frac{G_f^{-}}{l_{ch}}=\int\sigma^{-}d\epsilon^{-} \tag{22}$$

where G_f^{+} and G_f^{-} = tensile and compressive fracture energy, respectively; and l_{ch} =

characteristic length related to the dimension of the element mesh.

The tensile fracture energy of concrete is provided by several national codes, whereas the compressive fracture energy remains a controversial problem in structural engineering, especially for RC. This paper adopts the values recommended by Saritas and Filippou[20, 29].

3 Modeling of Compression-Softening Effect

3.1 Mechanism of Compression-Softening

As mentioned previously, the properties of RC are somewhat different from those of plain concrete because of the complex interaction between concrete and steel. Fig. 1 shows that RC can bear tensile stress even after cracking, and the corresponding tensile strain can develop to an order of magnitude such as $(10-20)\times10^{-3}$ due to the presence of reinforcement bars (tension-stiffening). This amount of tensile strain actually has no physical meaning for plain concrete because at such a time its strength already has decayed to zero. Meanwhile, the large amount of tensile strain will cause the softening of compressive strength in the orthogonal direction, i.e., compression-softening (Fig. 2).

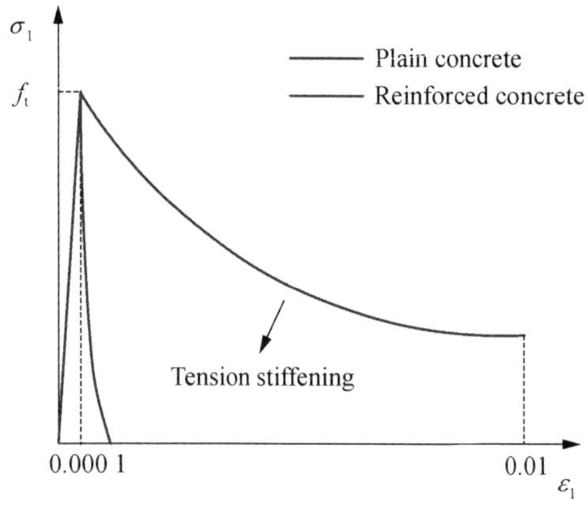

Fig. 1. (Color) Tensile stress-strain curves of plain concrete and RC

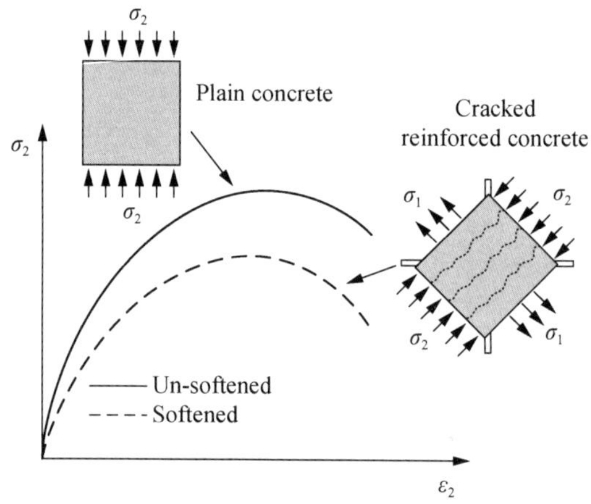

Fig. 2. Compression-softening effect of RC

This kind of softening effect is different from the tension-compression softening effect of plain concrete. Tension-compression softening of plain concrete means that the tension-compression stress state may decrease the tensile strength of plain concrete, which can be seen in the biaxial peak stress envelopes predicted by Kupfer et al.[31] (Fig. 3), and the typical failure mode of this case is one major tensile crack. Most concrete plasticity models already capture this kind of compression-softening effect by using a yield surface[13-14, 29, 32]. On the other hand, compression-softening of RC means the reduction of compressive strength of RC under the tension-compression stress state[4, 33], and the failure mode is several parallel cracks normal to the tensile principal direction. Fig. 4 shows the two different failure modes.

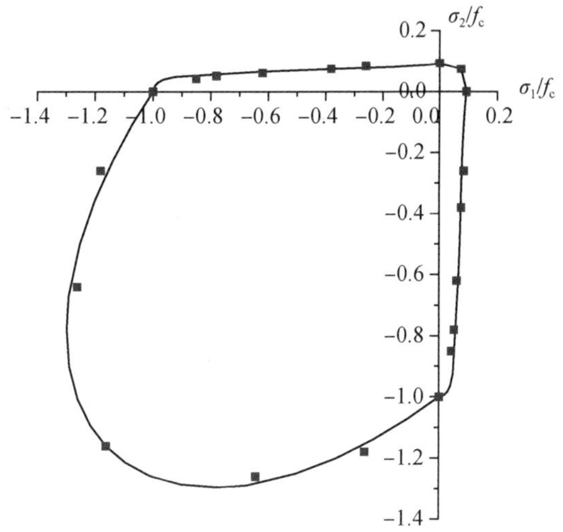

Fig. 3. (Color) Tension-compression softening effect of plain concrete

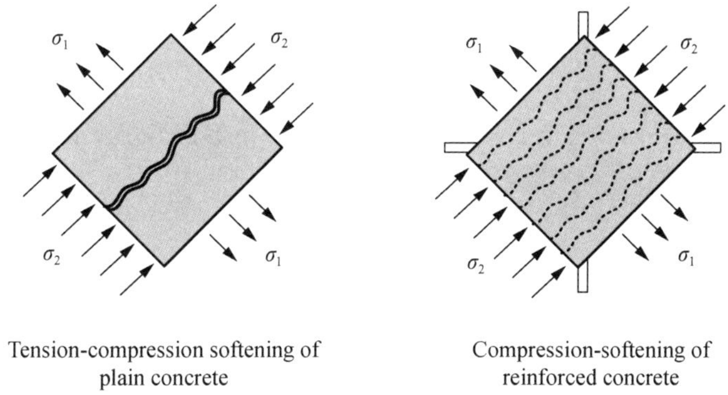

Fig. 4. Failure modes of tension-compression softening and compression-softening

3.2 Compression-Softening in MCFT and STM

In concrete constitutive models, tension-stiffening can be considered by modifying the descending branch of the tensile stress-strain relationship in phenomenological models (most smeared crack models), or by identifying the material parameters of the damage model according to certain experimental data. However, consideration of the compression-softening effect is relatively complicated because the principal compressive stress is related to both the principal compressive strain and the principal tensile strain. Robinson and Demorieux[34] found the compression-softening effect. Before that time, the stress-strain relationship obtained from standard uniaxial compressive test was used in the modeling of RC structures, which resulted in an overestimation of the shear behavior. Vecchio and Collins[35] first quantified the concept by introducing a softening coefficient, which was a breakthrough for solving shear problems of RC structures. By performing systematic experimental and theoretical investigations, Vecchio and Collins[33] established the MCFT, which considered the equilibrium condition and compatibility condition, as well as material constitutive laws. Hsu[4] developed the STM with a similar idea. Both MCFT and STM are smeared-crack models, and the STM can be further divided into two categories, the rotational angle (RA) model and the fixed angle (FA) model[36]. The main difference lies in the assumption of the direction of cracks. The significant achievement of MCFT, RA-STM, and FA-STM was the introduction of the softening coefficient to quantify the compression-softening effect. The established stress-strain relationships for RC in MCFT are, for the tensile part

$$
\begin{aligned}
\sigma_1 &= E_c \epsilon_1 \quad \epsilon_1 \leqslant \epsilon_{cr} \\
\sigma_1 &= \frac{f_t}{1+\sqrt{200\epsilon_1}} \quad \epsilon_1 > \epsilon_{cr}
\end{aligned}
\tag{23}
$$

and for the compressive part

$$
\sigma_2 = f_{cmax}\left[2\left(\frac{\epsilon_2}{\epsilon_0}\right)+\left(\frac{\epsilon_2}{\epsilon_0}\right)^2\right] \tag{24}
$$

$$
\beta_{MCFT} = f_{cmax}/f'_c = \frac{1}{0.8+170\epsilon_1} \leqslant 1 \tag{25}
$$

and the established stress-strain relationships for RC in STM are, for the tensile part

$$
\begin{aligned}
\sigma_1 &= E_c \epsilon_1 \quad \epsilon_1 \leqslant \epsilon_{cr} \\
\sigma_1 &= f_t\left(\frac{\epsilon_{cr}}{\epsilon_1}\right)^{0.4} \quad \epsilon_1 > \epsilon_{cr}
\end{aligned}
\tag{26}
$$

and for the compressive part

$$
\begin{aligned}
\sigma_2 &= \beta f'_c\left[2\left(\frac{\epsilon_2}{\beta\epsilon_0}\right)+\left(\frac{\epsilon_2}{\beta\epsilon_0}\right)^2\right] \quad \epsilon_2 \leqslant \beta\epsilon_0 \\
\sigma_2 &= \beta f'_c\left[1-\left(\frac{\epsilon_2/\epsilon_0-\beta}{2-\beta}\right)^2\right] \quad \epsilon_2 > \beta\epsilon_0
\end{aligned}
\tag{27}
$$

$$
\beta_{STM} = \frac{1}{\sqrt{1+400\epsilon_1}} \leqslant 1 \tag{28}
$$

where E_c = elastic modulus of concrete; σ_1/ϵ_1 = principal tensile stress/strain; σ_2/ϵ_2 = principal compressive stress/strain; f_t and ϵ_{cr} = tensile strength and the corresponding strain; f'_c and ϵ_0 = compressive strength and the corresponding strain; and β_{MCFT} and β_{STM} = softening coefficients of the respective models.

Obviously, the concrete models in the two theories have the same trend and shape, and differences may be caused by the fact that they are fitting from different experimental results. The softening coefficients proposed to address the compression-softening are also very similar. Fig. 5 plots the two coefficients versus principal tensile strain. Either coefficient can be used to account for the compression-softening effect in simulation.

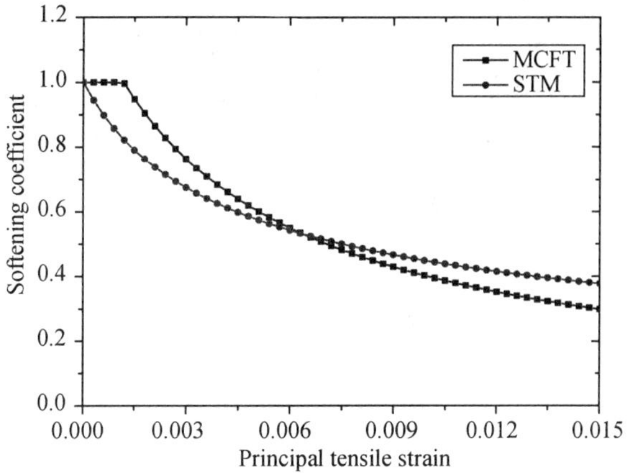

Fig. 5. (Color) Softening coefficients in MCFT and STM

3.3 Limitations of Previous Models for Analysis of RC Structures

In MCFT and STM, the behaviors of concrete at each principal direction are defined by the uniaxial stress-strain relations that are empirically derived from concrete experiments by fitting the test data. Although the models well reflect the shear mechanisms of RC (including tension-stiffening and compression-softening), they are uniaxial phenomenological models in nature and usually requires a large number of state variables and parameters to cover all possible responses of concrete, especially for cyclic loading cases. In addition, for multiaxial loading, some refinements (i.e., the Poisson effect, confinement, and so on) should be added to the original model[2], which will lose the simplicity and objectivity of the models[20, 37]. All such factors limit their future developments in engineering application from the computational point of view, especially for large-scale structures[15].

On the other hand, damage-plasticity models successfully reflect the behavior of plain concrete, but the compression-softening effect of RC (the ones in MCFT and STM) has not been accounted for until now. Fig. 6 shows the simulated compressive strength of a RC element under different principal tensile strains by the previous Wu-Li model. The model does not reflect the reduction of strength caused by transverse cracks of RC. Consequently, this paper includes compression-softening in the Wu-Li damage model for analysis of RC structures.

3.4 Softened Damage-Plasticity Model

In the Wu-Li model, the tensile part and the compressive part can be considered separately because of the positive and negative decomposition of the effective stress. The principal direction of the effective stress is defined in the projection tensors $\mathbb{P}^+$ and $\mathbb{P}^-$, and thus the damage-plasticity model is an equivalent form of the rotating angle models (MCFT and RA-STM)[24]. Furthermore, the energy equivalent strains, which bridge the gap between the uniaxial state and multiaxial states of concrete, make it convenient to implement the

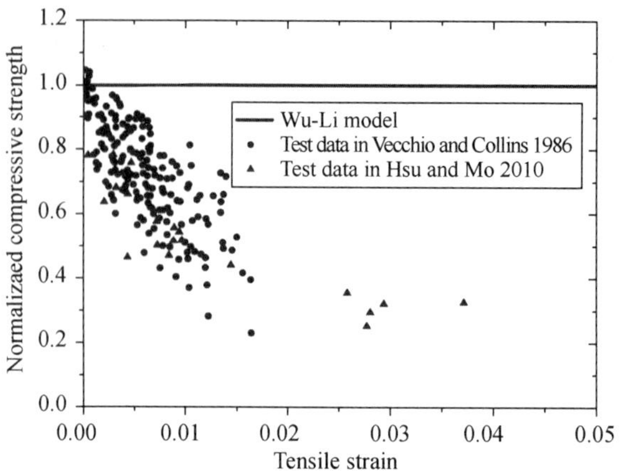

Fig. 6. (Color) Compression-softening by previous Wu-Li damage model

softening coefficient in the damage model for multiaxial loading conditions[38].

For a two-dimensional plane-stress situation, the damage-plasticity model in the compressive principal direction can be written as

$$\sigma_2 = (1-d^-)\bar{\sigma}_2 \tag{29}$$

Considering the compression-softening, a softening coefficient β should be introduced to Eq.(29)

$$\sigma_2^s = \beta(1-d^-)\bar{\sigma}_2 \tag{30}$$

where superscript s denotes the compression-softening effect.

The coefficient β_{STM} is adopted herein to represent the softening effect because the curve is smoother and more continuous than β_{MCFT}. This paper replaces the principal tensile strain ϵ_1 with the maximum tensile energy equivalent strain $\bar{\epsilon}_{max}^{e+}$ of the loading history, which makes it more convenient to calculate under the complex multiaxial stress state and accounts for the accumulated influence of compression-softening under reverse loading, and the monotonic character of the damage variable can be guaranteed, that is

$$\beta = \frac{1}{\sqrt{1+400\,\bar{\epsilon}_{max}^{e+}}} \tag{31}$$

Combining Eqs.(30) and (31), a new principal compressive stress expression can be derived as

$$\sigma_2^s = (1-d^{s-})\bar{\sigma}_2 = \beta(1-d^-)\bar{\sigma}_2 \tag{32}$$

where d^{s-} =softened damage variable, which has the form

$$d^{s-} = 1-\beta(1-d^-) \tag{33}$$

Using the constitutive law in Eq.(11), the Cauchy stress becomes

$$\boldsymbol{\sigma} = (1-d^+)\bar{\boldsymbol{\sigma}}^+ + (1-d^{s-})\bar{\boldsymbol{\sigma}}^- \tag{34}$$

Finally, the constitutive law can be reconstructed as

$$\begin{aligned}\boldsymbol{\sigma} &= (\mathbb{I} - \mathbb{D}^{s}) : \mathbb{E}_0 : (\boldsymbol{\epsilon} - \boldsymbol{\epsilon}^{p}) \\ \mathbb{D}^{s} &= d^{+}\,\mathbb{P}^{+} + d^{s-}\,\mathbb{P}^{-}\end{aligned} \tag{35}$$

where $\mathbb{D}^{s}$ = modified fourth-order damage tensor considering the compression-softening effect.

With Eq. (35), the new softened damage-plasticity model is proposed and yields the softening effect of compressive strength due to tensile strain(Fig. 7).

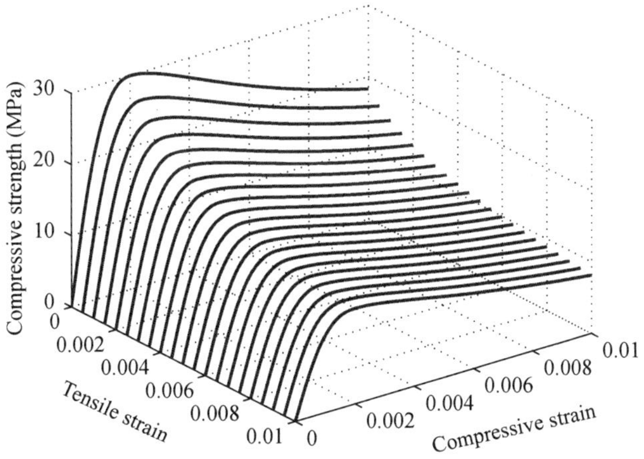

Fig. 7. **(Color) Softened stress-strain relationship by proposed model**

This kind of compression-softening effect means that the presence of transverse cracks in RC will increase the compressive damage extent in the orthogonal direction, and hence the introduction of the compression-softening effect in MCFT and STM into the damage-plasticity model can be done by reconstructing the damage surface shape, which can be achieved by modifying either the damage release rate [Y^{-} in Eq. (14)] or the damage evolution variable [d^{-} in Eq. (20)]. This paper uses the second method [Eqs. (32) and (33)] because the authors think it is more straightforward and physically meaningful, and the derivation is simple and easy to understand.

4 Computational Aspects

Finite element analysis usually requires an appropriate solution algorithm to compute the nodal displacements of the model, i.e., the Newton-Raphson (NR) family of methods. With the given strain increment at each material point calculated from the nodal displacements, the unknown stress needs to be updated according to the constitutive relationship. This section presents the implementation of the model and the recommended nonlinear solution strategy.

4.1 Numerical Implementation of Model

At each material point, the variables $\{\boldsymbol{\sigma}_{n+1}, \boldsymbol{\epsilon}^{p}_{n+1}, d^{\pm}_{n+1}\}$ at pseudotime t_{n+1} should be computed according to the variables $\{\bar{\boldsymbol{\sigma}}_n, \boldsymbol{\epsilon}_n^{\mathrm{p}}, d_n^{\pm}\}$ at previous pseudotime t_n and strain increment $\Delta\boldsymbol{\epsilon}$, and hence a numerical algorithm is needed to update the variable. This paper adopts an efficient and direct algorithm proposed by Faria et al.[12] and refined by Wu[26]. According to the operator-split concept[11, 14], the procedure consists of three parts: an elastic predictor, a plastic corrector, and a damage corrector. The detailed calculation is summarized as follows.

1. Elastic Predictor

First, update the strain and calculate the trial effective stress

$$\boldsymbol{\epsilon}_{n+1} = \boldsymbol{\epsilon}_n + \Delta\boldsymbol{\epsilon} \tag{36}$$

$$\bar{\boldsymbol{\sigma}}_{n+1}^{\mathrm{trial}} = \bar{\boldsymbol{\sigma}}_n + \mathbb{E}_0 : \Delta\boldsymbol{\epsilon} \tag{37}$$

The corresponding values of the damage energy release rates $Y_{n+1}^{\pm\mathrm{trial}}$ and the energy equivalent strains $\bar{\epsilon}_{n+1}^{\mathrm{e}\pm\mathrm{trial}}$ could also be obtained according to Eqs.(14) and (19).

2. Plastic Corrector

Using the empirical plastic model [Eqs.(17) and (18)], the real value of the variables at time t_{n+1} can be expressed as

$$\begin{aligned} \bar{\boldsymbol{\sigma}}_{n+1} &= \xi_{n+1}\bar{\boldsymbol{\sigma}}_{n+1}^{\mathrm{trial}}, \\ \boldsymbol{Y}_{n+1}^{\pm} &= \xi_{n+1}\boldsymbol{Y}_{n+1}^{\pm\mathrm{trial}}, \\ \bar{\epsilon}_{n+1}^{\mathrm{e}\pm} &= \xi_{n+1}\bar{\epsilon}_{n+1}^{\mathrm{e}\pm\mathrm{trial}} \end{aligned} \tag{38}$$

where ξ_{n+1} denotes the plastic modification factor, which has the following form[11, 14]

$$\xi_{n+1} = 1 - \frac{\xi^{\mathrm{p}}}{\|\bar{\boldsymbol{\sigma}}_{n+1}^{\mathrm{trial}}\|} E_0 \langle \boldsymbol{I}_{\bar{\boldsymbol{\sigma}}_{n+1}^{\mathrm{trial}}} : \Delta\boldsymbol{\epsilon} \rangle \tag{39}$$

where $\|\boldsymbol{x}\|$ = norm of tensor $\boldsymbol{x}$; and $\boldsymbol{I}_{\bar{\boldsymbol{\sigma}}_{n+1}^{\mathrm{trial}}}$ = trial unit effective stress tensor.

3. Damage Corrector

Compute the new damage variables with the energy equivalent strains $\bar{\epsilon}_{n+1}^{\mathrm{e}\pm}$ using Eq.(20)

$$d_{n+1}^{\pm} = g^{\pm}(\bar{\epsilon}_{n+1}^{\mathrm{e}\pm}) \tag{40}$$

Thus the Cauchy stress can be updated by Eq.(34)

$$\boldsymbol{\sigma}_{n+1} = (1 - d_{n+1}^{+})\bar{\boldsymbol{\sigma}}_{n+1}^{+} + (1 - d_{n+1}^{\mathrm{s}-})\bar{\boldsymbol{\sigma}}_{n+1}^{-} \tag{41}$$

There is no iteration in this implementation of the material model because an empirical plastic model is adopted, which also enables the application of the proposed model in the analysis of large-scale structures.

4.2 Nonlinear Solution Strategy

In nonlinear analysis, the full Newton-Raphson iterative method usually is adopted due to its quadratic convergence rate. However, for some practical problems, such as the coupling of damage and plasticity or asymmetry in material constitutive law, several numerical difficulties may arise in the full NR procedure. For the proposed damage model, the full

NR method is not suitable. The consistent tangential operator is too complicated to develop. Furthermore, the negative damage variable depends not only on the compressive part but also on the tensile part, which may cause asymmetry and concavity in the tangent stiffness matrix[39]. An alternative method is to use the secant stiffness algorithm to conduct the iterative procedure, i. e., the quasi-Newton method. Although the quasi-Newton method requires more iterations than the full NR method, it is numerically robust[39]. Consequently, this paper uses the secant stiffness method. Vecchio[40] adopted a similar approach for shear problem modeling of RC structures, and proved that the secant stiffness method is suitable and stable, even for cyclic loading cases. The iterative secant matrix can be derived as

$$\mathbb{K}_{sec} = \int_{\Omega} \boldsymbol{B}^{\mathrm{T}} \, \mathbb{S} \, \boldsymbol{B} \, \mathrm{d}\Omega \tag{42}$$

where $\boldsymbol{B}$ = assumed shape functions in a displacement-based finite element; Ω = entire problem domain; and $\mathbb{S}$ = secant stiffness tensor in the material level, originally defined as

$$\boldsymbol{\sigma} = \mathbb{S} : \boldsymbol{\epsilon} \tag{43}$$

However, for the damage-plasticity model in Eq. (35), the closed form cannot be easily obtained. In this case, the authors propose using the following Broyden-Fletcher-Goldfarb-Shannon(BFGS)[41] secant operator from a computational point of view

$$\mathbb{S}_{\mathrm{BFGS}} = \mathbb{E}_0 + \frac{\boldsymbol{\sigma} \otimes \boldsymbol{\sigma}}{\boldsymbol{\sigma} : \boldsymbol{\epsilon}} - \frac{(\mathbb{E}_0 : \boldsymbol{\epsilon}) \otimes (\boldsymbol{\epsilon} : \mathbb{E}_0)}{\boldsymbol{\epsilon} : \mathbb{E}_0 : \boldsymbol{\epsilon}} \tag{44}$$

The BFGS secant operator preserves the symmetry of $\mathbb{E}_0$ and has been proven to be well-conditioned in most cases[41].

4.3 Model Validation and Applications

The proposed element was implemented in *ABAQUS* through the subroutine user-defined material(UMAT)[42]. This section verifies the proposed model through a set of numerical examples. First the model is validated through the material tests for both plain and reinforced concrete. Then the RC shear panel tests are modeled to demonstrate the capacity of the proposed model over the previous Wu-Li damage model for shear behavior. Finally, some shear critical RC beams and RC shear wall are simulated to illustrate the performance of the model in engineering practice.

4.4 Material Level Tests

The model was validated through the uniaxial tensile and compressive tests of plain concrete[43]. The parameters used in the simulation were, for the tensile test, the elastic modulus E_c=35 000 MPa, tensile strength f_c^+=2.2 MPa, corresponding peak strain ϵ_c^+ = 0.000 1, and tensile fracture energy G_f^+=200 N/m; for compressive test, the parameters were the elastic modulus E_c = 35 000 MPa, compressive strength f_c^- = 39 MPa, peak strain ϵ_c^-=0.002, tensile fracture energy G_f^-=8 500 N/m, and plastic coefficient ξ_p=0.1.

Two kinds of meshes were used in the simulation to indicate the mesh sensitivity issue of the proposed model as shown in Fig. 8, in which the hatched part represents two weaker elements. Fig. 8 plots the numerical results against experimental data. The stress-strain curves predicted by the model agree well with the experimental curves, even in postpeak descending branches. Objective responses can be achieved regardless of the mesh type, which indicates that selecting appropriate descending parameters of the proposed model based on assumed fracture energies can avoid the mesh sensitivity issue.

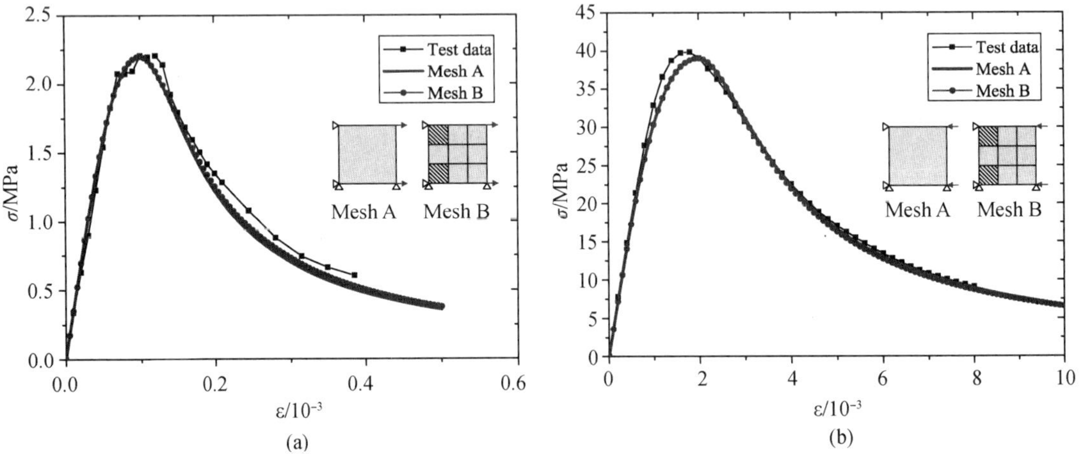

Fig. 8. (Color) Uniaxial tensile and compressive tests of plain concrete

The plain concrete biaxial compressive test conducted by Kupfer et al.[31] was also modeled. The material parameters were elastic modulus $E_c = 35\ 000$ MPa, compressive strength $f_c^- = 32$ MPa, peak strain $\epsilon_c^- = 0.002\ 4$, descending parameter $\alpha^- = 1.0$, and plastic coefficient $\xi_p = 0.1$. Fig. 9 shows the simulated results for specimens under three load conditions, $\sigma_2/\sigma_1 = -1/0$, $\sigma_2/\sigma_1 = -1/-0.52$, and $\sigma_2/\sigma_1 = -1/-1$, which matched the experimental results very well.

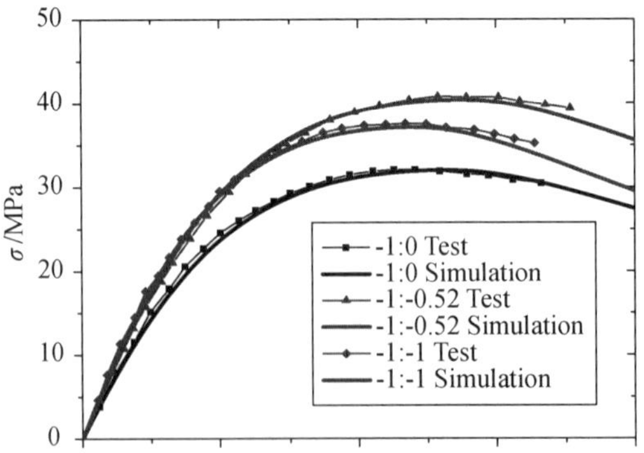

Fig. 9. (Color) Biaxial compressive test of plain concrete

The cyclic test of plain concrete by Karsan and Jirsa[44] was simulated as well. The modeling parameters were defined as follows: elastic modulus $E_c = 31\,000$ MPa, compressive strength $f_c^- = 27.6$ MPa, peak strain $\epsilon_c^- = 0.001\,8$, descending parameter $\alpha^- = 0.6$, and plastic coefficient $\xi_p = 0.08$. Fig. 10 plots the simulated result against the experimental data. The two curves were in good agreement with each other. The present model well reflected the strength softening, stiffness degradation, and residual deformation.
Fig. 11 shows the simulated compressive strengths of RC under different principal tensile strains calculated by the proposed model. The predicted compressive strengths were very close to those obtained from tests by Vecchio and Collins[1] and Hsu and Mo[45], and reflected the compression-softening effect.

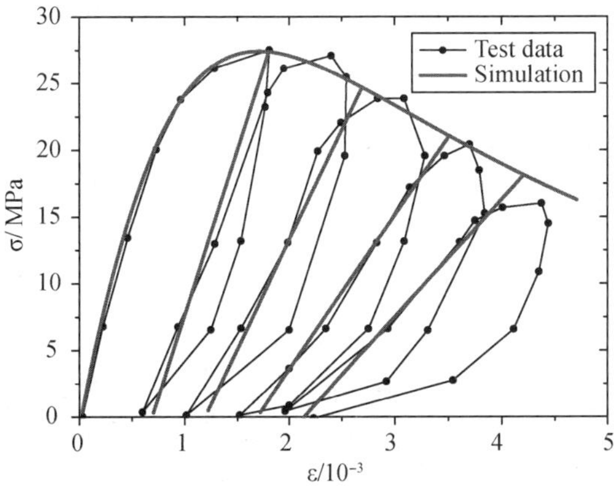

Fig. 10. (Color) Cyclic compressive test of plain concrete

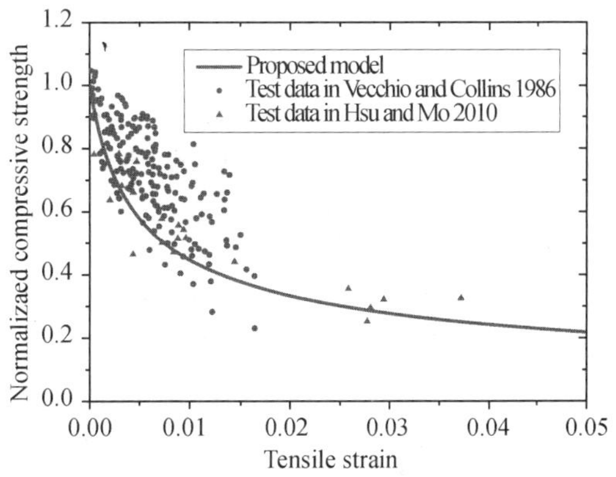

Fig. 11. (Color) Compression-softening of RC by proposed model

Fig. 12 shows the principal compressive stress-strain relationships of RC under different principal tensile strains obtained by STM, MCFT, the concrete damaged plasticity (CDP) model in ABAQUS, and the proposed model. The material properties were elastic modulus $E_c = 30\,000$ MPa, compressive strength $f_c^- = 28$ MPa, compressive peak strain $\epsilon_c^- = 0.001\,6$, tensile strength $f_c^+ = 2.8$ MPa, and tensile peak strain $\epsilon_c^+ = 0.000\,1$, and three load conditions, i.e., uniaxial, $\epsilon_2/\epsilon_1 = -1/0.3$, and $\epsilon_2/\epsilon_1 = -1/0.5$, were imposed. The curves were nearly the same for the CDP model and the proposed model in uniaxial cases, and the predicted compressive strength was the same as that in STM and MCFT. However, with principal tensile strains, the principal compressive strengths predicted by the proposed model still matched well with the STM and MCFT, which were fitted from experimental data and can be treated as the correct results, whereas the predictions of the CDP model were not consistent with the STM and MCFT. Although the CDP model can also consider the reduction of compressive strength due to transverse tensile strain, the results of the proposed model were better compared with the experimental results (STM and MCFT). As mentioned previously, this kind of compression-softening effect in damage-plasticity models can be achieved by modifying the yield function or the damage function according to the experimental results. This paper adopted the Wu-Li model as the framework; however, if the yield function in the CDP model is reconstructed based on STM and MCFT, the CDP model is also capable of representing the compression-softening effect.

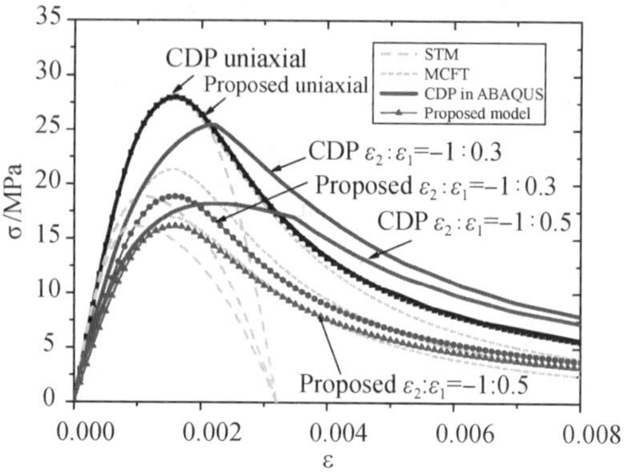

Fig. 12. (Color) Compressive stress-strain relationships under different principal tensile strains by different models

4.5 RC Shear Panel Tests

This section simulated a series of shear panel tests by Pang and Hsu[46]. The panels were 1.4 m×1.4 m×0.178 m and were subjected to pure shear(Fig. 13). Equal reinforcements were placed in orthogonal directions(45° to the x, y-coordinates). The reinforcing ratios were 0.77, 1.19, 1.79, and 2.98% for panels A1, A2, A3, and A4, respectively. The yield strength of the reinforcement bars was 460 MPa, and the compressive strength of the concrete was 42 MPa. The panels were simulated with only one four-node plane-stress element considering that the material and stress conditions were uniform throughout, and the reinforcement bars were modeled by a rebar layer, in which a perfect bond between concrete and steel was assumed. The parameters for concrete were elastic modulus E_c = 30 000 MPa, tensile strength f_c^+ = 3 MPa, peak tensile strain ϵ_c^+ = 0.000 1, tensile fracture energy G_f^+ = 200 N/m, descending parameter α^+ = 0.1, compressive strength f_c^- = 42 MPa, peak compressive strain ϵ_c^- = 0.002, compressive fracture energy G_f^- = 60 000 N/m, descending parameter α^- = 0.1, and plastic coefficient ξ_p = 0.1. The Menegotto-Pinto(M-P) constitutive model with isotropic hardening was used to model the behavior of the rebar layer with the parameters elastic modulus E_s = 200 000 MPa, yield strength f_y = 460 MPa, and hardening ratio 0.01. To maintain consistency with the secant algorithm, the uniaxial M-P model should also be cast in a secant way to provide the secant

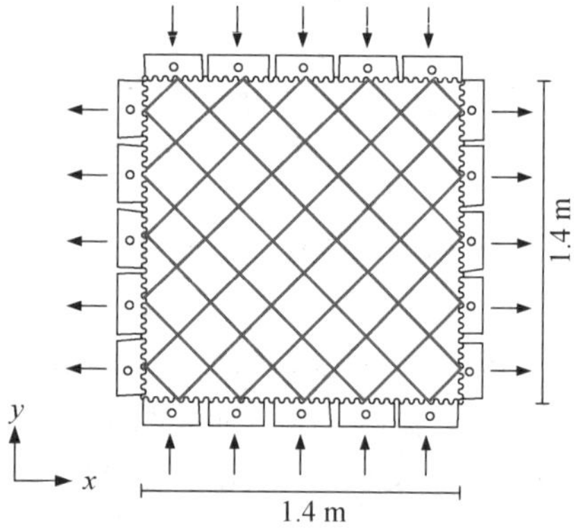

Fig. 13. (Color) RC shear panel tests

stiffness matrix for the nonlinear iteration procedure.

Table 1. Steel Properties of Bresler-Scordelis Beams

Bar type	Reinforcement			
	d/mm	f_y/MPa	f_u/MPa	E_s/MPa
No.2	6.4	325	430	190 000
No.4	12.7	345	542	201 000
No.9	28.7	555	933	218 000

Table 2. Concrete Properties of Bresler-Scordelis Beams

Beam number	Concrete	
	f_c'/MPa	f_t'/MPa
OA1	22.6	3.97
OA2	23.7	4.34
OA3	37.6	4.14
A1	24.1	3.86
A2	24.3	3.73
A3	35.1	4.34
B1	24.8	3.99
B2	23.2	3.76
B3	38.8	4.22
C1	29.6	4.22
C2	23.8	3.93
C3	35.1	3.86

Fig. 14 plots the simulated shear strain-shear stress curves against the experimental results. Numerical results agreed well with the experimental data, especially for the ultimate shear strength and the yielding of reinforcement bars. This example indicates that the typical behaviors of RC under pure shear can be described by the proposed model very well.

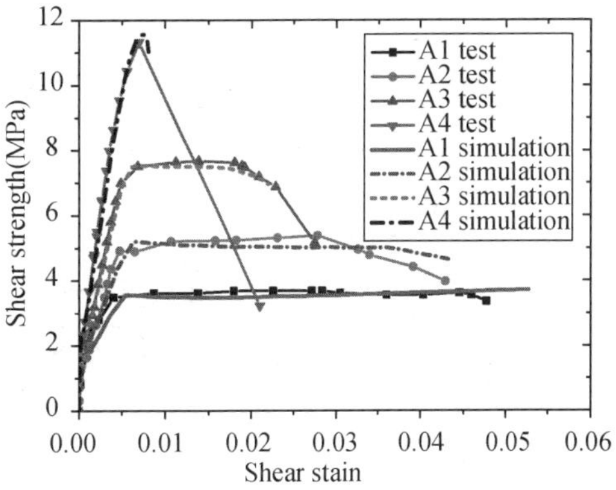

Fig. 14. (Color) Simulated and experimental shear stress-strain curves of the panels

Fig. 15 compares the results of the Wu-Li model(no compression-softening effect) and the proposed model. The results of Wu-Li damage model were higher than those of the proposed model, especially for the cases with high reinforcing ratios. The Wu-Li model cannot reflect the descending part of the panels (A2, A3, and A4) because it does not include the compression-softening.

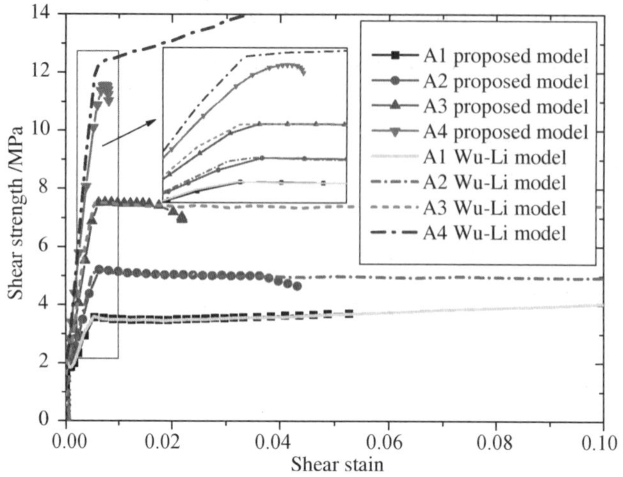

Fig. 15. (Color) Results of proposed model and Wu-Li model

4.6 Simply Supported RC Beam Tests

The simply supported RC beams tested by Bresler and Scordelis[47] were also considered as numerical examples. The test was designed to investigate the influence of different factors, e.g., shear reinforcement ratio, longitudinal reinforcement ratio, and shear span, on the behavior of the beams. Four series of three beams were tested, and each beam was heavily reinforced with bottom reinforcement but lightly reinforced with transverse steel to ensure shear-critical behavior. Fig. 16 shows the beam profiles and details of the cross sections. All the top longitudinal reinforcement used #4 bars, all the bottom longitudinal reinforcement used #9 bars, and the stirrups used #2 bars(Fig. 16).

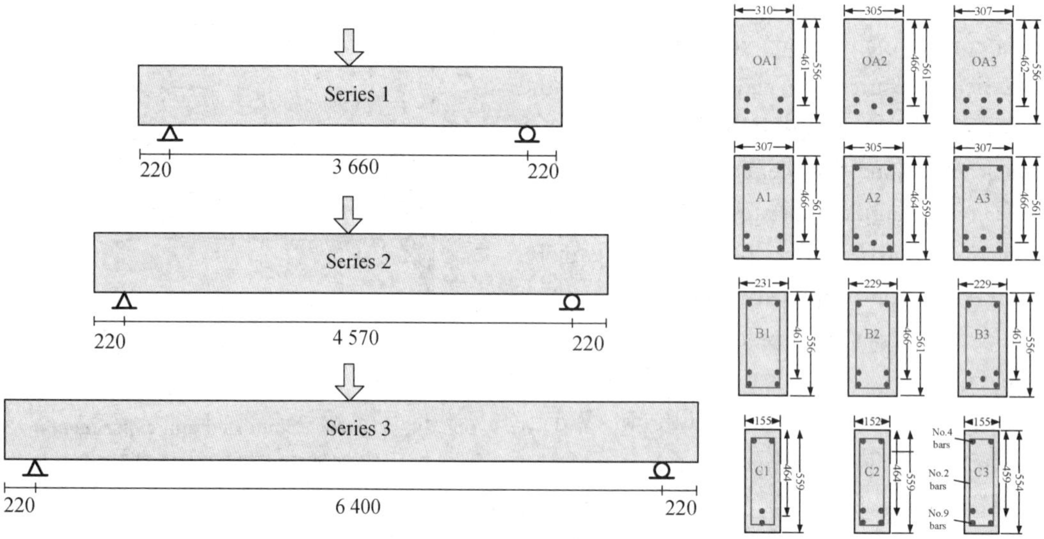

Fig. 16. (Color) Simply supported RC beam tests

In the simulation, a four-node plane-stress element was used to model the concrete and a two-node truss element was used to model the reinforcement. Reinforcement bars were embedded in concrete and there was no bond-slip. Tables 1 and 2 list the material properties of the tests, and Fig. 17 shows the finite meshes of the beams. The analysis terminated when convergence was not achieved, which also can be treated as the failure of the specimen.

Fig. 18 shows the load-deflection relationships at midspan computed by proposed model as well as the experimental results.

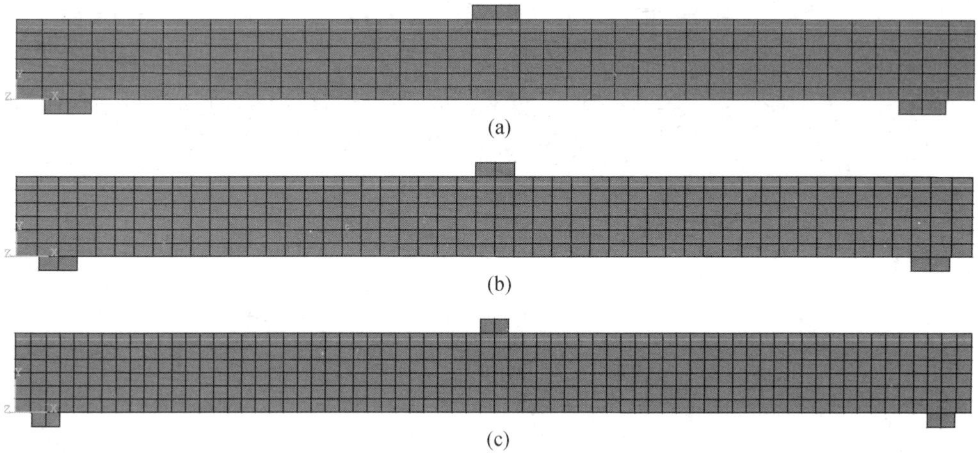

Fig. 17. (Color) Finite mesh of beams

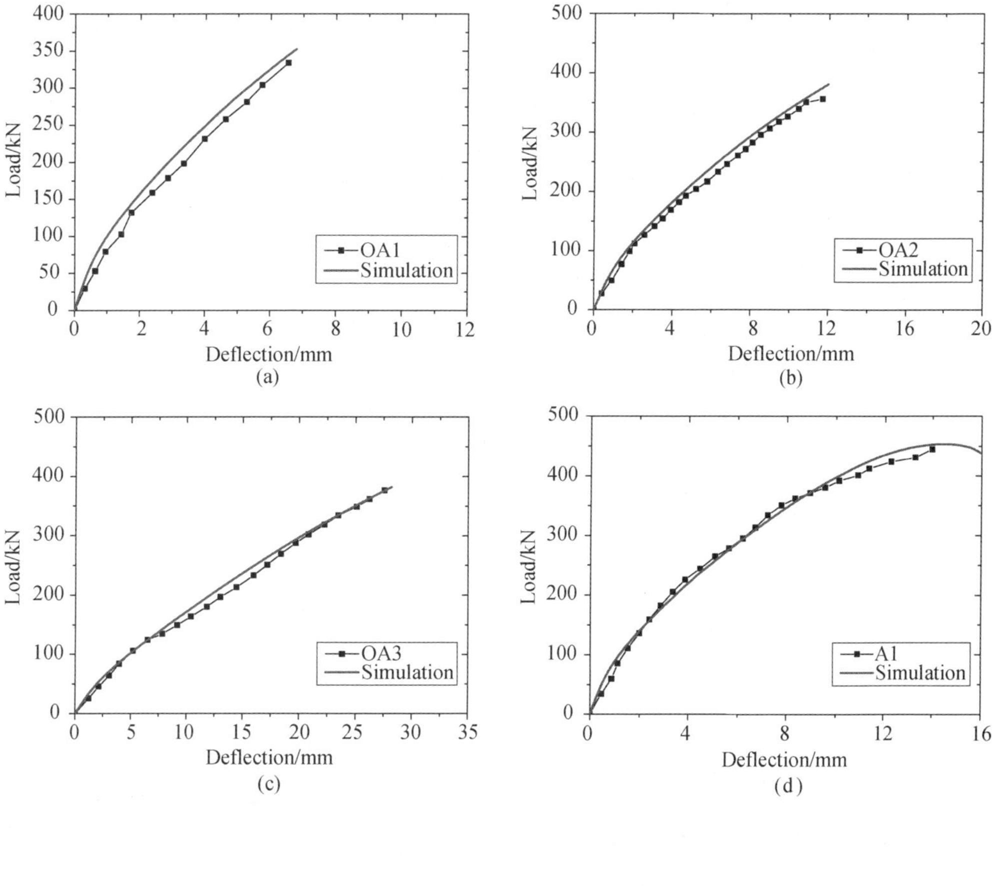

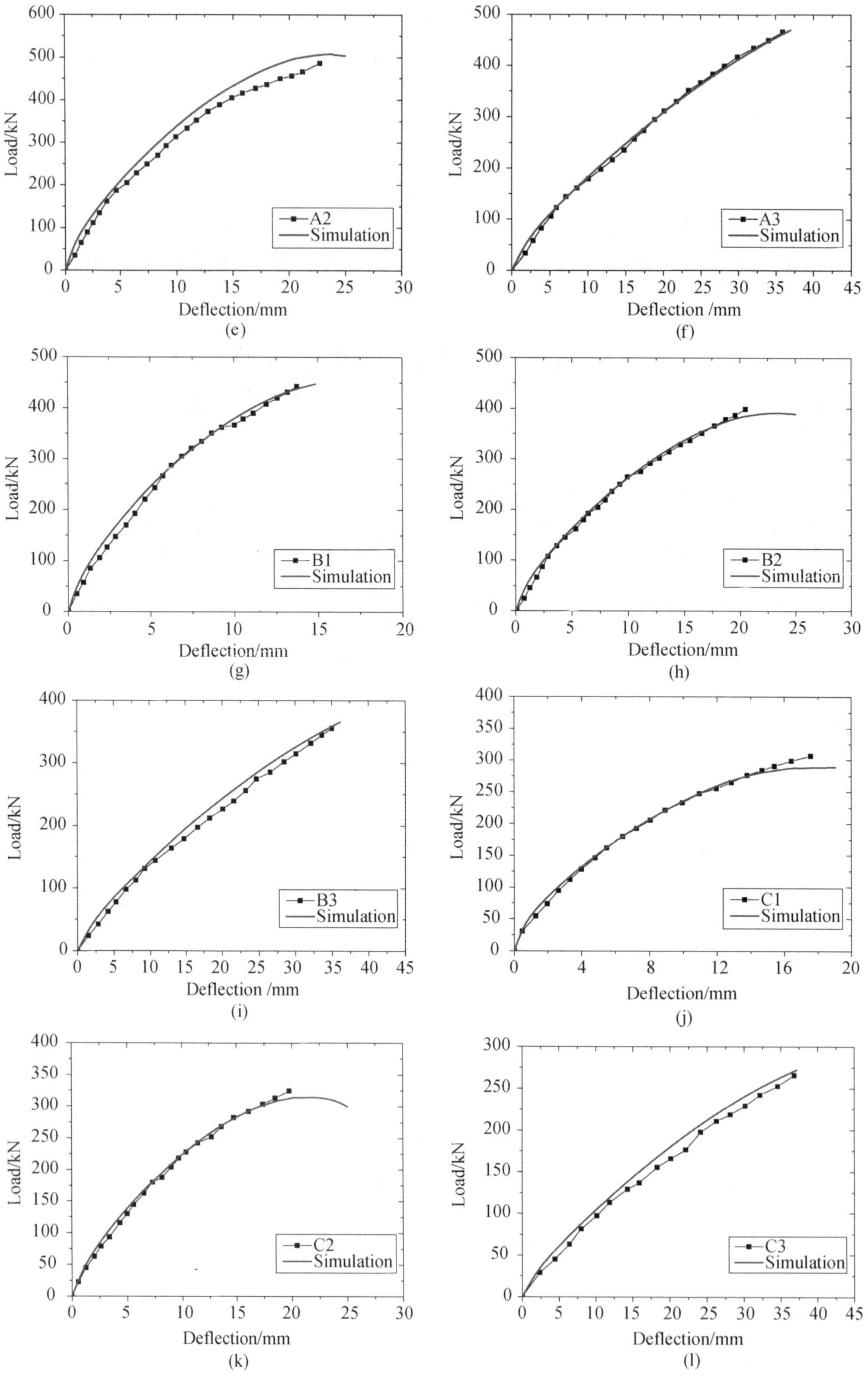

Fig. 18. (Color) Simulated and experimental load-deflection curves of beams

Excellent agreement was achieved for the beams in all cases. Table 3 compares the predicted ultimate loads and the corresponding midspan deflections with the experimental results. The ultimate load and the corresponding deflection were well reproduced by the model. The mean value of the ratio for the predicted peak load to experimental peak load was 1.01 and the corresponding coefficient of variation was 4.01%, and the mean value and coefficient of variation of the ratio for deflection were 1.04 and 3.36%, respectively.

Table 3. Comparison of Simulated and Experimental Results

Beam number	Capacity F_u		$F_{u,\text{model}}/F_{u,\text{test}}$	Deflection Δ_u		$\Delta_{u,\text{model}}/\Delta_{u,\text{test}}$
	Test/kN	Model/kN		Test/mm	Model/mm	
OA1	334	353	1.06	6.6	6.8	1.03
OA2	356	381	1.07	11.7	12.0	1.02
OA3	378	382	1.01	27.9	28.2	1.01
A1	467	454	0.97	14.2	14.3	1.00
A2	489	508	1.04	22.9	23.8	1.04
A3	467	470	1.01	35.8	37.0	1.03
B1	445	447	1.00	13.7	12.9	1.08
B2	400	391	0.98	20.8	23.1	1.11
B3	356	365	1.03	35.3	36.2	1.02
C1	311	289	0.93	17.8	19.1	1.07
C2	325	314	0.97	20.1	21.7	1.08
C3	269	272	1.01	36.8	37.1	1.01
Mean	—	—	1.01	—	—	1.04
Coefficient of variation/%	—	—	4.01	—	—	3.36

Fig. 19 compares the responses of Beams A1 and B1 predicted by the Wu-Li model and the proposed model. Similar conclusions as in the panel tests can be drawn, i.e., that the responses may be overestimated if compression-softening is not considered.

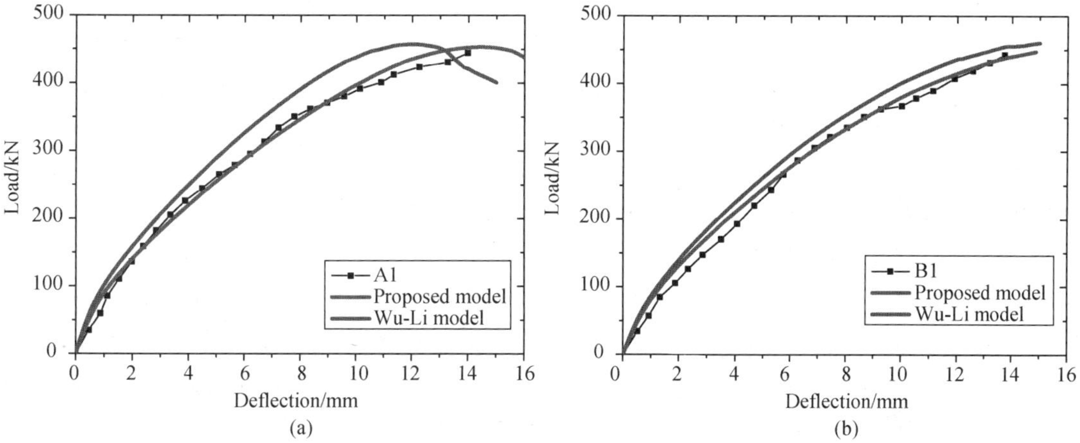

Fig. 19. (Color) Comparison of results of proposed model and Wu-Li model

4.7 RC Shear Wall Test

To demonstrate the ability of the proposed model for cyclic analysis of RC structures, specimen RW2 tested under cyclic loading by Thomsen and Wallace[48] was analyzed. The wall was in 1 : 4 scale, and Fig. 20 shows the geometrical and reinforcing details of the specimen. The compressive strength of concrete used in the specimen was 42.8 MPa on average, and the tensile and compressive fracture energies were set as 150 and 50 000 N/m, respectively. Table 4 lists the material properties for reinforcement. The experimental procedure included two stages: first, a constant axial force of $0.07A_g f_c'$ was applied on the top of the wall, and then the cyclic lateral load was imposed on the top of the wall.

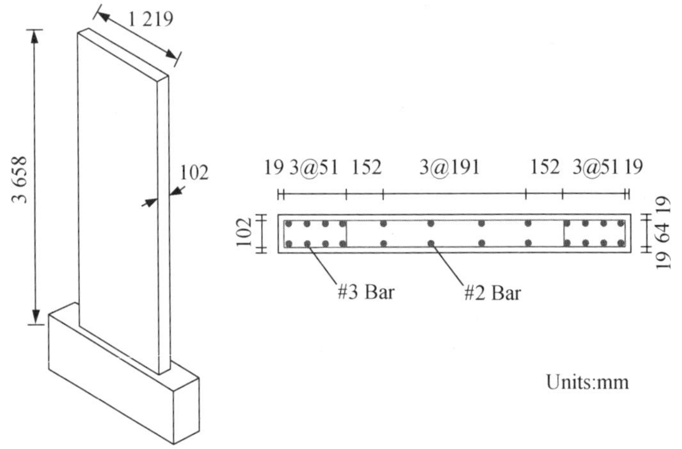

Fig. 20. (Color) RC shear wall test

Table 4. Properties of Reinforcement Bars in RW2

Bar type	Diameter d/mm	Yield strength f_y/MPa	Elastic modulus E_s/MPa
No.2	6.35	414	210 000
No.3	9.53	448	210 000
Smooth wire	4.75	448	210 000

Fig. 21 shows the established finite-element models; four-node plane-stress elements were used for the concrete and two-node truss elements were used for reinforcements without bond-slip. Fig. 22(a) compares the simulated top displacement versus lateral load curve of the specimen with the experimental result. The numerical result matched the experimental result very well. The peak strength, the residual displacement, and the energy dissipation were all reproduced by the proposed model. The pinching effect of the hysteretic loops can also be reflected. Furthermore, Fig. 22(b) compares the numerical results of the previous Wu-Li model and the proposed model; overestimation of the peak strength and energy dissipation were observed again, which indicates that the compression-softening effect of RC should be considered when modeling shear behavior of RC structures.

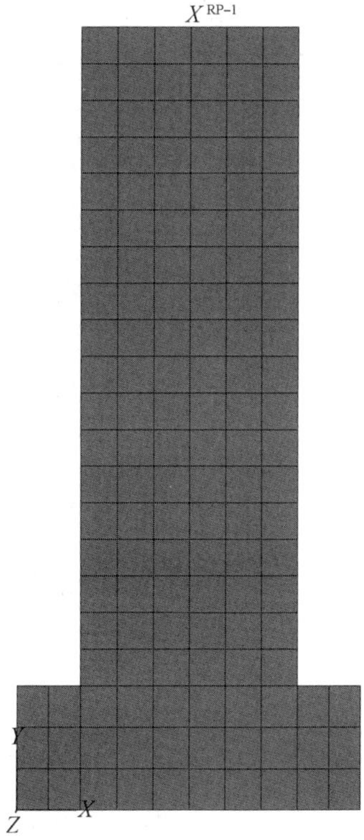

Fig. 21. (Color) Finite mesh of shear wall

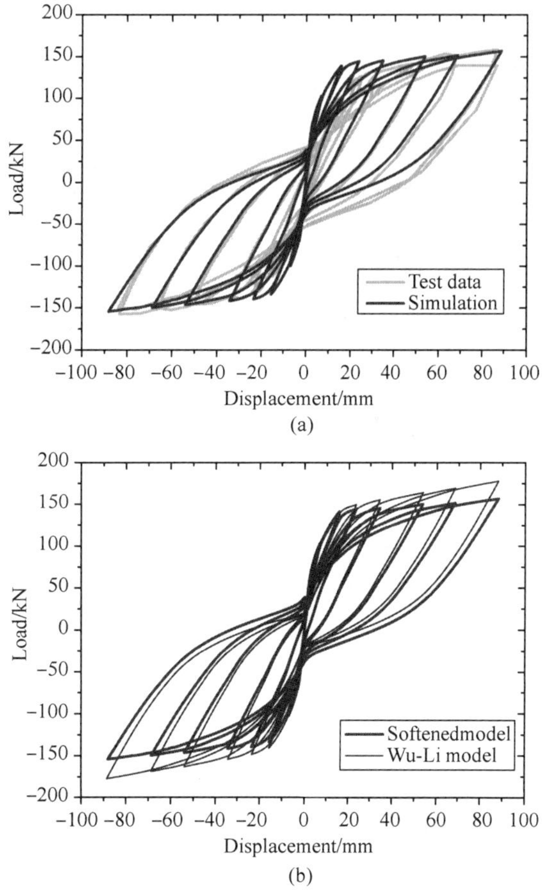

Fig. 22. (Color) Simulated and experimental load-displacement curves of RW2: (a) comparison of results of experiment and proposed model; (b) comparison of results of proposed model and Wu-Li model

Fig. 23 plots the vertical strain distributions of the base section at different drift levels obtained by the numerical method against the experimental results. The numerical results agreed with the experimental results very well. Fig. 24 shows the tensile damage contours of the specimen corresponding to different drift levels, which indicate the cracking patterns of the walls. From the figures, the damage features and the accumulation of damage in different directions can

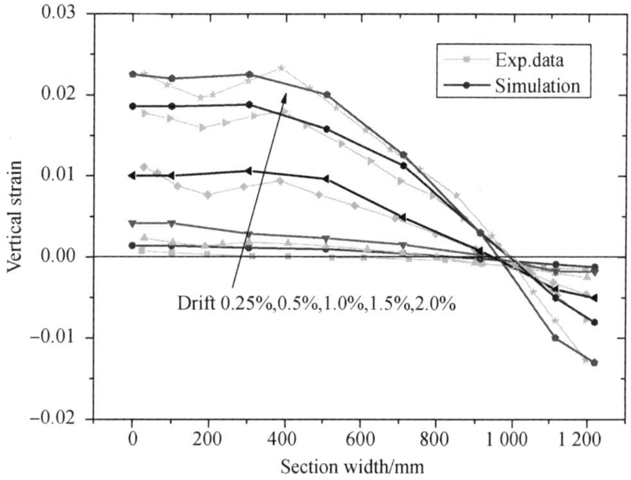

Fig. 23. (Color) Vertical strain distributions at different drift levels along base section width of RW2

be obtained. These results indicate that the proposed models and methods can not only predict realistic global responses, but can reproduce the local responses of the shear wall.

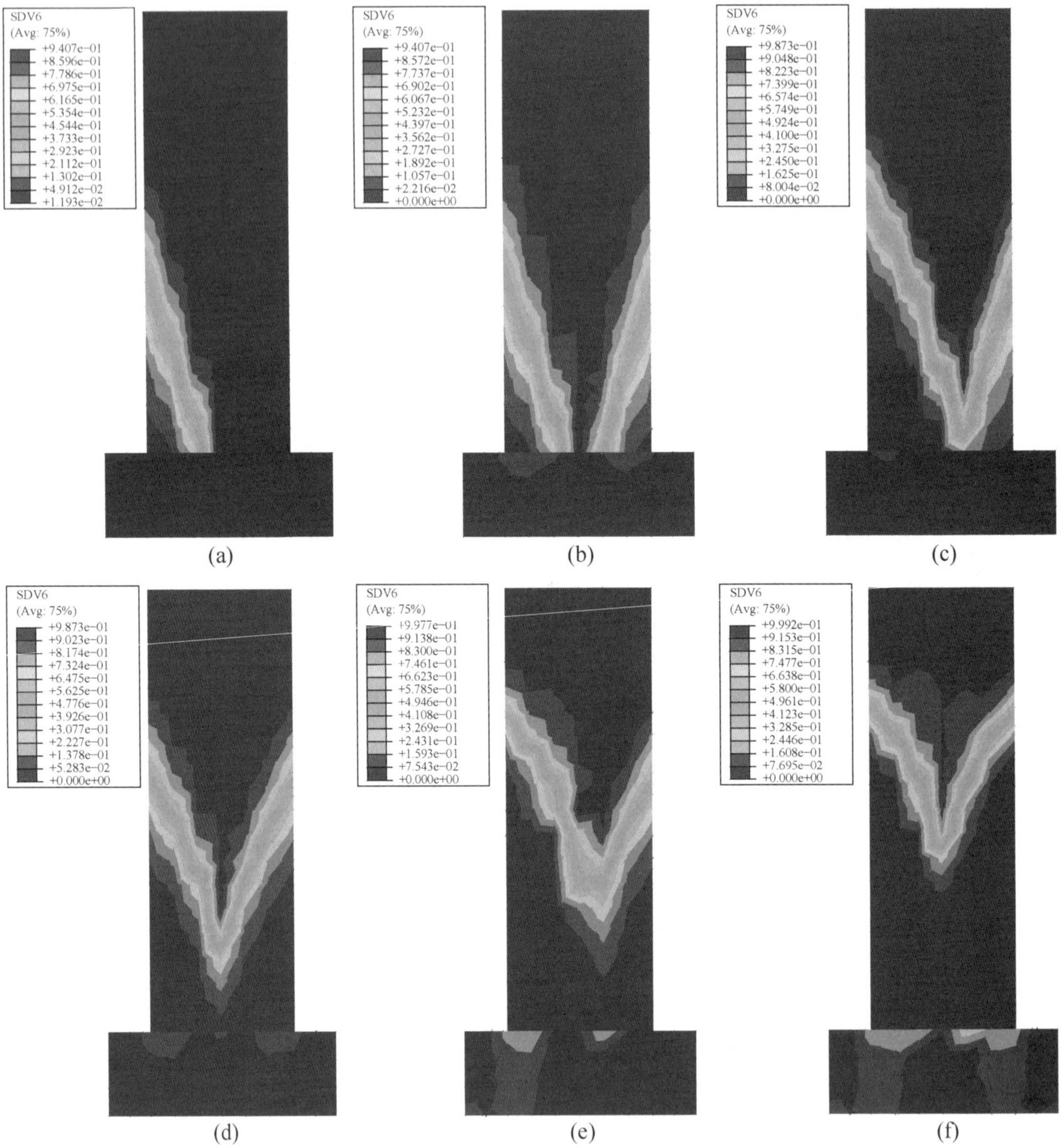

Fig. 24. **(Color) Tensile damage contours of RW2 at different drift levels**

5 Conclusion

This paper proposed a new damage-plasticity model to consider the compression-softening effect in MCFT and STM of cracked RC. The model was validated through several material tests and applied to simulate some typical members in RC structures. The following conclusions can be drawn:

(1) The properties of RC may be very different from that of plain concrete, especially when subjected to shear. The compressive strength in the compressive principal direction is reduced due to the presence of significant tensile strain in the orthogonal direction, which can be defined as the compression-softening effect and is a critical factor that affects the accuracy of the prediction of shear behavior.

(2) This paper introduced a damage-plasticity model for concrete as the basic framework, and the compressive damage evolution function was reconstructed through a softening coefficient developed by other phenomenological models, e.g., the MCFT and STM, to account for the compression-softening effect. Consequently, damage, plasticity, and compression softening were organized in a unified framework.

(3) Because of the complexity and asymmetry of the proposed model, the consistent tangent operator cannot be derived for the full NR method, and the paper adopted the quasi-Newton method (a secant stiffness algorithm) to solve the nonlinear finite-element equations. Although more iterations in the structural level may be needed compared with the full NR method, the present scheme is much more robust.

(4) Numerical simulations of a series of material tests, RC shear panels, RC simply supported beams, and RC shear wall indicated that the previous damage model may not reflect the properties of cracked RC, whereas the proposed model can well reproduce realistic responses of RC structures, even for shear behaviors. Moreover, the model can obtain not only the load-displacement curves but also the distribution of damages.

Acknowledgments

Financial supports from the National Natural Science Foundation of China (Grant Nos. 51708106, 51678439, and 51538010), the Natural Science Foundation of Jiangsu Province (Grant No. BK20170680), the National Key Research and Development Program of China (No. 2016YFC0701400), and the Fundamental Research Funds for the Central Universities (No. 2242017k30002) are greatly appreciated.

References

[1] VECCHIO F J, COLLINS M P. Compression response of cracked reinforced concrete[J]. Journal of Structural Engineering, 1993, 119(12): 3590-3610.

[2] SELBY R G, VECCHIO F J. A constitutive model for analysis of reinforced concrete solids[J]. Canadian Journal of Civil Engineering, 1997, 24(3): 460-470.

[3] BENTZ E C, VECCHIO F J, Collins M P. Simplified modified compression field theory for calculating shear strength of reinforced concrete elements[J]. ACI Structural Journal, 2006, 103(4): 614-624.

[4] HSU T T C. Softened truss model theory for shear and torsion[J]. Structural Journal, 1988, 85(6): 624-635.

[5] BELARBI A, HSU T T C. Constitutive laws of softened concrete in biaxial tension compression[J]. Structural Journal, 1995, 92(5): 562-573.

[6] HSU T T C, ZHU R R H. Softened membrane model for reinforced concrete elements in shear[J]. Structural Journal, 2002, 99(4): 460-469.

[7] WANG T, HSU T T C. Nonlinear finite element analysis of concrete structures using new constitutive models[J]. Computers & Structures, 2001, 79(32): 2781-2791.

[8] ROTS J G, BLAAUWENDRAAD J. Crack models for concrete, discrete or smeared? Fixed, multi-directional or rotating? [J]. Heron, 1989, 34 (1): 1-59.

[9] LOTFI H R, SHING P B. An appraisal of smeared crack models for masonry shear wall analysis[J]. Computers & Structures, 1991, 41(3): 413-425.

[10] MAEKAWA K, OKAMURA H, PIMANMAS A. Non-linear mechanics of reinforced concrete[M]. CRC Press, 2003.

[11] JU J W. On energy-based coupled elastoplastic damage theories: constitutive modeling and computational aspects[J]. International Journal of Solids and Structures, 1989, 25(7): 803-833.

[12] FARIA R, OLIVER J, CERVERA M. A strain-based plastic viscous-damage model for massive concrete structures[J]. International Journal of Solids and Structures, 1998, 35(14): 1533-1558.

[13] LEE J, FENVES G L. Plastic-damage model for cyclic loading of concrete structures[J]. Journal of Engineering Mechanics, 1998, 124(8): 892-900.

[14] WU J Y, LI J, FARIA R. An energy release rate-based plastic-damage model for concrete[J]. International Journal of Solids and Structures, 2006, 43(3-4): 583-612.

[15] TESSER L, FILIPPOU F C, TALLEDO D A, et al. Nonlinear analysis of R/C panels by a two parameter concrete damage model[C]//III ECCOMAS thematic conference on computational methods in structural dynamics and earthquake engineering. 2011: 25-28.

[16] SUIDAN M, SCHNOBRICH W C. Finite element analysis of reinforced concrete[J]. Journal of the Structural Division, 1973, 99(10): 2109-2122.

[17] CERVENKA V. Inelastic finite element analysis of reinforced concrete panels under in-plane loads[M]. University of Colorado at Boulder, 1970.

[18] COPE R J. Modeling of reinforced concrete behavior for finite element analysis of bridge slabs[J]. Numerical Method for Nonlinear Problems, 1980, 1: 457-470.

[19] GUPTA A K, AKBAR H. Cracking in reinforced concrete analysis[J]. Journal of structural Engineering, 1984, 110(8): 1735-1746.

[20] SARITAS A. Mixed formulation frame element for shear critical steel and reinforced concrete members[D]. University of California, Berkeley, 2006.

[21] STEVENS N J, UZUMERI S M, WILL G T. Constitutive model for reinforced concrete finite element analysis[J]. Structural Journal, 1991, 88(1): 49-59.

[22] LEMAITRE J. Evaluation of dissipation and damage in metals[C]//Proc. ICM Kyoto. 1971, 1.

[23] LI J, REN X D. Stochastic damage model for concrete based on energy equivalent strain[J]. International Journal of Solids and Structures, 2009, 46(11-12): 2407-2419.

[24] REN X D, ZENG S, LI J. A rate-dependent stochastic damage-plasticity model for quasi-brittle materials[J]. Computational Mechanics, 2015, 55(2): 267-285.

[25] LIANG J, REN X D, LI J. A competitive mechanism driven damage-plasticity model for fatigue behavior of concrete[J]. International Journal of Damage Mechanics, 2016, 25(3): 377-399.

[26] WU J Y. Damage energy release rate-based elastoplastic damageconstitutive model for concrete and its application to nonlinear analysisof structures[D]. Ph.D. thesis, Tongji Univ., Shanghai, China, 2004.

[27] FENG D, REN X, LI J. Stochastic damage hysteretic model for concrete based on micromechanical

approach[J]. International Journal of Non-Linear Mechanics, 2016, 83: 15-25.

[28] FENG D C, LI J. Stochastic nonlinear behavior of reinforced concrete frames. Ⅱ: Numerical simulation[J]. Journal of Structural Engineering, 2016, 142(3): 04015163.

[29] FENG D, KOLAY C, RICLES J M, et al. Collapse simulation of reinforced concrete frame structures[J]. The Structural Design of Tall and Special Buildings, 2016, 25(12): 578-601.

[29] SARITAS A, FILIPPOU F C. Analysis of RC walls with a mixed formulation frame finite element[J]. Computers and Concrete, 2013, 12(4): 519-536.

[30] BERTO L, SAETTA A, SCOTTA R, et al. A coupled damage model for RC structures: Proposal for a frost deterioration model and enhancement of mixed tension domain[J]. Construction and Building Materials, 2014, 65: 310-320.

[31] KUPFER H, HILSDORF H K, RUSCH H. Behavior of concrete under biaxial stresses[C]//Journal Proceedings. 1969, 66(8): 656-666.

[32] SARITAS A, FILIPPOU F C. Numerical integration of a class of 3d plastic-damage concrete models and condensation of 3d stress-strain relations for use in beam finite elements[J]. Engineering Structures, 2009, 31(10): 2327-2336.

[33] VECCHIO F J, COLLINS M P. The modified compression-field theory for reinforced concrete elements subjected to shear[J]. ACI J., 1986, 83(2): 219-231.

[34] ROBINSON J R. Essais de tractioncompression sur modeles d'ame de poutre en beton arme[J]. Institute de Recherches appliquees du Beton (IRABA), 1968.

[35] VECCHIO F. Stress-strain characteristics of reinforced concrete in pure shear, final report[C]// IABSE colloquium on advanced mechanics of reinforced concrete. International Association for Bridge and Structural Engineering, 1981: 211-225.

[36] PANG X B D, HSU T T C. Fixed angle softened truss model for reinforced concrete[J]. Structural Journal, 1996, 93(2): 196-208.

[37] TENG M. Plastic-damage model of lightweight concrete and normal weight concrete[D]. Ph.D. thesis, National Univ. of Singapore,Singapore, 2009.

[38] FENG D C, WU G, SUN Z Y, et al. A flexure-shear Timoshenko fiber beam element based on softened damage-plasticity model[J]. Engineering Structures, 2017, 140: 483-497.

[39] VECCHIO F J. Nonlinear finite element analysis of reinforced concrete membranes[J]. ACI Structural Journal, 1989, 86(1): 26-35.

[40] VECCHIO F J. Towards cyclic load modeling of reinforced concrete[J]. ACI Structural Journal, 1999, 96: 193-202.

[41] DENNIS JR J E, SCHNABEL R B. Numerical methods for unconstrained optimization and nonlinear equations[M]. Society for Industrial and Applied Mathematics, 1996.

[42] HIBBITT, KARLSSON, SORENSEN. ABAQUS/Standard User's Manual: Version 6.2. Vol. 1[M]. Hibbitt, Karlsson & Sorensen, 2001.

[43] ZENG S J. Dynamic experimental research and stochastic damage constitutive model for concrete[D]. Tongji Univ., Shanghai, China, 2012.

[44] KARSAN I D, JIRSA J O. Behavior of concrete under compressive loadings[J]. Journal of the Structural Division, 1969, 95(12): 2543-2564.

[45] HSU T T C, MO Y L. Unified theory of concrete structures[M]. John Wiley & Sons, 2010.

[46] PANG X B D, HSU T T C. Behavior of reinforced concrete membrane elements in shear[J]. Structural Journal, 1995, 92(6): 665-679.

[47] BRESLER B, SCORDELIS A C. Shear strength of reinforced concrete beams [C]//Journal Proceedings. 1963, 60(1): 51-74.

[48] THOMSEN IV J H, WALLACE J W. Displacement-based design of slender reinforced concrete structural walls—experimental verification[J]. Journal of Structural Engineering, 2004, 130(4): 618-630.

Part 3

Numerical Methods for Damage Mechanics

Semi-implicit algorithm for elastoplastic damage models involving energy integration①

Ji Zhang[1, 2], **Jie Li**[2]

1 *School of Civil Engineering and Mechanics, Huazhong University of Science and Technology, Wuhan 430074, China*

2 *School of Civil Engineering, Tongji University, Shanghai 200092, China*

Abstract This study aims to develop a semi-implicit constitutive integration algorithm for a class of elastoplastic damage models where calculation of damage energy release rates involves integration of free energy. The constitutive equations with energy integration are split into the elastic predictor, plastic corrector, and damage corrector. The plastic corrector is solved with an improved format of the semi-implicit spectral return mapping, which is characterized by constant flow direction and plastic moduli calculated at initial yield, enforcement of consistency at the end, and coordinate-independent formulation with an orthogonally similar stress tensor. The tangent stiffness consistent with the updating algorithm is derived. The algorithm is implemented with a recently proposed elastoplastic damage model for concrete, and several typical mechanical tests of reinforced concrete components are simulated. The present semi-implicit algorithm proves to achieve a balance between accuracy, stability, and efficiency compared with the implicit and explicit algorithms and calculate free energy accurately with small time steps.

1 Introduction

Due to its accuracy, fast convergence, and unconditional stability, the (fully) implicit backward Euler return mapping method[1] has become the most favorable one for numerically integrating inelastic constitutive models. In this method, quite large time steps can be used, hence far fewer global (finite element) iterations.

However, sufficiently small time steps are warranted to accurately calculate free energy for a class of elastoplastic damage models. In these constitutive theories, a damage energy release rate is derived from a Helmholtz free energy potential. Unlike the elastic free energy (i.e., strain energy), which is determined from the current state, the plastic free energy (a part of the plastic work[2]) has to be integrated along the plastic loading history[3-4].

With very small time steps and frequent iterations, the implicit return mapping method is too expensive computationally, especially for quasi-brittle material models, which usually employ complex hardening laws and flow rules. The extremely complex partial derivatives

① Originally published in *Advances in Materials Science and Engineering*, 2016(1): 1-9.

of the plastic moduli and flow direction with respect to the stress and hardening variables are required for the process of return mapping, and the inversion of certain matrices containing these derivatives needs to be carried out to construct the consistent tangent stiffness.

Return mapping can be made much simpler with the so-called semi-implicit method[5], where some quantities, including the plastic moduli and flow direction, are calculated at the beginning of a time step and remained unchanged throughout the process of return mapping, and some, including the plasticity consistency parameter, are determined at the end of the time step. For a single small time step, this approach is much easier to implement and more efficient than the implicit method; at the same time, the stress is return-mapped effectively onto the subsequent yield surface by enforcing the consistency condition, and the problem of drift in the explicit forward Euler method is overcome. Thus, the semi-implicit return mapping method is an ideal choice for the abovementioned plastic free energy integration.

It is the purpose of the present study to develop a semi-implicit constitutive integration algorithm for a class of elastoplastic damage models where calculation of damage energy release rates involves integration of free energy. It is intended that, with small time steps, free energy can be integrated accurately, and the numerical algorithm remains efficient and stable.

2 Constitutive Framework

The class of elastoplastic damage models under discussion are mainly for quasi-brittle materials, which exhibit distinct behavior in tension and compression. The general constitutive framework is presented as follows.

2.1 Overall State Description

Two scalars d^+ and d^-, the tensile and compressive damage variables, are used to describe a damage state as

$$\boldsymbol{\sigma} = (1-d^+)\bar{\boldsymbol{\sigma}}^+ + (1-d^-)\bar{\boldsymbol{\sigma}}^-, \tag{1}$$

where $\boldsymbol{\sigma}$ is the nominal (Cauchy) stress and $\bar{\boldsymbol{\sigma}}^\pm$ is the tensile/compressive part of the effective stress $\bar{\boldsymbol{\sigma}}$

$$\bar{\boldsymbol{\sigma}}^\pm = \pm\sum_i \langle \pm\bar{\sigma}_i \rangle \underset{\rightarrow i}{n} \otimes \underset{\rightarrow i}{n}, \tag{2}$$

in which $\bar{\sigma}_i$ are the principal effective stresses, $\underset{\rightarrow i}{n}$ is the orthonormal basis along the principal directions, and $\langle x \rangle = (|x| + x)/2$.

Elastoplasticity is formulated in the effective stress space

$$\bar{\boldsymbol{\sigma}} = \bar{\mathbb{C}} : (\boldsymbol{\varepsilon} - \boldsymbol{\varepsilon}^{\mathrm{p}}), \tag{3}$$

where $\bar{\mathbb{C}}$ is the effective elastic stiffness, $\boldsymbol{\varepsilon}$ is the strain, and $\boldsymbol{\varepsilon}^{\mathrm{p}}$ is the plastic strain.

2.2 Damage Evolution: Drive Chain

The tensile/compressive Helmholtz free energy $\psi^\pm$ is split into its elastic part $\psi^{e\pm}$ and plastic part $\psi^{p\pm}$, mapped into their effective values $\bar{\psi}^{e\pm}$ and $\bar{\psi}^{p\pm}$, and calculated as

$$\begin{aligned}\psi^{\pm} &= \psi^{e\pm} + \psi^{p\pm} = (1-d^{\pm})(\bar{\psi}^{e\pm} + \bar{\psi}^{p\pm}) \\ &= (1-d^{\pm})\left(\frac{1}{2}\bar{\boldsymbol{\sigma}}^{\pm} : \boldsymbol{\varepsilon}^{e} + \int_0^{\kappa^{\pm}} \bar{Z}^{\pm}\mathrm{d}\kappa^{\pm}\right),\end{aligned} \tag{4}$$

where $\boldsymbol{\varepsilon}^{e}$ is the elastic strain, $\bar{Z}^{\pm}$ is the tensile/compressive effective plastic energy storage rate[4], and $\kappa^{\pm}$ is the tensile/compressive hardening variable.

The tensile/compressive damage energy release rate is defined and calculated as

$$Y^{\pm} = -\frac{\partial \psi^{\pm}}{\partial d^{\pm}} = \frac{1}{2}\bar{\boldsymbol{\sigma}}^{\pm} : \boldsymbol{\varepsilon}^{e} + \int_0^{\kappa^{\pm}} \bar{Z}^{\pm}\mathrm{d}\kappa^{\pm}. \tag{5}$$

The tensile/compressive damage threshold is defined as

$$R^{\pm} = \max\{R_0^{\pm},\ \max_{\tau\in[0,\,t]} Y_{\tau}^{\pm}\}, \tag{6}$$

where $R_0^{\pm}$ is the initial tensile/compressive damage threshold, τ is any instant throughout the loading history, and t is the current instant.

Damage evolution is formulated as a function of $R_0^{\pm}$ and $R^{\pm}$

$$d^{\pm} = G^{\pm}(R_0^{\pm},\ R^{\pm}). \tag{7}$$

2.3 Plastic Evolution: Intertwined Components

Two scalars κ^{+} and κ^{-}, the tensile and compressive hardening variables, are used to characterize plasticity, which includes the yield criterion,

$$f^{p}(\bar{\boldsymbol{\sigma}},\ \kappa^{+},\ \kappa^{-}) = 0, \tag{8}$$

and the flow rule

$$\dot{\boldsymbol{\varepsilon}}^{p} = \dot{\lambda}\frac{\partial g^{p}(\bar{\boldsymbol{\sigma}},\ \kappa^{+},\ \kappa^{-})}{\partial \bar{\boldsymbol{\sigma}}}, \tag{9}$$

where λ is the plasticity consistency parameter and g^{p} is the plastic potential, and the evolution law for $\kappa^{\pm}$ is as follows

$$\dot{\kappa}^{\pm} = \dot{\lambda}h^{\pm}(\bar{\boldsymbol{\sigma}},\ \kappa^{+},\ \kappa^{-}), \tag{10}$$

where $h^{\pm}$ is the tensile/compressive plastic modulus.

Finally, with the Karush-Kuhn-Tucker loading/unloading conditions,

$$\begin{aligned} &f^{p} \leqslant 0, \\ &\dot{\lambda} \geqslant 0, \\ &\dot{\lambda} f^{p} = 0, \end{aligned} \tag{11}$$

and the consistency condition,

$$\dot{\lambda}\dot{f}^{p} = 0, \tag{12}$$

the plastic relation is complete.

3 Operator Split

In accordance with the concept and methods of product formulas and operator split[3, 6], the constitutive equations presented in the previous section are replaced here with the elastic, plastic, and damage parts in Table 1, which can serve as a mathematical foundation for developing first-order accurate algorithms.

Table 1. Split of rate constitutive equations

Total	=Elastic part	+Plastic part	+Damage part
$\dot{\boldsymbol{\varepsilon}} = \nabla^s \underset{\rightarrow}{\dot{u}} = \dfrac{\underset{\rightarrow}{\dot{u}} \otimes \underset{\rightarrow}{\nabla} + \underset{\rightarrow}{\nabla} \otimes \underset{\rightarrow}{\dot{u}}}{2}$	$\dot{\boldsymbol{\varepsilon}} = \nabla^s \underset{\rightarrow}{\dot{u}}$		
$\dot{\boldsymbol{\varepsilon}}^{p} = \dot{\lambda} \dfrac{\partial g^{p}}{\partial \bar{\boldsymbol{\sigma}}}$		$\dot{\boldsymbol{\varepsilon}}^{p} = \dot{\lambda} \dfrac{\partial g^{p}}{\partial \bar{\boldsymbol{\sigma}}}$	
$\dot{\kappa}^{\pm} = \dot{\lambda} h^{\pm}$		$\dot{\kappa}^{\pm} = \dot{\lambda} h^{\pm}$	
$\dot{\bar{\boldsymbol{\sigma}}} = \bar{\mathbb{C}} : (\dot{\boldsymbol{\varepsilon}} - \dot{\boldsymbol{\varepsilon}}^{p})$	$\dot{\bar{\boldsymbol{\sigma}}} = \bar{\mathbb{C}} : \dot{\boldsymbol{\varepsilon}}$	$\dot{\bar{\boldsymbol{\sigma}}} = -\bar{\mathbb{C}} : \dot{\boldsymbol{\varepsilon}}^{p}$	
$\dot{\bar{Z}}^{\pm} = \dot{\bar{\boldsymbol{\sigma}}}^{\pm} \dfrac{\partial \bar{Z}^{\pm}}{\partial \bar{\boldsymbol{\sigma}}}$		$\dot{\bar{Z}}^{\pm} = \left(\dot{\bar{\boldsymbol{\sigma}}}^{\pm} \Big\vert_{\text{elastic part}} + \dot{\bar{\boldsymbol{\sigma}}}^{\pm} \right) \dfrac{\partial \bar{Z}^{\pm}}{\partial \bar{\boldsymbol{\sigma}}}$	
$\dot{\bar{\psi}}^{e\pm} = \bar{\boldsymbol{\sigma}}^{\pm} : (\dot{\boldsymbol{\varepsilon}} - \dot{\boldsymbol{\varepsilon}}^{p})$	$\dot{\bar{\psi}}^{e\pm} = \bar{\boldsymbol{\sigma}}^{\pm} : \dot{\boldsymbol{\varepsilon}}$	$\dot{\bar{\psi}}^{e\pm} = -\bar{\boldsymbol{\sigma}}^{\pm} : \dot{\boldsymbol{\varepsilon}}^{p}$	
$\dot{\bar{\psi}}^{p\pm} = \bar{Z}^{\pm} \dot{\kappa}^{\pm}$		$\dot{\bar{\psi}}^{p\pm} = \bar{Z}^{\pm} \dot{\kappa}^{\pm}$	
$\dot{Y}^{\pm} = \dot{\bar{\psi}}^{e\pm} + \dot{\bar{\psi}}^{p\pm}$			$\dot{Y}^{\pm} = \dot{\bar{\psi}}^{e\pm} \vert_{\text{elastic part}} + \dot{\bar{\psi}}^{e\pm} \vert_{\text{plastic part}} + \dot{\bar{\psi}}^{p\pm} \vert_{\text{plastic part}}$
$\dot{R}^{\pm} = \begin{cases} \dot{Y}^{\pm}, & Y^{\pm} = R^{\pm},\ \dot{Y}^{\pm} > 0 \\ 0, & Y^{\pm} < R^{\pm} \text{ or } \dot{Y}^{\pm} \leqslant 0 \end{cases}$			$\dot{R}^{\pm} = \begin{cases} \dot{Y}^{\pm}, & Y^{\pm} = R^{\pm},\ \dot{Y}^{\pm} > 0 \\ 0, & Y^{\pm} < R^{\pm} \text{ or } \dot{Y}^{\pm} \leqslant 0 \end{cases}$
$\dot{d}^{\pm} = \dot{R}^{\pm} \dfrac{\partial G^{\pm}}{\partial R^{\pm}}$			$\dot{d}^{\pm} = \dot{R}^{\pm} \dfrac{\partial G^{\pm}}{\partial R^{\pm}}$
$\dot{\boldsymbol{\sigma}} = (1-d^{+})\dot{\bar{\boldsymbol{\sigma}}}^{+} + (1-d^{-})\dot{\bar{\boldsymbol{\sigma}}}^{-} - \dot{d}^{+}\bar{\boldsymbol{\sigma}}^{+} - \dot{d}^{-}\bar{\boldsymbol{\sigma}}^{-}$	$\dot{\boldsymbol{\sigma}} = (1-d^{+})\dot{\bar{\boldsymbol{\sigma}}}^{+} + (1-d^{-})\dot{\bar{\boldsymbol{\sigma}}}^{-}$	$\dot{\boldsymbol{\sigma}} = (1-d^{+})\dot{\bar{\boldsymbol{\sigma}}}^{+} + (1-d^{-})\dot{\bar{\boldsymbol{\sigma}}}^{-}$	$\dot{\boldsymbol{\sigma}} = -\dot{d}^{+}\bar{\boldsymbol{\sigma}}^{+} - \dot{d}^{-}\bar{\boldsymbol{\sigma}}^{-}$

These three parts are taken as split problems in a series: the first problem provides the initial condition for the second one, the second one for the third one, and the third one for the first one in the next time step; and all the split problems add up to the original elastoplastic damage problem. In this manner, the elastoplastic damage constitutive equations can be integrated incrementally over a sequence of time steps $[t_n,\ t_{n+1}]$, $n=0,\ 1,\ 2\ \cdots$

3.1 Elastic Predictor

The first problem, usually referred to as the elastic predictor, is simply the update of the strain

$$\boldsymbol{\varepsilon}_{n+1} = \boldsymbol{\varepsilon}_n + \nabla^s (\Delta \underset{\rightarrow}{u}_{n+1}) \tag{13}$$

and the effective stress

$$\bar{\boldsymbol{\sigma}}^{\text{trial}}_{n+1} = \bar{\mathbb{C}} : (\boldsymbol{\varepsilon}_{n+1} - \boldsymbol{\varepsilon}^{\text{p}}_n) \tag{14}$$

assuming purely elastic behavior.

3.2 Plastic Corrector

The second problem, the plastic corrector, is to relax the plastic strain increment $\Delta\boldsymbol{\varepsilon}^{\text{p}}_{n+1}$ and direct the excursive effective stress $\bar{\boldsymbol{\sigma}}^{\text{trial}}_{n+1}$ back to the updated yield surface:

$$\bar{\boldsymbol{\sigma}}_{n+1} = \bar{\boldsymbol{\sigma}}^{\text{trial}}_{n+1} - \bar{\mathbb{C}} : \Delta\boldsymbol{\varepsilon}^{\text{p}}_{n+1} \tag{15}$$

if $\bar{\boldsymbol{\sigma}}^{\text{trial}}_{n+1}$ violates the yield criterion

$$f^{\text{p}}(\bar{\boldsymbol{\sigma}}^{\text{trial}}_{n+1}, \kappa^+_n, \kappa^-_n) > 0. \tag{16}$$

Various types of algorithms have been developed for this problem, including the explicit forward Euler method, implicit backward Euler method, and semi-implicit method, which differ from one another in their specific (iterative) updating schemes for Eq. (15). In a broader sense, all these methods can be regarded as "return mapping"; many researchers mean only those that accurately enforce consistency by this popular term.

3.3 Damage Corrector

The third problem, the damage corrector, is to evaluate damage evolution

$$\boldsymbol{\sigma}_{n+1} = (1 - d^+_{n+1})\bar{\boldsymbol{\sigma}}^+_{n+1} + (1 - d^-_{n+1})\bar{\boldsymbol{\sigma}}^-_{n+1}. \tag{17}$$

Like the elastic predictor, it is also straightforward and involves no iteration. However, its formulation plays an important role in the construction of the algorithmic consistent tangent stiffness as demonstrated in the following Section 5.

4 Semi-Implicit Return Mapping

For the plastic corrector problem, the semi-implicit method[5] is adopted and improved in the present study in order to integrate free energy accurately and efficiently with small time steps $[t_n, t_{n+1}]$.

4.1 General Format

Different from the implicit backward Euler method, the flow direction $\partial g^{\text{p}}/\partial\bar{\boldsymbol{\sigma}}$ and plastic moduli h^+, h^- at the beginning of the elastoplastic loading range t_{n+e} are used in the whole elastoplastic range $[t_{n+e}, t_{n+1}]$, where $e \in [0, 1]$ is the relative length of the purely elastic range $[t_n, t_{n+e}]$, so that the iterations become much easier and simpler[7]. Different from the explicit forward Euler method, the consistency condition is enforced at the end of the step t_{n+1} so that drift no longer occurs.

The set of nonlinear Eqs. (8)—(10) are cast in a semi-implicit format of their residuals

$$
\begin{aligned}
f_{n+1}^{\mathrm{p}} &= f^{\mathrm{p}}(\bar{\boldsymbol{\sigma}}_{n+1}, \kappa_{n+1}^{+}, \kappa_{n+1}^{-}), \\
\mathbf{R}_{n+1}(\boldsymbol{\varepsilon}^{\mathrm{p}}) &= -\boldsymbol{\varepsilon}_{n+1}^{\mathrm{p}} + \boldsymbol{\varepsilon}_{n}^{\mathrm{p}} + \Delta\lambda_{n+1} \frac{\partial g_{n+e}^{\mathrm{p}}}{\partial \bar{\boldsymbol{\sigma}}_{n+e}} \\
R_{n+1}(\kappa^{\pm}) &= -\kappa_{n+1}^{\pm} + \kappa_{n}^{\pm} + \Delta\lambda_{n+1} h_{n+e}^{\pm},
\end{aligned} \tag{18}
$$

where

$$
\begin{aligned}
g_{n+e}^{\mathrm{p}} &= g^{\mathrm{p}}(\bar{\boldsymbol{\sigma}}_{n+e}, \kappa_{n}^{+}, \kappa_{n}^{-}), \\
h_{n+e}^{\pm} &= h^{\pm}(\bar{\boldsymbol{\sigma}}_{n+e}, \kappa_{n}^{+}, \kappa_{n}^{-}).
\end{aligned} \tag{19}
$$

To solve Eq. (18), the Newton-Raphson iterations

$$
\begin{aligned}
f_{n+1}^{\mathrm{p}(k)} + \frac{\partial f_{n+1}^{\mathrm{p}(k)}}{\partial \bar{\boldsymbol{\sigma}}_{n+1}^{(k)}} : \delta\bar{\boldsymbol{\sigma}}_{n+1}^{(k)} + \frac{\partial f_{n+1}^{\mathrm{p}(k)}}{\partial \kappa_{n+1}^{+(k)}} \delta\kappa_{n+1}^{+(k)} + \frac{\partial f_{n+1}^{\mathrm{p}(k)}}{\partial \kappa_{n+1}^{-(k)}} \delta\kappa_{n+1}^{-(k)} &= 0, \\
\mathbf{R}_{n+1}^{(k)}(\boldsymbol{\varepsilon}^{\mathrm{p}}) + \bar{\mathbb{C}}^{-1} : \delta\bar{\boldsymbol{\sigma}}_{n+1}^{(k)} + \frac{\partial g_{n+e}^{\mathrm{p}}}{\partial \bar{\boldsymbol{\sigma}}_{n+e}} \delta(\Delta\lambda)_{n+1}^{(k)} &= \mathbf{0}, \\
R_{n+1}^{(k)}(\kappa^{\pm}) - \delta\kappa_{n+1}^{\pm(k)} + h_{n+e}^{\pm} \delta(\Delta\lambda)_{n+1}^{(k)} &= 0
\end{aligned} \tag{20}
$$

are used, where the superscript (k) denotes the kth iteration within a time step, leading to

$$
\begin{aligned}
(\Delta\lambda)_{n+1}^{(k+1)} &= (\Delta\lambda)_{n+1}^{(k)} + \delta(\Delta\lambda)_{n+1}^{(k)}, \\
\kappa_{n+1}^{\pm(k+1)} &= \kappa_{n+1}^{\pm(k)} + \delta\kappa_{n+1}^{\pm(k)}, \\
\bar{\boldsymbol{\sigma}}_{n+1}^{(k+1)} &= \bar{\boldsymbol{\sigma}}_{n+1}^{(k)} + \delta\bar{\boldsymbol{\sigma}}_{n+1}^{(k)}, \\
\boldsymbol{\varepsilon}_{n+1}^{\mathrm{p}(k+1)} &= \boldsymbol{\varepsilon}_{n+1}^{\mathrm{p}(k)} - \bar{\mathbb{C}}^{-1} : \delta\bar{\boldsymbol{\sigma}}_{n+1}^{(k)},
\end{aligned} \tag{21}
$$

where $\delta(\Delta\lambda)_{n+1}^{(k)}$, $\delta\kappa_{n+1}^{\pm(k)}$ and $\delta\bar{\boldsymbol{\sigma}}_{n+1}^{(k)}$ are obtained by solving Eq. (20).

Remark 1. It is noted that when $e=0$, that is, the stress state is on the current yield surface at the beginning of the time step t_n, the above procedure reduces to the existing semi-implicit format[5].

4.2 Spectral Version

The plastic constitutive equations for quasi-brittle materials often involve terms of principal stresses. As a consequence, eigenvalue calculation needs to be performed for each iteration (k) in the return mapping process developed in the previous subsection, which substantially increases the computational cost. For these cases, the spectral version[8-9] of the return mapping methods is much more efficient. It is adopted and formulated rigorously in the present subsection.

The matrix $[\bar{\boldsymbol{\sigma}}_{n+1}^{\text{trial}}]$ corresponding to the tensor $\bar{\boldsymbol{\sigma}}_{n+1}^{\text{trial}}$ in the structural coordinate system is spectrally decomposed as

$$
[\bar{\boldsymbol{\sigma}}_{n+1}^{\text{trial}}] = [Q][\hat{\bar{\boldsymbol{\sigma}}}_{n+1}^{\text{trial}}][Q]^{\mathrm{T}}, \tag{22}
$$

where $[\mathbf{Q}] = [[\underset{\rightarrow 1}{n}][\underset{\rightarrow 2}{n}][\underset{\rightarrow 3}{n}]]$ and $[\hat{\bar{\boldsymbol{\sigma}}}_{n+1}^{\text{trial}}] = \mathrm{diag}\{\bar{\sigma}_1^{\text{trial}}, \bar{\sigma}_2^{\text{trial}}, \bar{\sigma}_3^{\text{trial}}\}$, which corresponds to $\bar{\boldsymbol{\sigma}}_{n+1}^{\text{trial}}$ in the principal coordinate system.

In this study, a new tensor $\hat{\bar{\boldsymbol{\sigma}}}^{\text{trial}}_{n+1}$ corresponding to the diagonal matrix $[\hat{\bar{\boldsymbol{\sigma}}}^{\text{trial}}_{n+1}]$ in the structural coordinate system is defined, and Eq.(22) is expressed in a tensor form

$$\bar{\boldsymbol{\sigma}}^{\text{trial}}_{n+1}=\mathbf{Q}\cdot\hat{\bar{\boldsymbol{\sigma}}}^{\text{trial}}_{n+1}\cdot\mathbf{Q}^{\mathrm{T}},\tag{23}$$

where $\mathbf{Q}$ is the tensor corresponding to $[\mathbf{Q}]$ in the structural coordinate system. Because $[\mathbf{Q}]$ is an orthogonal matrix, $\mathbf{Q}$ is an orthogonal tensor. By definition, the newly introduced tensor $\hat{\bar{\boldsymbol{\sigma}}}^{\text{trial}}_{n+1}$ and the traditional tensor $\bar{\boldsymbol{\sigma}}^{\text{trial}}_{n+1}$ are orthogonally similar.

It has been proved[8-9] that, for an isotropic material with coaxial plastic flow, that is, a plastic potential g^{p} exists such that (9) holds, $\bar{\boldsymbol{\sigma}}_{n+1}$ after return mapping in Eq.(15) has the identical eigenspace with $\bar{\boldsymbol{\sigma}}^{\text{trial}}_{n+1}$. Thus, $[\bar{\boldsymbol{\sigma}}_{n+1}]$ corresponding to $\bar{\boldsymbol{\sigma}}_{n+1}$ in the structural coordinate system can be spectrally decomposed with the same eigenvector matrix $[\mathbf{Q}]$ as

$$[\bar{\boldsymbol{\sigma}}_{n+1}]=[\mathbf{Q}][\hat{\bar{\boldsymbol{\sigma}}}_{n+1}][\mathbf{Q}]^{\mathrm{T}},\tag{24}$$

where $[\hat{\bar{\boldsymbol{\sigma}}}_{n+1}]=\operatorname{diag}\{\bar{\sigma}_1,\ \bar{\sigma}_2,\ \bar{\sigma}_3\}$, which corresponds to $\bar{\boldsymbol{\sigma}}_{n+1}$ in the principal coordinate system.

In a similar way, $\hat{\bar{\boldsymbol{\sigma}}}_{n+1}$ corresponding to $[\hat{\bar{\boldsymbol{\sigma}}}_{n+1}]$ in the structural coordinate system is defined, and Eq. (24) is expressed in a tensor form

$$\bar{\boldsymbol{\sigma}}_{n+1}=\mathbf{Q}\cdot\hat{\bar{\boldsymbol{\sigma}}}_{n+1}\cdot\mathbf{Q}^{\mathrm{T}},\tag{25}$$

which indicates that $\hat{\bar{\boldsymbol{\sigma}}}_{n+1}$ and $\bar{\boldsymbol{\sigma}}_{n+1}$ are orthogonally similar.

The return mapping procedure expressed by Eqs.(18)—(21) is carried out here on the orthogonally similar tensor $\hat{\bar{\boldsymbol{\sigma}}}$ as

$$\hat{f}^{\mathrm{p}}_{n+1}=\hat{f}^{\mathrm{p}}(\hat{\bar{\boldsymbol{\sigma}}}_{n+1},\ \kappa^{+}_{n+1},\ \kappa^{-}_{n+1})=f^{\mathrm{p}}(\bar{\boldsymbol{\sigma}}_{n+1},\ \kappa^{+}_{n+1},\ \kappa^{-}_{n+1}),$$

$$\mathbf{R}_{n+1}(\boldsymbol{\varepsilon}^{\mathrm{p}})=-\boldsymbol{\varepsilon}^{\mathrm{p}}_{n+1}+\boldsymbol{\varepsilon}^{\mathrm{p}}_{n}+\Delta\lambda_{n+1}\frac{\partial\hat{g}^{\mathrm{p}}_{n+e}}{\partial\hat{\bar{\boldsymbol{\sigma}}}_{n+e}},\tag{26}$$

$$R_{n+1}(\kappa^{\pm})=-\kappa^{\pm}_{n+1}+\kappa^{\pm}_{n}+\Delta\lambda_{n+1}\hat{h}^{\pm}_{n+e},$$

$$\begin{aligned}\hat{g}^{\mathrm{p}}_{n+e}&=\hat{g}^{\mathrm{p}}(\hat{\bar{\boldsymbol{\sigma}}}_{n+e},\ \kappa^{+}_{n},\ \kappa^{-}_{n})=g^{\mathrm{p}}(\bar{\boldsymbol{\sigma}}_{n+e},\ \kappa^{+}_{n},\ \kappa^{-}_{n}),\\ \hat{h}^{\pm}_{n+e}&=\hat{h}^{\pm}(\hat{\bar{\sigma}}_{n+e},\ \kappa^{+}_{n},\ \kappa^{-}_{n})=h^{\pm}(\bar{\boldsymbol{\sigma}}_{n+e},\ \kappa^{+}_{n},\ \kappa^{-}_{n}),\end{aligned}\tag{27}$$

$$\begin{aligned}&\hat{f}^{\mathrm{p}(k)}_{n+1}+\frac{\partial\hat{f}^{\mathrm{p}(k)}_{n+1}}{\partial\hat{\bar{\boldsymbol{\sigma}}}^{(k)}_{n+1}}:\delta\hat{\bar{\boldsymbol{\sigma}}}^{(k)}_{n+1}+\frac{\partial\hat{f}^{\mathrm{p}(k)}_{n+1}}{\partial\kappa^{+(k)}_{n+1}}\delta\kappa^{+(k)}_{n+1}+\frac{\partial\hat{f}^{\mathrm{p}(k)}_{n+1}}{\partial\kappa^{-(k)}_{n+1}}\delta\kappa^{-(k)}_{n+1}=0,\\ &\mathbf{R}^{(k)}_{n+1}(\boldsymbol{\varepsilon}^{\mathrm{p}})+\bar{\mathbb{C}}^{-1}:\delta\hat{\bar{\boldsymbol{\sigma}}}^{(k)}_{n+1}+\frac{\partial\hat{g}^{\mathrm{p}}_{n+e}}{\partial\hat{\bar{\boldsymbol{\sigma}}}_{n+e}}\delta(\Delta\lambda)^{(k)}_{n+1}=\mathbf{0},\\ &R^{(k)}_{n+1}(\kappa^{\pm})-\delta\kappa^{\pm(k)}_{n+1}+\hat{h}^{\pm}_{n+e}\delta(\Delta\lambda)^{(k)}_{n+1}=0,\end{aligned}\tag{28}$$

$$\begin{aligned}(\Delta\lambda)^{(k+1)}_{n+1}&=(\Delta\lambda)^{(k)}_{n+1}+\delta(\Delta\lambda)^{(k)}_{n+1},\\ \kappa^{\pm(k+1)}_{n+1}&=\kappa^{\pm(k)}_{n+1}+\delta\kappa^{\pm(k)}_{n+1},\\ \hat{\bar{\boldsymbol{\sigma}}}^{(k+1)}_{n+1}&=\hat{\bar{\boldsymbol{\sigma}}}^{(k)}_{n+1}+\delta\hat{\bar{\boldsymbol{\sigma}}}^{(k)}_{n+1},\\ \boldsymbol{\varepsilon}^{\mathrm{p}(k+1)}_{n+1}&=\boldsymbol{\varepsilon}^{\mathrm{p}(k)}_{n+1}-\bar{\mathbb{C}}^{-1}:\delta\hat{\bar{\boldsymbol{\sigma}}}^{(k)}_{n+1},\end{aligned}\tag{29}$$

where eigenvalue calculation needs to be performed only for the first iteration ($k=1$), and then $\bar{\sigma}_i$ are iterated upon with $\vec{n}_i$ fixed. The existence of the functions $\hat{f}^{\mathrm{p}}$, $\hat{g}^{\mathrm{p}}$, and $\hat{h}^{\pm}$ in Eqs.(26) and (27) is based on the assumption that the material is isotropic[10].
Remark 2. It can be seen from Eqs.(28) and (29) that the introduction of the orthogonally similar tensor $\hat{\bar{\boldsymbol{\sigma}}}$ is necessary for the coordinate-independent implementation of the double dot operation between higher-order tensors.

5 Consistent Tangent Stiffness

The differentiation with respect to the strain increment $\Delta\boldsymbol{\varepsilon}_{n+1}$ of the resulting $\bar{\boldsymbol{\sigma}}_{n+1}$ and $\boldsymbol{\sigma}_{n+1}$ gives, respectively, the tangent stiffness consistent with the updating algorithms in elastic-plastic parts and elastic-plastic-damage parts:

$$\begin{aligned}\bar{\mathbb{C}}_{n+1}^{\tan}&=\frac{\partial\bar{\boldsymbol{\sigma}}_{n+1}}{\partial\boldsymbol{\varepsilon}_{n+1}},\\ \mathbb{C}_{n+1}^{\tan}&=\frac{\partial\boldsymbol{\sigma}_{n+1}}{\partial\boldsymbol{\varepsilon}_{n+1}}\end{aligned}\tag{30}$$

which lead to

$$\mathbb{C}_{n+1}^{\tan}=\frac{\partial\boldsymbol{\sigma}_{n+1}}{\partial\bar{\boldsymbol{\sigma}}_{n+1}}:\bar{\mathbb{C}}_{n+1}^{\tan}.\tag{31}$$

On the basis of the continuum elastoplastic tangent stiffness expression

$$\bar{\mathbb{C}}^{\tan}=\bar{\mathbb{C}}-\frac{(\partial g^{\mathrm{p}}/\partial\bar{\boldsymbol{\sigma}}:\bar{\mathbb{C}})\otimes(\bar{\mathbb{C}}:\partial f^{\mathrm{p}}/\partial\bar{\boldsymbol{\sigma}})}{\partial g^{\mathrm{p}}/\partial\bar{\boldsymbol{\sigma}}:\bar{\mathbb{C}}:\partial f^{\mathrm{p}}/\partial\bar{\boldsymbol{\sigma}}-(\partial f^{\mathrm{p}}/\partial\kappa^{+})h^{+}-(\partial f^{\mathrm{p}}/\partial\kappa^{-})h^{-}}\tag{32}$$

replacing the flow direction and plastic moduli with their values at t_{n+e} and the other quantities at t_{n+1}, one obtains

$$\bar{\mathbb{C}}_{n+1}^{\tan}=\bar{\mathbb{C}}-\frac{(\partial g_{n+e}^{\mathrm{p}}/\partial\bar{\boldsymbol{\sigma}}_{n+e}:\bar{\mathbb{C}})\otimes(\bar{\mathbb{C}}:\partial f_{n+1}^{\mathrm{p}}/\partial\bar{\boldsymbol{\sigma}}_{n+1})}{\partial g_{n+e}^{\mathrm{p}}/\partial\bar{\boldsymbol{\sigma}}_{n+e}:\bar{\mathbb{C}}:\partial f_{n+1}^{\mathrm{p}}/\partial\bar{\boldsymbol{\sigma}}_{n+1}-(\partial f_{n+1}^{\mathrm{p}}/\partial\kappa_{n+1}^{+})h_{n+e}^{+}-(\partial f_{n+1}^{\mathrm{p}}/\partial\kappa_{n+1}^{-})h_{n+e}^{-}}.\tag{33}$$

$\partial\boldsymbol{\sigma}_{n+1}/\partial\bar{\boldsymbol{\sigma}}_{n+1}$ in Eq. (31) needs to be derived in accordance with specific damage formulation. From Eq.(1), one can obtain

$$\frac{\partial\boldsymbol{\sigma}}{\partial\bar{\boldsymbol{\sigma}}}=(1-d^{+})\frac{\partial\bar{\boldsymbol{\sigma}}^{+}}{\partial\bar{\boldsymbol{\sigma}}}+(1-d^{-})\frac{\partial\bar{\boldsymbol{\sigma}}^{-}}{\partial\bar{\boldsymbol{\sigma}}}-\bar{\boldsymbol{\sigma}}^{+}\otimes\frac{\partial d^{+}}{\partial\bar{\boldsymbol{\sigma}}}-\bar{\boldsymbol{\sigma}}^{-}\otimes\frac{\partial d^{-}}{\partial\bar{\boldsymbol{\sigma}}},\tag{34}$$

where

$$\begin{aligned}\frac{\partial\bar{\boldsymbol{\sigma}}^{+}}{\partial\bar{\boldsymbol{\sigma}}}&=\sum_{i}H(\bar{\sigma}_i)\vec{n}_i\otimes\vec{n}_i\otimes\vec{n}_i\otimes\vec{n}_i+2\sum_{i<j}\frac{\langle\bar{\sigma}_i\rangle-\langle\bar{\sigma}_j\rangle}{\bar{\sigma}_i-\bar{\sigma}_j}\operatorname{sym}(\vec{n}_i\otimes\vec{n}_j)\otimes\operatorname{sym}(\vec{n}_i\otimes\vec{n}_j),\\ \frac{\partial\bar{\boldsymbol{\sigma}}^{-}}{\partial\bar{\boldsymbol{\sigma}}}&=\sum_{i}H(-\bar{\sigma}_i)\vec{n}_i\otimes\vec{n}_i\otimes\vec{n}_i\otimes\vec{n}_i-2\sum_{i<j}\frac{\langle-\bar{\sigma}_i\rangle-\langle-\bar{\sigma}_j\rangle}{\bar{\sigma}_i-\bar{\sigma}_j}\operatorname{sym}(\vec{n}_i\otimes\vec{n}_j)\otimes\operatorname{sym}(\vec{n}_i\otimes\vec{n}_j)\end{aligned}\tag{35}$$

as derived in[11], and

$$\frac{\partial d^{\pm}}{\partial \bar{\boldsymbol{\sigma}}} = \frac{\partial d^{\pm}}{\partial R^{\pm}} \frac{\partial R^{\pm}}{\partial \bar{\boldsymbol{\sigma}}}. \tag{36}$$

In(36), $\partial d^{\pm}/\partial R^{\pm}$ depend on damage evolution functions and $\partial R^{\pm}/\partial \bar{\boldsymbol{\sigma}}$ on definitions of $Y^{\pm}$. If the formulation of $Z^{\pm}$ in[4] is adopted, the specific expressions read

$$\frac{\partial R^+}{\partial \bar{\boldsymbol{\sigma}}} = \frac{1}{2}\left(\bar{\boldsymbol{\sigma}} : \bar{\mathbb{C}}^{-1} : \frac{\partial \bar{\boldsymbol{\sigma}}^+}{\partial \bar{\boldsymbol{\sigma}}} + \bar{\boldsymbol{\sigma}}^+ : \bar{\mathbb{C}}^{-1}\right) + \frac{1}{w}\Delta\kappa^+ H(\bar{\sigma}_{\max})\,\underset{\to\max}{n}\otimes\underset{\to\max}{n},$$

$$\frac{\partial R^-}{\partial \bar{\boldsymbol{\sigma}}} = \frac{1}{2}\left(\bar{\boldsymbol{\sigma}} : \bar{\mathbb{C}}^{-1} : \frac{\partial \bar{\boldsymbol{\sigma}}^-}{\partial \bar{\boldsymbol{\sigma}}} + \bar{\boldsymbol{\sigma}}^- : \bar{\mathbb{C}}^{-1}\right) - \frac{1}{1-w}\Delta\kappa^- H(-\bar{\sigma}_{\min})\,\underset{\to\min}{n}\otimes\underset{\to\min}{n}, \tag{37}$$

where

$$w = \frac{\sum_i \langle \bar{\sigma}_i \rangle}{\sum_i |\bar{\sigma}_i|}. \tag{38}$$

6 Implementation

The complete algorithm of the present study is summarized as follows.
Summary of integration algorithm is as follows

$$\boldsymbol{\varepsilon}_{n+1} = \boldsymbol{\varepsilon}_n + \nabla^s(\Delta\underset{\to}{u}_{n+1})$$

$$\bar{\boldsymbol{\sigma}}_{n+1}^{\text{trial}} = \bar{\mathbb{C}} : (\boldsymbol{\varepsilon}_{n+1} - \boldsymbol{\varepsilon}_n^{\text{p}})$$

If $f^{\text{p}}(\bar{\boldsymbol{\sigma}}_{n+1}^{\text{trial}}, \kappa_n^+, \kappa_n^-) > \text{TOL}$

$$\Delta\lambda_{n+1}^{(k+1)} = \Delta\lambda_{n+1}^{(k)} + \delta(\Delta\lambda)_{n+1}^{(k)}$$

$$\boldsymbol{\varepsilon}_{n+1}^{\text{p}(k+1)} = \boldsymbol{\varepsilon}_n^{\text{p}} + \Delta\lambda_{n+1}^{(k+1)} \frac{\partial \hat{g}_{n+e}^{\text{p}}}{\partial \hat{\bar{\boldsymbol{\sigma}}}_{n+e}}$$

$$\kappa_{n+1}^{\pm(k+1)} = \kappa_n^{\pm} + \Delta\lambda_{n+1}^{(k+1)} \hat{h}_{n+e}^{\pm}$$

$$\hat{\bar{\boldsymbol{\sigma}}}_{n+1}^{(k+1)} = \hat{\bar{\boldsymbol{\sigma}}}_{n+1}^{\text{trial}} - \Delta\lambda_{n+1}^{(k+1)} \bar{\mathbb{C}} : \frac{\partial \hat{g}_{n+e}^{\text{p}}}{\partial \hat{\bar{\boldsymbol{\sigma}}}_{n+e}}$$

If $\hat{f}^{\text{p}}(\hat{\bar{\boldsymbol{\sigma}}}_{n+1}^{(k+1)}, \kappa_{n+1}^{+(k+1)}, \kappa_{n+1}^{-(k+1)}) \leqslant \text{TOL}$

$$\bar{\psi}_{n+1}^{\text{e}\pm} = \frac{1}{2}\bar{\boldsymbol{\sigma}}_{n+1}^{\pm} : (\boldsymbol{\varepsilon}_{n+1} - \boldsymbol{\varepsilon}_{n+1}^{\text{p}})$$

$$\bar{\psi}_{n+1}^{\text{p}\pm} = \bar{\psi}_n^{\text{p}\pm} + \frac{1}{2}(\bar{Z}_n^{\pm} + \bar{Z}_{n+1}^{\pm})\Delta\kappa_{n+1}^{\pm} \tag{39}$$

$$Y_{n+1}^{\pm} = \bar{\psi}_{n+1}^{\text{e}\pm} + \bar{\psi}_{n+1}^{\text{p}\pm}$$

$$R_{n+1}^{\pm} = \max\{R_n^{\pm}, Y_{n+1}^{\pm}\}$$

$$d_{n+1}^{\pm} = G^{\pm}(R_0^{\pm}, R_{n+1}^{\pm})$$

$$\boldsymbol{\sigma}_{n+1} = (1 - d_{n+1}^+)\bar{\boldsymbol{\sigma}}_{n+1}^+ + (1 - d_{n+1}^-)\bar{\boldsymbol{\sigma}}_{n+1}^-$$

$$\mathbb{C}_{n+1}^{\tan} = \left((1-d_{n+1}^+)\frac{\partial \bar{\boldsymbol{\sigma}}_{n+1}^+}{\partial \bar{\boldsymbol{\sigma}}_{n+1}} + (1-d_{n+1}^-)\frac{\partial \bar{\boldsymbol{\sigma}}_{n+1}^-}{\partial \bar{\boldsymbol{\sigma}}_{n+1}} - \frac{\partial d_{n+1}^+}{\partial R_{n+1}^+}\bar{\boldsymbol{\sigma}}_{n+1}^+\otimes\frac{\partial R_{n+1}^+}{\partial \bar{\boldsymbol{\sigma}}_{n+1}} - \frac{\partial d_{n+1}^-}{\partial R_{n+1}^-}\bar{\boldsymbol{\sigma}}_{n+1}^-\otimes\frac{\partial R_{n+1}^-}{\partial \bar{\boldsymbol{\sigma}}_{n+1}}\right) :$$

$$\left(\bar{\mathbb{C}} - \frac{(\partial g_{n+e}^{\text{p}}/\partial \bar{\boldsymbol{\sigma}}_{n+e} : \bar{\mathbb{C}}) \otimes (\bar{\mathbb{C}} : \partial f_{n+1}^{\text{p}}/\partial \bar{\boldsymbol{\sigma}}_{n+1})}{\partial g_{n+e}^{\text{p}}/\partial \bar{\boldsymbol{\sigma}}_{n+e} : \bar{\mathbb{C}} : \partial f_{n+1}^{\text{p}}/\partial \bar{\boldsymbol{\sigma}}_{n+1} - (\partial f_{n+1}^{\text{p}}/\partial \kappa_{n+1}^+)h_{n+e}^+ - (\partial f_{n+1}^{\text{p}}/\partial \kappa_{n+1}^-)h_{n+e}^-}\right).$$

As numerical examples, this algorithm is implemented with the specific elastoplastic damage model for concrete presented in[4]. Several typical mechanical tests of reinforced concrete components are simulated with the commercial finite element program Abaqus and its user material subroutine UMAT. Solid elements are used for concrete and truss elements for steel rebars, between which perfect bonding is assumed.

The uniaxial mechanical properties for concrete are first calibrated to fit the stress-strain curves of uniaxial tension and compression recommended in the Chinese code[12]. The tension stiffening effects due to the presence of reinforcement are accounted for by using the descending stress-strain curve by Stevens et al.[13] and the confining effects resulting from closely spaced stirrups by fitting the stress-strain curve for core concrete by Kent and Park[14] and Scott et al.[15]. Besides, the common values for the equibiaxial strength ratio $f_{b0}^-/f_0^- = 1.2$, Poisson's ratio $v = 0.2$, and dilation parameter $\alpha^p = 0.2$ are set for multiaxial behavior. For reinforcing steel, uniaxial elastoplastic models are employed.

6.1 Slab

For application in reinforced concrete slabs, the test by McNeice[16] [Fig. 1 (a)] is simulated. The two-way slab has a bottom reinforcement ratio of 0.85% in each direction. In the simulation, Young's modulus $E_s = 200\times10^3$ N/mm^2 and yield point $f_s = 345$ N/mm^2 are set for steel, and effective Young's modulus $\bar{E} = 35.1\times10^3$ N/mm^2, initial damage stresses in uniaxial tension and compression $f_0^+ = 3.17$ N/mm^2, $f_0^- = 21.0$ N/mm^2, and initial effective yield stresses in uniaxial tension and compression $\bar{f}_y^+ = 3.17$ N/mm^2, $\bar{f}_y^- = 37.92$ N/mm^2 for concrete. The calculated load-deflection curve is shown along with the test data in Fig. 1(b).

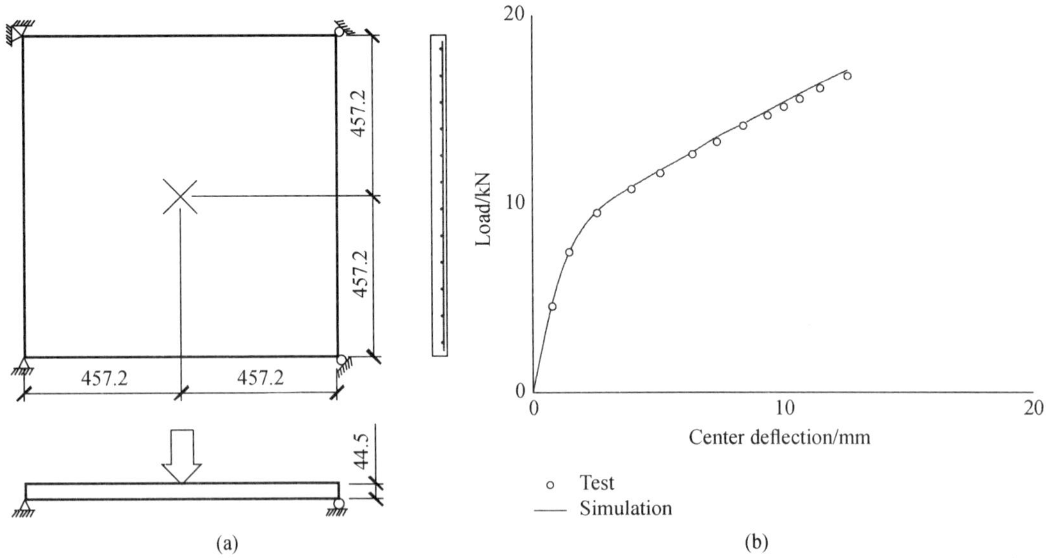

Fig. 1. Reinforced concrete slab under out-of-plane loading[16]. (a) Setup (dimensions in mm). (b) Results

For comparison, the implicit backward Euler return mapping algorithm is also implemented, and the calculated load-deflection curve is precisely the same as in Fig. 1(b). The times of calculation consumed with the semi-implicit and implicit methods are 179 s and 804 s, respectively, at 2 000 displacement steps.

6.2 Wall

For application in reinforced concrete walls, the tests by Lefas et al.[17] [Fig. 2(a)] are simulated. In one of the two selected tests(SW22 and SW24), a constant vertical load is held while the horizontal load increases. These loads are transferred through the top spreader beams, which are cast integrally with the walls. The vertical and horizontal reinforcement ratios in the walls are 2.49% and 0.82%, respectively, and the ratio of closed stirrups to confined concrete of 140 mm-wide concealed columns at both edges is

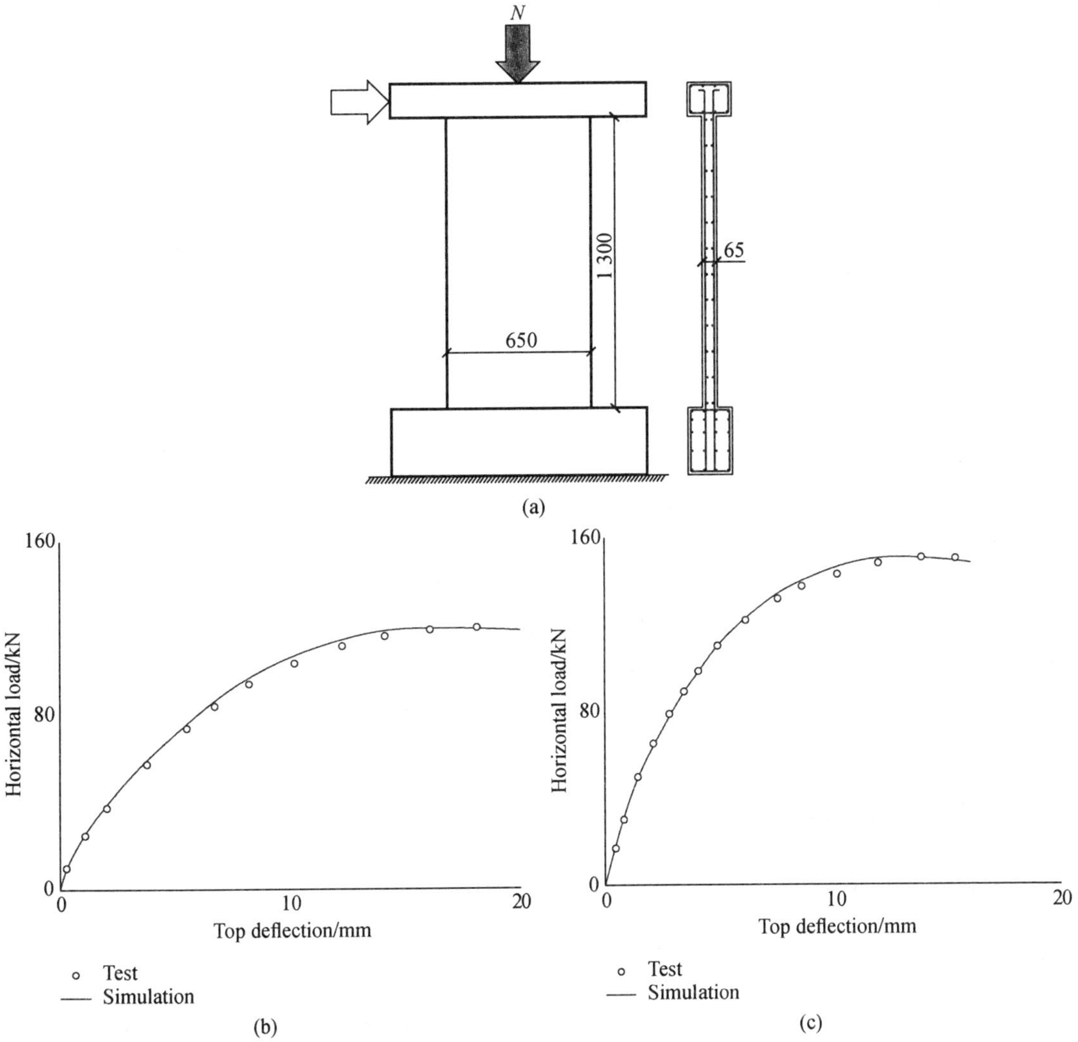

Fig. 2. Reinforced concrete walls under in-plane loading[17]. (a) Setup(dimensions in mm). (b) Results for SW24 ($N=0$, $f_{cu}=48.3$ N/mm^2). (c) Results for SW22($N=182$ kN, $f_{cu}=50.6$ N/mm^2)

0.52%. In the simulations, $E_s=210\times 10^3$ N/mm^2, $f_{s,v}=470$ N/mm^2 (vertical rebars), $f_{s,h}=520$ N/mm^2 (horizontal rebars), and $f_{s,s}=420$ N/mm^2 (stirrups) are set for steel, $\bar{E}=35.4\times 10^3$ N/mm^2, $f_0^+=3.33$ N/mm^2, $f_0^-=21.3$ N/mm^2, $\bar{f}_y^+=3.33$ N/mm^2, and $\bar{f}_y^-=38.5$ N/mm^2 for concrete of SW24, and $\bar{E}=36.7\times 10^3$ N/mm^2, $f_0^+=3.42$ N/mm^2, $f_0^-=22.0$ N/mm^2, $\bar{f}_y^+=3.42$ N/mm^2, and $\bar{f}_y^-=40.8$ N/mm^2 for concrete of SW22. The calculated load-deflection curves are displayed along with the test data in Figs. 2(b) and 2(c), respectively, where f_{cu} is the cubic strength of concrete.

For comparison, the implicit return mapping algorithm is also implemented, and the calculated load-deflection curves are precisely the same as in Figs. 2(b) and 2(c). The times of calculation consumed with the semi-implicit and implicit methods are 495 s/508 s and 1 870 s/1 922 s, respectively, at 2 000 displacement steps.

7 Conclusions

History-dependent energy calculation makes the quality of stability in large time steps of the popular implicit backward Euler return mapping algorithm no longer attractive for some inelastic constitutive models. More flexible and efficient integrating schemes are favored.

The semi-implicit constitutive integration algorithm developed for a class of elastoplastic damage models that involve integration of free energy in the present study achieves a balance between accuracy, stability, and efficiency. In comparison with the implicit backward Euler algorithm, it is by far less challenging mathematically and less time-consuming computationally. While it is less stable, the problem can be fixed with small time steps, which are necessary for free energy integration. Another drawback is the need to find the stress where the yield surface is first reached within a time step. In comparison with the explicit forward Euler algorithm, it is much more accurate and stable at the cost of iterations for consistency.

The splitting of operators and the elastic-plastic-damage updating algorithm on its basis presented in this paper are particularly formulated for damage energy release rates that are derived from elastoplastic free energy potentials. Nevertheless, the basic idea and method are also applicable to other types of constitutive models that include history-dependent calculation of certain quantities.

Competing Interests

The authors declare that they have no competing interests.

Acknowledgments

The financial support from the National Science Foundation of China-National Science Foundation (NSFC-NSF) Joint Project (Grant no. 51261120374) and NSFC Projects (Grant Nos. 51108336 and 51378377) is greatly appreciated.

References

[1] SIMO J C, HUGHES T J R. Computational Inelasticity[M]. New York: Springer, 1998.

[2] ZHANG J, LI J. Microelement formulation of free energy for quasi-brittle materials[J]. Journal of Engineering Mechanics, 2014, 140(8): 06014008.

[3] JU J W. On energy-based coupled elastoplastic damage theories: constitutive modeling and computational aspects[J]. International Journal of Solids and Structures, 1989, 25(7): 803-833.

[4] ZHANG J, LI J. Elastoplastic damage model for concrete based on consistent free energy potential[J]. Science China Technological Sciences, 2014, 57(11): 2278-2286.

[5] DUNNE F, PETRINIC N. Introduction to computational plasticity[M]. OUP Oxford, 2005.

[6] CHORIN A J, HUGHES T J R, MCCRACKEN M F, et al. Product formulas and numerical algorithms[J]. Communications on Pure and Applied Mathematics, 1978, 31(2): 205-256.

[7] ZHANG J, LI J. Construction of homogeneous loading functions for elastoplastic damage models for concrete[J]. Science China Physics, Mechanics and Astronomy, 2014, 57(3): 490-500.

[8] SIMO J C. Algorithms for static and dynamic multiplicative plasticity that preserve the classical return mapping schemes of the infinitesimal theory[J]. Computer Methods in Applied Mechanics and Engineering, 1992, 99(1): 61-112.

[9] LEE J, FENVES G L. A return-mapping algorithm for plastic-damage models: 3-D and plane stress formulation[J]. International Journal for Numerical Methods in Engineering, 2001, 50(2): 487-506.

[10] SHAMES I H. Elastic and inelastic stress analysis[M]. CRC Press, 1997.

[11] FARIA R, OLIVER J O O, RUIZ M C. On isotropic scalar damage models for the numerical analysis of concrete structures[M]. Barcelona, Spain: Centro Internacional de Métodos Numéricos en Ingeniería, 2000.

[12] Chinese National Standard. Code for design of concrete structures: GB 50010[R]. China Architecture & Building Press, Beijing, China, 2002.

[13] STEVENS N J, UZUMERI S M, WILL G T. Constitutive model for reinforced concrete finite element analysis[J]. Structural Journal, 1991, 88(1): 49-59.

[14] KENT D C, PARK R. Flexural members with confined concrete[J]. Journal of the Structural Division, 1971, 97(7): 1969-1990.

[15] SCOTT B D, PARK R, PRIESTLEY M J N. Stress-strain behavior of concrete confined by overlapping hoops at low and high strain rates[C]//Journal Proceedings. 1982, 79(1): 13-27.

[16] MCNEICE G M. Elastic-plastic bending analysis of plates and slabs by the finite element method[D]. University of London, 1967.

[17] LEFAS I D, KOTSOVOS M D, AMBRASEYS N N. Behavior of reinforced concrete structural walls: strength, deformation characteristics, and failure mechanism[J]. Structural Journal, 1990, 87(1): 23-31.

Implicit gradient delocalization method for force-based frame element①

De-Cheng Feng[1], **Xiaodan Ren**[2], **Jie Li**[3]

1 *Ph.D. Candidate, Dept. of Structural Engineering, Tongji Univ., Shanghai 200092, China.*
2 *Associate Professor, Dept. of Structural Engineering, Tongji Univ., Shanghai 200092, China.*
3 *Professor, Dept. of Structural Engineering and State Key Laboratory on Disaster Reduction in Civil Engineering, Tongji Univ., Shanghai 200092, China (corresponding author). E-mail: lijie@tongji.edu.cn*

Abstract In this study, a novel force-based frame element with delocalization is developed to address localization issues induced by material strain softening. A regularization procedure for section deformations is introduced on the basis of the concept of deformation gradient dependency. The internal length scale of the gradient-based regularization model is discussed and related to the plastic hinge length of the structural members. The weak form of the governing equation is derived, and a subscale numerical strategy is utilized within each element to implement the proposed model in the frame element. In addition, an adaptive scheme is developed for the RC members to determine the real-time gradient coefficient. Finally, two numerical examples are utilized and the results show that the proposed model can achieve an objective response for and excellent prediction of the nonlinear behavior of RC members.

Keywords Force-based frame element; Strain softening; Delocalization; Implicit gradient method; Plastic hinge; Analysis and computation

1 Introduction

The nonlinear analysis of RC frame structures has been an open research area for over 50 years. However, predicting the actual performance of RC structures under static and dynamic loads remains a challenge. Frame elements, which are extensively used to simulate RC frame structures, can be grouped into two categories, namely, lumped plasticity and distributed plasticity. Lumped plasticity[1-4] stems from the phenomenon in which plastic deformations of the frames frequently concentrate at the ends of beams or columns. Consequently, a common approach for such models is to introduce nonlinear springs at each end of an elastic element[2]. Despite its simplicity in application, lumped plasticity models are frequently inaccurate because the plastic zone length is ignored, which can be clearly observed in experiments. The limitations of the lumped plasticity approach have motivated the development of distributed plasticity models[5-6], which allow plastic

① Originally published in *Journal of Structural Engineering*, 2016, 142(2): 04015122.

deformations to occur anywhere within the element. Distributed plasticity models provide a more accurate description of the plastic behavior of structural members than lumped plasticity models. Moreover, the coupling effect between the axial force and moment can be considered by introducing the fiber section model in both methods[7].

Distributed plasticity models can be formulated in two ways. The first is the classical displacement-based method, which approximates the displacement field within the one-dimensional(1D) element by using finite-element(FE) shape functions, e.g., the cubic Hermite polynomials[8-9]. However, the displacement field may not be well approximated by FE shape functions because of deformation localization within the domain of a plastic hinge. Consequently, a refined mesh may be required to achieve a real nonlinear response from the 1D element. By contrast, the force-based approach approximates the internal force distribution without nodal loading, which is simpler than displacement distribution[7, 10-12]. Evidently, the primary advantage of the force-based element is that the equilibrium condition is strictly satisfied even within the range of strong nonlinear material responses[13]. Considering this feature, one element is sufficient to capture the linear and nonlinear performances of a structural member. Therefore, the total number of degrees of freedom can be reduced in the structural model and computational efficiency can be enhanced.

However, localization issues may occur when modeling strain-softening sectional behavior using a force-based element[14]. Deformations will localize at a single integration point of the element, which will lead to objectivity loss. The simulated response depends on the number of quadrature points. To date, a few delocalization/regularization techniques have been proposed for force-based element formulations to overcome this problem. Coleman and Spacone[14] first introduced a method to maintain objective results by modifying the uniaxial stress-strain relationships to ensure constant fracture energy. However, this method could only achieve objective responses at the global level, i.e., load-displacement curve. The local responses, i.e., moment-rotation curve, remained nonobjective. A family of plastic hinge integration methods[15-16] have also been proposed to address this problem; however, they appear to underestimate the postyield response for strain-hardening behavior. Then, two accurate methods[17-18], which are also on the basis of the scale of the integration weights, are established to model hardening and softening sectional behavior. The successful applications of nonlocal integral formulations to address this issue in a force-based element were also reported by Khaloo and Tariverdilo[19], Addessi and Ciampi[16], and Valipour and Foster[20]. These types of approaches propose the regularization of state variables in their integral form, in which a characteristic length is introduced by the integral weight function. However, other investigations[16, 18] have revealed that nonlocal integral formulations appear more suitable for cases in which element size is comparable to characteristic length, which is typically smaller than the length of the frame members. Therefore, the merits of the force-based element, from which coarse meshes are usually adopted, may be degraded.

The present study aims to construct a novel force-based frame element on the basis of the implicit gradient model to address localization issues. The work begins with the force-based element formulation. Then, section deformations are utilized as regularized state variables and the governing equation of regularization is derived by applying the Taylor expansion. Afterward, a subscale numerical strategy is developed to solve the governing equation and implemented in the framework of the force-based frame element. In particular, the gradient coefficient of RC frame members is correlated with the plastic hinge length to consider the development of the plastic zone during the response time history. Finally, numerical examples are utilized to verify the accuracy and efficiency of the new element. The prediction of the plastic hinge length of a structural member is also demonstrated.

2 Force-Based Element Formulation

In this study, a force-based element formulation[7] is on the basis of the framework of the two-dimensional (2D) Euler-Bernoulli beam theory without geometric nonlinearity. As shown in Fig. 1, the element has six degrees of freedom in the global reference system. By transforming the element displacements and forces from the global system to the local system and eliminating rigid modes, as shown in Fig. 1, the element deformations $\mathbf{q}$ and forces $\mathbf{Q}$ in the local reference system can be obtained and grouped into the following vectors

$$\mathbf{q}=[q_1 \quad q_2 \quad q_3]^{\mathrm{T}} \tag{1}$$

$$\mathbf{Q}=[Q_1 \quad Q_2 \quad Q_3]^{\mathrm{T}} \tag{2}$$

where q_1 = axial extension; q_2 and q_3 = nodal rotations; Q_1 = axial force; and Q_2 and Q_3 = bending moments of the end nodes.

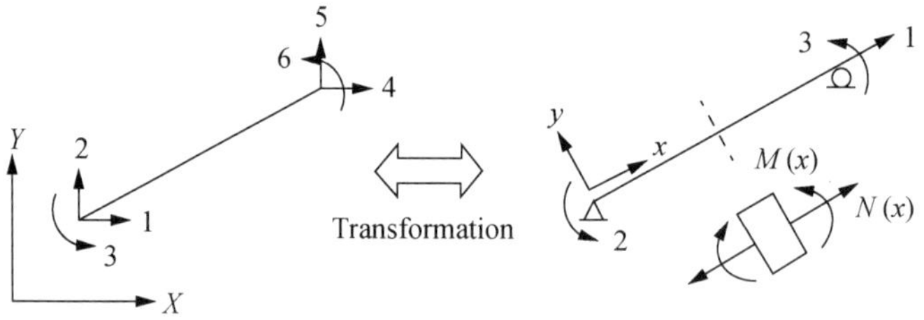

Fig. 1. Transformation from global coordinates to local coordinates

Ignoring shear deformations, the section deformation vector $\mathbf{d}(x)$ and the section force vector $\mathbf{D}(x)$ can be written as follows

$$\mathbf{d}(x)=[e(x) \quad \kappa(x)]^{\mathrm{T}} \tag{3}$$

$$\mathbf{D}(x)=[N(x) \quad M(x)]^{\mathrm{T}} \tag{4}$$

where $e(x)$ = axial strain along the reference axis; $\kappa(x)$ = curvature; $N(x)$ = axial force; and $M(x)$ = bending moment.

Equilibrium between section forces and element nodal forces in the strong form can be expressed as follows

$$\mathbf{D}(x)=\mathbf{b}(x)\mathbf{Q} \tag{5}$$

where matrix $\mathbf{b}(x)$ contains the force interpolation functions that define the constant axial force and the linear bending moment distribution along element length L as follows

$$\mathbf{b}(x)=\begin{bmatrix} 1 & 0 & 0 \\ 0 & \dfrac{x}{L}-1 & \dfrac{x}{L} \end{bmatrix} \tag{6}$$

The compatibility relationship between section deformations and element nodal deformations can be obtained by applying the principle of virtual force as follows

$$\mathbf{q}=\int_L \mathbf{b}(x)^{\mathrm{T}}\mathbf{d}(x)\mathrm{d}x \tag{7}$$

Subsequently, element flexibility is attained using the derivative of the compatibility equation as follows

$$\mathbf{F}=\frac{\partial \mathbf{q}}{\partial \mathbf{Q}}=\int_L \mathbf{b}(x)^{\mathrm{T}}\mathbf{f}_s(x)\mathbf{b}(x)\mathrm{d}x \tag{8}$$

where $\mathbf{f}_s$ denotes section flexibility, which is expressed as the inverse of the section stiffness as follows

$$\mathbf{f}_s=\mathbf{k}_s^{-1} \tag{9}$$

The fiber section model shown in Fig. 2 is frequently used to establish the section force-deformation relationship on the basis of the uniaxial stress-strain behavior of the materials[7]. In the 2D Euler-Bernoulli beam theory, the uniaxial strain of the fiber at height y of cross section x is expressed as follows

$$\varepsilon(x, y)=\mathbf{l}(y)\mathbf{d}(x) \tag{10}$$

where $\mathbf{l}(y)=[1 \quad -y]$ is a simple geometric vector. On the basis of the material constitutive law, the uniaxial stress $\sigma(y, x)$ and tangent modulus $E(x, y)$ of the fiber can be obtained. Then, the section-resisting forces $\mathbf{D}_{\mathrm{R}}(x)$ and the stiffness matrix $\mathbf{k}_s(x)$ are determined as follows

$$\mathbf{D}_{\mathrm{R}}(x)=\int_{A(x)} \mathbf{l}^{\mathrm{T}}(y)\sigma(x, y)\mathrm{d}A \tag{11}$$

$$\mathbf{k}_s(x)=\int_{A(x)} \mathbf{l}^{\mathrm{T}}(y)E(x, y)\mathbf{l}(y)\mathrm{d}A \tag{12}$$

where $A(x)$=area of the section. Integrations over the sections in Eqs.(11) and (12) are frequently computed using the 2D rectangular method, with the domain discretization shown in Fig. 2.

The compatibility relation expressed in Eq.(7) and the element flexibility expressed in Eq.(8) are calculated by numerical integration as follows

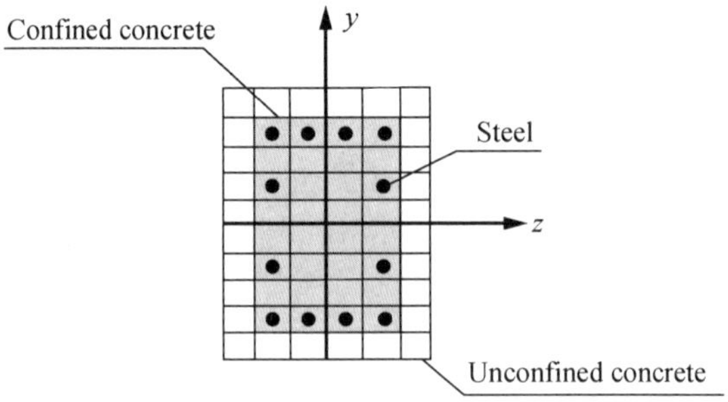

Fig. 2. Fiber section model

$$\mathbf{q}=\sum_{i=1}^{N_{\mathrm{p}}}\mathbf{b}(x_i)^{\mathrm{T}}\mathbf{d}(x_i)w_i \quad \mathbf{F}=\sum_{i=1}^{N_{\mathrm{p}}}\mathbf{b}(x_i)^{\mathrm{T}}\mathbf{f}_{\mathrm{s}}(x_i)\mathbf{b}(x_i)w_i \tag{13}$$

where N_{p} represents the number of integration points over the element length; x_i indicates the position of the ith integration point; and w_i denotes the corresponding weight. The common choice for Eq.(13) is the Gauss-Lobatto quadrature[21], in which sample points are placed at element ends, where the largest internal forces occur in the absence of distributed loads. The order of accuracy for the Gauss-Lobatto quadrature is proven to be $2N_{\mathrm{p}}-3$, such that at least three integration points are required for the quadratic terms in Eq.(8) for linear elastic problems. In practice, four to six integration points are typically used to obtain an accurate response under nonlinear material behavior[13]. Evidently, a merit of the force-based element is that the force interpolation matrix $\mathbf{b}(x)$ is always accurate for small deformation problems, regardless of the level of material nonlinearity. Such a feature enables the modeling of a structural member with a single element, which can significantly improve computation efficiency.

3 Localization Issues

The force-based formulation successfully describes the spreading of plasticity for an element with strain-hardening sectional behavior. In addition, an objective response can be achieved by increasing the number of integration points. However, when simulating strain-softening sectional behavior, a nonobjective response may occur, which is also referred to as a localization issue[14, 22]. Deformation tends to concentrate at the end integration point when modeling softening behavior. Thus, the plastic hinge length is implied by the integration weight of the end point, which is not an objective variable of the physical problem.

To address this problem, Coleman and Spacone[14] first developed a regularization technique that modifies the material stress-strain behavior on the basis of the invariability of fracture energy release within the element. However, modification is relatively inconvenient and only achieves objectivity at the global level. In fact, the main solution to this problem is scaling the element integration weights to match the prescribed plastic hinge length. Typical cases are the plastic hinge integration methods proposed by Scott and Fenves[15] and Addessi and Ciampi[16]. One of the drawbacks of these methods is that flexible responses for hardening sectional behavior are produced. Scott and Hamutçuoğlu[17] and Almeida et al.[18] have developed two integration schemes to ensure accurate results for hardening and softening behavior. The application of a nonlocal integral theory is an alternative method to address this problem[16, 19-20]. However, only a few studies that utilize the nonlocal gradient model to address objectivity loss have been published. The present work aims to establish an implicit gradient delocalization formulation for a force-based frame element.

4 Method for Delocalization

The nonlocal integral theory[23-25] defines an integral transformation to regularize/delocalize the fields of state variables. For a 1D frame element, the nonlocal variable $\bar{f}(x)$ of a certain section at x is the weighted average of the local variable $f(x)$ over the surrounding problem domain Ω[16, 19], which is expressed as follows

$$\bar{f}(x)=\frac{1}{V_r}\int_{\Omega}\alpha(x, \xi)f(\xi)\mathrm{d}\Omega \tag{14}$$

where $\alpha(x, \xi)$ = nonlocal weight function that only depends on the distance between the source point ξ and the receiver point x; and V_r = normalizing factor expressed as follows

$$V_r=\int_{\Omega}\alpha(x, \xi)\mathrm{d}\Omega \tag{15}$$

The weight function is typically defined by bell-shaped functions, e. g., the Gauss distribution is expressed as follows

$$\alpha(r)=\frac{1}{\sqrt{2\pi}\,l}\exp\left(-\frac{r^2}{2l^2}\right) \tag{16}$$

where $r=|x-\xi|$; and l = characteristic length that defines the nonlocal interaction region that contributes to the nonlocal variable.

Using Eq. (14), the element formulation becomes a double integral. The first integral is within the nonlocal interaction region to derive the nonlocal counterpart, whereas the second integral is along the entire element length to generate element information, as shown in Fig. 3. Notably, the nonlocal averaging concept is a simple but physically meaningful solution to localization problems. The nonlocal integral model only requires an additional integral process over a spatial neighborhood of each integration point. However, sufficient integration points should be arranged within the interaction region to obtain accurate results[20]. Notably, the interaction region (nonlocal region) is typically one order of magnitude lower than the common length of the frame members[18]; thus, increasing the number of integration points or refined mesh types is required.

Peerlings et al.[26-27] proposed an alternative approach to overcome the drawbacks of the nonlocal integral of each section by introducing a Taylor expansion to Eq. (14). The nonlocal counterpart can be expressed as follows

$$\bar{f}(x) - c\,\nabla^2 \bar{f}(x) = f(x) \tag{17}$$

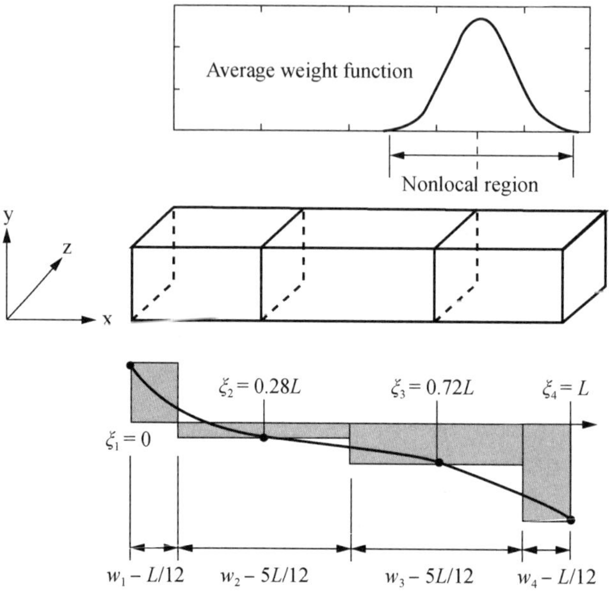

Fig. 3. Nonlocal integral model for force-based element

where ∇ denotes the Laplacian operator; and $c = l^2/2$ is the gradient coefficient determined using the weight function in Eq. (16)[27]. Eq. (17) is the so-called implicit gradient model that calculates the regularized variable implicitly by using the local variable $f(x)$ without derivatives. Thus, a low-order regularity of the local variable is required[26].

An additional boundary condition must be supplemented to determine the solution to Eq. (17). In general, the natural boundary condition is used as follows[26-29]

$$\nabla \bar{f} \cdot \mathbf{n} = 0 \tag{18}$$

where **n** denotes the unit that is normal to the element axis.

In fact, the implicit gradient model can be related to the nonlocal integral model by solving the partial differential equation, i.e., Eq. (17), using Green's function method[27]. Green's function can be defined as the analytical solution of Eq. (17), with the local variable $f(x)$ replaced by a Dirac function $\delta(x-\xi)$ and calculated as follows

$$G(x,\xi) - c\,\nabla^2 G(x,\xi) = \delta(x-\xi) \tag{19}$$

where $G(x,\xi)$ = Green's function. For the 1D case, Green's function is expressed as follows[30]

$$G(x,\xi) = \frac{1}{2\sqrt{c}} \exp\left(-\frac{|x-\xi|}{\sqrt{c}}\right) \tag{20}$$

Notably, the right side of Eq.(17) can be expressed as follows

$$f(x)=\int_{\Omega} f(\xi)\delta(x-\xi)\mathrm{d}\Omega \tag{21}$$

By substituting Eq.(19) into Eq.(21), the regularized variable can be rewritten as follows

$$\bar{f}(x)=\int_{\Omega} G(x, \xi)f(\xi)\mathrm{d}\Omega \tag{22}$$

Evidently, Eq.(22) has the same form as Eq.(14) for the nonlocal integral model, which indicates that the implicit gradient model is a special case of the nonlocal integral model, where the weight function $\alpha(x, \xi)$ is defined as Green's function $G(x, \xi)$. The normalized factor $V_r=1$ for Green's function is derived as follows[27]

$$V_r=\int_{\Omega} G(x, \xi)\mathrm{d}\Omega=\int_{\Omega}\delta(x-\xi)\mathrm{d}\Omega-c\int_{\Omega}\nabla^2 G(x, \xi)\mathrm{d}\Omega=1 \tag{23}$$

where the divergence theorem and Eq.(18) are used.

The advantage of the implicit gradient model over the integral model is that it can be easily implemented in the FE code because the gradients are only computed at each individual integration point. Meanwhile, interactions exist among different integration points for the nonlocal integral models. Considering that the combination of the implicit gradient model and the force-based frame element has not been extensively investigated, an implicit gradient regularized/delocalized force-based element was developed in this study.

5 Element Implementation

5.1 Nonlocal Variable

To set up an implicit gradient formulation, the nonlocal variable for further regularization should be initially decided, although the choice of the variable remains somewhat controversial. Jirásek[31] investigated the most commonly used nonlocal variables in integral-type models and revealed that the nonlocal variable should be carefully selected because stress locking might occur when certain variables are selected, such as damage inelastic stress, inelastic stress increment, and inelastic strain. Realistic results are obtained when nonlocal averaging is operated under a strain or damage energy release rate. Considering that the present study aims to achieve delocalization/regularization at the element level, section deformation $\mathbf{d}(x)$, which corresponds to strain in the material constitutive models, is selected as the nonlocal variable for regularization.

5.2 Numerical Scheme of the Governing Equation

A special numerical scheme should be implemented to solve the partial differential Eq.(17). With section deformation $\mathbf{d}$ defined as the nonlocal variable, the partial differential equation can be converted to the following expression

$$\bar{\mathbf{d}}-c\nabla^2\bar{\mathbf{d}}=\mathbf{d} \tag{24}$$

The weak form of Eq.(24) is expressed as follows

$$\int_L w\bar{\mathbf{d}}\mathrm{d}x - c\int_L w\,\nabla^2\bar{\mathbf{d}}\mathrm{d}x = \int_L w\mathbf{d}\mathrm{d}x \tag{25}$$

where w = trial function. By implementing the integration of parts, the second term on the left side of Eq.(25) can be rewritten as follows

$$c\int_L w\,\nabla^2\bar{\mathbf{d}}\mathrm{d}x = cw\,\nabla\,\bar{\mathbf{d}}\,|_L - c\int_L \nabla w\,\nabla\,\bar{\mathbf{d}}\mathrm{d}x \tag{26}$$

Substituting the natural boundary condition into Eq.(26) yields the following equation

$$\int_L w\bar{\mathbf{d}}\mathrm{d}x + c\int_L \nabla w\,\nabla\,\bar{\mathbf{d}}\mathrm{d}x = \int_L w\mathbf{d}\mathrm{d}x \tag{27}$$

and FE discretization is required to solve Eq.(27). By introducing a 1D FE displacement-based shape function, the local section deformation $\mathbf{d}$, nonlocal section deformation $\bar{\mathbf{d}}$, and trial function w can be expressed as follows

$$\mathbf{d} = \mathbf{N}(x)\mathbf{q} \quad \bar{\mathbf{d}} = \mathbf{N}(x)\bar{\mathbf{q}} \quad w = \mathbf{N}(x)h \tag{28}$$

where $\mathbf{q}$ and $\bar{\mathbf{q}}$ = local and nonlocal nodal deformations, respectively; h = variable related to the trial function w; and $\mathbf{N}(x)$ contains the interpolation function matrix. Notably, this discretization is separated from the original force-based formulation because the procedure only aims to solve Eq.(24) and obtain the nonlocal counterpart. The assembly of element stiffness still follows a force-based manner. Meanwhile, the interpolation functions only need to satisfy C^0 continuity requirements because the local variable has no Laplacian operator in the partial differential equation, i.e., Eq.(24).

Accordingly, the gradients of nonlocal section deformation and trial function are derived as follows

$$\nabla\,\bar{\mathbf{d}} = \nabla\,\mathbf{N}\bar{\mathbf{q}} = \mathbf{B}\bar{\mathbf{q}} \quad \nabla w = \nabla\,\mathbf{N}h = \mathbf{B}h \tag{29}$$

where $\mathbf{B} = \nabla\mathbf{N}$. Substituting Eqs.(28) and (29) into Eq.(27) yields the following expression

$$\mathbf{A}\bar{\mathbf{q}} = \mathbf{y} \tag{30}$$

where

$$\mathbf{A} = \int_L \mathbf{N}^{\mathrm{T}}\mathbf{N}\mathrm{d}x + c\int_L \mathbf{B}^{\mathrm{T}}\mathbf{B}\mathrm{d}x \tag{31}$$

$$\mathbf{y} = \int_L \mathbf{N}^{\mathrm{T}}\mathbf{d}\mathrm{d}x \tag{32}$$

Thus, the nonlocal section deformation is obtained as follows

$$\bar{\mathbf{d}} = \mathbf{N}\mathbf{A}^{-1}\mathbf{y} \tag{33}$$

For each local section deformation component d (axial strain or curvature), Eq.(33) can be reduced to the following expression

$$\bar{d} = \mathbf{N}\mathbf{A}^{-1}\mathbf{y} \tag{34}$$

A three-point Lagrangian interpolation technique can be adopted to approximate the relevant variable fields of the element as follows

$$\mathbf{N}(x)=[(2\zeta-1)(\zeta-1)\quad -4\zeta(\zeta-1)\quad \zeta(2\zeta-1)] \tag{35}$$

$$\mathbf{B}(x)=\frac{1}{L}[4\zeta-3\quad 4-8\zeta\quad 4\zeta-1] \tag{36}$$

where $\zeta=x/L$ is the dimensionless position along the element. Therefore, Eq.(31) can be rewritten as follows

$$\mathbf{A}=\begin{bmatrix} \frac{2}{15}L+c\frac{7}{3L} & \frac{1}{15}L-c\frac{8}{3L} & -\frac{1}{30}L+c\frac{1}{3L} \\ \frac{1}{15}L-c\frac{8}{3L} & \frac{8}{15}L+c\frac{16}{3L} & \frac{1}{15}L-c\frac{8}{3L} \\ -\frac{1}{30}L+c\frac{1}{3L} & \frac{1}{15}L-c\frac{8}{3L} & \frac{2}{15}L+c\frac{7}{3L} \end{bmatrix} \tag{37}$$

Numerical integration should be applied to calculate $\mathbf{y}$ in Eq.(34). The Gaussian quadrature is also adopted, as shown in Eq.(13), to yield the following expression

$$\mathbf{y}=\sum_{i=1}^{N_p}\mathbf{N}(x_i)^{\mathrm{T}}d(x_i)w_i \tag{38}$$

This implicit gradient process can be easily implanted in a computer code on the basis of the force-based element. Notably, the entire process occurs within the element; thus, data transfer among different elements need not be implemented.

5.3 Determining the Gradient Coefficient

As mentioned previously, the gradient coefficient c is of the dimension length squared that introduces an internal length scale into the gradient model, which is similar to the characteristic length l in the nonlocal integral models. In fact, equivalence between the implicit gradient model and the nonlocal integral model can be achieved by setting the weight function $\alpha(x,\xi)$ in the integral model as Green's function $G(x,\xi)$, as shown in Eq.(22)[32]. From the physical perspective, the gradient coefficient c in $G(x,\xi)$ and characteristic length l in Gauss's distribution, i.e., Eq.(16), determine the width of the nonlocal region, as shown in Fig. 4. Previous studies have revealed that the relationship between c and l is $c=l^2/2$ [26-27], and the characteristic length is typically equal to the plastic hinge length l_p in a beam element[20]; thus, the gradient coefficient can be expressed as follows

$$c=\frac{l_p^2}{2} \tag{39}$$

In existing studies[33], the plastic hinge length l_p is calculated by an empirical formula. Then, the gradient coefficient c can be computed. However, the plastic hinge length is not a fixed value but an evolving variable, which indicates that the coefficient c is not a constant but a time-dependent variable. In fact, the plastic hinge length can be easily evaluated at

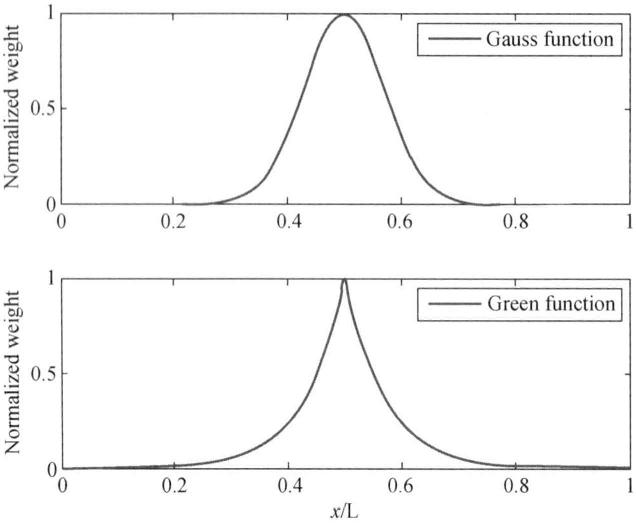

Fig. 4. Gauss distribution and Green's function

every loading step during the numerical analysis. The yielding of the reinforcement fiber or the crushing of the concrete fiber can be considered the index, which indicates the initiation of a plastic hinge. Before the index is reached, the plastic hinge length $l_p = 0$. If any reinforcement bar yields or concrete fiber crushes for the first time, then the corresponding moment M_p is recorded. Consequently, the plastic hinge length can be determined through linear interpolation, as shown in Fig. 5. Thus, the coefficient c can be derived using Eq.(39). Notably, the evolution of the plastic hinge is an irreversible process. Thus, l_p and c are monotonic increasing variables. Only when the value of the current step is larger than that of the previous step can the parameters be updated.

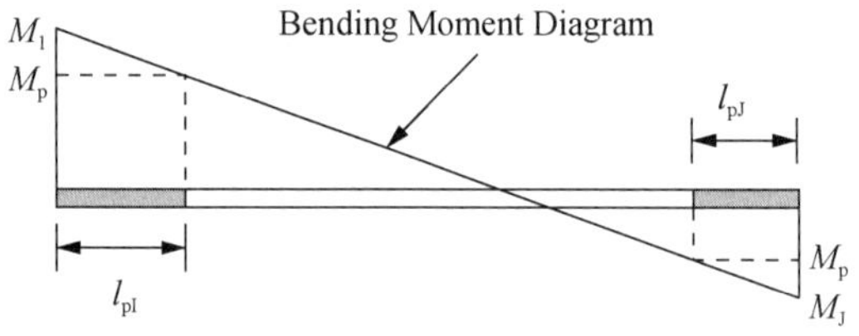

Fig. 5. Calculation of plastic hinge length

This implicit gradient formulation can be easily implemented by adding several modifications to the original force-based element formulation, as shown in Fig.6. Meanwhile, the flowchart of the nonlocal integral model is also included in Fig.6 for comparison. Notably, the element and section state determination shown in Fig.6 are on the basis of the solution procedure proposed by Spacone et al.[7], in which no distributed load is applied to the element. However, if the procedure proposed by Taucer et al.[11] is adopted, then the distributed load can also be considered. The main modifications of the proposed element involve each computing step. After local section deformations are

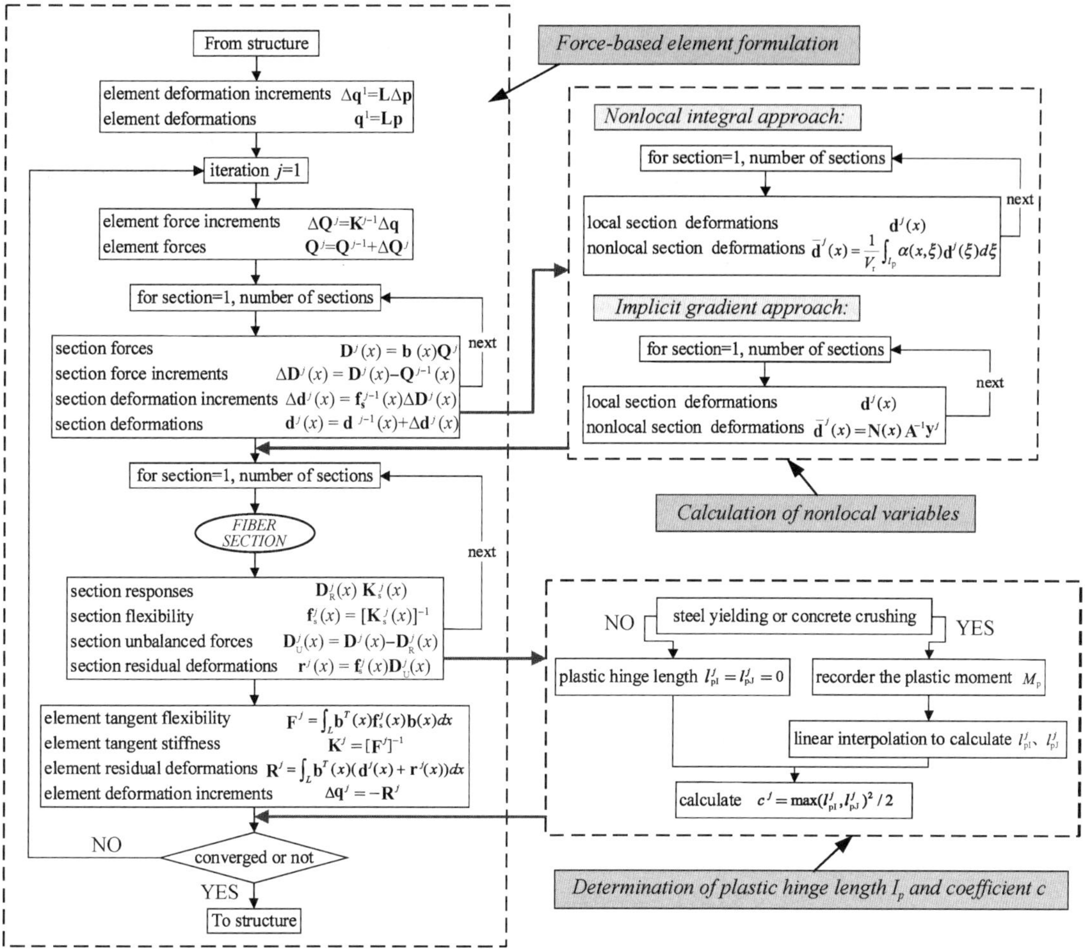

Fig. 6. Implementation of proposed element

obtained in iteration j, nonlocal deformations are calculated as follows

$$\bar{\mathbf{d}}^j(x) = \mathbf{N}\mathbf{A}^{-1}\mathbf{y}^j \tag{40}$$

$$\mathbf{y}^j = \int_L \mathbf{N}^{\mathrm{T}}\mathbf{d}^j \,\mathrm{d}x \tag{41}$$

Then, the sectional responses $\mathbf{D}_{\mathrm{R}}^j(x)$ and $\mathbf{k}_{\mathrm{s}}^j(x)$ are determined on the basis of $\bar{\mathbf{d}}^j(x)$. The real-time plastic hinge length at each end of the element is calculated through linear interpolation on the basis of current section internal forces, and thus, the gradient coefficient c^j is obtained using Eq. (39).

6 Numerical Examples

The element is implemented in ABAQUS[34] through the subroutine user-defined elements. This section presents two numerical examples to verify the proposed model. The comparative investigation of the nonlocal integral approach and the implicit gradient approach is conducted in the first example. The second example validates the performance of the element by comparing its prediction with the experimental results of a RC bridge pier under constant axial load and increasing top displacement.

6.1 Cantilever Beam

Fig. 7 shows a cantilever beam under increasing vertical displacement v placed at the free end. A bilinear moment-curvature relationship is adopted to describe the sectional behavior, where EI is elastic stiffness, M_y and φ_y are yield moment and curvature, respectively, and the modulus αEI denotes softening behavior. Setting $\alpha = -0.05$ and modeling the beam with a single force-based element, the global and local responses without delocalization/regularization are plotted in Fig. 8, where $\theta = ML/(2EI)$ is the rotation of the fixed section. Evidently, as the number of integration points increases, the postpeak response becomes increasingly brittle. Meanwhile, deformation localizes more to the first section.

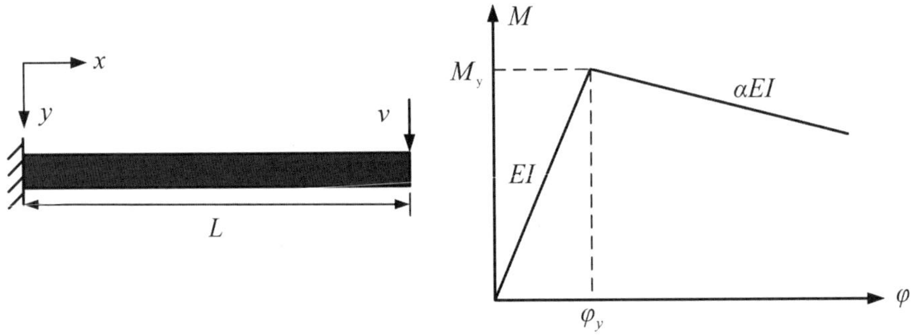

Fig. 7. Cantilever beam with softening behavior

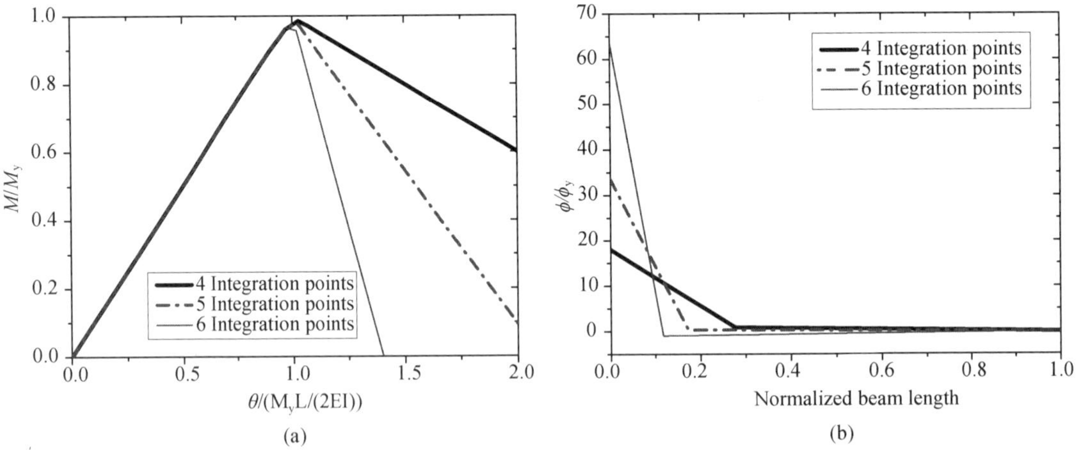

Fig. 8. Beam responses by force-based element: (a) moment-rotation curve; (b) curvature distribution

The nonlocal integral model and the implicit gradient model are used to simulate the beam. The gradient coefficient c and characteristic length l are set as constants in this example to highlight the difference between the two models. The characteristic length l is assumed as

$l = 0.15L$, and the gradient coefficient c is calculated as $c = l^2/2$.

The regularized results shown in Fig. 9 indicate that both approaches can provide converging results. In the implicit gradient approach, one element with five to seven integration points can obtain objective responses. However, in the nonlocal integral approach, a sufficient number of Gauss-Lobatto integration points should be within the nonlocal region to guarantee the accuracy of the nonlocal integral process, as shown in Fig. 10. Notably, an objective result can also be achieved for the nonlocal integral method as the number of integration points increases. Fig. 11 shows the corresponding curvature fields for both methods at the final step. Similarly, both regularized approaches provide objective predictions of the curvature distribution.

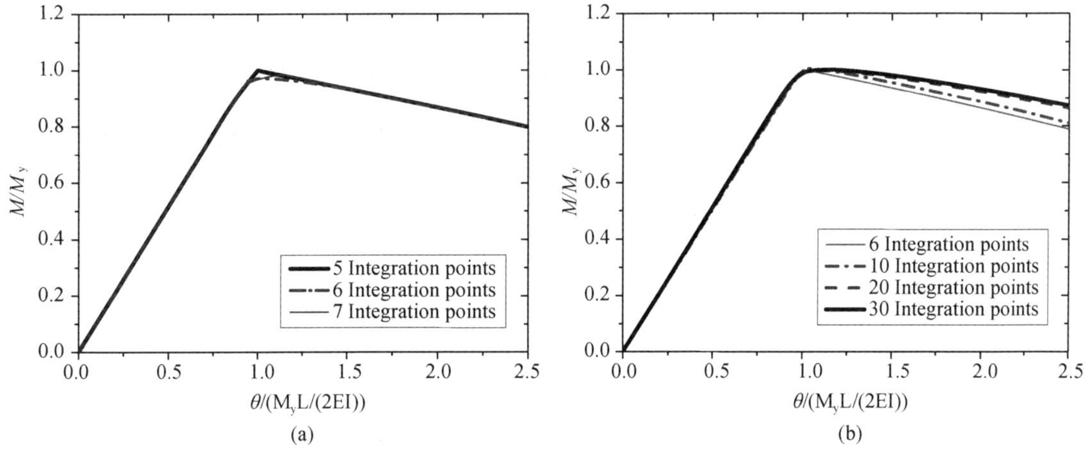

Fig. 9. Beam responses by delocalized element: (a) implicit gradient model; (b) nonlocal integral model

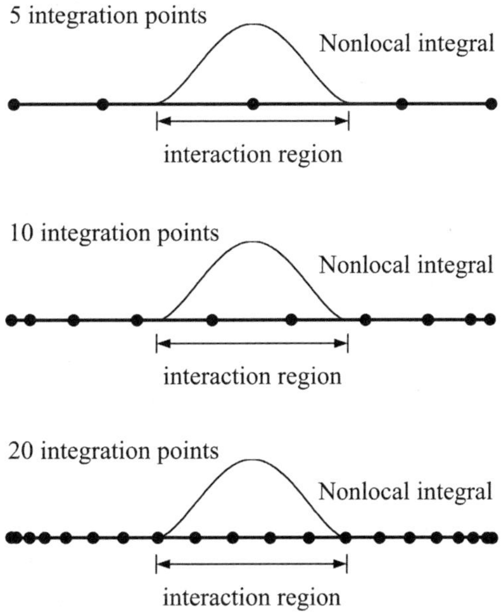

Fig. 10. Nonlocal integral model with different numbers of Gauss-Lobatto integration points

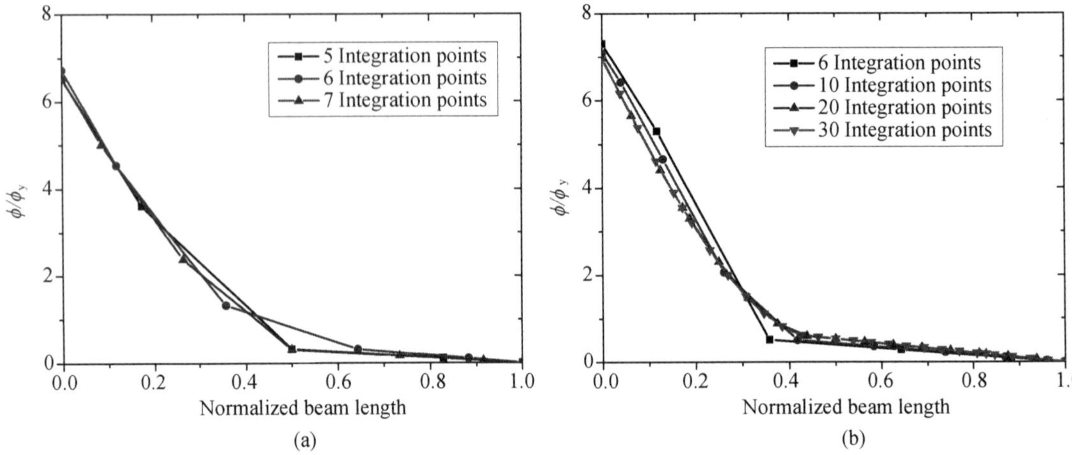

Fig. 11. Curvature distribution of beam: (a) implicit gradient model; (b) nonlocal integral model

The effect of the order of the interpolation function $\mathbf{N}(x)$ in Eq. (35) is also investigated. A total of three-, four-, and five-point Lagrangian interpolation techniques are used, and the numerical results are shown in Fig. 12. It can be concluded that the order of the interpolation function has a slight influence on the element response because the implicit gradient in Eq. (24) does not contain the Laplacian operator of the local variable. Thus, low-order regularity, particularly C^0 continuity requirements, of the local variable is required.

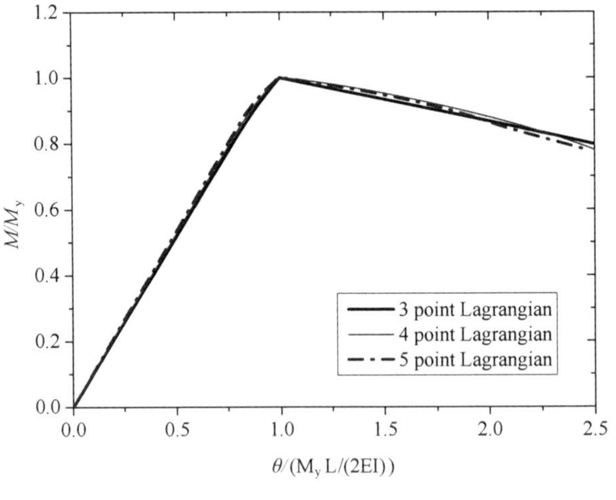

Fig. 12. Influence of interpolation function N

6.2 RC Bridge Pier

The RC bridge pier, i.e., Specimen 7 in the test of Tanaka and Park[35], is simulated and analyzed to verify the accuracy of the proposed delocalization model under strain-softening behavior. Fig. 13 shows the details of the pier. The compressive strength of concrete, f_c', is 32 MPa, and the strain corresponding to the compressive strength, ε_c, is 0.002 4.

Mander's model[36] is used to analyze the confining effect of transverse steel. When this model is used, the peak strength of the confined concrete, f'_{cc}, is 39 MPa at a strain, ε_{cc}, of 0.005 2, and the ultimate strain of the confined concrete, ε_{ccu}, is 0.024 8. For reinforcing steel, a bilinear stress-strain relationship is assumed with the elastic modulus E_s = 200 GPa, yield stress f_y = 510 MPa, and hardening ratio α = 0.01.

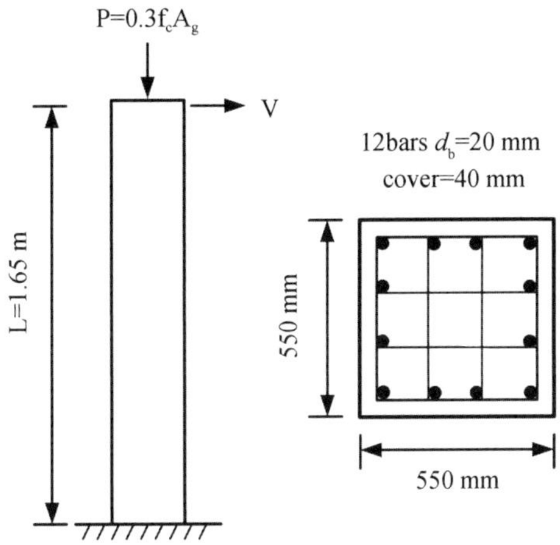

Fig. 13. Bridge pier test by Tanaka and Park(Specimen 7)

A single element is used to model the bridge pier. Fig. 14(a) shows the results of the force-based element without delocalization/regularization. The simulated responses decrease faster than the test results and demonstrate a high dependency on the number of integration points. This computational nonobjectivity originates from the rapidly increased plastic deformation in the localized region. The proposed delocalization model is used to establish an objective result. Fig. 14(b) shows the computed load-displacement responses. Evidently, the results match the test data well, and dependency on the number of integration points has been eliminated.

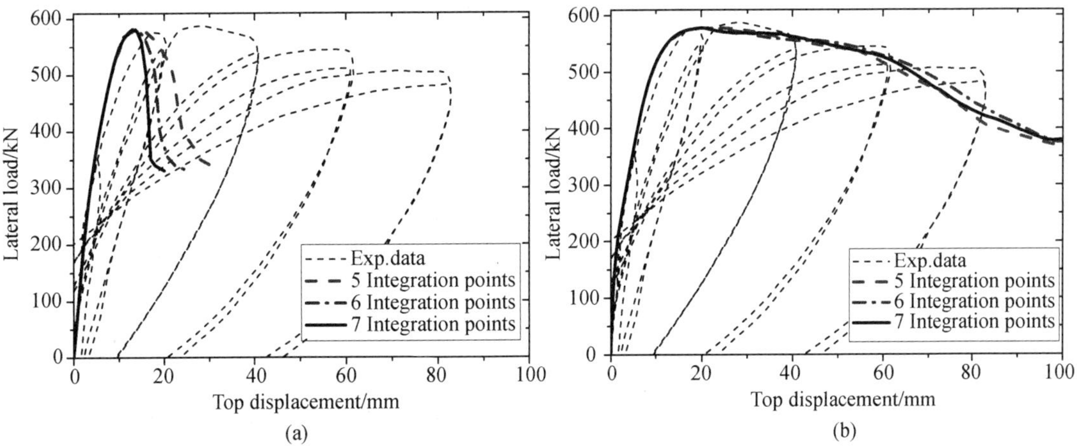

Fig. 14. Numerical results of bridge pier: (a) force-based element; (b) proposed model

Fig. 15 shows the evolution of the plastic hinge. Regardless of the number of integration points in the element, the base section reaches the plastic moment when the top displacement is 6.58 mm and the plastic hinge is fully formed at 18.58 mm. The eventual plastic hinge length is approximately 300 mm, which is close to the value (356 mm) calculated by the following empirical formula[33]

$$l_p = 0.08L + 0.022 f_y d_b \tag{42}$$

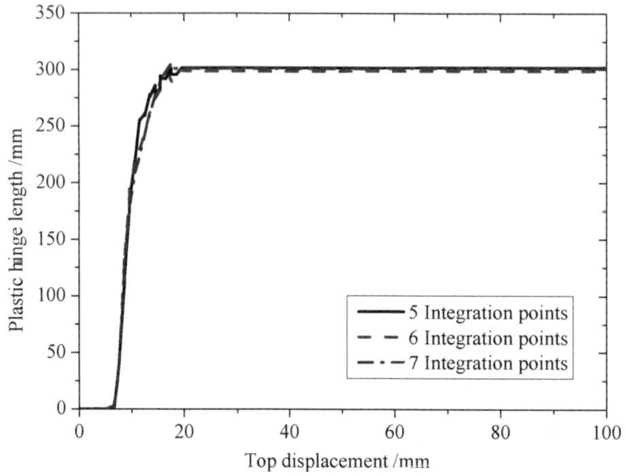

Fig. 15. Evolution of plastic hinge

where L = length of the beam or column; and f_y and d_b = strength and diameter of the longitudinal bar, respectively.

By contrast, Fig. 16 plots the moment and curvature distributions along the pier at different times. Notably, objective results are obtained before and after the plastic hinge is completely formed.

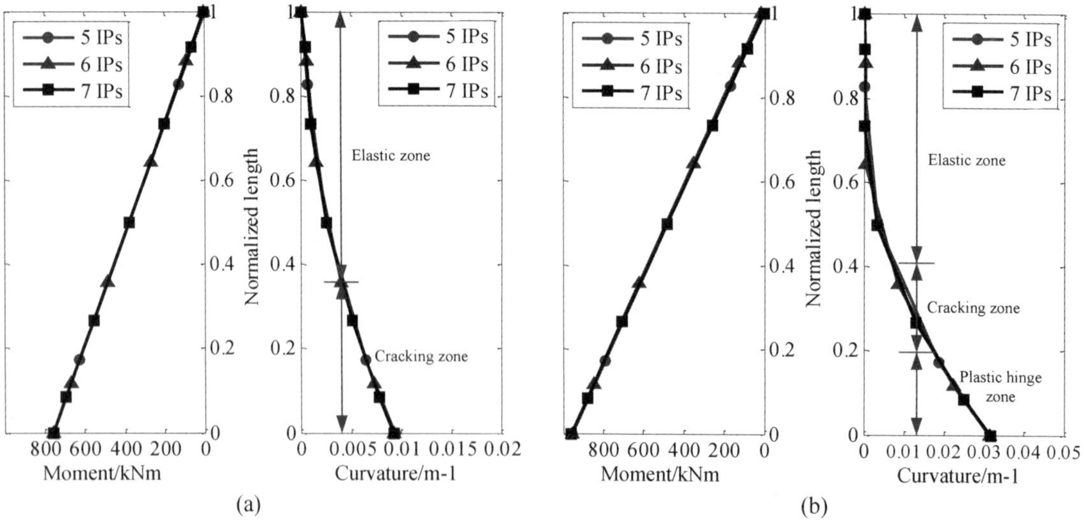

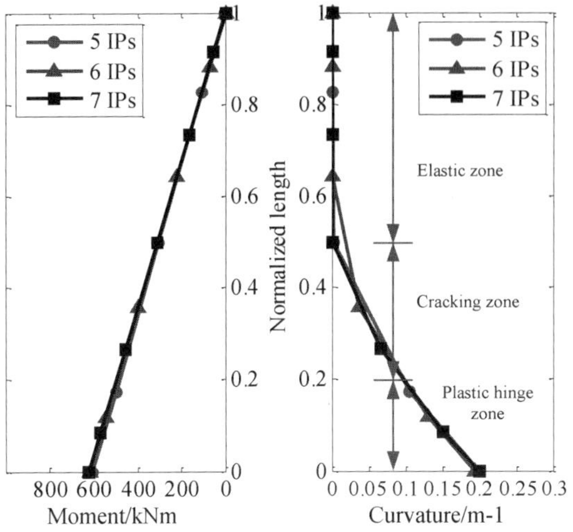

(c)

Fig. 16. Moment and curvature distributions along pier height: (a) top displacement=6.58 mm; (b) top displacement=18.58 mm; (c) top displacement=100 mm

7 Conclusion

In this study, a novel force-based frame element on the basis of the nonlocal implicit gradient delocalization formulation is developed to overcome the localization problem originating from strain-softening behavior. The section deformation is selected as the nonlocal variable and solved through a mixed formulation strategy. From the physical perspectives of localization issues, the gradient coefficient c in the model is related to the plastic hinge length. A new method is also developed to calculate the instantaneous plastic hinge length, depending on the element state at each loading step, and to update the gradient coefficient adaptively. The element can be implemented in the framework of the force-based element without any modifications. Using this model, the objective response for the strain-softening behavior can be attained at the global, i.e., load-displacement, and local, i.e., moment-curvature, levels. The plastic hinge evolution process can also be realized.

Acknowledgments

Financial support from the National Natural Science Foundation of China (Grant No. 91315301, 51261120374, and 51208374) is greatly appreciated.

References

[1] CLOUGH R W, BENUSKA K L, Wilson E L. Inelastic earthquake response of tall buildings[C]// Proc., 3rd World Conf. on Earthquake Engineering, New Zealand National Committee on Earthquake Engineering, Wellington, New Zealand, 1965.

[2] GIBERSON M F. The response of nonlinear multi-story structures subjected to earthquake excitation

[D]. California Institute of technology, 1967.
[3] HILMY S I, ABEL J F. Material and geometric nonlinear dynamic analysis of steel frames using computer graphics[J]. Computers & Structures, 1985, 21(4): 825-840.
[4] POWELL G H, CHEN P F S. 3D beam-column element with generalized plastic hinges[J]. Journal of Engineering Mechanics, 1986, 112(7): 627-641.
[5] SOLEIMANI D, POPOV E P, BERTERO V V. Nonlinear beam model for R/C frame analysis[C]// Electronic Computation. ASCE, 1979: 483-509.
[6] MEYER C, ROUFAIEL M S L, Arzoumanidis S G. Analysis of damaged concrete frames for cyclic loads[J]. Earthquake Engineering & Structural Dynamics, 1983, 11(2): 207-228.
[7] SPACONE E, FILIPPOU F C, TAUCER F F. Fibre beam-column model for nonlinear analysis of R/C frames: Part Ⅰ. Formulation[J]. Earthquake Engineering & Structural Dynamics, 1996, 25(7): 711-725.
[8] HELLESLAND J, SCORDELIS A. Analysis of RC bridge columns under imposed deformations[C]// IABSE Colloquium. the Netherlands: Delft, 1981: 545-559.
[9] MARI A R. Nonlinear geometric, material and time dependent analysis of three dimensional reinforced and prestressed concrete frames[D]. SESM Rep. 82-12, Dept. of Civil Engineering, Univ. of California, Berkeley, Berkeley, CA, 1984.
[10] CIAMPI V, CARLESIMO L. A nonlinear beam element for seismic analysis of structures[C]//8th European Conference on Earthquake Engineering. 1986, 3: 6.3.
[11] TAUCER F, SPACONE E, FILIPPOU F C. A fiber beam-column element for seismic response analysis of reinforced concrete structures[D]. EERC Rep. 91/17, Earthquake Engineering Research Center, College of Engineering, Univ. of California, Berkeley, Berkeley, CA, 1991.
[12] SPACONE E, FILIPPOU F C, TAUCER F F. Fibre beam-column model for nonlinear analysis of R/C frames: part Ⅱ. Applications[J]. Earthquake Engineering & Structural Dynamics, 1996, 25(7): 727-742.
[13] NEUENHOFER A, FILIPPOU F C. Evaluation of nonlinear frame finite-element models[J]. Journal of Structural Engineering, 1997, 123(7): 958-966.
[14] COLEMAN J, SPACONE E. Localization issues in force-based frame elements[J]. Journal of Structural Engineering, 2001, 127(11): 1257-1265.
[15] SCOTT M H, FENVES G L. Plastic hinge integration methods for force-based beam-column elements [J]. Journal of Structural Engineering, 2006, 132(2): 244-252.
[16] ADDESSI D, CIAMPI V. A regularized force-based beam element with a damage-plastic section constitutive law[J]. International Journal for Numerical Methods in Engineering, 2007, 70(5): 610-629.
[17] SCOTT M H, HAMUTÇUOǦLU O M. Numerically consistent regularization of force-based frame elements[J]. International Journal for Numerical Methods in Engineering, 2008, 76(10): 1612-1631.
[18] ALMEIDA J P, DAS S, PINHO R. Adaptive force-based frame element for regularized softening response[J]. Computers & Structures, 2012, 102: 1-13.
[19] KHALOO A R, TARIVERDILO S. Localization analysis of reinforced concrete members with softening behavior[J]. Journal of Structural Engineering, 2002, 128(9): 1148-1157.
[20] VALIPOUR H R, FOSTER S J. Nonlocal damage formulation for a flexibility-based frame element [J]. Journal of Structural Engineering, 2009, 135(10): 1213-1221.
[21] HILDEBRAND F B. Introduction to numerical analysis[M]. Courier Corporation, 1987.

[22] CALABRESE A, ALMEIDA J P, PINHO R. Numerical issues in distributed inelasticity modeling of RC frame elements for seismic analysis[J]. Journal of Earthquake Engineering, 2010, 14(S1): 38-68.

[23] ERINGEN A C, EDELEN D G B. On nonlocal elasticity[J]. International Journal of Engineering Science, 1972, 10(3): 233-248.

[24] BAŽANT Z P, OH B H. Crack band theory for fracture of concrete[J]. Matériaux et Construction, 1983, 16(3): 155-177.

[25] BAŽANT Z P, JIRÁSEK M. Nonlocal integral formulations of plasticity and damage: survey of progress[J]. Journal of Engineering Mechanics, 2002, 128(11): 1119-1149.

[26] PEERLINGS R H J, DE BORST R, Brekelmans W A M, et al. Gradient enhanced damage for quasibrittle materials[J]. International Journal for Numerical Methods in Engineering, 1996, 39(19): 3391-3403.

[27] PEERLINGS R H J, GEERS M G D, DE BORST R, et al. A critical comparison of nonlocal and gradient-enhanced softening continua[J]. International Journal of Solids and Structures, 2001, 38(44-45): 7723-7746.

[28] LASRY D, BELYTSCHKO T. Localization limiters in transient problems[J]. International Journal of Solids and Structures, 1988, 24(6): 581-597.

[29] MÜHLHAUS H B, ALFANTIS E C. A variational principle for gradient plasticity[J]. International Journal of Solids and Structures, 1991, 28(7): 845-857.

[30] ZAUDERER E. Partial differential equations of applied mathematics[M]. Wiley, New York, 2011.

[31] JIRASEK M. Nonlocal models for damage and fracture: comparison of approaches[J]. International Journal of Solids and Structures, 1998, 35(31-32): 4133-4145.

[32] TOVO R, LIVIERI P, BENVENUTI E. An implicit gradient type of static failure criterion for mixed-mode loading[J]. International Journal of Fracture, 2006, 141(3): 497-511.

[33] PAULAY T, PRIESTLEY M J N. Seismic design of reinforcedconcrete and masonry buildings[M]. Wiley, New York, 1992.

[34] HIBBETT, KARLSSON, SORENSEN. ABAQUS/standard: User'smanual, Hibbett, Karlsson and Sorensen, Pawtucket, RI, 1998.

[35] TANAKA H, PARK R. Effect of lateral confining reinforcement on the ductile behaviour of reinforced concrete columns[M]. Rep. No.90-2, Dept. of Civil Engineering, Univ. of Canterbury, Canterbury, U.K, 1990.

[36] MANDER J, PRIESTLEY M J N, PARK R. Theoretical Stress-Strain Model for Confined Concrete Columns[J]. Journal of Structural Engineering, 1804-1826.

Two-level consistent secant operators for cyclic loading of structures①

Xiaodan Ren[1], **Jie Li**[2]

1 *Associate Professor, College of Civil Engineering, Tongji Univ., 1239 Siping Rd., Shanghai 200092, China (corresponding author). Email: rxdtj@tongji.edu.cn*

2 *Professor, State Key Laboratory of Disaster Reduction in Civil Engineering, Tongji Univ., 1239 Siping Rd., Shanghai 200092, China.*

Abstract Two-level consistent secant operators are proposed for the nonlinear simulation of structures especially under cyclic loading. At the structural level, the quasi-Newton (Q-N) update of stiffness matrix is used to search the equilibrium of the overall system. At the material level, the general formulas of secant operators are proposed for the nonlinear material model. The consistency between the Q-N stiffness matrix at the structural level and the secant operator at the material level is proposed and considered as a good trade-off between efficiency and robustness. Applications are made to numerical examples that include strain-softening and cyclic loading. Significant improvements are demonstrated in relation to the full Newton methods and the inconsistent secant methods.

Keywords Stiffness matrix; Secant operator; Plasticity; Damage; Cyclic loading

1 Introduction

In recent years, the rapid development of innovative composite materials and engineering structures has presented systematic challenges for the analytical and computational methods. The simulation of engineering structures experiencing strong nonlinearities usually involve two aspects: the development of a representative numerical model and the solution of the developed nonlinear system. For the first aspect, numbers of breakthroughs have been made in the past 20 years, including the extended finite element method (XFEM) for crack propagation[1-4], meshfree/meshless methods for extremely large deformation[5-7], elaborate plasticity/damage models for complex material nonlinearities[8-12] and nonlocal/gradient methods for strain localization[13-16]. For the second aspect, the currently used methods for the solution of nonlinear equations in nonlinear structural analysis are more or less old-fashioned. They are primarily based on the concept of Newtonian iterative procedures. The full Newton procedure with the consistent tangent operator has a quadratic rate of (asymptotic) convergence[17-18]. Thus, it is the most widely used method in the solution of nonlinear equations because of its efficiency.

---

① Originally published in *Journal of Engineering Mechanics*, 2018, 144(8): 04018065.

Practical engineering structures may experience unstable nonlinearities, which may not be effectively solved by the full Newton procedure. Some examples are, the propagation of multiple cracks in brittle solids, the composite material with complex structures and material constitutive law with certain unsmoothness. For these cases, researchers consider the quasi-Newton (Q-N) methods, which have superlinear convergence and improved robustness. The cornerstone of the Q-N methods for nonlinear finite element (FE) solutions was set by Matthies and Strang[19]. They introduced the Broyden-Fletcher-Goldfarb-Shanno (BFGS)[20] update to the stiffness matrix $\mathbb{K}$ and developed the direct update formula for $\mathbb{K}^{-1}$. Subsequently, Bathe and Cimento[21] applied the Q-N method with BFGS update to implicit static and dynamic simulation of structures. This method has been adopted in nonlinear FE packages, such as ADINA and ABAQUS. However, Q-N methods are not frequently used in practice. For the stable structure problem, we usually prefer the full Newton method due to its quadratic convergence rate. For the irregular problem with structural instability, the current version of Q-N method does not seem to be robust enough to overcome most of the problems.

As is well known, the Newtonian methods (including the full Newton and the Q-N methods) are widely used not only in nonlinear equations but also in optimization problems. It could be indicated that the two classes of problems are of the same mathematical nature. At present, the full Newton method is more frequently used than the Q-N method to solve nonlinear systems developed by the FE method because of its efficiency. However, mathematicians seem to prefer the Q-N methods when solving optimization problems[22], especially for rather complicated problems with substantial nonlinearities and discontinuities. By learning this fact, we believe that the current version of Q-N methods for the solution of nonlinear FE equations still have substantial room for further improvement.

In the context of the full Newton procedure, Simo and Taylor[17] established the concept of the consistent tangent operator, which emphasized that the tangent modulus in the linearized problem must be obtained through consistent linearization of the response function resulting from the integration algorithm to preserve the quadratic rate of asymptotic convergence. According to our point of view, *consistent* indicates that the linearization at the structural level should be in accordance with the linearization at the material level. Thus, the full Newton procedure at the structural level should coordinate with the full linearization of the material model discretized by a certain integration scheme to achieve optimal efficiency. For the Q-N procedure, the concept of *consistency* still remains. In other words, the Q-N procedure at the structural level should coordinate with the Q-N linearization of the material model. Therefore, we propose the consistent secant operators for nonlinear stimulation of structures. The improvement of robustness for the structural simulation is observed within the proposed framework.

In the present work, we concentrate on the application of the proposed method to the cyclic loading problem of structures. On one hand, the cyclic loading problem plays a very important role in earthquake engineering. The seismic behaviors of structures are usually investigated by using cyclic loading tests. On the other hand, the cyclic loading tests are difficult to reproduce in numerical simulations. The materials usually experience severe damage, cracking, or local buckling, for which the commonly used full Newton method is likely to diverge. Although the arc-length methods[23-24] are a class of good techniques for monotonic loading problems with instabilities; they are seldom used in the cyclic loading problem which is displacement controlled loading in its nature. To sum up, the robustness and accuracy of the proposed consistent secant method may be well demonstrated based on the cyclic loading problem, which is also of special importance in engineering application.

2 Secant Operators in Different Levels

In the present study, the displacement based FE formulation of the static nonlinear structural problems is considered. The weak form is expressed as follows

$$\int_{\Omega} \delta\boldsymbol{\varepsilon} : \boldsymbol{\sigma}(\boldsymbol{\varepsilon})\mathrm{d}\Omega - \int_{\Omega} \delta\boldsymbol{u} \cdot \boldsymbol{b}\mathrm{d}\Omega - \int_{\Gamma_t} \delta\boldsymbol{u} \cdot \boldsymbol{t}\mathrm{d}\Gamma = 0 \tag{1}$$

where $\boldsymbol{\sigma}$, $\boldsymbol{\varepsilon}$, and $\boldsymbol{u}$ = stress, strain, and displacement, respectively; and $\boldsymbol{b}$ and $\boldsymbol{t}$ = body force and surface traction.

The displacement based FE approximation is expressed as follows

$$\begin{aligned} \boldsymbol{u} &= \boldsymbol{N}\tilde{u} \\ \boldsymbol{\varepsilon} &= \boldsymbol{B}\tilde{u} \end{aligned} \tag{2}$$

where $\boldsymbol{N}$ and $\boldsymbol{B}$ = shape functions for displacement and strain, respectively. Substituting Eq.(2) into Eq.(1) yields the nonlinear FE equations with the nodal displacement vector $\tilde{\boldsymbol{u}}$ as unknown. We have

$$\boldsymbol{P}(\tilde{\boldsymbol{u}}) = \boldsymbol{f} \tag{3}$$

where the inner force and external force vectors

$$\begin{aligned} \boldsymbol{P}(\tilde{\boldsymbol{u}}) &= \int_{\Omega} \boldsymbol{B}^{\mathrm{T}}\boldsymbol{\sigma}(\boldsymbol{B}\tilde{\boldsymbol{u}})\mathrm{d}\Omega = \mathbf{A}_{e=1}^{n_{\mathrm{el}}} \int_{\Omega_e} \boldsymbol{B}_e^{\mathrm{T}}\boldsymbol{\sigma}(\boldsymbol{B}_e\tilde{\boldsymbol{u}})\mathrm{d}\Omega \\ \boldsymbol{f} &= \int_{\Omega} \boldsymbol{N}^{\mathrm{T}}\boldsymbol{b}\mathrm{d}\Omega + \int_{\Gamma_t} \boldsymbol{N}^{\mathrm{T}}\boldsymbol{t}\mathrm{d}\Gamma = \mathbf{A}_{e=1}^{n_{\mathrm{el}}} \left(\int_{\Omega_e} \boldsymbol{N}_e^{\mathrm{T}}\boldsymbol{b}\mathrm{d}\Omega + \int_{\Gamma_t^e} \boldsymbol{N}_e^{\mathrm{T}}\boldsymbol{t}\mathrm{d}\Gamma\right) \end{aligned} \tag{4}$$

where n_{el} denotes the number of elements, and the subscript e denotes that the corresponding variables are defined at the element level. The vector expressions of $\boldsymbol{\sigma}$ and $\boldsymbol{\varepsilon}$ in the present paper follow the Voigt notation.

To solve the nonlinear FE Eq.(3), the following iterative schemes are often adopted.

$$\begin{aligned} \mathbf{K}^m \Delta\tilde{\boldsymbol{u}}^{m+1} &= \boldsymbol{f} - \boldsymbol{P}(\tilde{\boldsymbol{u}}^m) \\ \tilde{\boldsymbol{u}}^{m+1} &= \tilde{\boldsymbol{u}}^m + \Delta\tilde{\boldsymbol{u}}^{m+1} \end{aligned} \tag{5}$$

In the classic Newton-Raphson (N-R) method, the iterative matrix is defined as the following tangential stiffness matrix

$$\mathbf{K}_{\tan}^{m} = \left.\frac{\partial \boldsymbol{P}}{\partial \tilde{\boldsymbol{u}}}\right|_{\tilde{\boldsymbol{u}}^m} = \mathop{\mathbf{A}}_{e=1}^{n_{el}} \int_{\Omega_e} \boldsymbol{B}_e^{\mathrm{T}} \left[\frac{\partial \boldsymbol{\sigma}}{\partial \boldsymbol{\varepsilon}}\right]_{\tilde{\boldsymbol{u}}^m} \boldsymbol{B}_e \mathrm{d}\Omega \tag{6}$$

The N-R iterative matrix [Eq.(6)] offers the optimal (quadratic) convergence rate, but experiences numerical singularities for many cases including perfect plasticity and strain softening. The closed form of consistent tangential operator $\mathbb{T} = \frac{\partial \boldsymbol{\sigma}_m}{\partial \boldsymbol{\varepsilon}_m}$ [18] may not be easily developed for some commonly used constitutive models because of its complexity in relation to second-order derivatives to the evolutionary potential function.

2.1 Structural Level Based Secant Operator

Many other choices for the matrix iteration in Eq.(5) could also achieve good convergence. The Q-N method, which adopts the secant stiffness matrix as the iterative matrix in Eq.(5), is developed for more stable iterations of Eq.(5). To develop the secant stiffness matrix, an appropriate state prior to the current state of the structural system is defined as the reference state. Then the nonlinear FE equation of the reference state could be easily developed as follows

$$\boldsymbol{P}(\tilde{\boldsymbol{u}}_{\mathrm{ref}}) = \boldsymbol{f}_{\mathrm{ref}} \tag{7}$$

where the subscript *ref* denotes that the corresponding variables are defined for the reference state. The incremental equations between Eqs.(3) and (7) read

$$\boldsymbol{P}(\tilde{\boldsymbol{u}}) - \boldsymbol{P}(\tilde{\boldsymbol{u}}_{\mathrm{ref}}) = \boldsymbol{f} - \boldsymbol{f}_{\mathrm{ref}} \tag{8}$$

Define the secant stiffness matrix as follows

$$\boldsymbol{P}(\tilde{\boldsymbol{u}}) - \boldsymbol{P}(\tilde{\boldsymbol{u}}_{\mathrm{ref}}) = \mathbf{K}_{\mathrm{sec}}(\tilde{\boldsymbol{u}} - \tilde{\boldsymbol{u}}_{\mathrm{ref}}) \tag{9}$$

Substituting Eq.(9) into Eq.(8), we finally obtain the secant equation as follows

$$\mathbf{K}_{\mathrm{sec}}(\tilde{\boldsymbol{u}} - \tilde{\boldsymbol{u}}_{\mathrm{ref}}) = \boldsymbol{f} - \boldsymbol{f}_{\mathrm{ref}} \tag{10}$$

The secant [Eq.(10)] is usually considered as the standard definition of secant stiffness matrix. By adopting $\mathbf{K}_{\mathrm{sec}}$ as the iterative matrix in Eq.(5), the Quasi-Newton (Q-N) method could be built.

The secant equation [Eq.(10)] does not completely specify the secant matrix $\mathbf{K}_{\mathrm{sec}}$ when the dimension of the nonlinear Eq.(3) is larger than 1. In fact, if we define N as the dimension of the nonlinear system, an $N(N-1)$-dimensional subspace of matrix satisfies Eq.(10). Actually, several prominent studies (e.g., those by Broyden, Powell, and Dennis) were finished, and a series of successful methods to develop the secant matrices were proposed. The Broyden-Fletcher-Goldfarb-Shannon (BFGS)[20] update, which was introduced to solve the nonlinear FE equation by Matthies and Strang[19], has been proven to be one of the most efficient and robust methods. Based on the definition of the current state as the m-th state solved by the m-th iteration and the reference state as the state solved by the $(m-1)$-th iteration, the secant equation was

$$
\begin{aligned}
&\mathbf{K}_{\mathrm{sec}}^{m}\Delta\tilde{\boldsymbol{u}}^{m}=\Delta\boldsymbol{f}^{m}\\
&\Delta\tilde{\boldsymbol{u}}^{m}=\tilde{\boldsymbol{u}}^{m}-\tilde{\boldsymbol{u}}^{m-1}\\
&\Delta\boldsymbol{f}^{m}=\boldsymbol{f}^{m}-\boldsymbol{f}^{m-1}
\end{aligned}
\tag{11}
$$

Matthies and Strang introduced the BFGS update to develop the inverse secant matrix for each iteration as follows

$$
\begin{aligned}
&(\mathbf{K}_{\mathrm{sec}}^{m})^{-1}=\boldsymbol{\Pi}^{m}(\mathbf{K}_{\mathrm{sec}}^{m-1})^{-1}\boldsymbol{\Pi}^{m}+\rho_{m}\Delta\tilde{\boldsymbol{u}}^{m}(\Delta\tilde{\boldsymbol{u}}^{m})^{\mathrm{T}}\\
&\boldsymbol{\Pi}^{m}=[\boldsymbol{\Lambda}-\rho_{m}\Delta\boldsymbol{f}^{m}(\Delta\tilde{\boldsymbol{u}}^{m})^{\mathrm{T}}]
\end{aligned}
\tag{12}
$$

where $\boldsymbol{\Lambda}$ = identity matrix and

$$
\rho_{m}=(\Delta\tilde{\boldsymbol{u}}^{m})^{\mathrm{T}}\Delta\boldsymbol{f}^{m}
\tag{13}
$$

Although more iterations are usually needed for the Matthies-Strang (M-S) secant method than the N-R method, the former is still computationally effective because the inverse secant matrix is directly solved without performing the matrix inversion in each iteration. Furthermore, the M-S secant method is usually more stable than the N-R method because its resulting secant matrix is well-posed. One major difficulty of the M-S method is the selection of the initial matrix, $\mathbf{K}_{\mathrm{sec}}^{0}$. The proponents[19, 21] did not discuss this issue. ABAQUS[25] selected the tangential stiffness matrix $\mathbf{K}_{\mathrm{tan}}^{0}$ as the initial matrix, which may, at the same time, degrade the stability of the secant method. In addition, the bandwidth of the resulting secant matrix may be large if the update of Eq. (12) has been performed iteratively for numerous times without updating the initial matrix.

2.2 Material-Level-Based Secant Operator

According to the present study, an alternative way to develop the secant method is also based on the reference state. The reference state is defined on the domain and boundary (Ω_{ref}, Γ_{rcf}), and the current state is defined on the domain and boundary (Ω, Γ). For a small deformation problem, (Ω_{ref}, Γ_{ref}) and (Ω, Γ) are the same. For a large deformation problem, the geometric mapping between (Ω_{ref}, Γ_{ref}) and (Ω, Γ) should be employed. The boundary conditions and the solution of the governing equations are known for the reference state. For the current state, the boundary conditions are known, and the governing equations are to be solved.

For simplicity, consider the small deformation problem in the present work. Thus, the weak form of the governing equation defined on the reference state could be expressed as follows

$$
\int_{\Omega}\delta\boldsymbol{\varepsilon}:\boldsymbol{\sigma}(\boldsymbol{\varepsilon}_{\mathrm{ref}})\mathrm{d}\Omega-\int_{\Omega}\delta\boldsymbol{u}\cdot\boldsymbol{b}_{\mathrm{ref}}\mathrm{d}\Omega-\int_{\Gamma_t}\delta\boldsymbol{u}\cdot\boldsymbol{t}_{\mathrm{ref}}\mathrm{d}\Gamma=0
\tag{14}
$$

The incremental between Eqs. (1) and (14) gives

$$
\int_{\Omega}\delta\boldsymbol{\varepsilon}:[\boldsymbol{\sigma}(\boldsymbol{\varepsilon})-\boldsymbol{\sigma}(\boldsymbol{\varepsilon}_{\mathrm{ref}})]\mathrm{d}\Omega-\int_{\Omega}\delta\boldsymbol{u}\cdot(\boldsymbol{b}-\boldsymbol{b}_{\mathrm{ref}})\mathrm{d}\Omega-\int_{\Gamma_t}\delta\boldsymbol{u}\cdot(\boldsymbol{t}-\boldsymbol{t}_{\mathrm{ref}})\mathrm{d}\Gamma=0
\tag{15}
$$

The secant stiffness tensor bridging stress and strain can be defined as follows

$$\boldsymbol{\sigma} - \boldsymbol{\sigma}_{\text{ref}} = \mathbb{S} : (\boldsymbol{\varepsilon} - \boldsymbol{\varepsilon}_{\text{ref}}) \tag{16}$$

How to determine $\mathbb{S}$ will be discussed in the next section. Then, by using Eq.(16), Eq.(15) further reads

$$\int_{\Omega} \delta\boldsymbol{\varepsilon} : \mathbb{S} : (\boldsymbol{\varepsilon} - \boldsymbol{\varepsilon}_{\text{ref}})\mathrm{d}\Omega - \int_{\Omega} \delta\boldsymbol{u} \cdot (\boldsymbol{b} - \boldsymbol{b}_{\text{ref}})\mathrm{d}\Omega - \int_{\Gamma_t} \delta\boldsymbol{u} \cdot (\boldsymbol{t} - \boldsymbol{t}_{\text{ref}})\mathrm{d}\Gamma = 0 \tag{17}$$

By substituting the FE approximation Eq.(2) into Eq.(17), we obtain the same secant equation as shown in Eq.(10) with the secant matrix explicitly expressed by the secant stiffness tensor in the material level. The following is obtained

$$\mathbb{K}_{\text{sec}} = \int_{\Omega} \boldsymbol{B}^{\mathrm{T}}\, \mathbb{S}\, \boldsymbol{B}\mathrm{d}\Omega = \mathop{\mathbf{A}}_{e=1}^{n_{\text{el}}} \int_{\Omega_e} \boldsymbol{B}_e^{\mathrm{T}}\, \mathbb{S}\, \boldsymbol{B}_e \mathrm{d}\Omega \tag{18}$$

The matrix expression of the fourth-order tensor $\mathbb{S}$ also obeys the Voight rule. By giving $\mathbb{S}$ and using the secant matrix defined by Eq.(18), we can perform the iterative scheme of Eq.(5) and solve the nonlinear FE equations.

On one hand, the secant stiffness matrix $\mathbb{K}_{\text{sec}}$ defined by Eq.(18) carries more information from the material level than the structural-level-based matrix such as $\mathbf{K}_{\text{sec}}$ in Eq.(12). Thus, the material nonlinearity of the structure should be better described by the former. On the other hand, comparing with $\mathbf{K}_{\text{sec}}$ in Eq.(12), which is obtained from the linearization of the traction term, the complete information of structural force and geometry may be lacking for $\mathbb{K}_{\text{sec}}$ in Eq.(18).

2.3 Two-Level Strategy

To address the problems encountered in the schemes at different levels, and to develop economical numerical algorithms to solve the nonlinear FE equations, we propose a two-level update strategy for the secant operators. The idea is to start the iterative update of the secant matrix [Eq.(12)] at the structural level by using the material level based secant matrix [Eq.(18)]. In addition, the updates of Eq.(12) are restarted by the secant matrix [Eq.(18)] after a certain number of iterations, so that the bandwidth and the cumulative errors of the secant matrix could be reduced effectively.

We consider the incremental solution scheme of the nonlinear structural problem. The loading procedure is discretized by a series of time steps, and the solution is performed step by step. We assume that the n-th step has been solved, with the equilibrium achieved and all the variables determined. Then, by providing the loading conditions at the $(n+1)$-th step, we can solve the corresponding state variables, such as displacement, internal force, stress, and strain, through the proposed two-level scheme starting from the n-th step, as shown in Fig. 1. Moreover, the details of the proposed two-level strategy are shown in Fig. 2.

3 Secant Operators for Material Inelasticity

At each integration point of the structural FE model, the material stiffness matrix could be

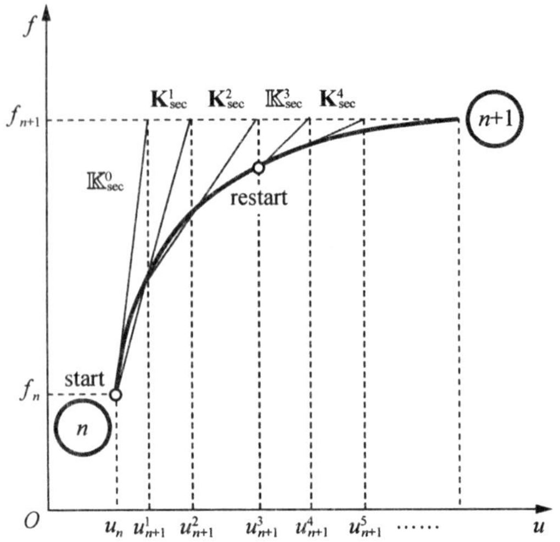

Fig. 1. Incremental moduli

begin

  Import $\tilde{\boldsymbol{u}}_n, f_n, \sigma_n, \varepsilon_n, f_{n+1}$;

  /* Initialize iteration */

  Calculate $\mathbb{S}^0$; /* to be developed in the next section */

  $(\mathbf{K}^0_{sec})^{-1} \leftarrow \left(\int_\Omega \boldsymbol{B}^T \mathbb{S}^0 \boldsymbol{B} d\Omega\right)^{-1}$;

  $\tilde{\boldsymbol{u}}^0_{n+1} \leftarrow \tilde{\boldsymbol{u}}_n, \boldsymbol{P}^0_{n+1} = f_n$;

  /* Iterative solution with maximum M iterations */

  for $m \leftarrow 1$ **to** M **do**

    $\Delta\tilde{\boldsymbol{u}}^m_{n+1} \leftarrow (\mathbf{K}^{m-1}_{sec})^{-1}(f_{n+1} - \boldsymbol{P}^{m-1}_{n+1})$;

    $\tilde{\boldsymbol{u}}^m_{n+1} \leftarrow \tilde{\boldsymbol{u}}^{m-1}_{n+1} + \Delta\tilde{\boldsymbol{u}}^m_{n+1}$;

    $\boldsymbol{P}^m_{n+1} \leftarrow \int_\Omega \boldsymbol{B}^T \sigma(\boldsymbol{B}\tilde{\boldsymbol{u}}^m_{n+1}) d\Omega$;

    $\Delta f^m_{n+1} \leftarrow \boldsymbol{P}^m_{n+1} - \boldsymbol{P}^{m-1}_{n+1}, \rho_m \leftarrow (\Delta\tilde{\boldsymbol{u}}^m_{n+1})^T \Delta f^m_{n+1}$;

    /* Error norm calculation */

    DispError $\leftarrow \frac{\|\tilde{\boldsymbol{u}}^m_{n+1} - \tilde{\boldsymbol{u}}^{m-1}_{n+1}\|_p}{\|\tilde{\boldsymbol{u}}^m_{n+1} - \tilde{\boldsymbol{u}}_n\|_p}$;

    ForceError $\leftarrow \frac{\|\boldsymbol{P}(\tilde{\boldsymbol{u}}^m_{n+1}) - f_{n+1}\|_p}{\|f_n - f_{n+1}\|_p}$;

    if DispError $\le \epsilon_d$ **AND** ForceError $\le \epsilon_f$ **then**

      $\tilde{\boldsymbol{u}}_{n+1} \leftarrow \tilde{\boldsymbol{u}}^m_{n+1}$;

      return with $\tilde{\boldsymbol{u}}_{n+1}$;

      else

        if $m{=}M$ **then**

          return with divergence message ;

        end

      end

    end

    /* Update secant matrix */

    if mod(m,mRestart)=0 **then**

      /* Restart secant matrix update every mRestart iterations */

      Calculate $\mathbb{S}^m_n$; /* to be developed in the next section */

      $(\mathbf{K}^m_{sec})^{-1} \leftarrow \left(\int_\Omega \boldsymbol{B}^T \mathbb{S}^m \boldsymbol{B} d\Omega\right)^{-1}$;

      else

        /* MS-BFGS update */

        $\mathbf{R}^m \leftarrow \left[\boldsymbol{\Lambda} - \rho_m \Delta f^m_{n+1} (\Delta\tilde{\boldsymbol{u}}^m_{n+1})^T\right]$;

        $(\mathbf{K}^m_{sec})^{-1} \leftarrow \mathbf{R}^m (\mathbf{K}^{m-1}_{sec})^{-1} \mathbf{R}^m + \rho_m \Delta\tilde{\boldsymbol{u}}^m_{n+1} (\Delta\tilde{\boldsymbol{u}}^m_{n+1})^T$;

      end

    end

  end

end

Fig. 2. Two-level algorithm

developed. Previously, the tangent modulus $\mathbb{T}=\left.\frac{\partial \boldsymbol{\sigma}}{\partial \boldsymbol{\varepsilon}}\right|_m$ was usually adopted. Then, Simo and Hughes[18] proposed the algorithm consistent tangential operator, $\mathbb{T}^{\mathrm{alg}}=\frac{\partial \boldsymbol{\sigma}_m}{\partial \boldsymbol{\varepsilon}_m}$, which preserves the quadratic rate of convergence at the structural level. In the present study, we propose another class of operators for material inelasticity considering the concept of secant operator. The expressions of secant operators for the plasticity and damage models are also developed.

3.1 Secant Operators

Based on the original definition of secant operators in Eq. (16), the commonly used expressions are as follows

(1) Origin operator

$$\boldsymbol{\sigma}=\mathbb{S}_{\mathrm{org}}:\boldsymbol{\varepsilon} \tag{19}$$

(2) Incremental operator

$$(\boldsymbol{\sigma}^m-\boldsymbol{\sigma}^{m-1})=\mathbb{S}_{\mathrm{inc}}^m:(\boldsymbol{\varepsilon}^m-\boldsymbol{\varepsilon}^{m-1}) \tag{20}$$

(3) Plasticity shifted operator

$$\boldsymbol{\sigma}=\mathbb{S}_{\mathrm{org}}:(\boldsymbol{\varepsilon}-\boldsymbol{\varepsilon}^{\mathrm{p}}) \tag{21}$$

The first way to determine the secant operator is to solve it in a closed form. The original operator may be directly solved through the continuum material inelastic model, and the incremental operator should be solved through the discretized material inelastic model. On the other hand, if the closed forms of the secant operators could not be solved easily, or the solved operators are ill-conditioned, some computational techniques are required. By applying the general form of the BFGS operator[20] to the numerical solution of stress-strain laws, the following BFGS secant operator could be developed

$$\mathbb{S}_{\mathrm{BFGS}}(\mathbb{S}_c,\Delta\boldsymbol{\sigma},\Delta\boldsymbol{\varepsilon})=\mathbb{S}_c+\frac{\Delta\boldsymbol{\sigma}\otimes\Delta\boldsymbol{\sigma}}{\Delta\boldsymbol{\sigma}:\Delta\boldsymbol{\varepsilon}}-\frac{(\mathbb{S}_c:\Delta\boldsymbol{\varepsilon})\otimes(\Delta\boldsymbol{\varepsilon}:\mathbb{S}_c)}{\Delta\boldsymbol{\varepsilon}:\mathbb{S}_c:\Delta\boldsymbol{\varepsilon}} \tag{22}$$

where $\mathbb{S}_c$=reference fourth-order stiffness tensor; $\Delta\boldsymbol{\sigma}$ and $\Delta\boldsymbol{\varepsilon}$ =stress and strain increments between current state and reference state. The BFGS secant operator reserves the symmetry of $\mathbb{S}_c$ and has been proven to be well-conditioned in most cases[20]. Considering two commonly used reference states, we obtain the following two BFGS operators

(1) BFGS-origin operator

$$\mathbb{S}_{\mathrm{BO}}=\mathbb{S}_{\mathrm{BFGS}}(\mathbb{E}_0,\boldsymbol{\sigma},\boldsymbol{\varepsilon}) \tag{23}$$

where $\mathbb{E}_0$=elastic stiffness tensor.

(2) BFGS-incremental operator

$$\mathbb{S}_{\mathrm{BI}}^m=\mathbb{S}_{\mathrm{BFGS}}(\mathbb{S}_{\mathrm{BI}}^{m-1},\boldsymbol{\sigma}^m-\boldsymbol{\sigma}^{m-1},\boldsymbol{\varepsilon}^m-\boldsymbol{\varepsilon}^{m-1}) \tag{24}$$

(3) BFGS operator with plastic shift

$$\mathbb{S}_{BP}=\mathbb{S}_{BFGS}(\mathbb{E}_0, \boldsymbol{\sigma}, \boldsymbol{\varepsilon}-\boldsymbol{\varepsilon}^{p}) \tag{25}$$

For nonsymmetrical systems, which may be derived from the nonassociate plasticity or anisotropic damage model, the moduli developed based on the Broyden update[26] are more appropriate. We consider the following good Broyden update.

$$\mathbb{S}_{GB}(\mathbb{S}_c, \Delta\boldsymbol{\sigma}, \Delta\boldsymbol{\varepsilon})=\mathbb{S}_c+\frac{(\Delta\boldsymbol{\sigma}-\mathbb{S}_c:\Delta\boldsymbol{\varepsilon})\otimes\Delta\boldsymbol{\varepsilon}}{\Delta\boldsymbol{\varepsilon}:\Delta\boldsymbol{\varepsilon}} \tag{26}$$

We can further develop

(1) good-Broyden-origin operator

$$\mathbb{S}_{GBO}=\mathbb{S}_{GB}(\mathbb{E}_0, \boldsymbol{\sigma}, \boldsymbol{\varepsilon}) \tag{27}$$

(2) good-Broyden-incremental operator

$$\mathbb{S}^{m}_{GBI}=\mathbb{S}^{GB}(\mathbb{S}^{m-1}_{BI}, \boldsymbol{\sigma}^{m}-\boldsymbol{\sigma}^{m-1}, \boldsymbol{\varepsilon}^{m}-\boldsymbol{\varepsilon}^{m-1}) \tag{28}$$

(3) good-Broyden operator with plastic shift

$$\mathbb{S}_{GBP}=\mathbb{S}_{GB}(\mathbb{E}_0, \boldsymbol{\sigma}, \boldsymbol{\varepsilon}-\boldsymbol{\varepsilon}^{p}) \tag{29}$$

3.2 Plastic Model

As one of the possible applications of the proposed secant method, the classic stress-based plastic model[27] includes the following equations:

(1) stress-strain relation

$$\boldsymbol{\sigma}=\mathbb{E}_0:(\boldsymbol{\varepsilon}-\boldsymbol{\varepsilon}_p) \tag{30}$$

(2) Yield function

$$f(\boldsymbol{\sigma}, \boldsymbol{q})=0 \tag{31}$$

(3) Plastic evolution

$$\begin{aligned}\dot{\boldsymbol{\varepsilon}}_p&=\dot{\lambda}\,\partial_{\boldsymbol{\sigma}}\tilde{f}(\boldsymbol{\sigma}, \boldsymbol{q})\\ \dot{\boldsymbol{\alpha}}&=\dot{\lambda}\,\partial_{\boldsymbol{q}}\tilde{f}(\boldsymbol{\sigma}, \boldsymbol{q})\\ \boldsymbol{q}&=-\boldsymbol{H}\cdot\dot{\boldsymbol{\alpha}}\end{aligned} \tag{32}$$

(4) Kuhn-Tucker condition

$$f\leqslant 0,\quad \dot{\lambda}\geqslant 0,\quad \dot{\lambda}\dot{f}=0 \tag{33}$$

where $\boldsymbol{\varepsilon}_p$ = plastic strain tensor; f and $\tilde{f}$ = yield function and evolution potential, respectively; $\boldsymbol{\alpha}$, $\boldsymbol{q}$, and $\boldsymbol{H}$ = internal hardening variable and back stress and hardening modulus, respectively. The operator of the partial derivative is defined as $\partial_x=\frac{\partial}{\partial_x}$. In the case of isotropic elasticity, we have the elastic stiffness tensor

$$\mathbb{E}_0=K(\mathbf{1}\otimes\mathbf{1})+2G\left[\mathbb{I}-\frac{1}{3}(\mathbf{1}\otimes\mathbf{1})\right]=K(\mathbf{1}\otimes\mathbf{1})+2G\,\mathbb{I}_{dev} \tag{34}$$

where K and G=bulk and shear moduli, respectively; and **1** and $\mathbb{I}$=identity tensors of the second and fourth orders, respectively.

The return-mapping algorithm suggests solving the material inelastic models in two steps. First is the elastic prediction. We freeze the inelastic evolution and calculate the trial stress as follows

$$\boldsymbol{\sigma}^{\text{trial}}=\mathbb{E}_0:(\boldsymbol{\varepsilon}^m-\boldsymbol{\varepsilon}_{\text{p}}^{m-1}) \tag{35}$$

Check the yield condition

$$f(\boldsymbol{\sigma}^{\text{trial}},\boldsymbol{q}^{m-1})=\begin{cases}\leqslant 0 & \text{elastic step}\\ >0 & \text{plastic step}\end{cases} \tag{36}$$

For the elastic step, the updates of the variables are as follows

$$\begin{aligned}\boldsymbol{\varepsilon}_{\text{p}}^m&=\boldsymbol{\varepsilon}_{\text{p}}^{m-1}\\ \boldsymbol{\alpha}^m&=\boldsymbol{\alpha}^{m-1}\\ \boldsymbol{q}^m&=\boldsymbol{q}^{m-1}\\ \boldsymbol{\sigma}^m&=\boldsymbol{\sigma}^{\text{trial}}\end{aligned} \tag{37}$$

And the secant stiffness tensor is adopted as the elastic stiffness tensor. One obtains

$$\mathbb{S}=\mathbb{E}_0 \tag{38}$$

For the plastic step, the backward Euler discretization is applied. Then, we have the following finite discretized system

$$\begin{aligned}&\boldsymbol{\varepsilon}_{\text{p}}^m-\boldsymbol{\varepsilon}_{\text{p}}^{m-1}=(\lambda^m-\lambda^{m-1})\partial_{\boldsymbol{\sigma}}\tilde{f}(\boldsymbol{\sigma}^m,\boldsymbol{q}^m)\\ &\boldsymbol{\alpha}^m-\boldsymbol{\alpha}^{m-1}=(\lambda^m-\lambda^{m-1})\partial_{\boldsymbol{q}}\tilde{f}(\boldsymbol{\sigma}^m,\boldsymbol{q}^m)\\ &\boldsymbol{q}^m-\boldsymbol{q}^{m-1}=-\boldsymbol{H}^m\cdot(\boldsymbol{\alpha}^m-\boldsymbol{\alpha}^{m-1})\\ &f(\boldsymbol{\sigma}^m,\boldsymbol{q}^m)=0\end{aligned} \tag{39}$$

We have four equations in Eq.(39) and four unknowns $\boldsymbol{\varepsilon}_{\text{p}}^m$, $\boldsymbol{\alpha}^m$, $\boldsymbol{q}^m$ and λ^m. By applying the Newton scheme to the nonlinear Eq.(39), the unknowns could be solved iteratively. Furthermore, the backward Euler discretization of the consistency condition reads as follows

$$\partial_{\boldsymbol{\sigma}}f^m:(\boldsymbol{\sigma}^m-\boldsymbol{\sigma}^{m-1})-\partial_{\boldsymbol{q}}f^m\cdot(\boldsymbol{q}^m-\boldsymbol{q}^{m-1})=0 \tag{40}$$

By combining Eqs.(30), (39) and (40), we further have

$$(\boldsymbol{\sigma}^m-\boldsymbol{\sigma}^{m-1})=\left\{\mathbb{E}_0-\frac{(\mathbb{E}_0:\partial_{\boldsymbol{\sigma}}f^m)\otimes(\partial_{\boldsymbol{\sigma}}\tilde{f}^m:\mathbb{E}_0)}{\partial_{\boldsymbol{\sigma}}f^m:\mathbb{E}_0:\partial_{\boldsymbol{\sigma}}\tilde{f}^m+\partial_{\boldsymbol{q}}f^m\cdot\boldsymbol{H}^m\cdot\partial_{\boldsymbol{q}}\tilde{f}^m}\right\}:(\boldsymbol{\varepsilon}^m-\boldsymbol{\varepsilon}^{m-1}) \tag{41}$$

Thus, we have the incremental secant moduli

$$\mathbb{S}_{\text{inc}}^m=\mathbb{E}_0-\frac{(\mathbb{E}_0:\partial_{\boldsymbol{\sigma}}f^m)\otimes(\partial_{\boldsymbol{\sigma}}\tilde{f}^m:\mathbb{E}_0)}{\partial_{\boldsymbol{\sigma}}f^m:\mathbb{E}_0:\partial_{\boldsymbol{\sigma}}\tilde{f}^m+\partial_{\boldsymbol{q}}f^m\cdot\boldsymbol{H}^m\cdot\partial_{\boldsymbol{q}}\tilde{f}^m}=\mathbb{C}_{\text{ep}}^m \tag{42}$$

Interestingly, the elastoplastic tangent modulus $\mathbb{C}_{\text{ep}}^m$ is reduced to the algorithmic incremental secant moduli $\mathbb{S}_{\text{inc}}^m$ within the framework of the return mapping scheme. Thus, if the full Newton method is applied to solve the nonlinear FE equation, the consistent tangent moduli[18] is required to reserve the quadratic convergence rate instead of the elastoplastic tangent moduli. And it is also easy to testify that the plasticity-shifted secant operator reduces to the elastic stiffness operator for the plasticity model.

Based on the aforementioned rate forms of the plastic model, the closed-form solution of the origin modulus $\mathbb{S}_{\text{org}}$ could not be explicitly solved. Meanwhile, Eq.(42) indicates that the closed form of the incremental secant modulus $\mathbb{S}_{\text{inc}}^{m}$ is expressed by the elastoplastic tangent modulus $\mathbb{C}_{\text{ep}}^{m}$. For the simple plastic model, such as von Mises plasticity, $\mathbb{C}_{\text{ep}}^{m}$ could be easily determined in the same way as $\mathbb{S}_{\text{inc}}^{m}$. Thus, $\mathbb{S}_{\text{inc}}^{m}$ and $\mathbb{C}_{\text{ep}}^{m}$ could be candidates of the undetermined material level secant tensor in the algorithm shown in Fig. 2. If some advanced plastic models that consider nonassociate plastic evolution or yield function with vertices are applied[28], $\mathbb{C}_{\text{ep}}^{m}$ may experience asymmetry or discontinuity. These properties often result in ill-condition and even divergence in the iterative solution of the equations. In this case, the proposed secant operators could be applied to the algorithm shown in Fig. 2. The BFGS family expressed in Eqs.(23)—(25) work effectively for the associate plasticity. For the nonassociate plasticity, the good Broyden family expressed in Eqs.(27)—(29) performs better because of its asymmetrical nature.

3.3 Damage Model

To simulate the material softening and its induced structural failure, damage models are often implemented at the integration points of the FE model. Over the past 30 years, developments have occurred, and the damage models are now widely adopted in studies or engineering designs. In the present paper, we recommend a biscalar elastic damage model, that is a simplified version of the elastoplastic damage model developed for quasi-brittle materials[10], to illustrate the possible application of the proposed secant method.

The damaged stress-strain relation is expressed as follows

$$\boldsymbol{\sigma}=(\mathbb{I}-\mathbb{D}):\mathbb{E}_0:\boldsymbol{\varepsilon}^{e},\quad \boldsymbol{\varepsilon}^{e}=(\boldsymbol{\varepsilon}-\boldsymbol{\varepsilon}^{p}) \tag{43}$$

For brittle materials like concrete and rocks, the damage evolutions under tension and compression are quite different. Thus, the decomposition of damage tensor is introduced as follows

$$\mathbb{D}=d^{+}\mathbb{P}^{+}+d^{-}\mathbb{P}^{-} \tag{44}$$

where d^{+} and d^{-} =tensile damage scalar and compressive damage scalar, respectively; $\mathbb{P}^{+}$ and $\mathbb{P}^{-}$=fourth-order projective tensors.

To develop the fourth-order damage tensor, the effective stress, which represents the stress of the undamaged materials, should be defined beforehand. We have

$$\bar{\boldsymbol{\sigma}}=\mathbb{E}_0:(\boldsymbol{\varepsilon}-\boldsymbol{\varepsilon}^{p}) \tag{45}$$

The projective tensors satisfy the following split of effective stress

$$\bar{\boldsymbol{\sigma}}=\bar{\boldsymbol{\sigma}}^{+}+\bar{\boldsymbol{\sigma}}^{-}$$

$$\bar{\boldsymbol{\sigma}}^{+}=\mathbb{P}^{+}:\bar{\boldsymbol{\sigma}},\quad \bar{\boldsymbol{\sigma}}^{-}=\mathbb{P}^{-}:\bar{\boldsymbol{\sigma}} \tag{46}$$

By introducing the eigendecomposition of effective stress, we further have the following expressions of projective tensors:

$$\begin{aligned}\mathbb{P}^{+}&=\sum_{i}H(\hat{\bar{\sigma}}_{i})\boldsymbol{p}_{i}\otimes\boldsymbol{p}_{i}\otimes\boldsymbol{p}_{i}\otimes\boldsymbol{p}_{i}\\ \mathbb{P}^{-}&=\mathbb{I}-\mathbb{P}^{+}\end{aligned} \tag{47}$$

where $\hat{\bar{\sigma}}_i$ and $\boldsymbol{p}_i$ = ith eigenvalue and eigenvector of the effective stress $\bar{\boldsymbol{\sigma}}$, respectively. The Heaviside function is

$$\mathrm{H}(x)=\begin{cases}0 & x<0\\ 1 & x\geqslant 0\end{cases} \tag{48}$$

According to Li and Ren[12], damage evolutions are driven by the energy equivalent strain as follows

$$d^{\pm}=g^{\pm}(\bar{\epsilon}^{\pm}) \tag{49}$$

where the superscript $\pm$ denotes that $+$ and $-$ are considered, respectively. The energy equivalent strains are

$$\begin{aligned}\bar{\epsilon}^{+}&=\sqrt{\frac{2Y^{+}}{E_0}}\\ \bar{\epsilon}^{-}&=\frac{1}{E_0(\alpha-1)}\sqrt{\frac{Y^{-}}{b_0}}\end{aligned} \tag{50}$$

and the damage release rates are

$$\begin{aligned}Y^{+}&=\frac{1}{2E_0}\left[\frac{2(1+\nu_0)}{3}3\bar{J}_2^{+}+\frac{1-2\nu_0}{3}(\bar{I}_1^{+})^2-\nu_0\bar{I}_1^{+}\bar{I}_1^{-}\right]\\ Y^{-}&=b_0(\alpha\bar{I}_1^{-}+\sqrt{3\bar{J}_2^{-}})^2\end{aligned} \tag{51}$$

where $\bar{\boldsymbol{I}}_1^{\pm}$ = first invariants of tensile and compressive effective stresses $\bar{\boldsymbol{\sigma}}^{\pm}$, respectively; and $\bar{\boldsymbol{J}}_2^{\pm}$ = second invariants of the deviatoric components of effective stresses. The parameter α is related to the biaxial strength. Thus, we have

$$\alpha=\frac{\dfrac{f_{\mathrm{bc}}}{f_{\mathrm{c}}}-1}{2\dfrac{f_{\mathrm{bc}}}{f_{\mathrm{c}}}-1} \tag{52}$$

where f_{c} and f_{bc} = compressive strengths of concrete under uniaxial and biaxial load conditions, respectively.

The experiential damage evolution functions are adopted to describe the progressive degradation of material integrity. We have

$$d^{\pm}=g^{\pm}(\bar{\epsilon}^{\pm})=\begin{cases}1-\dfrac{\rho^{\pm}n^{\pm}}{n^{\pm}-1+(x^{\pm})^{n^{\pm}}} & x^{\pm}\leqslant 1\\ 1-\dfrac{\rho^{\pm}}{\alpha_c^{\pm}(x^{\pm}-1)^2+x^{\pm}} & x^{\pm}>1\end{cases} \tag{53}$$

with the definitions of symbols

$$x^{\pm}=\frac{\bar{\epsilon}^{\pm}}{\varepsilon_c^{\pm}},\quad \rho^{\pm}=\frac{f_c^{\pm}}{E_c\varepsilon_c^{\pm}},\quad n^{\pm}=\frac{E_c\varepsilon_c^{\pm}}{E_0\varepsilon_c^{\pm}-f_c^{\pm}} \tag{54}$$

where E_c = Young's modulus of the undamaged material; $f_c^{\pm}$ and $\varepsilon_c^{\pm}$ = stress and strain corresponding to the stress-strain peak, respectively; and $\alpha_c^{\pm}$ = material parameters governing the descending part of the stress-strain curves.

The plastic strain $\boldsymbol{\varepsilon}^{p}$ could be solved by the theoretical model[10] and the empirical models. To simplify the computation of the proposed model without loss of too much accuracy, we consider the empirical model proposed by Faria et al.[29]. We have

$$\dot{\boldsymbol{\varepsilon}}^{p}=b^{p}\bar{\boldsymbol{\sigma}},\quad b^{p}=\xi^{p}E_0 H(\dot{d}^{-})\frac{\langle\boldsymbol{\epsilon}^{e}:\dot{\boldsymbol{\epsilon}}\rangle}{\bar{\boldsymbol{\sigma}}:\bar{\boldsymbol{\sigma}}} \tag{55}$$

where $\xi^{p}\geqslant 0$ = plastic coefficient governing the evolution of plastic strain; and $\langle\cdot\rangle$ = Macaulay brackets.

The aforementioned damage-plasticity model is developed in an implicit incremental form. Thus, the return-mapping is needed for its numerical solution. Whereas the stress of the current state has been solved, the secant operators required in the algorithm shown in Fig. 2 could be easily calculated based on Eqs.(23)—(25) and Eqs.(27)—(29). Comparing with the consistent tangent operators for damage-plasticity models developed in the literature[9-10] the proposed material level secant operators are extremely easy to form and code.

4 Numerical Examples

4.1 Reinforced Concrete Wall-Type Structure under Cyclic Shear

Consider the reinforced concrete (RC) shear wall under cyclic shear as the numerical example to show the performance of the proposed method in 2D. Systematic tests were performed by Mo et al.[30] at the University of Houston for framed shear walls under constant axial load at the top of each column and a reversed cyclic shear at the top beam. We take the specimen named FSW4 as our numerical example, as shown in Fig. 3. The dimensions of the wall panel are 914.4 mm×914.4 mm×76.2 mm. The cross section of the boundary beams and columns are 152.4 mm×152.4 mm. The steel ratio of the wall panel is 0.55% in each direction, and the longitudinal steel ratio for beam and column is 3.33%. The vertical load applied on the top of each column is 534 kN.

A FE model is developed based on the experimental arrangements. The wall panel is modeled by 10×10 layered shell elements. The beam and column is modeled by the beam-column elements. The shell elements share the same nodes with beam-column elements on the boundary of the wall panel. The damage-plasticity model is considered as the material model of concrete. The uniaxial compressive strength of concrete is tested to be 49. 5 MPa. Based on this value, the model parameters of concrete are characterized as shown in Table 1. The nonlinearity of steel is reproduced by the plastic model with isotropic linear hardening. The elastic module of steel is 180 GPa. The yielding stress of steel is 500 MPa, and the ultimate strength is 800 MPa at the strain of 0.2. Beyond the ultimate strain, the ductile damage of steel bar is considered. The linear decay of stress is considered following the increase of displacement. And the rupture strain of steel reinforcing bar is set to be 0.35.

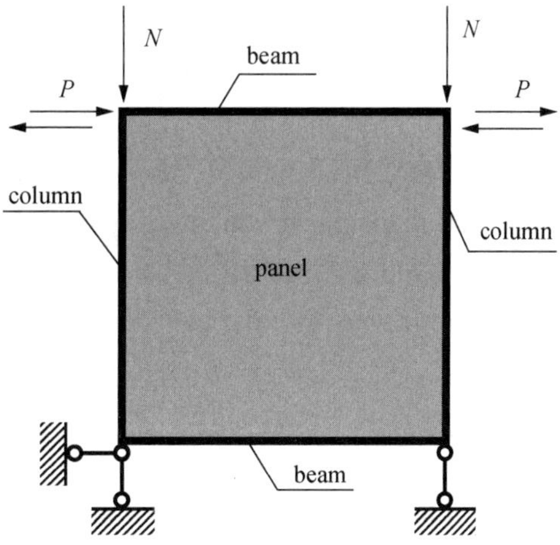

Fig. 3. RC wall-type structure

Table 1. Model parameters of concrete in the shear panel test

Material properties	Tension	Compression
Elasticity	E_0(35 GPa), ν(0.20)	
Biaxial strength increase	N/A	$\frac{f_{bc}}{f_c}$(1.16)
Damage evolution	f_c^+(4.0 MPa)	f_c^-(49.5 MPa)
	ε_c^+(150×10^{-6})	ε_c^-($1\,800\times10^{-6}$)
	α_c^+(0.06)	α_c^-(0.5)
Plasticity	N/A	ξ^p(0.5)

We have developed numbers of expressions for secant operators in the material level. The selection of secant operators for the structural simulation should be based on the governing mechanism of the problem and also certain trial runs. In the cyclic loading condition, the residual deformation usually plays an important role for the behaviors of structure. We performed three to five trial runs based on the BFGS operator and good-Broyden operator, and then adopted the BFGS secant operator shown in Eq.(25) with plastic shift due to its relatively better performance for this problem. The results are shown in Fig. 4 considering three types of strategies in the simulation. The two-level tangent operator is the full Newton method with consistent tangent operator. The tangent-secant operator is the Q-N method in the structural level with tangent operator in the material level. And the two-level secant operator is the Q-N method in the structural level with the secant operator in the material level, which is the consistent secant operator proposed by the present work.

According to Fig. 4(a), the two-level tangent operator and the tangent-secant operator diverge at nearly the same point, for which the concrete within the tensile domain is cracking. The required values of displacement increments are 10^{-10} and 10^{-7} mm, respectively. It is difficult to finish the simulations in a reasonable amount of time with such small displacement increments. The consistent secant operator converges during the whole loading process. The smallest required displacement increment is 10^{-3} mm throughout the simulation, which indicates the proposed method is very stable and effective.

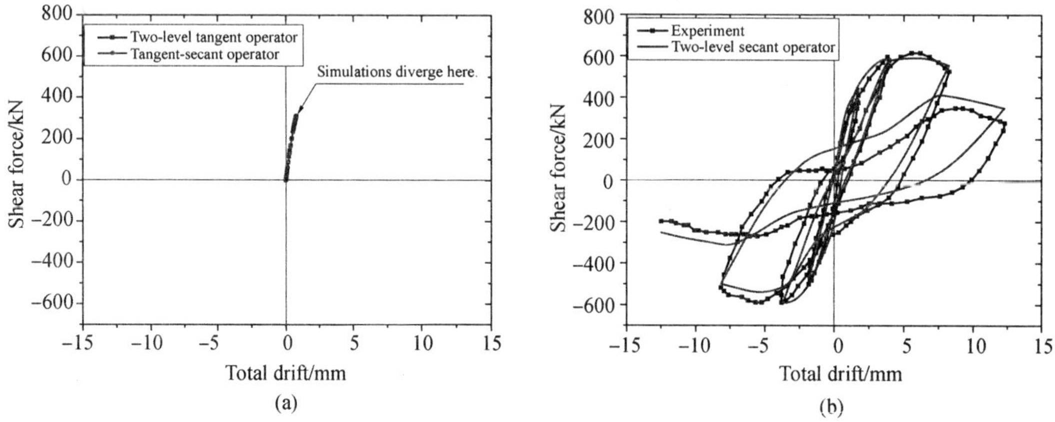

Fig. 4. Shear force versus drift displacement of wall-type structure

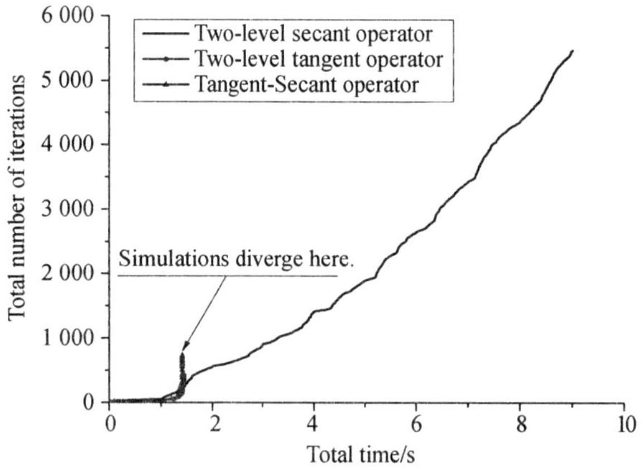

Fig. 5. Iteration count for RC panel simulations

The iteration count throughout the whole simulations is shown in Fig. 5. The horizontal ordinate is the pseudo time scale used in the numerical simulation. For each loading cycle, the pseudo time scale is set to be 1 s. The vertical ordinate is the total number of iterations counted from the starting to the current time. In the beginning, it is not surprisingly to see that the tangent operator needs less iterations to achieve convergence than the secant operator because the structure undergoes very weak damage in this stage. The reason is that the convergence of the full-Newton method is quadratic for well-posed problems, and the convergence of the quasi-Newton method is super-linear. As shown in Fig. 5, the situation is changed when the structure experiences cracking and damage. The number of iterations for the tangent operator goes up sharply, whereas the number of iterations for the consistent secant operator goes up steadily. If the required time increments become too small, the simulations are judged to be divergent.

One may see that the secant operator is more stable than the tangent operator especially for the problems undergoing serious nonlinearity. Fig. 5 also demonstrates that the concept of consistent secant operator guarantees a very stable and robust solution procedure for the whole simulation. Moreover, the computational burden for the secant operator is much lower than that of the tangent operator because the inversion of the structural Jacobian matrix is not needed for each iteration.

4.2 Reinforced Concrete Column under Lateral Deformation Reversals

The performance of the proposed method in 3D is verified by the hysterical behaviors of a reinforced concrete (RC) column, which was tested by Saatcioglu and Grira[31]. The dimensions of the RC column specimen (BG - 2) are shown in Fig. 6. The longitudinal steel bars are $8 \times \phi 20$ mm and the stirrups are $\phi 9.53$ mm at 76 mm. The constant vertical load applied on the top of the column is 1 782 kN. A 3D FE model is developed for the column structure(Fig. 7). Concrete is modeled by 3D *brick* elements with eight nodes and reduced integration. Steel reinforcing bars are modeled as truss elements embedded in concrete elements. The damage-plasticity model is also considered for concrete. With the tested uniaxial compressive strength 45 MPa, the material parameters for concrete in the column are characterized and shown in Table 2. The Young's module and the plastic module of steel are 180 and 3.6 GPa, respectively. The yielding stress of steel is 500 MPa.

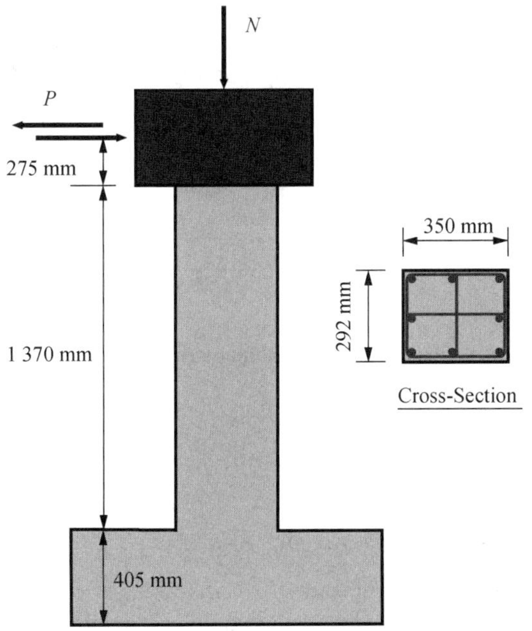

Fig. 6. RC column structure

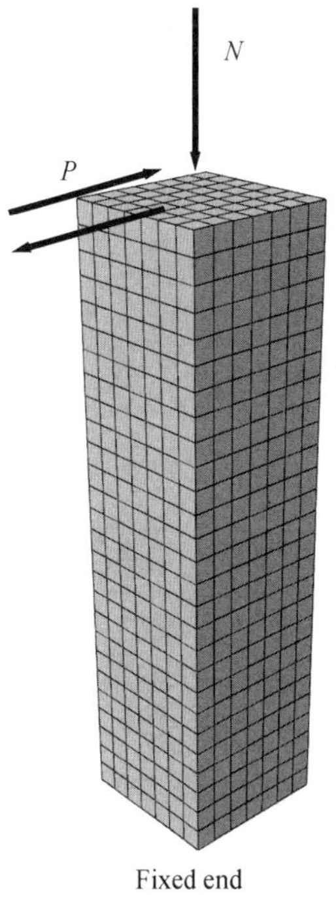

Fig. 7. FE model for column structure

Table 2. Model parameters of concrete in the column

Material properties	Tension	Compression
Elasticity	E_0(30 GPa), ν(0.20)	
Biaxial strength increase	N/A	$\frac{f_{bc}}{f_c}$(1.16)
Damage evolution	f_c^+(4.0 MPa)	f_c^-(45 MPa)
	ε_c^+(180×10^{-6})	ε_c^-($2\,000\times10^{-6}$)
	α_c^+(0.01)	α_c^-(0.06)
Plasticity	N/A	ξ^p(0.5)

Three to five trial runs revealed that the BFGS operator and the good-Broyden operator gave rather similar performance for this problem. Because the BFGS operator has been shown in the last example, we chose the other one, e.g., the good-Broyden operator with plastic shift, as the secant operator in the material level for this example. The results are shown in Fig. 8. The two-level tangent operators and the tangent-secant operators diverge after numbers of loading cycles. The consistent two-level secant operator is able to converge throughout all of the loading cycles. The minimum required displacement increment is 10^{-3} mm, which guarantees the simulation with reasonable cost and accuracy.

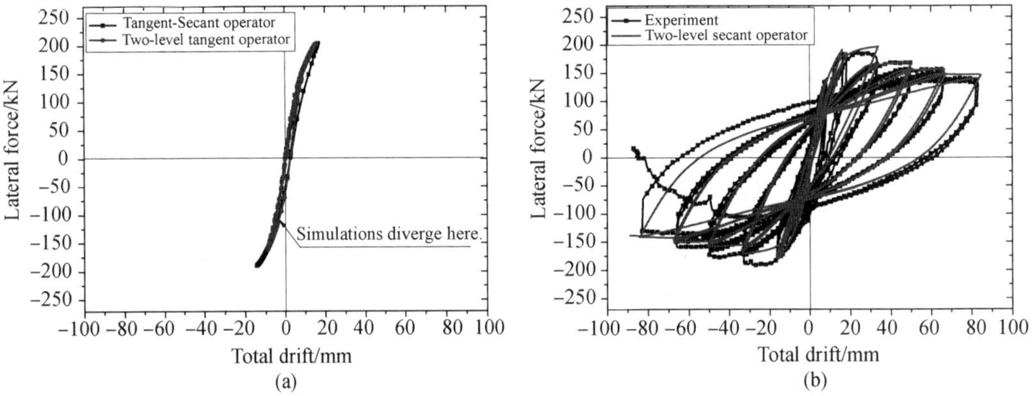

Fig. 8. Lateral force versus drift displacement of column structure

The iteration count for the simulations of the RC column is shown in Fig. 9. The horizontal ordinate is the pseudo time scale, and the vertical ordinate is the total number of iterations. For each loading cycle, the pseudo time scale is also set to be 1 s. The number of iterations increases steadily and smoothly following the increase of the time scale for the consistent secant operator. For the tangent operators, the simulations are unstable. Sometimes large time increments could be finished by very few iterations. Sometimes rather small time increments require a large number of iterations. Finally, the simulations based on tangent operators are judged to be divergent when the required time increments are too small. One may argue that the simulations based on tangent operators might survive and catch up with the secant operators if we change the convergence criterion more or less. But when we pay attention to the overall performance of the operators throughout the simulation, we could reach the conclusion that the consistent secant operator offers a robust and effective way for the simulation of structures experiencing serious nonlinearities.

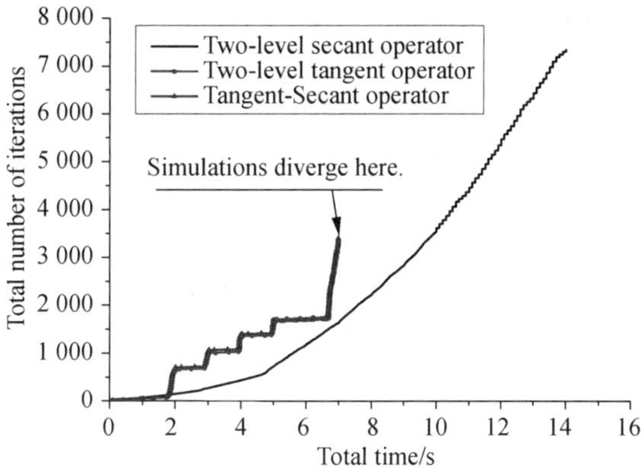

Fig. 9. Iteration count for RC column simulations

5 Conclusions

Based on the theoretical developments and numerical results, the following conclusions are obtained:

(1) This paper has presented a two-scale consistent secant strategy to solve nonlinear FE equations. The consistency between the Q-N update at the structural level and the secant operator at the material level could lead to a good trade-off between efficiency and stability.

(2) Regardless of the number of iterations at the structural level, the essential feature of the proposed secant operator at the material level is the ability to simulate the cyclic behaviors of structure with a rather complex constitutive law.

(3) The scalability of the proposed strategy, which is apparently better than the full-Newton algorithm, deserves further attention in the future study.

Acknowledgments

Financial supports from the National Science Foundation of China (Grant Nos. 51261120374, 51678439, and 51208374) and the Fundamental Research Funds for the Central Universities are sincerely appreciated. The authors also thank Dr. De-Cheng Feng at Southeast University for valuable suggestions and kind help.

References

[1] ALFAIATE J, SIMONE A, SLUYS L J. Non-homogeneous displacement jumps in strong embedded

discontinuities[J]. International Journal of Solids and Structures, 2003, 40(21): 5799-5817.

[2] MOËS N, DOLBOW J, BELYTSCHKO T. A finite element method for crack growth without remeshing[J]. International Journal for Numerical Methods in Engineering, 1999, 46(1): 131-150.

[3] DOLBOW J, MOËS N, BELYTSCHKO T. Discontinuous enrichment in finite elements with a partition of unity method[J]. Finite Elements in Analysis and Design, 2000, 36(3-4): 235-260.

[4] BELYTSCHKO T, MOËS N, USUI S, et al. Arbitrary discontinuities in finite elements[J]. International Journal for Numerical Methods in Engineering, 2001, 50(4): 993-1013.

[5] BELYTSCHKO T, LU Y Y, GU L. Element-free Galerkin methods[J]. International Journal for Numerical Methods in Engineering, 1994, 37(2): 229-256.

[6] LIU W K, JUN S, ZHANG Y F. Reproducing kernel particle methods[J]. International Journal for Numerical Methods in Fluids, 1995, 20(8-9): 1081-1106.

[7] CHEN J S, PAN C, WU C T, et al. Reproducing kernel particle methods for large deformation analysis of non-linear structures[J]. Computer Methods in Applied Mechanics and Engineering, 1996, 139(1-4): 195-227.

[8] JU J W. On energy-based coupled elastoplastic damage theories: constitutive modeling and computational aspects[J]. International Journal of Solids and Structures, 1989, 25(7): 803-833.

[9] LEE J, FENVES G L. Plastic-damage model for cyclic loading of concrete structures[J]. Journal of Engineering Mechanics, 1998, 124(8): 892-900.

[10] WU J Y, LI J, FARIA R. An energy release rate-based plastic-damage model for concrete[J]. International Journal of Solids and Structures, 2006, 43(3-4): 583-612.

[11] GRASSL P, JIRÁSEK M. Damage-plastic model for concrete failure[J]. International Journal of Solids and Structures, 2006, 43(22-23): 7166-7196.

[12] LI J, REN X D. Stochastic damage model for concrete based on energy equivalent strain[J]. International Journal of Solids and Structures, 2009, 46(11-12): 2407-2419.

[13] BAZANT Z P, PIJAUDIER-CABOT G. Nonlocal continuum damage, localization instability and convergence[J]. J. Appl. Mech. 1988, 55(2):287-293.

[14] PEERLINGS R H J, BORST R, Brekelmans W A M, et al. Some observations on localization in non-local and gradient damage models[J]. European Journal of Mechanics. A, Solids, 1996, 15(6): 937-953.

[15] BAŽANT Z P, JIRÁSEK M. Nonlocal integral formulations of plasticity and damage: survey of progress[J]. Journal of Engineering Mechanics, 2002, 128(11): 1119-1149.

[16] ASKES H, SLUYS L J. Explicit and implicit gradient series in damage mechanics[J]. European Journal of Mechanics-A/Solids, 2002, 21(3): 379-390.

[17] SIMO J C, TAYLOR R L. Consistent tangent operators for rate-independent elastoplasticity[J]. Computer methods in Applied Mechanics and Engineering, 1985, 48(1): 101-118.

[18] SIMO J C, HUGHES T. Computational inelasticity[M]. New York, NY: Springer, 1998.

[19] MATTHIES H, STRANG G. The solution of nonlinear finite element equations[J]. International Journal for Numerical Methods in Engineering, 1979, 14(11): 1613-1626.

[20] DENNIS J, SCHNABEL R. Numerical methods for unconstrained optimization and nonlinear equations [M]. Philadelphia, PA: Society for Industrial and Applied Mathematics, 1996.

[21] BATHE K J, ARTHUR P. CIMENTO. Some practical procedures for the solution of nonlinear finite element equations[J]. Computer Methods in Applied Mechanics and Engineering, 1980, 22(1): 59-85.

[22] NOCEDAL J, WRIGHT S J. Numerical optimization[M]. Berlin,Germany: Springer, 1999.

[23] RIKS E. An incremental approach to the solution of snapping and buckling problems[J]. International Journal of Solids and Structures, 1979, 15(7): 529-551.

[24] CRISFIELD M A. A fast incremental/iterative solution procedure that handles "snap-through"[M]// Computational Methods in Nonlinear Structural and Solid Mechanics. Pergamon, 1981: 55-62.

[25] SIMULIA. Abaqus documentation[M]. New York, NY: DassaultSystèmes, 2012.

[26] BROYDEN C G. Quasi-Newton methods and their application to function minimisation [J]. Mathematics of Computation, 1967, 21(99): 368-381.

[27] LUBLINER J. Plasticity theory[M]. New York, NY: Dover Publication, 2006.

[28] PAPANIKOLAOU V K, KAPPOS A J. 2007. Confinement-sensitive plasicity constitutive model for concrete in triaxial compression[J]. Int. J.Solids Struct, 2007, 44(21): 7021-7048.

[29] FARIA R, OLIVER J, CERVERA M. A strain-based plastic viscous-damage model for massive concrete structures[J]. International Journal of Solids and Structures, 1998, 35(14): 1533-1558.

[30] MO Y L, ZHONG J, HSU T T C. Seismic simulation of RC wall-type structures[J]. Engineering Structures, 2008, 30(11): 3167-3175.

[31] SAATCIOGLU M, GRIRA M. Confinement of reinforced concrete columns with welded reinforced grids[J]. Structural Journal, 1999, 96(1): 29-39.

Part 4

Application of Damage Mechanics

Stochastic nonlinear behavior of reinforced concrete frames. Ⅱ: numerical simulation①

De-Cheng Feng[1], **Jie Li**[2]

1 *Ph.D. Candidate, College of Civil Engineering, Tongji Univ., 1239 Siping Rd., Shanghai 200092, P.R. China. E-mail: aufdc@163.com*

2 *Professor, The State Key Laboratory on Disaster Reduction in Civil Engineering and College of Civil Engineering, Tongji Univ., 1239 Siping Rd., Shanghai 200092, P.R. China (corresponding author). E-mail: lijie@ tongji.edu.cn*

Abstract Experimental investigations on the stochastic nonlinear behavior of eight half-scale reinforced concrete (RC) frames with the same conditions in the companion paper indicates that the coupling effect between randomness and nonlinearity of concrete will cause a remarkable fluctuation in structural nonlinear responses. In the current paper, a validation program is presented through numerical simulation. Deterministic nonlinear analysis of test specimens is performed on the basis of a force-based beam-column element model. Probability density evolution method (PDEM) is introduced to analyze the stochastic response analysis of structures. Probability density functions of the stochastic structural responses of the specimens are provided at both the load-displacement level and internal-force level. Comparative studies between numerical and experimental results reveal that the numerical method in conjunction with PDEM can not only qualify the fundamental trend of stochastic nonlinear responses of structures but also generate the probabilistic details in structural damage evolution.

Keywords Reinforced concrete frame; Nonlinearity; Randomness; Force-based frame element; Probability density evolution method; Analysis and computation

1 Introduction

In the companion paper[1], a series of tests is conducted to experimentally investigate the stochastic nonlinear behavior of reinforced concrete (RC) frames. The study indicates that the stochastic nonlinear properties of concrete materials will result in the random processes of cracking and hinging. Significant fluctuations in the nonlinear responses of structures are observed. Accordingly, the coupling between randomness and nonlinearity of materials, especially concrete, should be carefully considered within the nonlinear analysis procedure of RC structures.

Nonlinear analysis of RC frame structures has been an active research area for more than 50 years, and significant progress has been made. Several celebrated models have been developed to simulate the nonlinear behavior of beam and/or column members under

① Originally published in *Journal of Structural Engineering*, 2016, 142(3): 04015163.

arbitrary loadings. These models include lumped plasticity models[2-5] that fix the plasticity of an element to the ends of elastic elements, as well as distributed plasticity models[6-7] that allow plastic deformations to occur at any location within the entire element. Distributed models can provide a refined description for frame members. Two schemes are typically developed to formulate distributed models. One is the conventional displacement-based method, which uses certain shape functions(e.g., cubic Hermitian polynomials) to approximate the displacement field along an element[8-9]. However, the displacement field in this case is highly complicated in the strong nonlinear phase, wherein a refined mesh type is usually needed to gain an accurate solution. The other is the force-based approach, which assumes the nominated formulation of force distribution in an element[10-11]. The advantage of this approach is that the equilibrium condition can be exclusively satisfied in a rigorous sense even when the element exhibits strong nonlinear material response. By taking this advantage, coarse meshcs are sufficient for simulation. Some other methods are also proposed to model the nonlinear behavior, even stochastic cracking, of RC[12-13], but the computational cost is relatively high for the simulation of total RC frames. Consequently, it can be concluded that the force-based element can reach the best tradeoff between numerical accuracy and computational efficiency for the present investigation.

As indicated in previous investigations, the randomness from material properties affects the structural behavior, which should be included in the nonlinear analysis of RC frames. Stochastic analysis of structures has gained increasing attention in the comprehensive capture of structural performances. The approaches available to attain the statistics of stochastic structural response can be broadly classified into three categories, namely, Monte Carlo simulation method[14], random perturbation technique[15], and orthogonal polynomial expansion method[16-17]. However, no approach with both accuracy and efficiency in the nonlinear analysis of stochastic structures has been developed. In addition, the previously-mentioned methods mainly aim to attain the second-order statistical quantities of structural responses, such as mean, standard deviation, and so forth, which are not sufficient to understand the comprehensive probabilistic information of structural responses. Li and Chen[18-19] proposed a family of probability density evolution method (PDEM), which is proven to be capable of obtaining the probability density function(PDF) of responses and its evolution at different levels. Integration with this method enables the fundamental trend of stochastic nonlinear responses of structures to be qualified and the probabilistic details during structural damage evolution to be obtained.

In this paper, a deterministic method for the numerical analysis of frame structures is presented. Material models that consider concrete-reinforcement interaction are introduced, and the force-based frame element is employed to simulate test specimens. The generalized probability density equation is derived, and its numerical algorithm is detailed. Stochastic response analysis of the testing models is performed using PDEM. Comparative studies of numerical and experimental results are conducted. PDEM reveals the stochastic evolution from material scale to structure scale, thereby achieving an in-depth understanding on the stochastic nonlinear behavior of RC frames.

2 Deterministic Analytical Model for RC Frames

2.1 Concrete

The uniaxial elasto-plastic damage model, developed by our group[20-22] and adopted by the Chinese design code[23], is employed to reproduce the nonlinear behavior of concrete herein. As shown in Fig. 1(a), the monotonic stress-strain relation in tension can be expressed as follows

$$\sigma_t=(1-d_t)E_c(\varepsilon_t-\varepsilon_t^p) \tag{1}$$

$$d_t=\begin{cases}1-\dfrac{\rho_t n_t}{n_t-1+x_t} & x_t\leqslant 1\\[2ex] 1-\dfrac{\rho_t}{\alpha_t(x_t-1)^2+x_t} & x_t>1\end{cases} \tag{2}$$

$$x_t=\frac{\varepsilon_t-\varepsilon_t^p}{\varepsilon_{t,r}},\quad \rho_t=\frac{f_{t,r}}{E_c\varepsilon_{t,r}},\quad n_t=\frac{1}{1-\rho_t} \tag{3}$$

where σ_t = uniaxial tensile stress; ε_t = uniaxial tensile strain; E_c = initial elastic modulus; d_t = tensile damage variable; α_t = tensile descending parameter that controls the shape of the descending curve; $f_{t,r}$ = tensile peak strength; $\varepsilon_{t,r}$ = strain corresponding to $f_{t,r}$; and ε_t^p = tensile plastic strain that is usually small and can be set as zero.

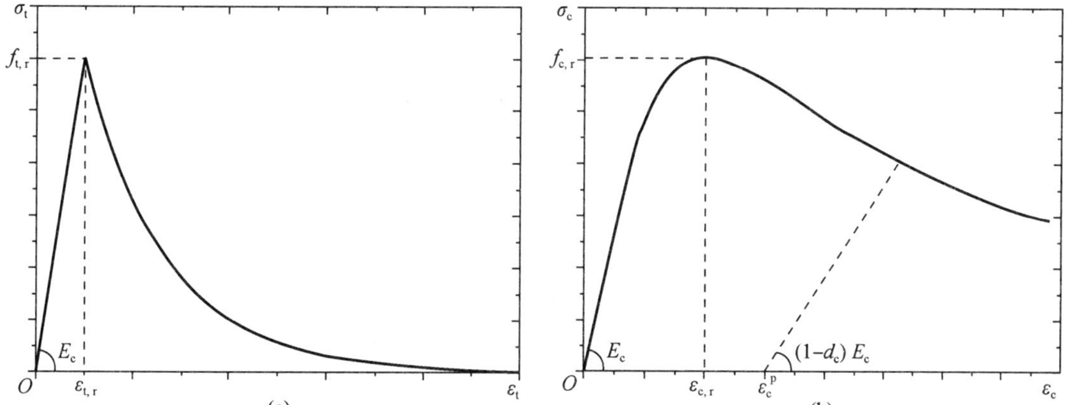

Fig. 1. Stress-strain relationship of concrete: (a) tension; (b) compression

The compressive stress-strain relationship illustrated in Fig. 1(b) is defined as

$$\sigma_c=(1-d_c)E_c(\varepsilon_c-\varepsilon_c^p) \tag{4}$$

$$d_c=\begin{cases}1-\dfrac{\rho_c n_c}{n_c-1+x_c} & x_c\leqslant 1\\[2ex] 1-\dfrac{\rho_c}{\alpha_c(x_c-1)^2+x_c} & x_c>1\end{cases} \tag{5}$$

$$x_c = \frac{\varepsilon_c - \varepsilon_c^p}{\varepsilon_{c,r}}, \quad \rho_c = \frac{f_{c,r}}{E_c \varepsilon_{c,r}}, \quad n_c = \frac{1}{1-\rho_c} \tag{6}$$

$$\varepsilon_c^p = \xi_p (d_c)^{n_p} \tag{7}$$

where σ_c = uniaxial compressive stress; ε_c = uniaxial compressive strain; d_c = compressive damage variable; α_c = compressive descending parameter; $f_{c,r}$ = compressive peak strength; $\varepsilon_{c,r}$ = strain corresponding to $f_{c,r}$; and ε_c^p = compressive plastic strain related to compressive damage d_c and plastic parameters ξ_p and n_p.

The previously-mentioned model is proposed only for plain concrete. The interaction between concrete and steel bars definitely contributes to the performance of an individual material. For concrete in tension, tension stiffening should be considered[24]. Tension stiffening represents the capacity bearing the tensile stress among crack surfaces and contributes to the flexural stiffness of RC members. Hence, the tensile stress-strain relation should be modified in case the tensile stress reduces gradually to zero as the tensile strain increases. The gradual reduction in tensile stress is often approximated by linear, multilinear, or exponential curves. Stenvens et al.[25] proposed the descending branch of the tensile stress-strain relationship based on a series of tests, as follows

$$\frac{\sigma}{f_{t,r}} = (1-\theta)\exp[\vartheta(\varepsilon_{t,r} - \varepsilon)] + \theta \tag{8}$$

where θ = parameter related to the ratio of reinforcement ρ_s and the diameter of steel bars d_b and is given by $\theta = 0.075\rho_s / d_b$; and ϑ = model parameter of the descending branch and is given by

$$\vartheta = 270/\sqrt{\theta}, \quad \text{and} \quad \vartheta \leqslant 1\ 000 \tag{9}$$

Eq.(8) describes the tension stiffening effect dependent on the reinforcement ratio and the diameter of steel bars. The results are shown in Fig. 2.

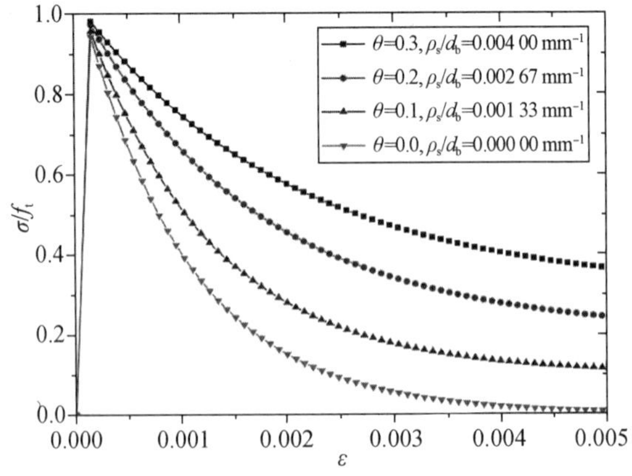

Fig. 2. Tension stiffening effect of concrete

For concrete in compression, the confinement of stirrups will clearly improve the strength and ductility of the confined concrete, as shown in Fig. 3. In this paper, Mander's model[26] is adopted to consider the effects of confinement. On the basis of this model, the key parameters are identified for the concrete damage model considering confinement. For the rectangular sections, the lateral confining stress on the concrete can be given by

$$f_{\mathrm{l}x}=\rho_x f_{\mathrm{yh}} \quad f_{\mathrm{l}y}=\rho_y f_{\mathrm{yh}} \tag{10}$$

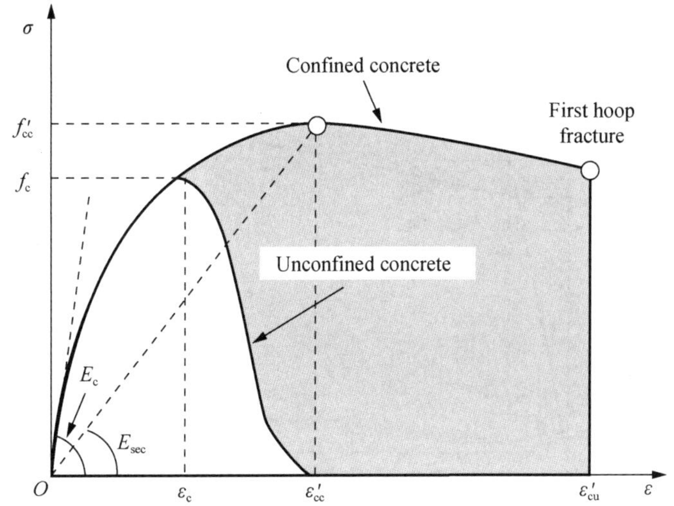

Fig. 3. Confined concrete model by Mander et al.[26] (adapted from Mander 1988, © ASCE)

where ρ_x, ρ_y = ratio of the volume of stirrups to the volume of core concrete in the x and y directions, respectively; and f_{yh} = yielding strength of stirrups. The effective lateral confining stress can be attained as follows

$$f'_{\mathrm{l}}=k_{\mathrm{e}} f_{\mathrm{l}} \tag{11}$$

$$k_{\mathrm{e}}=\frac{A_{\mathrm{e}}}{A_{\mathrm{cc}}}=\frac{\left[b_{\mathrm{c}}d_{\mathrm{c}}-\sum_{i}^{n}\frac{(W'_i)^2}{6}\right]\left(1-\frac{s'}{2b_{\mathrm{c}}}\right)\left(1-\frac{s'}{2d_{\mathrm{c}}}\right)}{b_{\mathrm{c}}d_{\mathrm{c}}(1-\rho_{\mathrm{cc}})} \tag{12}$$

where $b_{\mathrm{c}}(d_{\mathrm{c}})$ = width (height) of a section; w_i = clear distance among longitudinal bars; and ρ_{cc} = ratio of longitudinal bars to the area of concrete core. Finally, the confined strength ratio $R=f'_{\mathrm{cc}}/f_{\mathrm{c}}$ can be obtained according to the figure proposed by Mander et al.[26], and the corresponding peak strain is

$$\varepsilon'_{\mathrm{cc}}=\varepsilon_{\mathrm{c}}\left[1+5\left(\frac{f'_{\mathrm{cc}}}{f_{\mathrm{c}}}-1\right)\right] \tag{13}$$

2.2 Reinforcement Bars

Hsu and Mo[24] conducted systematic experiments to study the stress-strain relationship for steel bars embedded in concrete. Their results indicated that the nonlinear behavior of steel bars in concrete differs from that of bare bars. For steel bars embedded in concrete, the stress-strain curve is characterized by the relationship between average stress and average strain within a specific length that contains several cracks, other than the local stress-strain relationship at one point. As shown in Fig. 4, the yielding strength f'_y of the bars in concrete is lower than that of the bare bars because stress is averaged within a certain length. However, the postyield stage of the bars in concrete is close to the hardening region of the bare bars because the contribution of concrete can be neglected under large strain. Hence, a simplified bilinear model for steel bars embedded in concrete can be given by[24]

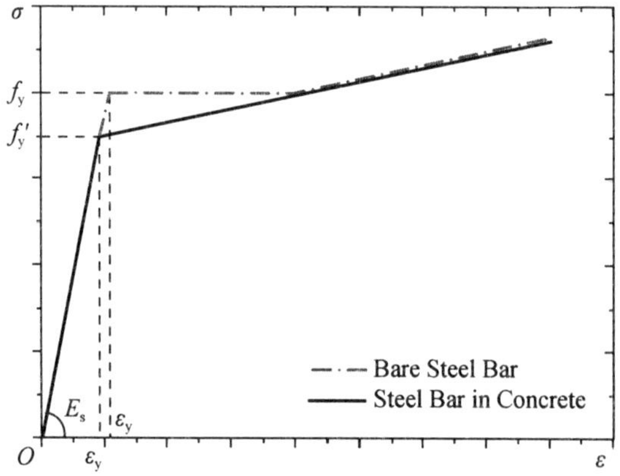

Fig. 4. Stress-strain relationship of reinforcement bars

$$f_s = E_s \varepsilon_s \quad \varepsilon_s \leqslant \varepsilon'_y \tag{14}$$

$$f_s = (0.91 - 2B) f_y + (0.02 + 0.25B) E_s \varepsilon_s \quad \varepsilon_s > \varepsilon'_y \tag{15}$$

In Eqs.(14) and (15)

$$\varepsilon'_y = f'_y / E_s \tag{16}$$

$$f'_y = (0.93 - 2B) f_y \tag{17}$$

$$B = \frac{1}{\rho}\left(\frac{f_{t,r}}{f_y}\right)^{1.5} \tag{18}$$

where f_s = stress of steel; ε_s = strain of steel; E_s = elastic modulus of steel; f_y/f'_y = yield stress of bare bars/bars in concrete; and ρ = reinforcement ratio. To simplify the numerical simulation, the stress-strain relation of steel bar in compression is modeled by similar equations in Eqs.(14)—(18).

2.3 Force-Based Fiber Frame Element

The force-based element[10] is used to simulate RC frames. The element is based on 2D Euler-Bernoulli theory, as shown in Fig. 5(a). In the formulation, the element forces and deformations are defined as vectors **Q** and **q** respectively. The sectional forces **D**(x) and deformations **d**(x) are expressed in the following vectors

$$\mathbf{D}(x) = \{N(x) \quad M(x)\}^{\mathrm{T}} \tag{19}$$

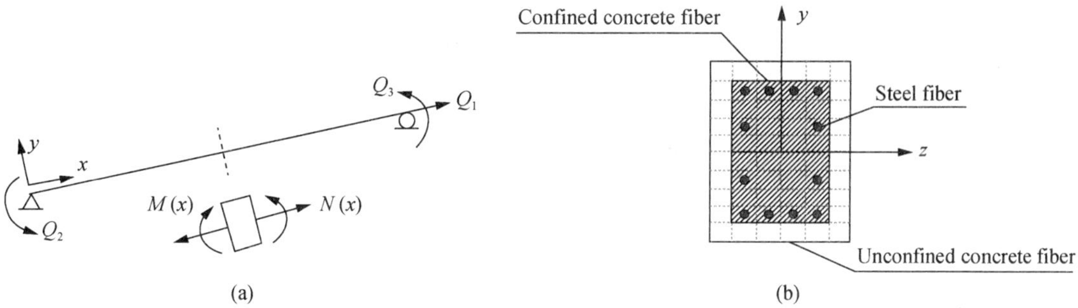

Fig. 5. Force-based fiber frame element: (a) force-based frame element; (b) fiber section model

$$\mathbf{d}(x)=\{e(x) \quad \kappa(x)\}^{\mathrm{T}} \tag{20}$$

where $N(x)$ = axial load; $M(x)$ = bending moment; $e(x)$ = axial strain; and $\kappa(x)$ = curvature.

Assuming the force distribution along the element length can express the equilibrium between sectional forces $\mathbf{D}(x)$ and element forces $\mathbf{Q}$ as

$$\mathbf{D}(x)=\mathbf{b}(x)\mathbf{Q} \tag{21}$$

where the matrix $\mathbf{b}(x)$ contains the force interpolation functions that define constant axial force and linear distribution of bending moment, i.e.

$$\mathbf{b}(x)=\begin{bmatrix} 1 & 0 & 0 \\ 0 & \dfrac{x}{L}-1 & \dfrac{x}{L} \end{bmatrix} \tag{22}$$

The compatibility relationship between sectional deformations $\mathbf{d}(x)$ and element deformations $\mathbf{q}$ can be obtained through applying the principle of virtual force

$$\mathbf{q}=\int_L \mathbf{b}(x)^{\mathrm{T}}\mathbf{d}(x)\mathrm{d}x \tag{23}$$

Subsequently, element flexibility is provided by the derivative of the compatibility equation

$$\mathbf{f}=\frac{\partial \mathbf{q}}{\partial \mathbf{Q}}=\int_L \mathbf{b}(x)^{\mathrm{T}}\mathbf{f}_s(x)\mathbf{b}(x)\mathrm{d}x \tag{24}$$

where $\mathbf{f}_{\mathrm{s}}$ = section flexibility, can be calculated by the inverse of the section stiffness, $\mathbf{f}_{\mathrm{s}}=\mathbf{k}_{\mathrm{s}}^{-1}$.

The fiber section model in Fig. 5(b) is often employed to establish the section force-deformation relationship based on the uniaxial stress-strain behavior of materials[10]. In a 2D case, the strain of a fiber can be expressed as

$$\varepsilon(x, y)=\mathbf{l}(y)\mathbf{d}(x) \tag{25}$$

where x = coordinate of a section; y = distance from the fiber to the center of a section; and $\mathbf{l}(y)=[1 \quad -y]$ = simple geometric vector.

The stress $\sigma(x, y)$ and tangent modulus $E(x, y)$ of the fiber can be obtained according to material constitutive law. The sectional resisting forces $\mathbf{D}_R(x)$ and stiffness matrix $\mathbf{k}_s(x)$ are then integrated

$$\mathbf{D}_R(x) = \int_{\mathbf{A}(x)} \mathbf{l}^T(y)\sigma(x, y)\mathrm{d}\mathbf{A} \tag{26}$$

$$\mathbf{k}_s(x) = \int_{\mathbf{A}(x)} \mathbf{l}^T(y)E(x, y)\mathbf{l}(y)\mathrm{d}\mathbf{A} \tag{27}$$

The integrations over section in Eqs.(26) and (27) are often computed by the 2D rectangle rule with the domain discretization shown in Fig. 5(b).

The compatibility in Eq.(23) and the element flexibility in Eq.(24) are evaluated using the following quadrature

$$\mathbf{q} = \sum_{i=1}^{N_p} \mathbf{b}(\xi_i)^T \mathbf{d}(\xi_i) w_i \quad \mathbf{f} = \sum_{i=1}^{N_p} \mathbf{b}(\xi_i)^T \mathbf{f}_s(\xi_i)\mathbf{b}(\xi_i) w_i \tag{28}$$

where N_p = number of integration points over the element length; and ξ_i and w_i = location and weight of the ith point respectively. The plastic hinge integration method is used in the formulation to overcome the localization issue[27]. The plastic hinge length l_p is calculated using the following empirical formula[28]

$$l_p = 0.08L + 0.022 f_y d_b \tag{29}$$

where L = length of the beam or column; f_y = yield strength of the longitudinal bar; and d_b = diameter of the bar.

3 Deterministic Analysis of Test Specimens

3.1 Finite-Element Model

The finite-element model for the test specimens is shown in Fig. 6. Each beam or column is simulated with a single force-based element. Based on Eq.(29), the plastic hinge length is

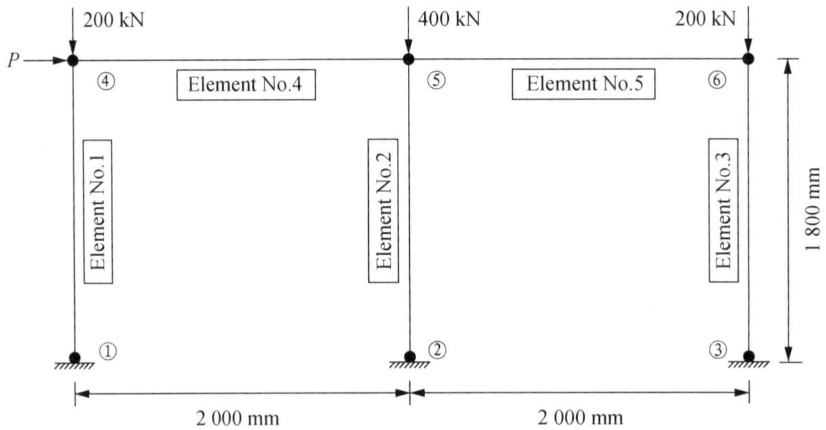

Fig. 6. Finite-element model of test specimen

calculated to be 310 mm for the beam and 295 mm for the column. The material properties used in the analysis are adopted in accordance with the test data in the preliminary study (mean value) and the previously-mentioned material models. The initial elastic modulus of concrete $E_c = 33\ 470$ MPa; the tensile properties are $f_{t,r} = 2.92$ MPa, $\varepsilon_{t,r} = 0.000\ 1$ MPa, and $\alpha_t = 0.1$; and the compressive properties are $f_{c,r} = 36.1$ MPa, $\varepsilon_{c,r} = 0.002$, and $\alpha_c = 3$ for unconfined concrete and $f'_{cc} = 39.7$ MPa, $\varepsilon'_{cc} = 0.002\ 4$, and $\alpha_c = 0.3$ for confined concrete. For steel, the yield strength is $f'_y = 500$ MPa in consideration of tension stiffening, and the elastic modulus is $E_s = 189\ 730$ MPa.

3.2 Simulated Results

The load-displacement curve simulated by deterministic analysis is compared with the test results(Fig. 7). At the load-displacement level, the analytical result agrees well with the test results. The simulated peak lateral load is 111.6 kN, and the mean value of the test results is 107.0 kN. The difference between the maximum test value (110.8 kN) and the analytical result is 0.7%, and the difference between the minimum test value (103.5 kN) and the analytical result is 7.3%. Hence, the deterministic analysis method can be used to estimate the overall structural responses because of the small variability observed in the global scale.

The historical diagrams of internal forces at column bases are shown in Fig. 8. Owing to the large scattering of internal forces, the results of conventional deterministic analysis can only capture the trend of responses. Nevertheless, the trend is not sufficient to assess structural safety and reliability. As listed in Table 1, a difference of more than 20% for the peak values of internal forces is observed. For example, the simulated peak value of shear force of the left column is 31.5 kN, and the maximum test result is 42.4 kN, which exhibits a difference of 34.9%. These results prove that the deterministic analysis method cannot obtain the redistribution and fluctuation of the internal forces. A comprehensive stochastic method should be introduced to study the randomness of the nonlinear behavior in depth.

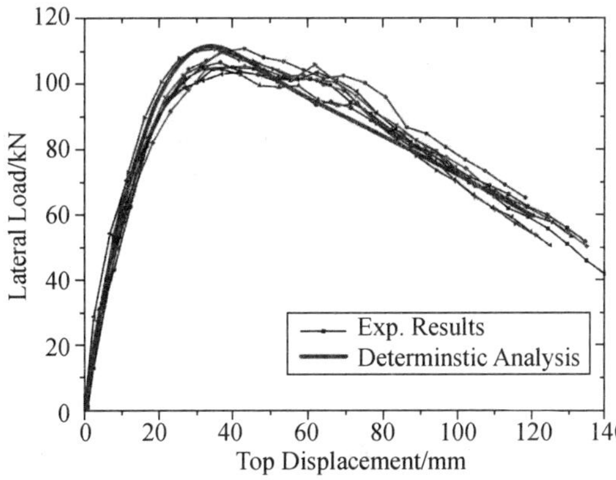

Fig. 7. Load-displacement curves by deterministic analysis

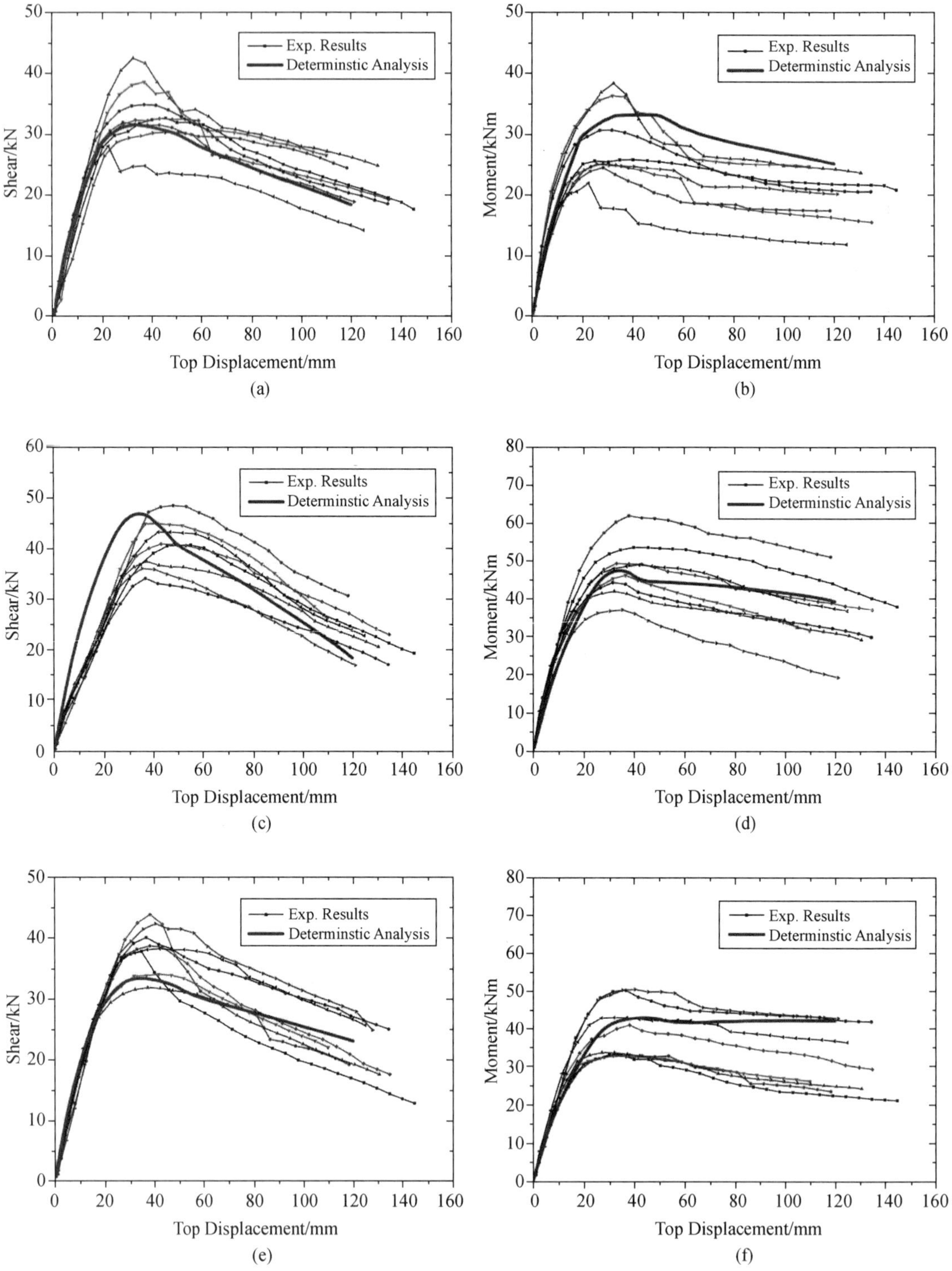

Fig. 8. Internal forces of columns by deterministic analysis: (a) shear of left column; (b) moment of left column; (c) shear of middle column; (d) moment of middle column; (e) shear of right column; (f) moment of right column

Table 1. Peak Internal Forces by Deterministic Analysis and Test Results

Details	Left column		Middle column		Right column	
	Shear /kN	Moment /kN·m	Shear /kN	Moment /kN·m	Shear /kN	Moment /kN·m
Test max F_{max}	42.4	38.2	48.4	61.9	43.9	50.6
Test min F_{min}	27.9	21.8	34.0	37.2	31.9	33.1
Analysis F_{ana}	31.5	33.1	46.8	47.5	33.4	43.2
$\|F_{ana}-F_{max}\|/F_{max}(\%)$	34.9	15.6	3.4	30.4	31.4	17.3
$\|F_{ana}-F_{min}\|/F_{min}(\%)$	11.2	34.1	27.2	21.7	4.5	23.2

4 Stochastic Analysis Method for RC Frames

4.1 Generalized Density Evolution Equation

A random system generally has the following expression

$$\mathbf{X}(\tau)=\mathbf{G}(\boldsymbol{\Theta},\tau) \quad \text{or} \quad X_l(\tau)=G_l(\boldsymbol{\Theta},\tau) \quad l=1,2,\cdots,m \tag{30}$$

where $\mathbf{X}$ = state vector of a structure; $\boldsymbol{\Theta}$ = random vector characterizing the randomness in material properties; τ = generalized time reflecting the evolution direction of a structure; and m = system dimension. System velocity can be expressed as

$$\dot{\mathbf{X}}(\tau)=\dot{\mathbf{H}}(\boldsymbol{\Theta},\tau) \quad \text{or} \quad \dot{X}_l(\tau)=\dot{H}_l(\boldsymbol{\Theta},\tau) \quad l=1,2,\cdots,m \tag{31}$$

Obviously, $\mathbf{H}=\partial\mathbf{G}/\partial\tau$.

In structural engineering, the quantities of interests are not only the displacement, velocity, and acceleration response of structures, but also the internal force, stress, and/or strain at some typical points. These quantities $\mathbf{Z}=(Z_1, Z_2, \cdots, Z_m)^{\mathrm{T}}$ can generally be determined by the state vector $\mathbf{X}$, i.e.

$$\dot{\mathbf{Z}}(\tau)=\psi[\dot{\mathbf{X}}(\tau)] \tag{32}$$

where $\psi[\cdot]$ = transfer operator. For instance, if $Z_l(\tau)$ represents the strain of concrete at a certain time τ, then $\psi[\cdot]$ will be an operator bridging strain and any other responses. Combining Eqs. (31) and (32) yields

$$\dot{\mathbf{Z}}(\tau)=\psi[\dot{\mathbf{H}}(\boldsymbol{\Theta},\tau)]=\mathbf{h}(\boldsymbol{\Theta},\tau) \tag{33}$$

For convenience, the stochastic system $[\mathbf{Z}(\tau), \boldsymbol{\Theta}]$ is included, and its joint PDF is denoted as $p_{\mathbf{Z}\boldsymbol{\Theta}}(\mathbf{z}, \boldsymbol{\theta}, \tau)$. The random event upon the stochastic system can be defined by $\{[\mathbf{Z}(\tau), \boldsymbol{\Theta}] \in \Omega_\tau \times \Omega_\theta\}$, where Ω_θ denotes an arbitrary subdomain of the distribution domain of $\boldsymbol{\Theta}$, and Ω_τ denotes the distribution domain of $\mathbf{Z}$ at time τ. Through a small time increment $\mathrm{d}\tau$, the random event will evolve into $\{[\mathbf{Z}(\tau+\mathrm{d}\tau), \boldsymbol{\Theta}] \in \Omega_{\tau+\mathrm{d}\tau} \times \Omega_\theta\}$. According to the principle of preservation of probability[19, 29], the following expression can be obtained:

$$\Pr\{[\mathbf{Z}(\tau), \boldsymbol{\Theta}] \in \Omega_\tau \times \Omega_\theta\}=\Pr\{[\mathbf{Z}(\tau+\mathrm{d}\tau), \boldsymbol{\Theta}] \in \Omega_{\tau+\mathrm{d}\tau} \times \Omega_\theta\} \tag{34}$$

Specifically

$$\int_{\Omega_\tau \times \Omega_\theta} p_{\mathbf{Z\Theta}}(\mathbf{z}, \boldsymbol{\theta}, \tau)\mathrm{d}\mathbf{z}\mathrm{d}\boldsymbol{\theta} = \int_{\Omega_{\tau+\mathrm{d}\tau} \times \Omega_\theta} p_{\mathbf{Z\Theta}}(\mathbf{z}, \boldsymbol{\theta}, \tau + \mathrm{d}\tau)\mathrm{d}\mathbf{z}\mathrm{d}\boldsymbol{\theta} \tag{35}$$

The domain $\Omega_{\tau+\mathrm{d}\tau}$ is attributed to the transition of Ω_τ and its boundary, i.e.

$$\Omega_{\tau+\mathrm{d}\tau} = \Omega_\tau + \int_{\partial\Omega_\tau} (\mathbf{v}_Z \mathrm{d}\tau) \cdot \mathbf{n}\mathrm{d}S = \Omega_\tau + \int_{\partial\Omega_\tau} [\mathbf{h}(\boldsymbol{\theta}, \tau)\mathrm{d}\tau] \cdot \mathbf{n}\mathrm{d}S \tag{36}$$

Substituting Eq. (36) into Eq. (35) leads to

$$\begin{aligned}
\int_{\Omega_\tau \times \Omega_\theta} p_{\mathbf{Z\Theta}}(\mathbf{z}, \boldsymbol{\theta}, \tau)\mathrm{d}\mathbf{z}\mathrm{d}\boldsymbol{\theta} &= \int_{\Omega_{\tau+\mathrm{d}\tau} \times \Omega_\theta} p_{\mathbf{Z\Theta}}(\mathbf{z}, \boldsymbol{\theta}, \tau + \mathrm{d}\tau)\mathrm{d}\mathbf{z}\mathrm{d}\boldsymbol{\theta} \\
&= \int_{\Omega_{\tau+\mathrm{d}\tau} \times \Omega_\theta} \left[p_{\mathbf{Z\Theta}}(\mathbf{z}, \boldsymbol{\theta}, \tau) + \frac{\partial p_{\mathbf{Z\Theta}}(\mathbf{z}, \boldsymbol{\theta}, \tau)}{\partial t} \tau \mathrm{d}\tau \right] \mathrm{d}\mathbf{z}\mathrm{d}\boldsymbol{\theta} \\
&= \int_{\Omega_\tau \times \Omega_\theta} \left[p_{\mathbf{Z\Theta}}(\mathbf{z}, \boldsymbol{\theta}, \tau) + \frac{\partial p_{\mathbf{Z\Theta}}(\mathbf{z}, \boldsymbol{\theta}, \tau)}{\partial \tau} \mathrm{d}\tau \right] \mathrm{d}\mathbf{z}\mathrm{d}\boldsymbol{\theta} + \\
&\quad \int_{\partial\Omega_\tau \times \Omega_\theta} \left[p_{\mathbf{Z\Theta}}(\mathbf{z}, \boldsymbol{\theta}, \tau) + \frac{\partial p_{\mathbf{Z\Theta}}(\mathbf{z}, \boldsymbol{\theta}, \tau)}{\partial \tau} \mathrm{d}\tau \right] [\mathbf{h}(\boldsymbol{\theta}, \tau)\mathrm{d}\tau] \cdot \mathbf{n}\mathrm{d}S\mathrm{d}\boldsymbol{\theta}
\end{aligned} \tag{37}$$

Subtracting $\int_{\Omega_\tau \times \Omega_\Theta} p_{\mathbf{Z\Theta}}(\mathbf{z}, \boldsymbol{\theta}, \tau)\mathrm{d}\mathbf{z}\mathrm{d}\boldsymbol{\theta}$ on both sides of Eq. (37) yields

$$\begin{aligned}
\int_{\Omega_\tau \times \Omega_\Theta} \left[\frac{\partial p_{\mathbf{Z\Theta}}(\mathbf{z}, \boldsymbol{\theta}, \tau)}{\partial \tau} \mathrm{d}\tau \right] \mathrm{d}\mathbf{z}\mathrm{d}\boldsymbol{\theta} = -\int_{\partial\Omega_\tau \times \Omega_\Theta} \left[p_{\mathbf{Z\Theta}}(\mathbf{z}, \boldsymbol{\theta}, \tau) + \frac{\partial p_{\mathbf{Z\Theta}}(\mathbf{z}, \boldsymbol{\theta}, \tau)}{\partial \tau} \mathrm{d}\tau \right] \times \\
[\mathbf{h}(\boldsymbol{\theta}, \tau)\mathrm{d}\tau] \cdot \mathbf{n}\mathrm{d}S\mathrm{d}\boldsymbol{\theta}
\end{aligned} \tag{38}$$

Applying the divergence theorem and omitting high-order terms will lead to

$$\int_{\Omega_\tau \times \Omega_\Theta} \left[\frac{\partial p_{\mathbf{Z\Theta}}(\mathbf{z}, \boldsymbol{\theta}, \tau)}{\partial \tau} \mathrm{d}\tau \right] \mathrm{d}\mathbf{z}\mathrm{d}\boldsymbol{\theta} = -\int_{\Omega_\tau \times \Omega_\Theta} \left[\sum_{l=1}^{m} \frac{\partial [p_{\mathbf{Z\Theta}}(\mathbf{z}, \boldsymbol{\theta}, \tau) h_l(\boldsymbol{\theta}, \tau)\mathrm{d}\tau]}{\partial z_l} \right] \mathrm{d}\mathbf{z}\mathrm{d}\boldsymbol{\theta} \tag{39}$$

In accordance with the arbitrariness of $\Omega_\tau \times \Omega_\theta$

$$\frac{\partial p_{\mathbf{Z\Theta}}(\mathbf{z}, \boldsymbol{\theta}, \tau)}{\partial \tau} + \sum_{l=1}^{m} h_l(\boldsymbol{\theta}, \tau) \frac{\partial p_{\mathbf{Z\Theta}}(\mathbf{z}, \boldsymbol{\theta}, \tau)}{\partial z_l} = 0 \tag{40}$$

or in an alternative form

$$\frac{\partial p_{\mathbf{Z\Theta}}(\mathbf{z}, \boldsymbol{\theta}, \tau)}{\partial \tau} + \sum_{l=1}^{m} \dot{Z}_l(\boldsymbol{\theta}, \tau) \frac{\partial p_{\mathbf{Z\Theta}}(\mathbf{z}, \boldsymbol{\theta}, \tau)}{\partial z_l} = 0 \tag{41}$$

Eqs. (40) and (41) are so-called generalized density evolution equations. Introducing the initial and boundary conditions

$$p_{\mathbf{Z\Theta}}(\mathbf{z}, \boldsymbol{\theta}, \tau)\Big|_{\tau=\tau_0} = \delta(\mathbf{z} - \mathbf{z}_0) p_{\boldsymbol{\Theta}}(\boldsymbol{\theta}) \tag{42}$$

$$p_{\mathbf{Z\Theta}}(\mathbf{z}, \boldsymbol{\theta}, \tau)\Big|_{z_l \to \pm\infty} = 0 \quad l = 1, 2, \cdots, m \tag{43}$$

where z_0 = initial value of $\mathbf{Z}$; and $\delta(\cdot)$ = Dirac function. By using numerical schemes, the generalized density evolution equation can be solved and the joint PDF $p_{Z\Theta}(\mathbf{z}, \boldsymbol{\theta}, \tau)$ can be obtained. The joint PDF of $Z(\tau)$ is given by integration as follows:

$$p_Z(\mathbf{z}, \tau)=\int_{\Omega_\Theta} p_{Z\Theta}(\mathbf{z}, \boldsymbol{\theta}, \tau)\mathrm{d}\boldsymbol{\theta} \tag{44}$$

4.2 Analytical Solution for Structural Responses of RC Frames

The generalized density evolution equation[Eq.(41)], provides an opportunity to calculate the probability density evolution process of structural responses. In fact, for RC frame structures, the probability density evolution equation for a response in the 1D domain can be written as

$$\frac{\partial p_{z\Theta}(z_l, \boldsymbol{\theta}, \tau)}{\partial \tau}+\frac{\partial Z_1}{\partial \tau}\frac{\partial p_{z\Theta}(z_l, \boldsymbol{\theta}, \tau)}{\partial z_l}=0 \tag{45}$$

where Z_1=response of concern, e.g., the internal force of a specific section. The calculation of $\partial Z_1/\partial\tau$ depends on a specific type of response and the structural loading mechanism.

For RC frames, the static governing equation for the entire structure is

$$\mathbf{K}\Delta\mathbf{U}=\Delta\mathbf{P} \tag{46}$$

where $\mathbf{K}$ = global stiffness matrix of the structure; $\Delta\mathbf{U}$ = displacement increment of the structure; and $\Delta\mathbf{P}$=load increment of the structure. The generalized time τ depends on the specific loading mechanism of the structure. In case of force loading, τ denotes the loading increment factor; in case of displacement loading, τ denotes the loading parameter of the characteristic displacement. For simplicity, the proportional displacement control is considered in the paper, namely

$$\mathbf{U}=\mathbf{U}_0 f(\tau) \quad \Delta\mathbf{U}=\mathbf{U}_0\dot{f} \tag{47}$$

where $\mathbf{U}_0$ =initial displacements of the structure; $f(\tau)$ =time-dependent loading function; and $\dot{f}$ =derivative of $f(\tau)$.

In consideration of the internal force $\mathbf{D}(x)$ of a certain section of a frame, according to the formulation of the force-based element in section "Force-Based Fiber Frame Element," the following relations are obtained in incremental format

$$\Delta\mathbf{D}(x)=\mathbf{b}(x)\Delta\mathbf{Q} \tag{48}$$

$$\Delta\mathbf{d}(x)=\mathbf{f}_s(x)\Delta\mathbf{D}(x)=\mathbf{f}_s(x)\mathbf{b}(x)\Delta\mathbf{Q} \tag{49}$$

$$\mathbf{q}=\int_L \mathbf{b}^{\mathrm{T}}(x)\mathbf{d}(x)\mathrm{d}x \tag{50}$$

$$\mathbf{f}=\frac{\partial\mathbf{q}}{\partial\mathbf{Q}}=\int_L \mathbf{b}^{\mathrm{T}}(x)\mathbf{f}_s(x)\mathbf{b}(x)\mathrm{d}x \tag{51}$$

$$\mathbf{k}=(\mathbf{f})^{-1} \tag{52}$$

where $\mathbf{k}$ =element stiffness matrix.

Consequently, it can be obtained

$$\frac{\partial Z_l}{\partial \tau}=\frac{\partial \mathbf{D}(x)}{\partial \tau}=\frac{\partial \mathbf{D}(x)}{\partial \mathbf{Q}}\frac{\partial \mathbf{Q}}{\partial \mathbf{q}}\frac{\partial \mathbf{q}}{\partial \tau}=\mathbf{b}(x)\mathbf{k}\frac{\partial \mathbf{q}}{\partial \tau} \tag{53}$$

The increment of element deformation $\Delta\mathbf{q}$ depends on the structural displacement increment $\Delta\mathbf{U}$. Thus

$$\Delta\mathbf{q}=\mathbf{TC}\Delta\mathbf{U} \tag{54}$$

where $\mathbf{C}$ = element location matrix; and $\mathbf{T}$ = coordinate transform matrix.

Hence

$$\frac{\partial \mathbf{q}}{\partial \tau}=\mathbf{TC}\frac{\partial \mathbf{U}}{\partial \tau}=\mathbf{TCU}_0\dot{f} \tag{55}$$

Substitution of Eq.(55) to Eq.(53) obtains

$$\frac{\partial Z_l}{\partial \tau}=\mathbf{b}(x)\mathbf{kTCU}_0\dot{f} \tag{56}$$

The details of the iteration algorithm to calculate the element stiffness $\mathbf{k}$ can be seen in Spacone et al.[10].

4.3 Numerical Algorithm

The analytical solution of $\dot{Z}_l$ can be obtained in seldom cases, for which the function $f(\tau)$ is simple, and the analytical form of $\dot{f}$ can be obtained. However, for a general structural system, the analytical solution of $\dot{\mathbf{Z}}(\boldsymbol{\theta}, \tau)$ cannot be derived; thus, the generalized density evolution Eq.(40) is analytically intractable for most problems. Nevertheless, the equation can be solved through a numerical algorithm. The procedure to solve the problem can generally be performed in the following steps[19, 30]:

(1) Step 1: A set of discrete representative points is selected in the basic distribution domain $\Omega_{\boldsymbol{\Theta}}$ of random parameters, and the point set is defined as

$$\boldsymbol{\theta}_q=(\boldsymbol{\theta}_{1,q}, \boldsymbol{\theta}_{2,q}, \cdots, \boldsymbol{\theta}_{s,q});\quad q=1, 2, \cdots, n_{\text{sel}} \tag{57}$$

where s = number of random parameters; and n_{sel} = total number of the selected points. The assigned probability of each point in the point set is computed through

$$P_q=\int_{V_q} p_{\boldsymbol{\Theta}}(\boldsymbol{\theta})\mathrm{d}\boldsymbol{\theta} \tag{58}$$

where V_q = representative volume of each point.

(2) Step 2: For a given $\boldsymbol{\theta}_q$ obtained from Step 1, the deterministic analysis of the structure is conducted, and the generalized velocity $\dot{Z}_l(\boldsymbol{\theta}_q, \tau_k)$ of the response according to Eq.(33) is obtained, where $\tau_k = k\cdot\Delta\tau$, and $\Delta\tau$ is the time step.

(3) Step 3: With the representative point set and assigned probability in Step 1, as well as the velocity of response in Step 2, the generalized probability density evolution Eq.(41) can be written in the following discrete manner:

$$\frac{\partial p_{\mathbf{Z\Theta}}(\mathbf{z}, \boldsymbol{\theta}_q, \tau)}{\partial \tau} + \sum_{l=1}^{m} \dot{Z}_l(\boldsymbol{\theta}_q, \tau) \frac{\partial p_{\mathbf{Z\Theta}}(\mathbf{z}, \boldsymbol{\theta}_q, \tau)}{\partial z_l} = 0 \tag{59}$$

This equation can be solved under the initial and boundary conditions using the finite-difference method to obtain $p_{\mathbf{Z\Theta}}(\mathbf{z}, \boldsymbol{\theta}_q, \tau)$.

(4) Step 4: Taking the numerical integration with respect to $\boldsymbol{\theta}_q$ gives the PDF as follows:

$$p_{\mathbf{Z}}(\mathbf{z}, \tau) = \sum_{q=1}^{n_{\text{sel}}} p_{\mathbf{Z\Theta}}(\mathbf{z}, \boldsymbol{\theta}_q, \tau) \tag{60}$$

The numerical algorithm of PDEM involves combining a series of deterministic analyses and solving the generalized density evolution equation. The point selection in Step 1 is important because it is directly related to the efficiency of PDEM. The suitable point selecting strategy can be found in, e.g., Li and Chen[19] and Chen and Zhang[31].

4.4 Total Variation Diminishing (TVD) Difference Scheme

In principle, any difference schemes with behaviors satisfying the requirements of convergence, consistency, and stability can be employed in Step 3[19]. The computing experiences suggest that the Lax-Wendroff finite difference scheme can offer a reliable result with second-order accuracy. The shortcomings of the Lax-Wendroff algorithm are the possible unphysical negative values of PDF and the algorithmic oscillations around unsmooth points. An effective way to overcome the shortcomings of the Lax-Wendroff scheme is to introduce numerical damping to suppress the dispersion of the scheme. In this work, a TVD format, referred to as the TVD scheme, is adopted by imposing a flux limiter to the Lax-Wendroff scheme[32].

On the basis of the Lax-Wendroff scheme, the TVD scheme of Eq.(40) can be constructed by adding a flux limiter, i.e.

$$\begin{aligned} p_{j,k+1} = p_{j,k} - r_L \Big[& \frac{1}{2}(g_k + |g_k|)(p_{j,k} - p_{j-1,k}) \Big] + \\ & \frac{1}{2}(g_k - |g_k|)(p_{j+1,k} - p_{j,k}) - \\ & \frac{1}{2}(1 - |r_L g_k|) r_L g_k \, | \, [\psi(r^+_{j+1/2}, r^-_{j+1/2})(p_{j+1,k} - p_{j,k}) - \\ & \psi(r^+_{j-1/2}, r^-_{j-1/2})(p_{j,k} - p_{j-1,k})] \end{aligned} \tag{61}$$

where $p_{j,k} = p_{\mathbf{Z\Theta}}(\mathbf{z}_i, \boldsymbol{\theta}_q, \tau_k)$; $r_L = \Delta\tau/\Delta x$ = lattice ratio; $\psi(r^+, r^-)$ = flux limiter; and

$$\begin{aligned} r^+_{j+1/2} &= \frac{p_{j+2,k} - p_{j+1,k}}{p_{j+1,k} - p_{j,k}} \quad & r^-_{j+1/2} &= \frac{p_{j,k} - p_{j-1,k}}{p_{j+1,k} - p_{j,k}} \\ r^+_{j-1/2} &= \frac{p_{j+1,k} - p_{j,k}}{p_{j,k} - p_{j-1,k}} \quad & r^-_{j-1/2} &= \frac{p_{j-1,k} - p_{j-2,k}}{p_{j,k} - p_{j-1,k}} \end{aligned} \tag{62}$$

$$g_k = \frac{1}{2}[\dot{\mathbf{Z}}(\boldsymbol{\theta}_q, \tau_{k-1}) + \dot{\mathbf{Z}}(\boldsymbol{\theta}_q, \tau_k)] \tag{63}$$

A variety of candidates exists for the flux limiter. According to the computational experience, the Roe-Sweby flux limiter with small dissipation is an ideal choice, i.e.

$$\psi_{sb}(r^{-}) = \max[0, \min(2r^{-}, 1), \min(r^{-}, 2)] \tag{64}$$

To account for the adaptive difference direction, the following expression is used:

$$\psi(r^{+}, r^{-}) = H(-g_k)\psi_{sb}(r^{+}) + H(g_k)\psi_{sb}(r^{-}) \tag{65}$$

where $H(\cdot)$ is the Heaviside function, i.e.

$$H(x) = \begin{cases} 1 & x \geqslant 0 \\ 0 & \text{otherwise} \end{cases} \tag{66}$$

Previous investigations have proven that the TVD scheme does work well in the probability density evolution analysis of most problems[19].

5 Stochastic Response Analysis of Test Specimens

5.1 Random Parameters

Random parameters should be specified first to conduct the stochastic response analysis. In case of a relatively large axial compressive ratio(0.167 for the side column and 0.333 for the middle column), the response of the specimens is governed by the properties of steel bars and the compression properties of concrete. Thus, randomness is involved in the parameters of the steel reinforcing bar model and the compressive part of the concrete model, namely, yield strength f_y' and elastic modulus E_s in the reinforcement model, and compressive strength $f_{c,r}$, corresponding strain $\varepsilon_{c,r}$, and descending parameter α_c in the concrete model. According to the rules mentioned previously, confined concrete can be modeled by the parameters calculated in unconfined concrete. The values of the other parameters are adopted from the deterministic analysis.

The probability density distribution of the parameters is assumed to be the lognormal distribution, which is usually adopted to describe concrete property. The mean values of the parameters and the coefficient of variation (COV) are listed in Table 2 according to the results of material tests. Given that only six specimens are tested in the material test of concrete, the COV of the concrete parameters is adopted according to existing experiments[33] and research[34] to avoid offset estimation. A total of five random variables are involved, and the point-selection scheme based on generalized F-discrepancy[31] is employed to generate 128 representative points. These random variables are assumed to be uncorrelated. In fact, the current analytical framework can consider the correlations among random variables, however, the physical background and sufficient testing data are lacking for the qualification of correlations in existing literature. Thus, the uncorrelated case is chosen to demonstrate the capacity of the proposed method. The step-forwards in this direction will be discussed in the forthcoming works. The assumed distribution in comparison with the stress-strain curves for concrete under compression is shown in Fig. 9, which demonstrates that the selected distribution can capture the results of the material tests.

Table 2. Probabilistic Information of the Material Parameters

Parameter	Distribution	Mean	COV
$f_{c,r}$	Lognormal	36.1 MPa	0.15
$\varepsilon_{c,r}$	Lognormal	0.002	0.15
α_c	Lognormal	3.0	0.15
f_y'	Lognormal	500 MPa	0.1
E_s	Lognormal	189 730 MPa	0.05

Note: COV=coefficient of variation.

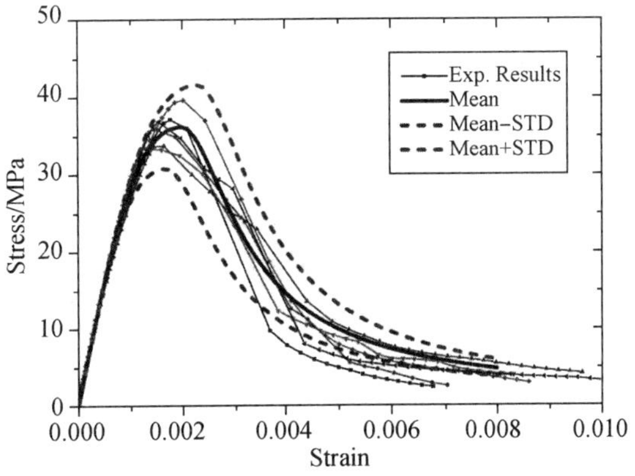

Fig. 9. Stress-strain curves for concrete by assumed distribution and test results

5.2 Load-Displacement Response

The stochastic responses of the frames are obtained through PDEM. The load-displacement response by PDEM, together with the experimental data, is plotted in Fig. 10. The test data of the eight frames are covered by the mean value plus/minus one time of the standard deviation of the numerical results. PDEM can secure the stochastic nonlinear behavior in the sense of the second-order statistical quantities. The instantaneous PDF of the lateral load against the top drift can also be obtained. Fig. 11 shows the PDFs of the lateral load at three different time points. In the initial loading stage (2 mm), with most parts of the structure at linear elastic, PDF is of a rather peak value and a narrow distributed domain. Hence, the random variation is relatively small. After the plastic hinge occurs (22 mm) and the peak strength is reached (39 mm), the influence of the randomness of material property on structural behavior becomes obvious. Consequently, the probability density distributes in a wide range. The PDF distribution shape turns to be irregular and different from the distribution functions assigned to the random parameters. A complex process of random evolution from the material scale to the structural scale is thus needed. The transfer and flow of the probability within this process will lead to stochastic damage evolutions and cause a variety of structural behavior and failure modes. As a result, the PDF curve is no longer smooth, and multiple peaks occur.

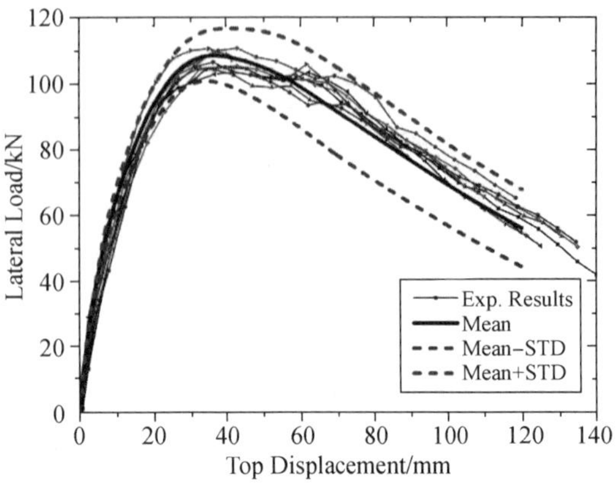

Fig. 10. Load-displacement curves by PDEM and test results

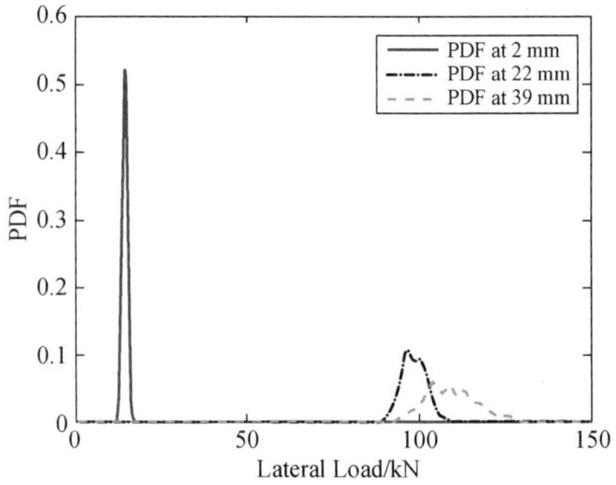

Fig. 11. PDFs at typical time instants

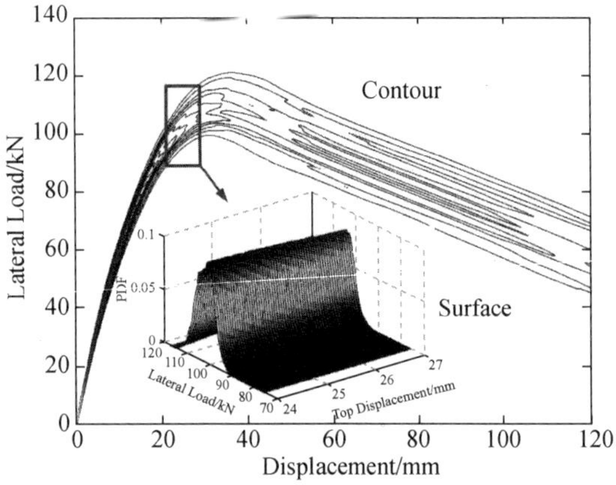

Fig. 12. PDF evolution surface and the corresponding contour

As shown in Fig. 12, PDF surfaces/contours can be constructed based on the PDF curves of the lateral load at different top displacements. Probability density varies against time. With increasing loading displacement, the peak value of PDF decreases and the distribution width grows. The corresponding contour is similar to water flowing through a river. However, the flow in the river behaves not as the stationary process, and the intensive part is related to the peak of the mountain. The width of the river naturally indicates the interval over which the response distributes. The entire evolution process can help to understand the probabilistic information of a stochastic structure deeply. This characteristic is also the advantage of PDEM against other stochastic analysis methods.

5.3 Internal Forces

As mentioned previously, the internal forces are directly relevant to the safety of a structure. However, the preliminary study indicates that the randomness in material causes fluctuation of internal forces. Consequently, the deterministic analysis method is not adequate to handle the realistic response of the frames. The mean value and standard deviation of internal forces of the frame members by PDEM against the test results are shown in Fig. 13. An acceptable agreement is obtained between the experimental and numerical results. The test data are nearly covered by the interval of mean ± 2 times standard deviation by PDEM, except for the moment of the left column. The variability calculated by PDEM seems to under predict the variability observed in some cases, i.e.,

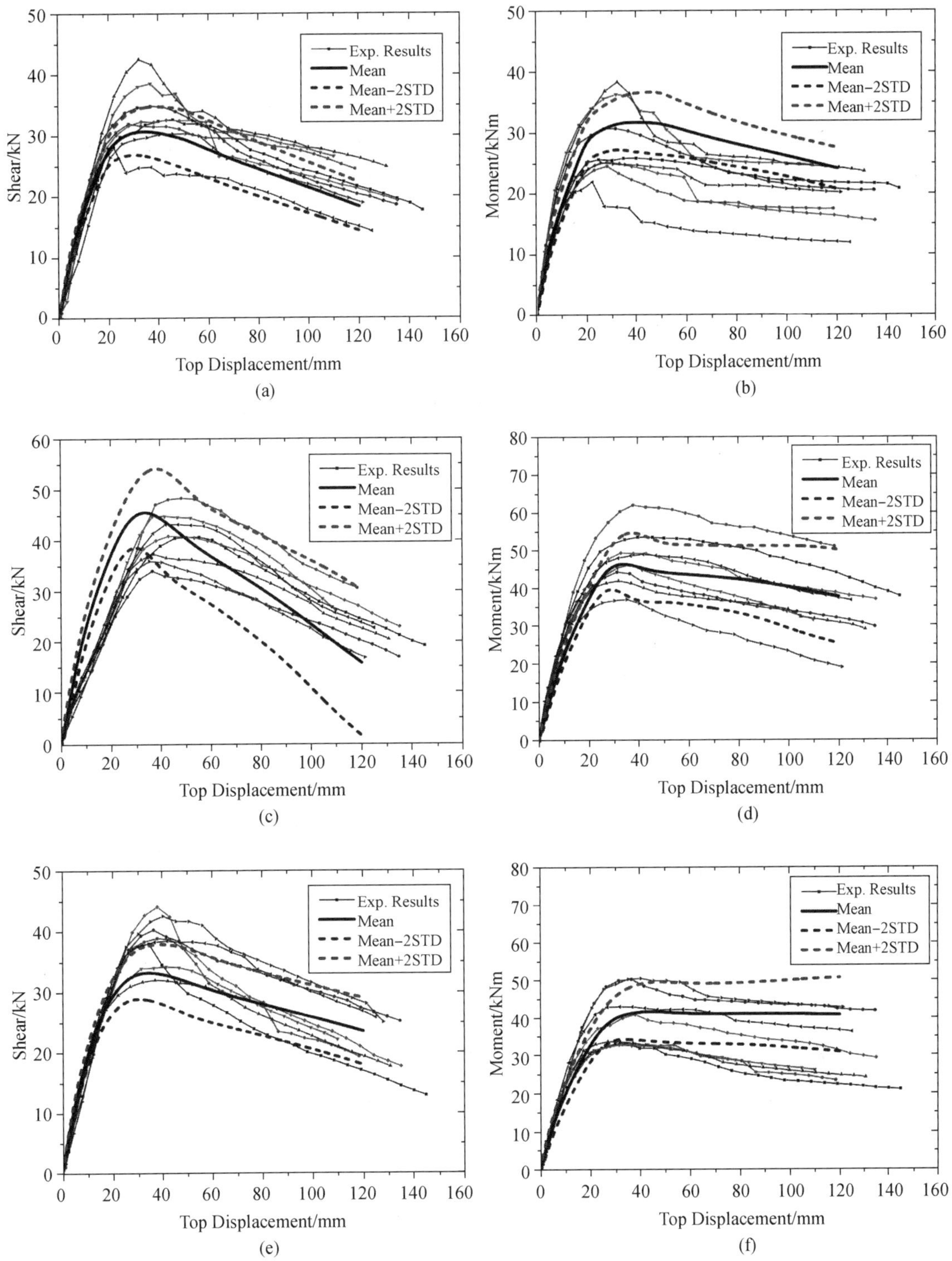

Fig. 13. Internal forces of columns by PDEM and test results: (a) shear of left column; (b) moment of left column; (c) shear of middle column; (d) moment of middle column; (e) shear of right column; (f) moment of right column

moments of the middle and right columns. Possible reasons may be as follows: some other uncertainties in structures are not included in the analysis, the assumed COV of the parameters is relatively small in comparison with the materials in the test specimen, the correlation of the random variables, or the accuracy of the deterministic analysis. The results presented in this paper partly verify that PDEM can reveal the influence of the randomness of materials on structural behavior to some extent.

The probabilistic information of internal forces is presented in Figs. 14 and 15, where the similar situation relevant to the load-displacement response is observed. In the initial loading stage, the variability of the internal forces is relatively small. Along with the process of loading, the variability grows and the PDFs change from a single-peak curve to a double-peak curve. This phenomenon indicates that different failure modes may occur. Figs. 14 and 15 imply that the structural response is a complex stochastic evolutionary process rather than a stationary process. PDEM can achieve a comprehensive representation of the stochastic evolution in structure.

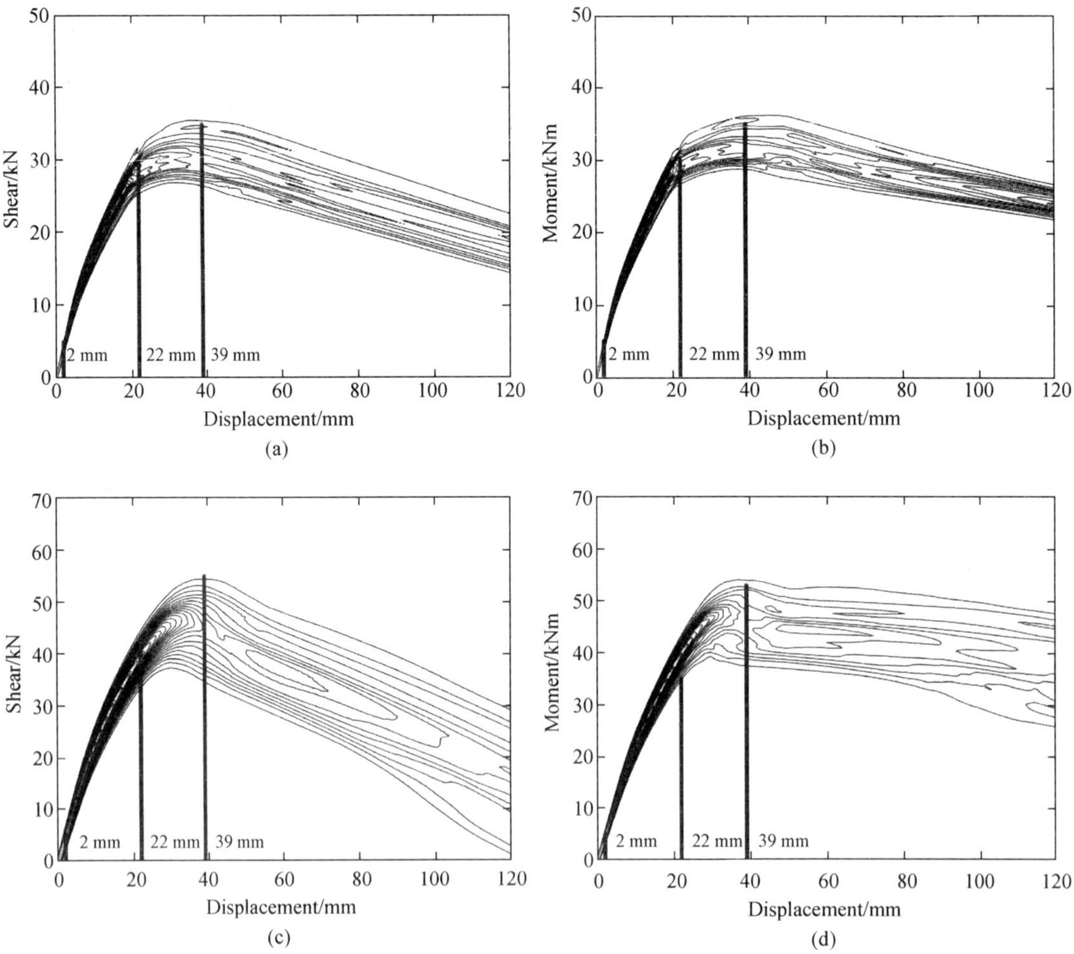

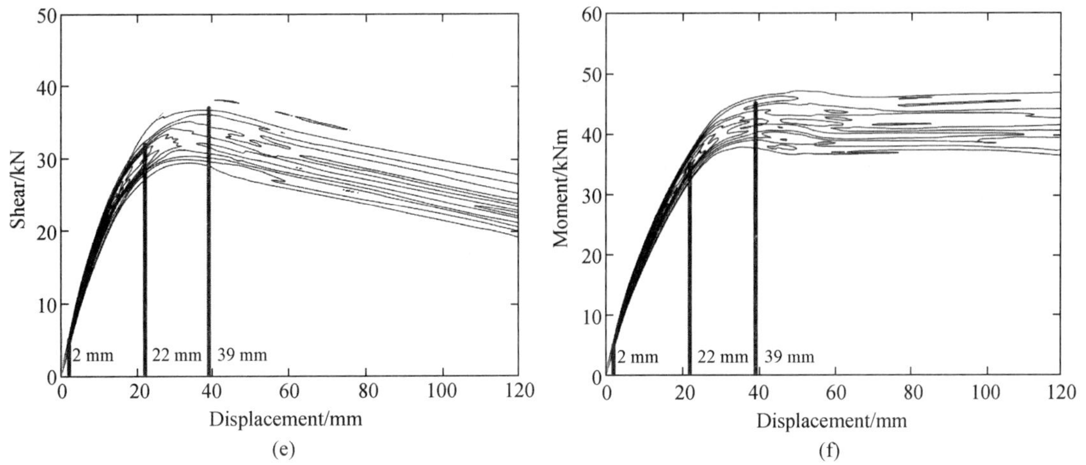

Fig. 14. PDF contours of the internal forces: (a) shear of left column; (b) moment of left column; (c) shear of middle column; (d) moment of middle column; (e) shear of right column; (f) moment of right column

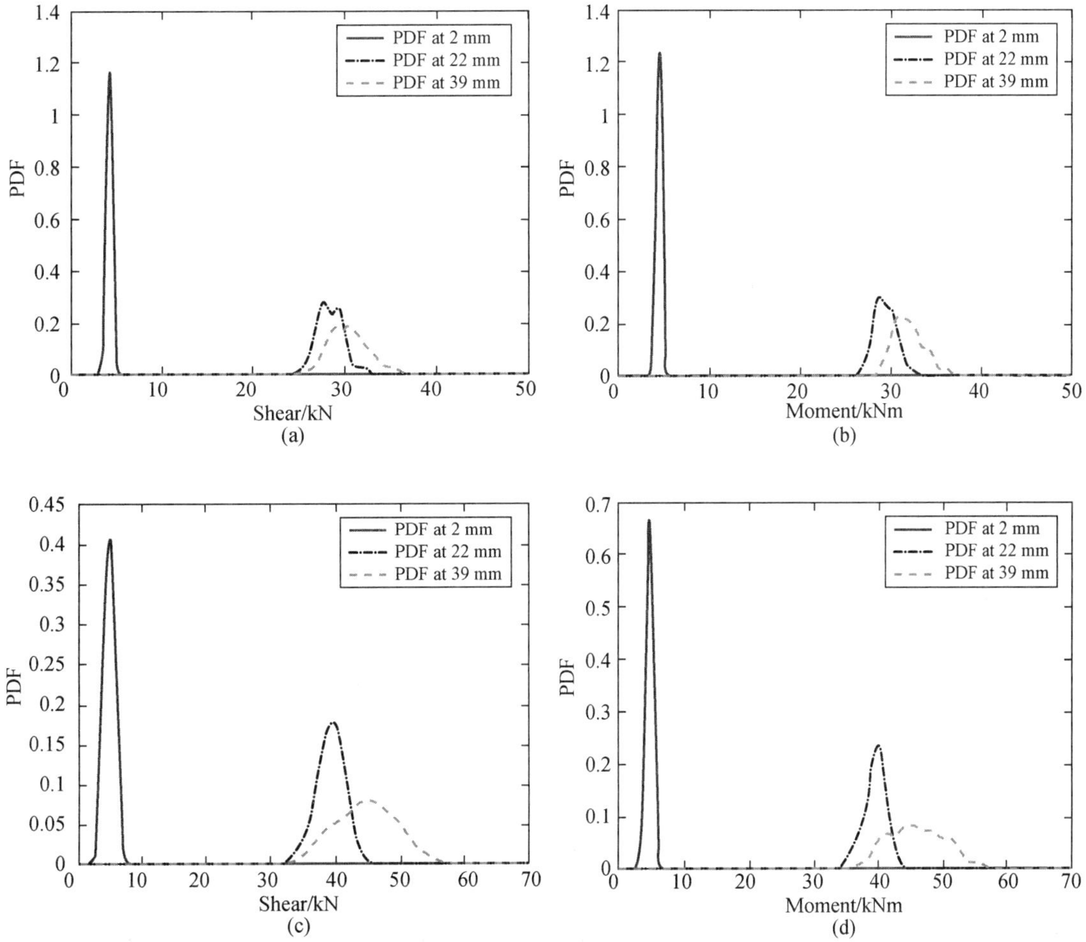

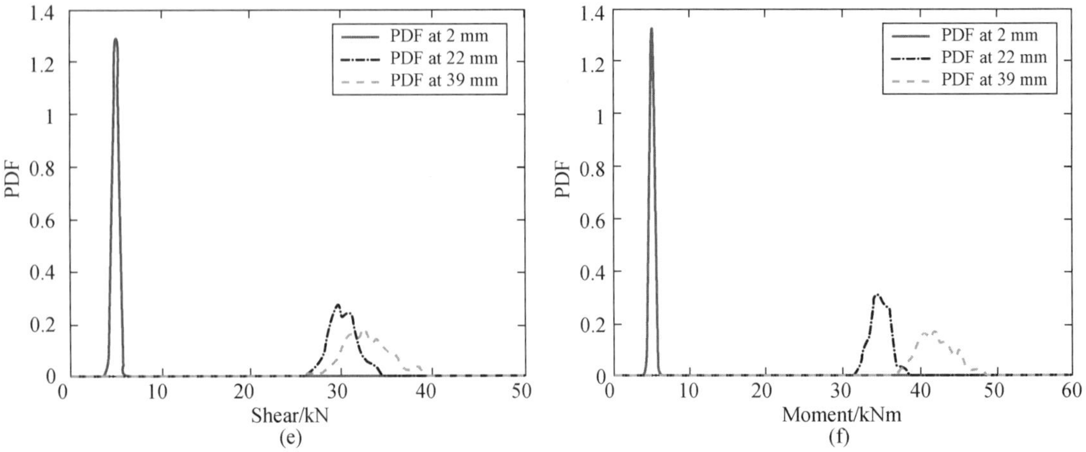

Fig. 15. PDFs of the internal forces at typical time instants: (a) shear of left column; (b) moment of left column; (c) shear of middle column; (d) moment of middle column; (e) shear of right column; (f) moment of right column

In structural design, the structural performance may deviate from the design goal, thereby resulting in unexpected failure modes if the influence of the stochastic evolution on the structure is neglected. On the contrary, the enhanced design of structures can be achieved if the stochastic evolution is handled in a logical manner, such as PDEM through quantificational optimization. This method provides a new perspective on structural performance control so the structure behaves as the expected failure mechanism with a certain probability. The coupling between randomness and nonlinearity should be appropriately included in structural design.

6 Conclusions

Numerical simulations of test specimens, which are experimentally investigated in the preliminary study, are performed through PDEM. On the basis of the results, a few conclusions are made as follows:

(1) Conventional deterministic analysis can be used to estimate the lateral load and top displacement of a structure because a small variability exists in such levels. However, in the internal force level, deterministic analysis cannot predict the actual response of an individual member because of the remarkable fluctuation and redistribution caused by the coupling between the randomness and nonlinearity of materials.

(2) The well-developed PDEM is introduced to conduct the stochastic analysis of concrete structures. The second-order quantities and PDFs of the responses can be readily derived. Thus, PDEM is a good approach to reveal scientifically the natural properties from material to structure.

(3) The results by PDEM seem to underestimate the variability of the responses measured

from the tests. Such underestimation indicates that some other factors affect the stochastic behavior of structures, i.e., other uncertainties in the structures, the distribution and correlation of random variables, the COV of the random variables, and so on. In-depth investigation on the coupling effect of randomness and nonlinearity inherent in the structural behavior should be conducted.

(4) Stochastic evolution in the global and local levels should be included in structural design and analysis to guarantee structural safety. PDEM can attain the reliability of structural performance or manipulate the structure to behave as the expected failure mechanism with a specific probability.

Acknowledgments

Financial supports from the National Natural Science Foundation of China (Grant Nos. 51261120374 and 91315301 - 01) are greatly appreciated. The authors would like to acknowledge the kind help of Building Structure Laboratory at Tongji University to accomplish the experiments.

References

[1] LI J, FENG D C, GAO X, et al. Stochastic nonlinear behavior of reinforced concrete frames. Ⅰ: Experimental investigation[J]. Journal of Structural Engineering, 2016, 142(3): 04015162.

[2] CLOUGH R W, Benuska K L, Wilson E L. Inelastic earthquake response of tall buildings[C]// Proceedings, Third World Conference on Earthquake Engineering, New Zealand. 1965, 11.

[3] GIBERSON M F. The response of nonlinear multi-story structures subjected to earthquake excitation [D]. California Institute of technology, 1967.

[4] HILMY S I, ABEL J F. Material and geometric nonlinear dynamic analysis of steel frames using computer graphics[J]. Computers & Structures, 1985, 21(4): 825-840.

[5] POWELL G H, CHEN P F S. 3D beam-column element with generalized plastic hinges[J]. Journal of Engineering Mechanics, 1986, 112(7): 627-641.

[6] SOLEIMANI D, POPOV E P, BERTERO V V. Nonlinear beam model for R/C frame analysis[C]// Electronic Computation. ASCE, 1979: 483-509.

[7] MEYER C, ROUFAIEL M S L, ARZOUMANIDIS S G. Analysis of damaged concrete frames for cyclic loads[J]. Earthquake Engineering & Structural Dynamics, 1983, 11(2): 207-228.

[8] HELLESLAND J, SCORDELIS A. Analysis of RC bridge columns under imposed deformations[C]// IABSE colloquium. The Netherlands: Delft, 1981: 545-559.

[9] MARI A R. Nonlinear geometric, material and time dependent analysis of three dimensional reinforced and prestressed concrete frames[D]. Department of Civil Engineering, University of California, 1984.

[10] SPACONE E, FILIPPOU F C, TAUCER F F. Fibre beam-column model for non-linear analysis of R/C frames: Part Ⅰ. Formulation[J]. Earthquake Engineering & Structural Dynamics, 1996, 25(7): 711-725.

[11] SPACONE E, FILIPPOU F C, TAUCER F F. Fibre beam-column model for non-linear analysis of R/C frames: part Ⅱ. Applications[J]. Earthquake Engineering & Structural Dynamics, 1996, 25(7): 727-742.

[12] CUSATIS G, MENCARELLI A, PELESSONE D, et al. Lattice discrete particle model (LDPM) for failure behavior of concrete. Ⅱ: Calibration and validation[J]. Cement and Concrete Composites,

2011, 33(9): 891-905.

[13] CUSATIS G, PELESSONE D, MENCARELLI A. Lattice discrete particle model (LDPM) for failure behavior of concrete. Ⅰ: Theory[J]. Cement and Concrete Composites, 2011, 33(9): 881-890.

[14] SHINOZUKA M. Monte Carlo solution of structural dynamics[J]. Computers & Structures, 1972, 2 (5-6): 855-874.

[15] KLEIBER M, HIEN T D. The stochastic finite element method: basic perturbation technique and computer implementation[M]. Joh Wiley & Sons, 1992.

[16] GHANEM R, SPANOS P D. Polynomial chaos in stochastic finite elements[J]. J. Appl. Mech., 1990, 57(1), 197-202.

[17] LI J. Stochastic structural systems: Analysis and modeling[M]. Beijing: Science Press, 1996.

[18] LI J, CHEN J B. The probability density evolution method for dynamic response analysis of non-linear stochastic structures[J]. International Journal for Numerical Methods in Engineering, 2006, 65(6): 882-903.

[19] LI J, CHEN J. Stochastic dynamics of structures[M]. New York: Wiley, 2009.

[20] WU J Y, LI J, FARIA R. An energy release rate-based plastic-damage model for concrete[J]. International Journal of Solids and Structures, 2006, 43(3-4): 583-612.

[21] REN X D. Multi-scale based stochastic damage constitutive theory for concrete[D]. Shanghai: Tongji Univ., 2010.

[22] ZENG S J. Dynamic experimental research and stochastic damage constitutive model for concrete[D]. Shanghai: Tongji Univ., 2012.

[23] Ministry of Housing and Urban-Rural Development of the People's Republic of China. Code for design of concrete structures: GB 50010—2010[R]. Beijing, 2010.

[24] HSU T T C, MO Y L. Unified theory of concrete structures[M]. John Wiley & Sons, 2010.

[25] STEVENS N J, UZUMERI S M, WILL G T. Constitutive model for reinforced concrete finite element analysis[J]. Structural Journal, 1991, 88(1): 49-59.

[26] MANDER J, PRIESTLEY M J N, PARK R. Theoretical Stress-Strain Model for Confined Concrete Columns[J]. Journal of Structural Engineering, 1804-1826.

[27] SCOTT M H, FENVES G L. Plastic hinge integration methods for force-based beam-column elements[J]. Journal of Structural Engineering, 2006, 132(2): 244-252.

[28] PAULAY T, PRIESTLEY M J N. Seismic design of reinforced concrete and masonry buildings[M]. New York, John Wiley&Sons Inc, 1992.

[29] LI J, CHEN J. The principle of preservation of probability and the generalized density evolution equation[J]. Structural Safety, 2008, 30(1): 65-77.

[30] LI J, CHEN J, SUN W, et al. Advances of the probability density evolution method for nonlinear stochastic systems[J]. Probabilistic Engineering Mechanics, 2012, 28: 132-142.

[31] Chen J, Zhang S. Improving point selection in cubature by a new discrepancy[J]. SIAM Journal on Scientific Computing, 2013, 35(5): A2121-A2149.

[32] Chen J B, Li J. Dynamic response and reliability analysis of non-linear stochastic structures[J]. Probabilistic Engineering Mechanics, 2005, 20(1): 33-44.

[33] REN X D, YANG W, ZHOU Y, et al. Behavior of high-performance concrete under uniaxial and biaxial loading[J]. ACI materials Journal, 2008, 105(6): 548.

[34] Hamutçuoğlu O M, Scott M H. Finite element reliability analysis of bridge girders considering moment-shear interaction[J]. Structural Safety, 2009, 31(5): 356-362.

Multiscale stochastic structural analysis toward reliability assessment for large complex reinforced concrete structures[①]

Hao Zhou,[1] **Jie Li,**[1,2] **, Xiaodan Ren**[1]

1 School of Civil Engineering, Tongji University, Shanghai, China

2 State Key Laboratory of Disaster Reduction in Civil Engineering, Tongji University, Shanghai, China

Abstract This paper focuses on a multiscale methodology to model and analyze large high-rise buildings subject to disastrous dynamic excitations. Starting from the material randomness and the nonlinear behavior of concrete, a mesoscopic stochastic damage model (SDM) is recommended in which the fracture strain of concrete at the microlevel is modeled as a Gaussian random field. By integrating the SDM and the refined structural elements into the finite element analysis, the structural dynamic responses can be comprehensively investigated using the explicit integration algorithm to solve the dynamic equations. To represent the probability information of structural responses, the probability density evolution method (PDEM) is employed. Also, the randomness propagation across different levels can be readily addressed via PDEM. The absorbing boundary condition corresponding to the failure criterion of structures is introduced to assess the dynamic reliability. As a case study, the stochastic dynamic analysis and the reliability assessment are illustratively carried out in terms of a prototype reinforced concrete structure. The simulated results show that the randomness of concrete materials plays a critical role in the stochastic response and dynamic reliability of reinforced concrete structures.

Keywords Stochastic damage model; Nonlinear; Randomness propagation; Finite element analysis; Probability density evolution method; Absorbing boundary condition

1 Introduction

As computers are becoming more and more powerful, there has been an increasing emphasis on the elaborate modeling of structures in order to reduce costs and improve structural safety. Many engineers believe that structures may be modeled to arbitrary accuracy merely by refining the finite element mesh. Actually, the finite element method (FEM) is able to model the geometry of structures accurately, but uncertainties that exist in structures, such as the random material strength, could hardly be reflected through this approach. To ensure that structures do not fail during their intended design life periods, stochastic structural analysis that incorporates the uncertain aspects should be implemented.

---

① Originally published in *International Journal for Multiscale Computational Engineering*, 2016, 14(3): 303-321.

Concrete materials are widely used in civil engineering with nonlinearity and randomness as their two essential characteristics. The nonlinear behaviors of concrete mainly include the different properties under tension and compression, the unilateral effect, the stiffness degradation, the residual strain after unloading, the strain stiffening in tension, stress softening in compression, etc. While due to the intrinsic heterogeneity of mesostructures and ingredients of concrete material, the above macro-mechanical properties would also have a conspicuous variability. Experiences indicate that the tested stress-strain curves are quite different even for concrete specimens prepared based on the same mixture ratios and the same curing conditions[1]. Furthermore, recent investigations have indicated that the randomness of concrete at the material level is able to deviate the nonlinear responses dramatically and change the nonlinear behaviors of concrete structures subject to disastrous excitations. In other words, the mechanical performance of concrete relies on the coupling of the nonlinearity and the randomness[2].

To describe the nonlinearity of concrete, a series of constitutive models based on the formulation of continuum damage mechanics (CDM) have been developed over the past decades[3-8]. By introducing the framework of thermodynamics and its energy-based representations, the damage and the plasticity could be well organized in a class of unified theories. Currently, some of these models have been used as standard tools for the nonlinear analysis of concrete structures. However, the empirical damage evolution functions are widely adopted in most of the CDM-based material models, in which many empirical parameters without clear physical senses are required. Moreover, neither the influence of randomness nor the coupling effect between the nonlinearity and the randomness could be rationally reflected by the CDM-based models.

In recent years, the stochastic damage model (SDM) based on the micro-mechanical approach makes it feasible to reasonably represent the nonlinearity and the randomness as well as their coupling effect at the macrolevel of concrete[1-2, 9]. In addition, the SDM is capable of capturing the uncertain relationships between micro-and macroscales of concrete. The prototype of SDM is a parallel element model at the microlevel. It is assumed that the stiffness of all the elements is identical whereas the rupture strengths are independent and identically distributed (i. i. d.) random variables[9]. The monotonic load-displacement behavior of brittle materials at microscale can then be calculated through the stochastic averaging approaches. In history, Krajcinovic and Fanella[10] first introduced the conventional parallel fiber bundle model to the modeling of random damage behavior of concrete, although their results were restricted to the mean value evolution of damage. Kandarpa et al.[11] investigated the model in detail and derived the standard deviation of damage evolution. Li and Zhang[9] defined the fracture strain of the element as a 1D random field, based on which the uniaxial SDM for concrete was developed. Li and Ren[2] further extended the SDM from uniaxial to multiaxial damage-plasticity framework based on the idea of energy equivalent strain.

On the other hand, the approach to uncertain propagation in structural system has also experienced a long-run development. To attain the statistics of expected responses in the stochastic structural system, the statistical approaches such as the Monte Carlo method[12], the nonstatistical approaches such as the random perturbation method[13], the orthogonal polynomials expansion method[14], etc., have been developed during the past few decades. However, when it comes to the complicated nonlinear stochastic structures, especially considering the coupling effect of nonlinearity and randomness, the above-mentioned approaches seem to be powerless. Fortunately, by integrating the random event description of the principle of preservation of probability with the uncoupled physical equation, a family of probability density evolution methods (PDEMs) has been developed in recent years[15]. In this method, a completely uncoupled 1D governing partial differential equation is derived, which is capable of capturing the instantaneous probability density function (PDF) and its evolution of any response of the nonlinear stochastic structures. Actually, the PDEM provides a feasible way to address the propagation of randomness from micro- to meso- and then to macroscale in the spatial domain and describe the fluctuation effect induced by the development of nonlinearity in the temporal domain. Furthermore, it is demonstrated that the PDEM is applicable for the stochastic response analysis of multiple-degree-of-freedom(MDOF) nonlinear systems with balanced accuracy and efficiency[15].

By consolidating the SDM and the PDEM as well as the refined FEM, a multiscale analysis framework toward the stochastic dynamic response and the reliability assessment for large complex reinforced concrete (RC) structures is developed in this work. Of which the SDM is capable of capturing the uncertain relationships among multiple scales. While the PDEM can handle the issue of randomness propagation across different scales. As an example of an engineering application for the developed methodology, the nonlinear analysis of a prototype high-rise building structure is implemented. Results show that the randomness of concrete material notably affects the reliability of RC structures.

2 Multiscale Analytical Framework for RC Structures

2.1 Stochastic Damage Constitutive Model of Concrete

2.1.1 Physical Stochastic Damage Evolution Law of Concrete

According to Li and Zhang[9], a representative volume element(RVE) of structures in uniaxial loading condition can be idealized as a series of microelements joined in parallel (Fig. 1). The elements are linked with rigid bars on the ends so that each of them bears uniform deformation during the loading process. The individual element represents the micro-properties of the material while the bundle describes the response of the RVE.

Let N denote the total number of microelements. The spatial coordinate of the ith element is $i/N(i=1, 2, \cdots, N)$. The stress-strain relationship is considered as the perfect elasto-brittle type with random fracture strain Δ_i (Fig. 2).

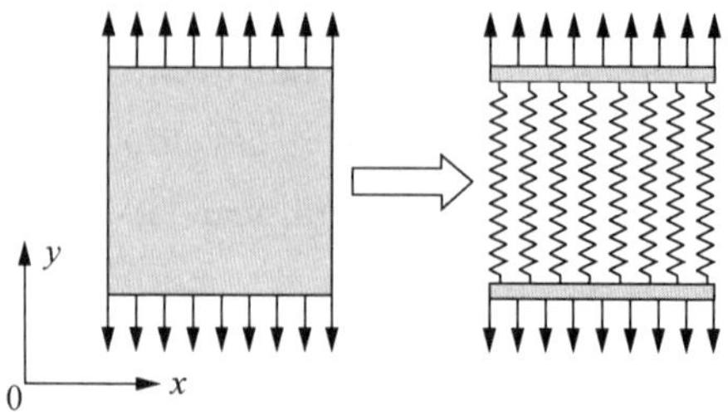

Fig. 1. Parallel element model

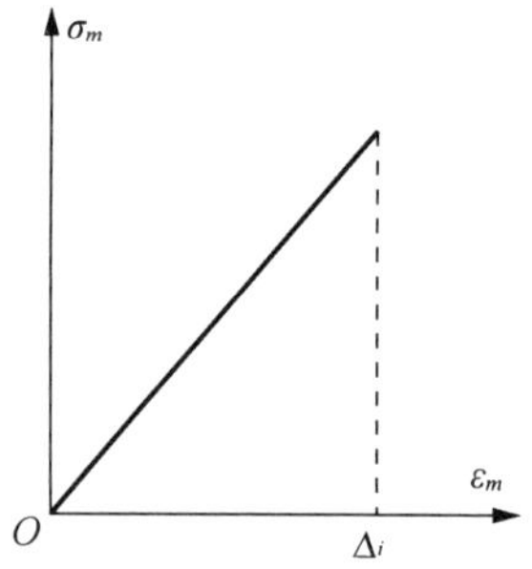

Fig. 2. Stress-strain relationships of microelements

According to the classic definition of damage by Robotnov[16], we have

$$d=\frac{A_{\mathrm{d}}}{A} \tag{1}$$

where A_{d} and A are the cracked area and the initial undamaged area, respectively.

By introducing the random distribution of fracture strain of microelements, the damage variable for the parallel element model can be defined as

$$D(\varepsilon^{\mathrm{e}})=\frac{1}{N}\sum_{i=1}^{N}H(\varepsilon^{\mathrm{e}}-\Delta_i) \tag{2}$$

where ε^{e} is the elastic strain of the element, Δ_i is the random fracture strain, and $H(\cdot)$ is the Heaviside function, which is defined as

$$H(x)=\begin{cases}0 & x\leqslant 0\\ 1 & x>0\end{cases} \tag{3}$$

Taking the limit of Eq. (2) as N approaches infinity and accounting for the definition of stochastic integral, one obtains

$$D(\varepsilon^{\mathrm{e}})=\int_0^1 H[\varepsilon^{\mathrm{e}}-\Delta(x)]\mathrm{d}x \tag{4}$$

where $\Delta(x)$ is the 1D microfracture strain random field, and x denotes the spatial coordinate of the microelement.

It is indicated that Eq. (4) is actually a stochastic damage evolution law. Suppose $\Delta(x)$ is a homogenous random field with the first-order and the second-order PDFs as follows

$$f(\Delta;\ x)=f(\Delta) \tag{5}$$

$$f(\Delta_1,\ \Delta_2;\ x_1,\ x_2)=f(\Delta_1,\ \Delta_2;\ |x_1-x_2|) \tag{6}$$

Let $\varphi(x)=H[\varepsilon^{\mathrm{e}}-\Delta(x)]$. Obviously, $\varphi(x)$ obeys the (0, 1) distribution. Thus, performing expectation operator on Eq. (4), we obtain the expectation of the damage scalar $D(\varepsilon^{\mathrm{e}})$,

$$\mu_{\mathrm{D}}(\varepsilon^{\mathrm{e}})=E\left[\int_{0}^{1}H[\varepsilon^{\mathrm{e}}-\Delta(x)]\mathrm{d}x\right]=F(\varepsilon^{\mathrm{e}}) \tag{7}$$

The variance of $D(\varepsilon^{\mathrm{e}})$ reads

$$V_{\mathrm{D}}^{2}(\varepsilon^{\mathrm{e}})=E[D(\varepsilon^{\mathrm{e}})]^{2}-[\mu_{D}(\varepsilon^{\mathrm{e}})]^{2}=2\int_{0}^{1}(1-\eta)F(\varepsilon^{\mathrm{e}},\varepsilon^{\mathrm{e}};\eta)\mathrm{d}\eta-[F(\varepsilon^{\mathrm{e}})]^{2} \tag{8}$$

where $\eta=|x_2-x_1|$; $F(\varepsilon^{\mathrm{e}})$ and $F(\varepsilon^{\mathrm{e}},\varepsilon^{\mathrm{e}};\eta)$ are the first-order and the second-order cumulative distribution function (CDF) of $\Delta(x)$, respectively.

Let μ_{Δ} and σ_{Δ} be the first-order expectation and the standard deviation of $\Delta(x)$, respectively. Given that $\Delta(x)$ follows the lognormal distribution with the first-order parameters(λ,ζ), the relations among parameters can be given as

$$\lambda=E[\ln\Delta(x)]=\ln\left(\frac{\mu_{\Delta}}{\sqrt{1+\sigma_{\Delta}^{2}/\mu_{\Delta}^{2}}}\right) \tag{9}$$

$$\zeta^{2}=\mathrm{Var}[\ln\Delta(x)]=\ln(1+\sigma_{\Delta}^{2}/\mu_{\Delta}^{2}) \tag{10}$$

Accordingly, the first-order CDF of $\Delta(x)$ can be given by

$$F(\varepsilon^{\mathrm{e}})=\Phi\left[\frac{\ln\varepsilon^{\mathrm{e}}-\lambda}{\zeta}\right]=\Phi(\alpha) \tag{11}$$

where $\Phi(\cdot)$ is the CDF of the standard normal distribution.

If $\Delta(x)$ is an independent random field, then its variance function is

$$V_{\mathrm{D}}^{2}(\varepsilon^{\mathrm{e}})=2\int_{0}^{1}(1-\eta)[F(\varepsilon^{\mathrm{e}})]^{2}\mathrm{d}\eta-[F(\varepsilon^{\mathrm{e}})]^{2}=0 \tag{12}$$

This means that ignoring the spatial correlation of the microfracture strain field will lead to a deterministic damage evolution model. Therefore, the spatial correlation of $\Delta(x)$ should be taken into account.

Consider the exponential autocorrelation coefficient function for $Z(x)=\ln\Delta(x)$ as follows

$$\rho_{Z}(\eta)=\exp(-\varsigma\cdot\eta) \tag{13}$$

where ζ is a correlation scale parameter. Then the second-order CDF of $\Delta(x)$ is

$$F(\varepsilon^{\mathrm{e}},\varepsilon^{\mathrm{e}};\eta)=\Phi\left[\frac{\ln\varepsilon^{\mathrm{e}}-\lambda}{\zeta},\frac{\ln\varepsilon^{\mathrm{e}}-\lambda}{\zeta}\mid\rho_{Z}\right]=\Phi(\alpha,\alpha\mid\rho_{Z}) \tag{14}$$

where $\Phi(y_1,y_2\mid\rho)$ is the standard expression of a second-order CDF for 2D normal distribution, which is defined as a double integral of its second-order PDF. In addition, the double integral function $\Phi(\alpha,\alpha\mid\rho_{Z})$ could be reduced to a single integral expression by introducing[17]

$$\Phi(\alpha,\alpha\mid\rho_{Z})=\Phi(\alpha)-\frac{1}{\pi}\int_{0}^{\beta}\frac{1}{1+t^{2}}\exp\left[-\frac{\alpha^{2}}{2}(1+t^{2})\right]\mathrm{d}t \tag{15}$$

where the upper limit of the integral reads

$$\beta=\sqrt{\frac{1-\rho_Z}{1+\rho_Z}} \tag{16}$$

Since the performance of concrete depends much on the loading condition, two groups of parameters are introduced to describe the damage evolutions under tension and compression, respectively. In this paper, $(\lambda^+, \zeta^+, \omega^+)$ denotes the tensile material parameters triple related to the evolution of D^+, whereas $(\lambda^-, \zeta^-, \omega^-)$ denotes the compressive material parameters triple related to the evolution of D^-.

2.1.2 Multidimensional Damaged Plasticity Constitutive Model

Within the framework of the CDM[8], the concept of the energy equivalent strain is proposed by Li and Ren[2] to bridge the gap between the uniaxial and the multiaxial constitutive models. Based on the damage energy release rates (DERRs) and the damage consistent condition, the energy equivalent strain is expressed as

$$\begin{cases} \varepsilon_{eq}^{e+}=\sqrt{\dfrac{2Y^+}{E_0}} \\ \varepsilon_{eq}^{e-}=\dfrac{1}{(1-\alpha)E_0}\sqrt{\dfrac{Y^-}{b_0}} \end{cases} \tag{17}$$

where E_0 is the Young's modulus of the initial undamaged material; b_0 is a material parameter; Y^+, Y^- are the tensile and shear DERR, respectively (for details, see Wu et al.[8] and Li and Ren[2]); and the parameter α is related to the biaxial strength increase as

$$\alpha=\frac{f_{bc}/f_c-1}{2f_{bc}/f_c-1} \tag{18}$$

where f_c and f_{bc} are the uniaxial compressive strength and the biaxial compressive strength, respectively.

Substituting Eq. (17) into Eq. (4), the multidimensional damage evolution functions for plain concrete are established as

$$D^{\pm}(\varepsilon_{eq}^{e\pm})=\int_0^1 H[\varepsilon_{eq}^{e\pm}-\Delta^{\pm}(x)]dx \tag{19}$$

Then the constitutive law of concrete can be described as[1]

$$\sigma=(1-D^+)\bar{\sigma}^+ + (1-D^-)\bar{\sigma}^- = (\mathbb{I}-\mathbb{D}):\bar{\sigma}=(\mathbb{I}-\mathbb{D}):\mathbb{E}_0:(\varepsilon-\varepsilon^p) \tag{20}$$

where $\bar{\sigma}$ denotes the effective stress tensor, $\mathbb{E}_0$ denotes the initial undamaged elastic stiffness, $\mathbb{I}$ is a fourth-order unit tensor, and $\mathbb{D}$ is a fourth-order damage tensor.

The evolution of plastic strain can be calculated according to the following empirical formula[18]

$$\dot{\varepsilon}^{p\pm}=f_p^{\pm}\dot{\varepsilon}^{e\pm}=H[\dot{D}^{\pm}]\xi_p^{\pm}(D^{\pm})^{n_p^{\pm}}\dot{\varepsilon}^{e\pm} \tag{21}$$

where scalars $f_p^{\pm}$ are considered as the function of the corresponding damage variable $D^{\pm}$;

$\xi_p^{\pm}$ and $n_p^{\pm}$ are material parameters fitted by the experimental data.

2.1.3 Generation of Random Field of Microscopic Fracture Strain

To obtain the stress-strain relationships in the sense of random samples and ensemble samples, we need to model the random field of fracture strain by numerical simulations. As one of the most effective ways, the stochastic harmonic function (SHF) method[19] can be used to generate the random field of the microscopic fracture strain $\Delta(x)$ as

$$\Delta(x)=\sum_{i=1}^{N} A(\omega_i)\cos(\omega_i x+\varphi_i) \tag{22}$$

where ω_i and φ_i are random frequencies and random phase angles, respectively. The amplitudes $A(\omega_i)$ are functions of ω_i, and N is the number of the components.

In the case φ_i are independent uniformly distributed random variables over $[0, 2\pi]$, the correlation function of $\Delta(x)$ is

$$R_\Delta(\tau)=\frac{1}{2}\sum_{i=1}^{N} E[A^2(\omega_i)\cos(\omega_i\tau)] \tag{23}$$

where $\tau=x_2-x_1$ and $E(\cdot)$ denotes an expectation operator.

By Fourier transform of $R_\Delta(\tau)$, the power spectral density (PSD) function can be given as

$$S_\Delta(\omega)=\int_{-\infty}^{\infty} R_\Delta(\tau)e^{-i\omega\tau}\mathrm{d}\tau \tag{24}$$

Suppose that the ω_i are mutually independent random variables and each ω_i obeys a uniform distribution over $[\omega_{i-1}, \omega_i]$, i.e., the PDF of ω_i in corresponding interval reads

$$p_{\omega_i}(\omega)=\begin{cases}\dfrac{1}{\omega_i-\omega_{i-1}}=\dfrac{1}{\delta\omega_i}, & \text{for } \omega\in(\omega_{i-1}, \omega_i)\\ 0, & \text{otherwise}\end{cases} \tag{25}$$

Then the associated random variables $A(\omega_i)$ can be given by

$$A(\omega_i)=\sqrt{\frac{2S_\Delta(\omega_i)\delta\omega_i}{\pi}} \tag{26}$$

If a target PSD $S_\Delta(\omega)$ is prescribed by adopting the autocorrelation function described in Eq.(13) for a stochastic process $\Delta(x)$, the samples of stress-strain curve of concrete could be generated by the following steps:

(1) Specify the lower and upper cutoff frequencies ω_l and ω_u, and partition the interval $[\omega_l, \omega_u]$ into $[\omega_{i-1}, \omega_i]$.

(2) Generate samples of $\Delta(x)$ by Eqs.(22)—(26) providing that the involved random variables ω_i and φ_i follow the above described distributions.

(3) Generate the stress-strain curves with the samples of $\Delta(x)$ in accordance with the theory of SDM. Each random field $\Delta(x)$ corresponds to one stress-strain curve of concrete.

It has been demonstrated that containing only several components of SHFs are adequate for applications to nonlinear stochastic dynamics. That is, with a small number of random variables, the SHF method can work well for the nonlinear problems of stochastic dynamics[19].

To illustrate the above procedures, the stochastic damage constitutive curves of C30 concrete are generated using the SHF representation and the theory of SDM, as shown in Fig. 3. In which, totally eight basic random variables with four components of SHFs are involved during the simulations. And these constitutive curves will be used for the subsequent analyses.

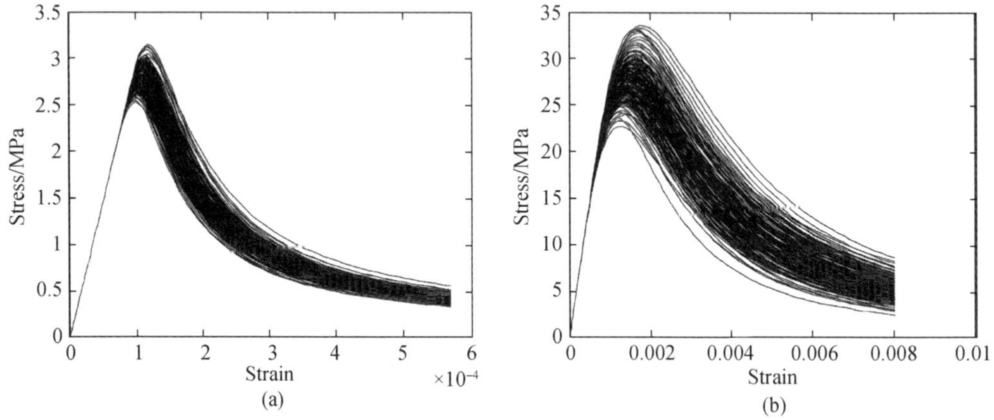

Fig. 3. Stress-strain relationship samples of concrete(C30): (a) uniaxial tension, (b) uniaxial compression

2.2 Refined Structural Element Models

In general, two kinds of components, i.e., the 1D frame members and 2D floor slab and shear wall members, are mainly involved in complex RC frame-shear wall structures. During the structural finite element analysis, the fiber beam-column element and the multilayer shell element are widely adopted to respectively model the two types of structural members with fair efficiency and accuracy. Integrating the SDM of concrete into the following refined structural elements, the damage and fracture of concrete material as well as the nonlinear whole-process response of structures can be comprehensively described via the refined finite element analysis.

2.2.1 Fiber Beam-Column Element

The fiber beam-column element is capable of well tracing the highly nonlinear behavior of RC members under cyclic loadings[20-22]. As shown in Fig. 4, the cross section of an element is divided into several fibers. Each individual fiber has a uniaxial material constitutive law, where different fibers can have different constitutive laws. Hence the confinement effect induced by stirrups can be considered. The fiber element is suitable for cross sections of arbitrary shape and can account for the coupling effect of axial force and bending moment.

With regard to the confined concrete in a fiber beam-column element, it is well known that the strength and ductility will be significantly enhanced compared with the plain concrete. Hence the damage evolution function of the confined concrete subjected to uniaxial

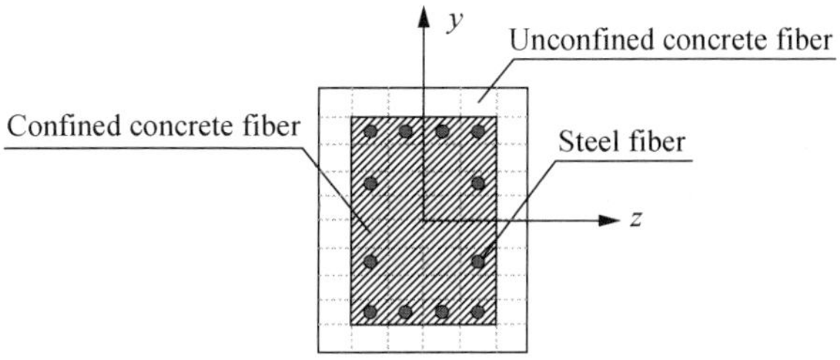

Fig. 4. Fiber beam-column model

compressive loading should be revised as

$$D_{con}^{-}(\varepsilon_{eq}^{e-})=\int_{0}^{1}H[\gamma\varepsilon_{eq}^{e-}-\Delta^{-}(x)]dx \tag{27}$$

where γ is a reduction coefficient, which is considered as a function of the plastic strain as

$$\gamma=1-\left(\frac{\psi\varepsilon^{p}}{\varepsilon^{p}+\vartheta/100}\right)^{\kappa} \tag{28}$$

where ψ, ϑ, κ are the model parameters that can be identified by test results.

The skeleton curve of the stress-strain relationship used for the steel reinforcement bars is based on the model proposed by Hsu and Mo[23], as shown in Fig. 5. The simplified bilinear model for a steel reinforcement bar embedded in concrete can be considered as

$$f_{s}=\begin{cases}E_{s}\varepsilon_{s}, & \text{for } \varepsilon_{s}\leqslant\varepsilon_{y}' \\ (0.91-2B)f_{y}+(0.02+0.25B)E_{s}\varepsilon_{s}, & \text{for } \varepsilon_{s}>\varepsilon_{y}'\end{cases} \tag{29}$$

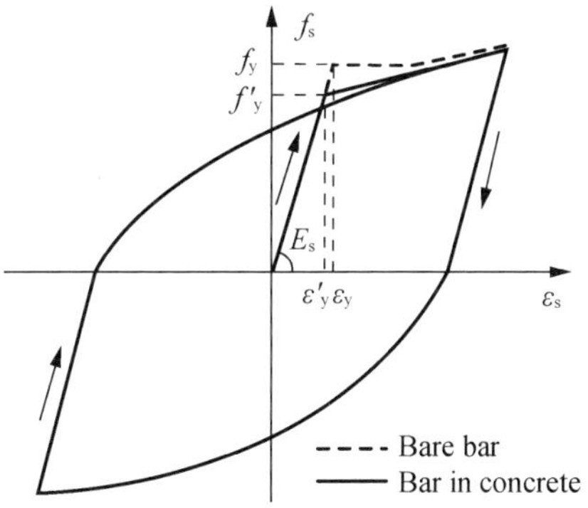

Fig. 5. Stress-strain relationship of steel reinforcement bar

where

$$\varepsilon_{y}'=\frac{f_{y}'}{E_{s}} \tag{30}$$

$$f_{y}'=(0.93-2B)f_{y} \tag{31}$$

$$B=\frac{1}{\rho}\left(\frac{f_{t,r}}{f_{y}}\right)^{1.5} \tag{32}$$

where f_s, ε_s, and E_s are the stress, strain, and elastic modulus of the steel bar, respectively; f_y/f_y' is the yield stress ratio of a bare bar to a bar embedded in concrete; ε_y' is the strain at yield stress; ρ is the reinforcement ratio; and $f_{t,r}$ is the tensile strength of concrete.

In order to account for the Bauschinger effect, the model proposed by Menegotto and Pinto[24] is adopted to simulate the unloading and reloading paths.

2.2.2 Multilayer Shell Element

Analogous to the fiber beam-column element, the shell element used for shear walls and floor slabs is made up of membrane element and shell element by a combination type[25-26]. Of which a number of layers with different thicknesses and different material properties (i.e., concrete layers or rebar layers) are included, as shown in Fig. 6. The reinforcement bars are smeared into one or more layers according to their physical locations and directions. And the layer thickness is determined depending on the reinforcement ratio. The multilayer shell element is capable of simulating the coupled in-plane/out-of-plane bending as well as in-plane direct shear and the coupled bending-shear behavior of RC shear walls and floor slabs.

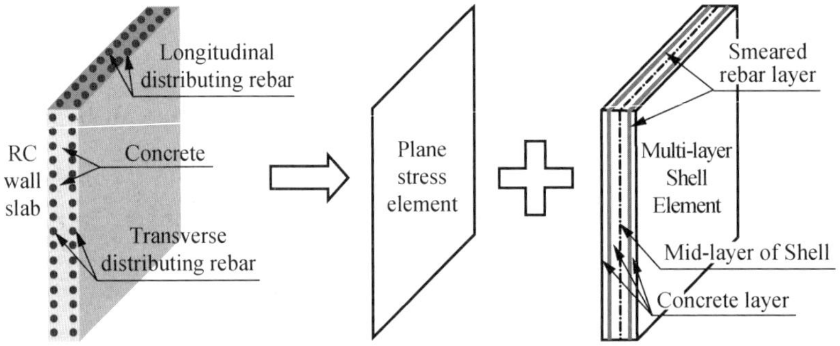

Fig. 6. Diagram of multilayer shell element

2.3 Numerical Implementation for the Nonlinear Dynamic Analysis

Although theoretical results for specific situations indicate that unconditional stability of an implicit method holds at least for certain nonlinear systems, no proof of unconditional stability exists that encompasses the wide range of conditions found in practice[27]. Besides, considering the stiffness degradation, strength softening, and irreversible deformation of concrete, convergence problems will be frequently encountered during the highly nonlinear analysis using an implicit Newmark integral algorithm. To possibly avoid such issues on stability and convergence, the central difference method, which is the most popular one among the several explicit integration algorithms, is adopted herein to solve the dynamical equations.

One of the notable advantages of the central difference method lies in that no iterative solving procedures are involved in the update scheme. However, the time step used must be limited by a critical value such that the solution will not grow unboundedly. A stable time step for a linear problem is given by[28]

$$\Delta t \leqslant \alpha \Delta t_{\text{crit}}, \quad \Delta t_{\text{crit}} = \frac{2}{\omega_{\max}} \leqslant \min_{e} \frac{l_e}{c_e} \tag{33}$$

where $\omega_{\max}$ is the maximum frequency of the linearized system; l_e is a characteristic length of element e; c_e is the current wave speed in element e; α is a reduction factor, and a good choice for α is $0.8 \leqslant \alpha \leqslant 0.98$.

In the case of nonlinear analysis of a dynamic system, the stability of the system can be detected by an energy balance check that requires that[29]

$$| W_{\text{kin}} + W_{\text{int}} + W_{\text{ext}} | \leqslant \delta \cdot \max(W_{\text{kin}}, W_{\text{int}}, W_{\text{ext}}) \tag{34}$$

where δ is a small tolerance, generally on the order of 10^{-2}; W_{int}, W_{ext}, and W_{kin} are the internal energy, external energy, and kinetic energy of the system (for more details, refer to Belytschko et al.[27]).

2.4 Multiscale Structural Analysis Framework

Through the above discussions, a three-scale analytical framework for RC structures can be established, arriving here. From the random fracture of microelements in a RVE, to the mesoscopic constitutive relationship of concrete using SDM, and then to the structural dynamic responses if integrated with the refined finite element analysis, both the micro-properties of concrete material and the macroscopic responses of RC structures can be roundly investigated within the framework. Different from the CDM-based models, the damage evolution law of concrete in SDM has clear physical meaning. It is the random fracture of microelements that constitutes the nonlinear behaviors of concrete, e.g., the stiffness degradation, the strength softening, etc.

From another point of view, the SDM seizes the essence of the uncertainty relationships of mechanical properties among different scales of concrete. Thus, the coupling effect of nonlinearity and randomness could be rationally represented in the above multiscale analytical framework for RC structures.

3 Randomness Propagation and Dynamic Reliability

In fact, the above analytical framework has dealt with the issue of randomness propagation across spatial scales. If integrated with the general stochastic dynamic analysis procedures, the framework could also theoretically work well in describing the random fluctuation effect of structural responses with the damage evolution and the development of nonlinearity in the temporal domain. However, when it comes to the stochastic dynamic responses of complex nonlinear RC structures in real engineering applications, the computational efficiency usually becomes the primary consideration during the stochastic analyses. To this end, the traditional approaches like the Monte Carlo method, the perturbation method, and the orthogonal expansion method all seem to be unpractical. Whereas the PDEM developed in the past few years could adequately address the above issue. With PDEM, the instantaneous probability density function (PDF) and its evolution of any response of the nonlinear stochastic structures could be well captured. Moreover, it provides a feasible way to simultaneously address the propagation of randomness from micro to meso and then to macroscale in the spatial domain and describe the fluctuation effect of structural responses in the temporal domain with balanced accuracy and efficiency.

3.1 Probability Density Evolution Method

In general, considering a MDOF nonlinear stochastic dynamical system, the equations of motion can be uniformly expressed as

$$\mathrm{M}(\Theta)\ddot{\mathrm{u}}(t)+\mathrm{C}(\Theta)\dot{\mathrm{u}}(t)+\mathrm{f}^{\mathrm{int}}(\Theta,\ \mathrm{u}(t))=\mathrm{f}^{\mathrm{ext}}(\Theta,\ \mathrm{u}(t)) \tag{35}$$

where $\Theta=(\Theta_1,\ \Theta_2,\ \cdots,\ \Theta_s)$ are the random parameters involved in both the physical properties of the system and the external dynamic excitations, in which s is the total number of the basic random variables. M, C are the mass and damping matrices of the structure system, and $\mathrm{f}^{\mathrm{int}}(\cdot)$, $\mathrm{f}^{\mathrm{ext}}(\cdot)$ are the restoring force and external dynamic excitation vector, respectively. Here, $\ddot{\mathrm{u}}(t)$, $\dot{\mathrm{u}}(t)$, $\mathrm{u}(t)$ are the vectors of nodal accelerations, velocities, and displacements of the system, respectively. This is the equation to be resolved in which all the randomness from the initial conditions, excitations, and system parameters are involved and exposed in a unified manner.

Suppose that $Z=(Z_1,\ Z_2,\ \cdots,\ Z_m)^{\mathrm{T}}$ denotes all the physical quantities of interest in the system(e.g., the displacements, velocities, stress, internal forces, etc.), then considering Eq.(35) the augmented system $(Z,\ \Theta)$ is probability preserved because all the random factors are involved. According to the random event description of the principle of preservation of probability, the joint PDF of the augmented state vector $(Z,\ \Theta)$ satisfies the governing partial differential equation[15]

$$\frac{\partial p_{Z\Theta}(z,\ \theta,\ t)}{\partial t}+\dot{Z}(\theta,\ t)\frac{\partial p_{Z\Theta}(z,\ \theta,\ t)}{\partial z}=0 \tag{36}$$

where $\dot{Z}(\theta,\ t)$ is the velocity of the response for a prescribed θ.

It is easy to specify the initial condition of Eq.(36), i.e.,

$$p_{Z\Theta}(z,\ \theta,\ t)\Big|_{t=0}=\delta(z-z_0)p_{\Theta}(\theta) \tag{37}$$

where z_0 is the deterministic initial value.

Eq.(36) is referred to as a generalized density evolution equation(GDEE), which reveals the intrinsic connections between a stochastic dynamic system and its deterministic counterpart. Using the GDEE and a physical equation such as Eq.(35), the PDF of a physical quantity of interest can be easily obtained once its velocity is known since varying of the PDF results from varying of the state of the quantity.

Actually, solving the initial-value problem of Eqs.(36) and (37) with Eq.(35), the instantaneous PDF of $Z(t)$ can be obtained by

$$p_Z(z, t) = \int_{\Omega_\Theta} p_{Z\Theta}(z, \theta, t)\mathrm{d}\theta \tag{38}$$

where Ω_Θ is the distribution domain of Θ.

3.2 Dynamic Reliability Assessment

The dynamic reliability for the first passage problem can be resolved by an initial-boundary-value partial differential equation problem instead of the widely used level-crossing-process-based approach[30].

The first passage reliability of a general stochastic dynamical system is defined by

$$R(t) = \Pr\{Z(\tau) \in \Omega_s, \tau \in [0, t]\} \tag{39}$$

where $\Pr\{\cdot\}$ denotes the probability of the random events, and Ω_s is the safe domain. From Eq. (39), the total probability of the random events that are always in the safe domain over the time interval $[0, t]$ constitutes the dynamic reliability. This means that the random events as well as their "adherent" probabilities will "disappear" once entering the failure domain. In other words, the probability density transits one-way outside the boundary[30]. Based on this, an absorbing boundary condition should be imposed on Eq. (36), i.e.,

$$p_{Z\Theta}(z, \theta, t) = 0, \quad z \in \Omega_f \tag{40}$$

where Ω_f is the failure domain, $\Omega_f \cap \Omega_s = \emptyset$, $\Omega_f \cup \Omega_s = \Omega$, and Ω is the response space of Z. Solving the simultaneous Eqs. (36), (37), and (40), we can obtain the "remaining" joint PDF $\breve{p}_{Z\Theta}(z, \theta, t)$. Then the remaining PDF reads

$$\breve{p}_Z(z, t) = \int_{\Omega_\Theta} \breve{p}_{Z\Theta}(z, \theta, t)\mathrm{d}\theta \tag{41}$$

and the reliability yields

$$R(t) = \int_{\Omega_s} \breve{p}_Z(z, t)\mathrm{d}z \tag{42}$$

In the case of the symmetric double boundary, the reliability equation, Eq. (42), converts to

$$R(t) = \int_{-z_B}^{z_B} \breve{p}_Z(z, t)\mathrm{d}z \tag{43}$$

where z_B is the value of the symmetric boundary. Definitely, Eq. (43) will convert to the dynamic response analysis problem while $z_B \to \infty$.

3.3 Numerical Implementation for the GDEE and the Dynamic Reliability

As is pointed out above, for a general dynamical system, the explicit expression of $\dot{Z}(\theta, t)$ is usually unknown, which makes Eq. (36) analytically intractable for most problems. Nevertheless, Eqs. (36)—(38) can be numerically solved expediently by the following steps:

Step 1 Select a representative point set in the domain Ω_Θ and denote it as $\theta_q = (\theta_{1,q}, \theta_{2,q}, \cdots, \theta_{s,q})$ $(q = 1, 2, \cdots, N)$, where N is the total number of the selected points. Simultaneously, determine the corresponding assigned probability P_q.

Step 2 For the prescribed $\Theta = \theta_q$, solve the deterministic ordinary differential equation (35) to evaluate the value of $\dot{Z}(\theta_q, t_m)$, where $t_m = m\Delta t$ $(m = 0, 1, 2, \cdots)$, and Δt is the time step.

Step 3 Employ θ_q in place of θ, and $\dot{Z}(\theta_q, t_m)$ in place of $\dot{Z}(\theta, t)$ in Eq.(36). Solve the partial differential equation(36) with the finite difference method to obtain the discretized value $p_{Z\Theta}(z_j, \theta_q, t_k)$, where $x_j = j\Delta x(j = 0, \pm 1, \pm 2, \cdots)$, Δx is the space step, $t_k = k \cdot \Delta\hat{t}$ $(k = 0, 1, 2, \cdots)$, and $\Delta\hat{t}$ is the time step in the finite difference method.

Step 4 Carry out the numerical integration in Eq.(38) to obtain the numerical solution of the PDF of structural responses, and in Eq.(43) to evaluate the dynamic reliability.

In Step 1, special techniques are needed in the strategy of selecting representative points in Ω_Θ. For instance, the number theoretical method based algorithm developed by Li and Chen[31] is employed in the subsequent numerical example. In Step 3, the total variation diminishing(TVD) schemes are appropriate in dynamic response analysis while a hybrid difference scheme is preferred in evaluation of the extreme value distribution and the reliability[30].

From the above steps, it is seen that a traditional deterministic dynamic response analysis process is embedded in the procedure. An additional finite difference method is then employed. This makes it convenient to use the existing commercial software. For the sake of clarity, Fig. 7 shows the basic implementation process of the above-proposed methodology for the stochastic structural analysis as well as the reliability assessment. Additionally, the

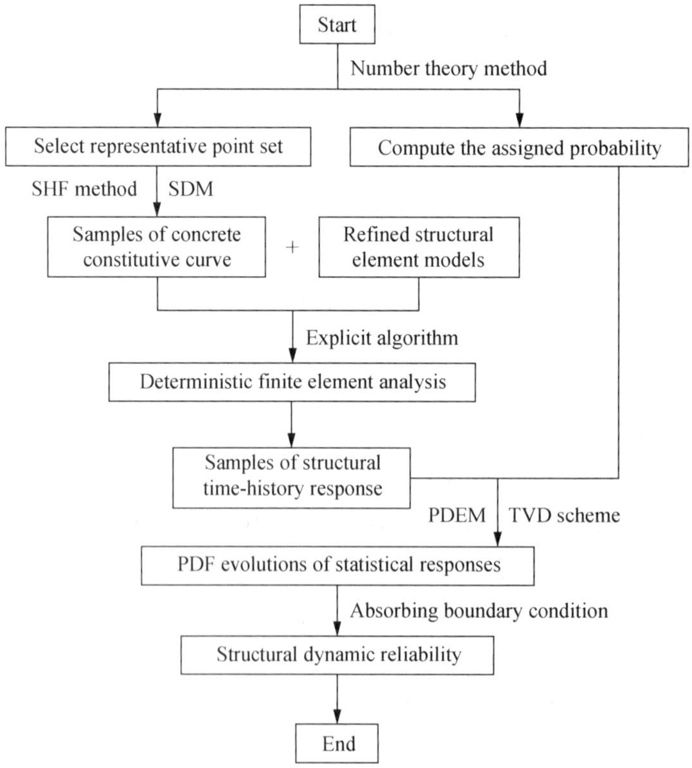

Fig. 7. Procedure for the stochastic structural analysis and the reliability assessment

stochastic nonlinear analysis of a prototype high-rise building structure is carried out to illustrate the overall procedure in the following. The deterministic numerical analyses are performed by the finite element package ABAQUS, in which the constitutive models for concrete and steel materials are implemented with VUMAT and Python script.

4 Numerical Example

4.1 Description of Model Structure

The Pudong Broadcast & TV Center (hereinafter referred to as BTC) is an 18-story large high-rise building, which is located in Pudong New Area, Shanghai, China. The main format of the lateral-force-resisting system of BTC is a RC frame-shear wall structure. And it is designed in accordance with the Chinese building design codes[32-33] with an intended working life of 50 years. The seismic precautionary intensity is categorized as 7.0(the peak ground acceleration for a 10% exceedance probability in 50 years is $0.10g$, where g is the acceleration of gravity) and the site class of type III(soil shear wave velocity 150 m/s $<$ $v_s \leqslant$ 250 m/s) is assigned.

The overall configuration of the BTC structure is shown in Fig. 8. There is a shear wall core tube with an elevator shaft located close to either side of the structure; whereas, the frame structure is used for structural connections of two core tubes and other areas of the structure. The plane size of the ground floor is 38 m×36 m. The structural total height is 86.4 m. Story heights of the structure are 6.3 m for the first two floors, 5.5 m for the third to fourth floors, 4.6 m for the fifth to sixteenth floors, and 4.2 m and 3.4 m for the last two floors. The building structure is vertically divided into three areas, i.e., stories one to seven are area Ⅰ, stories eight to 13 are area Ⅱ, and stories 14 to 18 are area Ⅲ(Fig. 8). The thicknesses of the core walls are 0.4 m, 0.4 m, and 0.3 m for the three areas from bottom to top and the corresponding strength grades of concrete are C50, C40, and C40, respectively. The concrete used for the floor slabs, beams, and columns are of grades C30, C40, and C60, respectively. Steel specifications of HRB335, HRB400, and

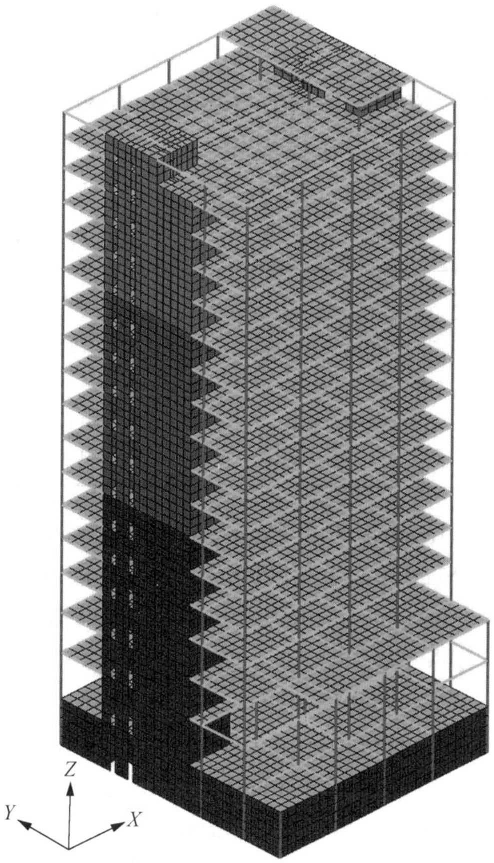

Fig. 8. Isometric view of the BTC structure model

HRB500 are used for the longitudinal reinforcement bars and HPB300 for all the stirrups. The above material classification grades are all in accordance with the Chinese design code for concrete structures[32].

In the finite element model of BTC structure, the fiber beam-column element and the multilayer shell element are adopted to model the frame members and the floor slab and shear wall members, respectively. In total, 53 372 elements are developed. The first three natural periods of the structure are, in order, 1.75 s, 1.16 s, and 0.76 s.

4.2 Earthquake Input

The ground motion recorded at the USC 90081 Carson-Water St station in the January 17, 1994 Northridge earthquake is selected as the deterministic earthquake input to the BTC structure along the X-axis after the gravity analysis is performed. To acquire a significant nonlinear behavior during the simulation, the peak ground acceleration (PGA) of the ground motion is scaled to $1.0g$, which is much larger than the maximal considered earthquake for the design of the BTC structure (i.e., PGA $=0.22g$). The acceleration time history after amplification and the corresponding response spectrum of Northridge ground motion are illustrated in Figs. 9 and 10, respectively. The 5% Rayleigh damping ratio is adopted in the numerical model. The seismic time history analysis is implemented by the aforementioned explicit dynamic integration algorithm with a time step of 10^{-6} s.

4.3 Stochastic Structural Analysis

Ignoring the randomness of steel, the mechanical properties of reinforcement bars used in the BTC structure are listed in Table 1.

Table 1. Material properties of steel

	Symbol(unit)	HPB300	HRB335	HRB400	HRB500
Elasticity modulus	E_s/GPa	210	200	200	200
Poisson's ratio	μ_s	0.3	0.3	0.3	0.3
Yield strength	f_y/MPa	300	335	400	500
Ultimate strength	f_u/MPa	420	455	540	630

In the structural finite element model, the stochastic damage constitutive model described in Section 2.1 is adopted to simulate the concrete material. The property parameters of concrete excerpted from the Chinese design code[32] or identified by test results are listed in Table 2, in which a symbol with a superscript "+" denotes a tensile parameter while "−" for a compressive one. By employing the SHF method, the random constitutive relationships are derived regarding the two groups of parameters (λ^+, ζ^+, ω^+) and (λ^-, ζ^-, ω^-) of the stochastic damage evolution laws in the mesolevel of concrete. It should be noted that here the parameters for plastic model, stirrup confinement effect model, and other aspects involved in SDM are all considered as deterministic variables in view of their insensitivity to the constitutive curves.

Table 2. Material parameters of concrete

	Symbol(unit)	C30	C40	C50	C60
Elasticity modulus	E_0/MPa	30 000	32 500	34 500	36 000
Poisson's ratio	μ_0	0.2	0.2	0.2	0.2
Biaxial strength increase	f_{bc}/f_c	1.16	1.16	1.16	1.16
Mean value	λ^+	4.894 4	4.941 4	5.060 3	5.085 4
	λ^-	7.402 6	7.631 9	7.713 6	7.795 1
Standard deviation	ζ^+	0.442 5	0.418 8	0.407 3	0.391 7
	ζ^-	0.695 8	0.617 5	0.568 0	0.519 7
Correlation function	ω^+	40	45	50	55
	ω^-	70	75	80	85
Plasticity	ξ_p^+	0.250 9	0.296 4	0.334 7	0.382 1
	n_p^+	0.430 2	0.415 7	0.368 5	0.311 5
	ξ_p^-	0.314 1	0.366 9	0.418 5	0.463 0
	n_p^-	0.549 3	0.503 1	0.448 6	0.376 1
Confinement effect	ψ	0.872 1	0.871 9	0.871 6	0.871 2
	ν	0.434 7	0.466 2	0.492 2	0.520 2
	κ	0.555 9	0.629 5	0.699 6	0.726 0

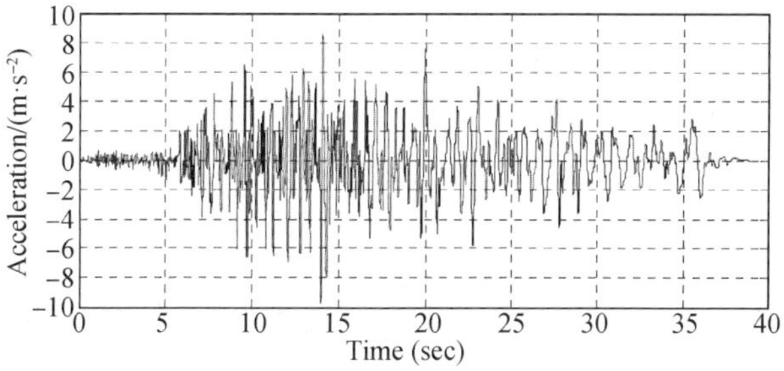

Fig. 9. Acceleration time history (1.0*g*)

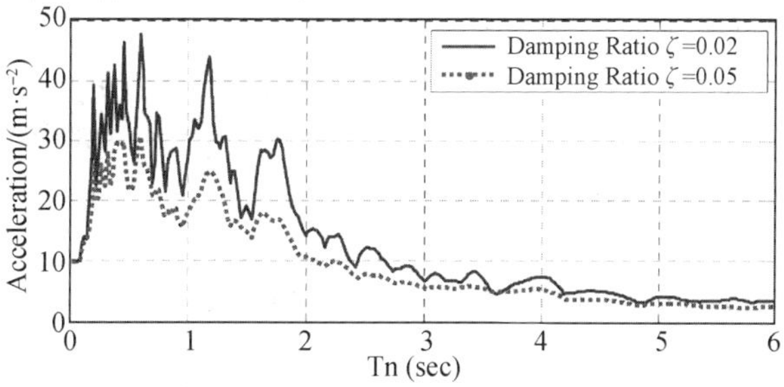

Fig. 10. Acceleration response spectrum (1.0*g*)

During the stochastic analyses of structures, a total of 200 representative points in terms of 32 basic random variables including 16 random frequencies and 16 random phase angles in the SHF representations are intelligently selected. And the corresponding assigned probabilities are computed according to the number theoretical method based algorithm[31].

After a series of deterministic seismic response analysis, many concerned physical quantities such as stress, strain, damage, reaction forces, displacements, velocities and accelerations, etc., of the BTC structure are obtained. Accordingly, we can get the probability density information varying with time and further the dynamic reliability of structure via the PDEM. Since the final purpose in this stage is the reliability evaluation, we further focus on the responses that most likely result in the macrofailure and destruction of the structures, i.e., the interstory drift ratio, which is also an important index in design specifications. Here, the interstory drift ratio is defined as the interstory drift for a particular floor normalized by the story height.

From the analysis of interstory drifts, it is found that the most dangerous story locates on the sixteenth floor. Then, the mean value and standard deviation (Std. D) of interstory drift ratios of the sixteenth floor changing over time are computed by the PDEM with two different difference schemes, i.e., the Lax-Wendroff (L-W) scheme and the TVD scheme (for details, refer to Li and Chen[15]). The results are pictured in Fig. 11, which shows good agreement. Thus, the time step chosen in the PDEM is deemed to be reasonable. Also, we can see that the random fluctuation of the structural response increases gradually with the evolution of nonlinearity. This fluctuation effect is inconspicuous at the initial elastic stage or when it is subjected to a small nonlinearity, e.g., the first 10 s. Then, the randomness is quickly enlarged, induced by the development of nonlinearity to some extent, and begins to fluctuate with the external loading-unloading conditions.

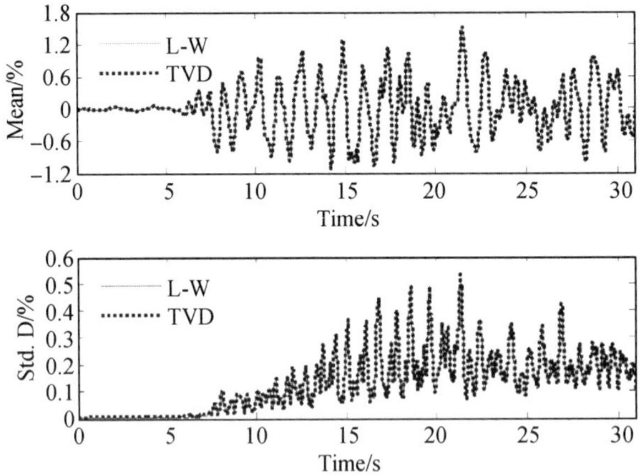

Fig. 11. The mean and standard deviation (Std.D) of the interstory drift ratios

Typical PDFs at three specified time instants are plotted in Fig. 12. As is seen, each PDF curve has two or more peaks, which is much different from the well-known regular distributions. Fig. 13 pictures the PDF surface varying with time, and the corresponding contour to it is shown in Fig. 14, which seem like a mountain and a river stretching to the distance, respectively.

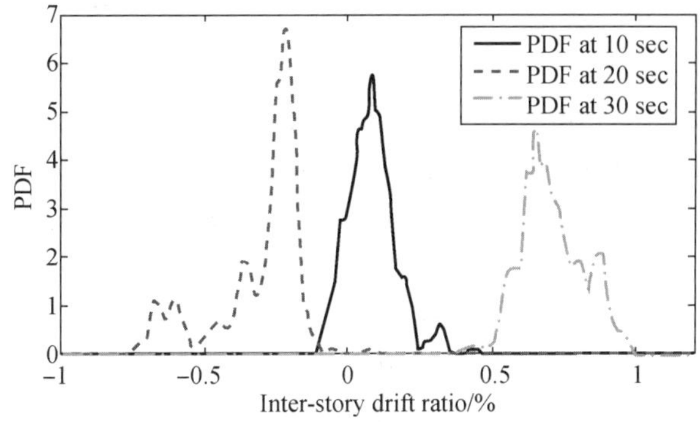

Fig. 12. The PDFs at three typical instants of time

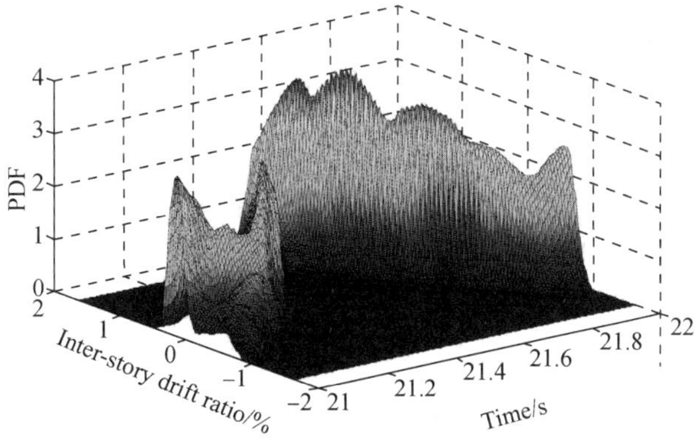

Fig. 13. The PDF surface evolving against time

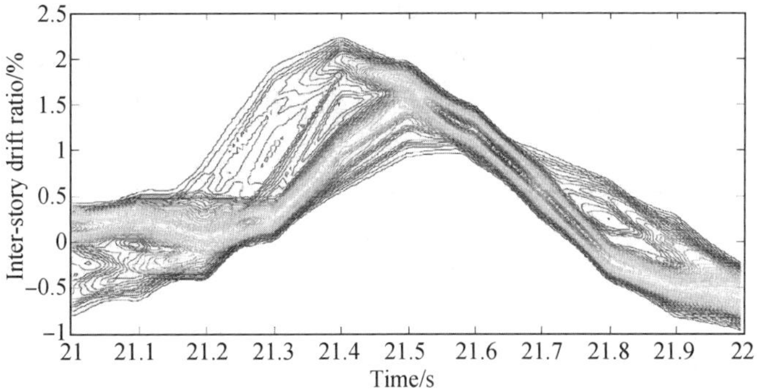

Fig. 14. The contour to the PDF surface

As an epitome of the idea of physical stochastic system, the above phenomena and features of the probabilistic information are analogous to that discussed in the previous literature[15]. Herein, the whole structure can be regarded as a physical system. When the randomness transits through the system, the probability structure will be notably changed, stirring up a fluctuation of the structural responses even though the system itself does not change. Moreover, the fluctuation effect will be aggravated due to the inevitable change of the system properties with the damage evolution and the development of nonlinearity. In this example, it is the random damage evolution in material level of concrete that leads to the dramatic fluctuation of the macroscopic nonlinear structural behaviors. Thus, it is shown that the coupling effect of nonlinearity and randomness in the stochastic dynamic system could be comprehensively studied with the presented method.

From the above results and discussions, on the one hand, it is indicated that the structural response process is a complex stochastic damage evolution process and should be investigated from the development process of the nonlinearity. On the other hand, the visible evolution of PDFs involves plentiful information valuable in capturing the fluctuation effect of the stochastic dynamical system, while the fluctuation is of paramount value in probing the global performance of the nonlinear system and is also very important to better understand the mechanism of structural dynamic reliability. In this regard, the above integrated methodology provides a new perspective to fully and efficiently investigate the nonlinear behaviors of complex RC structures.

4.4 Dynamic Reliability

On account of the obvious consistency of the appearance position of maximum interstory drift ratios for the BTC structure, a single failure mode is considered herein. The dynamic reliability is defined by the sixteenth interstory drift ratio with a symmetrical double boundary, which reads

$$R(t)=\Pr\{|Z_{\text{drift}}(\tau)|\leqslant z_{\text{B}},\ \tau\in[0,t]\}\tag{44}$$

With Eq.(44) and the PDEM, the dynamic reliabilities of the BTC structure are evaluated. The reliabilities varying with time are pictured in Fig. 15. It is seen that the reliabilities grow with the increase of assumed thresholds. Also, the reliabilities decline in a ladder shape rather than smoothly with time increasing. This might mean that the level-crossing process of the response of stochastic structural analysis is not a Poisson or Markov process.

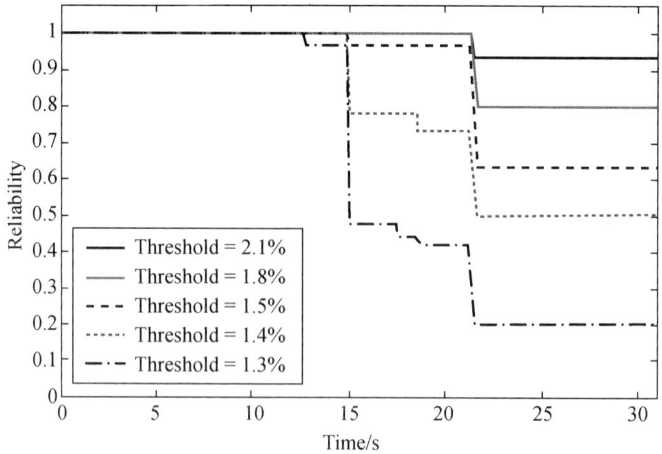

Fig. 15. The time-variant reliabilities of the BTC structure with different failure thresholds

5 Summary

A multiscale stochastic analytical framework for large complex RC structures under strong earthquake excitations is proposed. The parametric uncertainties of concrete materials are quantified based on the stochastic damage model. By integrating the model into the refined finite element analysis, the nonlinear responses of a structural dynamic system in various scales could be fully represented, while the probability density evolution method is employed to address the issue of randomness propagation in both the spatial and the temporal domain. Within the framework, the random fluctuation effect of structural responses induced by the random damage evolution of concrete at the microlevel can be investigated in depth. Further, by introducing an absorbing boundary condition in accordance with the structural failure criterion, the global performance regarding of the dynamic reliability of a general structure can be systematically assessed. To illustrate the overall procedure, the stochastic structural analysis and the reliability evaluation are carried out with respect to a prototype high-rise RC structure. Results show that the randomness of concrete material remarkably affects the seismic response of structures and should be properly deliberated for their intended design life period by structural engineers. It is also indicated that the proposed methodology provides a new perspective to fully and efficiently investigate the nonlinear behaviors of complex RC structures.

Acknowledgments

Financial supports from the National Natural Science Foundation of China (Grants No. 51261120374 and No. 51208374) are gratefully appreciated.

References

[1] LI J, WU J Y, CHEN J B. Stochastic Damage Mechanics of Concrete Structure[M]. Beijing: Science Press, 2014.

[2] LI J, REN X D. Stochastic damage model for concrete based on energy equivalent strain[J]. International Journal of Solids and Structures, 2009, 46(11-12): 2407-2419.

[3] SIMO J C, JU J W. Strain-and stress-based continuum damage models—Ⅰ. Formulation[J]. International Journal of Solids and Structures, 1987, 23(7): 821-840.

[4] JU J W. On energy-based coupled elastoplastic damage theories: constitutive modeling and computational aspects[J]. International Journal of Solids and Structures, 1989, 25(7): 803-833.

[5] LUBLINER J, OLIVER J, OLLER S, et al. A plastic-damage model for concrete[J]. International Journal of Solids and Structures, 1989, 25(3): 299-326.

[6] FARIA R, OLIVER J, CERVERA M. A strain-based plastic viscous-damage model for massive concrete structures[J]. International Journal of Solids and Structures, 1998, 35(14): 1533-1558.

[7] LEE J, FENVES G L. Plastic-damage model for cyclic loading of concrete structures[J]. Journal of Engineering Mechanics, 1998, 124(8): 892-900.

[8] WU J Y, LI J, FARIA R. An energy release rate-based plastic-damage model for concrete[J]. International Journal of Solids and Structures, 2006, 43(3-4): 583-612.

[9] LI J, ZHANG Q Y. The stochastic damage constitutive law of concrete[J]. Journal of Tongji University. (Nature Science Edition), 2001, 29(10): 1135-1141.

[10] KRAJCINOVIC D, FANELLA D. A micromechanical damage model for concrete[J]. Engineering Fracture Mechanics, 1986, 25(5-6): 585-596.

[11] KANDARPA S, KIRKNER D J, SPENCER Jr B F. Stochastic damage model for brittle materials subjected to monotonic loading[J]. Journal of Engineering Mechanics, 1996, 122(8): 788-795.

[12] SHINOZUKA M. Monte Carlo solution of structural dynamics[J]. Computers & Structures, 1972, 2 (5-6): 855-874.

[13] KLEIBER M, HIEN T D. The stochastic finite element method: basic perturbation technique and computer implementation[M]. John Wiley & Sons, 1992.

[14] GHANEM R, SPANOS P D. Polynomial chaos in stochastic finite elements[J]. ASME J. Appl. Mech. 1990, 57(1):197-202.

[15] LI J, CHEN J B. Stochastic dynamics of structures[M]. John Wiley & Sons, 2009.

[16] RABORNO V Y N. Creep rupture[M]//Applied Mechanics. Internationa l Union of Theoretical and Applied Mechanics. Berlin: Springer, 1969.

[17] ZHANG X T, FANG K T. Introduction to Multivariate Statistical Analysis[M]. Beijing: Science Press, 1982.

[18] REN X D, ZENG S, LI J. A rate-dependent stochastic damage-plasticity model for quasibrittle materials[J]. Computational Mechanics, 2015, 55(2): 267-285.

[19] CHEN J B, SUN W L, LI J, et al. Stochastic harmonic function representation of stochastic processes [J]. Journal of Applied Mechanics, 2013, 80(1).

[20] SPACONE E, CIAMPI V, FILIPPOU F C. Mixed formulation of nonlinear beam finite element[J]. Computers & Structures, 1996, 58(1): 71-83.

[21] SPACONE E FILIPPOU F C, TAUCER F. Fiber beam-column model for nonlinear analysis of R/C frames: part Ⅰ formulation[J]. Earthquake Engineering and Structural Dynamics, 1996, 25(7): 711-725.

[22] NEUENHOFER A, FILIPPOU F C. Evaluation of nonlinear frame finiteelement models[J]. Journal of Structural Engineering, 1997, 123(7): 958-966.

[23] HSU T T C, MO Y L. Unified theory of concrete structures[M]. John Wiley & Sons, 2010.

[24] MENEGOTTO M, PINTO P E. Method of analysis for cyclically loaded reinforced concrete plane frames including changes ingeometry and nonelastic behavior of elements under combined normal force and bending[M]. IABSE Symposium on the Resistanceand Ultimate Deformability of Structures Acted on by Well-Defined Repeated Loads, Lisbon, 1973.

[25] HAND F R, PECKNOLD D A, SCHNOBRICH W C. Nonlinear layered analysis of RC plates and shells[J]. Journal of the Structural Division, 1973, 99(7): 1491-1505.

[26] LIN C S, SCORDELIS A C. Nonlinear analysis of RC shells of general form[J]. Journal of the Structural Division, 1975, 101(3): 523-538.

[27] BELYTSCHKO T, LIU W K, MORAN B. Nonlinear finite-elements for continua and structures[M]. John Wiley & Sons, Hoboken, 2000.

[28] HUGHES T J R. Analysis of transient algorithms with particular reference to stability behavior[J]. Computational methods for transient analysis (A 84 - 29160 12 - 64). Amsterdam, North-Holland,

1983: 67-155.

[29] BELYTSCHKO T. Overview of Semidiscretization and Time Integration Procedures [M]// Computational Methods for Transient Analysis. BELYTSCHKO T, HUGHES T J R. North-Holland Publishing, Amsterdam, 1983: 1-65.

[30] CHEN J B, LI J. Dynamic response and reliability analysis of nonlinear stochastic structures[J]. Probabilistic Engineering Mechanics, 2005, 20(1): 33-44.

[31] LI J, CHEN J B. The number theoretical method in response analysis of nonlinear stochastic structures [J]. Computational Mechanics, 2007, 39(6): 693-708.

[32] Ministry of construction of the People's Republic of China. Code for Design of Concrete Structures: GB 50010—2010[S]. Beijing: China Architecture & Buildings Press, 2010.

[33] Ministry of construction of the People's Republic of China. Code for Seismic Design of Buildings: GB 50011—2010[S]. Beijing: China Architecture & Buildings Press, 2010.

Incremental dynamic analysis of seismic collapse of super-tall building structures①

Tiancan Huang[1], **Xiaodan Ren**[1], **Jie Li**[1,2]

1 School of Civil Engineering, Tongji University, 200092, Shanghai, China

2 The State Key Laboratory on Disaster Reduction in Civil Engineering, Tongji University, Shanghai, China

Abstract In the present work, the incremental dynamic analysis is performed for the super-tall building structure of ZhongNan Center, which is more than 700 m high and is currently under construction in Suzhou, China. The Chi-Chi earthquake record is used as the ground motion input. The peak ground accelerations are scaled to five levels including 0.2g, 0.4g, 0.8g, 1.4g, and 2.0g. The plastic damage model is adopted to represent the material failure. The finite element model is developed by using fiber beam elements for beam-columns and laminated shell elements for shear walls and slabs. Under the attack of strong seismic ground motions, the super tall structure experiences overall collapse, which is well captured by the developed finite element model. More importantly, the collapse patterns of the super tall structure are rather different under the excitations of earthquake records with same wave pattern but different peak accelerations.

Keywords Collapse pattern; Damage model; Finite element method; Incremental dynamic analysis; Structural collapse; Super-tall building

1 Introduction

With the rapid developments of modern materials and construction techniques, super-tall buildings have been experiencing incredibly rapid development all over the world. However, the safety of super-tall buildings has not been well investigated and understood. It has been reported[1-3] that many high-tall buildings, which were designed and constructed in accordance with modern seismic design principles, have partially or totally collapsed during earthquakes (Fig. 1).

(a) (b) (c)

Fig. 1. Collapse of modern buildings (a) 1995 Kobe, Japan earthquake; (b) 1999 Chi-Chi, Taiwan earthquake; and (c) 2010 Offshore Maule, Chile earthquake

① Originally published in *The Structural Design of Tall and Special Buildings*, 2017, 26(16): e1370.

As is well known, one of the essential objectives of the performance-based design is to prevent the structures from overall collapse. Structures subjected to earthquake excitations usually experience a series of stages including the elastic stage, the damaged stage, the partial collapse stage, and the total collapse stage. Thus, the elaborate methods to assess structural damage and collapse under strong earthquake ground motions are required to implement the performance-based design procedure. In recent years, the nonlinear static procedure and the direct dynamic time integration method have been widely used.[4]

The nonlinear static procedure, also known as the push over analysis, was introduced in Federal Emergency Management Agency.[5] Some design codes such as ATC40 and seismic design code of China (GB 50011) also include this approach. It is recommended for structures that are roughly deformed in the pattern of their first natural modes of vibration under seismic excitations. Static pushover analysis has been widely used due to its convenience and simplicity in application. On the other hand, it is based on the incorrect assumptions that the nonlinear response of a structure can be related to the response of an equivalent single degree-of-freedom system and that the distribution of the equivalent lateral forces remains constant over the height of the structure during the entire duration of the structural response.[4] The direct dynamic time integration[6] is an accurate way to simulate the behaviors of structures under earthquake excitations. Both the structural behaviors and the earthquake records are modeled accurately. Therefore, it could be presumed that an accurate structural response is obtained. However, this method is unable to provide the seismic performance of structures in an explicit way and the additional post-processing of simulated results should be introduced.

In recent years, incremental dynamic analyses (IDA) have emerged as a powerful means to study the overall behavior of structures, by which the response of structures from elastic stage to damage and finally to overall collapse could be investigated.[7] An incremental dynamic analysis involves a series of nonlinear dynamic analyses in which the intensity of the ground motion selected for the collapse investigation is incrementally increased until the overall collapse of the structure is reached. The concept of IDA was proposed as early as 1977 by Bertero.[8] In the milestone work of Vamvatsikos and Cornell,[9] the details of IDA were well described. Later, the enhanced techniques for IDA were also proposed.[10] It was also recommended by FEMA (2000a)[7] to estimate the seismic performance of structure. With the rapid development of computational techniques and facilities, we are capable of applying the IDA to super-tall building structures, defined on average as over 500 m high, and made up of hundreds of thousands of structural elements.

The present paper aims at assessing the anticollapse capacities of super-tall structures under earthquake ground motions based on IDA analysis. In order to accurately obtain the nonlinear responses of super tall building structures subjected to earthquake excitations, a finite element model is developed by using the finite element package ABAQUS with plasticity damage model implemented as user's defined subroutine. In order to perform

IDA, a set of ground motion records were chosen with their Peak ground acceleration (PGA) scaled to five levels. The patterns of structural collapse under different levels of ground motions are obtained and discussed.

2 Description of The Super-Tall Building

Zhongnan Center Tower, a 729 m high office tower, is located in the west bund of JinJi Lake, Suzhou, China.* The tower has 137 stories and the structural height reaches 598 m (Fig. 2). The aspect ratio of height to width is 8.7, and it is designed to be the tallest building in China thus far. The building is divided into nine zones along its vertical direction (Fig. 3). With a square base of 61 m by 61 m, the structure is symmetric and gradually shrinking along the height direction(Fig.4). The primary structural components include core tube, mega columns, coupling beams, outriggers, belt trusses, and floor slabs (Fig. 5).

Fig. 2. Zhongnan Center Tower

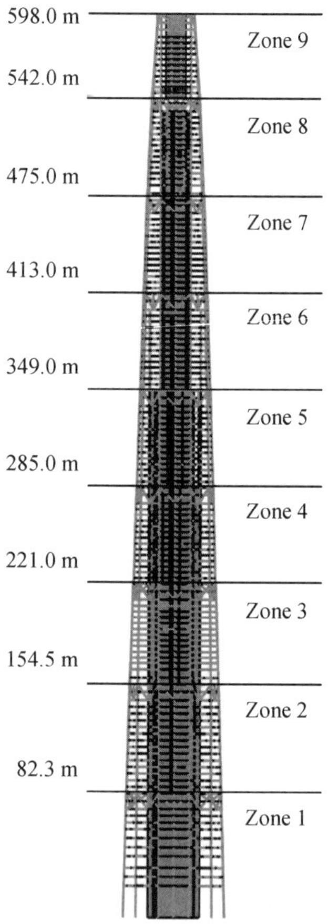

Fig. 3. The partition of elevation

The core tube is designed as square tabular structures, with 4×4 tubes from bottom to 349 m and 2×2 tubes in the upper levels. The maximum thickness of the inner tube is 1.2 m in bottom, and it is reduced to 0.5 m for the top level. The maximum thickness of

* The structural design considered in the present study is not the finalized design of ZhongNan Center Tower. The improved design has been finished partially based on the results and findings of the present work. The contents of the present work are only for research purposes.

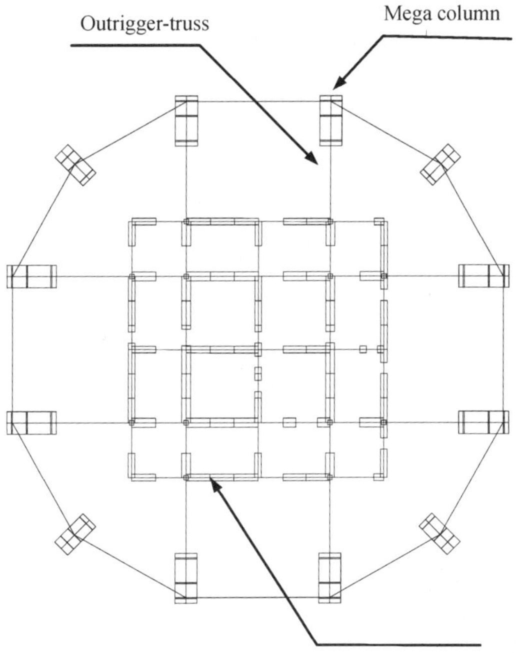

Fig. 4. Structural plan for regular levels

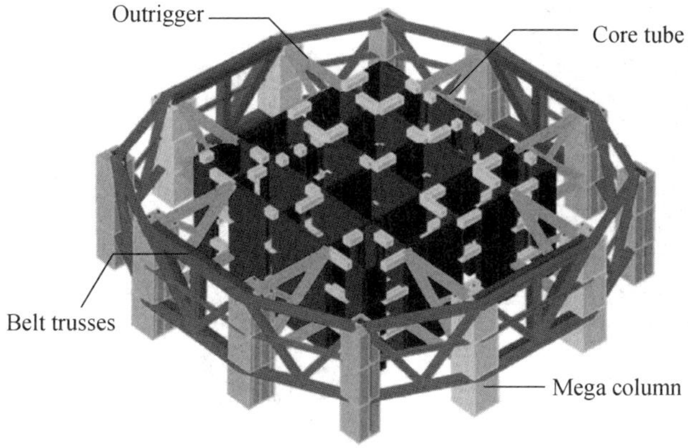

Fig. 5. Truss system connecting core tube and mega columns

the outer tube is 0.9 m in bottom, and it is reduced to 0.45 m for the top level.

The mega columns are rectangular, varying from 3.8 m×4.9 m to 1.9 m×1.9 m. There are 6 outriggers connecting the core tube and mega columns located in different levels of the structure. According to the structural plan, the core tube bears a majority of the horizontal shear and part of the overturning moment. With the help of outrigger trusses, the mega columns provide substantial bending stiffness.

According to the seismic design code of China[11], the tower is located in Suzhou with the seismic design intensity categorized as 7. The site soil in Suzhou is specified as type-Ⅲ and the characteristic period of the site soil is 0.57 s.

3 Development of The Structural Model

3.1 3D finite element model

As shown in Fig. 6, a three-dimensional finite element model is developed based on ABAQUS platform. The structural components of the tower, including core tubes, mega columns, outriggers, belt trusses, and floor slabs, are modeled by appropriate finite elements. The fiber beam-column element is used to model the beam-columns (Fig. 7), and the multilayer shell element is used for the shear walls (Fig. 8). As for the floor slabs, the regular shell elements (SR4) were used. The refined mesh was developed to ensure the accuracy of numerical simulation (Table 1).

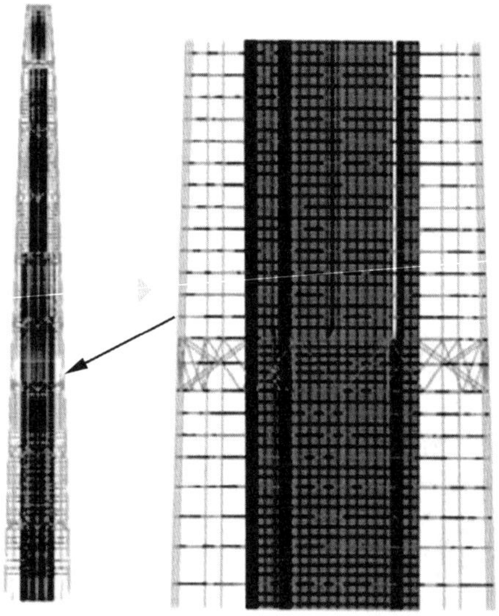

Fig. 6. Finite element mesh

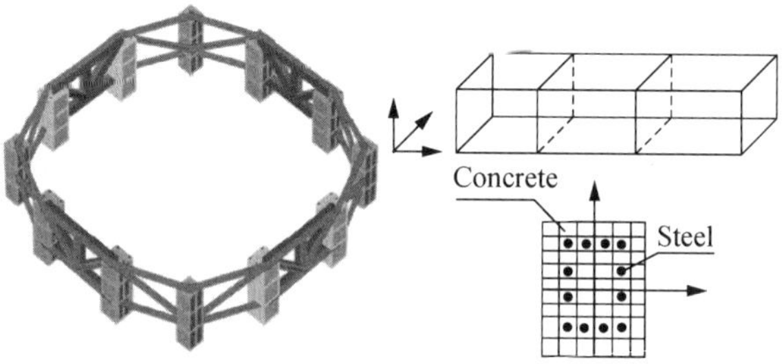

Fig. 7. Fiber beam-column element

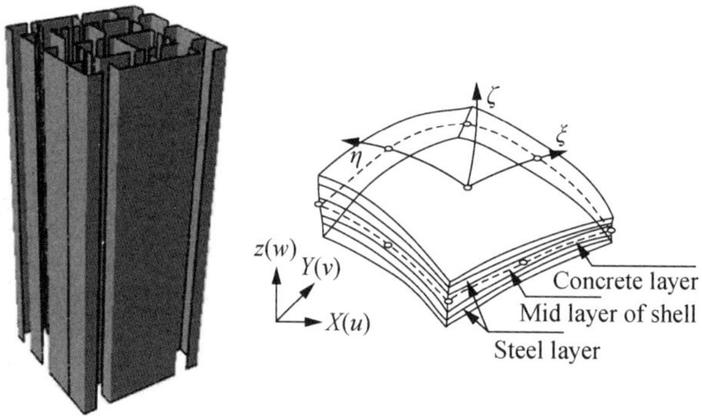

Fig. 8. Mutilayer shell element

Table 1. Summary of finite elements

Number of Nodes	225 632
Number of fiber beam-column elements	53 822
Number of multilayer shell elements	27 021
Number of SR4 elements	62 277

The materials considered in the structural design include ① concrete used in core tube (C60), mega columns (C70, C60, C50) and floor slabs (C35); ② steel tube used in outrigger-trusses (Q235) and steel reinforcing bars in core tube (HRB335), mega column and floor slabs. The constitutive models describing nonlinear behaviors and dynamic failures of materials are briefly summarized in the following subsections.

3.2 Plastic-damage concrete model

According to the principle of strain equivalence[12-13], the effective stress tensor can be represented as

$$\bar{\boldsymbol{\sigma}} = \mathbb{E}_0 : \boldsymbol{\varepsilon}^{e} = \mathbb{E}_0 : (\boldsymbol{\varepsilon} - \boldsymbol{\varepsilon}^{p}), \tag{1}$$

where $\mathbb{E}_0$ is forth-order initial elastic tensor; $\boldsymbol{\varepsilon}$, $\boldsymbol{\varepsilon}^{e}$, and $\boldsymbol{\varepsilon}^{p}$ are second-order tensors, denoting the strain tensor and its elastic and plastic components, respectively.

To account for the unilateral effect, the effective strain tensor is decomposed as follows:

$$\bar{\boldsymbol{\sigma}} = \bar{\boldsymbol{\sigma}}^{+} + \bar{\boldsymbol{\sigma}}^{-}, \tag{2}$$

$$\bar{\boldsymbol{\sigma}}^{+} = \mathbf{P}^{+} : \bar{\boldsymbol{\sigma}}, \tag{3}$$

$$\bar{\boldsymbol{\sigma}}^{-} = \bar{\boldsymbol{\sigma}} - \bar{\boldsymbol{\sigma}}^{+} = \mathbf{P}^{-} : \bar{\boldsymbol{\sigma}}, \tag{4}$$

where $\mathbf{P}^{+}$ and $\mathbf{P}^{-}$ are rank-four stress projection tensors in the following form

$$\mathbf{P}^{+} = \sum_i H(\hat{\bar{\sigma}}_i)(\mathbf{P}_i \otimes \mathbf{P}_i \otimes \mathbf{P}_i \otimes \mathbf{P}_i), \tag{5}$$

$$\mathbf{P}^{-} = \mathbf{I} - \mathbf{P}^{+} \tag{6}$$

and $\mathbf{I}$ is the forth-order identity tensor; $\hat{\bar{\sigma}}_i$ is the ith the eigenvalue of the effective stress tensor; $\mathbf{P}_i$ is the ith normalized eigenvector corresponding to $\hat{\bar{\sigma}}_i$; $H(\cdot)$ is the Heaviside function defined as follows

$$H(x) = \begin{cases} 1, & x > 0 \\ 0, & x \leqslant 0 \end{cases}. \tag{7}$$

Based on the principle of thermodynamics, we could obtain the final form for the constitutive law, which is expressed by Cauchy stress tensor[14]

$$\boldsymbol{\sigma} = (\mathbf{I} - \mathbf{D}) : \bar{\boldsymbol{\sigma}} = (\mathbf{I} - \mathbf{D}) : \mathbb{E}_0 : (\boldsymbol{\varepsilon} - \boldsymbol{\varepsilon}^{p}), \tag{8}$$

where the fourth-order damage tensor $\mathbf{D}$ is expressed as

$$\mathbf{D} = d^{+}\mathbf{P}^{+} + d^{-}\mathbf{P}^{-}. \tag{9}$$

It should be noted that Eq. (8) is the standard relation between the Cauchy stress tensor $\boldsymbol{\sigma}$ and the effective stress $\bar{\boldsymbol{\sigma}}$ in Continuum Damage Mechanics(CDM), according to the "strain equivalence" assumption[12,15].

The evolutionary law of damage scalars[16] is expressed as

$$\begin{cases} d^{+} = f_d^{+}(\varepsilon_{eq}^{+}) \\ d^{-} = f_d^{-}(\varepsilon_{eq}^{-}) \end{cases}, \tag{10}$$

where the energy equivalent strains are

$$\begin{cases} \varepsilon_{eq}^{+} = \dfrac{\sqrt{Y^{+}}}{E_0} \\ \varepsilon_{eq}^{-} = \dfrac{\sqrt{Y^{-}}}{(\alpha - 1)E_0} \end{cases}, \tag{11}$$

And the expressions of damage energy release rate[17] are

$$\begin{cases} Y^{+}=\dfrac{2(1+\upsilon)}{3}3\bar{J}_2^{+}+\dfrac{1-2\upsilon}{3}\bar{I}_1^{+}\bar{I}_1^{+}-\upsilon\bar{I}_1^{+}\bar{I}_1^{-} \\ Y^{-}=(\alpha\bar{I}_1^{-}+\sqrt{3\bar{J}_2^{-}})^2 \end{cases}, \tag{12}$$

where $\bar{I}_1^{+}$, $\bar{I}_1^{-}$, $\bar{J}_2^{+}$, and $\bar{J}_2^{-}$ are invariants of $\bar{\boldsymbol{\sigma}}^{+}$ and $\bar{\boldsymbol{\sigma}}^{-}$, respectively; υ is Piosson's ratio; and α is material parameter related to biaxial compressive strength increase of concrete. Based on the Mander's model[18] and the Guo's model[19], the damage evolutionary functions are adopted as follows

$$f_{\mathrm{d}}^{\pm}(\varepsilon)=\begin{cases} 1-\dfrac{\rho^{\pm}n^{\pm}}{n^{\pm}-1+(x^{\pm})^{n^{\pm}}} & x^{\pm}\leqslant 1 \\ 1-\dfrac{\rho^{\pm}}{\alpha_{\mathrm{d}}^{\pm}(x^{\pm}-1)^2+x^{\pm}} & x^{\pm}\geqslant 1 \end{cases}, \tag{13}$$

$$x^{\pm}=\frac{\varepsilon}{\varepsilon^{\pm}},\quad \rho^{\pm}=\frac{f^{\pm}}{E_0\varepsilon^{\pm}},\quad n^{\pm}=\frac{E_0\varepsilon^{\pm}}{E_0\varepsilon^{\pm}-f^{\pm}}, \tag{14}$$

where $f^{\pm}$ denotes the uniaxial tensile strength and the uniaxial compressive strength, as appropriate; $\varepsilon^{\pm}$ is the strain corresponding to the peak value of the stress; and $\alpha_{\mathrm{d}}^{\pm}$ is the parameter governs the slope of the descending part of stress-strain curve.

According to literature[12], the plastic evolutions could be developed based on the concept of operator splitting. In the present work, we turn to the empirical plastic evolutionary model[20]. By analogy of decomposition of damage tensor, we introduce the decomposition of plastic strain as follows

$$\boldsymbol{\varepsilon}^{\mathrm{p}}=\boldsymbol{\varepsilon}^{\mathrm{p}^{+}}+\boldsymbol{\varepsilon}^{\mathrm{p}^{-}}, \tag{15}$$

where $\boldsymbol{\varepsilon}^{\mathrm{p}^{+}}$ and $\boldsymbol{\varepsilon}^{\mathrm{p}^{-}}$ are plastic strains under tension and compression, respectively. Considering the coupling between damage and plasticity, the evolutions of plastic strains are expressed as follows

$$\dot{\boldsymbol{\varepsilon}}^{\mathrm{p}^{+}}=H(\dot{d}^{+})\xi_{\mathrm{p}}^{+}(d^{+})^{n_{\mathrm{p}}^{+}}\mathbb{E}_0^{-1}:\dot{\bar{\boldsymbol{\sigma}}}^{+},\ \dot{\boldsymbol{\varepsilon}}^{\mathrm{p}^{-}}=H(\dot{d}^{-})\xi_{\mathrm{p}}^{-}(d^{-})^{n_{\mathrm{p}}^{-}}\mathbb{E}_0^{-1}:\dot{\bar{\boldsymbol{\sigma}}}^{-}, \tag{16}$$

where ξ_{p}^{+}, ξ_{p}^{-}, n_{p}^{+} and n_{p}^{-} are material parameters related to plasticity.

Thus far, the framework of damage-plasticity model could be used for the numerical simulation of structures.

3.3 Steel constitutive law

A standard elastoplastic strain-hardening model has been used to simulate the nonlinear behavior of steel tubes and steel reinforcing bars under cyclic loading. The fundamental stress-strain relation is

$$\boldsymbol{\sigma}=\mathbb{E}_0:(\boldsymbol{\varepsilon}-\boldsymbol{\varepsilon}^{\mathrm{p}}). \tag{17}$$

The evolutionary law of plastic strain is

$$\begin{cases} \dot{\boldsymbol{\varepsilon}}^{p}=\dot{\lambda}\partial_{\sigma}f(\boldsymbol{\sigma}) \\ F(\boldsymbol{\sigma},\alpha^{p})=f(\boldsymbol{\sigma})-k\alpha^{p} \\ F\geqslant 0,\quad \dot{\lambda}\leqslant 0,\quad \dot{\lambda}F=0 \end{cases}, \tag{18}$$

where k is the hardening modulus and λ is the evolutionary coefficient. The von Mises type yield function is

$$f(\boldsymbol{\sigma})=\sqrt{J_2}, \tag{19}$$

where J_2 is the second invariant of deviatoric stress tensor **s**. The equivalent plastic strain is in the following form:

$$\dot{\alpha}^{p}=\sqrt{\frac{2}{3}\dot{\boldsymbol{\varepsilon}}^{p}:\dot{\boldsymbol{\varepsilon}}^{p}}. \tag{20}$$

The uniaxial stress-strain relation of steel is shown in Fig. 9.

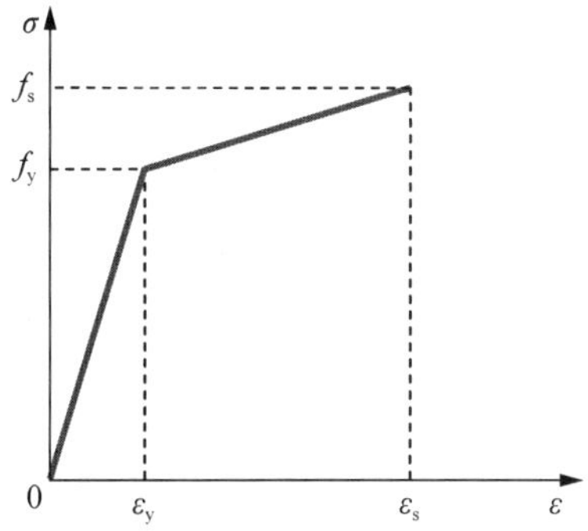

Fig. 9. Uniaxial stress-strain relation of steel

3.4 Central difference method for time integration

To simulate the collapse of structures, material damage and softening should be taken into account, and the contacts between elements and nodes should be reproduced properly. The system with those strong nonlinearities could hardly be solved by implicit solver, which is developed based on the Newton-Raphson method due to illness of system Jacobian. Thus, in the present work, Abaqus/Explicit is adopted, which has been used and verified for a wide variety of simulations including crash, blast, material failure, and impact applications.

The explicit dynamic analysis procedure in Abaqus/Explicit is based upon the implementation of the central difference method[21].

Consider the nth time increment as follows:

$$\Delta t^{n+1/2}=t^{n+1}-t^{n},\quad t^{n+1/2}=\frac{1}{2}(t^{n+1}+t^{n}),\quad \Delta t^{n}=t^{n+1/2}-t^{n-1/2}. \tag{21}$$

Denote the vectors of displacement, velocity, and acceleration by **u**, **v**, and **a**. The central difference formula for the velocity could be expressed as follows

$$\dot{\mathbf{u}}^{n+1/2}=\mathbf{v}^{n+1/2}=\frac{\mathbf{u}^{n+1}-\mathbf{u}^{n}}{t^{n+1}-t^{n}}=\frac{1}{\Delta t^{n+1/2}}(\mathbf{u}^{n+1}-\mathbf{u}^{n}). \tag{22}$$

This difference formula can be converted to an integration formula, as follows

$$\mathbf{u}^{n+1}=\mathbf{u}^{n}+\Delta t^{n+1/2}\mathbf{v}^{n+1/2}. \tag{23}$$

The velocity at the nth time step is defined as follows

$$\begin{aligned}\mathbf{v}^{n}&=\frac{1}{2}(\mathbf{v}^{n+1/2}+\mathbf{v}^{n-1/2})\\&=\frac{1}{2}\left[\frac{1}{\Delta t^{n+1/2}}(\mathbf{u}^{n+1}-\mathbf{u}^{n})-\frac{1}{\Delta t^{n-1/2}}(\mathbf{u}^{n}-\mathbf{u}^{n-1})\right].\end{aligned} \tag{24}$$

For the equal time steps, the formula of velocity reduces to

$$\mathbf{v}^{n}=\frac{\mathbf{u}^{n+1}-2\mathbf{u}^{n}+\mathbf{u}^{n-1}}{2\Delta t^{n}}. \tag{25}$$

The acceleration is

$$\ddot{\mathbf{u}}^{n}=\mathbf{a}^{n}=\frac{\mathbf{v}^{n+1/2}-\mathbf{v}^{n-1/2}}{t^{n+1/2}-t^{n-1/2}}=\frac{\Delta t^{n-1/2}(\mathbf{u}^{n+1}-\mathbf{u}^{n})-\Delta t^{n+1/2}(\mathbf{u}^{n}-\mathbf{u}^{n-1})}{\Delta t^{n}\Delta t^{n+1/2}\Delta t^{n-1/2}}. \tag{26}$$

For the case of equal time steps, the reduced form of acceleration could be obtained as follows

$$\ddot{\mathbf{u}}^{n}=\mathbf{a}^{n}=\frac{\mathbf{u}^{n+1}-2\mathbf{u}^{n}+\mathbf{u}^{n-1}}{(\Delta t^{n})^{2}}. \tag{27}$$

The integration formula of acceleration is

$$\mathbf{v}^{n+1/2}=\mathbf{v}^{n-1/2}+\Delta t^{n}\mathbf{a}^{n}. \tag{28}$$

The system of structural dynamics could be expressed by the following second order equations

$$\mathbf{M}\mathbf{a}^{n}+\mathbf{C}\mathbf{v}^{n}+\mathbf{f}_{\text{int}}(\mathbf{u}^{n})=\mathbf{f}_{\text{ext}}^{n}, \tag{29}$$

where $\mathbf{M}$ and $\mathbf{C}$ are mass and damping matrices; and $\mathbf{f}_{\text{int}}$ and $\mathbf{f}_{\text{ext}}$ are internal and external forces. Substituting Eqs. (25) and (27) into Eq. (29) gives the explicit integration formula as follows

$$\left[\frac{\mathbf{M}}{(\Delta t^{n})^{2}}+\frac{\mathbf{C}}{2\Delta t^{n}}\right]\mathbf{u}^{n+1}=\mathbf{f}_{\text{ext}}^{n}-\mathbf{f}_{\text{int}}(\mathbf{u}^{n})+\left[\frac{\mathbf{M}}{(\Delta t^{n})^{2}}+\frac{\mathbf{C}}{2\Delta t^{n}}\right](2\mathbf{u}^{n}-\mathbf{u}^{n-1}). \tag{30}$$

According to Eq. (30) the displacement at the $(n+1)$th step could be explicitly updated by $\mathbf{u}^{n}$ without any matrix interactions and inversions. Thus, material damage and softening cannot enhance the difficulties to solve the structural system. It is also very easy to be parallelized so that the collapse of large-scale structures could be simulated by taking advantage of large-scale computers or servers.

On the other hand, the central difference method is conditionally stable. The stability limit for the dynamic system without damping is given in terms of the highest eigenvalue ω_{max} as follows

$$\Delta t_{stable} \leqslant \frac{2}{\omega_{max}}. \tag{31}$$

In Abaqus/Explicit, a small amount of damping is usually introduced to suppress the high frequency oscillations. With damping the stable time increment is given by

$$\Delta t_{stable} \leqslant \frac{2}{\omega_{max}(\sqrt{1+\xi^2}-\xi)}, \tag{32}$$

where ξ is the fraction of critical damping in the highest natural mode. The stable time increment defined by Eq. (31) or Eq. (32) is usually rather small ($10^{-7} \sim 10^{-5}$ s) for regular structural simulations compared with the total time length of earthquakes ($10^1 \sim 10^2$ s). Thus, usually a great number of time increments is required for the full process simulation of structural collapse. That is the price for the explicit method for which the Jacobian of system is neglected.

4 Description of Single Seismic Record Incremental Dynamic Analysis

The seismic collapse of structures is by its essence a dynamic stability problem, which is closely related to both the structural performance and the dynamic excitations. IDA is a parametric analysis that estimates structural performance under seismic loads. It can reflect the dynamic response process of structures, which are subjected to different levels of ground motion while structural responses are shifted from the elastic state to the nonlinear state until collapse. IDA not only can evaluate the anticollapse capacity of structure, but also can grasp the seismic performance of structure.

This paper develops the IDA process specifically for the structure of Zhongnan Center, a super-tall building. The key steps of implementation are as follows:

(1) Select a ground motion record and determine the intensity measure.

(2) Define the scale factor and obtain a suit of ground motion records of different intensities.

(3) Calculate the dynamic response of the structure subjected to the ground motion records.

(4) Assess seismic performance and anticollapse capacity of structure-based on the simulated results.

The ground motion of Chi-Chi, Taiwan 1999 is adopted for this study. The earthquake was of a Richter magnitude of 7.62 with moderate epicentral distance of 35.39 km. The recorded PGA was 0.103 1g, peak ground velocity was 43.44 cm/s, and permanent ground deformation was 33.27 cm. Fig. 10 shows the ground motion record scaled to 1.0g and the response spectra is shown in Fig. 11. PGA was adopted as the Monotonic Scalable Ground Motion Intensity Measure, which was scaled to five levels including 0.2g, 0.4g, 0.8g, 1.4g, and 2.0g.

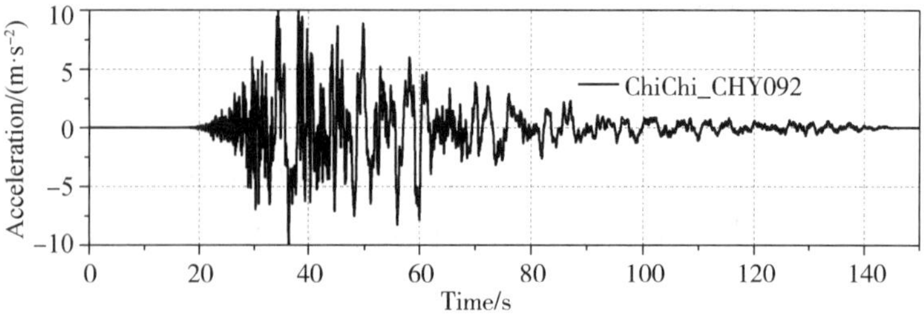

Fig. 10. **Scaled acceleration of Chi-Chi earthquake(1.0*g*)**

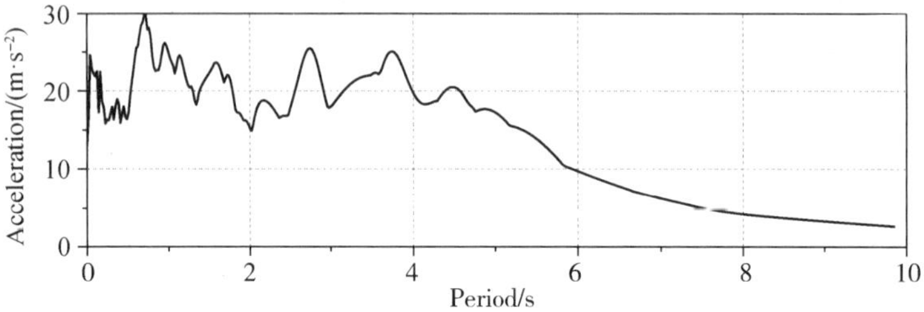

Fig. 11. **Scaled acceleration spectra of Chi-Chi earthquake(5% damping ratio)**

5 Results and Discussions

5.1 Basis dynamic characteristics of the super-tall building

The linear perturbation analysis is performed to calculate the natural periods of the super-tall building structure. Using the Lanczos algorithm, the first 50 natural modes are calculated. The first three natural modes (Fig. 12) and 10 free natural periods are presented (Table 2).

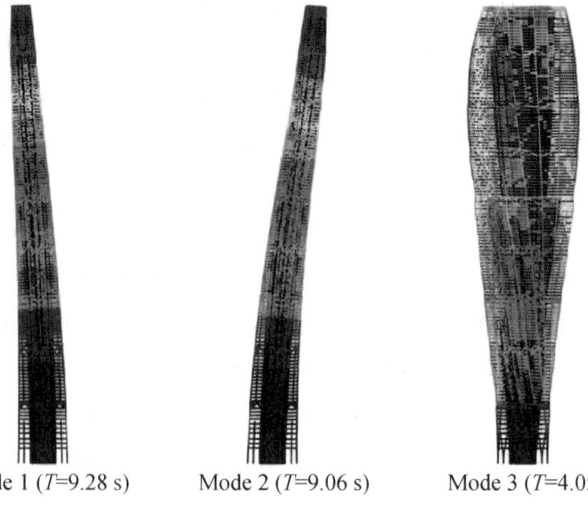

Fig. 12. **The first three mode shapes of model**

Table 2. The natural periods of structure

T1	T2	T3	T4	T5	T6	T7	T8	T9	T10
9.28 s	9.06 s	4.05 s	3.77 s	3.65 s	2.80 s	2.14 s	2.08 s	2.02 s	1.39 s

5.2 Patterns of collapse

The super-tall structure collapses in rather different patterns under different levels of seismic excitations. They are briefly summarized as follows

Case 1. PGA=$0.2g$

Figs. 13 and 14 show the collapse process of the structure subject to the ground motion with scaled PGA to be 3.92 m/s^2 ($0.2g$). The damage of the structure occurs in the top and the middle part where the thickness and section of the core tube are changed. The local outrigger trusses and belt trusses start to yield in Zone 9 at 35 s as parts of the coupling beams have failed. At 42.5 s, the bottom of the core tube starts to crack and the damage of shear wall in Zone 8 and 9 propagates gradually. At 50 s the cracks penetrate the core tube in Zone 9, and the core tube finally ruptures at the height of 542 m.

(a) t=27.5 s (b) t=35 s (c) t=42.5 s (d) t=50 s (e) t=65 s

Fig. 13. Collapse process of the structure

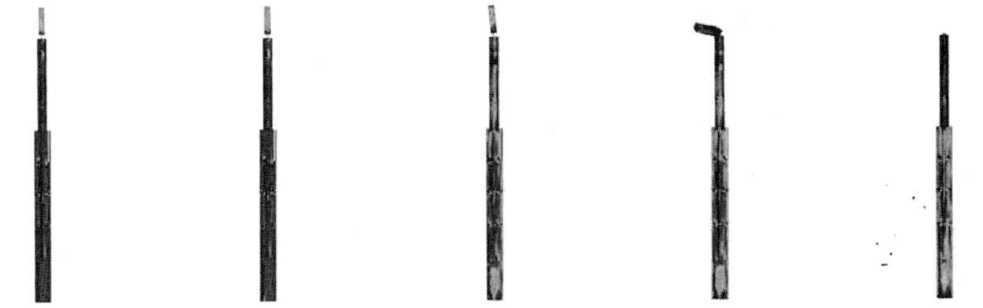

(a) t=27.5 s (b) t=35 s (c) t=42.5 s (d) t=50 s (e) t=65 s

Fig. 14. Tensile damage evolution of core-tube

Case 2. PGA=$0.4g$

In Case 2(Figs. 15 and 16), the initial damage of the structure also occurs in the top and the middle part where the section of the core tube changes. More coupling beam and belt trusses have failed in Zone 8 and Zone 9 until 35 s compared to Case 1; in the meantime, the bottom of the core tube starts to crack. Until 38 s, most of the belt trusses at the height of the 542 m region have failed, and the structure in Zone 9 has seriously inclined. The shear walls in Zone 6 and 8 appear to be damaged at 42.5 s. At 50 s the cracks cut through the shear wall in Zone 6 so that the structure in Zones 6～8 collapses.

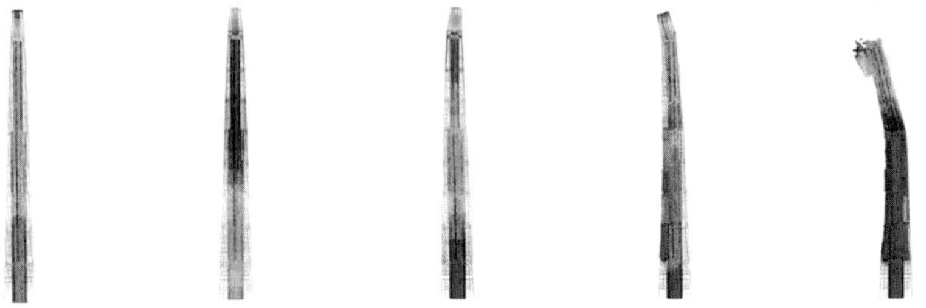

(a) t=27.5 s (b) t=35 s (c) t=38 s (d) t=42.5 s (e) t=50 s

Fig. 15. Collapse process of the structure

(a) t=27.5 s (b) t=35 s (c) t=38 s (d) t=42.5 s (e) t=50 s

Fig. 16. Tensile damage evolution of core-tube

Case 3. PGA=0.8g

In Case 3(Figs. 17 and 18), the initial damage of the structure also occurs in the top and the middle part where the section of the core tube changes. Up till to 35 s the core tube in Zone 1 and 9 has been damaged more seriously compared to Case 1 and Case 2; in the meanwhile, many coupling beams, outrigger trusses, and belt trusses in the region have failed. In the time of 38 s, the damage of structure develops rather rapidly, and the shear wall appears to be seriously damaged in Zone 3, Zone 6, and Zone 9. At 42.5 s the cracks cut through the core tube at different sections at the heights of 221 m, 349 m, and 542 m, simultaneously. Then the whole structure collapses rapidly due to impacts of dropping structural components.

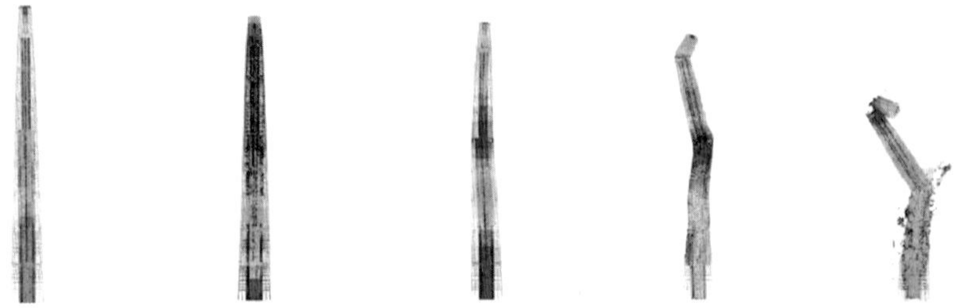

(a) t=27.5 s (b) t=35 s (c) t=38 s (d) t=42.5 s (e) t=47 s

Fig. 17. Tensile damage evolution of core-tube

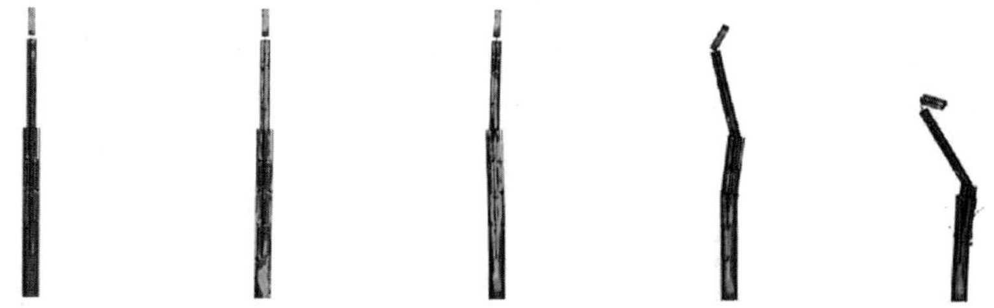

(a) t=27.5 s (b) t=35 s (c) t=38 s (d) t=42.5 s (e) t=47 s

Fig. 18. Collapse process of the structure

Case 4. PGA$=1.4g$

In Case 4(in Figs. 19 and 20), the damage of the structure also initiates in the top and the middle part where the section of the core tube changes. Later on, different phenomena are shown. At 35 s the damage of core tube is mainly concentrated in Zones 6~9. Especially in Zone 9, most of the structural components fail, and the structure experiences significant tilt. At 37.5 s, the damage in core tube develops in a rather rapid way to form the damage zones. Finally, at the time of 40 s, the crack cuts through the core tube at the heights of 151 m and 349 m, and the structure collapses.

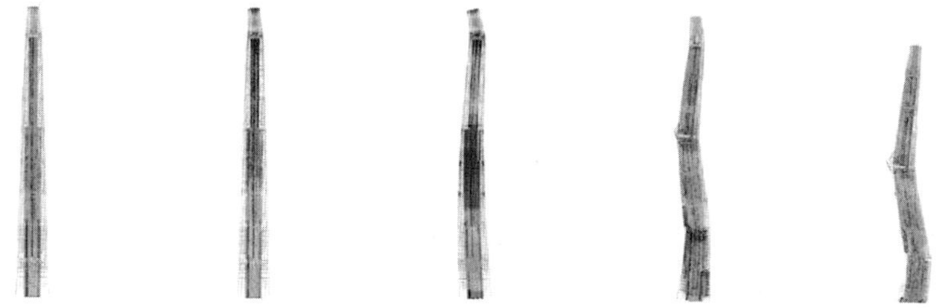

(a) t=27.5 s (b) t=35 s (c) t=37.5 s (d) t=40 s (e) t=42.5 s

Fig. 19. Tensile damage evolution of core-tube

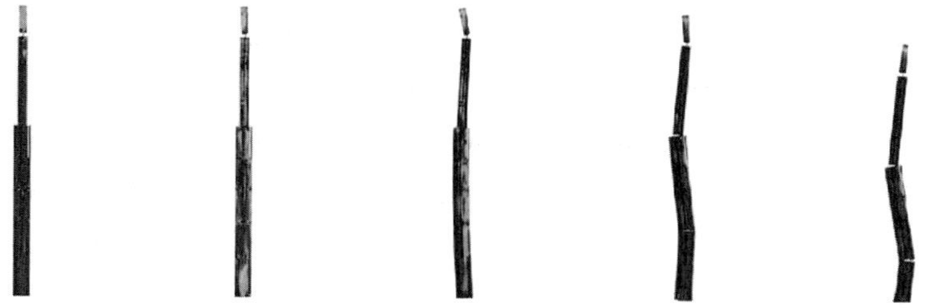

(a) t=27.5 s (b) t=35 s (c) t=37.5 s (d) t=40 s (e) t=42.5 s

Fig. 20. Collapse process of the structure

Case 5. PGA$=2.0g$

In Case 5(Figs. 21 and 22), the PGA reaches $2.0g$. The initial damage of the structure also occurred in the top and the middle part where the section of shear wall tube changes. Later, the structural damage distribution shows different features from other cases. At 35 s, the bottom of core tube starts to crack and form an evident damage zone. After that, governing cracks and damage bands rapidly penetrate the bottom of the core tube due to the fatal earthquake excitations. The core tube finally ruptures at the bottom, and the structure collapses at 38 s.

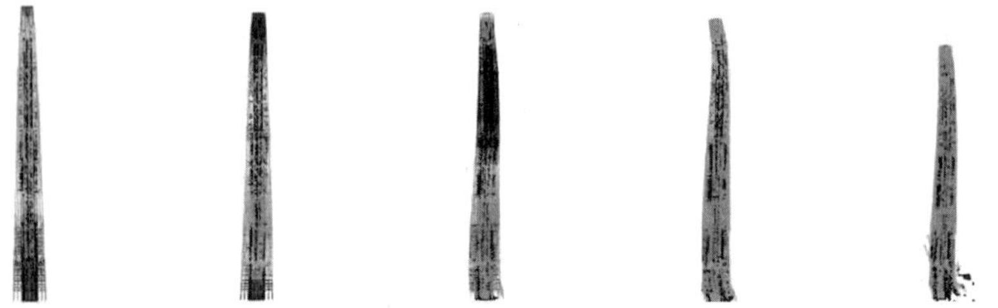

(a) t=27.5 s (b) t=35 s (c) t=36.5 s (d) t=38 s (e) t=41 s

Fig. 21. Tensile damage evolution of core-tube

(a) t=27.5 s (b) t=35 s (c) t=36.5 s (d) t=38 s (e) t=41 s

Fig. 22. Tensile damage evolution of core-tube

It is indicated by the analytical results of IDA that the damage of the super-tall building structure initiates in the similar regions, which are mainly within upper region and the middle region whereas the section of shear wall tube changes. The reason is that in the early stage of earthquake, the super-tall building structure is governed by the elastic deformation, for which the stresses concentrate in the regions with changes of stiffness. Then with the increase of structural deformation, the P-Δ effect gradually governs the development of the structural damage. As the peak intensity of ground acceleration increases, the structural deformation increases. In the meanwhile, the bending moment of the total structure in the bottom part also increases faster than the top part due to the P-Δ effect. Thus, the damage concentration zones of structure gradually goes downward. On the other hand, the super-tall building structure usually has a series of transfer region, which the internal forces are transferred between internal tube and outer columns by trusses. Thus, the inertial tube would experience sudden change of overall bending moment within transfer region. This will significantly affect the development process of structural damage evolution. As a result, when subjected to earthquakes with the same waveform but different PGAs, the super-tall building structure experiences rather different patterns of damage evolution. Finally, the structure collapses in quite different schemas.

6 Conclusions

According to the IDA-based seismic collapse simulation and analysis of super-tall building structure in the present paper, several conclusions could be drawn:

(1) To numerically reproduce the seismic collapse of super tall structures, a number of essential factors should be carefully considered, e. g., material softening under severe deformation modeled by damage model, deformation of element modeled by proper structural elements, contact between dropping elements, and illness of the system determined by explicit algorithm.

(2) Incremental dynamic analysis gives a comprehensive assessment of not only the regular behavior but also the collapse of structures under seismic loads. As a dynamic equivalence of static pushover analysis, the super-tall structure could be dynamically pushed over to its collapse in rather different patterns.

(3) The initiation of structural damage is governed by the weakness of structures, including the transition stories of shear wall thickness, sectional shape, and truss arrangement. The subsequent damage evolutions are closed related to the real-time waveforms and intensities of the corresponding earthquake record. Although the sets of ground motion records adopted for IDA are of the same waveform, their local waveforms and intensities are quite different corresponding to the critical moments including damage initiation, unstable damage propagation, and structural collapse. Thus, the final patterns of collapse for the super tall structures are quite different.

(4) In the design of super-tall building structures, at least multiple collapse patterns should be considered, even for the ground motion record with similar waveforms but different intensities.

Acknowledgements

Financial supports from the National Science Foundation of China are gratefully appreciated (GNs:51261120374, 91315301, and 51678439).

References

[1] HUANG S C, SKOKAN M J. Collapse of the Tungshing building during the 1999 Chi-Chi earthquake in Taiwan[C]//Proc., 7th US National Conf. on Earthquake Engineering. 2002.

[2] LEW M, NAEIM F, CARPENTER L D, et al. The significance of the 27 February 2010 offshore Maule, Chile earthquake[J]. The Structural Design of Tall and Special Buildings, 2010, 19(8): 826-837.

[3] NAKASHIMA M, INOUE K, TADA M. Classification of damage to steel buildings observed in the 1995 Hyogoken-Nanbu earthquake[J]. Engineering Structures, 1998, 20(4-6): 271-281.

[4] VILLAVERDE R. Methods to assess the seismic collapse capacity of building structures: State of the art[J]. Journal of Structural Engineering, 2007, 133(1): 57-66.

[5] Building Seismic Safety Council. Nehrp Guidelines for the Seismic Rehabilitation of Buildings: FEMA Publication 273[S]. Washington, D.C.: Federal Emergency Management Agency, 1997: 2-12.
[6] CLOUGH R W, PENZIEN J. Dynamics of structures[M]. New York: McGraw Hill, 1975.
[7] Building Seismic Safety Council. Recommended seismic design criteria for new steel moment frame buildings: FEMA Publication 350[S]. Washington, D.C.: Federal Emergency Management Agency, 2000.
[8] BERTERO V V. Strength and deformation capacities of buildings under extreme environments[J]. Structural Engineering and Structural Mechanics, 1977, 53(1): 29-79.
[9] VAMVATSIKOS D, CORNELL C A. Incremental dynamic analysis[J]. Earthquake Engineering & Structural Dynamics, 2002, 31(3): 491-514.
[10] VAMVATSIKOS D, CORNELL C A. Applied incremental dynamic analysis[J]. Earthquake Spectra, 2004, 20(2): 523-553.
[11] Ministry of construction of the People's Republic of China. Code for Seismic Design of Buildings: GB 50011—2010[S]. Beijing: China Architecture & Buildings Press, 2010.
[12] JU J W. Energy-based coupled elastoplastic damage models at finite strains[J]. Journal of Engineering Mechanics, 1989, 115(11): 2507-2525.
[13] LEMAITRE J. Evalution of dissipation and damage in metals submitted to dynamic loading[C]// Proceedings of International Conference of Mechanical Behavior of Materials. 1971.
[14] LI J, WU J Y. Energy-based CDM model for nonlinear analysis of confined concrete[C]//ACI SP-238, In: Proc. of the Symposium on Confined Concrete, Changsha, China. 2004.
[15] LEMAITRE J. A continuous damage mechanics model for ductile fracture[J]. Journal of Engineering Materials & Technology, 1985, 107(107): 83-89.
[16] LI J, REN X D. Stochastic damage model for concrete based on energy equivalent strain[J]. International Journal of Solids and Structures, 2009, 46(11-12): 2407-2419.
[17] WU J Y, LI J, FARIA R. An energy release rate-based plastic-damage model for concrete[J]. International Journal of Solids and Structures, 2006, 43(3-4): 583-612.
[18] MANDER J B, PRIESTLEY M J N, PARK R. Theoretical stress-strain model for confined concrete [J]. Journal of Structural Engineering, 1988, 114(8): 1804-1826.
[19] GUO Z, SHI X. Reinforced concrete theory and analyse[M]. Beijing: Tsinghua University Press, 2003.
[20] Ren X D, Zeng S, Li J. A rate-dependent stochastic damage-plasticity model for quasi-brittle materials [J]. Computational Mechanics, 2015, 55(2): 267-285.
[21] BELYTSCHKO T, LIU W K, MORAN B, et al. Nonlinear finite elements for continua and structures [M]. John wiley & sons, 2014.

Stochastic analysis of fatigue of concrete bridges①

Ruofan Gao[1,2], **Jie Li**[1,3], **Alfredo H.-S. Ang**[2]

1 *School of Civil Engineering, Tongji University, Shanghai, China*
2 *Department of Civil and Environmental Engineering, University of California, Irvine, California, USA*
3 *School of Civil Engineering and the State Key Laboratory on Disaster Reduction in Civil Engineering, Tongji University, Shanghai, China*

Abstract A novel method is proposed in this work for the assessment of the remaining fatigue life and fatigue reliability of concrete bridges subjected to random loads. The fatigue reliability of a bridge is a function of the fatigue damage accumulation; a stochastic fatigue damage model (SFDM) with physical mechanism is introduced for deriving the fatigue damage process. In order to implement the probabilistic analysis, based on the probability density evolution method (PDEM), the generalised density evolution equation (GDEE) for the remaining fatigue life is developed. Finally, a prestressed concrete continuous beam bridge located in China is illustrated. The random fatigue load acting on the bridge is modelled as the compound Poisson process, and the simulation of the random load uses the stochastic harmonic function (SHF) method. To simplify the reliability analysis, an equivalent constant-amplitude (ECA) load process is introduced based on energy equivalence. By employing SFDM, the finite element analysis of the bridge under the fatigue loading is performed. Then, the fatigue damage accumulation process of the bridge under the fatigue loading is obtained. Through solving the probability density evolution equation for the remaining fatigue life, the probability density functions (PDFs) of the remaining fatigue life evolving with time is obtained. The fatigue reliability is then calculated by integrating the PDF of the corresponding remaining life.
Keywords Fatigue reliability; Concrete bridges; Stochastic fatigue damage model; Probability density evolution method

1 Introduction

Concrete bridges are always subjected to random fatigue loads (such as vehicle load, wind load, etc.) during their service life. Fatigue damage can be expected to occur in these concrete bridges under random loads. With the accumulation of damage, the remaining life of a concrete bridge as well as its reliability will be reduced. Thus, it is important to examine the fatigue behaviour and assess the fatigue reliability of concrete bridges under random loads.

A review of previous studies suggests that there are two methods commonly employed to

① Originally published in *Structure and Infrastructure Engineering*, 2019, 15(7): 925-939.

analyse the stochastic fatigue behaviour of concrete bridges. The first is the traditional S-N curve method[1-5]. This method aims at obtaining the S-N-P curves based on extensive experimental data and evaluating the probability of fatigue failure based on established fatigue damage laws; this method provides a simple way for studying the random fatigue of metallic structures. However, the S-N-P curve method is entirely empirical and not easy to apply for other materials. Another method for the probabilistic analysis of concrete fatigue is the fracture mechanics approach[6-10]. This approach focuses on modifying the well-known Paris' law in order to derive the suitable fatigue crack propagation laws for concrete. The randomness of external loadings is considered to predict the crack length and estimate the fatigue life and reliability of concrete structures. The physical mechanism of this approach is explicit, but it cannot present fatigue damage state before the formation of the macro crack, especially for materials such as concrete with a large number of micro defects.

An analysis method for the stochastic fatigue of concrete bridges is proposed. Based on the stochastic damage mechanics (SDM)[11-14] and the probability density evolution method (PDEM)[15-20] developed in recent years, the resulting concrete constitutive model can accurately reflect the concrete damage mechanisms and present the whole damage evolution process. Li and Zhang proposed the stochastic damage constitutive model for concrete under monotonous loading[14]. Ding and Li extended this stochastic damage constitutive model to fatigue[21-23]. Li and Chen[19] have developed the PDEM for stochastic analysis.

Based on the SDM and PDEM, this paper developed a universal probabilistic analysis method for assessing the remaining fatigue life and reliability of concrete bridges under random fatigue loads. As the fatigue life and reliability are dependent on the damage accumulation, a stochastic fatigue damage constitutive model for concrete is introduced to perform the finite element analysis and derive the cumulative damage process of concrete bridges. The PDEM is then applied to the probabilistic analysis of fatigue of concrete bridges, through which a universal probability density evolution equation (PDEE) for the remaining fatigue life of concrete bridge is obtained.

Finally, the method is illustrated for a prestressed concrete continuous beam bridge. The random fatigue load is described by the compound Poisson process and simulated by the stochastic harmonic function(SHF) method[24-26]. By performing the finite element analysis on the bridge under fatigue loads, the damage evolution processes are derived. After that, through solving the PDEE for the remaining fatigue life, the probability density functions (PDFs) of the remaining life of the bridge as well as the fatigue reliability are obtained.

2 Constitutive model with stochastic fatigue damage

In order to estimate the damage accumulation of concrete bridges under fatigue loads, a stochastic fatigue damage constitutive model for concrete with physical mechanism is introduced[21].

2.1 Elastoplastic damage constitutive relationship

Based on the framework of continuum damage mechanics, the damage constitutive model can be described by[27]

$$\boldsymbol{\sigma} = (\mathbb{I} - \mathbb{D}) : \mathbb{S}_0 : (\boldsymbol{\varepsilon} - \boldsymbol{\varepsilon}^{\mathrm{p}}) \tag{1}$$

where $\boldsymbol{\sigma}$ is the second order stress tensor, $\mathbb{D}$ is the fourth order damage tensor, $\mathbb{I}$ is the fourth order identity tensor, $\mathbb{S}_0$ is the fourth order initial elastic stiffness tensor, $\boldsymbol{\varepsilon}$ is the second order strain tensor, $\boldsymbol{\varepsilon}^{\mathrm{p}}$ is the plastic components of the total strain tensor.
Under the undamaged state, the effective stress tensor $\bar{\boldsymbol{\sigma}}$ is defined as

$$\bar{\boldsymbol{\sigma}} = \mathbb{S}_0 : (\boldsymbol{\varepsilon} - \boldsymbol{\varepsilon}^{\mathrm{p}}) \tag{2}$$

To consider the tensile and shear damage separately, the effective stress tensor $\bar{\boldsymbol{\sigma}}$ is decomposed into tensile and compressive components $\bar{\boldsymbol{\sigma}}^+$, $\bar{\boldsymbol{\sigma}}^-$ [28-30]

$$\bar{\boldsymbol{\sigma}} = \bar{\boldsymbol{\sigma}}^+ + \bar{\boldsymbol{\sigma}}^-,\ \bar{\boldsymbol{\sigma}}^+ = \mathbb{P}^+ : \bar{\boldsymbol{\sigma}},\ \bar{\boldsymbol{\sigma}}^- = \mathbb{P}^- : \bar{\boldsymbol{\sigma}} \tag{3}$$

where the fourth order projective tensors are formulated as

$$\begin{aligned} \mathbb{P}^+ &= \sum_i H(\bar{\sigma}_i)\mathbf{n}^{(i)} \otimes \mathbf{n}^{(i)} \otimes \mathbf{n}^{(i)} \otimes \mathbf{n}^{(i)} \\ \mathbb{P}^- &= \mathbb{I} - \mathbb{P}^+ \end{aligned} \tag{4}$$

in which $\mathbb{I}$ denotes the fourth order identity tensor; $\bar{\sigma}_i$ is the ith eigenvalue of the effective stress tensor $\bar{\boldsymbol{\sigma}}$; $\mathbf{n}^{(i)}$ is the ith eigenvector of the effective stress tensor $\bar{\boldsymbol{\sigma}}$. The Heaviside function is defined by

$$H(x) = \begin{cases} 0 & \text{if } x \leqslant 0 \\ 1 & \text{if } x > 0 \end{cases} \tag{5}$$

The damage tensor $\mathbb{D}$ is simplified by adopting the tensile and the compressive damage variables d^+, d^- as follows[30]

$$\mathbb{D} = d^+\mathbb{P}^+ + d^-\mathbb{P}^+ \tag{6}$$

Substituting Eqs.(3) and (6) into Eq. (1) yields

$$\boldsymbol{\sigma} = (1 - d^+)\bar{\boldsymbol{\sigma}}^+ + (1 - d^-)\bar{\boldsymbol{\sigma}}^- \tag{7}$$

2.2 Evolution model for plastic strain

For the development of plasticity in concrete, the evolution of plastic strain is considered to be mainly caused by the slippage on the crack surface of each scale and is dependent on the evolution of damage[2]. Considering the coupling of plasticity and damage, a practical plastic evolution model is proposed by Li and Ren, et al., expressed as follows[2, 27]

$$\dot{\boldsymbol{\varepsilon}}^{\mathrm{p}} = f_p(\mathbb{D})\dot{\boldsymbol{\varepsilon}}^{\mathrm{e}} \tag{8}$$

in which f_p denotes the plastic evolution function.
Because of the difference in mechanical behaviour of concrete under tension and compression, the plastic strain is also decomposed into tensile and compressive components

with the following form[27]

$$\begin{cases} \dot{\varepsilon}^{p} = \dot{\varepsilon}^{p+} + \dot{\varepsilon}^{p-} \\ \dot{\varepsilon}^{p+} = f_p^{+}(d^{+})\dot{\varepsilon}^{e+} \\ \dot{\varepsilon}^{p-} = f_p^{-}(d^{-})\dot{\varepsilon}^{e-} \end{cases} \tag{9}$$

where $f_p^{\pm}$ are functions of the corresponding damage variables $d^{\pm}$, derived using experimental data[27]

$$f_p^{\pm}(d^{\pm}) = \xi_p^{\pm}(d^{\pm})^{n_p^{\pm}} \tag{10}$$

where $\xi_p^{\pm}$ and $n_p^{\pm}$ are plastic parameters related to the material.

2.3 Evolution law for fatigue damage

Based on the research at the micro-level mechanism, a mesoscopic stochastic fracture model for concrete is developed to describe the physical evolution process of damage[21, 31-32]. Considering the classic definition of damage and the random distribution of the fracture strain, the damage variable in this model is defined as[13, 33]

$$d = \int_0^1 H[\varepsilon - \Delta(x)]\mathrm{d}x \tag{11}$$

where $\Delta(x)$ is the 1-D micro fracture strain random field; x is the spatial coordinate of the micro-spring.

Eq. (11) is only applicable for the damage evolution under static loading. From the perspective of energy dissipation, the mesoscopic stochastic fracture model may be extended to the fatigue case. Thus, the damage law for concrete under fatigue loading is developed by Ding and Li[21-23], given as follows

$$d^{\pm} = \int_0^1 H[E_f^{\pm} - E_s^{\pm}]\mathrm{d}x \tag{12}$$

where $E_s^{\pm}$ is the elastic energy, expressed as

$$E_s^{\pm} = \frac{1}{2}E_0[\Delta^{\pm}(x)]^2 \tag{13}$$

with $E_f^{\pm}$ denoting the cumulated energy induced by fatigue loads. By introducing the rate process theory and fatigue disturbance effect[21, 34-35], the fatigue energy dissipation $E_f^{\pm}$ is obtained as follows[21]

$$E_f^{\pm} = \int_0^t C_0 \mathrm{e}^{-Q_0/kT}\mathrm{e}^{-\kappa d}(Y^{\pm})^2\left(\frac{Y^{\pm}}{\Gamma \cdot \mathrm{e}^{-\chi|\Delta\delta|}}\right)^{\omega}\mathrm{d}t \tag{14}$$

in which C_0 represents the material constant; Q_0 denotes the energy barrier for the fracture of the particle bonds in the micro crack tip; k, T are Boltzmann constant and absolute temperature, respectively; $Y^{\pm}$ are the macro tensile and compressive damage energy release rate[30]; $\mathrm{e}^{-\chi|\Delta\delta|}$ denotes the oscillation effect of the homogenised surface energy due to repeated loading with χ denoting the material constant; κ is the material parameter used to consider the shielding effect of the multi cracking behaviour; Γ is the homogenised surface energy density of the micro crack; $\tilde{\omega}$ is the spatial scale transfer parameter. For more details, see Ref.[21 - 23] and Ref.[36 - 37].

3 Probabilistic method for analysis of remaining fatigue life

The basic principle of the PDEM[16-18] is briefly described below, and then applied to the stochastic fatigue analysis.

3.1 The generalised PDEM

According to the principle of preservation of probability[18], the following generalised density evolution equation (GDEE) is

$$\frac{\partial p_{\mathbf{X\Theta}}(\mathbf{x}, \boldsymbol{\theta}, t)}{\partial t} + \sum_{j=1}^{n_d} \dot{H}_j(\boldsymbol{\theta}, t) \frac{\partial p_{\mathbf{X\Theta}}(\mathbf{x}, \boldsymbol{\theta}, t)}{\partial x_j} = 0 \tag{15}$$

where $\mathbf{X} = (X_1, X_2, \cdots, X_{n_d})$ represents the state vector; $\boldsymbol{\Theta} = (\Theta_1, \Theta_2, \cdots, \Theta_{n_\Theta})$ is the random vector, which characterises the randomness involved in the system; t is the generalised time; and n_d is the dimension of the system.

The corresponding initial condition of Eq.(15) can be expressed as follow

$$p_{\mathbf{Z\Theta}}(\mathbf{x}, \boldsymbol{\theta}, t)\Big|_{t=t_0} = \delta(\mathbf{x} - \mathbf{x}_0) p_{\boldsymbol{\Theta}}(\boldsymbol{\theta}) \tag{16}$$

where $\mathbf{x}_0$ is the initial value vector of the system; $\delta(\cdot)$ denotes the Dirac function. By solving the generalised density evolution equation, the joint PDF, $p_{\mathbf{X\Theta}}(\mathbf{x}, \boldsymbol{\theta}, t)$, can be obtained, and then, the PDF of $\mathbf{X}(t)$ can be evaluated by the following integration

$$p_{\mathbf{X}}(\mathbf{x}, t) = \int_{\Omega_0} p_{\mathbf{X\theta}}(\mathbf{x}, \boldsymbol{\theta}, t) \mathrm{d}\boldsymbol{\theta} \tag{17}$$

The specific process of solving the GDEEs is summarised with the following steps:

Step 1. Select the representative discretised points $\boldsymbol{\theta}_q$ in the probability-assigned space Ω_θ based on the point selection method[38], denoted as

$$\boldsymbol{\theta}_q = (\theta_{q,1}, \theta_{q,2}, \cdots, \theta_{q,s}); \; q = 1, 2, \cdots, n_{\mathrm{sel}} \tag{18}$$

where n_{sel} is the total number of the selected points. Simultaneously, the assigned probability of each point is determined by the following equation

$$P_q = \int_{V_q} p_{\boldsymbol{\Theta}}(\theta) \mathrm{d}\theta; \; q = 1, 2, \cdots, n_{\mathrm{sel}} \tag{19}$$

Step 2. Obtain the generalised velocity $\dot{H}_j(\boldsymbol{\theta}_q, t)$ by solving the related physical equation for the given $\boldsymbol{\theta}_q$.

Step 3. Substitute the generalised velocity into the generalised density evolution equation [Eq.(15)], and solve Eq.(15) using the finite difference method based on the following initial conditions[19-20]

$$p_{\mathbf{X\Theta}}(\boldsymbol{x}, \boldsymbol{\theta}_q, t)\Big|_{t=t_0} = \delta(\boldsymbol{x} - \boldsymbol{x}_0)P_q \tag{20}$$

Step 4. Sum up the numerical results obtained from Step 3 $p_{\mathbf{X\Theta}}(x, \boldsymbol{\theta}_q, t)$ to derive the PDF of $X(t)$, expressed as follows

$$p_{\mathbf{X}}(\boldsymbol{x}, t) = \sum_{q=1}^{n_{\mathrm{sel}}} p_{\mathbf{X\Theta}}(\boldsymbol{x}, \boldsymbol{\theta}_q, t) \tag{21}$$

3.2 Probability density evolution equation for remaining fatigue life

In this paper, the randomness of the vehicle load acting on the bridge is considered. The random vehicle load model is described by the filtered compound Poisson process, expressed as follows[39]

$$L(t) = \sum_{n=0}^{N_L(t)} \xi_n I(t; \tau_n) \tag{22}$$

$$I(t; \tau_n) = \begin{cases} 1 & t \in \tau_n \\ 0 & t \notin \tau_n \end{cases} \tag{23}$$

where $\{L(t), t \geqslant 0\}$ denotes the stochastic process of vehicle load; $\{N_L(t), t \geqslant 0\}$ denotes the Poisson process that describes the arrival times of the vehicle load; $\{\xi_n, n=1, 2, \cdots\}$ denote the weights of the vehicle loads, which are independent and identically distributed. Accordingly, the time intervals $\{\widetilde{T}_n, n=1, 2, \cdots\}$ between two adjacent vehicle loads follow the exponential distribution. Here, W_n denotes the occurrence time of the nth vehicle load. The relationship between W_n and $\widetilde{T}_n$ is expressed by the following equation

$$W_n = \widetilde{T}_1 + \widetilde{T}_2 + \cdots + \widetilde{T}_n, \ n=1, 2, \cdots \tag{24}$$

As seen from Eq. (24), the occurrence time W_n changes randomly with the occurrence number of vehicle loads n. This means the occurrence number n is a function of the current time t, and changes randomly with t, expressed as

$$n = n(t) \tag{25}$$

Fig. 1 illustrates some samples of curves of the occurrence number W_n changing with the time t. It can be seen from Fig. 1 that the relationship between the occurrence number of vehicle load and time is randomised. Here, the remaining fatigue life N_{rc} is introduced, and defined as

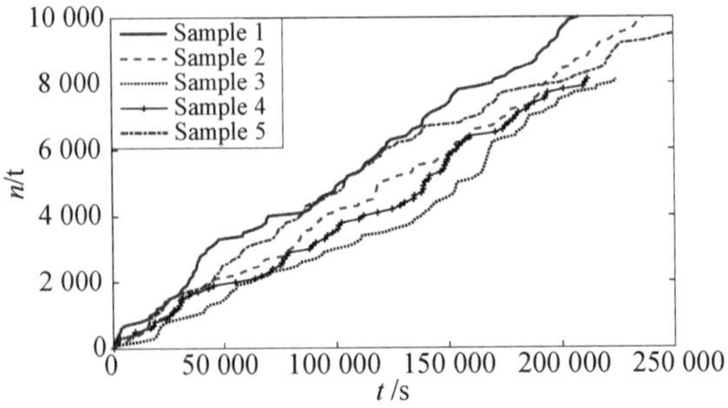

Fig. 1. Random relationship between the occurrence number and the time

$$N_{rc}(t) = N_{lc} - n(t) \tag{26}$$

where N_{lc} is the total fatigue life. For a specific sample, the fatigue life N_{lc} is determined and will not change with the time t. Fig. 2 presents the definition of the remaining fatigue life. Based on the above analysis, the generalised density evolution equation for the remaining fatigue life can be obtained as follows

$$\frac{\partial p_{N_{rc}t}(N_{rc}, \mathbf{\Theta}, t)}{\partial t} + \frac{\partial N_{rc}(\mathbf{\Theta}, t)}{\partial t} \cdot \frac{\partial p_{N_{rc}t}(N_{rc}, \mathbf{\Theta}, t)}{\partial N_{rc}} = 0 \tag{27}$$

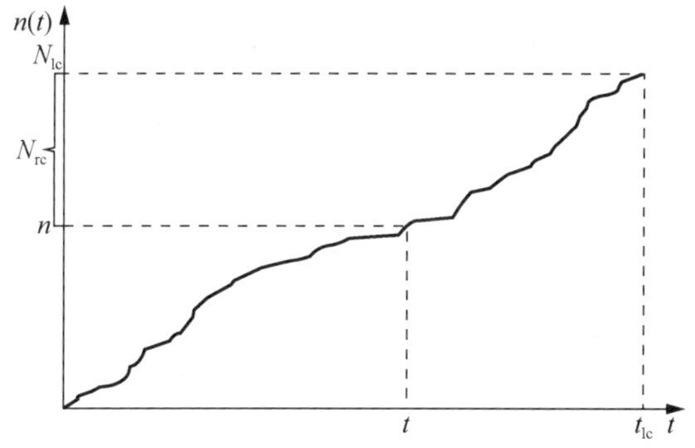

Fig. 2. Definition diagram of remaining fatigue life

4 Stochastic fatigue analysis of a prestressed concrete bridge

The newly developed stochastic fatigue analysis method means that the PDEM is combined with the SFDM to obtain the probabilistic information of concrete bridges under random fatigue loads including the PDF of the remaining fatigue life and the fatigue reliability. As the basis of the stochastic fatigue analysis, the SFDM is adopted to acquire the real damage evolution process of the concrete bridge under random loads. Based on the damage development derived by SFDM, the universal GDEE of the remaining fatigue life obtained by applying the PDEM [Eq.(27)] is solved to get the PDF of the remaining fatigue life as well as the fatigue reliability.

A prestressed concrete continuous beam bridge, namely the Xiwei bridge, is examined to demonstrate the application of the newly developed stochastic fatigue analysis method.

4.1 Bridge description

The Xiwei Bridge is located in Fujian Province of China, crossing the Dazhang river. The overall length of this bridge is 308.5 m. The Bridge consists of two units, each of three-span prestressed continuous beams with span lengths of 30 m-30 m-30 m. The length of the bridge is 90 m and the width is 19 m. The superstructure of the bridge is a three-span prestressed concrete continuous T-shaped beam. The piers of the substructure are columns supported by pile foundation. The heights of the piers are 4.14 m, 6.97 m, and 15.53 m, respectively. Fig. 3 displays the overall layout of the Xiwei bridge.

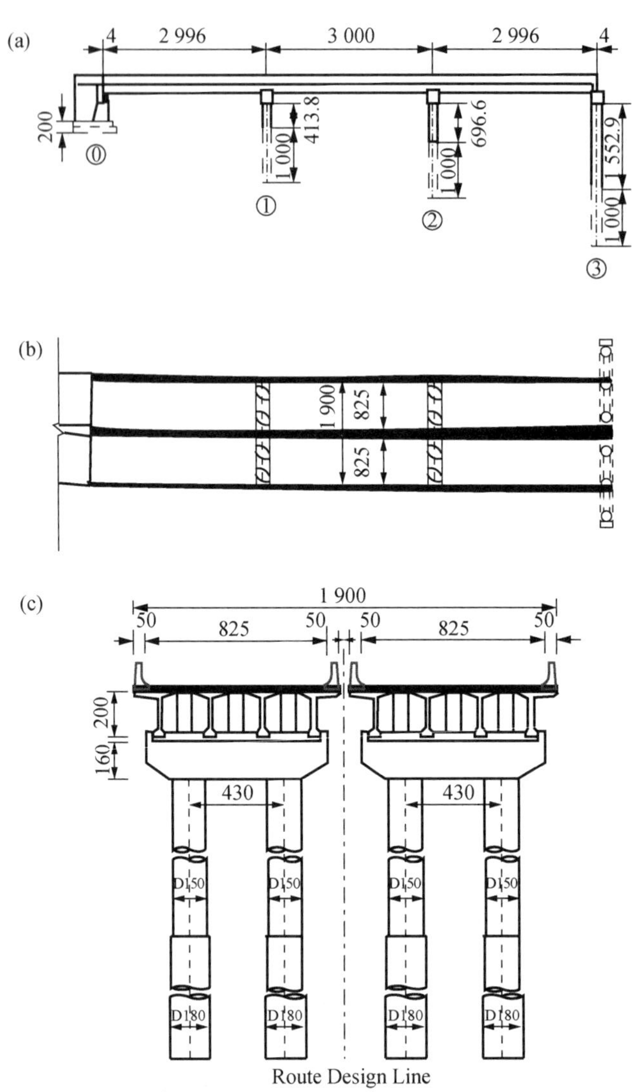

Fig. 3. Drawings of the first structure of Xiwei bridge (units:cm): (a) elevation; (b) plan; (c) section

4.1.1 Cross section of Xiwei bridge

The cross section of the prestressed concrete continuous beam of the Xiwei bridge is T shape. The width of the flange, web and horseshoe is 1.7 m, 0.2 m and 0.5 m, respectively. The reinforcement of the beam is given in Fig. 4. The prestressed steel bars are tensioned in three batches, and their position is presented in Fig. 5. The piers are circular shape, with diameters of 1.5 m. Fig. 6 shows the reinforcement layout of the piers, while the parameters in Figure 6 are listed in Table 1.

4.1.2 Material properties

By the definition of the Chinese concrete structures' design code, the concrete strength of the bridge beams and decks is C50, and that of the piers is C30. The parameter values for concrete are listed in Table 2. For the steel bars, there are two types: HPB300 and

HRB400. The type of steel bars with the diameter greater than 12 mm is HRB400, and all the rest is HPB300. The parameter values for the steel bars are presented in Table 3. The prestressed steel bundles are composed of low relaxation steel strands with grade 270, the parameter values of which are shown in Table 4.

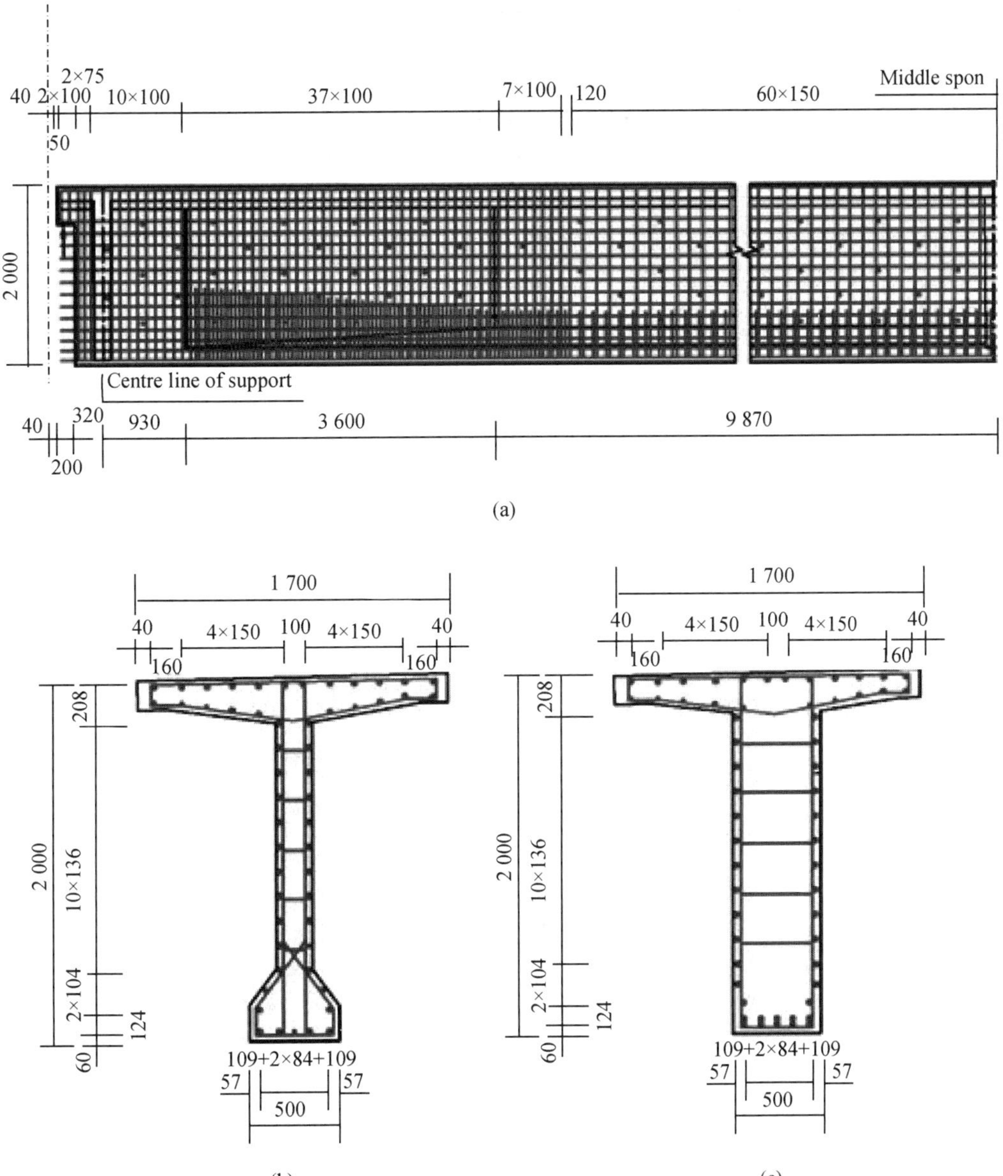

Fig. 4. **Reinforcement of prestressed continuous beam (units: mm) (a) elevation; (b) section of mid-span; (b) section at support**

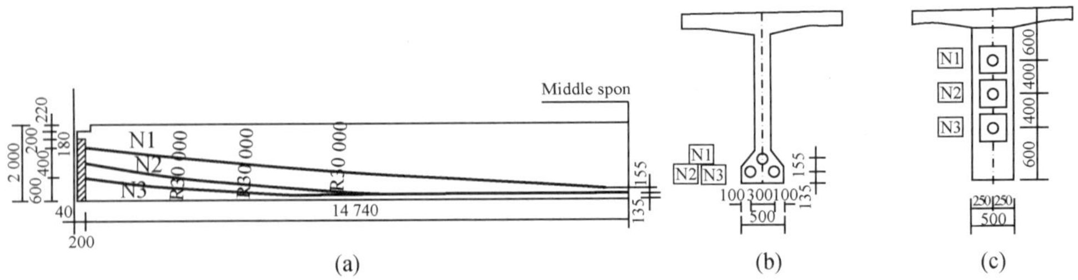

Fig. 5. Prestressed reinforcement of beam(units: mm): (a) elevation; (b) section of mid-span; (c) section at support

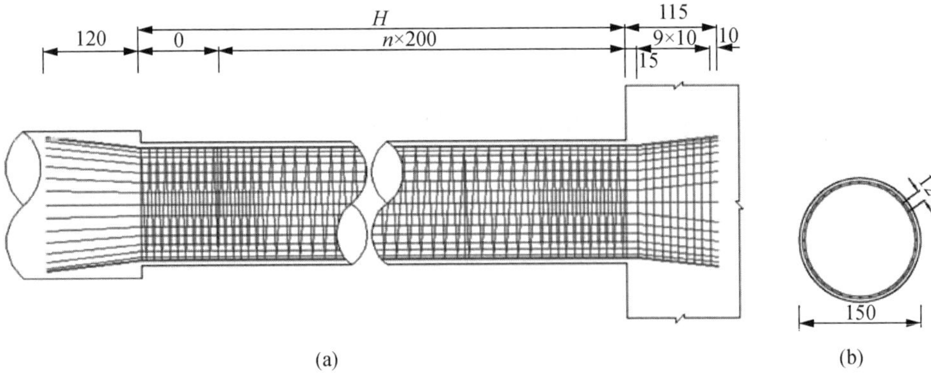

Fig. 6. Reinforcement of piers (units: cm): (a) elevation; (b) section

Table 1. Parameters of reinforcement of piers

Pier number	H/m	a/m	n
1	4.138	0	2
2	6.966	0	3

Table 2. Material parameter values for concrete

Strength level	Density /(kN · m^{-3})	Elastic modulus /MPa	Tensile strength /MPa	Compressive strength/MPa	Poisson's ratio
C50	24	35 000	3	35	0.167
C30		30 000	1.5	18	

Table 3. Material parameter values for steel bars

Types	Density /(kN · m^{-3})	Elastic modulus /MPa	Yield strength /MPa	Poisson's ratio
HPB300	78	210 000	250	0.3
HRB400		200 000	330	

Table 4. Material parameter values for prestressed steel bars

Diameter /mm	Density /(kN · m^{-3})	Elastic modulus /MPa	Yield strength /MPa	Tension control stress/MPa	Poisson's ratio
15.2	78	195 000	1 860	1 395	0.3

4.2 Finite element model

4.2.1 Model description

The finite element analysis of the Xiwei bridge uses the ABAQUS software. The Xiwei bridge is symmetric in the transverse direction, and thus, the finite element model of one half of the bridge as built, is seen in Fig. 7. The concrete continuous beams and piers use three-dimensional beam elements while the concrete decks are modelled with shell elements. The combined finite element method is applied to model the steel bars[40]. The total number of finite elements is 17 408.

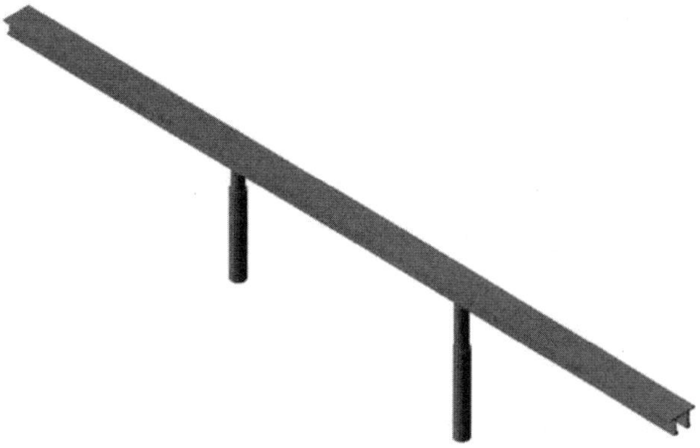

Fig. 7. Finite element model of the first structure of Xiwei bridge

4.2.2 Constitutive relations of materials

In the finite element analysis, the stochastic fatigue damage constitutive model for concrete as introduced in Section 3 is used. Here, the tensile plasticity evolution is neglected[30]. The model parameters are presented in Table 5[21, 41].

Table 5. Parameters in concrete fatigue damage constitutive model

Stress conditions	κ	$\Gamma/(\mathrm{J}\cdot\mathrm{m}^{-3})$	$C_0/(\mathrm{J}^{-1}\cdot\mathrm{m}\cdot\mathrm{s}^{-1})$	χ	$\bar{\omega}$	ξ_p	n_p
Tensile stress	25	0.155	1.2×10^{-3}	0.006	9	—	—
Compressive stress			1.52×10^{-14}	0.036	8.5	2.5	8

Because the stress levels of the longitudinal steel bars are high, the damage development of the longitudinal steel bars is taken into consideration. Based on the damage evolution of ductile material, which proposed by Lemaitre[42] and the approach for modifying the damage evolution equation of ductile material proposed by Peerlings[43], Ding recommends the following equation to consider the damage of the steel bars[21]

$$\dot{d}_s = \frac{\sigma_f^2}{2E_s^2 S}\left[\frac{2}{3}(1+\nu_s) + 3(1-2\nu_s)\left(\frac{\sigma}{3\sigma_f}\right)^2\right] e^{\alpha_s d_s} \mid \dot{\sigma} \mid \tag{28}$$

where E_s is the initial elastic modulus of steel bars; ν_s is the Poisson's ratio of steel bars; σ_f, S and α_s are material constants, and their values are taken as: $\sigma_f = 100$ MPa, $S = 2.4$ MPa, and $\alpha = 10$.

4.3 Loading condition

4.3.1 Dead load

The dead load of the bridge is divided into two parts; namely, the first can be estimated from the material density listed in Tables 2—4. The dead load intensity of the second part including the bridge deck pavement, sidewalk slab, etc., is calculated to be 18.2 kN/m.

4.3.2 Live load

The live load acting on the bridge refers to the vehicle load. As stated above, the random vehicle load model is described by the filtered compound Poisson process. The parameters of the vehicle load are evaluated based on the data collected from more than 60 000 vehicles under general traffic condition. By using the K-S test method, the vehicle data is examined for five types of probability distribution, including normal, lognormal, Gumbel, Weibull and Gamma. The distribution parameters are obtained by the maximum likelihood method[39]. The test results are shown in Table 6.

Table 6. Statistical information of vehicle load

Vehicle data	Distribution type	Parameters	
Weight/t ξ_n	Lognormal	$\mu_\xi = 1.667$	$\sigma_\xi = 0.816$
Time interval/s $\widetilde{T}_n$	Exponential	$\lambda\widetilde{T} = 0.039$	

The simulation of the compound Poisson process of the vehicle load is based on the SHFs[25-26]. The process of simulation of the random vehicle load is presented in Fig. 8. It can be seen from the simulation process that the basic random variables of the stochastic system are the random frequencies $\boldsymbol{\omega} = \{\omega_i, i=1, 2, \cdots, N_Y\}$ and random phase angles $\boldsymbol{\phi} = \{\varphi_i, i=1, 2, \cdots, N_Y\}$ of SHFs.

To simplify the stochastic fatigue analysis, an equivalent constant-amplitude(ECA) loading method[44] for random vehicle load based on the concept of energy equivalence is introduced to convert the random vehicle load to the ECA load. In this method, the equivalent load ratio R_{eq} is considered to be the mean value of the load ratios R_j of random loading by the logarithmic form, formulated as follows

$$\ln(R_{eq}) = \frac{\sum_{j=1}^{n}[\ln(R_j)]}{n} \tag{29}$$

Moreover, the equivalent load amplitude ΔF_{eq} is obtained by the following equation

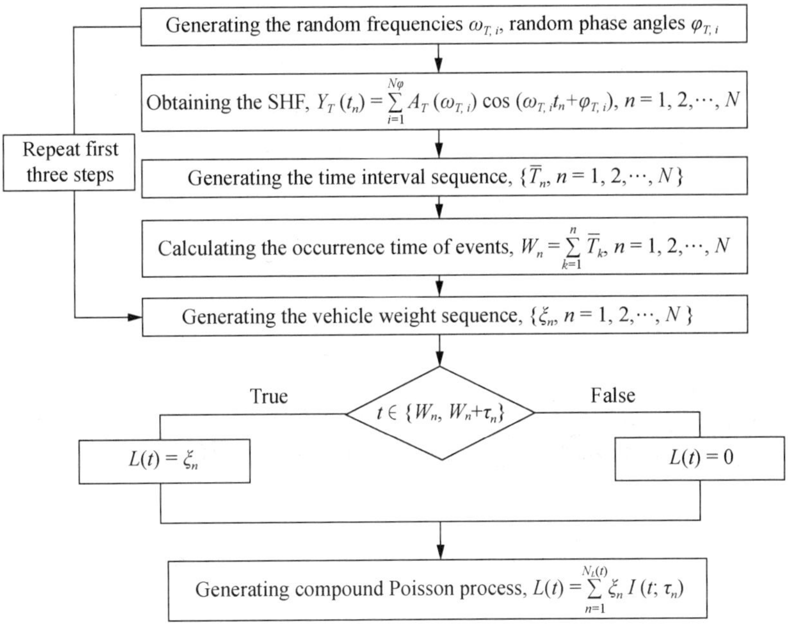

Fig. 8. Flow chart for generation of compound Poisson process based on SHFs

$$\Delta F_{\mathrm{eq}}=\left[\frac{\sum_{j=1}^{n}A_j(\Delta F_j)^{s+4}}{A_{\mathrm{eq}}\sum_{j=1}^{n}(\Delta F_j)^s}\right]^{\frac{1}{4}} \tag{30}$$

where ΔF_j is the load amplitude of the jth load cycle in the random loading; s is the material parameter, and according to Research by Le and Bazant[36], $s=8$; A_j and A_{eq} are expressed as follows

$$A_j=\frac{1+\sum_{i=1}^{4}(\bar{c}R_j)^i}{(1-\bar{c}R_j)^4},\quad A_{\mathrm{eq}}=\frac{1+\sum_{i=1}^{4}(\bar{c}R_{\mathrm{eq}})^i}{(1-\bar{c}R_{\mathrm{eq}})^4} \tag{31}$$

In addition, the number of equivalent cycles n_{eq} is given as follows

$$n_{\mathrm{eq}}=\frac{\sum_j^n(\Delta F_j)^s}{(\Delta F_{\mathrm{eq}})^s} \tag{32}$$

According to the ECA load method, the energy dissipation caused by random loading E is equal to that induced by ECA loading E_{eq}, formulated as:

$$E=E_{\mathrm{eq}} \tag{33}$$

$$E=\beta t_{\mathrm{c}}\sum_{j=1}^{n}A_j(\Delta F_j)^{s+4} \tag{34}$$

$$E_{\mathrm{eq}}=\beta t_{\mathrm{c}}n_{\mathrm{eq}}A_{\mathrm{eq}}(\Delta F_{\mathrm{eq}})^{s+4} \tag{35}$$

where the parameter β is related to the material property, and the detailed information is seen in Research by Gao and Li[44] and Le and Bažant[36]; t_{c} denotes the duration time per load cycle.

Based on the simulation method for the vehicle load, the random vehicle load process is generated. A typical sample of the random vehicle load process is presented in Fig. 9. By employing the ECA load method, the random vehicle load is converted to the ECA cyclic load. Fig. 10 depicts the corresponding ECA load for the sample of the random vehicle load of Fig. 9.

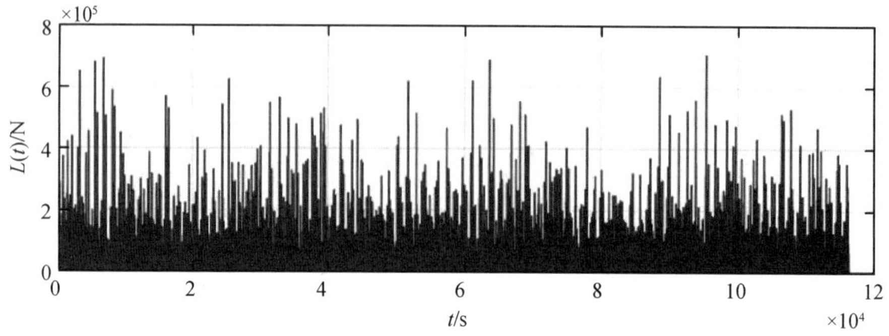

Fig. 9. A sample of the random vehicle load process

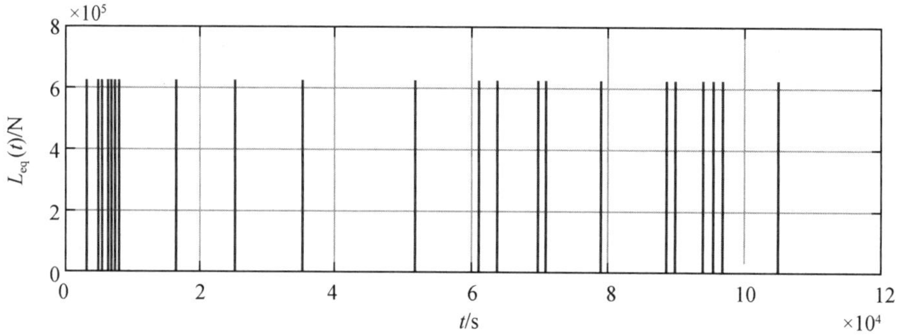

Fig. 10. The ECA load for the random vehicle load

It is shown below under Special Note that the fatigue damage accumulation and fatigue reliability under the random vehicle load is virtually the same as that under the corresponding ECA load. For the prestressed concrete continuous beam bridge, the vehicle load is loaded at the midspans of the adjacent two spans based on the principle of the most unfavourable loading position, shown in Fig. 11.

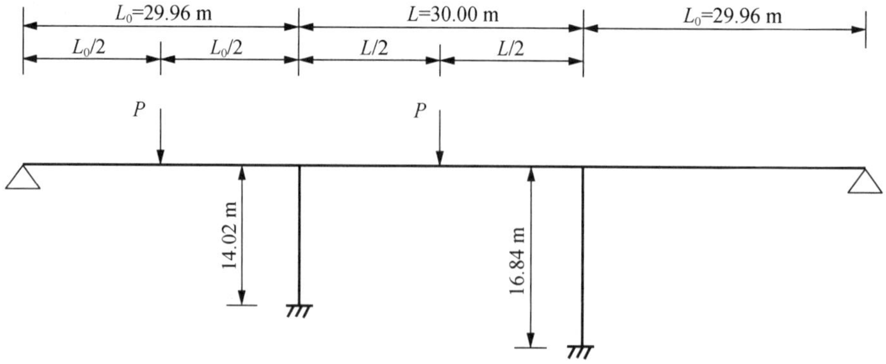

Fig. 11. Loading position of vehicle load

4.4 Structural fatigue criterion

In general, the concrete crush in compressive zone near the middle support is considered as the failure criterion for the prestressed concrete continuous beam bridge under the fatigue loading. Thus, the compressive damage in compressive zone is selected as the failure threshold, and when the cumulative damage exceeds the fatigue damage threshold, the material has failed. Considering that structural failure is usually not caused by the destruction of one material element, the damage concentration zone is introduced[41]. The elements in the damage concentration zone must satisfy the following: the principal tensile strain within the compressive zone exceeds the tensile ultimate strain of concrete, and simultaneously, the accumulated damage exceeds the fatigue failure threshold, as expressed below

$$\{e_i \in ED \mid \varepsilon_{pr}^i \geqslant \varepsilon_t \cap d^i \geqslant d_{thr},\ i=1,\ 2,\ \cdots\} \tag{36}$$

where e_i denotes the ith concrete element; ED is the set of elements constituting the damage concentration zone; ε_{pr}^i is the principal tensile strain of the ith concrete element; ε_t is the tensile ultimate strain of concrete; d^i is the accumulated damage of ith element; d_{thr} is the fatigue damage threshold.

According to the typical three-stage law of the fatigue damage development of concrete, Stage Ⅲ is an abrupt damage evolving phase, in which the damage growth rate is steep and the value of damage rises sharply. Thus, the fatigue life of the beginning of Stage Ⅲ is nearly the same as that of the end of Stage Ⅲ. For convenience of numerical simulation, the theoretical damage threshold is reduced to a proper extent, and the damage value of the beginning of Stage Ⅲ is set as the fatigue damage threshold, that is, $d_{thr}=0.7$.

4.5 Application of stochastic fatigue analysis

Based on the above finite element model, the stochastic fatigue analysis of the Xiwei bridge is performed. As mentioned before, the basic random variables are the frequencies $\boldsymbol{\omega}$ and phase angles $\boldsymbol{\varphi}$ of the SHFs. In this analysis, 20 random frequencies $\boldsymbol{\omega}=\{\omega_i,\ i=1,\ 2,\ \cdots,\ 20\}$ and 20 random phase angles $\boldsymbol{\varphi}=\{\varphi_i,\ i=1,\ 2,\ \cdots,\ 20\}$ are used to generate the random vehicle load. Therefore, the random variables in the stochastic analysis of the Xiwei bridge are composed of two parts:

(1) The frequencies of the SHFs: $\boldsymbol{\omega}=\{\omega_i,\ i=1,\ 2,\ \cdots,\ 20\}$;

(2) The phase angles of the SHFs: $\boldsymbol{\varphi}=\{\varphi_i,\ i=1,\ 2,\ \cdots,\ 20\}$.

Then, 308 representative samples of vehicle load, $n_{sel}=308$, are generated by applying the Sobol point set together with GF-discrepancy[38]. The whole process of the stochastic fatigue analysis of the Xiwei bridge is summarised in the flow chart of Fig. 12.

In the finite element analysis of the bridge, the fatigue acceleration algorithm is adopted together with the stochastic fatigue damage constitutive model to reduce computing time[21, 41]. The acceleration algorithm consists of the combination of three parts; namely, the ABAQUS finite element software, Python and Fortran programming languages. The

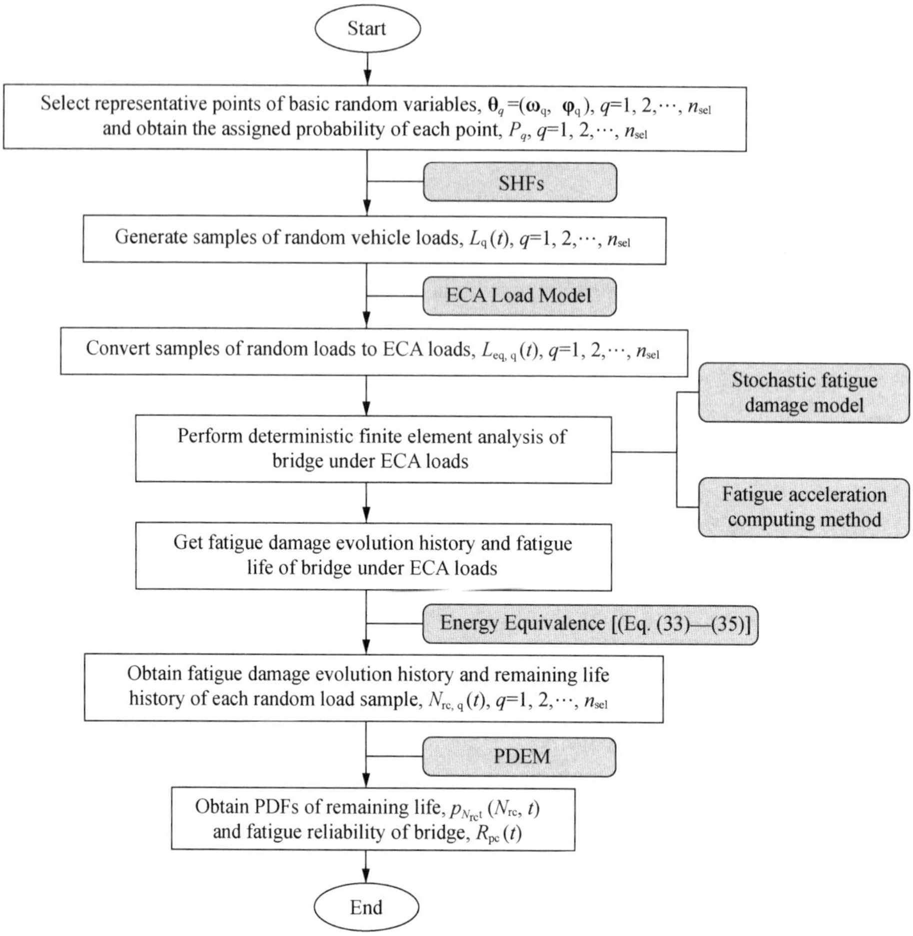

Fig. 12. Flow chart of stochastic fatigue analysis of Xiwei bridge

specific procedures of this algorithm are shown in Fig. 13.

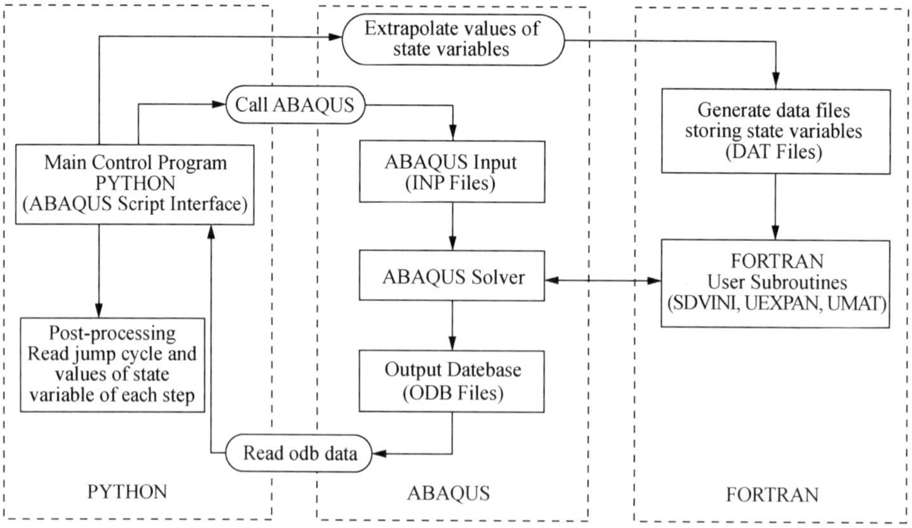

Fig. 13. Flow chart of fatigue acceleration algorithm

4.6 Results of analysis

According to the above analysis, the fatigue effects of the bridge decks and the main beams are the important members of the bridge. The fatigue failure of the main beams, which are the main load-bearing components, will cause the collapse of the whole bridge and are of primary concern. Fig. 14 shows the fatigue damage evolution process of prestressed concrete continuous beams under a typical random load sample, where "SDV3" denotes the tensile damage state of the concrete beams and "SDV4" indicates the compressive damage state. In addition, t is the service time of the Xiwei bridge, and n represents the number of the random vehicle load cycles.

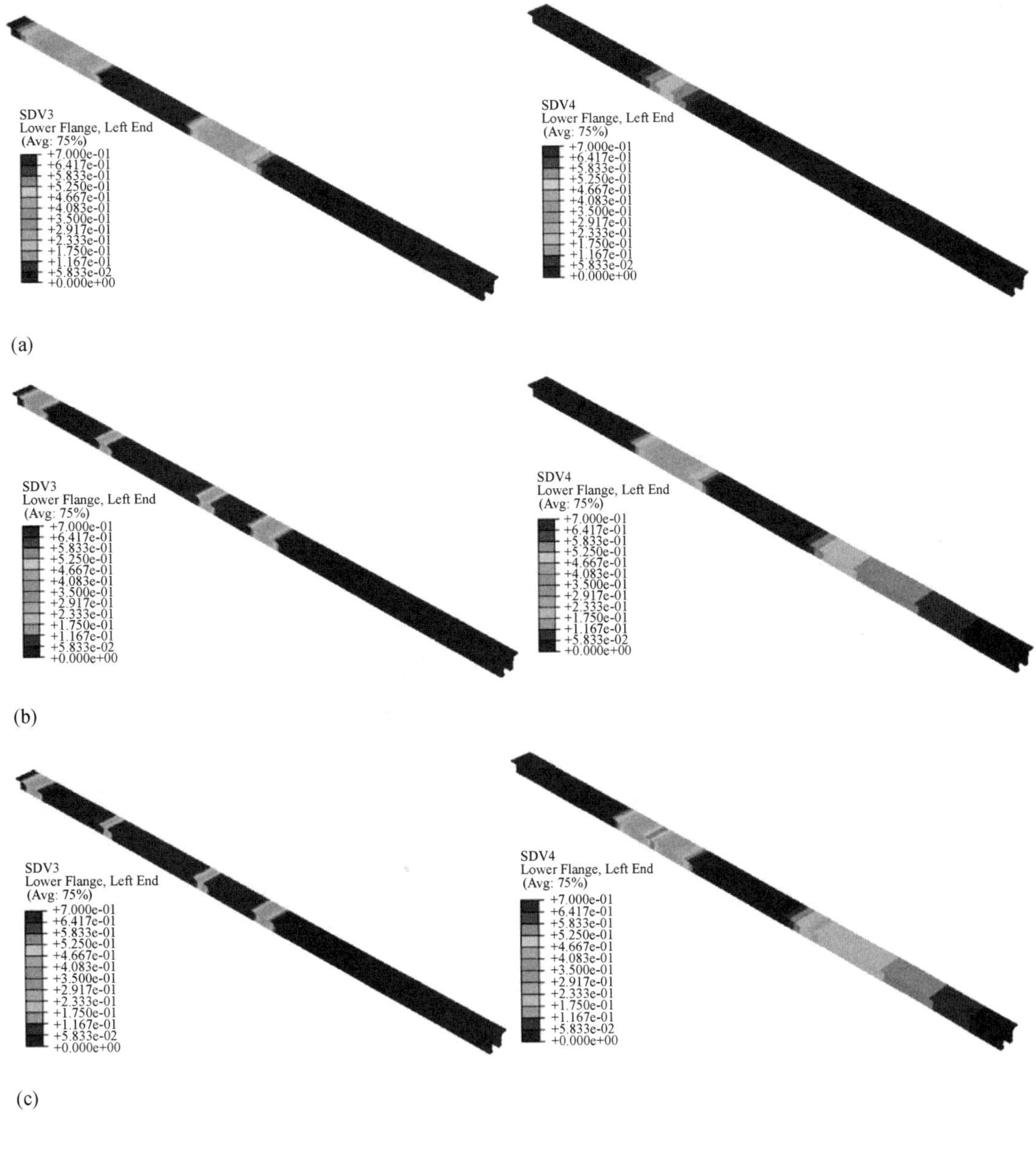

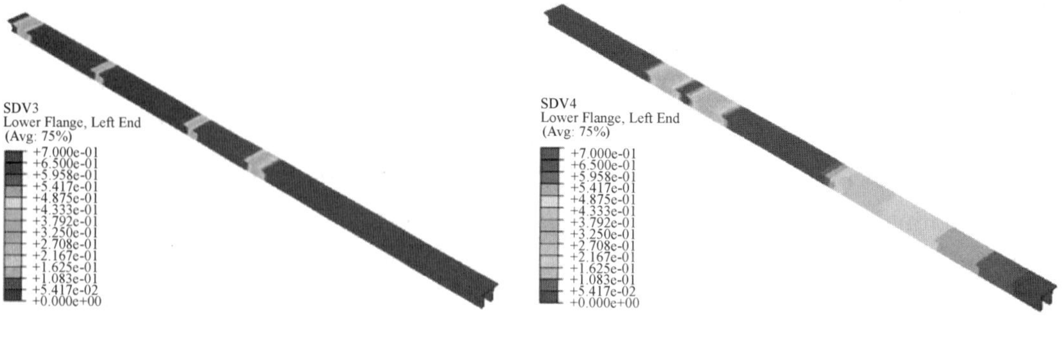

(d)

Fig. 14. Fatigue damage evolution process of the prestressed concrete continuous beams under the random vehicle load: (a) t=10th year, n=7 429 000 cycles; (b) t=24th year, n=25 143 000 cycles; (c) t=86th year, n=88 034 500 cycles; (d) t=113th year, n=115 175 000 cycles

From Fig.14, the tensile damage zone of the concrete continuous beams is mainly concentrated in the middle of the first and second spans, whereas the compressive damage is distributed in the first middle support area. With the increase of the vehicle load cycles, the compressive damage of the first middle support zone gradually reaches the damage threshold, which indicates that the shear compressive failure of the concrete beams occurs. The results are shown in Figs. 15-17. Fig. 15 presents the calculated probability density evolution surface of the remaining fatigue life of the Xiwei bridge. It is obvious from Fig. 15 that the PDF of the remaining fatigue life evolves against the time due to the randomness of the vehicle load. In order to learn the randomness more clearly, the PDFs of the remaining fatigue life at three specific instants of time(0th year, 50th year and 105th year) are shown in Fig. 16. And the PDF of the total fatigue life N_{lc} of the structure is the same as the remaining fatigue life N_{rc} at 0th year.

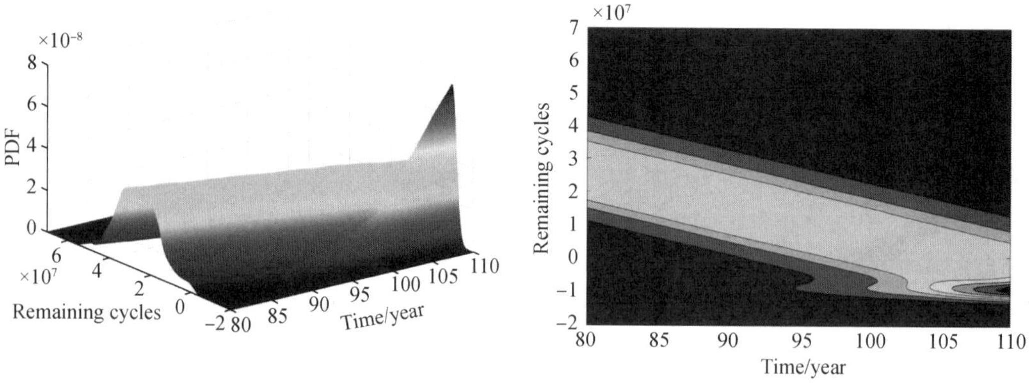

(a) (b)

Fig. 15. The probability density evolution results of remaining fatigue life of the bridge: (a) PDF evolving against time; (b) Contour to the PDF surface

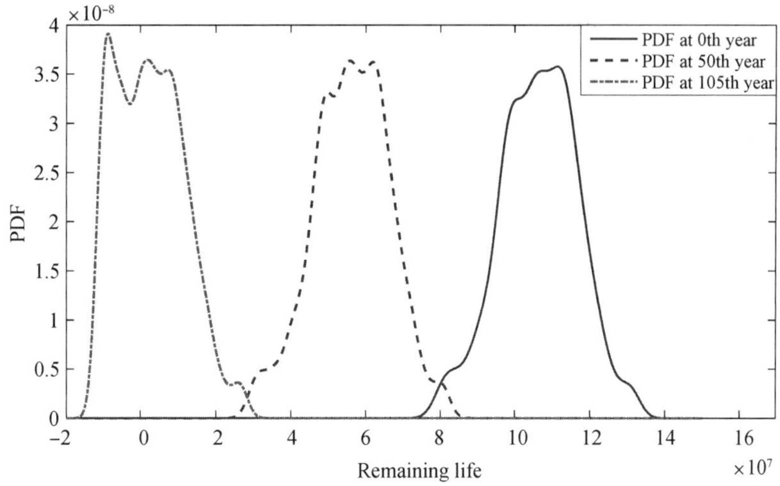

Fig. 16. PDFs at three specific instants of time

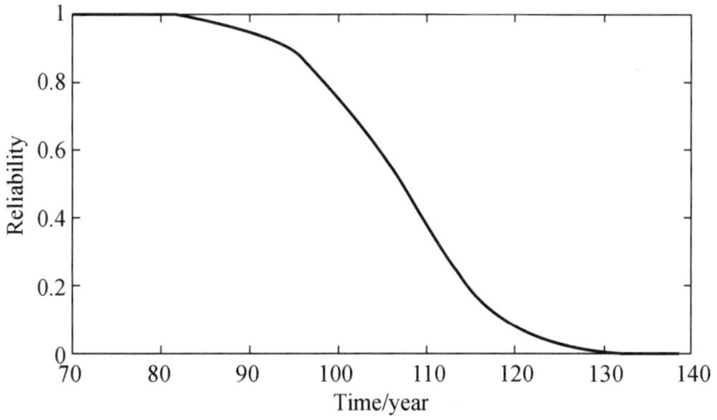

Fig. 17. The fatigue reliability of Xiwei bridge

Based on the definition of the remaining fatigue life, one can show that when the remaining fatigue life N_{rc} is greater than zero, the structure is reliable; and otherwise, the structure is considered to have failed. Hence, the fatigue reliability $R_{pc}(t)$ may be calculated by integrating the joint PDF $p_{N_{rc}t}(N_{rc}, t)$:

$$R_{pc}(t) = P_r(N_{rc} > 0) = \int_0^{+\infty} p_{N_{rc}t}(N_{rc}, t)\mathrm{d}N_{rc} \tag{37}$$

The calculated result of the fatigue reliability is shown in Fig. 17. The figure shows that the structure of the Xiwei bridge is in the reliable state before its service time reaches 80 years. However, its reliability starts to decrease after 80 years of service time, and approaches zero as time reaches 140 years. The fatigue reliabilities of the bridge at different service time are summarised in Table 7.

Table 7. Fatigue reliability values of Xiwei bridge at different service years

Service time/years	Reliability
80	1.000 0
90	0.950 7
100	0.753 5
110	0.390 8
120	0.084 8
130	0.009 5

Special note-On accuracy of fatigue reliability of the bridge

The results for the Xiwei bridge presented in Figs. 15—17 and Table 7 were calculated based on the ECA loadings. This is based on the implicit assumption that the fatigue reliability of the bridge, at any service life, under the ECA loading is the same(or close to) that under the corresponding random loading. To show that this is a valid assumption, the fatigue reliability of a much simplified problem-that of a simply supported R/C beam-is evaluated in place of the Xiwei bridge. Calculations for the reliability of the bridge(e.g., for a service life of 50 years) would require enormous calculation time.

The finite element simulation for fatigue analysis for a concrete beam under the random vehicle load with a loading period of 3 days is examined to show that the fatigue damage and reliability under the random vehicle load are almost the same as that under the ECA load. The reinforced concrete beam is 5.0 m long, 0.5 m high and 0.3 m wide, simply supported on a span of 4.6 m and loaded at the middle point, as shown in Figure 18. The compressive strength, tensile strength and Young's modulus of concrete are 44.8 MPa, 3.5 MPa and 3.25×10^4 MPa, respectively. The finite element model of the concrete beam is based on ABAQUS, as shown in Figure 19. The concrete element is in plane stress, and the element for the steel bars is the truss element. The bond connection between the concrete and reinforcement elements is considered by using the 'embedded' interaction type in ABAQUS.

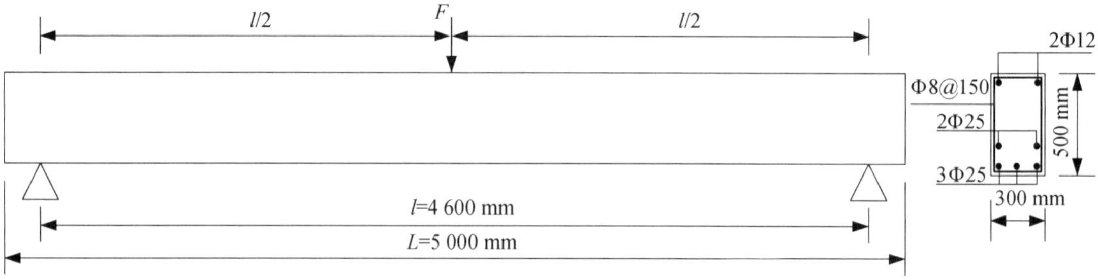

Fig. 18. Size and loading of concrete beam

Fig. 19. Finite element model of concrete beam

On the basis of the finite element model of the reinforced concrete beam, the validation steps are described as follows:

Step 1. Select the representative points for the basic random variables of the vehicle load, $\boldsymbol{\theta}_q=\{\boldsymbol{\omega}_q, \boldsymbol{\varphi}_q\}$, $q=1, 2, \cdots, 308$; see Section 4.5 for details.

Step 2. Generate the samples of the random vehicle load with the load period of 3 days, $L_{\mathrm{q}}(t)$, $q=1, 2, \cdots, 308$, and convert the samples to the corresponding ECA load, $L_{\mathrm{eq,\,q}}(t)$, $q=1, 2, \cdots, 308$. The detailed information is given in Section 4.3 and 4.5.

Step 3. Subject the samples of the random vehicle load to concrete beam. For each given sample, perform the finite element numerical analysis in ABAQUS based on the fatigue damage constitutive model.

Step 4. Obtain the cumulative damage of the concrete beam under the samples of the random vehicle load.

Step 5. Solve the following PDEE by using the finite difference method to obtain the PDF of the damage, $p_{d_{\mathrm{c}}t}$

$$\frac{\partial p_{d_{\mathrm{c}}t}(d_{\mathrm{c}}, \boldsymbol{\Theta}, t)}{\partial t}+\dot{d}_{\mathrm{c}}\frac{\partial p_{d_{\mathrm{c}}t}(d_{\mathrm{c}}, \boldsymbol{\Theta}, t)}{\partial d_{\mathrm{c}}} \tag{38}$$

where d_{c} is the compressive damage at the top of the midpoint of the beam.

Step 6. Obtain the fatigue reliability of the concrete beam at the end of the third day under the random vehicle load, R_{rand}, by integrating the solved PDF at the end of the 3rd day, $p_{d_{\mathrm{c}}}$, over the interval between zero and the damage threshold, expressed as follows

$$R_{\mathrm{rand}}=\int_0^{d_{\mathrm{thr}}} p_{d_{\mathrm{c}}}\,\mathrm{d}d_{\mathrm{c}} \tag{39}$$

where d_{thr} is the damage threshold, set equal to 0.7.

Step 7. Subject the ECA load in Step 2 to the concrete beam. Repeat Steps 3—6, and obtain the corresponding PDF of d_{c}, $p_{d_{\mathrm{c}},\mathrm{eq}}$, as well as the reliability, R_{ECA}, of the concrete beam under the ECA load.

Step 8. Compare the PDFs of the damage and reliability of the beam at the end of the third day under the random load and the ECA load.

The results of the above validation process are shown in Figs. 20—23 and Table 8. Figs. 20—22 present the compressive damage of the concrete beam at the end of the third day for several samples of the random vehicle load and their corresponding ECA load. From these figures, it can be seen that the damages under the random load are almost the same as those under the corresponding ECA load. The PDFs of the compressive damage at the top of the mid-section of the beam at the end of the third day under both the random load and the ECA load are displayed in Fig. 23, showing that the PDF under the random load agrees well with that under the ECA load. The calculated reliability of the concrete beam at the end of third day under the random load as well as the ECA load is given in Table 8. The relative error of the reliability is only 0.100 8%, which proves the reliability under the random load is nearly the same as that under the ECA load. Therefore, it is reasonable to conclude that the fatigue damage and reliability under the random load are almost the same as that under the ECA load.

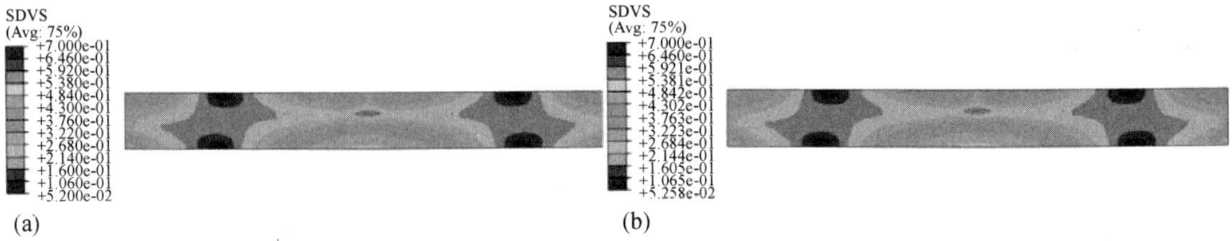

Fig. 20. Compressive damage distribution of concrete beam at the end of the 3rd day under sample 10: (a) random load; (b) ECA load

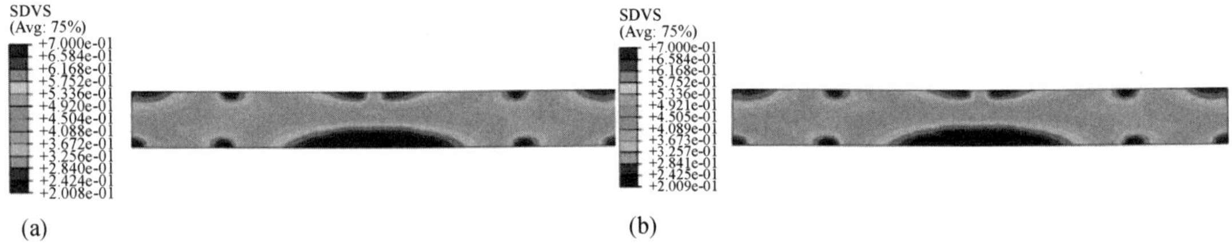

Fig. 21. Compressive damage distribution of concrete beam at the end of the 3rd day under sample 28: (a) random load; (b) ECA load

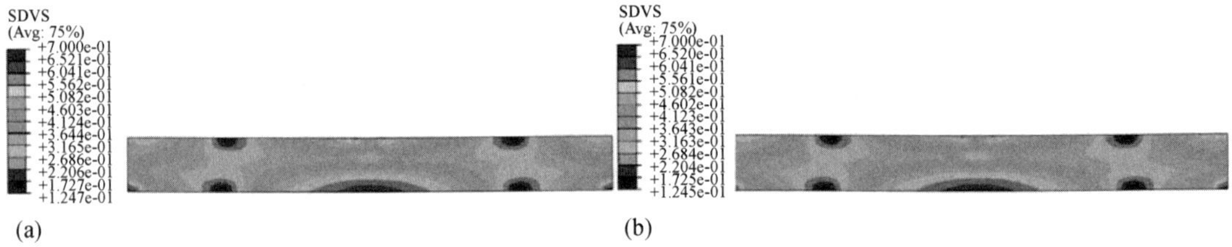

Fig. 22. Compressive damage distribution of concrete beam at the end of the 3rd day under sample 165: (a) random load; (b) ECA load

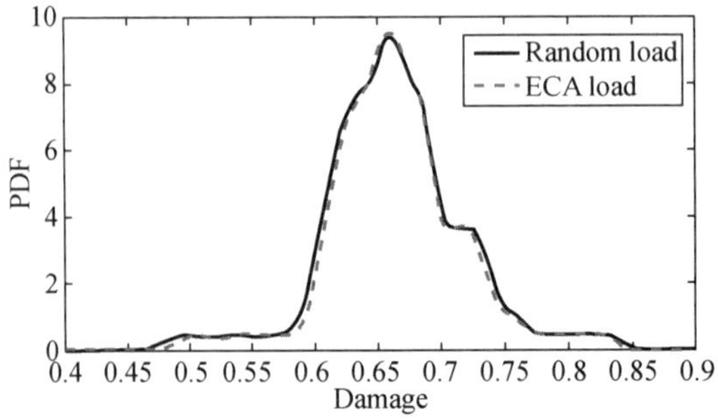

Fig. 23. PDFs of compressive damage at the end of the 3rd day under both random load and ECA load

Table 8. Fatigue reliability of concrete beam at the end of the third day under both random load and ECA load

	Reliability
Random load R_{rand}	0.793 6
ECA load R_{ECA}	0.794 4
Relative error	0.100 8%

5 Conclusion

A new stochastic analysis method for the remaining fatigue life and reliability of concrete bridges under random loads is presented. According to this method, the following developments for the stochastic fatigue analysis of concrete bridges are summarised:

(1) The fatigue damage accumulation process of the concrete bridge under random load is obtained based on the SFDM.

(2) The PDEE for concrete structures is effective for fatigue analysis.

(3) By combining the SFDM with PDEM, the PDF of the remaining fatigue life of a concrete bridge evolving against time is derived. Its time-dependent fatigue reliability is evaluated by integrating the PDF of the remaining fatigue life.

In the present work, for the sake of simplicity, the randomness of the loads, namely, the vehicle loads, is considered, and the variability of the materials' properties is ignored. For the next work plan, both the randomness of the loads and that of the materials' properties will be taken into consideration.

Disclosure statement

No potential conflict of interest was reported by the authors.

Funding

This study was performed at the University of California, Irvine during the one-year that Ms Ruofan Gao spent as Visiting Scholar performing her doctoral dissertation for her Ph.D. degree at the Tongji University in Shanghai, China. Her research was supported by the Chinese Scholarship Committee. This work was supported by National Natural Science Foundation of China [Grant No. 51538010], [Grant No. 51261120374].

References

[1] D'ANGELO L, NUSSBAUMER A. Reliability based fatigue assessment of existing motorway bridge [J]. Structural Safety, 2015, 57: 35-42.

[2] HOLMEN. Fatigue of concrete by constant and variable amplitude loading[J]. Special Publication, 1982, 75: 71-110.

[3] KWON K, FRANGOPOL D M. Bridge fatigue reliability assessment using probability density functions of equivalent stress range based on field monitoring data[J]. International Journal of Fatigue, 2010, 32(8): 1221-1232.

[4] LUO Y, YAN D H, YUAN M, et al. Probabilistic modeling of fatigue damage in orthotropic steel

bridge decks under stochastic traffic loadings[J]. Journal of Highway and Transportation Research and Development, 2017, 11(3): 62-70.

[5] OH B H. Fatigue life distributions of concrete for various stress levels[J]. Materials Journal, 1991, 88(2): 122-128.

[6] LEANDER J, ZAMIRI AKHLAGHI F, AL-EMRANI M. Fatigue reliability assessment of welded bridge details using probabilistic fracture mechanics[C]//IABSE Congress Stockholm 2016. IABSE c/o ETH Hönggerberg, 2016.

[7] ROM S, AGERSKOV H. Fatigue in aluminum highway bridges under random loading [J]. International Journal of Applied Science and Technology, 2014, 4(6): 95-107.

[8] SAIN T, KISHEN J C. Probabilistic assessment of fatigue crack growth in concrete[J]. International Journal of Fatigue, 2008, 30(12): 2156-2164.

[9] SLOWIK V, PLIZZARI G A, SAOUMA V E. Fracture of concrete under variable amplitude fatigue loading[J]. Materials Journal, 1996, 93(3): 272-283.

[10] XI Y, BAZANT Z P. Analysis of crack propagation in concrete structures by Markov chain model and r curve method[C]//Proceedings of the 1996 7th Specialty Conference on Probabilistic Mechanics and Structural Reliability. ASCE, 1996: 358-361.

[11] KANDARPA S, KIRKNER D J, SPENCER JR B F. Stochastic damage model for brittle materials subjected to monotonic loading[J]. Journal of Engineering Mechanics, 1996, 122(8): 788-795.

[12] LI J, Ren X D. Stochastic damage model for concrete based on energy equivalent strain [J]. International Journal of Solids and Structures, 2009, 46(11-12): 2407-2419.

[13] LI J, WU J Y, CHEN J B. Stochastic damage mechanics of concrete structures[M]. Beijing: Science Press, 2014.

[14] LI J, ZHANG Q Y. Study of stochastic damage constitutive relationship for concrete material[J]. Journal of Tongji University(Nature Science Edition), 2001, 29(10): 1135-1141.

[15] CHEN J B, LI J. A note on the principle of preservation of probability and probability density evolution equation[J]. Probabilistic Engineering Mechanics, 2009, 24(1): 51-59.

[16] LI J, CHEN J B. Probability density evolution method for dynamic response analysis of structures with uncertain parameters[J]. Computational Mechanics, 2004, 34(5): 400-409.

[17] LI J, CHEN J B. The number theoretical method in response analysis of nonlinear stochastic structures [J]. Computational Mechanics, 2007, 39(6): 693-708.

[18] LI J, CHEN J B. The principle of preservation of probability and the generalized density evolution equation[J]. Structural Safety, 2008, 30(1): 65-77.

[19] LI J, CHEN J B. Stochastic dynamics of structures[M]. John Wiley & Sons, 2009.

[20] LI J, CHEN J B, SUN W L, et al. Advances of the probability density evolution method for nonlinear stochastic systems[J]. Probabilistic Engineering Mechanics, 2012, 28: 132-142.

[21] DING Z D. The fatigue constitutive relation of concrete and stochastic fatigue response of concrete structure[D]. Shanghai: Tongji University, 2015.

[22] DING Z D, LI J. Modelling of fatigue damage of concretewith stochastic character[C]//IALCCE 2014 Conference, Tokyo, Japan, 2014.

[23] DING Z D, LI J. The fatigue constitutive model of concrete based on micro-meso mechanics[J]. Chinese Journal of Theoretical and Applied Mechanics, 2014, 46(6): 911-919.

[24] CHEN J B, LI J. Stochastic harmonic function and spectral representations[J]. Chinese Journal of Theoretical and Applied Mechanics, 2011, 43(3): 505-513.

[25] CHEN J B, SUN W L, LI J, et al. Stochastic harmonic function representation of stochastic processes

[J]. Journal of Applied Mechanics, 2013, 80(1): 011001.

[26] GAO R F, LI J. Simulation of compound Poisson process based on stochastic harmonic function[J]. Journal of Tongji University(Natural Science Edition), 2017, 45(12): 1731-1738.

[27] REN X D, ZENG S J, LI J. A rate-dependent stochastic damage-plasticity model for quasi-brittle materials[J]. Computational Mechanics, 2015, 55(2): 267-285.

[28] FARIA R, OLIVER J, CERVERA M. A strain-based plastic viscous-damage model for massive concrete structures[J]. International Journal of Solids and Structures, 1998, 35(14): 1533-1558.

[29] JU J W. On energy-based coupled elastoplastic damage theories: constitutive modeling and computational aspects[J]. International Journal of Solids and structures, 1989, 25(7): 803-833.

[30] WU J Y, LI J, FARIA R. An energy release rate-based plastic-damage model for concrete[J]. International Journal of Solids and Structures, 2006, 43(3-4): 583-612.

[31] LI J. Recent research progress on the stochastic damage constitutional law of concrete[J]. Journal of Southeast University(Natural Science Edition), 2002, 32(5): 750-755.

[32] ZENG S J. Dynamic experimental research and stochastic damage constitutive model for concrete[D]. Shanghai: Tongji University, 2012.

[33] ZHOU H, LI J, REN X D. Multiscale stochastic structural analysis toward reliability assessment for large complex reinforced concrete structures[J]. International Journal for Multiscale Computational Engineering, 2016, 14(3): 303-321.

[34] KRAUSZ A S, KRAUSZ K. Fracture kinetics of crack growth[M]. Springer Science & Business Media, 1988.

[35] MAALI A, COHEN-BOUHACINA T, COUTURIER G, et al. Oscillatory dissipation of a simple confined liquid[J]. Physical Review Letters, 2006, 96(8): 086105.

[36] LE J L, BAŽANT Z P, BAZANT M Z. Unified nano-mechanics based probabilistic theory of quasibrittle and brittle structures: Ⅰ. Strength, static crack growth, lifetime and scaling[J]. Journal of the Mechanics and Physics of Solids, 2011, 59(7): 1291-1321.

[37] LE J L, BAŽANT Z P. Unified nano-mechanics based probabilistic theory of quasibrittle and brittle structures: Ⅱ. Fatigue crack growth, lifetime and scaling[J]. Journal of the Mechanics and Physics of Solids, 2011, 59(7): 1322-1337.

[38] CHEN J B, ZHANG S. Improving point selection in cubature by a new discrepancy[J]. SIAM Journal on Scientific Computing, 2013, 35(5): A2121-A2149.

[39] "The Research of Vehicle Load of Highway and Bridge" Research Team. The research of vehicle load of highway and bridge[R]. Highway, 1997, 3: 8-12.

[40] GE D, WAN J, PAN P, et al. Comparative study on simulation methods of steel bar in fiber beam eliement model[J]. Building Structure, 2013, 43(2): 60-62.

[41] LIANG J S. The study of fatigue damage constitutive relation of concrete and stochastic fatigue reliability analysis on concrete structures[D]. Shanghai: Tongji University, 2017.

[42] LEMAITRE J, DESMORAT R. Engineering damage mechanics: ductile, creep, fatigue and brittle failures[M]. Springer Science & Business Media, 2005.

[43] PEERLINGS R H, BREKELMANS W M, DE BORST R, et al. Gradient-enhanced damage modelling of high-cycle fatigue[J]. International Journal for Numerical Methods in Engineering, 2000, 49(12): 1547-1569.

[44] GAO R F, LI J. Equivalent constant-amplitude fatigue loads based on the energy equivalence principle [J]. Advances in Structural Engineering, 2018.